- Textile antennas and sensors (Section 13.11) *NEW*
- RFID (Section 13.12) *NEW*
- Commercial EM software—FEKO (Section 14.7) *NEW*
- COMSOL Multiphysics (Section 14.8) *NEW*
- CST Microwave Studio (Section 14.9) *NEW*

PHYSICAL CONSTANTS

Quantity (Units)	Symbol	Best Experimental Value*	Approximate Value for Problem Work
Permittivity of free space (F/m)	ε_o	8.854×10^{-12}	$\dfrac{10^{-9}}{36\pi}$
Permeability of free space (H/m)	μ_o	$4\pi \times 10^{-7}$	12.6×10^{-7}
Intrinsic impedance of free space (Ω)	η_o	376.6	120π
Speed of light in vacuum (m/s)	c	2.998×10^8	3×10^8
Electron charge (C)	e	-1.6022×10^{-19}	-1.6×10^{-19}
Electron mass (kg)	m_e	9.1093×10^{-31}	9.1×10^{-31}
Proton mass (kg)	m_p	1.6726×10^{-27}	1.67×10^{-27}
Neutron mass (kg)	m_n	1.6749×10^{-27}	1.67×10^{-27}
Boltzmann constant (J/K)	κ	1.38065×10^{-23}	1.38×10^{-23}
Avogadro number (/kg-mole)	N	6.0221×10^{23}	6×10^{23}
Planck constant (J $\cdot$ s)	h	6.626×10^{-34}	6.62×10^{-34}
Acceleration due to gravity (m/s²)	g	9.80665	9.8
Universal constant of gravitation N (m/kg)²	G	6.673×10^{-11}	6.66×10^{-11}
Electron-volt (J)	eV	1.602176×10^{-19}	1.6×10^{-19}

* Values recommended by CODATA (Committee on Data for Science and Technology, Paris).

ELEMENTS OF
ELECTROMAGNETICS

THE OXFORD SERIES
IN ELECTRICAL AND COMPUTER ENGINEERING

Adel S. Sedra, Series Editor

ELEMENTS OF ELECTROMAGNETICS

SEVENTH EDITION

MATTHEW N. O. SADIKU
Prairie View A&M University

New York • Oxford
OXFORD UNIVERSITY PRESS

Oxford University Press is a department of the University of Oxford. It furthers the University's objective of excellence in research, scholarship, and education by publishing worldwide. Oxford is a registered trade mark of Oxford University Press in the UK and certain other countries.

Published in the United States of America by Oxford University Press
198 Madison Avenue, New York, NY 10016, United States of America.

For titles covered by Section 112 of the US Higher Education Opportunity Act, please visit www.oup.com/us/he for the latest information about pricing and alternate formats.

Library of Congress Cataloging-in-Publication Data
Names: Sadiku, Matthew N. O., author.
Title: Elements of Electromagnetics / Matthew N.O. Sadiku, Prairie View A&M University.
Description: Seventh edition. | New York, NY, United States of America : Oxford University Press,
   [2018] | Series: The Oxford series in electrical and computer engineering
Identifiers: LCCN 2017046497 | ISBN 9780190698614 (hardcover)
Subjects: LCSH: Electromagnetism.
Classification: LCC QC760 .S23 2018 | DDC 537—dc23 LC record available at
   https://lccn.loc.gov/2017046497

9 8 7 6
Printed in Canada by Marquis

To my wife, Kikelomo

BRIEF TABLE OF CONTENTS

CONTENTS

† Indicates sections that may be skipped, explained briefly, or assigned as homework if the text is covered in one semester.

PART 2:   ELECTROSTATICS

4   ELECTROSTATIC FIELDS   111

5   ELECTRIC FIELDS IN MATERIAL SPACE   177

PREFACE

This new edition is intended to provide an introduction to engineering electromagnetics (EM) at the junior or senior level. Although the new edition improves on the previous editions, the core of the subject of EM has not changed. The fundamental objective of the first edition has been retained: to present EM concepts in a clearer and more interesting manner than other texts. This objective is achieved in the following ways:

1. To avoid complicating matters by covering EM and mathematical concepts simultaneously, vector analysis is covered at the beginning of the text and applied gradually. This approach avoids breaking in repeatedly with more background on vector analysis, thereby creating discontinuity in the flow of thought. It also separates mathematical theorems from physical concepts and makes it easier for the student to grasp the generality of those theorems. Vector analysis is the backbone of the mathematical formulation of EM problems.

2. Each chapter opens either with a historical profile of some electromagnetic pioneers or with a discussion of a modern topic related to the chapter. The chapter starts with a brief introduction that serves as a guide to the whole chapter and also links the chapter to the rest of the book. The introduction helps the students see the need for the chapter and how it relates to the previous chapter. Key points are emphasized to draw the reader's attention. A brief summary of the major concepts is discussed toward the end of the chapter.

3. To ensure that students clearly get the gist of the matter, key terms are defined and highlighted. Important formulas are boxed to help students identify essential formulas.

4. Each chapter includes a reasonable amount of solved examples. Since the examples are part of the text, they are clearly explained without asking the reader to fill in missing steps. In writing out the solution, we aim for clarity rather than efficiency. Thoroughly worked out examples give students confidence to solve problems themselves and to learn to apply concepts, which is an integral part of engineering education. Each illustrative example is followed by a problem in the form of a Practice Exercise, with the answer provided.

5. At the end of each chapter are ten review questions in the form of multiple-choice objective items. Open-ended questions, although they are intended to be thought-provoking, are ignored by most students. Objective review questions with answers immediately following them provide encouragement for students to do the problems and gain immediate feedback. A large number of problems are provided and are presented in the same order as the material in the main text. Approximately 20 to 25 percent of the problems in this edition have been replaced. Problems of intermediate difficulty are identified by a single asterisk; the most difficult problems are marked with a double asterisk. Enough problems are provided to allow the instructor to choose some as examples and assign some as homework problems. Answers to odd-numbered problems are provided in Appendix E.

6. Since most practical applications involve time-varying fields, six chapters are devoted to such fields. However, static fields are given proper emphasis because they are special cases of dynamic fields. Ignorance of electrostatics is no longer acceptable

because there are large industries, such as copier and computer peripheral manufacturing, that rely on a clear understanding of electrostatics.

7. The last section in each chapter is devoted to applications of the concepts covered in the chapter. This helps students see how concepts apply to real-life situations.

8. The last chapter covers numerical methods with practical applications and MATLAB programs. This chapter is of paramount importance because most practical problems are only solvable using numerical techniques. Since MATLAB is used throughout the book, an introduction to MATLAB is provided in Appendix C.

9. Over 130 illustrative examples and 300 figures are given in the text. Some additional learning aids such as basic mathematical formulas and identities are included in Appendix A. Another guide is a special note to students, which follows this preface.

NEW TO THE SIXTH EDITION

- Five new Application Notes designed to explain the real-world connections between the concepts discussed in the text.
- A revised Math Assessment test, for instructors to gauge their students' mathematical knowledge and preparedness for the course.
- New and updated end-of-chapter problems.

Solutions to the end-of-chapter problems and the Math Assessment, as well as PowerPoint slides of all figures in the text, can be found at the Oxford University Press Ancillary Resource Center.

Students and professors can view Application Notes from previous editions of the text on the book's companion website www.oup.com/us/sadiku.

Although this book is intended to be self-explanatory and useful for self-instruction, the personal contact that is always needed in teaching is not forgotten. The actual choice of course topics, as well as emphasis, depends on the preference of the individual instructor. For example, an instructor who feels that too much space is devoted to vector analysis or static fields may skip some of the materials; however, the students may use them as reference. Also, having covered Chapters 1 to 3, it is possible to explore Chapters 9 to 14. Instructors who disagree with the vector-calculus-first approach may proceed with Chapters 1 and 2, then skip to Chapter 4, and refer to Chapter 3 as needed. Enough material is covered for two-semester courses. If the text is to be covered in one semester, covering Chapters 1 to 9 is recommended; some sections may be skipped, explained briefly, or assigned as homework. Sections marked with the dagger sign (†) may be in this category.

ACKNOWLEDGMENTS

I thank Dr. Sudarshan Nelatury of Penn State University for providing the new Application Notes and the Math Assessment test. It would not be possible to prepare this edition without the efforts of Executive Editor Dan Kaveney, Associate Editor Christine Mahon, Assistant Editor Megan Carlson, Marketing Manager David Jurman, Marketing Assistant Colleen Rowe, Production Editor Claudia Dukeshire, and Designer Michele Laseau at Oxford University Press, as well as Susan Brown and Betty Pessagno.

I thank the reviewers who provided helpful feedback for this edition:

Mohammadreza (Reza) Barzegaran
Lamar University

Sudarshan Nelatury
Penn State Erie

Sharif M. A. Bhuiyan
Tuskegee University

Sima Noghanian
University of North Dakota

Muhammad Dawood
New Mexico State University

Vladimir Rakov
University of Florida

Robert Gauthier
Carleton University

Lisa Shatz
Suffolk University

Jesmin Khan
Tuskegee University

Kyle Sundqvist
Texas A&M University

Edwin Marengo
Northeastern University

Lili H. Tabrizi
California State University, Los Angeles

Perambur S. Neelakanta
Florida Atlantic University

I also offer thanks to those who reviewed previous editions of the text:

Yinchao Chen
University of South Carolina

Douglas T. Petkie
Wright State University

Satinderpaul Singh Devgan
Tennessee State University

James E. Richie
Marquette University

Dentcho Angelov Genov
Louisiana Tech University

Elena Semouchkina
Michigan Technological University

Scott Grenquist
Wentworth Institute of Technology

Barry Spielman
Washington University

Xiaomin Jin
Cal Poly State University, San Luis Obispo

Murat Tanik
University of Alabama–Birmingham

Jaeyoun Kim
Iowa State University

Erdem Topsakal
Mississippi State University

Caicheng Lu
University of Kentucky

Charles R. Westgate Sr.
SUNY–Binghamton

Perambur S. Neelakantaswamy
Florida Atlantic University

Weldon J. Wilson
University of Central Oklahoma

Kurt E. Oughstun
University of Vermont

Yan Zhang
University of Oklahoma

I am grateful to Dr. Kendall Harris, dean of the College of Engineering at Prairie View A&M University, and Dr. Pamela Obiomon, head of the Department of Electrical and

Computer Engineering, for their constant support. I would like to express my gratitude to Dr. Vichate Ungvichian, at Florida Atlantic University, for pointing out some errors. I acknowledge Emmanuel Shadare for help with the figures. A well-deserved expression of appreciation goes to my wife and our children for their constant support and prayer.

I owe special thanks for those professors and students who have used earlier editions of the book. Please keep sending those errors directly to the publisher or to me at sadiku@ieee.org.

—Matthew N.O. Sadiku
Prairie View, Texas

A NOTE TO THE STUDENT

Electromagnetic theory is generally regarded by students as one of the most difficult courses in physics or the electrical engineering curriculum. But this conception may be proved wrong if you take some precautions. From experience, the following ideas are provided to help you perform to the best of your ability with the aid of this textbook:

1. Pay particular attention to Part 1 on vector analysis, the mathematical tool for this course. Without a clear understanding of this section, you may have problems with the rest of the book.

2. Do not attempt to memorize too many formulas. Memorize only the basic ones, which are usually boxed, and try to derive others from these. Try to understand how formulas are related. There is nothing like a general formula for solving all problems. Each formula has limitations owing to the assumptions made in obtaining it. Be aware of those assumptions and use the formula accordingly.

3. Try to identify the key words or terms in a given definition or law. Knowing the meaning of these key words is essential for proper application of the definition or law.

4. Attempt to solve as many problems as you can. Practice is the best way to gain skill. The best way to understand the formulas and assimilate the material is by solving problems. It is recommended that you solve at least the problems in the Practice Exercise immediately following each illustrative example. Sketch a diagram illustrating the problem before attempting to solve it mathematically. Sketching the diagram not only makes the problem easier to solve, but also helps you understand the problem by simplifying and organizing your thinking process. Note that unless otherwise stated, all distances are in meters. For example $(2, -1, 5)$ actually means $(2 \text{ m}, -1 \text{ m}, 5 \text{ m})$.

You may use MATLAB to do number crunching and plotting. A brief introduction to MATLAB is provided in Appendix C.

A list of the powers of 10 and Greek letters commonly used throughout this text is provided in the tables located on the inside cover. Important formulas in calculus, vectors, and complex analysis are provided in Appendix A. Answers to odd-numbered problems are in Appendix E.

Matthew N. O. Sadiku received his BSc degree in 1978 from Ahmadu Bello University, Zaria, Nigeria, and his MSc and PhD degrees from Tennessee Technological University, Cookeville, Tennessee, in 1982 and 1984, respectively. From 1984 to 1988, he was an assistant professor at Florida Atlantic University, Boca Raton, Florida, where he did graduate work in computer science. From 1988 to 2000, he was at Temple University, Philadelphia, Pennsylvania, where he became a full professor. From 2000 to 2002, he was with Lucent/Avaya, Holmdel, New Jersey, as a system engineer and with Boeing Satellite Systems, Los Angeles, California, as a senior scientist. He is currently a professor of electrical and computer engineering at Prairie View A&M University, Prairie View, Texas.

He is the author of over 370 professional papers and over 70 books, including *Elements of Electromagnetics* (Oxford University Press, 7th ed., 2018), *Fundamentals of Electric Circuits* (McGraw-Hill, 6th ed., 2017, with C. Alexander), *Computational Electromagnetics with MATLAB* (CRC, 4th ed., 2018), *Metropolitan Area Networks* (CRC Press, 1995), and *Principles of Modern Communication Systems* (Cambridge University Press, 2017, with S. O. Agbo). In addition to the engineering books, he has written Christian books including *Secrets of Successful Marriages, How to Discover God's Will for Your Life*, and commentaries on all the books of the New Testament Bible. Some of his books have been translated into French, Korean, Chinese (and Chinese Long Form in Taiwan), Italian, Portuguese, and Spanish.

He was the recipient of the 2000 McGraw-Hill/Jacob Millman Award for outstanding contributions in the field of electrical engineering. He was also the recipient of Regents Professor award for 2012–2013 by the Texas A&M University System. He is a registered professional engineer and a fellow of the Institute of Electrical and Electronics Engineers (IEEE) "for contributions to computational electromagnetics and engineering education." He was the IEEE Region 2 Student Activities Committee Chairman. He was an associate editor for IEEE Transactions on Education. He is also a member of the Association for Computing Machinery (ACM) and the American Society of Engineering Education (ASEE). His current research interests are in the areas of computational electromagnetics, computer networks, and engineering education. His works can be found in his autobiography, *My Life and Work* (Trafford Publishing, 2017) or on his website, www.matthewsadiku.com. He currently resides with his wife Kikelomo in Hockley, Texas. He can be reached via email at sadiku@ieee.org.

MATH ASSESSMENT

1.1 Let θ be the angle between the vectors $\mathbf{A}$ and $\mathbf{B}$. What can be said about θ if (i) $|\mathbf{A} + \mathbf{B}| < |\mathbf{A} - \mathbf{B}|$, (ii) $|\mathbf{A} + \mathbf{B}| = |\mathbf{A} - \mathbf{B}|$, (iii) $|\mathbf{A} + \mathbf{B}| > |\mathbf{A} - \mathbf{B}|$?

1.2 Two sides of a parallelogram $ABCD$ denoted as $\mathbf{p} = 5\mathbf{a}_x$ and $\mathbf{q} = 3\mathbf{a}_x + 4\mathbf{a}_y$ are shown in Figure MA-1 Let the diagonals intersect at O and make an angle α. Find the coordinates of O and the magnitude of α. Based on the value of α, what can we call $ABCD$?

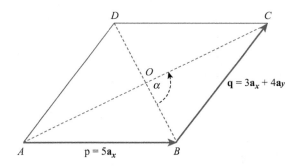

FIGURE MA-1 Parallelogram *ABCD*.

1.3 What is the distance R between the two points $A(3, 5, 1)$ and $B(5, 7, 2)$? Also find its reciprocal, $\dfrac{1}{R}$.

1.4 What is the distance vector $\mathbf{R}_{AB}$ from $A(3, 7, 1)$ to $B(8, 19, 2)$ and a unit vector $\mathbf{a}_{AB}$ in the direction of $\mathbf{R}_{AB}$?

1.5 Find the interval of values x takes so that a unit vector $\mathbf{u}$ satisfies $|(x - 2)\mathbf{u}| < |3\mathbf{u}|$.

1.6 There are four charges in space at four points A, B, C, and D, each 1 m from *every* other. You are asked to make a selection of coordinates for these charges. How do you place them in space and select their coordinates? There is no unique way.

1.7 A man driving a car starts at point O, drives in the following pattern

> 15 km northeast to point A,
> 20 km southwest to point B,
> 25 km north to C,
> 10 km southeast to D,
> 15 km west to E, and stops.

How far is he from his starting point, and in what direction?

1.8 A unit vector $\mathbf{a}_n$ makes angles α, β, and γ with the x-, y-, and z-axes, respectively. Express $\mathbf{a}_n$ in the rectangular coordinate system. Also express a nonunit vector $\overrightarrow{OP}$ of length ℓ parallel to $\mathbf{a}_n$.

1.9 Three vectors $\mathbf{p}$, $\mathbf{q}$, and $\mathbf{r}$ sum to a zero vector and have the magnitude of 10, 11, and 15, respectively. Determine the value of $\mathbf{p} \cdot \mathbf{q} + \mathbf{q} \cdot \mathbf{r} + \mathbf{r} \cdot \mathbf{p}$.

1.10 An experiment revealed that the point $Q(x', y', z')$ is 4 m from $P(2, 1, 4)$ and that the vector $\overrightarrow{QP}$ makes $45.5225°$, $59.4003°$, and $60°$ with the x-, y-, and z-axes, respectively. Determine the location of Q.

1.11 In a certain frame of reference with x-, y-, and z-axes, imagine the first octant to be a room with a door. Suppose that the height of the door is h and its width is ρ. The top-right corner P of the door when it is shut has the rectangular coordinates $(\rho, 0, h)$. Now if the door is turned by angle ϕ, so we can enter the room, what are the coordinates of P? What is the length of its diagonal $r = \overline{OP}$ in terms of ρ and z? Suppose the vector $\overrightarrow{OP}$ makes an angle θ with the z-axis; express ρ and h in terms of r and θ.

1.12 Consider two vectors $\mathbf{p} = \overrightarrow{OP}$ and $\mathbf{q} = \overrightarrow{OQ}$ in Figure MA-2. Express the vector $\overrightarrow{GR}$ in terms of $\mathbf{p}$ and $\mathbf{q}$. Assume that $\angle ORQ = 90°$.

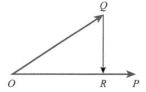

FIGURE MA-2 Orthogonal projection of one vector over another.

1.13 Consider the equations of two planes:

$$3x - 2y - z = 8$$
$$2x + y + 4z = 3$$

Let them intersect along the straight line ℓ. Obtain the coordinates of the points where ℓ meets the xy– and the yz–planes. Also determine the angle between ℓ and the xz-plane.

1.14 Given two vectors $\mathbf{p} = \mathbf{a}_x + \mathbf{a}_y$ and $\mathbf{q} = \mathbf{a}_y + \mathbf{a}_z$ of equal length, find a third vector $\mathbf{r}$ such that it has the same length and the angle between any two of them is $60°$.

1.15 Given $\mathbf{A} = 2xy\,\mathbf{a}_x + 3zy\,\mathbf{a}_y + 5z\,\mathbf{a}_z$ and $\mathbf{B} = \sin x\,\mathbf{a}_x + 2y\,\mathbf{a}_y + 5y\,\mathbf{a}_z$, find (i) $\nabla \cdot \mathbf{A}$, (ii) $\nabla \times \mathbf{A}$, (iii) $\nabla \cdot \nabla \times \mathbf{A}$, and (iv) $\nabla \cdot (\mathbf{A} \times \mathbf{B})$.

2.1 A triangular plate of base $b = 5$ and height $h = 4$ shown in Figure MA-3 is uniformly charged with a uniform surface charge density $\rho_s = 10$ C/m^2. You are to cut a rectangular piece so that maximum amount of charge is taken out. What should be the dimensions x and y of the rectangle? What is the magnitude of the charge extracted out?

2.2 Consider two fixed points $F_1(-c, 0)$ and $F_2(c, 0)$ in the xy-plane. Show that the locus of a point $P(x, y)$ that satisfies the constraint that the sum $PF_1 + PF_2$ remains constant and is equal to $2a$ is an ellipse. The equipotential loci due to a uniform line charge of length $2c$ are family of ellipses in the plane containing the charge. This problem helps in proving it.

2.3 Show that the ordinary angle subtended by a closed curve lying in a plane at a point P is 2π radians if P is enclosed by the curve and zero if not.

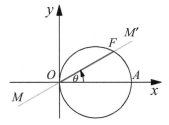

FIGURE MA-3 A rectangular piece cut out from a triangular plate.

2.4 Show that the solid angle subtended by a closed surface at a point P is 4π steradians if P is enclosed by the closed surface and zero if not.

2.5 The electrostatic potential $V(r)$ is known to obey the equation $V(r) = 2V(2r)$ with the boundary condition $V(5) = 3$ volts. Determine $V(15)$.

2.6 Evaluate the indefinite integrals (i) $\int \csc \theta \, d\theta$ and (ii) $\int \sec \theta \, d\theta$. Ignore the arbitrary constant.

2.7 A liquid drop is in the form of an ellipsoid $\dfrac{x^2}{a^2} + \dfrac{y^2}{b^2} + \dfrac{z^2}{c^2} = 1$ shown in Figure MA-4 and is filled with a charge of nonuniform density $\rho_v = x^2$ C/m^3. Find the total charge in the drop.

2.8 Two families of curves are said to be orthogonal to each other if they intersect at $90°$. Given a family $y^2 = cx^3$, find the equation for orthogonal trajectories and plot three to four members of each on the same graph.

2.9 Consider a vector given by $\mathbf{E} = (4xy + z)\mathbf{a}_x + 2x^2\mathbf{a}_y + x\mathbf{a}_z$. Find the line integral from $A(3, 7, 1)$ to $B(8, 9, 2)$ by   (i) evaluating the line integral $V_{AB} = -\displaystyle\int_A^B \mathbf{E} \cdot \mathbf{dl}$ along the line joining A to B and   (ii) evaluating $\left\{ -\displaystyle\int_A^C \mathbf{E} \cdot \mathbf{dl} - \int_C^D \mathbf{E} \cdot \mathbf{dl} - \int_D^B \mathbf{E} \cdot \mathbf{dl} \right\}$, where the stopovers C and D are $C(8, 7, 1)$ and $D(8, 9, 1)$.

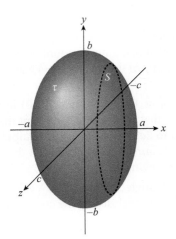

FIGURE MA-4 A non uniformly charged liquid drop.

2.10 Find the trigonometric Fourier series of a function $f(x) = x + x^2$ defined over the interval $-\pi < x < \pi$.

2.11 In a certain electrostatic system, there are found an infinite set of image point charges. The field intensity at a point may be written as

$$E = A \sum_{m=1}^{\infty} \sum_{n=m}^{\infty} \frac{(-1)^{(n-1)}}{n^2}$$

Simplify the double summation.

Hint: Integrating the following series term by term and substituting $x = 1$ helps in finding the result.

$$\frac{1}{1 + x} = 1 - x + x^2 - x^3 + \cdots + - \cdots$$

2.12 Solve the differential equation

$$\frac{d^2 V(x)}{dx^2} = \frac{k}{\sqrt{V(x)}}$$

subject to the boundary conditions $\left.\dfrac{dV}{dx}\right|_{x=0} = 0$ and $V(0) = 0$. Assume that k is a constant.

3.1 The location of a moving charge is given by the time-varying radius vector $\mathbf{r} = 2 \cos t\, \mathbf{a}_x + 2 \sin t\, \mathbf{a}_y + 3t\mathbf{a}_z$. Describe the trajectory of motion. Find the velocity and acceleration vectors at any instant t. In particular, indicate their directions at the specific instants $t = 0$ and $t = \pi/2$. Find their magnitudes at any instant.

3.2 The magnetic field strength $H(z)$ at a point on the z-axis shown in Figure MA-5 is proportional to the sum of cosine of angles and is given by $H = k(\cos \theta_1 + \cos \theta_2)$. Find $H(0)$. Also show that if $a \ll \ell$, $H\,(\pm\ell) \approx \dfrac{1}{2}H(0)$. This helps in finding the magnetic field along the axis of a long solenoid.

3.3 Suppose it is suggested that $\mathbf{B} = r(\mathbf{k} \times \mathbf{r})$ is the magnetic flux density vector, where $\mathbf{k}$ is a constant vector and $\mathbf{r} = r\mathbf{a}_r$ verify if it is solenoidal.

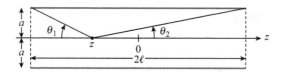

FIGURE MA-5 Toward finding magnetic field along the axis of a solenoid.

3.4 Evaluate the line integral $\oint_C \dfrac{(x+y)dx + (x-y)dy}{x^2 + y^2}$ where C is the circle $x^2 + y^2 = a^2$ of constant radius a.

3.5 Evaluate the line integral $\oint_C \dfrac{xdy - ydx}{x^2 + y^2}$ where C is a closed curve (i) encircling the origin n times, (ii) not enclosing the origin.

4.1 Show that $\nabla \cdot \nabla \times \mathbf{A} = 0$.

4.2 Show that $\nabla \times \nabla \psi = \mathbf{0}$.

4.3 Given that the imaginary unit is $j = \sqrt{-1}$ and that $x = j^j$, could the value of x be real? If so, is it unique? Can x have one value in the interval $(100, 120)$?

4.4 Show that $\nabla \cdot \mathbf{A} \times \mathbf{B} = \mathbf{B} \cdot \nabla \times \mathbf{A} - \mathbf{A} \cdot \nabla \times \mathbf{B}$.

4.5 Use *De Moivre's theorem* to prove that $\cos 3\theta = \cos^3 \theta - 3 \cos \theta \sin^2\theta$:

4.6 Determine $\sqrt{j}$.

4.7 Determine $\sqrt{j}$ using the Euler formula.

4.8 Find the phasors for the following field quantities:

(a) $E_x(z, t) = E_o \cos(\omega t - \beta z + \phi)$ (V/m)

(b) $E_y(z, t) = 100e^{-3z} \cos(\omega t - 5z + \pi/4)$ (V/m)

(c) $H_x(z, t) = H_o \cos(\omega t + \beta z)$ (A/m)

(d) $H_y(z, t) = 120\pi e^{-5z} \cos(\omega t + \beta z + \phi_h)$ (A/m)

4.9 Find the instantaneous time domain sinusoidal functions corresponding to the following phasors:

(a) $E_x(z) = E_o e^{j\beta z}$ (V/m)

(b) $E_y(z) = 100e^{-3z}e^{-j5z}$ (V/m)

(c) $I_s(z) = 5 + j4$ (A)

(d) $V_s(z) = j10e^{j\pi/3}$ (V)

4.10 Write the phasor expression $\tilde{I}$ for the following current using a cosine reference.

(a) $i(t) = I_o \cos(\omega t - \pi/6)$

(b) $i(t) = I_o \sin(\omega t + \pi/3)$

4.11 In a certain resonant cavity, the resonant modes are described by a triplet of nonnegative integers m, n, and p. Find possible solutions under the inequality constraints,

$$mn + np + pm \neq 0$$

$$\frac{13}{16} \leq \frac{m^2}{4} + \frac{n^2}{9} + p^2 \leq \frac{5}{4}$$

4.12 A voltage source $V(t) = 100 \cos(6\pi 10^9 t - 45°)$ (V) is connected to a series RLC circuit, as shown in Figure MA-6. Given $R = 10\,M\Omega$, $C = 100\,pF$, and $L = 1\,H$, use phasor notation to find the following:

(a) $i(t)$

(b) $V_c(t)$, the voltage across the capacitor

4.13 (i) Show that the locus of the points $P(x, y)$ obeying the equation

$$x^2 + y^2 + 2gx + 2fy + c = 0$$

represents a circle. (ii) Express the coordinates of the center and the radius. Use the following equations of circles to find the centers and radii.

$$x^2 + y^2 + 8x - 4y + 11 = 0$$
$$x^2 + y^2 - 10x - 6y + 9 = 0$$
$$225x^2 + 225y^2 + 90x - 300y + 28 = 0$$

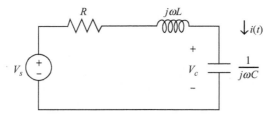

FIGURE MA-6 A series RLC circuit for Problem 4.12.

4.14 Recall the vector identity $\nabla \times \psi \mathbf{A} \equiv \psi \nabla \times \mathbf{A} + \nabla \psi \times \mathbf{A}$, where ψ is a scalar function and $\mathbf{A}$ is a vector point function. Suppose $\mathbf{A} = A_z \mathbf{a}_z$, where $A_z = \dfrac{e^{-jkr}}{r}$ and k is a constant. Simplify $\nabla \times \mathbf{A}$.

4.15 Between two points A and B on the brink of a circular water pond, a transmission line has to be run. It costs twice the money per meter length to install the cable through the water compared to installation on the edge. One might take the cable (a) completely around the arc on the surrounding land or (b) straight through in the water or (c) partly on the arc and for the remaining, straight in the water. (i) What path costs the maximum money? (ii) Suggest an arrangement that minimizes the cost. With some numerical values, plot the cost function.

4.16 Show the following series expansion assuming $|x| < 1$:

$$\frac{1}{(1 - x)^2} = 1 + 2x + 3x^2 + 4x^3 + \cdots$$

VECTOR ANALYSIS

CODES OF ETHICS

Engineering is a profession that makes significant contributions to the economic and social well-being of people all over the world. As members of this important profession, engineers are expected to exhibit the highest standards of honesty and integrity. Unfortunately, the engineering curriculum is so crowded that there is no room for a course on ethics in most schools. Although there are over 850 codes of ethics for different professions all over the world, the code of ethics of the Institute of Electrical and Electronics Engineers (IEEE) is presented here to give students a flavor of the importance of ethics in engineering professions.

We, the members of the IEEE, in recognition of the importance of our technologies in affecting the quality of life throughout the world, and in accepting a personal obligation to our profession, its members and the communities we serve, do hereby commit ourselves to the highest ethical and professional conduct and agree:

1. to accept responsibility in making engineering decisions consistent with the safety, health, and welfare of the public, and to disclose promptly factors that might endanger the public or the environment;
2. to avoid real or perceived conflicts of interest whenever possible, and to disclose them to affected parties when they do exist;
3. to be honest and realistic in stating claims or estimates based on available data;
4. to reject bribery in all its forms;
5. to improve the understanding of technology, its appropriate application, and potential consequences;
6. to maintain and improve our technical competence and to undertake technological tasks for others only if qualified by training or experience, or after full disclosure of pertinent limitations;
7. to seek, accept, and offer honest criticism of technical work, to acknowledge and correct errors, and to credit properly the contributions of others;
8. to treat fairly all persons regardless of such factors as race, religion, gender, disability, age, or national origin;
9. to avoid injuring others, their property, reputation, or employment by false or malicious action;
10. to assist colleagues and co-workers in their professional development and to support them in following this code of ethics.

—Courtesy of IEEE

CHAPTER 1

VECTOR ALGEBRA

Books are the quietest and most constant friends; they are the most accessible and wisest of counselors, and most patient of teachers.

—CHARLES W. ELLIOT

1.1 INTRODUCTION

Electromagnetics (EM) may be regarded as the study of the interactions between electric charges at rest and in motion. It entails the analysis, synthesis, physical interpretation, and application of electric and magnetic fields.

> **Electromagnetics (EM)** is a branch of physics or electrical engineering in which electric and magnetic phenomena are studied.

EM principles find applications in various allied disciplines such as microwaves, antennas, electric machines, satellite communications, bioelectromagnetics, plasmas, nuclear research, fiber optics, electromagnetic interference and compatibility, electromechanical energy conversion, radar meteorology, and remote sensing.[1,2] In physical medicine, for example, EM power, in the form either of shortwaves or microwaves, is used to heat deep tissues and to stimulate certain physiological responses in order to relieve certain pathological conditions. EM fields are used in induction heaters for melting, forging, annealing, surface hardening, and soldering operations. Dielectric heating equipment uses shortwaves to join or seal thin sheets of plastic materials. EM energy offers many new and exciting possibilities in agriculture. It is used, for example, to change vegetable taste by reducing acidity.

EM devices include transformers, electric relays, radio/TV, telephones, electric motors, transmission lines, waveguides, antennas, optical fibers, radars, and lasers. The design of these devices requires thorough knowledge of the laws and principles of EM.

---

[1]For numerous applications of electrostatics, see J. M. Crowley, *Fundamentals of Applied Electrostatics*. New York: John Wiley & Sons, 1986.

[2]For other areas of applications of EM, see, for example, D. Teplitz, ed., *Electromagnetism: Paths to Research*. New York: Plenum Press, 1982.

†1.2 A PREVIEW OF THE BOOK

The subject of electromagnetic phenomena in this book can be summarized in Maxwell's equations:

$$\nabla \cdot \mathbf{D} = \rho_v \qquad (1.1)$$

$$\nabla \cdot \mathbf{B} = 0 \qquad (1.2)$$

$$\nabla \times \mathbf{E} = -\frac{\partial \mathbf{B}}{\partial t} \qquad (1.3)$$

$$\nabla \times \mathbf{H} = \mathbf{J} + \frac{\partial \mathbf{D}}{\partial t} \qquad (1.4)$$

where ∇ = the vector differential operator
$\mathbf{D}$ = the electric flux density
$\mathbf{B}$ = the magnetic flux density
$\mathbf{E}$ = the electric field intensity
$\mathbf{H}$ = the magnetic field intensity
ρ_v = the volume charge density
$\mathbf{J}$ = the current density

Maxwell based these equations on previously known results, both experimental and theoretical. A quick look at these equations shows that we shall be dealing with vector quantities. It is consequently logical that we spend some time in Part 1 examining the mathematical tools required for this course. The derivation of eqs. (1.1) to (1.4) for time-invariant conditions and the physical significance of the quantities $\mathbf{D}$, $\mathbf{B}$, $\mathbf{E}$, $\mathbf{H}$, $\mathbf{J}$, and ρ_v will be our aim in Parts 2 and 3. In Part 4, we shall reexamine the equations for time-varying situations and apply them in our study of practical EM devices such as transmission lines, waveguides, antennas, fiber optics, and radar systems.

1.3 SCALARS AND VECTORS

Vector analysis is a mathematical tool with which electromagnetic concepts are most conveniently expressed and best comprehended. We must learn its rules and techniques before we can confidently apply it. Since most students taking this course have little exposure to vector analysis, considerable attention is given to it in this and the next two chapters.[3] This chapter introduces the basic concepts of vector algebra in Cartesian coordinates only. The next chapter builds on this and extends to other coordinate systems.

A quantity can be either a scalar or a vector. A scalar is a quantity that is completely specified by its magnitude.

---

†Indicates sections that may be skipped, explained briefly, or assigned as homework if the text is covered in one semester.
[3]The reader who feels no need for review of vector algebra can skip to the next chapter.

A **scalar** is a quantity that has only magnitude.

Quantities such as time, mass, distance, temperature, entropy, electric potential, and population are scalars. A vector has not only magnitude, but direction in space.

A **vector** is a quantity that is described by both magnitude and direction.

Vector quantities include velocity, force, momentum, acceleration displacement, and electric field intensity. Another class of physical quantities is called *tensors,* of which scalars and vectors are special cases. For most of the time, we shall be concerned with scalars and vectors.[4]
To distinguish between a scalar and a vector it is customary to represent a vector by a letter with an arrow on top of it, such as $\vec{A}$ and $\vec{B}$, or by a letter in boldface type such as **A** and **B**. A scalar is represented simply by a letter—for example, A, B, U, and V.
EM theory is essentially a study of some particular fields.

A **field** is a function that specifies a particular quantity everywhere in a region.

A field may indicate variation of a quantity throughout space and perhaps with time. If the quantity is scalar (or vector), the field is said to be a scalar (or vector) field. Examples of scalar fields are temperature distribution in a building, sound intensity in a theater, electric potential in a region, and refractive index of a stratified medium. The gravitational force on a body in space and the velocity of raindrops in the atmosphere are examples of vector fields.

1.4 UNIT VECTOR

A vector **A** has both magnitude and direction. The *magnitude* of **A** is a scalar written as A or $|\mathbf{A}|$. A *unit vector* $\mathbf{a}_A$ along **A** is defined as a vector whose magnitude is unity (i.e., 1) and its direction is along **A**; that is,

$$\mathbf{a}_A = \frac{\mathbf{A}}{|\mathbf{A}|} = \frac{\mathbf{A}}{A} \tag{1.5}$$

Note that $|\mathbf{a}_A| = 1$. Thus we may write **A** as

$$\mathbf{A} = A\mathbf{a}_A \tag{1.6}$$

which completely specifies **A** in terms of its magnitude A and its direction $\mathbf{a}_A$.
A vector **A** in Cartesian (or rectangular) coordinates may be represented as

$$(A_x, A_y, A_z) \quad \text{or} \quad A_x\mathbf{a}_x + A_y\mathbf{a}_y + A_z\mathbf{a}_z \tag{1.7}$$

---

[4]For an elementary treatment of tensors, see, for example, A. I. Borisenko and I. E. Tarapor, *Vector and Tensor Analysis with Applications.* New York: Dover, 1979.

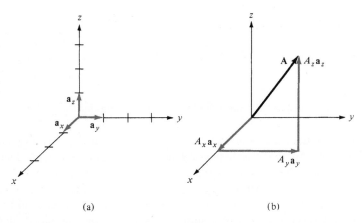

FIGURE 1.1 (**a**) Unit vectors $\mathbf{a}_x$, $\mathbf{a}_y$, and $\mathbf{a}_z$, (**b**) components of **A** along $\mathbf{a}_x$, $\mathbf{a}_y$, and $\mathbf{a}_z$.

where A_x, A_y, and A_z are called the *components of* **A** in the *x*-, *y*-, and *z*-directions, respectively; $\mathbf{a}_x$, $\mathbf{a}_y$, and $\mathbf{a}_z$ are unit vectors in the *x*-, *y*-, and *z*-directions, respectively. For example, $\mathbf{a}_x$ is a dimensionless vector of magnitude one in the direction of the increase of the *x*-axis. The unit vectors $\mathbf{a}_x$, $\mathbf{a}_y$, and $\mathbf{a}_z$ are illustrated in Figure 1.1(a), and the components of **A** along the coordinate axes are shown in Figure 1.1(b). The magnitude of vector **A** is given by

$$A = \sqrt{A_x^2 + A_y^2 + A_z^2} \qquad (1.8)$$

and the unit vector along **A** is given by

$$\mathbf{a}_A = \frac{A_x\mathbf{a}_x + A_y\mathbf{a}_y + A_z\mathbf{a}_z}{\sqrt{A_x^2 + A_y^2 + A_z^2}} \qquad (1.9)$$

1.5 VECTOR ADDITION AND SUBTRACTION

Two vectors **A** and **B** can be added together to give another vector **C**; that is,

$$\mathbf{C} = \mathbf{A} + \mathbf{B} \qquad (1.10)$$

The vector addition is carried out component by component. Thus, if $\mathbf{A} = (A_x, A_y, A_z)$ and $\mathbf{B} = (B_x, B_y, B_z)$.

$$\mathbf{C} = (A_x + B_x)\mathbf{a}_x + (A_y + B_y)\mathbf{a}_y + (A_z + B_z)\mathbf{a}_z \qquad (1.11)$$

Vector subtraction is similarly carried out as

$$\begin{aligned} \mathbf{D} = \mathbf{A} - \mathbf{B} &= \mathbf{A} + (-\mathbf{B}) \\ &= (A_x - B_x)\mathbf{a}_x + (A_y - B_y)\mathbf{a}_y + (A_z - B_z)\mathbf{a}_z \end{aligned} \qquad (1.12)$$

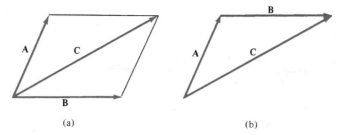

FIGURE 1.2 Vector addition **C** = **A** + **B**: (**a**) parallelogram rule, (**b**) head-to-tail rule.

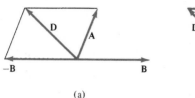

FIGURE 1.3 Vector subtraction **D** = **A** − **B**: (**a**) parallelogram rule, (**b**) head-to-tail rule.

(a)   (b)

Graphically, vector addition and subtraction are obtained by either the parallelogram rule or the head-to-tail rule as portrayed in Figures 1.2 and 1.3, respectively.

The three basic laws of algebra obeyed by any given vectors **A**, **B**, and **C** are summarized as follows:

Law	Addition	Multiplication
Commutative	$\mathbf{A} + \mathbf{B} = \mathbf{B} + \mathbf{A}$	$k\mathbf{A} = \mathbf{A}k$
Associative	$\mathbf{A} + (\mathbf{B} + \mathbf{C}) = (\mathbf{A} + \mathbf{B}) + \mathbf{C}$	$k(\ell\mathbf{A}) = (k\ell)\mathbf{A}$
Distributive	$k(\mathbf{A} + \mathbf{B}) = k\mathbf{A} + k\mathbf{B}$	

where k and ℓ are scalars. Multiplication of a vector with another vector will be discussed in Section 1.7.

1.6 POSITION AND DISTANCE VECTORS

A point P in Cartesian coordinates may be represented by (x, y, z).

The **position vector** $\mathbf{r}_P$ (or **radius vector**) of point P is defined as the directed distance from the origin O to P; that is,

$$\mathbf{r}_P = OP = x\mathbf{a}_x + y\mathbf{a}_y + z\mathbf{a}_z \tag{1.13}$$

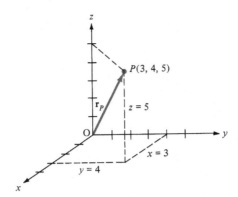

FIGURE 1.4 Illustration of position vector $\mathbf{r}_P = 3\mathbf{a}_x + 4\mathbf{a}_y = 5\mathbf{a}_z$.

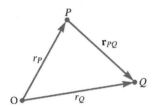

FIGURE 1.5 Distance vector $\mathbf{r}_{PQ}$.

The position vector of point P is useful in defining its position in space. Point $(3, 4, 5)$, for example, and its position vector $3\mathbf{a}_x + 4\mathbf{a}_y + 5\mathbf{a}_z$ are shown in Figure 1.4.

> The **distance vector** is the displacement from one point to another.

If two points P and Q are given by (x_P, y_P, z_P) and (x_Q, y_Q, z_Q), the *distance vector* (or *separation vector*) is the displacement from P to Q as shown in Figure 1.5; that is,

$$
\begin{aligned}
\mathbf{r}_{PQ} &= \mathbf{r}_Q - \mathbf{r}_P \\
&= (x_Q - x_P)\mathbf{a}_x + (y_Q - y_P)\mathbf{a}_y + (z_Q - z_P)\mathbf{a}_z
\end{aligned}
\tag{1.14}
$$

The difference between a point P and a vector $\mathbf{A}$ should be noted. Though both P and $\mathbf{A}$ may be represented in the same manner as (x, y, z) and (A_x, A_y, A_z), respectively, the point P is not a vector; only its position vector $\mathbf{r}_P$ is a vector. Vector $\mathbf{A}$ may depend on point P, however. For example, if $\mathbf{A} = 2xy\mathbf{a}_x + y^2\mathbf{a}_y - xz^2\mathbf{a}_z$ and P is $(2, -1, 4)$, then $\mathbf{A}$ at P would be $-4\mathbf{a}_x + \mathbf{a}_y - 32\mathbf{a}_z$. A vector field is said to be *constant* or *uniform* if it does not depend on space variables x, y, and z. For example, vector $\mathbf{B} = 3\mathbf{a}_x - 2\mathbf{a}_y + 10\mathbf{a}_z$ is a uniform vector while vector $\mathbf{A} = 2xy\mathbf{a}_x + y^2\mathbf{a}_y - xz^2\mathbf{a}_z$ is not uniform because $\mathbf{B}$ is the same everywhere, whereas $\mathbf{A}$ varies from point to point.

EXAMPLE 1.1

If $\mathbf{A} = 10\mathbf{a}_x - 4\mathbf{a}_y + 6\mathbf{a}_z$ and $\mathbf{B} = 2\mathbf{a}_x + \mathbf{a}_y$, find (a) the component of $\mathbf{A}$ along $\mathbf{a}_y$, (b) the magnitude of $3\mathbf{A} - \mathbf{B}$, (c) a unit vector along $\mathbf{A} + 2\mathbf{B}$.

Solution:

(a) The component of **A** along $\mathbf{a}_y$ is $A_y = -4$.

(b) $3\mathbf{A} - \mathbf{B} = 3(10, -4, 6) - (2, 1, 0)$

$$= (30, -12, 18) - (2, 1, 0)$$

$$= (28, -13, 18)$$

Hence,

$$|3\mathbf{A} - \mathbf{B}| = \sqrt{28^2 + (-13)^2 + (18)^2} = \sqrt{1277}$$

$$= 35.74$$

(c) Let $\mathbf{C} = \mathbf{A} + 2\mathbf{B} = (10, -4, 6) + (4, 2, 0) = (14, -2, 6)$.

A unit vector along **C** is

$$\mathbf{a}_c = \frac{\mathbf{C}}{|\mathbf{C}|} = \frac{(14, -2, 6)}{\sqrt{14^2 + (-2)^2 + 6^2}}$$

or

$$\mathbf{a}_c = 0.9113\mathbf{a}_x - 0.1302\mathbf{a}_y + 0.3906\mathbf{a}_z$$

Note that $|\mathbf{a}_c| = 1$ as expected.

PRACTICE EXERCISE 1.1

Given vectors $\mathbf{A} = \mathbf{a}_x + 3\mathbf{a}_z$ and $\mathbf{B} = 5\mathbf{a}_x + 2\mathbf{a}_y - 6\mathbf{a}_z$, determine

(a) $|\mathbf{A} + \mathbf{B}|$

(b) $5\mathbf{A} - \mathbf{B}$

(c) The component of **A** along $\mathbf{a}_y$

(d) A unit vector parallel to $3\mathbf{A} + \mathbf{B}$

Answer: (a) 7, (b) $(0, -2, 21)$, (c) 0, (d) $\pm(0.9117, 0.2279, 0.3419)$.

EXAMPLE 1.2

Points P and Q are located at $(0, 2, 4)$ and $(-3, 1, 5)$. Calculate

(a) The position of vector $\mathbf{r}_P$

(b) The distance vector from P to Q

(c) The distance between P and Q

(d) A vector parallel to PQ with magnitude of 10

Solution:

(a) $\mathbf{r}_P = 0\mathbf{a}_x + 2\mathbf{a}_y + 4\mathbf{a}_z = 2\mathbf{a}_y + 4\mathbf{a}_z$

(b) $\mathbf{r}_{PQ} = \mathbf{r}_Q - \mathbf{r}_P = (-3, 1, 5) - (0, 2, 4) = (-3, -1, 1)$

or $\mathbf{r}_{PQ} = -3\mathbf{a}_x - \mathbf{a}_y + \mathbf{a}_z$

(c) Since $\mathbf{r}_{PQ}$ is the distance vector from P to Q, the distance between P and Q is the magnitude of this vector; that is,

$$d = |\mathbf{r}_{PQ}| = \sqrt{9 + 1 + 1} = 3.317$$

Alternatively:

$$d = \sqrt{(x_Q - x_P)^2 + (y_Q - y_P)^2 + (z_Q - z_P)^2}$$
$$= \sqrt{9 + 1 + 1} = 3.317$$

(d) Let the required vector be $\mathbf{A}$, then

$$\mathbf{A} = A\mathbf{a}_A$$

where $A = 10$ is the magnitude of $\mathbf{A}$. Since $\mathbf{A}$ is parallel to PQ, it must have the same unit vector as $\mathbf{r}_{PQ}$ or $\mathbf{r}_{QP}$. Hence,

$$\mathbf{a}_A = \pm \frac{\mathbf{r}_{PQ}}{|\mathbf{r}_{PQ}|} = \pm \frac{(-3, -1, 1)}{3.317}$$

and

$$\mathbf{A} = \pm \frac{10(-3, -1, 1)}{3.317} = \pm(-9.045\mathbf{a}_x - 3.015\mathbf{a}_y + 3.015\mathbf{a}_z)$$

PRACTICE EXERCISE 1.2

Given points $P(1, -3, 5)$, $Q(2, 4, 6)$, and $R(0, 3, 8)$, find (a) the position vectors of P and R, (b) the distance vector $\mathbf{r}_{QR}$, (c) the distance between Q and R.

Answer: (a) $\mathbf{a}_x - 3\mathbf{a}_y + 5\mathbf{a}_z$, $3\mathbf{a}_x + 8\mathbf{a}_z$, (b) $-2\mathbf{a}_x - \mathbf{a}_y + 2\mathbf{a}_z$, (c) 3.

EXAMPLE 1.3

A river flows southeast at 10 km/hr and a boat floats upon it with its bow pointed in the direction of travel. A man walks upon the deck at 2 km/hr in a direction to the right and perpendicular to the direction of the boat's movement. Find the velocity of the man with respect to the earth.

Solution:

Consider Figure 1.6 as illustrating the problem. The velocity of the boat is

$$\mathbf{u}_b = 10(\cos 45° \, \mathbf{a}_x - \sin 45° \, \mathbf{a}_y)$$
$$= 7.071\mathbf{a}_x - 7.071\mathbf{a}_y \text{ km/hr}$$

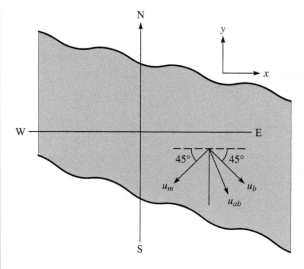

FIGURE 1.6 For Example 1.3.

The velocity of the man with respect to the boat (relative velocity) is

$$\mathbf{u}_m = 2(-\cos 45° \,\mathbf{a}_x - \sin 45° \,\mathbf{a}_y)$$
$$= -1.414\mathbf{a}_x - 1.414\mathbf{a}_y \text{ km/hr}$$

Thus the absolute velocity of the man is

$$\mathbf{u}_{ab} = \mathbf{u}_m + \mathbf{u}_b = 5.657\mathbf{a}_x - 8.485\mathbf{a}_y$$
$$|\mathbf{u}_{ab}| = 10.2\underline{/-56.3°}$$

that is, 10.2 km/hr at 56.3° south of east.

PRACTICE EXERCISE 1.3

An airplane has a ground speed of 350 km/hr in the direction due west. If there is a wind blowing northwest at 40 km/hr, calculate the true air speed and heading of the airplane.

Answer: 379.3 km/hr, 4.275° north of west.

1.7 VECTOR MULTIPLICATION

When two vectors **A** and **B** are multiplied, the result is either a scalar or a vector depending on how they are multiplied. Thus there are two types of vector multiplication:

1. Scalar (or dot) product: $\mathbf{A} \cdot \mathbf{B}$
2. Vector (or cross) product: $\mathbf{A} \times \mathbf{B}$

Multiplication of three vectors **A**, **B**, and **C** can result in either:

3. Scalar triple product: $\mathbf{A} \cdot (\mathbf{B} \times \mathbf{C})$

or

4. Vector triple product: $\mathbf{A} \times (\mathbf{B} \times \mathbf{C})$

A. Dot Product

The **dot product** of two vectors **A** and **B**, written as **A** · **B**, is defined geometrically as the product of the magnitudes of **A** and **B** and the cosine of the smaller angle between them when they are drawn tail to tail.

Thus,

$$\mathbf{A} \cdot \mathbf{B} = AB \cos \theta_{AB} \qquad (1.15)$$

where θ_{AB} is the *smaller* angle between **A** and **B**. The result of **A** · **B** is called either the *scalar product* because it is scalar or the *dot product* due to the dot sign. If $\mathbf{A} = (A_x, A_y, A_z)$ and $\mathbf{B} = (B_x, B_y, B_z)$, then

$$\mathbf{A} \cdot \mathbf{B} = A_x B_x + A_y B_y + A_z B_z \qquad (1.16)$$

which is obtained by multiplying **A** and **B** component by component. Two vectors **A** and **B** are said to be *orthogonal* (or perpendicular) with each other if **A** · **B** = 0.

Note that dot product obeys the following:

(i) *Commutative law:*

$$\mathbf{A} \cdot \mathbf{B} = \mathbf{B} \cdot \mathbf{A} \qquad (1.17)$$

(ii) *Distributive law:*

$$\mathbf{A} \cdot (\mathbf{B} + \mathbf{C}) = \mathbf{A} \cdot \mathbf{B} + \mathbf{A} \cdot \mathbf{C} \qquad (1.18)$$

(iii)

$$\mathbf{A} \cdot \mathbf{A} = |\mathbf{A}|^2 = A^2 \qquad (1.19)$$

Also note that

$$\mathbf{a}_x \cdot \mathbf{a}_y = \mathbf{a}_y \cdot \mathbf{a}_z = \mathbf{a}_z \cdot \mathbf{a}_x = 0 \qquad (1.20a)$$

$$\mathbf{a}_x \cdot \mathbf{a}_x = \mathbf{a}_y \cdot \mathbf{a}_y = \mathbf{a}_z \cdot \mathbf{a}_z = 1 \qquad (1.20b)$$

It is easy to prove the identities in eqs. (1.17) to (1.20) by applying eq. (1.15) or (1.16).

If **A** · **B** = 0, the two vectors **A** and **B** are orthogonal or perpendicular.

B. Cross Product

> The **cross product** of two vectors **A** and **B**, written as **A** $\times$ **B**, is a vector quantity whose magnitude is the area of the parallelogram formed by **A** and **B** (see Figure 1.7) and is in the direction of advance of a right-handed screw as A is turned into B.

Thus,

$$\mathbf{A} \times \mathbf{B} = AB \sin \theta_{AB}\mathbf{a}_n \tag{1.21}$$

where $\mathbf{a}_n$ is a unit vector normal to the plane containing **A** and **B**. The direction of $\mathbf{a}_n$ is taken as the direction of the right thumb when the fingers of the right hand rotate from **A** to **B** as shown in Figure 1.8(a). Alternatively, the direction of $\mathbf{a}_n$ is taken as that of the advance of a right-handed screw as **A** is turned into **B** as shown in Figure 1.8(b).

The vector multiplication of eq. (1.21) is called *cross product* owing to the cross sign; it is also called *vector product* because the result is a vector. If $\mathbf{A} = (A_x, A_y, A_z)$ and $\mathbf{B} = (B_x, B_y, B_z)$, then

$$\mathbf{A} \times \mathbf{B} = \begin{vmatrix} \mathbf{a}_x & \mathbf{a}_y & \mathbf{a}_z \\ A_x & A_y & A_z \\ B_x & B_y & B_z \end{vmatrix} \tag{1.22a}$$

$$= (A_yB_z - A_zB_y)\mathbf{a}_x + (A_zB_x - A_xB_z)\mathbf{a}_y + (A_xB_y - A_yB_x)\mathbf{a}_z \tag{1.22b}$$

which is obtained by "crossing" terms in cyclic permutation, hence the name "cross product."

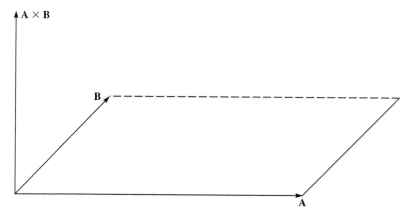

FIGURE 1.7 The cross product of **A** and **B** is a vector with magnitude equal to the area of the parallelogram and direction as indicated.

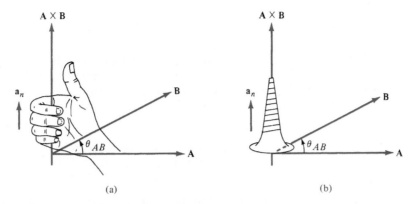

FIGURE 1.8 Direction of **A** × **B** and **a**$_n$ using (a) the right-hand rule and (b) the right-handed-screw rule.

Note that the cross product has the following basic properties:

(i) It is not commutative:

$$\mathbf{A} \times \mathbf{B} \neq \mathbf{B} \times \mathbf{A} \qquad (1.23a)$$

It is anticommutative:

$$\mathbf{A} \times \mathbf{B} = -\mathbf{B} \times \mathbf{A} \qquad (1.23b)$$

(ii) It is not associative:

$$\mathbf{A} \times (\mathbf{B} \times \mathbf{C}) \neq (\mathbf{A} \times \mathbf{B}) \times \mathbf{C} \qquad (1.24)$$

(iii) It is distributive:

$$\mathbf{A} \times (\mathbf{B} + \mathbf{C}) = \mathbf{A} \times \mathbf{B} + \mathbf{A} \times \mathbf{C} \qquad (1.25)$$

(iv) Scaling:

$$k\mathbf{A} \times \mathbf{B} = \mathbf{A} \times k\mathbf{B} = k(\mathbf{A} \times \mathbf{B}) \qquad (1.26)$$

(v)

$$\mathbf{A} \times \mathbf{A} = 0 \qquad (1.27)$$

Also note that

$$\begin{aligned}
\mathbf{a}_x \times \mathbf{a}_y &= \mathbf{a}_z \\
\mathbf{a}_y \times \mathbf{a}_z &= \mathbf{a}_x \\
\mathbf{a}_z \times \mathbf{a}_x &= \mathbf{a}_y
\end{aligned} \qquad (1.28)$$

which are obtained in cyclic permutation and illustrated in Figure 1.9. The identities in eqs. (1.23) to (1.28) are easily verified by using eq. (1.21) or (1.22). It should be noted that in obtaining **a**$_n$, we have used the right-hand or right-handed-screw rule because we want to

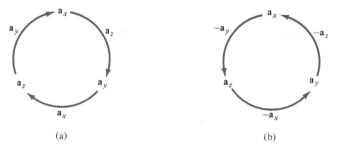

FIGURE 1.9 Cross product using cyclic permutation. (**a**) Moving clockwise leads to positive results. (**b**) Moving counterclockwise leads to negative results.

be consistent with our coordinate system illustrated in Figure 1.1, which is right-handed. A right-handed coordinate system is one in which the right-hand rule is satisfied: that is, $\mathbf{a}_x \times \mathbf{a}_y = \mathbf{a}_z$ is obeyed. In a left-handed system, we follow the left-hand or left-handed screw rule and $\mathbf{a}_x \times \mathbf{a}_y = -\mathbf{a}_z$ is satisfied. Throughout this book, we shall stick to right-handed coordinate systems.

Just as multiplication of two vectors gives a scalar or vector result, multiplication of three vectors **A**, **B**, and **C** gives a scalar or vector result, depending on how the vectors are multiplied. Thus we have a scalar or vector triple product.

C. Scalar Triple Product

Given three vectors **A**, **B**, and **C**, we define the scalar triple product as

$$\boxed{\mathbf{A} \cdot (\mathbf{B} \times \mathbf{C}) = \mathbf{B} \cdot (\mathbf{C} \times \mathbf{A}) = \mathbf{C} \cdot (\mathbf{A} \times \mathbf{B})} \tag{1.29}$$

obtained in cyclic permutation. If $\mathbf{A} = (A_x, A_y, A_z)$, $\mathbf{B} = (B_x, B_y, B_z)$, and $\mathbf{C} = (C_x, C_y, C_z)$, then $\mathbf{A} \cdot (\mathbf{B} \times \mathbf{C})$ is the volume of a parallelepiped having **A**, **B**, and **C** as edges and is easily obtained by finding the determinant of the 3×3 matrix formed by **A**, **B**, and **C**; that is,

$$\mathbf{A} \cdot (\mathbf{B} \times \mathbf{C}) = \begin{vmatrix} A_x & A_y & A_z \\ B_x & B_y & B_z \\ C_x & C_y & C_z \end{vmatrix} \tag{1.30}$$

Since the result of this vector multiplication is scalar, eq. (1.29) or (1.30) is called the *scalar triple product*.

D. Vector Triple Product

For vectors **A**, **B**, and **C**, we define the vector triple product as

$$\boxed{\mathbf{A} \times (\mathbf{B} \times \mathbf{C}) = \mathbf{B}(\mathbf{A} \cdot \mathbf{C}) - \mathbf{C}(\mathbf{A} \cdot \mathbf{B})} \tag{1.31}$$

which may be remembered as the "bac-cab" rule. It should be noted that

$$(\mathbf{A} \cdot \mathbf{B})\mathbf{C} \neq \mathbf{A}(\mathbf{B} \cdot \mathbf{C}) \tag{1.32}$$

but

$$(\mathbf{A} \cdot \mathbf{B})\mathbf{C} = \mathbf{C}(\mathbf{A} \cdot \mathbf{B}) \tag{1.33}$$

1.8 COMPONENTS OF A VECTOR

A direct application of scalar product is its use in determining the projection (or component) of a vector in a given direction. The projection can be scalar or vector. Given a vector **A**, we define the *scalar component* A_B of **A** along vector **B** as [see Figure 1.10(a)]

$$A_B = A \cos \theta_{AB} = |\mathbf{A}||\mathbf{a}_B| \cos \theta_{AB}$$

or

$$\boxed{A_B = \mathbf{A} \cdot \mathbf{a}_B} \tag{1.34}$$

The *vector component* $\mathbf{A}_B$ of **A** along **B** is simply the scalar component in eq. (1.34) multiplied by a unit vector along **B**; that is,

$$\boxed{\mathbf{A}_B = A_B\mathbf{a}_B = (\mathbf{A} \cdot \mathbf{a}_B)\mathbf{a}_B} \tag{1.35}$$

Both the scalar and vector components of **A** are illustrated in Figure 1.10. Notice from Figure 1.10(b) that the vector can be resolved into two orthogonal components: one component $\mathbf{A}_B$ parallel to **B**, another $(\mathbf{A} - \mathbf{A}_B)$ perpendicular to **B**. In fact, our Cartesian representation of a vector is essentially resolving the vector into three mutually orthogonal components as in Figure 1.1(b).

We have considered addition, subtraction, and multiplication of vectors. However, division of vectors **A/B** has not been considered because it is undefined except when **A** and **B** are parallel so that $\mathbf{A} = k\mathbf{B}$, where k is a constant. Differentiation and integration of vectors will be considered in Chapter 3.

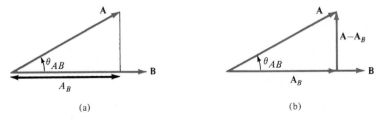

(a)　　　　　　　　　　　　　(b)

FIGURE 1.10 Components of **A** along **B**: (**a**) scalar component A_B, (**b**) vector component A_B.

EXAMPLE 1.4

Given vectors $\mathbf{A} = 3\mathbf{a}_x + 4\mathbf{a}_y + \mathbf{a}_z$ and $\mathbf{B} = 2\mathbf{a}_y - 5\mathbf{a}_z$, find the angle between $\mathbf{A}$ and $\mathbf{B}$.

Solution:

The angle θ_{AB} can be found by using either dot product or cross product.

$$\mathbf{A} \cdot \mathbf{B} = (3, 4, 1) \cdot (0, 2, -5)$$
$$= 0 + 8 - 5 = 3$$
$$|\mathbf{A}| = \sqrt{3^2 + 4^2 + 1^2} = \sqrt{26}$$
$$|\mathbf{B}| = \sqrt{0^2 + 2^2 + (-5)^2} = \sqrt{29}$$
$$\cos\theta_{AB} = \frac{\mathbf{A} \cdot \mathbf{B}}{|\mathbf{A}||\mathbf{B}|} = \frac{3}{\sqrt{(26)(29)}} = 0.1092$$
$$\theta_{AB} = \cos^{-1} 0.1092 = 83.73°$$

Alternatively:

$$\mathbf{A} \times \mathbf{B} = \begin{vmatrix} \mathbf{a}_x & \mathbf{a}_y & \mathbf{a}_z \\ 3 & 4 & 1 \\ 0 & 2 & -5 \end{vmatrix}$$
$$= (-20 - 2)\mathbf{a}_x + (0 + 15)\mathbf{a}_y + (6 - 0)\mathbf{a}_z$$
$$= (-22, 15, 6)$$
$$|\mathbf{A} \times \mathbf{B}| = \sqrt{(-22)^2 + 15^2 + 6^2} = \sqrt{745}$$
$$\sin\theta_{AB} = \frac{|\mathbf{A} \times \mathbf{B}|}{|\mathbf{A}||\mathbf{B}|} = \frac{\sqrt{745}}{\sqrt{(26)(29)}} = 0.994$$
$$\theta_{AB} = \sin^{-1} 0.994 = 83.73°$$

PRACTICE EXERCISE 1.4

If $\mathbf{A} = \mathbf{a}_x + 3\mathbf{a}_z$ and $\mathbf{B} = 5\mathbf{a}_x + 2\mathbf{a}_y - 6\mathbf{a}_z$, find θ_{AB}.

Answer: 120.6°.

EXAMPLE 1.5

Three field quantities are given by

$$\mathbf{P} = 2\mathbf{a}_x - \mathbf{a}_z$$
$$\mathbf{Q} = 2\mathbf{a}_x - \mathbf{a}_y + 2\mathbf{a}_z$$
$$\mathbf{R} = 2\mathbf{a}_x - 3\mathbf{a}_y + \mathbf{a}_z$$

Determine

(a) $(\mathbf{P} + \mathbf{Q}) \times (\mathbf{P} - \mathbf{Q})$

(b) $\mathbf{Q} \cdot \mathbf{R} \times \mathbf{P}$

(c) $\mathbf{P} \cdot \mathbf{Q} \times \mathbf{R}$

(d) $\sin \theta_{QR}$

(e) $\mathbf{P} \times (\mathbf{Q} \times \mathbf{R})$

(f) A unit vector perpendicular to both $\mathbf{Q}$ and $\mathbf{R}$

(g) The component of $\mathbf{P}$ along $\mathbf{Q}$

Solution:

(a)

$$
\begin{aligned}
(\mathbf{P} + \mathbf{Q}) \times (\mathbf{P} - \mathbf{Q}) &= \mathbf{P} \times (\mathbf{P} - \mathbf{Q}) + \mathbf{Q} \times (\mathbf{P} - \mathbf{Q}) \\
&= \mathbf{P} \times \mathbf{P} - \mathbf{P} \times \mathbf{Q} + \mathbf{Q} \times \mathbf{P} - \mathbf{Q} \times \mathbf{Q} \\
&= 0 + \mathbf{Q} \times \mathbf{P} + \mathbf{Q} \times \mathbf{P} - 0 \\
&= 2\mathbf{Q} \times \mathbf{P} \\
&= 2 \begin{vmatrix} \mathbf{a}_x & \mathbf{a}_y & \mathbf{a}_z \\ 2 & -1 & 2 \\ 2 & 0 & -1 \end{vmatrix} \\
&= 2(1 - 0)\, \mathbf{a}_x + 2(4 + 2)\, \mathbf{a}_y + 2(0 + 2)\, \mathbf{a}_z \\
&= 2\mathbf{a}_x + 12\mathbf{a}_y + 4\mathbf{a}_z
\end{aligned}
$$

(b) The only way $\mathbf{Q} \cdot \mathbf{R} \times \mathbf{P}$ makes sense is

$$
\begin{aligned}
\mathbf{Q} \cdot (\mathbf{R} \times \mathbf{P}) &= (2, -1, 2) \cdot \begin{vmatrix} \mathbf{a}_x & \mathbf{a}_y & \mathbf{a}_z \\ 2 & -3 & 1 \\ 2 & 0 & -1 \end{vmatrix} \\
&= (2, -1, 2) \cdot (3, 4, 6) \\
&= 6 - 4 + 12 = 14
\end{aligned}
$$

Alternatively:

$$
\mathbf{Q} \cdot (\mathbf{R} \times \mathbf{P}) = \begin{vmatrix} 2 & -1 & 2 \\ 2 & -3 & 1 \\ 2 & 0 & -1 \end{vmatrix}
$$

To find the determinant of a 3 × 3 matrix, we repeat the first two rows and cross multiply; when the cross multiplication is from right to left, the result should be negated as shown diagrammatically here. This technique of finding a determinant applies only to a 3 × 3 matrix. Hence,

$$
\mathbf{Q} \cdot (\mathbf{R} \times \mathbf{P}) = \begin{vmatrix} 2 & -1 & 2 \\ 2 & -3 & 1 \\ 2 & 0 & -1 \\ 2 & -1 & 2 \\ 2 & -3 & 1 \end{vmatrix}
\begin{matrix} \\ \\ + \\ + \\ + \\ + \end{matrix}
$$

$$
= +6 + 0 - 2 + 12 - 0 - 2
$$

$$
= 14
$$

as obtained before.

(c) From eq. (1.29)

$$\mathbf{P} \cdot (\mathbf{Q} \times \mathbf{R}) = \mathbf{Q} \cdot (\mathbf{R} \times \mathbf{P}) = 14$$

or

$$\mathbf{P} \cdot (\mathbf{Q} \times \mathbf{R}) = (2, 0, -1) \cdot (5, 2, -4)$$
$$= 10 + 0 + 4$$
$$= 14$$

(d)
$$\sin \theta_{QR} = \frac{|\mathbf{Q} \times \mathbf{R}|}{|\mathbf{Q}||\mathbf{R}|} = \frac{(5, 2, -4)|}{|(2, -1, 2)||(2, -3, 1)|}$$
$$= \frac{\sqrt{45}}{3\sqrt{14}} = \frac{\sqrt{5}}{\sqrt{14}} = 0.5976$$

(e)
$$\mathbf{P} \times (\mathbf{Q} \times \mathbf{R}) = (2, 0, -1) \times (5, 2, -4)$$
$$= (2, 3, 4)$$

Alternatively, using the bac-cab rule,

$$\mathbf{P} \times (\mathbf{Q} \times \mathbf{R}) = \mathbf{Q}(\mathbf{P} \cdot \mathbf{R}) - \mathbf{R}(\mathbf{P} \cdot \mathbf{Q})$$
$$= (2, -1, 2)(4 + 0 - 1) - (2, -3, 1)(4 + 0 - 2)$$
$$= (2, 3, 4)$$

(f) A unit vector perpendicular to both $\mathbf{Q}$ and $\mathbf{R}$ is given by

$$\mathbf{a} = \frac{\pm \mathbf{Q} \times \mathbf{R}}{|\mathbf{Q} \times \mathbf{R}|} = \frac{\pm (5, 2, -4)}{\sqrt{45}}$$
$$= \pm(0.745, 0.298, -0.596)$$

Note that $|\mathbf{a}| = 1$, $\mathbf{a} \cdot \mathbf{Q} = 0 = \mathbf{a} \cdot \mathbf{R}$. Any of these can be used to check $\mathbf{a}$.

(g) The component of $\mathbf{P}$ along $\mathbf{Q}$ is

$$\mathbf{P}_Q = |\mathbf{P}| \cos \theta_{PQ} \mathbf{a}_Q$$
$$= (\mathbf{P} \cdot \mathbf{a}_Q)\mathbf{a}_Q = \frac{(\mathbf{P} \cdot \mathbf{Q})\mathbf{Q}}{|\mathbf{Q}|^2}$$
$$= \frac{(4 + 0 - 2)(2, -1, 2)}{(4 + 1 + 4)} = \frac{2}{9}(2, -1, 2)$$
$$= 0.4444\mathbf{a}_x - 0.2222\mathbf{a}_y + 0.4444\mathbf{a}$$

PRACTICE EXERCISE 1.5

Let $\mathbf{E} = 3\mathbf{a}_y + 4\mathbf{a}_z$ and $\mathbf{F} = 4\mathbf{a}_x - 10\mathbf{a}_y + 5\mathbf{a}_z$.

(a) Find the component of $\mathbf{E}$ along $\mathbf{F}$.
(b) Determine a unit vector perpendicular to both $\mathbf{E}$ and $\mathbf{F}$.

Answer: (a) $(-0.2837, 0.7092, -0.3546)$, (b) $\pm(0.9398, 0.2734, -0.205)$.

EXAMPLE 1.6

Derive the cosine formula

$$a^2 = b^2 + c^2 - 2bc \cos A$$

and the sine formula

$$\frac{\sin A}{a} = \frac{\sin B}{b} = \frac{\sin C}{c}$$

using dot product and cross product, respectively.

Solution:

Consider a triangle as shown in Figure 1.11. From the figure, we notice that

$$\mathbf{a} + \mathbf{b} + \mathbf{c} = 0$$

that is,

$$\mathbf{b} + \mathbf{c} = -\mathbf{a}$$

Hence,

$$
\begin{aligned}
a^2 = \mathbf{a} \cdot \mathbf{a} &= (\mathbf{b} + \mathbf{c}) \cdot (\mathbf{b} + \mathbf{c}) \\
&= \mathbf{b} \cdot \mathbf{b} + \mathbf{c} \cdot \mathbf{c} + 2\mathbf{b} \cdot \mathbf{c} \\
a^2 &= b^2 + c^2 - 2bc \cos A
\end{aligned}
$$

where $(\pi - A)$ is the angle between $\mathbf{b}$ and $\mathbf{c}$.

The area of a triangle is half of the product of its height and base. Hence,

$$\left| \frac{1}{2}\mathbf{a} \times \mathbf{b} \right| = \left| \frac{1}{2}\mathbf{b} \times \mathbf{c} \right| = \left| \frac{1}{2}\mathbf{c} \times \mathbf{a} \right|$$

$$ab \sin C = bc \sin A = ca \sin B$$

Dividing through by abc gives

$$\frac{\sin A}{a} = \frac{\sin B}{b} = \frac{\sin C}{c}$$

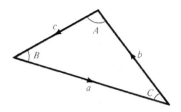

FIGURE 1.11 For Example 1.6.

PRACTICE EXERCISE 1.6

Show that vectors $\mathbf{a} = (4, 0, -1)$, $\mathbf{b} = (1, 3, 4)$, and $\mathbf{c} = (-5, -3, -3)$ form the sides of a triangle. Is this a right angle triangle? Calculate the area of the triangle.

Answer: Yes, 10.5.

EXAMPLE 1.7

Show that points $P_1(5, 2, -4)$, $P_2(1, 1, 2)$, and $P_3(-3, 0, 8)$ all lie on a straight line. Determine the shortest distance between the line and point $P_4(3, -1, 0)$.

Solution:
The distance vector $\mathbf{r}_{P_1P_2}$ is given by

$$\mathbf{r}_{P_1P_2} = \mathbf{r}_{P_2} - \mathbf{r}_{P_1} = (1, 1, 2) - (5, 2, -4)$$
$$= (-4, -1, 6)$$

Similarly,

$$\mathbf{r}_{P_1P_3} = \mathbf{r}_{P_3} - \mathbf{r}_{P_1} = (-3, 0, 8) - (5, 2, -4)$$
$$= (-8, -2, 12)$$

$$\mathbf{r}_{P_1P_4} = \mathbf{r}_{P_4} - \mathbf{r}_{P_1} = (3, -1, 0) - (5, 2, -4)$$
$$= (-2, -3, 4)$$

$$\mathbf{r}_{P_1P_2} \times \mathbf{r}_{P_1P_3} = \begin{vmatrix} \mathbf{a}_x & \mathbf{a}_y & \mathbf{a}_z \\ -4 & -1 & 6 \\ -8 & -2 & 12 \end{vmatrix}$$
$$= (0, 0, 0)$$

showing that the angle between $\mathbf{r}_{P_1P_2}$ and $\mathbf{r}_{P_1P_3}$ is zero $(\sin \theta = 0)$. This implies that P_1, P_2, and P_3 lie on a straight line.

Alternatively, the vector equation of the straight line is easily determined from Figure 1.12(a). For any point P on the line joining P_1 and P_2

$$\mathbf{r}_{P_1P} = \lambda \mathbf{r}_{P_1P_2}$$

where λ is a constant. Hence the position vector $\mathbf{r}_P$ of the point P must satisfy

$$\mathbf{r}_P - \mathbf{r}_{P_1} = \lambda (\mathbf{r}_{P_2} - \mathbf{r}_{P_1})$$

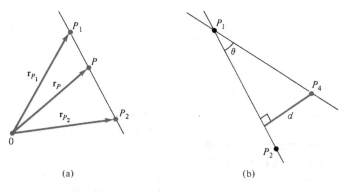

(a)   (b)

FIGURE 1.12 For Example 1.7.

that is,

$$\mathbf{r}_P = \mathbf{r}_{P_1} + \lambda(\mathbf{r}_{P_2} - \mathbf{r}_{P_1})$$
$$= (5, 2, -4) - \lambda(4, 1, -6)$$
$$\mathbf{r}_P = (5 - 4\lambda, 2 - \lambda, -4 + 6\lambda)$$

This is the vector equation of the straight line joining P_1 and P_2. If P_3 is on this line, the position vector of P_3 must satisfy the equation; $\mathbf{r}_3$ does satisfy the equation when $\lambda = 2$.

The shortest distance between the line and point $P_4(3, -1, 0)$ is the perpendicular distance from the point to the line. From Figure 1.12(b), it is clear that

$$d = r_{P_1P_4} \sin \theta = |\mathbf{r}_{P_1P_4} \times \mathbf{a}_{P_1P_2}|$$
$$= \frac{|(-2, -3, 4) \times (-4, -1, 6)|}{|(-4, -1, 6)|}$$
$$= \frac{\sqrt{312}}{\sqrt{53}} = 2.426$$

Any point on the line may be used as a reference point. Thus, instead of using P_1 as a reference point, we could use P_3. If $\angle P_4 P_3 P_2 = \theta'$, then

$$d = |\mathbf{r}_{P_3P_4}| \sin \theta' = |\mathbf{r}_{P_3P_4} \times \mathbf{a}_{P_3P_2}|$$

PRACTICE EXERCISE 1.7

If P_1 is $(1, 2, -3)$ and P_2 is $(-4, 0, 5)$, find

(a) The distance P_1P_2

(b) The vector equation of the line P_1P_2

(c) The shortest distance between the line P_1P_2 and point P_3 $(7, -1, 2)$

Answer: (a) 9.644, (b) $(1 - 5\lambda)\mathbf{a}_x + 2(1 - \lambda)\mathbf{a}_y + (8\lambda - 3)\mathbf{a}_z$, (c) 8.2.

MATLAB 1.1

```
% This script allows the user to input two vectors and
% then compute their dot product, cross product, sum,
% and difference
clear
vA = input('Enter vector A in the format [x y z]... \n >  ');
if isempty(vA); vA = [0 0 0]; end    % if the input is
                   % entered incorrectly set the vector to 0
vB = input('Enter vector B in the format [x y z]... \n >  ');
if isempty(vB); vB = [0 0 0]; end
disp('Magnitude of A:')
disp(norm(vA))                       % norm finds the magnitude of a
                                     % multi-dimensional vector
disp('Magnitude of B:')
disp(norm(vB))
disp('Unit vector in direction of A:')
disp(vA/norm(vA))                    % unit vector is the vector
                                     % divided by its magnitude
disp('Unit vector in direction of B:')
disp(vB/norm(vB))
disp('Sum A+B:')
disp(vA+vB)
disp('Difference A-B:')
disp(vA-vB)
disp('Dot product (A . B):')
disp(dot(vA,vB))          % dot takes the dot product of vectors
disp('Cross product (A x B):')
disp(cross(vA,vB))        % cross takes cross product of vectors
```

SUMMARY

1. A field is a function that specifies a quantity in space. For example, $\mathbf{A}(x, y, z)$ is a vector field, whereas $V(x, y, z)$ is a scalar field.
2. A vector $\mathbf{A}$ is uniquely specified by its magnitude and a unit vector along it, that is, $\mathbf{A} = A\mathbf{a}_A$.
3. Multiplying two vectors $\mathbf{A}$ and $\mathbf{B}$ results in either a scalar $\mathbf{A} \cdot \mathbf{B} = AB \cos \theta_{AB}$ or a vector $\mathbf{A} \times \mathbf{B} = AB \sin \theta_{AB} \, \mathbf{a}_n$. Multiplying three vectors $\mathbf{A}$, $\mathbf{B}$, and $\mathbf{C}$ yields a scalar $\mathbf{A} \cdot (\mathbf{B} \times \mathbf{C})$ or a vector $\mathbf{A} \times (\mathbf{B} \times \mathbf{C})$.
4. The scalar projection (or component) of vector $\mathbf{A}$ onto $\mathbf{B}$ is $A_B = \mathbf{A} \cdot \mathbf{a}_B$, whereas vector projection of $\mathbf{A}$ onto $\mathbf{B}$ is $\mathbf{A}_B = A_B \mathbf{a}_B$.
5. The MATLAB commands dot(A,B) and cross(A,B) are used for dot and cross products, respectively.

1.1 Tell which of the following quantities is not a vector: (a) force, (b) momentum, (c) acceleration, (d) work, (e) weight.

1.2 Which of the following is not a scalar field?
   (a) Displacement of a mosquito in space
   (b) Light intensity in a drawing room
   (c) Temperature distribution in your classroom
   (d) Atmospheric pressure in a given region
   (e) Humidity of a city

1.3 Of the rectangular coordinate systems shown in Figure 1.13, which are not right handed?

1.4 Which of these is correct?
   (a) $\mathbf{A} \times \mathbf{A} = |\mathbf{A}|^2$
   (b) $\mathbf{A} \times \mathbf{B} + \mathbf{B} \times \mathbf{A} = 0$
   (c) $\mathbf{A} \cdot \mathbf{B} \cdot \mathbf{C} = \mathbf{B} \cdot \mathbf{C} \cdot \mathbf{A}$
   (d) $\mathbf{a}_x \cdot \mathbf{a}_y = \mathbf{a}_z$
   (e) $\mathbf{a}_k = \mathbf{a}_x - \mathbf{a}_y$, where $\mathbf{a}_k$ is a unit vector

1.5 Which of the following identities is not valid?
   (a) $\mathbf{a}(\mathbf{b} + \mathbf{c}) = \mathbf{ab} + \mathbf{bc}$
   (b) $\mathbf{a} \times (\mathbf{b} + \mathbf{c}) = \mathbf{a} \times \mathbf{b} + \mathbf{a} \times \mathbf{c}$
   (c) $\mathbf{a} \cdot \mathbf{b} = \mathbf{b} \cdot \mathbf{a}$
   (d) $\mathbf{c} \cdot (\mathbf{a} \times \mathbf{b}) = -\mathbf{b} \cdot (\mathbf{a} \times \mathbf{c})$
   (e) $\mathbf{a}_A \cdot \mathbf{a}_B = \cos \theta_{AB}$

1.6 Which of the following statements are meaningless?
   (a) $\mathbf{A} \cdot \mathbf{B} + 2\mathbf{A} = 0$
   (b) $\mathbf{A} \cdot \mathbf{B} + 5 = 2\mathbf{A}$
   (c) $\mathbf{A}(\mathbf{A} + \mathbf{B}) + 2 = 0$
   (d) $\mathbf{A} \cdot \mathbf{A} + \mathbf{B} \cdot \mathbf{B} = 0$

1.7 Let $\mathbf{F} = 2\mathbf{a}_x - 6\mathbf{a}_y + 10\mathbf{a}_z$ and $\mathbf{G} = \mathbf{a}_x + G_y\mathbf{a}_y + 5\mathbf{a}_z$. If $\mathbf{F}$ and $\mathbf{G}$ have the same unit vector, G_y is
   (a) 6
   (b) −3
   (c) 0
   (d) 6

1.8 Given that $\mathbf{A} = \mathbf{a}_x + \alpha\mathbf{a}_y + \mathbf{a}_z$ and $\mathbf{B} = \alpha\mathbf{a}_x + \mathbf{a}_y + \mathbf{a}_z$, if **A and B** are normal to each other, α is
   (a) −2
   (b) −1/2
   (c) 0
   (d) 1
   (e) 2

1.9 The component of $6\mathbf{a}_x + 2\mathbf{a}_y - 3\mathbf{a}_z$ along $3\mathbf{a}_x - 4\mathbf{a}_y$ is
   (a) $-12\mathbf{a}_x - 9\mathbf{a}_y - 3\mathbf{a}_z$
   (b) $30\mathbf{a}_x - 40\mathbf{a}_y$
   (c) 10/7
   (d) 2
   (e) 10

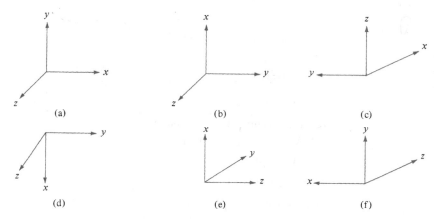

FIGURE 1.13 For Review Question 1.3.

1.10 Given $\mathbf{A} = -6\mathbf{a}_x + 3\mathbf{a}_y + 2\mathbf{a}_z$, the projection of $\mathbf{A}$ along $\mathbf{a}_y$ is

(a) -12 (d) 7

(b) -4 (e) 12

(c) 3

Answers: 1.1d, 1.2a, 1.3b,e, 1.4b, 1.5a, 1.6a,b,c, 1.7b, 1.8b, 1.9d, 1.10c.

PROBLEMS

Section 1.4—Unit Vector

1.1 Determine the unit vector along the direction OP, where O is the origin and P is point $(4, -5, 1)$.

1.2 Points $A(4, -6, 2)$, $B(-2, 0, 3)$, and $C(10, 1, -7)$ form a triangle. Show that $\mathbf{r}_{AB} + \mathbf{r}_{BC} + \mathbf{r}_{CA} = 0$.

Sections 1.5–1.7—Vector Addition, Subtraction, and Multiplication

1.3 If $\mathbf{A} = 4\mathbf{a}_x - 2\mathbf{a}_y + 6\mathbf{a}_z$ and $\mathbf{B} = 12\mathbf{a}_x + 18\mathbf{a}_y - 8\mathbf{a}_z$, determine:

(a) $\mathbf{A} - 3\mathbf{B}$

(b) $(2\mathbf{A} + 5\mathbf{B})/|\mathbf{B}|$

(c) $\mathbf{a}_x \times \mathbf{A}$

(d) $(\mathbf{B} \times \mathbf{a}_x) \cdot \mathbf{a}_y$

1.4 Let vectors $\mathbf{A} = 10\mathbf{a}_x - 6\mathbf{a}_y + 8\mathbf{a}_z$ and $\mathbf{B} = \mathbf{a}_x + 2\mathbf{a}_z$. Find: (a) $\mathbf{A} \cdot \mathbf{B}$, (b) $\mathbf{A} \times \mathbf{B}$, (c) $2\mathbf{A} - 3\mathbf{B}$.

1.5 Let $\mathbf{A} = -2\mathbf{a}_x + 5\mathbf{a}_y + \mathbf{a}_z$, $\mathbf{B} = \mathbf{a}_x + 3\mathbf{a}_z$, and $\mathbf{C} = 4\mathbf{a}_x - 6\mathbf{a}_y + 10\mathbf{a}_z$.

(a) Determine $\mathbf{A} - \mathbf{B} + \mathbf{C}$

(b) Find $\mathbf{A} \cdot (\mathbf{B} \times \mathbf{C})$

(c) Calculate the angle between $\mathbf{A}$ and $\mathbf{B}$

1.6 Let $\mathbf{A} = \mathbf{a}_x - \mathbf{a}_z$, $\mathbf{B} = \mathbf{a}_x + \mathbf{a}_y + \mathbf{a}_z$, $\mathbf{C} = \mathbf{a}_y + 2\mathbf{a}_z$, find:

(a) $\mathbf{A} \cdot (\mathbf{B} \times \mathbf{C})$

(b) $(\mathbf{A} \times \mathbf{B}) \cdot \mathbf{C}$

(c) $\mathbf{A} \times (\mathbf{B} \times \mathbf{C})$

(d) $(\mathbf{A} \times \mathbf{B}) \times \mathbf{C}$

1.7 Given that the position vectors of points T and S are $4\mathbf{a}_x + 6\mathbf{a}_y - \mathbf{a}_z$ and $10\mathbf{a}_x + 12\mathbf{a}_y + 8\mathbf{a}_z$, respectively, find: (a) the coordinates of T and S, (b) the distance vector from T to S, (c) the distance between T and S.

1.8 Let $\mathbf{A} = 4\mathbf{a}_x + 2\mathbf{a}_y - \mathbf{a}_z$ and $\mathbf{B} = \alpha\mathbf{a}_x + \beta\mathbf{a}_y + 3\mathbf{a}_z$

(a) If $\mathbf{A}$ and $\mathbf{B}$ are parallel, find α and β

(b) If $\mathbf{A}$ and $\mathbf{B}$ are perpendicular, find α and β

1.9 Let $\mathbf{A} = 10\mathbf{a}_x + 5\mathbf{a}_y - 2\mathbf{a}_z$. Find: (a) $\mathbf{A} \times \mathbf{a}_y$, (b) $\mathbf{A} \cdot \mathbf{a}_z$, (c) the angle between $\mathbf{A}$ and $\mathbf{a}_z$.

1.10 (a) Show that

$$(\mathbf{A} \cdot \mathbf{B})^2 + |\mathbf{A} \times \mathbf{B}|^2 = (AB)^2$$

(b) Show that

$$\mathbf{a}_x = \frac{\mathbf{a}_y \times \mathbf{a}_z}{\mathbf{a}_x \cdot \mathbf{a}_y \times \mathbf{a}_z}, \quad \mathbf{a}_y = \frac{\mathbf{a}_z \times \mathbf{a}_x}{\mathbf{a}_x \cdot \mathbf{a}_y \times \mathbf{a}_z}, \quad \mathbf{a}_z = \frac{\mathbf{a}_x \times \mathbf{a}_y}{\mathbf{a}_x \cdot \mathbf{a}_y \times \mathbf{a}_z}$$

1.11 Given that

$$\mathbf{P} = 2\mathbf{a}_x - \mathbf{a}_y - 2\mathbf{a}_z$$
$$\mathbf{Q} = 4\mathbf{a}_x + 3\mathbf{a}_y + 2\mathbf{a}_z$$
$$\mathbf{R} = -\mathbf{a}_x + \mathbf{a}_y + 2\mathbf{a}_z$$

find: (a) $|\mathbf{P} + \mathbf{Q} - \mathbf{R}|$, (b) $\mathbf{P} \cdot \mathbf{Q} \times \mathbf{R}$, (c) $\mathbf{Q} \times \mathbf{P} \cdot \mathbf{R}$, (d) $(\mathbf{P} \times \mathbf{Q}) \cdot (\mathbf{Q} \times \mathbf{R})$, (e) $(\mathbf{P} \times \mathbf{Q}) \times (\mathbf{Q} \times \mathbf{R})$, (f) $\cos \theta_{PR}$, (g) $\sin \theta_{PQ}$.

1.12 If $\mathbf{A} = 4\mathbf{a}_x - 6\mathbf{a}_y + \mathbf{a}_z$ and $\mathbf{B} = 2\mathbf{a}_x + 5\mathbf{a}_z$, find:

(a) $\mathbf{A} \cdot \mathbf{B} + 2|\mathbf{B}|^2$

(b) a unit vector perpendicular to both $\mathbf{A}$ and $\mathbf{B}$

1.13 Determine the dot product, cross product, and angle between
$$\mathbf{P} = 2\mathbf{a}_x - 6\mathbf{a}_y + 5\mathbf{a}_z \qquad \text{and} \qquad \mathbf{Q} = 3\mathbf{a}_y + \mathbf{a}_z$$

1.14 Prove that vectors $\mathbf{P} = 2\mathbf{a}_x + 4\mathbf{a}_y - 6\mathbf{a}_z$ and $\mathbf{Q} = 5\mathbf{a}_x + 2\mathbf{a}_y - 3\mathbf{a}_z$ are orthogonal vectors.

1.15 Simplify the following expressions:

(a) $\mathbf{A} \times (\mathbf{A} \times \mathbf{B})$

(b) $\mathbf{A} \times [\mathbf{A} \times (\mathbf{A} \times \mathbf{B})]$

1.16 A right angle triangle has its corners located at $P_1(5, -3, 1)$, $P_2(1, -2, 4)$, and $P_3(3, 3, 5)$. (a) Which corner is a right angle? (b) Calculate the area of the triangle.

1.17 Points P, Q, and R are located at $(-1, 4, 8)$, $(2, -1, 3)$, and $(-1, 2, 3)$, respectively. Determine (a) the distance between P and Q, (b) the distance vector from P to R, (c) the angle between QP and QR, (d) the area of triangle PQR, (e) the perimeter of triangle PQR.

1.18 Two points $P(2, 4, -1)$ and $Q(12, 16, 9)$ form a straight line. Calculate the time taken for a sonar signal traveling at 300 m/s to get from the origin to the midpoint of PQ.

1.19 Find the area of the parallelogram formed by the vectors $\mathbf{D} = 4\mathbf{a}_x + \mathbf{a}_y + 5\mathbf{a}_z$ and $\mathbf{E} = -\mathbf{a}_x + 2\mathbf{a}_y + 3\mathbf{a}_z$.

*1.20 (a) Prove that $\mathbf{P} = \cos\theta_1\mathbf{a}_x + \sin\theta_1\mathbf{a}_y$ and $\mathbf{Q} = \cos\theta_2\mathbf{a}_x + \sin\theta_2\mathbf{a}_y$ are unit vectors in the xy-plane, respectively, making angles θ_1 and θ_2 with the x-axis.

(b) By means of dot product, obtain the formula for $\cos(\theta_2 - \theta_1)$. By similarly formulating $\mathbf{P}$ and $\mathbf{Q}$, obtain the formula for $\cos(\theta_2 + \theta_1)$.

(c) If θ is the angle between $\mathbf{P}$ and $\mathbf{Q}$, find $\frac{1}{2}|\mathbf{P} - \mathbf{Q}|$ in terms of θ.

1.21 Consider a rigid body rotating with a constant angular velocity ω radians per second about a fixed axis through O as in Figure 1.14. Let $\mathbf{r}$ be the distance vector from O to P, the position of a particle in the body. The magnitude of the velocity $\mathbf{u}$ of the body at P is $|\mathbf{u}| = d|\omega| = |\mathbf{r}|\sin\theta\,|\omega|$ or $\mathbf{u} = \boldsymbol{\omega} \times \mathbf{r}$. If the rigid body is rotating at 3 rad/s about an axis parallel to $\mathbf{a}_x - 2\mathbf{a}_y + 2\mathbf{a}_z$ and passing through point $(2, -3, 1)$, determine the velocity of the body at $(1, 3, 4)$.

1.22 A cube of side 1 m has one corner placed at the origin. Determine the angle between the diagonals of the cube.

1.23 Given vectors $\mathbf{T} = 2\mathbf{a}_x - 6\mathbf{a}_y + 3\mathbf{a}_z$ and $\mathbf{S} = \mathbf{a}_x + 2\mathbf{a}_y + \mathbf{a}_z$, find (a) the scalar projection of $\mathbf{T}$ on $\mathbf{S}$, (b) the vector projection of $\mathbf{S}$ on $\mathbf{T}$, (c) the smaller angle between $\mathbf{T}$ and $\mathbf{S}$.

---

*Single asterisks indicate problems of intermediate difficulty.

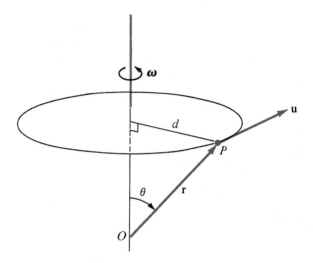

FIGURE 1.14 For Problem 1.21.

Section 1.8—Components of a Vector

1.24 Given two vectors **A** and **B**, show that the vector component of **A** perpendicular to **B** is

$$\mathbf{C} = \mathbf{A} - \frac{\mathbf{A} \cdot \mathbf{B}}{\mathbf{B} \cdot \mathbf{B}}\mathbf{B}$$

1.25 Let $\mathbf{A} = 20\mathbf{a}_x + 15\mathbf{a}_y - 10\mathbf{a}_z$ and $\mathbf{B} = \mathbf{a}_x + \mathbf{a}_y$. Find: (a) $\mathbf{A} \cdot \mathbf{B}$, (b) $\mathbf{A} \times \mathbf{B}$, (c) the component of **A** along **B**.

1.26 Figure 1.15 shows that **A** makes specific angles with respect to each axis. For $\mathbf{A} = 2\mathbf{a}_x - 4\mathbf{a}_y + 6\mathbf{a}_z$, find the direction angles α, β, and γ.

1.27 If $\mathbf{H} = 2xy\mathbf{a}_x - (x + z)\mathbf{a}_y + z^2\mathbf{a}_z$, find:
  (a) A unit vector parallel to **H** at $P(1, 3, -2)$
  (b) The equation of the surface on which $|\mathbf{H}| = 10$

1.28 Let $\mathbf{P} = 2\mathbf{a}_x - 4\mathbf{a}_y + \mathbf{a}_z$ and $\mathbf{Q} = \mathbf{a}_x + 2\mathbf{a}_y$. Find **R** which has magnitude 4 and is perpendicular to both **P** and **Q**.

1.29 Let $\mathbf{G} = x^2\mathbf{a}_x - y\mathbf{a}_y + 2z\mathbf{a}_z$ and $\mathbf{H} = yz\mathbf{a}_x + 3\mathbf{a}_y - xz\mathbf{a}_z$. At point $(1, -2, 3)$, (a) calculate the magnitude of **G** and **H**, (b) determine $\mathbf{G} \cdot \mathbf{H}$, (c) find the angle between **G** and **H**.

1.30 A vector field is given by $\mathbf{H} = 10yz^2\mathbf{a}_x - 8xyz\mathbf{a}_y + 12y^2\mathbf{a}_z$
  (a) Evaluate **H** at $P(-1, 2, 4)$
  (b) Find the component of **H** along $\mathbf{a}_x - \mathbf{a}_y$ at P.

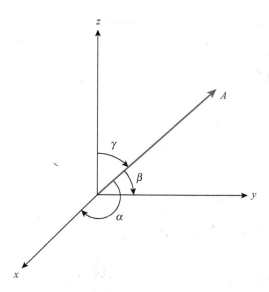

FIGURE 1.15 For Problem 1.26.

1.31 E and F are vector fields given by $\mathbf{E} = 2x\mathbf{a}_x + \mathbf{a}_y + yz\mathbf{a}_z$ and $\mathbf{F} = xy\mathbf{a}_x - y^2\mathbf{a}_y + xyz\mathbf{a}_z$. Determine:

(a) $|\mathbf{E}|$ at $(1, 2, 3)$

(b) The component of E along F at $(1, 2, 3)$

(c) A vector perpendicular to both E and F at $(0, 1, -3)$ whose magnitude is unity

1.32 Given two vector fields

$$\mathbf{D} = yz\mathbf{a}_x + xz\mathbf{a}_y + xy\mathbf{a}_z \quad \text{and} \quad \mathbf{E} = 5xy\mathbf{a}_x + 6(x^2 + 3)\mathbf{a}_y + 8z^2\mathbf{a}_z$$

(a) Evaluate $\mathbf{C} = \mathbf{D} + \mathbf{E}$ at point $P(-1, 2, 4)$. (b) Find the angle C makes with the x-axis at P.

ENHANCING YOUR SKILLS AND CAREER

The Accreditation Board for Engineering and Technology (ABET) establishes eleven criteria for accrediting engineering, technology, and computer science programs. The criteria are as follows:

A. Ability to apply mathematics science and engineering principles
B. Ability to design and conduct experiments and interpret data
C. Ability to design a system, component, or process to meet desired needs
D. Ability to function on multidisciplinary teams
E. Ability to identify, formulate, and solve engineering problems
F. Ability to understand professional and ethical responsibility
G. Ability to communicate effectively
H. Ability to understand the impact of engineering solutions in a global context
I. Ability to recognize the need for and to engage in lifelong learning
J. Ability to know of contemporary issues
K. Ability to use the techniques, skills, and modern engineering tools necessary for engineering practice

Criterion A applies directly to electromagnetics. As students, you are expected to study mathematics, science, and engineering with the purpose of being able to apply that knowledge to the solution of engineering problems. The skill needed here is the ability to apply the fundamentals of EM in solving a problem. The best approach is to attempt as many problems as you can. This will help you to understand how to use formulas and assimilate the material. Keep nearly all your basic mathematics, science, and engineering textbooks. You may need to consult them from time to time.

CHAPTER 2

COORDINATE SYSTEMS AND TRANSFORMATION

History teaches us that man learns nothing from history.

—HEGEL

2.1 INTRODUCTION

In general, the physical quantities we shall be dealing with in EM are functions of space and time. In order to describe the spatial variations of the quantities, we must be able to define all points uniquely in space in a suitable manner. This requires using an appropriate coordinate system.

A point or vector can be represented in any curvilinear coordinate system, which may be orthogonal or nonorthogonal.

An **orthogonal system** is one in which the coordinate surfaces are mutually perpendicular.

Nonorthogonal systems are hard to work with, and they are of little or no practical use. Examples of orthogonal coordinate systems include the Cartesian (or rectangular), the circular cylindrical, the spherical, the elliptic cylindrical, the parabolic cylindrical, the conical, the prolate spheroidal, the oblate spheroidal, and the ellipsoidal.[1] A considerable amount of work and time may be saved by choosing a coordinate system that best fits a given problem. A hard problem in one coordinate system may turn out to be easy in another system.

In this text, we shall restrict ourselves to the three best-known coordinate systems: the Cartesian, the circular cylindrical, and the spherical. Although we have considered the Cartesian system in Chapter 1, we shall consider it in detail in this chapter. We should bear in mind that the concepts covered in Chapter 1 and demonstrated in Cartesian coordinates are equally applicable to other systems of coordinates. For example, the procedure for finding the dot or cross product of two vectors in a cylindrical system is the same as that used in the Cartesian system in Chapter 1.

---

[1] For an introductory treatment of these coordinate systems, see M. R. Spiegel and J. Liu, *Mathematical Handbook of Formulas and Tables.* New York: McGraw-Hill, 2nd ed., 1999, pp. 126–130.

Sometimes, it is necessary to transform points and vectors from one coordinate system to another. The techniques for doing this will be presented and illustrated with examples.

2.2 CARTESIAN COORDINATES (*x*, *y*, *z*)

As mentioned in Chapter 1, a point P can be represented as (x, y, z) as illustrated in Figure 1.1. The ranges of the coordinate variables x, y, and z are

$$-\infty < x < \infty$$
$$-\infty < y < \infty \tag{2.1}$$
$$-\infty < z < \infty$$

A vector **A** in Cartesian (otherwise known as rectangular) coordinates can be written as

$$(A_x, A_y, A_z) \quad \text{or} \quad A_x\mathbf{a}_x + A_y\mathbf{a}_y + A_z\mathbf{a}_z \tag{2.2}$$

where $\mathbf{a}_x$, $\mathbf{a}_y$, and $\mathbf{a}_z$ are unit vectors along the *x*-, *y*-, and *z*-directions as shown in Figure 1.1. The coordinate system may be either right-handed or left-handed. See Figure 1.13. It is customary to use the right-handed system.

2.3 CIRCULAR CYLINDRICAL COORDINATES (ρ, ϕ, *z*)

The circular cylindrical coordinate system is very convenient whenever we are dealing with problems having cylindrical symmetry, such as dealing with a coaxial transmission line.

A point P in cylindrical coordinates is represented as (ρ, ϕ, z) and is as shown in Figure 2.1. Observe Figure 2.1 closely and note how we define each space variable: ρ is the

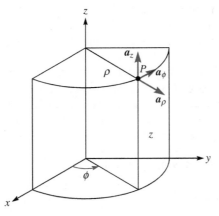

FIGURE 2.1 Point *P* and unit vectors in the cylindrical coordinate system.

radius of the cylinder passing through P or the radial distance from the z-axis; ϕ, called the *azimuthal* angle, is measured from the x-axis in the xy-plane; and z is the same as in the Cartesian system. The ranges of the variables are

$$0 \le \rho < \infty$$
$$0 \le \phi < 2\pi \tag{2.3}$$
$$-\infty < z < \infty$$

A vector **A** in cylindrical coordinates can be written as

$$(A_\rho, A_\phi, A_z) \quad \text{or} \quad A_\rho \mathbf{a}_\rho + A_\phi \mathbf{a}_\phi + A_z \mathbf{a}_z \tag{2.4}$$

where $\mathbf{a}_\rho$, $\mathbf{a}_\phi$, and $\mathbf{a}_z$ are unit vectors in the ρ-, ϕ-, and z-directions as illustrated in Figure 2.1. Note that $\mathbf{a}_\phi$ is not in degrees; it assumes the units of **A**. For example, if a force of 10 N acts on a particle in a circular motion, the force may be represented as $\mathbf{F} = 10\mathbf{a}_\phi$ N. In this case, $\mathbf{a}_\phi$ is in newtons.

The magnitude of **A** is

$$|\mathbf{A}| = (A_\rho^2 + A_\phi^2 + A_z^2)^{1/2} \tag{2.5}$$

Notice that the unit vectors $\mathbf{a}_\rho$, $\mathbf{a}_\phi$, and $\mathbf{a}_z$ are mutually perpendicular because our coordinate system is orthogonal; $\mathbf{a}_\rho$ points in the direction of increasing ρ, $\mathbf{a}_\phi$ in the direction of increasing ϕ, and $\mathbf{a}_z$ in the positive z-direction. Thus,

$$\mathbf{a}_\rho \cdot \mathbf{a}_\rho = \mathbf{a}_\phi \cdot \mathbf{a}_\phi = \mathbf{a}_z \cdot \mathbf{a}_z = 1 \tag{2.6a}$$
$$\mathbf{a}_\rho \cdot \mathbf{a}_\phi = \mathbf{a}_\phi \cdot \mathbf{a}_z = \mathbf{a}_z \cdot \mathbf{a}_\rho = 0 \tag{2.6b}$$
$$\mathbf{a}_\rho \times \mathbf{a}_\phi = \mathbf{a}_z \tag{2.6c}$$
$$\mathbf{a}_\phi \times \mathbf{a}_z = \mathbf{a}_\rho \tag{2.6d}$$
$$\mathbf{a}_z \times \mathbf{a}_\rho = \mathbf{a}_\phi \tag{2.6e}$$

where eqs. (2.6c) to (2.6e) are obtained in cyclic permutation (see Figure 1.9). They also show that the system is right-handed, following the cyclic ordering $\rho \to \phi \to z \to \rho \to \phi \to \dots$.

The relationships between the variables (x, y, z) of the Cartesian coordinate system and those of the cylindrical system (ρ, ϕ, z) are easily obtained from Figure 2.2 as

$$\rho = \sqrt{x^2 + y^2}, \quad \phi = \tan^{-1}\frac{y}{x}, \quad z = z \tag{2.7}$$

or

$$x = \rho \cos \phi, \quad y = \rho \sin \phi, \quad z = z \tag{2.8}$$

Whereas eq. (2.7) is for transforming a point from Cartesian (x, y, z) to cylindrical (ρ, ϕ, z) coordinates, eq. (2.8) is for $(\rho, \phi, z) \to (x, y, z)$ transformation.

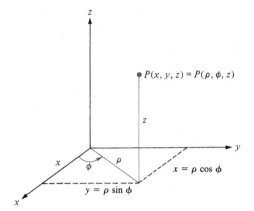

FIGURE 2.2 Relationship between (x, y, z) and (ρ, ϕ, z).

The relationships between $(\mathbf{a}_x, \mathbf{a}_y, \mathbf{a}_z)$ and $(\mathbf{a}_\rho, \mathbf{a}_\phi, \mathbf{a}_z)$ are obtained geometrically from Figure 2.3:

$$\mathbf{a}_x = \cos \phi \, \mathbf{a}_\rho - \sin \phi \, \mathbf{a}_\phi$$
$$\mathbf{a}_y = \sin \phi \, \mathbf{a}_\rho + \cos \phi \, \mathbf{a}_\phi \qquad (2.9)$$
$$\mathbf{a}_z = \mathbf{a}_z$$

or

$$\mathbf{a}_\rho = \cos \phi \, \mathbf{a}_x + \sin \phi \, \mathbf{a}_y$$
$$\mathbf{a}_\phi = -\sin \phi \, \mathbf{a}_x + \cos \phi \, \mathbf{a}_y \qquad (2.10)$$
$$\mathbf{a}_z = \mathbf{a}_z$$

Finally, the relationships between (A_x, A_y, A_z) and (A_ρ, A_ϕ, A_z) are obtained by simply substituting eq. (2.9) into eq. (2.2) and collecting terms. Thus,

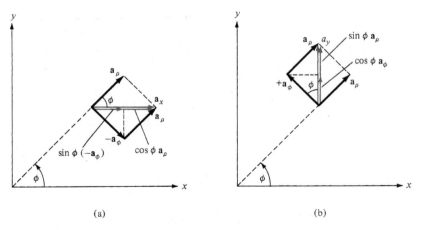

(a)   (b)

FIGURE 2.3 Unit vector transformation: **(a)** cylindrical components of $\mathbf{a}_x$, **(b)** cylindrical components of $\mathbf{a}_y$.

$$\mathbf{A} = (A_x \cos \phi + A_y \sin \phi)\mathbf{a}_\rho + (-A_x \sin \phi + A_y \cos \phi)\mathbf{a}_\phi + A_z\mathbf{a}_z \qquad (2.11)$$

or

$$
\begin{aligned}
A_\rho &= A_x \cos \phi + A_y \sin \phi \\
A_\phi &= -A_x \sin \phi + A_y \cos \phi \\
A_z &= A_z
\end{aligned}
\qquad (2.12)
$$

In matrix form, we write the transformation of vector $\mathbf{A}$ from (A_x, A_y, A_z) to (A_ρ, A_ϕ, A_z) as

$$
\begin{bmatrix} A_\rho \\ A_\phi \\ A_z \end{bmatrix} =
\begin{bmatrix} \cos \phi & \sin \phi & 0 \\ -\sin \phi & \cos \phi & 0 \\ 0 & 0 & 1 \end{bmatrix}
\begin{bmatrix} A_x \\ A_y \\ A_z \end{bmatrix}
\qquad (2.13)
$$

The inverse of the transformation $(A_\rho, A_\phi, A_z) \rightarrow (A_x, A_y, A_z)$ is obtained as

$$
\begin{bmatrix} A_x \\ A_y \\ A_z \end{bmatrix} =
\begin{bmatrix} \cos \phi & \sin \phi & 0 \\ -\sin \phi & \cos \phi & 0 \\ 0 & 0 & 1 \end{bmatrix}^{-1}
\begin{bmatrix} A_\rho \\ A_\phi \\ A_z \end{bmatrix}
\qquad (2.14)
$$

or directly from eqs. (2.4) and (2.10). Thus,

$$
\begin{bmatrix} A_x \\ A_y \\ A_z \end{bmatrix} =
\begin{bmatrix} \cos \phi & -\sin \phi & 0 \\ \sin \phi & \cos \phi & 0 \\ 0 & 0 & 1 \end{bmatrix}
\begin{bmatrix} A_\rho \\ A_\phi \\ A_z \end{bmatrix}
\qquad (2.15)
$$

An alternative way of obtaining eq. (2.13) or (2.15) is by using the dot product. For example,

$$
\begin{bmatrix} A_x \\ A_y \\ A_z \end{bmatrix} =
\begin{bmatrix} \mathbf{a}_x \cdot \mathbf{a}_\rho & \mathbf{a}_x \cdot \mathbf{a}_\phi & \mathbf{a}_x \cdot \mathbf{a}_z \\ \mathbf{a}_y \cdot \mathbf{a}_\rho & \mathbf{a}_y \cdot \mathbf{a}_\phi & \mathbf{a}_y \cdot \mathbf{a}_z \\ \mathbf{a}_z \cdot \mathbf{a}_\rho & \mathbf{a}_z \cdot \mathbf{a}_\phi & \mathbf{a}_z \cdot \mathbf{a}_z \end{bmatrix}
\begin{bmatrix} A_\rho \\ A_\phi \\ A_z \end{bmatrix}
\qquad (2.16)
$$

The derivation of this is left as an exercise.

Keep in mind that eqs. (2.7) and (2.8) are for point-to-point transformation, while eqs. (2.13) and (2.15) are for vector-to-vector transformation.

2.4 SPHERICAL COORDINATES (r, θ, ϕ)

Although cylindrical coordinates are covered in calculus texts, the spherical coordinates are rarely covered. The spherical coordinate system is most appropriate when one is dealing with problems having a degree of spherical symmetry. A point P can be represented as (r, θ, ϕ) and is illustrated in Figure 2.4. From Figure 2.4, we notice that r is defined as the distance from the origin to point P or the radius of a sphere centered at the origin and passing through P; θ (called the *colatitude*) is the angle between the z-axis and the position vector of P; and ϕ is measured from the x-axis (the same azimuthal angle in cylindrical coordinates). According to these definitions, the ranges of the variables are

$$
\begin{aligned}
0 &\le r < \infty \\
0 &\le \theta \le \pi \\
0 &\le \phi < 2\pi
\end{aligned}
\tag{2.17}
$$

A vector **A** in spherical coordinates may be written as

$$(A_r, A_\theta, A_\phi) \quad \text{or} \quad A_r \mathbf{a}_r + A_\theta \mathbf{a}_\theta + A_\phi \mathbf{a}_\phi \tag{2.18}$$

where $\mathbf{a}_r$, $\mathbf{a}_\theta$, and $\mathbf{a}_\phi$ are unit vectors along the r-, θ-, and ϕ-directions. The magnitude of **A** is

$$|\mathbf{A}| = (A_r^2 + A_\theta^2 + A_\phi^2)^{1/2} \tag{2.19}$$

The unit vectors $\mathbf{a}_r$, $\mathbf{a}_\theta$, and $\mathbf{a}_\phi$ are mutually orthogonal, $\mathbf{a}_r$ being directed along the radius or in the direction of increasing r, $\mathbf{a}_\theta$ in the direction of increasing θ, and $\mathbf{a}_\phi$ in the direction of increasing ϕ. Thus,

$$
\begin{aligned}
\mathbf{a}_r \cdot \mathbf{a}_r &= \mathbf{a}_\theta \cdot \mathbf{a}_\theta = \mathbf{a}_\phi \cdot \mathbf{a}_\phi = 1 \\
\mathbf{a}_r \cdot \mathbf{a}_\theta &= \mathbf{a}_\theta \cdot \mathbf{a}_\phi = \mathbf{a}_\phi \cdot \mathbf{a}_r = 0 \\
\mathbf{a}_r \times \mathbf{a}_\theta &= \mathbf{a}_\phi \\
\mathbf{a}_\theta \times \mathbf{a}_\phi &= \mathbf{a}_r \\
\mathbf{a}_\phi \times \mathbf{a}_r &= \mathbf{a}_\theta
\end{aligned}
\tag{2.20}
$$

FIGURE 2.4 Point P and unit vectors in spherical coordinates.

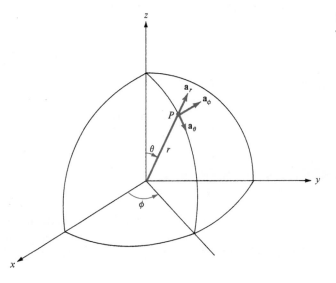

Equation (2.20) shows that the coordinate system is orthogonal and right-handed.

The space variables (x, y, z) in Cartesian coordinates can be related to variables (r, θ, ϕ) of a spherical coordinate system. From Figure 2.5 it is easy to notice that

$$r = \sqrt{x^2 + y^2 + z^2}, \quad \theta = \tan^{-1}\frac{\sqrt{x^2 + y^2}}{z}, \quad \phi = \tan^{-1}\frac{y}{x} \qquad (2.21)$$

or

$$x = r \sin \theta \cos \phi, \quad y = r \sin \theta \sin \phi, \quad z = r \cos \theta \qquad (2.22)$$

In eq. (2.21), we have $(x, y, z) \rightarrow (r, \theta, \phi)$ point transformation and in eq. (2.22), it is $(r, \theta, \phi) \rightarrow (x, y, z)$ point transformation.

The unit vectors $\mathbf{a}_x$, $\mathbf{a}_y$, $\mathbf{a}_z$ and $\mathbf{a}_r$, $\mathbf{a}_\theta$, $\mathbf{a}_\phi$ are related as follows:

$$\begin{aligned}
\mathbf{a}_x &= \sin \theta \cos \phi \, \mathbf{a}_r + \cos \theta \cos \phi \, \mathbf{a}_\theta - \sin \phi \, \mathbf{a}_\phi \\
\mathbf{a}_y &= \sin \theta \sin \phi \, \mathbf{a}_r + \cos \theta \sin \phi \, \mathbf{a}_\theta + \cos \phi \, \mathbf{a}_\phi \\
\mathbf{a}_z &= \cos \theta \, \mathbf{a}_r - \sin \theta \, \mathbf{a}_\theta
\end{aligned} \qquad (2.23)$$

or

$$\begin{aligned}
\mathbf{a}_r &= \sin \theta \cos \phi \, \mathbf{a}_x + \sin \theta \sin \phi \, \mathbf{a}_y + \cos \theta \, \mathbf{a}_z \\
\mathbf{a}_\theta &= \cos \theta \cos \phi \, \mathbf{a}_x + \cos \theta \sin \phi \, \mathbf{a}_y - \sin \theta \, \mathbf{a}_z \\
\mathbf{a}_\phi &= -\sin \phi \, \mathbf{a}_x + \cos \phi \, \mathbf{a}_y
\end{aligned} \qquad (2.24)$$

The components of vector $\mathbf{A} = (A_x, A_y, A_z)$ and $\mathbf{A} = (A_r, A_\theta, A_\phi)$ are related by substituting eq. (2.23) into eq. (2.2) and collecting terms. Thus,

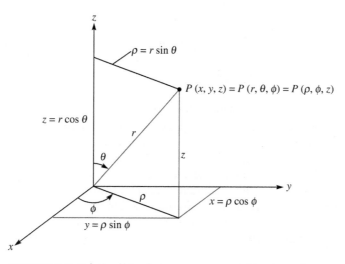

FIGURE 2.5 Relationships between space variables (x, y, z), (r, θ, ϕ), and $(\rho, \phi, z,)$.

$$\mathbf{A} = (A_x \sin\theta\cos\phi + A_y\sin\theta\sin\phi + A_z\cos\theta)\mathbf{a}_r + (A_x\cos\theta\cos\phi \\ + A_y\cos\theta\sin\phi - A_z\sin\theta)\mathbf{a}_\theta + (-A_x\sin\phi + A_y\cos\phi)\mathbf{a}_\phi \quad (2.25)$$

and from this, we obtain

$$
\begin{aligned}
A_r &= A_x\sin\theta\cos\phi + A_y\sin\theta\sin\phi + A_z\cos\theta \\
A_\theta &= A_x\cos\theta\cos\phi + A_y\cos\theta\sin\phi - A_z\sin\theta \\
A_\phi &= -A_x\sin\phi + A_y\cos\phi
\end{aligned}
\quad (2.26)
$$

In matrix form, the $(A_x, A_y, A_z) \rightarrow (A_r, A_\theta, A_\phi)$ vector transformation is performed according to

$$
\begin{bmatrix} A_r \\ A_\theta \\ A_\phi \end{bmatrix} =
\begin{bmatrix}
\sin\theta\cos\phi & \sin\theta\sin\phi & \cos\theta \\
\cos\theta\cos\phi & \cos\theta\sin\phi & -\sin\theta \\
-\sin\phi & \cos\phi & 0
\end{bmatrix}
\begin{bmatrix} A_x \\ A_y \\ A_z \end{bmatrix}
\quad (2.27)
$$

The inverse transformation $(A_r, A_\theta, A_\phi) \rightarrow (A_x, A_y, A_z)$ is similarly obtained, or we obtain it from eq. (2.23). Thus,

$$
\begin{bmatrix} A_x \\ A_y \\ A_z \end{bmatrix} =
\begin{bmatrix}
\sin\theta\cos\phi & \cos\theta\cos\phi & -\sin\phi \\
\sin\theta\sin\phi & \cos\theta\sin\phi & \cos\phi \\
\cos\theta & -\sin\theta & 0
\end{bmatrix}
\begin{bmatrix} A_r \\ A_\theta \\ A_\phi \end{bmatrix}
\quad (2.28)
$$

Alternatively, we may obtain eqs. (2.27) and (2.28) by using the dot product. For example,

$$
\begin{bmatrix} A_r \\ A_\theta \\ A_\phi \end{bmatrix} =
\begin{bmatrix}
\mathbf{a}_r\cdot\mathbf{a}_x & \mathbf{a}_r\cdot\mathbf{a}_y & \mathbf{a}_r\cdot\mathbf{a}_z \\
\mathbf{a}_\theta\cdot\mathbf{a}_x & \mathbf{a}_\theta\cdot\mathbf{a}_y & \mathbf{a}_\theta\cdot\mathbf{a}_z \\
\mathbf{a}_\phi\cdot\mathbf{a}_x & \mathbf{a}_\phi\cdot\mathbf{a}_y & \mathbf{a}_\phi\cdot\mathbf{a}_z
\end{bmatrix}
\begin{bmatrix} A_x \\ A_y \\ A_z \end{bmatrix}
\quad (2.29)
$$

For the sake of completeness, it may be instructive to obtain the point or vector transformation relationships between cylindrical and spherical coordinates. We shall use Figures 2.5 and 2.6 (where ϕ is held constant, since it is common to both systems). This will be left as an exercise (see Problem 2.16). Note that in a point or vector transformation, the point or vector has not changed; it is only expressed differently. Thus, for example, the magnitude of a vector will remain the same after the transformation, and this may serve as a way of checking the result of the transformation.

The distance between two points is usually necessary in EM theory. The distance d between two points with position vectors $\mathbf{r}_1$ and $\mathbf{r}_2$ is generally given by

$$d = |\mathbf{r}_2 - \mathbf{r}_1| \quad (2.30)$$

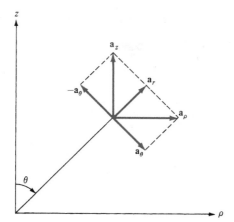

or

$$d^2 = (x_2 - x_1)^2 + (y_2 - y_1)^2 + (z_2 - z_1)^2 \text{ (Cartesian)} \qquad (2.31)$$

$$d^2 = \rho_2^2 + \rho_1^2 - 2\rho_1\rho_2 \cos(\phi_2 - \phi_1) + (z_2 - z_1)^2 \text{ (cylindrical)} \qquad (2.32)$$

$$\begin{aligned} d^2 = r_2^2 + r_1^2 &- 2r_1r_2 \cos\theta_2 \cos\theta_1 \\ &- 2r_1r_2 \sin\theta_2 \sin\theta_1 \cos(\phi_2 - \phi_1) \text{ (spherical)} \end{aligned} \qquad (2.33)$$

EXAMPLE 2.1

Given point $P(-2, 6, 3)$ and vector $\mathbf{A} = y\mathbf{a}_x + (x + z)\mathbf{a}_y$, express P and $\mathbf{A}$ in cylindrical and spherical coordinates. Evaluate $\mathbf{A}$ at P in the Cartesian, cylindrical, and spherical systems.

Solution:
At point P: $x = -2, y = 6, z = 3$. Hence,

$$\rho = \sqrt{x^2 + y^2} = \sqrt{4 + 36} = 6.32$$

$$\phi = \tan^{-1}\frac{y}{x} = \tan^{-1}\frac{6}{-2} = 108.43°$$

$$z = 3$$

$$r = \sqrt{x^2 + y^2 + z^2} = \sqrt{4 + 36 + 9} = 7$$

$$\theta = \tan^{-1}\frac{\sqrt{x^2 + y^2}}{z} = \tan^{-1}\frac{\sqrt{40}}{3} = 64.62°$$

Thus,

$$P(-2, 6, 3) = P(6.32, 108.43°, 3) = P(7, 64.62°, 108.43°)$$

In the Cartesian system, $\mathbf{A}$ at P is

$$\mathbf{A} = 6\mathbf{a}_x + \mathbf{a}_y$$

For vector **A**, $A_x = y$, $A_y = x + z$, $A_z = 0$. Hence, in the cylindrical system

$$\begin{bmatrix} A_\rho \\ A_\phi \\ A_z \end{bmatrix} = \begin{bmatrix} \cos\phi & \sin\phi & 0 \\ -\sin\phi & \cos\phi & 0 \\ 0 & 0 & 1 \end{bmatrix} \begin{bmatrix} y \\ x + z \\ 0 \end{bmatrix}$$

or

$$A_\rho = y\cos\phi + (x + z)\sin\phi$$
$$A_\phi = -y\sin\phi + (x + z)\cos\phi$$
$$A_z = 0$$

But $x = \rho\cos\phi$, $y = \rho\sin\phi$, and substituting these yields

$$\mathbf{A} = (A_\rho, A_\phi, A_z) = [\rho\cos\phi\sin\phi + (\rho\cos\phi + z)\sin\phi]\mathbf{a}_\rho$$
$$+ [-\rho\sin^2\phi + (\rho\cos\phi + z)\cos\phi]\mathbf{a}_\phi$$

At P

$$\rho = \sqrt{40}, \quad \tan\phi = \frac{6}{-2}$$

Hence,

$$\cos\phi = \frac{-2}{\sqrt{40}}, \quad \sin\phi = \frac{6}{\sqrt{40}}$$

$$\mathbf{A} = \left[\sqrt{40} \cdot \frac{-2}{\sqrt{40}} \cdot \frac{6}{\sqrt{40}} + \left(\sqrt{40} \cdot \frac{-2}{\sqrt{40}} + 3 \right) \cdot \frac{6}{\sqrt{40}} \right]\mathbf{a}_\rho$$
$$+ \left[-\sqrt{40} \cdot \frac{36}{40} + \left(\sqrt{40} \cdot \frac{-2}{\sqrt{40}} + 3 \right) \cdot \frac{-2}{\sqrt{40}} \right]\mathbf{a}_\phi$$

$$= \frac{-6}{\sqrt{40}}\mathbf{a}_\rho - \frac{38}{\sqrt{40}}\mathbf{a}_\phi = -0.9487\mathbf{a}_\rho - 6.008\mathbf{a}_\phi$$

Similarly, in the spherical system

$$\begin{bmatrix} A_r \\ A_\theta \\ A_\phi \end{bmatrix} = \begin{bmatrix} \sin\theta\cos\phi & \sin\theta\sin\phi & \cos\theta \\ \cos\theta\cos\phi & \cos\theta\sin\phi & -\sin\theta \\ -\sin\phi & \cos\phi & 0 \end{bmatrix} \begin{bmatrix} y \\ x + z \\ 0 \end{bmatrix}$$

or

$$A_r = y\sin\theta\cos\phi + (x + z)\sin\theta\sin\phi$$
$$A_\theta = y\cos\theta\cos\phi + (x + z)\cos\theta\sin\phi$$

$$A_\phi = -y \sin \phi + (x + z) \cos \phi$$

But $x = r \sin \theta \cos \phi$, $y = r \sin \theta \sin \phi$, and $z = r \cos \theta$. Substituting these yields

$$\begin{aligned}
\mathbf{A} &= (A_r, A_\theta, A_\phi) \\
&= r[\sin^2 \theta \cos \phi \sin \phi + (\sin \theta \cos \phi + \cos \theta) \sin \theta \sin \phi]\mathbf{a}_r \\
&\quad + r[\sin \theta \cos \theta \sin \phi \cos \phi + (\sin \theta \cos \phi + \cos \theta) \cos \theta \sin \phi]\mathbf{a}_\theta \\
&\quad + r[-\sin \theta \sin^2 \phi + (\sin \theta \cos \phi + \cos \theta) \cos \phi]\mathbf{a}_\phi
\end{aligned}$$

At P

$$r = 7, \quad \tan \phi = \frac{6}{-2}, \quad \tan \theta = \frac{\sqrt{40}}{3}$$

Hence,

$$\cos \phi = \frac{-2}{\sqrt{40}}, \quad \sin \phi = \frac{6}{\sqrt{40}}, \quad \cos \theta = \frac{3}{7}, \quad \sin \theta = \frac{\sqrt{40}}{7}$$

$$\begin{aligned}
\mathbf{A} &= 7 \cdot \left[\frac{40}{49} \cdot \frac{-2}{\sqrt{40}} \cdot \frac{6}{\sqrt{40}} + \left(\frac{\sqrt{40}}{7} \cdot \frac{-2}{\sqrt{40}} + \frac{3}{7} \right) \cdot \frac{\sqrt{40}}{7} \cdot \frac{6}{\sqrt{40}} \right] \mathbf{a}_r \\
&\quad + 7 \cdot \left[\frac{\sqrt{40}}{7} \cdot \frac{3}{7} \cdot \frac{6}{\sqrt{40}} \cdot \frac{-2}{\sqrt{40}} + \left(\frac{\sqrt{40}}{7} \cdot \frac{-2}{\sqrt{40}} + \frac{3}{7} \right) \cdot \frac{3}{7} \cdot \frac{6}{\sqrt{40}} \right] \mathbf{a}_\theta \\
&\quad + 7 \cdot \left[\frac{-\sqrt{40}}{7} \cdot \frac{36}{40} + \left(\frac{\sqrt{40}}{7} \cdot \frac{-2}{\sqrt{40}} + \frac{3}{7} \right) \cdot \frac{-2}{\sqrt{40}} \right] \mathbf{a}_\phi \\
&= \frac{-6}{7} \mathbf{a}_r - \frac{18}{7\sqrt{40}} \mathbf{a}_\theta - \frac{38}{\sqrt{40}} \mathbf{a}_\phi \\
&= -0.8571\mathbf{a}_r - 0.4066\mathbf{a}_\theta - 6.008\mathbf{a}_\phi
\end{aligned}$$

Note that $|\mathbf{A}|$ is the same in the three systems; that is,

$$|\mathbf{A}(x, y, z)| = |\mathbf{A}(\rho, \phi, z)| = |\mathbf{A}(r, \theta, \phi)| = 6.083$$

PRACTICE EXERCISE 2.1

(a) Convert points $P(1, 3, 5)$, $T(0, -4, 3)$, and $S(-3, -4, -10)$ from Cartesian to cylindrical and spherical coordinates.

(b) Transform vector

$$\mathbf{Q} = \frac{\sqrt{x^2 + y^2}\,\mathbf{a}_x}{\sqrt{x^2 + y^2 + z^2}} - \frac{yz\mathbf{a}_z}{\sqrt{x^2 + y^2 + z^2}}$$

to cylindrical and spherical coordinates.

(c) Evaluate $\mathbf{Q}$ at T in the three coordinate systems.

Answer: (a) $P(3.162, 71.56°, 5)$, $P(5.916, 32.31°, 71.56°)$, $T(4, 270°, 3)$,
$T(5, 53.13°, 270°)$, $S(5, 233.1°, -10)$, $S(11.18, 153.43°, 233.1°)$.

(b) $\dfrac{\rho}{\sqrt{\rho^2 + z^2}} (\cos\phi\, \mathbf{a}_\rho - \sin\phi\, \mathbf{a}_\phi - z\sin\phi\, \mathbf{a}_z)$, $\sin\theta(\sin\theta\cos\phi - $

$r\cos^2\theta\sin\phi)\mathbf{a}_r + \sin\theta\cos\theta(\cos\phi + r\sin\theta\sin\phi)\mathbf{a}_\theta - \sin\theta\sin\phi\, \mathbf{a}_\phi$.

(c) $0.8\mathbf{a}_x + 2.4\mathbf{a}_z, 0.8\mathbf{a}_\phi + 2.4\mathbf{a}_z, 1.44\mathbf{a}_r - 1.92\mathbf{a}_\theta + 0.8\mathbf{a}_\phi$.

EXAMPLE 2.2

Express the vector

$$\mathbf{B} = \frac{10}{r}\mathbf{a}_r + r\cos\theta\, \mathbf{a}_\theta + \mathbf{a}_\phi$$

in Cartesian and cylindrical coordinates. Find $\mathbf{B}(-3, 4, 0)$ and $\mathbf{B}(5, \pi/2, -2)$.

Solution:

Using eq. (2.28):

$$\begin{bmatrix} B_x \\ B_y \\ B_z \end{bmatrix} = \begin{bmatrix} \sin\theta\cos\phi & \cos\theta\cos\phi & -\sin\phi \\ \sin\theta\sin\phi & \cos\theta\sin\phi & \cos\phi \\ \cos\theta & -\sin\theta & 0 \end{bmatrix} \begin{bmatrix} \dfrac{10}{r} \\ r\cos\theta \\ 1 \end{bmatrix}$$

or

$$B_x = \frac{10}{r}\sin\theta\cos\phi + r\cos^2\theta\cos\phi - \sin\phi$$

$$B_y = \frac{10}{r}\sin\theta\sin\phi + r\cos^2\theta\sin\phi + \cos\phi$$

$$B_z = \frac{10}{r}\cos\theta - r\cos\theta\sin\theta$$

But $r = \sqrt{x^2 + y^2 + z^2}$, $\theta = \tan^{-1}\dfrac{\sqrt{x^2 + y^2}}{z}$, and $\phi = \tan^{-1}\dfrac{y}{x}$

Hence,

$$\sin\theta = \frac{\rho}{r} = \frac{\sqrt{x^2 + y^2}}{\sqrt{x^2 + y^2 + z^2}}, \quad \cos\theta = \frac{z}{r} = \frac{z}{\sqrt{x^2 + y^2 + z^2}}$$

$$\sin\phi = \frac{y}{\rho} = \frac{y}{\sqrt{x^2 + y^2}}, \quad \cos\phi = \frac{x}{\rho} = \frac{x}{\sqrt{x^2 + y^2}}$$

Substituting all these gives

$$B_x = \frac{10\sqrt{x^2 + y^2}}{(x^2 + y^2 + z^2)} \cdot \frac{x}{\sqrt{x^2 + y^2}} + \frac{\sqrt{x^2 + y^2 + z^2}}{(x^2 + y^2 + z^2)} \cdot \frac{z^2 x}{\sqrt{x^2 + y^2}} - \frac{y}{\sqrt{x^2 + y^2}}$$

$$= \frac{10x}{x^2 + y^2 + z^2} + \frac{xz^2}{\sqrt{(x^2 + y^2)(x^2 + y^2 + z^2)}} - \frac{y}{\sqrt{(x^2 + y^2)}}$$

$$B_y = \frac{10\sqrt{x^2 + y^2}}{(x^2 + y^2 + z^2)} \cdot \frac{y}{\sqrt{x^2 + y^2}} + \frac{\sqrt{x^2 + y^2 + z^2}}{x^2 + y^2 + z^2} \cdot \frac{z^2 y}{\sqrt{x^2 + y^2}} + \frac{x}{\sqrt{x^2 + y^2}}$$

$$= \frac{10y}{x^2 + y^2 + z^2} + \frac{yz^2}{\sqrt{(x^2 + y^2)(x^2 + y^2 + z^2)}} + \frac{x}{\sqrt{x^2 + y^2}}$$

$$B_z = \frac{10z}{x^2 + y^2 + z^2} - \frac{z\sqrt{x^2 + y^2}}{\sqrt{x^2 + y^2 + z^2}}$$

$$\mathbf{B} = B_x \mathbf{a}_x + B_y \mathbf{a}_y + B_z \mathbf{a}_z$$

where B_x, B_y, and B_z are as just given.

At $(-3, 4, 0)$, $x = -3$, $y = 4$, and $z = 0$, so

$$B_x = -\frac{30}{25} + 0 - \frac{4}{5} = -2$$

$$B_y = \frac{40}{25} + 0 - \frac{3}{5} = 1$$

$$B_z = 0 - 0 = 0$$

Thus,

$$\mathbf{B} = -2\mathbf{a}_x + \mathbf{a}_y$$

For spherical to cylindrical vector transformation (see Problem 2.16),

$$\begin{bmatrix} B_\rho \\ B_\phi \\ B_z \end{bmatrix} = \begin{bmatrix} \sin\theta & \cos\theta & 0 \\ 0 & 0 & 1 \\ \cos\theta & -\sin\theta & 0 \end{bmatrix} \begin{bmatrix} \dfrac{10}{r} \\ r\cos\theta \\ 1 \end{bmatrix}$$

or

$$B_\rho = \frac{10}{r}\sin\theta + r\cos^2\theta$$

$$B_\phi = 1$$

$$B_z = \frac{10}{r}\cos\theta - r\sin\theta\cos\theta$$

But $r = \sqrt{\rho^2 + z^2}$ and $\theta = \tan^{-1}\dfrac{\rho}{z}$

Thus,

$$\sin\theta = \frac{\rho}{\sqrt{\rho^2 + z^2}}, \quad \cos\theta = \frac{z}{\sqrt{\rho^2 + z^2}}$$

$$B_\rho = \frac{10\rho}{\rho^2 + z^2} + \sqrt{\rho^2 + z^2} \cdot \frac{z^2}{\rho^2 + z^2}$$

$$B_z = \frac{10z}{\rho^2 + z^2} - \sqrt{\rho^2 + z^2} \cdot \frac{\rho z}{\rho^2 + z^2}$$

Hence,

$$\mathbf{B} = \left(\frac{10\rho}{\rho^2 + z^2} + \frac{z^2}{\sqrt{\rho^2 + z^2}}\right)\mathbf{a}_\rho + \mathbf{a}_\phi + \left(\frac{10z}{\rho^2 + z^2} - \frac{\rho z}{\sqrt{\rho^2 + z^2}}\right)\mathbf{a}_z$$

At $(5, \pi/2, -2)$, $\rho = 5$, $\phi = \pi/2$, and $z = -2$, so

$$\mathbf{B} = \left(\frac{50}{29} + \frac{4}{\sqrt{29}}\right)\mathbf{a}_\rho + \mathbf{a}_\phi + \left(\frac{-20}{29} + \frac{10}{\sqrt{29}}\right)\mathbf{a}_z$$

$$= 2.467\mathbf{a}_\rho + \mathbf{a}_\phi + 1.167\mathbf{a}_z$$

Note that at $(-3, 4, 0)$,

$$|\mathbf{B}(x, y, z)| = |\mathbf{B}(\rho, \phi, z)| = |\mathbf{B}(r, \theta, \phi)| = 2.907$$

This may be used to check the correctness of the result whenever possible.

PRACTICE EXERCISE 2.2

Express the following vectors in Cartesian coordinates:

(a) $\mathbf{A} = \rho z \sin\phi\, \mathbf{a}_\rho + 3\rho \cos\phi\, \mathbf{a}_\phi + \rho \cos\phi \sin\phi\, \mathbf{a}_z$

(b) $\mathbf{B} = r^2 \mathbf{a}_r + \sin\theta\, \mathbf{a}_\phi$

Answer: (a) $\mathbf{A} = \dfrac{1}{\sqrt{x^2 + y^2}}[(xyz - 3xy)\mathbf{a}_x + (zy^2 + 3x^2)\mathbf{a}_y + xy\mathbf{a}_z]$.

(b) $\mathbf{B} = \dfrac{1}{\sqrt{x^2 + y^2 + z^2}}\{[x(x^2 + y^2 + z^2) - y]\mathbf{a}_x +$

$[y(x^2 + y^2 + z^2) + x]\mathbf{a}_y + z(x^2 + y^2 + z^2)\mathbf{a}_z\}$.

2.5 CONSTANT-COORDINATE SURFACES

Surfaces in Cartesian, cylindrical, or spherical coordinate systems are easily generated by keeping one of the coordinate variables constant and allowing the other two to vary. In the

Cartesian system, if we keep x constant and allow y and z to vary, an infinite plane is generated. Thus we could have infinite planes

$$x = \text{constant}$$
$$y = \text{constant} \qquad (2.34)$$
$$z = \text{constant}$$

which are perpendicular to the x-, y-, and z-axes, respectively, as shown in Figure 2.7. The intersection of two planes is a line. For example,

$$x = \text{constant}, \quad y = \text{constant} \qquad (2.35)$$

is the line RPQ parallel to the z-axis. The intersection of three planes is a point. For example,

$$x = \text{constant}, \quad y = \text{constant}, \quad z = \text{constant} \qquad (2.36)$$

is the point $P(x, y, z)$. Thus we may define point P as the intersection of three orthogonal infinite planes. If P is $(1, -5, 3)$, then P is the intersection of planes $x = 1$, $y = -5$, and $z = 3$.

Orthogonal surfaces in cylindrical coordinates can likewise be generated. The surfaces

$$\rho = \text{constant}$$
$$\phi = \text{constant} \qquad (2.37)$$
$$z = \text{constant}$$

are illustrated in Figure 2.8, where it is easy to observe that $\rho = $ constant is a circular cylinder, $\phi = $ constant is a semi-infinite plane with its edge along the z-axis, and $z = $ constant is the same infinite plane as in a Cartesian system. Where two surfaces meet is either a line or a circle. Thus,

$$z = \text{constant}, \quad \rho = \text{constant} \qquad (2.38)$$

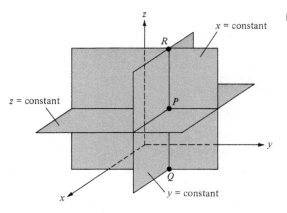

FIGURE 2.7 Constant x, y, and z surfaces.

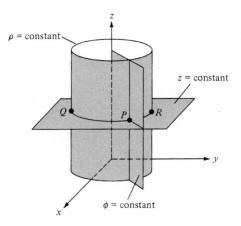

FIGURE 2.8 Constant ρ, ϕ, and z surfaces.

is a circle QPR of radius ρ, whereas z = constant, ϕ = constant is a semi-infinite line. A point is an intersection of the three surfaces in eq. (2.37). Thus,

$$\rho = 2, \quad \phi = 60°, \quad z = 5 \tag{2.39}$$

is the point $P(2, 60°, 5)$.

The orthogonal nature of the spherical coordinate system is evident by considering the three surfaces

$$\begin{aligned} r &= \text{constant} \\ \theta &= \text{constant} \\ \phi &= \text{constant} \end{aligned} \tag{2.40}$$

which are shown in Figure 2.9, where we notice that r = constant is a sphere of radius r with its center at the origin; θ = constant is a circular cone with the z-axis as its axis and the origin as its vertex; ϕ = constant is the semi-infinite plane as in a cylindrical system. A line is formed by the intersection of two surfaces. For example,

$$r = \text{constant}, \quad \phi = \text{constant} \tag{2.41}$$

FIGURE 2.9 Constant r, θ, and ϕ surfaces.

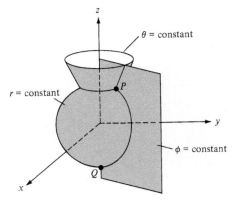

is a semicircle passing through Q and P. The intersection of three surfaces gives a point. Thus,

$$r = 5, \quad \theta = 30°, \quad \phi = 60° \tag{2.42}$$

is the point $P(5, 30°, 60°)$. We notice that in general, a point in three-dimensional space can be identified as the intersection of three mutually orthogonal surfaces. Also, a unit normal vector to the surface $n = $ constant is $\pm \mathbf{a}_n$, where n is x, y, z, ρ, ϕ, r, or θ. For example, to the plane $x = 5$, a unit normal vector is $\pm \mathbf{a}_x$ and to the plane $\phi = 20°$, a unit normal vector is $\pm \mathbf{a}_\phi$.

EXAMPLE 2.3

Two uniform vector fields are given by $\mathbf{E} = -5\mathbf{a}_\rho + 10\mathbf{a}_\phi + 3\mathbf{a}_z$ and $\mathbf{F} = \mathbf{a}_\rho + 2\mathbf{a}_\phi - 6\mathbf{a}_z$. Calculate

(a) $|\mathbf{E} \times \mathbf{F}|$
(b) The vector component of $\mathbf{E}$ at $P(5, \pi/2, 3)$ parallel to the line $x = 2, z = 3$
(c) The angle that $\mathbf{E}$ makes with the surface $z = 3$ at P

Solution:

(a) $\mathbf{E} \times \mathbf{F} = \begin{vmatrix} \mathbf{a}_\rho & \mathbf{a}_\phi & \mathbf{a}_z \\ -5 & 10 & 3 \\ 1 & 2 & -6 \end{vmatrix}$

$$= (-60 - 6)\mathbf{a}_\rho + (3 - 30)\mathbf{a}_\phi + (-10 - 10)\mathbf{a}_z$$
$$= (-66, -27, -20)$$
$$|\mathbf{E} \times \mathbf{F}| = \sqrt{66^2 + 27^2 + 20^2} = 74.06$$

(b) Line $x = 2, z = 3$ is parallel to the y-axis, so the component of $\mathbf{E}$ parallel to the given line is

$$(\mathbf{E} \cdot \mathbf{a}_y)\mathbf{a}_y$$

But at $P(5, \pi/2, 3)$

$$\mathbf{a}_y = \sin \phi \, \mathbf{a}_\rho + \cos \phi \, \mathbf{a}_\phi$$
$$= \sin \pi/2 \, \mathbf{a}_\rho + \cos \pi/2 \, \mathbf{a}_\phi = \mathbf{a}_\rho$$

Therefore,

$$(\mathbf{E} \cdot \mathbf{a}_y)\mathbf{a}_y = (\mathbf{E} \cdot \mathbf{a}_\rho)\mathbf{a}_\rho = -5\mathbf{a}_\rho \quad (\text{or} -5\mathbf{a}_y)$$

(c) Since the z-axis is normal to the surface $z = 3$, we can use the dot product to find the angle between the z-axis and $\mathbf{E}$, as shown in Figure 2.10:

$$\mathbf{E} \cdot \mathbf{a}_z = |E|(1) \cos \theta_{Ez} \rightarrow 3 = \sqrt{134} \cos \theta_{Ez}$$

$$\cos \theta_{Ez} = \frac{3}{\sqrt{134}} = 0.2592 \rightarrow \theta_{Ez} = 74.98°$$

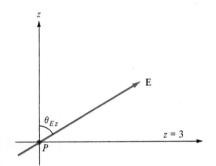

FIGURE 2.10 For Example 2.3(c).

Hence, the angle between $z = 3$ and **E** is

$$90° - \theta_{Ez} = 15.02°$$

PRACTICE EXERCISE 2.3

Given the vector field

$$\mathbf{H} = \rho z \cos \phi \, \mathbf{a}_\rho + e^{-2} \sin \frac{\phi}{2} \, \mathbf{a}_\phi + \rho^2 \mathbf{a}_z$$

at point $(1, \pi/3, 0)$, find

(a) $\mathbf{H} \cdot \mathbf{a}_x$

(b) $\mathbf{H} \times \mathbf{a}_\theta$

(c) The vector component of **H** normal to surface $\rho = 1$

(d) The scalar component of **H** tangential to the plane $z = 0$

Answer: (a) -0.0586, (b) $-0.06767 \, \mathbf{a}_\rho$, (c) $0 \, \mathbf{a}_\rho$, (d) 0.06767.

EXAMPLE 2.4

Given a vector field

$$\mathbf{D} = r \sin \phi \, \mathbf{a}_r - \frac{1}{r} \sin \theta \cos \phi \, \mathbf{a}_\theta + r^2 \mathbf{a}_\phi$$

determine

(a) **D** at $P(10, 150°, 330°)$

(b) The component of **D** tangential to the spherical surface $r = 10$ at P

(c) A unit vector at P perpendicular to **D** and tangential to the cone $\theta = 150°$

Solution:

(a) At P, $r = 10$, $\theta = 150°$, and $\phi = 330°$. Hence,

$$\mathbf{D} = 10 \sin 330° \, \mathbf{a}_r - \frac{1}{10} \sin 150° \cos 330° \, \mathbf{a}_\theta + 100 \, \mathbf{a}_\phi = (-5, -0.043, 100)$$

(b) Any vector $\mathbf{D}$ can always be resolved into two orthogonal components:

$$\mathbf{D} = \mathbf{D}_t + \mathbf{D}_n$$

where $\mathbf{D}_t$ is tangential to a given surface and $\mathbf{D}_n$ is normal to it. In our case, since $\mathbf{a}_r$ is normal to the surface $r = 10$,

$$\mathbf{D}_n = r \sin \phi \, \mathbf{a}_r = -5\mathbf{a}_r$$

Hence,

$$\mathbf{D}_t = \mathbf{D} - \mathbf{D}_n = -0.043\mathbf{a}_\theta + 100\mathbf{a}_\phi$$

(c) A vector at P perpendicular to $\mathbf{D}$ and tangential to the cone $\theta = 150°$ is the same as the vector perpendicular to both $\mathbf{D}$ and $\mathbf{a}_\theta$. Hence,

$$\mathbf{D} \times \mathbf{a}_\theta = \begin{vmatrix} \mathbf{a}_r & \mathbf{a}_\theta & \mathbf{a}_\phi \\ -5 & -0.043 & 100 \\ 0 & 1 & 0 \end{vmatrix}$$

$$= -100\mathbf{a}_r - 5\mathbf{a}_\phi$$

A unit vector along this is

$$\mathbf{a} = \frac{-100\mathbf{a}_r - 5\mathbf{a}_\phi}{\sqrt{100^2 + 5^2}} = -0.9988\mathbf{a}_r - 0.0499\mathbf{a}_\phi$$

PRACTICE EXERCISE 2.4

If $\mathbf{A} = 3\mathbf{a}_r + 2\mathbf{a}_\theta - 6\mathbf{a}_\phi$ and $\mathbf{B} = 4\mathbf{a}_r + 3\mathbf{a}_\phi$, determine

(a) $\mathbf{A} \cdot \mathbf{B}$

(b) $|\mathbf{A} \times \mathbf{B}|$

(c) The vector component of $\mathbf{A}$ along $\mathbf{a}_z$ at $(1, \pi/3, 5\pi/4)$

Answer: (a) -6, (b) 34.48, (c) $-0.116\mathbf{a}_r + 0.201\mathbf{a}_\theta$.

MATLAB 2.1

```
% This script allows the user to input a coordinate in either
% rectangular, cylindrical, or spherical coordinates and
% retrieve the answer in the other coordinate systems
clear
% prompt the user for the coordinate system
disp('Enter the coordinate system of the input coordinate');
coord_sys = input(' (r, c, or s)... \n > ','s');
% if user entered something other than "r" "c" or "s"
% set default as "r"
if isempty(coord_sys); coord_sys = 'r'; end
if coord_sys == 'r';
    % prompt the user for the coordinate
    disp('Enter the rectangular coordinate in the ');
    crd = input('format [x y z]... \n > ');
    % check input to see if empty and set to 0 if so
    if isempty(crd); crd = [0 0 0]; end
    disp('Cylindrical coordinates [rho phi(rad) z]:')
    % display the result... the [ ] and enclose a
    % three-dimensional vector
    disp([sqrt(crd(1)^2+crd(2)^2) atan2(crd(2),crd(1)) crd(3)])
    disp('Spherical coordinates [r phi(rad) theta(rad]:')
    disp([norm(crd) atan2(crd(2),crd(1)) acos(crd(3)/
    norm(crd))])
elseif coord_sys == 'c';   % if not r but c execute this block
    disp('Enter the cylindrical coordinate in the format');
    crd = input(' [\rho \phi z]... \n > ');
    % check input to see if empty and set to 0 if so
    if isempty(crd); crd = [0 0 0]; end
    disp('Rectangular coordinates [x y z]:')
    disp([crd(1)*cos(crd(2)) crd(1)*sin(crd(2)) crd(3)])
    disp('Spherical coordinates [r phi(rad) theta(rad]:')
    disp([sqrt(crd(1)^2+crd(3)^2) crd(2) crd(3)*cos(crd(3))])
else coord_sys == 's';   % if not r nor c but s execute this block
    disp('Enter the spherical coordinate in the');
    crd = input('format [\rho \phi \theta]... \n > ');
    if isempty(crd); crd = [0 0 0]; end
    disp('Rectangular coordinates [x y z]:')
    disp([crd(1)*cos(crd(2))*sin(crd(3)) ...
    crd(1)*sin(crd(2))*sin(crd(3)) crd(1)*cos(crd(3))])
    disp('Cylindrical coordinates [r phi(rad) theta(rad]:')
    disp([crd(1)*sin(crd(3)) crd(2) crd(1)*cos(crd(3))])
end
```

MATLAB 2.1

```
% This script allows the user to input a non-variable vector
% in rectangular coordinates and obtain the cylindrical, or
% spherical components. The user must also enter the point
% location where this transformation occurs; the result
```

```
% depends on the vector's observation point
clear
% prompt the user for the vectors and check to see if entered
% properly, else set to 0
disp('Enter the rectangular vector (in the ');
v = input(' format [x y z])... \n >  ');
if isempty(v); v = [0 0 0]; end
disp('Enter the location of the vector (in the ');
p = input(' format [x y z])... \n >  ');
if isempty(p); p = [0 0 0]; end
disp('Cylindrical components [rho phi(rad) z]:')
phi = atan2(p(2),p(1));
% Create the transformation matrix
cyl_p=[cos(phi) sin(phi) 0; ...    % The ellipses allow a single
                                   % command over multiple lines
        -sin(phi) cos(phi) 0; ...
          0 0 1];
disp((cyl_p*v')')    % the '  denotes a transpose from a row
                     % vector to a column vector
                   % The second transpose converts the column
                 % vector back to a row vector
disp('Spherical components [r phi(rad) theta(rad]:')
phi = atan2(p(3),sqrt(p(1)^2+p(2)^2));
theta = atan2(p(2),p(1));
% Create the transformation matrix
sph_p=[sin(theta)*cos(phi) sin(theta)*sin(phi) cos(theta); ...
          cos(theta)*cos(phi) cos(theta)*sin(phi) -sin(theta);...
          -sin(phi) cos(phi) 0];
disp((sph_p*v')')
```

SUMMARY

1. The three common coordinate systems we shall use throughout the text are the Cartesian (or rectangular), the circular cylindrical, and the spherical.
2. A point P is represented as $P(x, y, z)$, $P(\rho, \phi, z)$, and $P(r, \theta, \phi)$ in the Cartesian, cylindrical, and spherical systems, respectively. A vector field A is represented as (A_x, A_y, A_z) or $A_x\mathbf{a}_x + A_y\mathbf{a}_y + A_z\mathbf{a}_z$ in the Cartesian system, as (A_ρ, A_ϕ, A_z) or $A_\rho\mathbf{a}_\rho + A_\phi\mathbf{a}_\phi + A_z\mathbf{a}_z$ in the cylindrical system, and as (A_r, A_θ, A_ϕ) or $A_r\mathbf{a}_r + A_\theta\mathbf{a}_\theta + A_\phi\mathbf{a}_\phi$ in the spherical system. It is preferable that mathematical operations (addition, subtraction, product, etc.) be performed in the same coordinate system. Thus, point and vector transformations should be performed whenever necessary. A summary of point and vector transformations is given in Table 2.1.
3. Fixing one space variable defines a surface; fixing two defines a line; fixing three defines a point.
4. A unit normal vector to surface $n = $ constant is $\pm\mathbf{a}_n$.

TABLE 2.1 Relationships between Rectangular, Cylindrical, and Spherical Coordinates

Rectangular to Cylindrical

Variable change
$$\begin{cases} x = \rho \cos \phi \\ y = \rho \sin \phi \\ z = z \end{cases}$$

Component change
$$\begin{cases} A_\rho = A_x \cos \phi + A_y \sin \phi \\ A_\phi = -A_x \sin \phi + A_y \cos \phi \\ A_z = A_z \end{cases}$$

Cylindrical to Rectangular

Variable change
$$\begin{cases} \rho = \sqrt{x^2 + y^2} \\ \phi = \tan^{-1}\left(\dfrac{y}{x}\right) \\ z = z \end{cases} \begin{cases} \sin \phi = \dfrac{y}{\sqrt{x^2 + y^2}} \\ \cos \phi = \dfrac{x}{\sqrt{x^2 + y^2}} \end{cases}$$

Component change
$$\begin{cases} A_x = A_\rho \dfrac{x}{\sqrt{x^2 + y^2}} - A_\phi \dfrac{y}{\sqrt{x^2 + y^2}} \\ A_y = A_\rho \dfrac{y}{\sqrt{x^2 + y^2}} + A_\phi \dfrac{x}{\sqrt{x^2 + y^2}} \\ A_z = A_z \end{cases}$$

Rectangular to Spherical

Variable change
$$\begin{cases} x = r \sin \theta \cos \phi \\ y = r \sin \theta \sin \phi \\ z = r \cos \theta \end{cases}$$

Component change
$$\begin{cases} A_r = A_x \sin \theta \cos \phi + A_y \sin \theta \sin \phi \\ \qquad + A_z \cos \theta \\ A_\theta = A_x \cos \theta \cos \phi + A_y \cos \theta \sin \phi \\ \qquad - A_z \sin \theta \\ A_\phi = -A_x \sin \phi + A_y \cos \phi \end{cases}$$

Spherical to Rectangular

Variable change
$$\begin{cases} r = \sqrt{x^2 + y^2 + z^2} \\ \theta = \cos^{-1}\dfrac{z}{\sqrt{x^2 + y^2 + z^2}} \\ \phi = \tan^{-1}\left(\dfrac{y}{x}\right) \end{cases}$$
$$\begin{cases} \cos \theta = \dfrac{z}{\sqrt{x^2 + y^2 + z^2}} \\ \sin \theta = \dfrac{\sqrt{x^2 + y^2}}{\sqrt{x^2 + y^2 + z^2}} \end{cases}$$
$$\begin{cases} \cos \phi = \dfrac{x}{\sqrt{x^2 + y^2}} \\ \sin \phi = \dfrac{y}{\sqrt{x^2 + y^2}} \end{cases}$$

Component change
$$\begin{cases} A_x = \dfrac{A_r x}{\sqrt{x^2 + y^2 + z^2}} + \dfrac{A_\theta xz}{\sqrt{(x^2 + y^2)(x^2 + y^2 + z^2)}} - \dfrac{A_\phi y}{\sqrt{x^2 + y^2}} \\ A_y = \dfrac{A_r y}{\sqrt{x^2 + y^2 + z^2}} + \dfrac{A_\theta yz}{\sqrt{(x^2 + y^2)(x^2 + y^2 + z^2)}} + \dfrac{A_\phi x}{\sqrt{x^2 + y^2}} \\ A_z = \dfrac{A_r z}{\sqrt{x^2 + y^2 + z^2}} - \dfrac{A_\theta \sqrt{x^2 + y^2}}{\sqrt{x^2 + y^2 + z^2}} \end{cases}$$

Adopted with permission from G. F. Miner, *Lines and Electromagnetic Fields for Engineers*. New York: Oxford Univ. Press, 1996, p. 263.

REVIEW QUESTIONS

2.1 The ranges of θ and ϕ as given by eq. (2.17) are not the only possible ones. The following are all alternative ranges of θ and ϕ, except

(a) $0 \le \theta < 2\pi, 0 \le \phi \le \pi$

(b) $0 \le \theta < 2\pi, 0 \le \phi < 2\pi$

(c) $-\pi \le \theta \le \pi, 0 \le \phi \le \pi$

(d) $-\pi/2 \le \theta \le \pi/2, 0 \le \phi < 2\pi$

(e) $0 \le \theta \le \pi, -\pi \le \phi < \pi$

(f) $-\pi \le \theta < \pi, -\pi \le \phi < \pi$

2.2 At Cartesian point $(-3, 4, -1)$, which of these is incorrect?

(a) $\rho = -5$

(b) $r = \sqrt{26}$

(c) $\theta = \tan^{-1}\dfrac{5}{-1}$

(d) $\phi = \tan^{-1}\dfrac{4}{-3}$

2.3 Which of these is not valid at point $(0, 4, 0)$?

(a) $\mathbf{a}_\phi = -\mathbf{a}_x$

(b) $\mathbf{a}_\theta = -\mathbf{a}_z$

(c) $\mathbf{a}_r = 4\mathbf{a}_y$

(d) $\mathbf{a}_\rho = \mathbf{a}_y$

2.4 A unit normal vector to the cone $\theta = 30°$ is:

(a) $\mathbf{a}_r$

(b) $\mathbf{a}_\theta$

(c) $\mathbf{a}_\phi$

(d) none of these

2.5 At every point in space, $\mathbf{a}_\phi \cdot \mathbf{a}_\theta = 1$.

(a) True

(b) False

2.6 If $\mathbf{H} = 4\mathbf{a}_\rho - 3\mathbf{a}_\phi + 5\mathbf{a}_z$, at $(1, \pi/2, 0)$ the component of $\mathbf{H}$ parallel to surface $\rho = 1$ is

(a) $4\mathbf{a}_\rho$

(b) $5\mathbf{a}_z$

(c) $-3\mathbf{a}_\phi$

(d) $-3\mathbf{a}_\phi + 5\mathbf{a}_z$

(e) $5\mathbf{a}_\phi + 3\mathbf{a}_z$

2.7 Given $\mathbf{G} = 20\mathbf{a}_r + 50\mathbf{a}_\theta + 40\mathbf{a}_\phi$, at $(1, \pi/2, \pi/6)$ the component of $\mathbf{G}$ perpendicular to surface $\theta = \pi/2$ is

(a) $20\mathbf{a}_r$

(b) $50\mathbf{a}_\theta$

(c) $40\mathbf{a}_\phi$

(d) $20\mathbf{a}_r + 40\mathbf{a}_\theta$

(e) $-40\mathbf{a}_r + 20\mathbf{a}_\phi$

2.8 Where surfaces $\rho = 2$ and $z = 1$ intersect is

(a) an infinite plane

(b) a semi-infinite plane

(c) a circle

(d) a cylinder

(e) a cone

2.9 Match the items in the list at the left with those in the list at the right. Each answer can be used once, more than once, or not at all.

(a) $\theta = \pi/4$

(b) $\phi = 2\pi/3$

(c) $x = -10$

(d) $r = 1, \theta = \pi/3, \phi = \pi/2$

(e) $\rho = 5$

(f) $\rho = 3, \phi = 5\pi/3$

(g) $\rho = 10, z = 1$

(i)   infinite plane

(ii)   semi-infinite plane

(iii)   circle

(iv)   semicircle

(v)   straight line

(vi)   cone

(vii)   cylinder

(h) $r = 4, \phi = \pi/6$ (viii) sphere

(i) $r = 5, \theta = \pi/3$ (ix)   cube

(x)   point

2.10 A wedge is described by $z = 0$, $30° < \phi < 60°$. Which of the following is incorrect?

(a) The wedge lies in the xy-plane.

(b) It is infinitely long.

(c) On the wedge, $0 < \rho < \infty$.

(d) A unit normal to the wedge is $\pm \mathbf{a}_z$.

(e) The wedge includes neither the x-axis nor the y-axis.

Answers: 2.1b,f, 2.2a, 2.3c, 2.4b, 2.5b, 2.6d, 2.7b, 2.8c, 2.9a-(vi), b-(ii), c-(i), d-(x), e-(vii), f-(v), g-(iii), h-(iv), i-(iii), 2.10b.

PROBLEMS

Sections 2.3 and 2.4—Cylindrical and Spherical Coordinates

2.1 Convert the following Cartesian points to cylindrical and spherical coordinates:

(a) $P(2, 5, 1)$

(b) $Q(-3, 4, 0)$

(c) $R(6, 2, -4)$

2.2 Express the following points in Cartesian coordinates:

(a) $P_1(2, 30°, 5)$

(b) $P_2(1, 90°, -3)$

(c) $P_3(10, \pi/4, \pi/3)$

(d) $P_4(4, 30°, 60°)$

2.3 The rectangular coordinates at point P are $(x = 2, y = 6, z = -4)$. (a) What are its cylindrical coordinates? (b) What are its spherical coordinates?

2.4 The cylindrical coordinates of point Q are $\rho = 5$, $\phi = 120°$, $z = 1$. Express Q as rectangular and spherical coordinates.

2.5 Given point $T(10, 60°, 30°)$ in spherical coordinates, express T in Cartesian and cylindrical coordinates.

2.6 (a) If $V = xz - xy + yz$, express V in cylindrical coordinates.

(b) If $U = x^2 + 2y^2 + 3z^2$, express U in spherical coordinates.

2.7 Convert the following vectors to cylindrical and spherical systems:

(a) $\mathbf{F} = \dfrac{x\mathbf{a}_x + y\mathbf{a}_y + 4\mathbf{a}_z}{\sqrt{x^2 + y^2 + z^2}}$

(b) $\mathbf{G} = (x^2 + y^2)\left[\dfrac{x\mathbf{a}_x}{\sqrt{x^2 + y^2 + z^2}} + \dfrac{y\mathbf{a}_y}{\sqrt{x^2 + y^2 + z^2}} + \dfrac{z\mathbf{a}_z}{\sqrt{x^2 + y^2 + z^2}}\right]$

2.8 Let $\mathbf{B} = \sqrt{x^2 + y^2}\,\mathbf{a}_x + \dfrac{y}{\sqrt{x^2 + y^2}}\,\mathbf{a}_y + z\mathbf{a}_z$. Transform $\mathbf{B}$ to cylindrical coordinates.

2.9 Given vector $\mathbf{A} = 2\mathbf{a}_\rho + 3\mathbf{a}_\phi + 4\mathbf{a}_z$, convert $\mathbf{A}$ into Cartesian coordinates at point $(2, \pi/2, -1)$.

2.10 Express the following vectors in rectangular coordinates:

(a) $\mathbf{A} = \rho \sin \phi\, \mathbf{a}_\rho + \rho \cos \phi\, \mathbf{a}_\phi - 2z\, \mathbf{a}_z$

(b) $\mathbf{B} = 4r \cos \phi\, \mathbf{a}_r + r\, \mathbf{a}_\theta$

2.11 Given the vector field $\mathbf{F} = \dfrac{4\mathbf{a}_r}{r^2}$, express F in rectangular coordinates.

2.12 If $B = r \sin \theta \mathbf{a}_r - r^2 \cos \phi \mathbf{a}_\phi$, (a) find $\mathbf{B}$ at $(2, \pi/2, 3\pi/2)$, (b) convert $\mathbf{B}$ to Cartersian coordinates.

2.13 Let $\mathbf{B} = x\mathbf{a}_z$. Express $\mathbf{B}$ in

(a) cylindrical coordinates,

(b) spherical coordinates.

2.14 Prove the following:

(a) $\mathbf{a}_x \times \mathbf{a}_\rho = \cos \phi$
$\mathbf{a}_x \times \mathbf{a}_\phi = -\sin \phi$
$\mathbf{a}_y \times \mathbf{a}_\rho = \sin \phi$
$\mathbf{a}_y \times \mathbf{a}_\phi = \cos\phi$

(b) $\mathbf{a}_x \times \mathbf{a}_r = \sin \theta \cos \phi$
$\mathbf{a}_x \times \mathbf{a}_\theta = \cos \theta \cos \phi$
$\mathbf{a}_y \times \mathbf{a}_r = \sin \theta \sin \phi$

(c) $\mathbf{a}_y \times \mathbf{a}_\theta = \cos \theta \sin \phi$
$\mathbf{a}_z \times \mathbf{a}_r = \cos \theta$
$\mathbf{a}_z \times \mathbf{a}_\theta = -\sin \theta$

2.15 Prove the following expressions:

(a) $\mathbf{a}_\rho \times \mathbf{a}_\phi = \mathbf{a}_z$
$\mathbf{a}_z \times \mathbf{a}_\rho = \mathbf{a}_\phi$
$\mathbf{a}_\phi \times \mathbf{a}_z = \mathbf{a}_\rho$

(b) $\mathbf{a}_r \times \mathbf{a}_\phi = \mathbf{a}_\phi$
$\mathbf{a}_z \times \mathbf{a}_\rho = \mathbf{a}_\theta$
$\mathbf{a}_\theta \times \mathbf{a}_\phi = \mathbf{a}_r$

2.16 (a) Show that point transformation between cylindrical and spherical coordinates is obtained using

$$r = \sqrt{\rho^2 + z^2}, \quad \theta = \tan^{-1}\dfrac{\rho}{z}, \quad \phi = \phi$$

or

$$\rho = r\sin\theta, \quad z = r\cos\theta, \quad \phi = \phi$$

(b) Show that vector transformation between cylindrical and spherical coordinates is obtained using

$$\begin{bmatrix} A_r \\ A_\theta \\ A_\phi \end{bmatrix} = \begin{bmatrix} \sin\theta & 0 & \cos\theta \\ \cos\theta & 0 & -\sin\theta \\ 0 & 1 & 0 \end{bmatrix} \begin{bmatrix} A_\rho \\ A_\phi \\ A_z \end{bmatrix}$$

or

$$\begin{bmatrix} A_\rho \\ A_\phi \\ A_z \end{bmatrix} = \begin{bmatrix} \sin\theta & \cos\theta & 0 \\ 0 & 0 & 1 \\ \cos\theta & -\sin\theta & 0 \end{bmatrix} \begin{bmatrix} A_r \\ A_\theta \\ A_\phi \end{bmatrix}$$

(*Hint:* Make use of Figures 2.5 and 2.6.)

2.17 At point $P(2,0,-1)$, calculate the value of the following dot products:

(a) $\mathbf{a}_\rho \cdot \mathbf{a}_x$, (b)$\mathbf{a}_\phi \cdot \mathbf{a}_y$, (c)$\mathbf{a}_r \cdot \mathbf{a}_z$

2.18 Show that the vector fields

$$\mathbf{A} = \rho\sin\phi\,\mathbf{a}_\rho + \rho\cos\phi\,\mathbf{a}_\phi + \rho\mathbf{a}_z$$
$$\mathbf{B} = \rho\sin\phi\,\mathbf{a}_\rho + \rho\cos\phi\,\mathbf{a}_\phi - \rho\mathbf{a}_z$$

are perpendicular to each other at any point.

2.19 Given that $\mathbf{A} = 3\mathbf{a}_\rho + 2\mathbf{a}_\phi + \mathbf{a}_z$ and $\mathbf{B} = 5\mathbf{a}_\rho - 8\mathbf{a}_z$, find:

(a) $\mathbf{A} + \mathbf{B}$,   (b) $\mathbf{A} \cdot \mathbf{B}$,   (c) $\mathbf{A} \times \mathbf{B}$,   (d) the angle between $\mathbf{A}$ and $\mathbf{B}$.

2.20 Given that $\mathbf{G} = 3\rho\mathbf{a}_\rho + \rho\cos\phi\,\mathbf{a}_\phi - z^2\mathbf{a}_z$, find the component of $\mathbf{G}$ along $\mathbf{a}_x$ at point $Q(3,-4,6)$.

2.21 Let $\mathbf{G} = yz\mathbf{a}_x + xz\mathbf{a}_y + xy\mathbf{a}_z$. Transform $\mathbf{G}$ to cylindrical coordinates.

2.22 The transformation $(A_\rho, A_\phi, A_z) \rightarrow (A_x, A_y, A_z)$ in eq. (2.15) is not complete. Complete it by expressing $\cos\phi$ and $\sin\phi$ in terms of x, y, and z. Do the same thing to the transformation $(A_r, A_\theta, A_\phi) \rightarrow (A_x, A_y, A_z)$ in eq. (2.28).

2.23 In Practice Exercise 2.2, express $\mathbf{A}$ in spherical and $\mathbf{B}$ in cylindrical coordinates. Evaluate $\mathbf{A}$ at $(10, \pi/2, 3\pi/4)$ and $\mathbf{B}$ at $(2, \pi/6, 1)$.

2.24 Calculate the distance between the following pairs of points:

(a) $(2, 1, 5)$ and $(6, -1, 2)$
(b) $(3, \pi/2, -1)$ and $(5, 3\pi/2, 5)$
(c) $(10, \pi/4, 3\pi/4)$ and $(5, \pi/6, 7\pi/4)$

2.25 Calculate the distance between points $P(4, 30°, 0°)$ and $Q(6, 90°, 180°)$.

2.26 At point $(0, 4, -1)$, express $\mathbf{a}_\rho$ and $\mathbf{a}_\phi$ in Cartesian coordinates.

2.27 Let $\mathbf{A} = (2z - \sin\phi)\mathbf{a}_\rho + (4\rho + 2\cos\phi)\mathbf{a}_\phi - 3\rho z\mathbf{a}_z$ and $\mathbf{B} = \rho\cos\phi\mathbf{a}_\rho + \sin\phi\mathbf{a}_\phi + \mathbf{a}_z$.

    (a) Find the minimum angle between $\mathbf{A}$ and $\mathbf{B}$ at $(1, 60°, -1)$.

    (b) Determine a unit vector normal to both $\mathbf{A}$ and $\mathbf{B}$ at $(1, 90°, 0)$.

2.28 Given vectors $\mathbf{A} = 2\mathbf{a}_x + 4\mathbf{a}_y + 10\mathbf{a}_z$ and $\mathbf{B} = -5\mathbf{a}_\rho + \mathbf{a}_\phi - 3\mathbf{a}_z$, find

    (a) $\mathbf{A} + \mathbf{B}$ at $P(0, 2, -5)$

    (b) The angle between $\mathbf{A}$ and $\mathbf{B}$ at P

    (c) The scalar component of $\mathbf{A}$ along $\mathbf{B}$ at P

2.29 Given that $\mathbf{B} = \rho^2 \sin\phi\mathbf{a}_\rho + (z - 1)\cos\phi\mathbf{a}_\phi + z^2\mathbf{a}_z$, find $\mathbf{B} \cdot \mathbf{a}_x$ at $(4, \pi/4, -1)$.

2.30 A vector field in "mixed" coordinate variables is given by

$$\mathbf{G} = \frac{x\cos\phi}{\rho}\mathbf{a}_x + \frac{2yz}{\rho^2}\mathbf{a}_y + \left(1 - \frac{x^2}{\rho^2}\right)\mathbf{a}_z$$

Express $\mathbf{G}$ completely in the spherical system.

Section 2.5—Constant-Coordinate Surfaces

2.31 Describe the intersection of the following surfaces:

    (a) $x = 2, \quad y = 5$

    (b) $x = 2, \quad y = -1, \quad z = 10$

    (c) $r = 10, \quad \theta = 30°$

    (d) $\rho = 5, \quad \phi = 40°$

    (e) $\phi = 60°, z = 10$

    (f) $r = 5, \quad \phi = 90°$

2.32 If $\mathbf{J} = r\sin\theta\cos\phi\,\mathbf{a}_r - \cos 2\theta\sin\phi\,\mathbf{a}_\theta + \tan\dfrac{\theta}{2}\ln r\,\mathbf{a}_\phi$ at $T(2, \pi/2, 3\pi/2)$, determine the vector component of $\mathbf{J}$ that is:

    (a) Parallel to $\mathbf{a}_z$

    (b) Normal to surface $\phi = 3\pi/2$

    (c) Tangential to the spherical surface $r = 2$

    (d) Parallel to the line $y = -2, z = 0$

2.33 If $\mathbf{H} = \rho^2\cos\phi\mathbf{a}_\rho - \rho\sin\phi\mathbf{a}_\phi$, find $\mathbf{H} \cdot \mathbf{a}_x$ at point $P(2, 60°, -1)$.

2.34 If $\mathbf{r} = x\mathbf{a}_x + y\mathbf{a}_y + z\mathbf{a}_z$, describe the surface defined by:

    (a) $\mathbf{r} \cdot \mathbf{a}_x + \mathbf{r} \cdot \mathbf{a}_y = 5$

    (b) $|\mathbf{r} \times \mathbf{a}_z| = 10$

George Gabriel Stokes (1819–1903), mathematician and physicist, was one of Ireland's preeminent scientists of all time. He made significant contributions to the fields of fluid dynamics, optics, and mathematical physics.

Born in Sligo, Ireland, as the youngest son of the Reverend Gabriel Stokes, George Stokes was a religious man. In one of his books, he detailed his view of God and his relationship to the world.

Although Stokes's basic field was physics, his most important contribution was in fluid mechanics, where he described the motion of viscous fluids. These equations are known today as the Navier–Stokes equations and are considered fundamental equations. Stokes was an applied mathematician working in physics, and like many of his predecessors, he branched out into other areas while continuing to develop his own specialty. His mathematical and physical papers were published in five volumes. Several discoveries were named for him. For example, the Stokes's theorem, to be discussed in this chapter, reduced selected surface integrals to line integrals.

Carl Friedrich Gauss (1777–1855), German mathematician, astronomer, and physicist, is considered to be one of the leading mathematicians of all time because of his wide range of contributions.

Born in Brunswick, Germany, as the only son of uneducated parents, Gauss was a prodigy of astounding depth. Gauss taught himself reading and arithmetic by the age of 3. Recognizing the youth's talent, the Duke of Brunswick in 1792 provided him with a stipend to allow him to pursue his education. Before his 25th birthday, he was already famous for his work in mathematics and astronomy. At the age of 30 he went to Göttingen to become director of the observatory. From there, he worked for 47 years until his death at almost age 78. He found no fellow mathematical collaborators and worked alone for most of his life, engaging in an amazingly rich scientific activity. He carried on intensive empirical and theoretical research in many branches of science, including observational astronomy, celestial mechanics, surveying, geodesy, capillarity, geomagnetism, electromagnetism, actuarial science, and optics. In 1833 he constructed the first telegraph. He published over 150 works and did important work in almost every area of mathematics. For this reason, he is sometimes called the "prince of mathematics." Among the discoveries of C. F. Gauss are the method of least squares, Gaussian distribution, Gaussian quadrature, the divergence theorem (to be discussed in this chapter), Gauss's law (to be discussed in Chapter 4), the Gauss–Markov theorem, and Gauss–Jordan elimination. Gauss was deeply religious and conservative. He dominated the mathematical community during and after his lifetime.

CHAPTER 3

VECTOR CALCULUS

This nation was founded by men of many nations and background. It was founded on the principle that all men are created equal, and that the rights of every man are diminished when the rights of one man are threatened.

—JOHN F. KENNEDY

3.1 INTRODUCTION

Chapter 1 has focused mainly on vector addition, subtraction, and multiplication in Cartesian coordinates, and Chapter 2 extended all these to other coordinate systems. This chapter deals with vector calculus—integration and differentiation of vectors.

The concepts introduced in this chapter provide a convenient language for expressing certain fundamental ideas in electromagnetics or mathematics in general. A student may feel uneasy about these concepts at first—not seeing what they are "good for." Such a student is advised to concentrate simply on learning the mathematical techniques and to wait for their applications in subsequent chapters.

3.2 DIFFERENTIAL LENGTH, AREA, AND VOLUME

Differential elements in length, area, and volume are useful in vector calculus. They are defined in the Cartesian, cylindrical, and spherical coordinate systems.

A. Cartesian Coordinate Systems

From Figure 3.1, we notice that the differential displacement $d\mathbf{l}$ at point S is the vector from point $S(x, y, z)$ to point $B(x + dx, y + dy, z + dz)$.

1. Differential displacement is given by

$$d\mathbf{l} = dx\,\mathbf{a}_x + dy\,\mathbf{a}_y + dz\,\mathbf{a}_z \qquad (3.1)$$

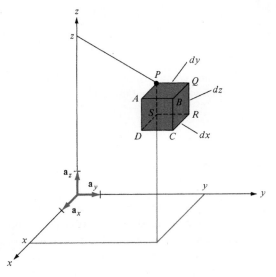

FIGURE 3.1 Differential elements in the right-handed Cartesian coordinate system.

2. Differential normal surface area is given by

$$dS = \begin{matrix} dy\,dz\,\mathbf{a}_x \\ dx\,dz\,\mathbf{a}_y \\ dx\,dy\,\mathbf{a}_z \end{matrix} \qquad (3.2)$$

and illustrated in Figure 3.2.

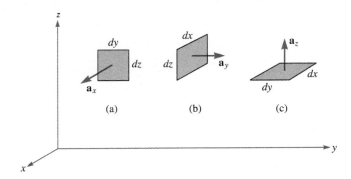

FIGURE 3.2 Differential normal surface areas in Cartesian coordinates:
(**a**) $dS = dy\,dz\,\mathbf{a}_x$, (**b**) $dS = dx\,dz\,\mathbf{a}_y$, (**c**) $dS = dx\,dy\,\mathbf{a}_z$.

3. Differential volume is given by

$$dv = dx\,dy\,dz \qquad (3.3)$$

These differential elements are very important as they will be referred to throughout the book. The student is encouraged not to memorize them, but to learn how to derive them from Figures 3.1 and 3.2. Notice from eqs. (3.1) to (3.3) that $d\mathbf{l}$ and $d\mathbf{S}$ are vectors, whereas dv is a scalar. Observe from Figure 3.1 that if we move from point P to Q (or Q to P), for example, $d\mathbf{l} = dy\,\mathbf{a}_y$ because we are moving in the y-direction, and if we move from Q to S (or S to Q), $d\mathbf{l} = dy\,\mathbf{a}_y + dz\,\mathbf{a}_z$ because we have to move dy along y, dz along z, and $dx = 0$ (no movement along x). Similarly, to move from D to Q (or Q to D) would mean that $d\mathbf{l} = dx\,\mathbf{a}_x + dy\,\mathbf{a}_y + dz\,\mathbf{a}_z$.

The way $d\mathbf{S}$ is defined is important. The differential surface (or area) element $d\mathbf{S}$ may generally be defined as

$$d\mathbf{S} = dS\,\mathbf{a}_n \tag{3.4}$$

where dS is the area of the surface element and $\mathbf{a}_n$ is a unit vector normal to the surface dS (and directed away from the volume if dS is part of the surface describing a volume). If we consider surface $ABCD$ in Figure 3.1, for example, $d\mathbf{S} = dy\,dz\,\mathbf{a}_x$, whereas for surface $PQRS$, $d\mathbf{S} = -dy\,dz\,\mathbf{a}_x$ because $\mathbf{a}_n = -\mathbf{a}_x$ is normal to $PQRS$.

What we have to remember at all times about differential elements is $d\mathbf{l}$ and how to get $d\mathbf{S}$ and dv from it. When $d\mathbf{l}$ is remembered, $d\mathbf{S}$ and dv can easily be found. For example, $d\mathbf{S}$ along $\mathbf{a}_x$ can be obtained from $d\mathbf{l}$ in eq. (3.1) by multiplying the components of $d\mathbf{l}$ along $\mathbf{a}_y$ and $\mathbf{a}_z$; that is, $dy\,dz\,\mathbf{a}_x$. Similarly, $d\mathbf{S}$ along $\mathbf{a}_z$ is the product of the components of $d\mathbf{l}$ along $\mathbf{a}_x$ and $\mathbf{a}_y$; that is, $dx\,dy\,\mathbf{a}_z$. Also, dv can be obtained from $d\mathbf{l}$ as the product of the three components of $d\mathbf{l}$, that is, $dx\,dy\,dz$. The idea developed here for Cartesian coordinates will now be extended to other coordinate systems.

B. Cylindrical Coordinate Systems

From Figure 3.3, the differential elements in cylindrical coordinates can be found as follows:

1. Differential displacement is given by

$$\boxed{d\mathbf{l} = d\rho\,\mathbf{a}_\rho + \rho\,d\phi\,\mathbf{a}_\phi + dz\,\mathbf{a}_z} \tag{3.5}$$

2. Differential normal surface area is given by

$$\boxed{\begin{aligned} d\mathbf{S} &= \rho\,d\phi\,dz\,\mathbf{a}_\rho \\ &\quad\ \ d\rho\,dz\,\mathbf{a}_\phi \\ &\quad\ \ \rho\,d\rho\,d\phi\,\mathbf{a}_z \end{aligned}} \tag{3.6}$$

and illustrated in Figure 3.4.
3. Differential volume is given by

$$\boxed{dv = \rho\,d\rho\,d\phi\,dz} \tag{3.7}$$

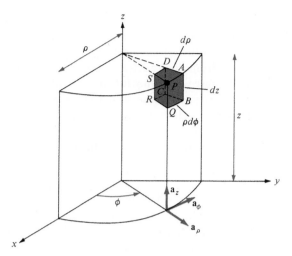

FIGURE 3.3 Differential elements in cylindrical coordinates.

As mentioned in the preceding section on Cartesian coordinates, we need only remember $d\mathbf{l}$; $d\mathbf{S}$ and dv can easily be obtained from $d\mathbf{l}$. For example, $d\mathbf{S}$ along $\mathbf{a}_z$ is the product of the components of $d\mathbf{l}$ along $\mathbf{a}_\rho$ and $\mathbf{a}_\phi$, that is, $d\rho\,\rho\,d\phi\,\mathbf{a}_z$. Also, dv is the product of the three components of $d\mathbf{l}$, that is, $d\rho\,\rho\,d\phi\,dz$.

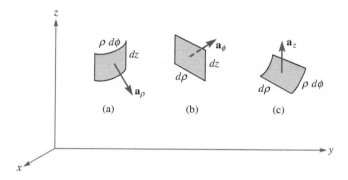

FIGURE 3.4 Differential normal surface areas in cylindrical coordinates: (a) $d\mathbf{S} = \rho\,d\phi\,dz\,\mathbf{a}_\rho$, (b) $d\mathbf{S} = d\rho\,dz\,\mathbf{a}_\phi$, (c) $d\mathbf{S} = \rho\,d\rho\,d\phi\,\mathbf{a}_z$.

C. Spherical Coordinate Systems

From Figure 3.5, the differential elements in spherical coordinates can be found as follows:

1. The differential displacement is

$$d\mathbf{l} = dr\,\mathbf{a}_r + r\,d\theta\,\mathbf{a}_\theta + r\sin\theta\,d\phi\,\mathbf{a}_\phi \qquad (3.8)$$

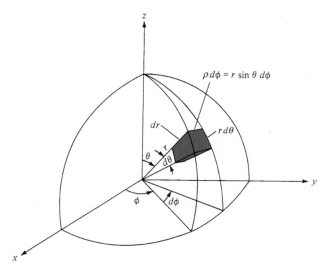

FIGURE 3.5 Differential elements in the spherical coordinate system.

2. The differential normal surface area is

$$dS = r^2 \sin \theta \, d\theta \, d\phi \, \mathbf{a}_r$$
$$r \sin \theta \, dr \, d\phi \, \mathbf{a}_\theta \qquad (3.9)$$
$$r \, dr \, d\theta \, \mathbf{a}_\phi$$

and illustrated in Figure 3.6.

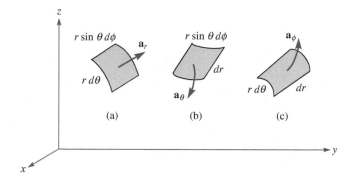

FIGURE 3.6 Differential normal surface areas in spherical coordinates:
(**a**) $dS = r^2 \sin \theta \, d\theta \, d\phi \, \mathbf{a}_r$, (**b**) $dS = r \sin \theta \, dr \, d\phi \, \mathbf{a}_\theta$, (**c**) $dS = r \, dr \, d\theta \, \mathbf{a}_\phi$.

3. The differential volume is

$$dv = r^2 \sin \theta \, dr \, d\theta \, d\phi \qquad (3.10)$$

Again, we need to remember only $d\mathbf{l}$, from which $d\mathbf{S}$ and dv are easily obtained. For example, $d\mathbf{S}$ along $\mathbf{a}_\theta$ is obtained as the product of the components of $d\mathbf{l}$ along $\mathbf{a}_r$ and $\mathbf{a}_\theta$, that is, $dr \cdot r \sin\theta \, d\phi$; dv is the product of the three components of $d\mathbf{l}$, that is, $dr \cdot r \, d\theta \cdot r \sin\theta \, d\phi$.

EXAMPLE 3.1

Consider the object shown in Figure 3.7. Calculate

(a) The length BC

(b) The length CD

(c) The surface area $ABCD$

(d) The surface area ABO

(e) The surface area $AOFD$

(f) The volume $ABDCFO$

Solution:

Although points A, B, C, and D are given in Cartesian coordinates, it is obvious that the object has cylindrical symmetry. Hence, we solve the problem in cylindrical coordinates. The points are transformed from Cartesian to cylindrical coordinates as follows:

$$A(5, 0, 0) \ \rightarrow \ A(5, 0°, 0)$$

$$B(0, 5, 0) \ \rightarrow \ B\left(5, \frac{\pi}{2}, 0\right)$$

$$C(0, 5, 10) \ \rightarrow \ C\left(5, \frac{\pi}{2}, 10\right)$$

$$D(5, 0, 10) \ \rightarrow \ D(5, 0°, 10)$$

(a) Along BC, $dl = dz$; hence,

$$BC = \int_L dl = \int_0^{10} dz = 10$$

(b) Along CD, $dl = \rho \, d\phi$ and $\rho = 5$, so

$$CD = \int_0^{\pi/2} \rho \, d\phi = 5 \phi \Big|_0^{\pi/2} = 2.5\pi$$

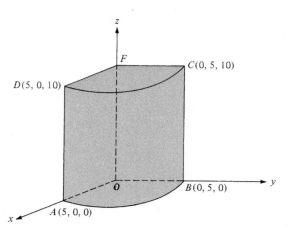

FIGURE 3.7 For Example 3.1.

(c) For *ABCD*, $dS = \rho \, d\phi \, dz$, $\rho = 5$. Hence,

$$\text{area } ABCD = \int_S dS = \int_{\phi=0}^{\pi/2}\int_{z=0}^{10} \rho \, d\phi \, dz = 5 \int_0^{\pi/2} d\phi \int_0^{10} dz \Bigg|_{\rho=5} = 25\pi$$

(d) For *ABO*, $dS = \rho \, d\phi \, d\rho$ and $z = 0°$, so

$$\text{area } ABO = \int_{\phi=0}^{\pi/2}\int_{\rho=0}^{5} \rho \, d\phi \, d\rho = \int_0^{\pi/2} d\phi \int_0^5 \rho \, d\rho = 6.25\pi$$

(e) For *AOFD*, $dS = d\rho \, dz$ and $\phi = 0°$, so

$$\text{area } AOFD = \int_{\rho=0}^{5}\int_{z=0}^{10} d\rho \, dz = 50$$

(f) For volume *ABDCFO*, $dv = \rho \, d\phi \, dz \, d\rho$. Hence,

$$v = \int_v dv = \int_{\rho=0}^{5}\int_{\phi=0}^{\pi/2}\int_{z=0}^{10} \rho \, d\phi \, dz \, d\rho = \int_0^{10} dz \int_0^{\pi/2} d\phi \int_0^5 \rho \, d\rho = 62.5\pi$$

PRACTICE EXERCISE 3.1

Refer to Figure 3.8; disregard the differential lengths and imagine that the object is part of a spherical shell. It may be described as $3 \le r \le 5$, $60° \le \theta \le 90°$, $45° \le \phi \le 60°$ where surface $r = 3$ is the same as *AEHD*, surface $\theta = 60°$ is *AEFB*, and surface $\phi = 45°$ is *ABCD*. Calculate

(a) The arc length *DH*
(b) The arc length *FG*
(c) The surface area *AEHD*

(d) The surface area *ABDC*
(e) The volume of the object

Answer: (a) 0.7854, (b) 2.618, (c) 1.179, (d) 4.189, (e) 4.276.

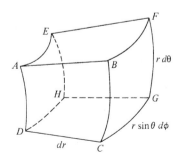

FIGURE 3.8 For Practice Exercise 3.1 (and also Review Question 3.3).

3.3 LINE, SURFACE, AND VOLUME INTEGRALS

The familiar concept of integration will now be extended to cases in which the integrand involves a vector. By "line" we mean the path along a curve in space. We shall use terms such as *line, curve,* and *contour* interchangeably.

> The **line integral** $\int_L \mathbf{A} \cdot d\mathbf{l}$ is the integral of the tangential component of **A** along curve *L*.

Given a vector field **A** and a curve *L*, we define the integral

$$\int_L \mathbf{A} \cdot d\mathbf{l} = \int_a^b |\mathbf{A}| \cos \theta \, dl \tag{3.11}$$

as the *line integral* of **A** around *L* (see Figure 3.9). If the path of integration is a closed curve such as *abca* in Figure 3.9, eq. (3.11) becomes a closed contour integral

$$\oint_L \mathbf{A} \cdot d\mathbf{l} \tag{3.12}$$

which is called the *circulation* of **A** around *L*. A common example of a line integral is the work done on a particle. In this case **A** is the force **F** and

$$\int_P^Q F \cdot d\mathbf{l} = \int_{X_P}^{X_Q} F_x \, dx + \int_{Y_P}^{Y_Q} F_y \, dy + \int_{Z_P}^{Z_Q} F_z \, dz$$

Given a vector field **A**, continuous in a region containing the smooth surface *S*, we define the *surface integral* or the *flux* of **A** through *S* (see Figure 3.10) as

$$\Psi = \int_S |\mathbf{A}| \cos \theta \, dS = \int_S \mathbf{A} \cdot \mathbf{a}_n \, dS$$

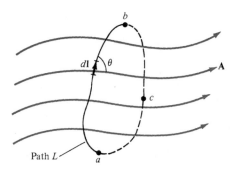

FIGURE 3.9 Path of integration of vector field **A**.

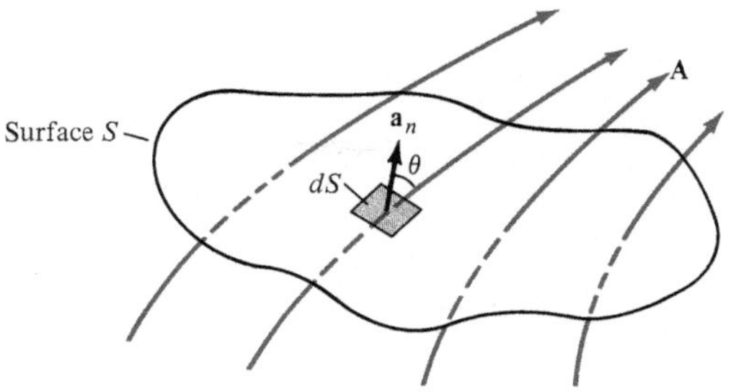

FIGURE 3.10 The flux of a vector field **A** through surface *S*.

or simply

$$\Psi = \int_S \mathbf{A} \cdot d\mathbf{S}$$ (3.13)

where, at any point on *S*, $\mathbf{a}_n$ is the unit normal to *S*. For a closed surface (defining a volume), eq. (3.13) becomes

$$\Psi = \oint_S \mathbf{A} \cdot d\mathbf{S}$$ (3.14)

which is referred to as the *net outward flux of* **A** from *S*. Notice that a closed path defines an open surface, whereas a closed surface defines a volume (see Figures 3.12 and 3.17).
      We define the integral

$$\int_v \rho_v \, dv$$ (3.15)

as the *volume integral* of the scalar ρ_v over the volume *v*. The physical meaning of a line, surface, or volume integral depends on the nature of the physical quantity represented by **A** or ρ_v. Note that $d\mathbf{l}$, $d\mathbf{S}$, and dv are all as defined in Section 3.2.

| EXAMPLE 3.2 |

Given that $\mathbf{F} = x^2\mathbf{a}_x - xz\mathbf{a}_y - y^2\mathbf{a}_z$, calculate the circulation of **F** around the (closed) path shown in Figure 3.11.

Solution:

The circulation of **F** around path *L* is given by

$$\oint_L \mathbf{F} \cdot d\mathbf{l} = \left(\int_① + \int_② + \int_③ + \int_④ \right) \mathbf{F} \cdot d\mathbf{l}$$

where the path is broken into segments numbered 1 to 4 as shown in Figure 3.11.
      For segment ①, $y = 0 = z$

$$\mathbf{F} = x^2\mathbf{a}_x - xz\mathbf{a}_y - y^2\mathbf{a}_z, \quad d\mathbf{l} = dx \, \mathbf{a}_x$$

Notice that $d\mathbf{l}$ is always taken as along $+\mathbf{a}_x$ so that the direction on segment ① is taken care of by the limits of integration. Also, since $d\mathbf{l}$ is in the $\mathbf{a}_x$-direction, only the $\mathbf{a}_x$ component of vector **F** will be integrated, owing to the definition of the dot product. Thus,

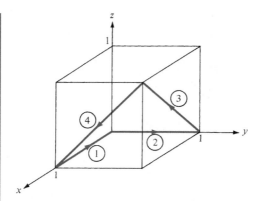

FIGURE 3.11 For Example 3.2.

$$\int_1 \mathbf{F} \cdot d\mathbf{l} = \int_1^0 x^2 dx = \frac{x^3}{3}\bigg|_1^0 = -\frac{1}{3}$$

For segment ②, $x = 0 = z$, $\mathbf{F} = x^2\mathbf{a}_x - xz\mathbf{a}_y - \mathbf{a}_z$, $d\mathbf{l} = dy\,\mathbf{a}_y$, $\mathbf{F} \cdot d\mathbf{l} = 0$. Hence,

$$\int_{②} \mathbf{F} \cdot d\mathbf{l} = 0$$

For segment ③, $y = 1$, $\mathbf{F} = x^2\mathbf{a}_x - xz\mathbf{a}_y - \mathbf{a}_z$, and $d\mathbf{l} = dx\,\mathbf{a}_x + dz\,\mathbf{a}_z$, so

$$\int_{③} \mathbf{F} \cdot d\mathbf{l} = \int (x^2 dx - dz)$$

But on ③, $z = x$; that is, $dx = dz$. Hence,

$$\int_{③} \mathbf{F} \cdot d\mathbf{l} = \int_0^1 (x^2 - 1)\, dx = \frac{x^3}{3} - x \bigg|_0^1 = -\frac{2}{3}$$

For segment ④, $x = 1$, so $\mathbf{F} = \mathbf{a}_x - za_y - y^2\mathbf{a}_z$, and $d\mathbf{l} = dy\,\mathbf{a}_y + dz\,\mathbf{a}_z$. Hence,

$$\int_{④} \mathbf{F} \cdot d\mathbf{l} = \int (-z\,dy - y^2 dz)$$

But on ④, $z = y$; that is, $dz = dy$, so

$$\int_{④} \mathbf{F} \cdot d\mathbf{l} = \int_1^0 (-y - y^2)\, dy = -\frac{y^2}{2} - \frac{y^3}{3}\bigg|_1^0 = \frac{5}{6}$$

By putting all these together, we obtain

$$\oint_L \mathbf{F} \cdot d\mathbf{l} = -\frac{1}{3} + 0 - \frac{2}{3} + \frac{5}{6} = -\frac{1}{6}$$

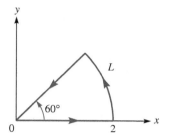

FIGURE 3.12 For Practice Exercise 3.2, L is a closed path.

PRACTICE EXERCISE 3.2

Calculate the circulation of

$$\mathbf{A} = \rho \cos \phi \, \mathbf{a}_\rho + z \sin \phi \, \mathbf{a}_z$$

around the edge L of the wedge defined by $0 \le \rho \le 2, 0 \le \phi \le 60°, z = 0$ and shown in Figure 3.12.

Answer: 1.

3.4 DEL OPERATOR

The del operator, written ∇, is the vector differential operator. In Cartesian coordinates,

$$\nabla = \frac{\partial}{\partial x} \mathbf{a}_x + \frac{\partial}{\partial y} \mathbf{a}_y + \frac{\partial}{\partial z} \mathbf{a}_z \qquad (3.16)$$

This vector differential operator, otherwise known as the *gradient operator,* is not a vector in itself, but when it operates on a scalar function, for example, a vector ensues. The operator is useful in defining

1. The gradient of a scalar V, written as ∇V
2. The divergence of a vector $\mathbf{A}$, written as $\nabla \cdot \mathbf{A}$
3. The curl of a vector $\mathbf{A}$, written as $\nabla \times \mathbf{A}$
4. The Laplacian of a scalar V, written as $\nabla^2 V$

Each of these will be defined in detail in the subsequent sections. Before we do that, it is appropriate to obtain expressions for the del operator ∇ in cylindrical and spherical coordinates. This is easily done by using the transformation formulas of Sections 2.3 and 2.4.

To obtain ∇ in terms of ρ, ϕ, and z, we recall from eq. (2.7) that[1]

$$\rho = \sqrt{x^2 + y^2}, \quad \tan \phi = \frac{y}{x}$$

Hence

$$\frac{\partial}{\partial x} = \cos \phi \frac{\partial}{\partial \rho} - \frac{\sin \phi}{\rho} \frac{\partial}{\partial \phi} \tag{3.17}$$

$$\frac{\partial}{\partial y} = \sin \phi \frac{\partial}{\partial \rho} + \frac{\cos \phi}{\rho} \frac{\partial}{\partial \phi} \tag{3.18}$$

Substituting eqs. (3.17) and (3.18) into eq. (3.16) and making use of eq. (2.9), we obtain ∇ in cylindrical coordinates as

$$\boxed{\nabla = \mathbf{a}_\rho \frac{\partial}{\partial \rho} + \mathbf{a}_\phi \frac{1}{\rho} \frac{\partial}{\partial \phi} + \mathbf{a}_z \frac{\partial}{\partial z}} \tag{3.19}$$

Similarly, to obtain ∇ in terms of r, θ, and ϕ, we use

$$r = \sqrt{x^2 + y^2 + z^2}, \quad \tan \theta = \frac{\sqrt{x^2 + y^2}}{z}, \quad \tan \phi = \frac{y}{x}$$

to obtain

$$\frac{\partial}{\partial x} = \sin \theta \cos \phi \frac{\partial}{\partial r} + \frac{\cos \theta \cos \phi}{r} \frac{\partial}{\partial \theta} - \frac{\sin \phi}{\rho} \frac{\partial}{\partial \phi} \tag{3.20}$$

$$\frac{\partial}{\partial y} = \sin \theta \sin \phi \frac{\partial}{\partial r} + \frac{\cos \theta \sin \phi}{r} \frac{\partial}{\partial \theta} + \frac{\cos \phi}{\rho} \frac{\partial}{\partial \phi} \tag{3.21}$$

$$\frac{\partial}{\partial z} = \cos \theta \frac{\partial}{\partial r} - \frac{\sin \theta}{r} \frac{\partial}{\partial \theta} \tag{3.22}$$

Substituting eqs. (3.20) to (3.22) into eq. (3.16) and using eq. (2.23) results in ∇ in spherical coordinates:

$$\boxed{\nabla = \mathbf{a}_r \frac{\partial}{\partial r} + \mathbf{a}_\theta \frac{1}{r} \frac{\partial}{\partial \theta} + \mathbf{a}_\phi \frac{1}{r \sin \theta} \frac{\partial}{\partial \phi}} \tag{3.23}$$

Notice that in eqs. (3.19) and (3.23), the unit vectors are placed to the left of the differential operators because the unit vectors depend on the angles.

---

[1] A more general way of deriving ∇, $\nabla \cdot \mathbf{A}$, $\nabla \times \mathbf{A}$, ∇V, and $\nabla^2 V$ is by using the curvilinear coordinates. See, for example, M. R. Spiegel, *Vector Analysis and an Introduction to Tensor Analysis*. New York: McGraw-Hill, 1959, pp. 135–165.

3.5 GRADIENT OF A SCALAR

The gradient of a scalar field at any point is the maximum rate of change of the field at that point.

> The **gradient** of a scalar field V is a vector that represents both the magnitude and the direction of the maximum space rate of increase of V.

A mathematical expression for the gradient can be obtained by evaluating the difference in the field dV between points P_1 and P_2 of Figure 3.13, where V_1, V_2, and V_3 are contours on which V is constant. From calculus,

$$dV = \frac{\partial V}{\partial x} dx + \frac{\partial V}{\partial y} dy + \frac{\partial V}{\partial z} dz$$
$$= \left(\frac{\partial V}{\partial x} \mathbf{a}_x + \frac{\partial V}{\partial y} \mathbf{a}_y + \frac{\partial V}{\partial z} \mathbf{a}_z \right) \cdot (dx\, \mathbf{a}_x + dy\, \mathbf{a}_y + dz\, \mathbf{a}_z) \tag{3.24}$$

For convenience, let

$$\mathbf{G} = \frac{\partial V}{\partial x} \mathbf{a}_x + \frac{\partial V}{\partial y} \mathbf{a}_y + \frac{\partial V}{\partial z} \mathbf{a}_z \tag{3.25}$$

Then

$$dV = \mathbf{G} \cdot d\mathbf{l} = G \cos \theta \, dl$$

or

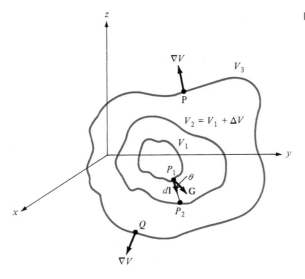

FIGURE 3.13 Gradient of a scalar.

$$\frac{dV}{dl} = G \cos \theta \qquad (3.26)$$

where $d\mathbf{l}$ is the differential displacement from P_1 to P_2 and θ is the angle between $\mathbf{G}$ and $d\mathbf{l}$. From eq. (3.26), we notice that dV/dl is a maximum when $\theta = 0$, that is, when $d\mathbf{l}$ is in the direction of $\mathbf{G}$. Hence,

$$\left.\frac{dV}{dl}\right|_{\text{max}} = \frac{dV}{dn} = G \qquad (3.27)$$

where dV/dn is the normal derivative. Thus G has its magnitude and direction as those of the maximum rate of change of V. By definition, $\mathbf{G}$ is the gradient of V. Therefore:

$$\text{grad } V = \nabla V = \frac{\partial V}{\partial x}\mathbf{a}_x + \frac{\partial V}{\partial y}\mathbf{a}_y + \frac{\partial V}{\partial z}\mathbf{a}_z \qquad (3.28)$$

By using eq. (3.28) in conjunction with eqs. (3.16), (3.19), and (3.23), the gradient of V can be expressed in Cartesian, cylindrical, and spherical coordinates. For Cartesian coordinates

$$\nabla V = \frac{\partial V}{\partial x}\mathbf{a}_x + \frac{\partial V}{\partial y}\mathbf{a}_y + \frac{\partial V}{\partial z}\mathbf{a}_z$$

for cylindrical coordinates,

$$\nabla V = \frac{\partial V}{\partial \rho}\mathbf{a}_\rho + \frac{1}{\rho}\frac{\partial V}{\partial \phi}\mathbf{a}_\phi + \frac{\partial V}{\partial z}\mathbf{a}_z \qquad (3.29)$$

and for spherical coordinates,

$$\nabla V = \frac{\partial V}{\partial r}\mathbf{a}_r + \frac{1}{r}\frac{\partial V}{\partial \theta}\mathbf{a}_\theta + \frac{1}{r \sin \theta}\frac{\partial V}{\partial \phi}\mathbf{a}_\phi \qquad (3.30)$$

The following computation formulas on gradient, which are easily proved, should be noted:

(i) $\nabla(V + U) = \nabla V + \nabla U$ \qquad\qquad (3.31a)

(ii) $\nabla(VU) = V\nabla U + U\nabla V$ \qquad\qquad (3.31b)

(iii) $\nabla\left[\dfrac{V}{U}\right] = \dfrac{U\nabla V - V\nabla U}{U^2}$ \qquad\qquad (3.31c)

(iv) $\nabla V^n = nV^{n-1}\nabla V$ \qquad\qquad (3.31d)

where U and V are scalars and n is an integer.

Also take note of the following fundamental properties of the gradient of a scalar field V:

1. The magnitude of ∇V equals the maximum rate of change in V per unit distance.
2. ∇V points in the direction of the maximum rate of change in V.

3. ∇V at any point is perpendicular to the constant V surface that passes through that point (see points P and Q in Figure 3.13).
4. The projection (or component) of ∇V in the direction of a unit vector $\mathbf{a}$ is $\nabla V \cdot \mathbf{a}$ and is called the *directional derivative* of V along $\mathbf{a}$. This is the rate of change of V in the direction of $\mathbf{a}$. For example, dV/dl in eq. (3.26) is the directional derivative of V along $P_1 P_2$ in Figure 3.13. Thus the gradient of a scalar function V provides us with both the direction in which V changes most rapidly and the magnitude of the maximum directional derivative of V.
5. If $\mathbf{A} = \nabla V$, V is said to be the scalar potential of $\mathbf{A}$.

EXAMPLE 3.3

Find the gradient of the following scalar fields:

(a) $V = e^{-z} \sin 2x \cosh y$
(b) $U = \rho^2 z \cos 2\phi$
(c) $W = 10r \sin^2 \theta \cos \phi$

Solution:

(a) $\nabla V = \dfrac{\partial V}{\partial x} \mathbf{a}_x + \dfrac{\partial V}{\partial y} \mathbf{a}_y + \dfrac{\partial V}{\partial z} \mathbf{a}_z$

$= 2e^{-z} \cos 2x \cosh y\, \mathbf{a}_x + e^{-z} \sin 2x \sinh y\, \mathbf{a}_y - e^{-z} \sin 2x \cosh y\, \mathbf{a}_z$

(b) $\nabla U = \dfrac{\partial U}{\partial \rho} \mathbf{a}_\rho + \dfrac{1}{\rho} \dfrac{\partial U}{\partial \phi} \mathbf{a}_\phi + \dfrac{\partial U}{\partial z} \mathbf{a}_z$

$= 2\rho z \cos 2\phi\, \mathbf{a}_\rho - 2\rho z \sin 2\phi\, \mathbf{a}_\phi + \rho^2 \cos 2\phi\, \mathbf{a}_z$

(c) $\nabla W = \dfrac{\partial W}{\partial r} \mathbf{a}_r + \dfrac{1}{r} \dfrac{\partial W}{\partial \theta} \mathbf{a}_\theta + \dfrac{1}{r \sin \theta} \dfrac{\partial W}{\partial \phi} \mathbf{a}_\phi$

$= 10 \sin^2 \theta \cos \phi\, \mathbf{a}_r + 10 \sin 2\theta \cos \phi\, \mathbf{a}_\theta - 10 \sin \theta \sin \phi\, \mathbf{a}_\phi$

PRACTICE EXERCISE 3.3

Determine the gradient of the following scalar fields:

(a) $U = x^2 y + xyz$
(b) $V = \rho z \sin \phi + z^2 \cos^2 \phi + \rho^2$
(c) $f = \cos \theta \sin \phi \ln r + r^2 \phi$

Answer: (a) $y(2x + z)\mathbf{a}_x + x(x + z)\mathbf{a}_y + xy\mathbf{a}_z$

(b) $(z \sin \phi + 2\rho)\mathbf{a}_\rho + \left(z \cos \phi - \dfrac{z^2}{\rho} \sin 2\phi \right)\mathbf{a}_\phi +$

$(\rho \sin \phi + 2z \cos^2 \phi)\mathbf{a}_z$

(c) $\left(\dfrac{\cos \theta \sin \phi}{r} + 2r\phi \right)\mathbf{a}_r - \dfrac{\sin \theta \sin \phi}{r} \ln r\, \mathbf{a}_\theta +$

$\left(\dfrac{\cot \theta}{r} \cos \phi \ln r + r \csc \theta \right)\mathbf{a}_\phi$

EXAMPLE 3.4

Given $W = x^2y^2 + xyz$, compute ∇W and the directional derivative dW/dl in the direction $3\mathbf{a}_x + 4\mathbf{a}_y + 12\mathbf{a}_z$ at $(2, -1, 0)$.

Solution:

$$\nabla W = \frac{\partial W}{\partial x}\mathbf{a}_x + \frac{\partial W}{\partial y}\mathbf{a}_y + \frac{\partial W}{\partial z}\mathbf{a}_z$$

$$= (2xy^2 + yz)\mathbf{a}_x + (2x^2y + xz)\mathbf{a}_y + (xy)\mathbf{a}_z$$

At $(2, -1, 0)$: $\nabla W = 4\mathbf{a}_x - 8\mathbf{a}_y - 2\mathbf{a}_z$
Hence,

$$\frac{dW}{dl} = \nabla W \cdot \mathbf{a}_l = (4, -8, -2) \cdot \frac{(3, 4, 12)}{13} = -\frac{44}{13}$$

PRACTICE EXERCISE 3.4

Given $\Phi = xy + yz + xz$, find gradient Φ at point $(1, 2, 3)$ and the directional derivative of Φ at the same point in the direction toward point $(3, 4, 4)$.

Answer: $5\mathbf{a}_x + 4\mathbf{a}_y + 3\mathbf{a}_z$, 7.

EXAMPLE 3.5

Find the angle at which line $x = y = 2z$ intersects the ellipsoid $x^2 + y^2 + 2z^2 = 10$.

Solution:

Let the line and the ellipsoid meet at angle ψ as shown in Figure 3.14. On line $x = y = 2z$, for two unit increments along z, there is a unit increment along x and a unit increment along y. Thus, the line can be represented by

$$\mathbf{r}(\lambda) = 2\lambda\,\mathbf{a}_x + 2\lambda\,\mathbf{a}_y + \lambda\,\mathbf{a}_z$$

where λ is a parameter. Where the line and the ellipsoid meet,

$$(2\lambda)^2 + (2\lambda)^2 + 2\lambda^2 = 10 \rightarrow \lambda = \pm 1$$

Taking $\lambda = 1$ (for the moment), the point of intersection is $(x, y, z) = (2, 2, 1)$. At this point, $\mathbf{r} = 2\mathbf{a}_x + 2\mathbf{a}_y + \mathbf{a}_z$.

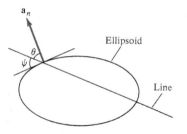

FIGURE 3.14 For Example 3.5; plane of intersection of a line with an ellipsoid.

The surface of the ellipsoid is defined by

$$f(x, y, z) = x^2 + y^2 + 2z^2 - 10$$

The gradient of f is

$$\nabla f = 2x\mathbf{a}_x + 2y\mathbf{a}_y + 4z\mathbf{a}_z$$

At $(2, 2, 1)$, $\nabla f = 4\mathbf{a}_x + 4\mathbf{a}_y + 4\mathbf{a}_z$. Hence, a unit vector normal to the ellipsoid at the point of intersection is

$$\mathbf{a}_n = \pm\frac{\nabla f}{|\nabla f|} = \pm\frac{\mathbf{a}_x + \mathbf{a}_y + \mathbf{a}_z}{\sqrt{3}}$$

Taking the positive sign (for the moment), the angle between $\mathbf{a}_n$ and $\mathbf{r}$ is given by

$$\cos\theta = \frac{\mathbf{a}_n \cdot \mathbf{r}}{|\mathbf{a}_n \cdot \mathbf{r}|} = \frac{2 + 2 + 1}{\sqrt{3}\sqrt{9}} = \frac{5}{3\sqrt{3}} = \sin\psi$$

Hence, $\psi = 74.21°$. Because we had choices of $+$ or $-$ for λ and $\mathbf{a}_n$, there are actually four possible angles, given by $\sin\psi = \pm 5/(3\sqrt{3})$.

PRACTICE EXERCISE 3.5

Calculate the angle between the normals to the surfaces $x^2y + z = 3$ and $x \log z - y^2 = -4$ at the point of intersection $(-1, 2, 1)$.

Answer: 73.4°.

3.6 DIVERGENCE OF A VECTOR AND DIVERGENCE THEOREM

From Section 3.3, we have noticed that the net outflow of the flux of a vector field $\mathbf{A}$ from a closed surface S is obtained from the integral $\oint \mathbf{A} \cdot d\mathbf{S}$. We now define the divergence of $\mathbf{A}$ as the net outward flow of flux per unit volume over a closed incremental surface.

The **divergence** of $\mathbf{A}$ at a given point P is the *outward* flux per unit volume as the volume shrinks about P.

Hence,

$$\text{div } \mathbf{A} = \nabla \cdot \mathbf{A} = \lim_{\Delta v \to 0} \frac{\oint_S \mathbf{A} \cdot d\mathbf{S}}{\Delta v} \tag{3.32}$$

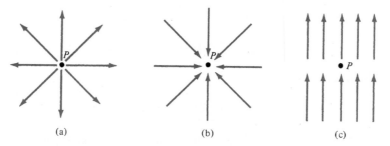

FIGURE 3.15 Illustration of the divergence of a vector field at P: (**a**) positive divergence, (**b**) negative divergence, (**c**) zero divergence.

where Δv is the volume enclosed by the closed surface S in which P is located. Physically, we may regard the divergence of the vector field $\mathbf{A}$ at a given point as a measure of how much the field diverges or emanates or originates from that point. Figure 3.15(a) shows that the divergence of a vector field at point P is positive because the vector diverges (or spreads out) at P. In Figure 3.15(b) a vector field has negative divergence (or convergence) at P, and in Figure 3.15(c) a vector field has zero divergence at P. The divergence of a vector field can also be viewed as simply the limit of the field's source strength per unit volume (or source density); it is positive at a *source* point in the field, and negative at a *sink* point, or zero where there is neither sink nor source.

We can obtain an expression for $\nabla \cdot \mathbf{A}$ in Cartesian coordinates from the definition in eq. (3.32). Suppose we wish to evaluate the divergence of a vector field $\mathbf{A}$ at point $P(x_o, y_o, z_o)$; we let the point be enclosed by a differential volume as in Figure 3.16. The surface integral in eq. (3.32) is obtained from

$$\oint_S \mathbf{A} \cdot d\mathbf{S} = \left(\iint_{\text{front}} + \iint_{\text{back}} + \iint_{\text{left}} + \iint_{\text{right}} + \iint_{\text{top}} + \iint_{\text{bottom}} \right) \mathbf{A} \cdot d\mathbf{S} \quad (3.33)$$

A three-dimensional Taylor series expansion of A_x about P is

$$A_x(x, y, z) = A_x(x_o, y_o, z_o) + (x - x_o) \frac{\partial A_x}{\partial x}\bigg|_P + (y - y_o) \frac{\partial A_x}{\partial y}\bigg|_P$$

$$+ (z - z_o) \frac{\partial A_x}{\partial z}\bigg|_P + \text{higher-order terms} \quad (3.34)$$

For the front side, $x = x_o + dx/2$ and $d\mathbf{S} = dy\, dz\, \mathbf{a}_x$. Then,

$$\iint_{\text{front}} \mathbf{A} \cdot d\mathbf{S} = dy\, dz \left[A_x(x_o, y_o, z_o) + \frac{dx}{2} \frac{\partial A_x}{\partial x}\bigg|_P \right] + \text{higher-order terms}$$

For the back side, $x = x_o - dx/2$ and $d\mathbf{S} = dy\, dz(-\mathbf{a}_x)$. Then,

$$\iint_{\text{back}} \mathbf{A} \cdot d\mathbf{S} = -dy\, dz \left[A_x(x_o, y_o, z_o) - \frac{dx}{2} \frac{\partial A_x}{\partial x}\bigg|_P \right] + \text{higher-order terms}$$

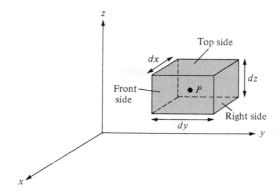

FIGURE 3.16 Evaluation of $\nabla \cdot \mathbf{A}$ at point $P(x_o, y_o, z_o)$.

Hence,

$$\iint_{\text{front}} \mathbf{A} \cdot d\mathbf{S} + \iint_{\text{back}} \mathbf{A} \cdot d\mathbf{S} = dx\, dy\, dz \left.\frac{\partial A_x}{\partial x}\right|_P + \text{higher-order terms} \qquad (3.35)$$

By taking similar steps, we obtain

$$\iint_{\text{left}} \mathbf{A} \cdot d\mathbf{S} + \iint_{\text{right}} \mathbf{A} \cdot d\mathbf{S} = dx\, dy\, dz \left.\frac{\partial A_y}{\partial y}\right|_P + \text{higher-order terms} \qquad (3.36)$$

and

$$\iint_{\text{top}} \mathbf{A} \cdot d\mathbf{S} + \iint_{\text{bottom}} \mathbf{A} \cdot d\mathbf{S} = dx\, dy\, dz \left.\frac{\partial A_z}{\partial z}\right|_P + \text{higher-order terms} \qquad (3.37)$$

Substituting eqs. (3.35) to (3.37) into eq. (3.33) and noting that $\Delta v = dx\, dy\, dz$, we get

$$\lim_{\Delta v \to 0} \frac{\oint_S \mathbf{A} \cdot d\mathbf{S}}{\Delta v} = \left. \left(\frac{\partial A_x}{\partial x} + \frac{\partial A_y}{\partial y} + \frac{\partial A_z}{\partial z} \right) \right|_{\text{at } P} \qquad (3.38)$$

because the higher-order terms will vanish as $\Delta v \to 0$. Thus, the divergence of $\mathbf{A}$ at point $P(x_o, y_o, z_o)$ in a Cartesian system is given by

$$\boxed{\nabla \cdot \mathbf{A} = \frac{\partial A_x}{\partial x} + \frac{\partial A_y}{\partial y} + \frac{\partial A_z}{\partial z}} \qquad (3.39)$$

Similar expressions for $\nabla \cdot \mathbf{A}$ in other coordinate systems can be obtained directly from eq. (3.32) or by transforming eq. (3.39) into the appropriate coordinate system. In cylindrical coordinates, substituting eqs. (2.15), (3.17), and (3.18) into eq. (3.39) yields

$$\boxed{\nabla \cdot \mathbf{A} = \frac{1}{\rho} \frac{\partial}{\partial \rho} (\rho A_\rho) + \frac{1}{\rho} \frac{\partial A_\phi}{\partial \phi} + \frac{\partial A_z}{\partial z}} \qquad (3.40)$$

Substituting eqs. (2.28) and (3.20) to (3.22) into eq. (3.39), we obtain the divergence of **A** in spherical coordinates as

$$\nabla \cdot \mathbf{A} = \frac{1}{r^2} \frac{\partial}{\partial r} (r^2 A_r) + \frac{1}{r \sin \theta} \frac{\partial}{\partial \theta} (A_\theta \sin \theta) + \frac{1}{r \sin \theta} \frac{\partial A_\phi}{\partial \phi} \qquad (3.41)$$

Note the following properties of the divergence of a vector field:

1. It produces a scalar field (because scalar product is involved).
2. $\nabla \cdot (\mathbf{A} + \mathbf{B}) = \nabla \cdot \mathbf{A} + \nabla \cdot \mathbf{B}$
3. $\nabla \cdot (V\mathbf{A}) = V\nabla \cdot \mathbf{A} + \mathbf{A} \cdot \nabla V$

From the definition of the divergence of **A** in eq. (3.32), it is not difficult to expect that

$$\oint_S \mathbf{A} \cdot d\mathbf{S} = \int_v \nabla \cdot \mathbf{A} \, dv \qquad (3.42)$$

This is called the *divergence theorem,* otherwise known as the *Gauss–Otrogradsky theorem.*

> The **divergence theorem** states that the total outward flux of a vector field **A** through the *closed* surface S is the same as the volume integral of the divergence of **A**.

To prove the divergence theorem, subdivide volume v into a large number of small cells. If the kth cell has volume Δv_k and is bounded by surface S_k

$$\oint_S A \cdot d\mathbf{S} = \sum_k \oint_{S_k} A \cdot d\mathbf{S} = \sum_k \frac{\oint_{S_k} \mathbf{A} \cdot d\mathbf{S}}{\Delta v_k} \Delta v_k \qquad (3.43)$$

Since the outward flux to one cell is inward to some neighboring cells, there is cancellation on every interior surface, so the sum of the surface integrals over the S_k's is the same as the surface integral over the surface S. Taking the limit of the right-hand side of eq. (3.43) and incorporating eq. (3.32) gives

$$\oint_S \mathbf{A} \cdot d\mathbf{S} = \int_v \nabla \cdot \mathbf{A} \, dv \qquad (3.44)$$

which is the divergence theorem. The theorem applies to any volume v bounded by the closed surface S such as that shown in Figure 3.17 provided that **A** and $\nabla \cdot \mathbf{A}$ are continuous in the region. With a little experience, one comes to understand that the volume integral on the right-hand side of eq. (3.42) is easier to evaluate than the surface integral(s) on the left-hand side of the equation. For this reason, to determine the flux

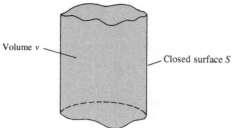

FIGURE 3.17 Volume *v* enclosed by surface *S*.

Volume *v*

Closed surface *S*

of **A** through a closed surface, we simply find the right-hand side of eq. (3.42) instead of the left-hand side of the equation.

EXAMPLE 3.6

Determine the divergence of these vector fields:

(a) $\mathbf{P} = x^2yz\,\mathbf{a}_x + xz\,\mathbf{a}_z$

(b) $\mathbf{Q} = \rho \sin \phi \, \mathbf{a}_\rho + \rho^2 z \, \mathbf{a}_\phi + z \cos \phi \, \mathbf{a}_z$

(c) $\mathbf{T} = \dfrac{1}{r^2} \cos \theta \, \mathbf{a}_r + r \sin \theta \cos \phi \, \mathbf{a}_\theta + \cos \theta \, \mathbf{a}_\phi$

Solution:

(a) $\nabla \cdot \mathbf{P} = \dfrac{\partial}{\partial x}P_x + \dfrac{\partial}{\partial y}P_y + \dfrac{\partial}{\partial z}P_z$

$\qquad = \dfrac{\partial}{\partial x}(x^2yz) + \dfrac{\partial}{\partial y}(0) + \dfrac{\partial}{\partial z}(xz)$

$\qquad = 2xyz + x$

(b) $\nabla \cdot \mathbf{Q} = \dfrac{1}{\rho}\dfrac{\partial}{\partial \rho}(\rho Q_\rho) + \dfrac{1}{\rho}\dfrac{\partial}{\partial \phi}Q_\phi + \dfrac{\partial}{\partial z}Q_z$

$\qquad = \dfrac{1}{\rho}\dfrac{\partial}{\partial \rho}(\rho^2 \sin \phi) + \dfrac{1}{\rho}\dfrac{\partial}{\partial \phi}(\rho^2 z) + \dfrac{\partial}{\partial z}(z \cos \phi)$

$\qquad = 2 \sin \phi + \cos \phi$

(c) $\nabla \cdot \mathbf{T} = \dfrac{1}{r^2}\dfrac{\partial}{\partial r}(r^2 T_r) + \dfrac{1}{r \sin \theta}\dfrac{\partial}{\partial \theta}(T_\theta \sin \theta) + \dfrac{1}{r \sin \theta}\dfrac{\partial}{\partial \phi}(T_\phi)$

$\qquad = \dfrac{1}{r^2}\dfrac{\partial}{\partial r}(\cos \theta) + \dfrac{1}{r \sin \theta}\dfrac{\partial}{\partial \theta}(r \sin^2 \theta \cos \phi) + \dfrac{1}{r \sin \theta}\dfrac{\partial}{\partial \phi}(\cos \theta)$

$\qquad = 0 + \dfrac{1}{r \sin \theta}2r \sin \theta \cos \theta \cos \phi + 0$

$\qquad = 2 \cos \theta \cos \phi$

PRACTICE EXERCISE 3.6

Determine the divergence of the following vector fields and evaluate them at the specified points.

(a) $\mathbf{A} = yz\mathbf{a}_x + 4xy\mathbf{a}_y + y\mathbf{a}_z$ at $(1, -2, 3)$
(b) $\mathbf{B} = \rho z \sin \phi\, \mathbf{a}_\rho + 3\rho z^2 \cos \phi\, \mathbf{a}_\phi$ at $(5, \pi/2, 1)$
(c) $\mathbf{C} = 2r \cos \theta \cos \phi\, \mathbf{a}_r + r^{1/2}\mathbf{a}_\phi$ at $(1, \pi/6, \pi/3)$

Answer: (a) $4x$, 4, (b) $(2 - 3z)z \sin \phi$, -1, (c) $6 \cos \theta \cos \phi$, 2.598.

EXAMPLE 3.7

If $\mathbf{G}(r) = 10e^{-2z}(\rho\mathbf{a}_\rho + \mathbf{a}_z)$, determine the flux of $\mathbf{G}$ out of the entire surface of the cylinder $\rho = 1, 0 \le z \le 1$. Confirm the result by using the divergence theorem.

Solution:

If Ψ is the flux of $\mathbf{G}$ through the given surface, shown in Figure 3.18, then

$$\Psi = \oint_S \mathbf{G} \cdot d\mathbf{S} = \Psi_t + \Psi_b + \Psi_s$$

where Ψ_t, Ψ_b, and Ψ_s are the fluxes through the top, bottom, and sides (curved surface) of the cylinder as in Figure 3.18.

For Ψ_t, $z = 1$, $d\mathbf{S} = \rho \, d\rho \, d\phi \, \mathbf{a}_z$. Hence,

$$\Psi_t = \iint \mathbf{G} \cdot d\mathbf{S} = \int_{\rho=0}^{1} \int_{\phi=0}^{2\pi} 10e^{-2}\rho \, d\rho \, d\phi = 10e^{-2}(2\pi) \left.\frac{\rho^2}{2}\right|_0^1$$

$$= 10\pi e^{-2}$$

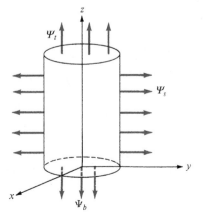

FIGURE 3.18 For Example 3.7.

For Ψ_b, $z = 0$ and $dS = \rho \, d\rho \, d\phi (-\mathbf{a}_z)$. Hence,

$$\Psi_b = \int_b \mathbf{G} \cdot d\mathbf{S} = \int_{\rho=0}^{1} \int_{\phi=0}^{2\pi} 10e^0 \rho \, d\rho \, d\phi = -10(2\pi) \left.\frac{\rho^2}{2}\right|_0^1$$

$$= -10\pi$$

For Ψ_s, $\rho = 1$ and $dS = \rho \, dz \, d\phi \, \mathbf{a}_\rho$. Hence,

$$\Psi_s = \int_s \mathbf{G} \cdot d\mathbf{S} = \int_{z=0}^{1} \int_{\phi=0}^{2\pi} 10e^{-2z}\rho^2 \, dz \, d\phi = 10(1)^2(2\pi) \left.\frac{e^{-2z}}{-2}\right|_0^1$$

$$= 10\pi(1 - e^{-2})$$

Thus,

$$\Psi = \Psi_t + \Psi_b + \Psi_s = 10\pi e^{-2} - 10\pi + 10\pi(1 - e^{-2}) = 0$$

Alternatively, since S is a closed surface, we can apply the divergence theorem:

$$\Psi = \oint_S \mathbf{G} \cdot d\mathbf{S} = \int_v (\nabla \cdot \mathbf{G}) \, dv$$

But

$$\nabla \cdot \mathbf{G} = \frac{1}{\rho}\frac{\partial}{\partial \rho}(\rho G_\rho) + \frac{1}{\rho}\frac{\partial}{\partial \phi} G_\phi + \frac{\partial}{\partial z} G_z$$

$$= \frac{1}{\rho}\frac{\partial}{\partial \rho}(\rho^2 10e^{-2z}) - 20e^{-2z}$$

$$= \frac{1}{\rho}(20\rho e^{-2z}) - 20e^{-2z} = 0$$

showing that **G** has no outward flux. Hence,

$$\Psi = \int_v (\nabla \cdot \mathbf{G}) \, dv = 0$$

PRACTICE EXERCISE 3.7

Determine the flux of $\mathbf{D} = \rho^2 \cos^2 \phi \, \mathbf{a}_\rho + z \sin \phi \, \mathbf{a}_\phi$ over the closed surface of the cylinder $0 \le z \le 1$, $\rho = 4$. Verify the divergence theorem for this case.

Answer: 64π.

3.7 CURL OF A VECTOR AND STOKES'S THEOREM

In Section 3.3, we defined the circulation of a vector field **A** around a closed path L as the integral $\oint_L \mathbf{A} \cdot d\mathbf{l}$.

> The **curl** of **A** is an axial (or rotational) vector whose magnitude is the maximum circulation of **A** per unit area as the area tends to zero and whose direction is the normal direction of the area when the area is oriented to make the circulation maximum.[2]

That is,

$$\text{curl } \mathbf{A} = \nabla \times \mathbf{A} = \left(\lim_{\Delta S \to 0} \frac{\oint_L \mathbf{A} \cdot d\mathbf{l}}{\Delta S} \right)_{\max} \mathbf{a}_n \tag{3.45}$$

where the area ΔS is bounded by the curve L and $\mathbf{a}_n$ is the unit vector normal to the surface ΔS and is determined by using the right-hand rule.

To obtain an expression for $\nabla \times \mathbf{A}$ from the definition in eq. (3.45), consider the differential area in the yz-plane as in Figure 3.19. The line integral in eq. (3.45) is obtained as

$$\oint_L \mathbf{A} \cdot d\mathbf{l} = \left(\int_{ab} + \int_{bc} + \int_{cd} + \int_{da} \right) \mathbf{A} \cdot d\mathbf{l} \tag{3.46}$$

We expand the field components in a Taylor series expansion about the center point $P(x_o, y_o, z_o)$ as in eq. (3.34) and evaluate eq. (3.46). On side ab, $d\mathbf{l} = dy \, \mathbf{a}_y$ and $z = z_o - dz/2$, so

$$\int_{ab} \mathbf{A} \cdot d\mathbf{l} = dy \left[A_y(x_o, y_o, z_o) - \frac{dz}{2} \frac{\partial A_y}{\partial z} \bigg|_P \right] \tag{3.47}$$

On side bc, $d\mathbf{l} = dz \, \mathbf{a}_z$ and $y = y_o + dy/2$, so

$$\int_{bc} \mathbf{A} \cdot d\mathbf{l} = dz \left[A_z(x_o, y_o, z_o) + \frac{dy}{2} \frac{\partial A_z}{\partial y} \bigg|_P \right] \tag{3.48}$$

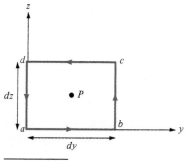

FIGURE 3.19 Contour used in evaluating the x-component of $\nabla \times \mathbf{A}$ at point $P(x_o, y_o, z_o)$.

---

[2] Because of its rotational nature, some authors use rot **A** instead of curl **A**.

On side cd, $d\mathbf{l} = dy\, \mathbf{a}_y$ and $z = z_o + dz/2$, so

$$\int_{cd} \mathbf{A} \cdot d\mathbf{l} = -dy \left[A_y(x_o, y_o, z_o) + \frac{dz}{2} \left. \frac{\partial A_y}{\partial z} \right|_P \right] \tag{3.49}$$

On side da, $d\mathbf{l} = dz\, \mathbf{a}_z$ and $y = y_o - dy/2$, so

$$\int_{da} \mathbf{A} \cdot d\mathbf{l} = -dz \left[A_z(x_o, y_o, z_o) - \frac{dy}{2} \left. \frac{\partial A_z}{\partial y} \right|_P \right] \tag{3.50}$$

Substituting eqs. (3.47) to (3.50) into eq. (3.46) and noting that $\Delta S = dy\, dz$, we have

$$\lim_{\Delta S \to 0} \oint_L \frac{\mathbf{A} \cdot d\mathbf{l}}{\Delta S} = \frac{\partial A_z}{\partial y} - \frac{\partial A_y}{\partial z}$$

or

$$(\text{curl } \mathbf{A})_x = \frac{\partial A_z}{\partial y} - \frac{\partial A_y}{\partial z} \tag{3.51}$$

The y- and x-components of the curl of $\mathbf{A}$ can be found in the same way. We obtain

$$(\text{curl } \mathbf{A})_y = \frac{\partial A_x}{\partial z} - \frac{\partial A_z}{\partial x} \tag{3.52a}$$

$$(\text{curl } \mathbf{A})_z = \frac{\partial A_y}{\partial x} - \frac{\partial A_x}{\partial y} \tag{3.52b}$$

The definition of $\nabla \times \mathbf{A}$ in eq. (3.45) is independent of the coordinate system. In Cartesian coordinates the curl of $\mathbf{A}$ is easily found using

$$\nabla \times \mathbf{A} = \begin{vmatrix} \mathbf{a}_x & \mathbf{a}_y & \mathbf{a}_z \\ \dfrac{\partial}{\partial x} & \dfrac{\partial}{\partial y} & \dfrac{\partial}{\partial z} \\ A_x & A_y & A_z \end{vmatrix} \tag{3.53}$$

or

$$\nabla \times \mathbf{A} = \left[\frac{\partial A_z}{\partial y} - \frac{\partial A_y}{\partial z} \right] \mathbf{a}_x + \left[\frac{\partial A_x}{\partial z} - \frac{\partial A_z}{\partial x} \right] \mathbf{a}_y + \left[\frac{\partial A_y}{\partial x} - \frac{\partial A_x}{\partial y} \right] \mathbf{a}_z \tag{3.54}$$

By transforming eq. (3.54) using point and vector transformation techniques used in Chapter 2, we obtain the curl of $\mathbf{A}$ in cylindrical coordinates as

$$
\nabla \times \mathbf{A} = \frac{1}{\rho}
\begin{vmatrix}
\mathbf{a}_\rho & \rho\, \mathbf{a}_\phi & \mathbf{a}_z \\
\dfrac{\partial}{\partial \rho} & \dfrac{\partial}{\partial \phi} & \dfrac{\partial}{\partial z} \\
A_\rho & \rho A_\phi & A_z
\end{vmatrix}
$$

or

$$
\nabla \times \mathbf{A} = \left[\frac{1}{\rho} \frac{\partial A_z}{\partial \phi} - \frac{\partial A_\phi}{\partial z} \right] \mathbf{a}_\rho + \left[\frac{\partial A_\rho}{\partial z} - \frac{\partial A_z}{\partial \rho} \right] \mathbf{a}_\phi \\
+ \frac{1}{\rho} \left[\frac{\partial (\rho A_\phi)}{\partial \rho} - \frac{\partial A_\rho}{\partial \phi} \right] \mathbf{a}_z
\tag{3.55}
$$

and in spherical coordinates as

$$
\nabla \times \mathbf{A} = \frac{1}{r^2 \sin \theta}
\begin{vmatrix}
\mathbf{a}_r & r\, \mathbf{a}_\theta & r \sin \theta\, \mathbf{a}_\phi \\
\dfrac{\partial}{\partial r} & \dfrac{\partial}{\partial \theta} & \dfrac{\partial}{\partial \phi} \\
A_r & r A_\theta & r \sin \theta\, A_\phi
\end{vmatrix}
$$

or

$$
\nabla \times \mathbf{A} = \frac{1}{r \sin \theta} \left[\frac{\partial (A_\phi \sin \theta)}{\partial \theta} - \frac{\partial A_\theta}{\partial \phi} \right] \mathbf{a}_r \\
+ \frac{1}{r} \left[\frac{1}{\sin \theta} \frac{\partial A_r}{\partial \phi} - \frac{\partial (r A_\phi)}{\partial r} \right] \mathbf{a}_\theta + \frac{1}{r} \left[\frac{\partial (r A_\theta)}{\partial r} - \frac{\partial A_r}{\partial \theta} \right] \mathbf{a}_\phi
\tag{3.56}
$$

Note the following properties of the curl:

1. The curl of a vector field is another vector field.
2. $\nabla \times (\mathbf{A} + \mathbf{B}) = \nabla \times \mathbf{A} + \nabla \times \mathbf{B}$
3. $\nabla \times (\mathbf{A} \times \mathbf{B}) = \mathbf{A}(\nabla \cdot \mathbf{B}) - \mathbf{B}(\nabla \cdot \mathbf{A}) + (\mathbf{B} \cdot \nabla)\mathbf{A} - (\mathbf{A} \cdot \nabla)\mathbf{B}$
4. $\nabla \times (V\mathbf{A}) = V\nabla \times \mathbf{A} + \nabla V \times \mathbf{A}$
5. The divergence of the curl of a vector field vanishes; that is, $\nabla \cdot (\nabla \times \mathbf{A}) = 0$.
6. The curl of the gradient of a scalar field vanishes; that is, $\nabla \times \nabla V = \mathbf{0}$ or $\nabla \times \nabla = \mathbf{0}$.

Other properties of the curl are given in Appendix A.10.

The physical significance of the curl of a vector field is evident in eq. (3.45); the curl provides the maximum value of the circulation of the field per unit area (or circulation density) and indicates the direction along which this maximum value occurs. The curl of a vector field $\mathbf{A}$ at a point P may be regarded as a measure of the circulation or how much the field curls around P. For example, Figure 3.20(a) shows that the curl of a vector field around P is directed out of the page. Figure 3.20(b) shows a vector field with zero curl.

FIGURE 3.20 Illustration of a curl: (**a**) curl at P points out of the page, (**b**) curl at P is zero.

(a)   (b)

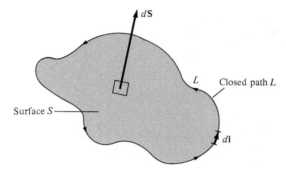

FIGURE 3.21 Determining the sense of $d\mathbf{l}$ and $d\mathbf{S}$ involved in Stokes's theorem.

Also, from the definition of the curl of **A** in eq. (3.45), we may expect that

$$\oint_L \mathbf{A} \cdot d\mathbf{l} = \int_S (\nabla \times \mathbf{A}) \cdot d\mathbf{S} \qquad (3.57)$$

This is called *Stokes's theorem*.

> **Stokes's theorem** states that the circulation of a vector field **A** around a (closed) path L is equal to the surface integral of the curl of **A** over the open surface S bounded by L (see Figure 3.21), provided **A** and $\nabla \times \mathbf{A}$ are continuous on S.

The proof of Stokes's theorem is similar to that of the divergence theorem. The surface S is subdivided into a large number of cells as in Figure 3.22. If the kth cell has surface area ΔS_k and is bounded by path L_k,

$$\oint_L \mathbf{A} \cdot d\mathbf{l} = \sum_k \oint_{L_k} \mathbf{A} \cdot d\mathbf{l} = \sum_k \frac{\oint_{L_k} \mathbf{A} \cdot d\mathbf{l}}{\Delta S_k} \Delta S_k \qquad (3.58)$$

As shown in Figure 3.22, there is cancellation on every interior path, so the sum of the line integrals around the L_k's is the same as the line integral around the bounding curve L. Therefore, taking the limit of the right-hand side of eq. (3.58) as $\Delta S_k \rightarrow 0$ and incorporating eq. (3.45) leads to

$$\oint_L \mathbf{A} \cdot d\mathbf{l} = \int_S (\nabla \times \mathbf{A}) \cdot d\mathbf{S}$$

which is Stokes's theorem.

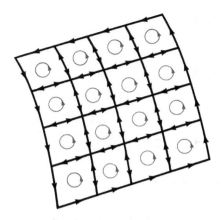

FIGURE 3.22 Illustration of Stokes's theorem.

The direction of $d\mathbf{l}$ and $d\mathbf{S}$ in eq. (3.57) must be chosen using the right-hand rule or right-handed-screw rule. Using the right-hand rule, if we let the fingers point in the direction of $d\mathbf{l}$, the thumb will indicate the direction of $d\mathbf{S}$ (see Figure 3.21). Note that whereas the divergence theorem relates a surface integral to a volume integral, Stokes's theorem relates a line integral (circulation) to suface integral.

EXAMPLE 3.8

Determine the curl of each of the vector fields in Example 3.6.

Solution:

(a) $\nabla \times \mathbf{P} = \left(\dfrac{\partial P_z}{\partial y} - \dfrac{\partial P_y}{\partial z} \right) \mathbf{a}_x + \left(\dfrac{\partial P_x}{\partial z} - \dfrac{\partial P_z}{\partial x} \right) \mathbf{a}_y + \left(\dfrac{\partial P_y}{\partial x} - \dfrac{\partial P_x}{\partial y} \right) \mathbf{a}_z$

$= (0 - 0)\mathbf{a}_x + (x^2 y - z)\mathbf{a}_y + (0 - x^2 z)\mathbf{a}_z$

$= (x^2 y - z)\mathbf{a}_y - x^2 z \mathbf{a}_z$

(b) $\nabla \times \mathbf{Q} = \left[\dfrac{1}{\rho} \dfrac{\partial Q_z}{\partial \phi} - \dfrac{\partial Q_\phi}{\partial z} \right] \mathbf{a}_\rho + \left[\dfrac{\partial Q_\rho}{\partial z} - \dfrac{\partial Q_z}{\partial \rho} \right] \mathbf{a}_\phi + \dfrac{1}{\rho} \left[\dfrac{\partial}{\partial \rho} (\rho Q_\phi) - \dfrac{\partial Q_\rho}{\partial \phi} \right] \mathbf{a}_z$

$= \left(\dfrac{-z}{\rho} \sin \phi - \rho^2 \right) \mathbf{a}_\rho + (0 - 0)\mathbf{a}_\phi + \dfrac{1}{\rho} (3\rho^2 z - \rho \cos \phi)\mathbf{a}_z$

$= -\dfrac{1}{\rho} (z \sin \phi + \rho^3)\mathbf{a}_\rho + (3\rho z - \cos \phi)\mathbf{a}_z$

(c) $\nabla \times \mathbf{T} = \dfrac{1}{r \sin \theta} \left[\dfrac{\partial}{\partial \theta} (T_\phi \sin \theta) - \dfrac{\partial}{\partial \phi} T_\theta \right] \mathbf{a}_r$

$+ \dfrac{1}{r} \left[\dfrac{1}{\sin \theta} \dfrac{\partial}{\partial \phi} T_r - \dfrac{\partial}{\partial r} (rT_\phi) \right] \mathbf{a}_\theta + \dfrac{1}{r} \left[\dfrac{\partial}{\partial r} (rT_\theta) - \dfrac{\partial}{\partial \theta} T_r \right] \mathbf{a}_\phi$

$= \dfrac{1}{r \sin \theta} \left[\dfrac{\partial}{\partial \theta} (\cos \theta \sin \theta) - \dfrac{\partial}{\partial \phi} (r \sin \theta \cos \phi) \right] \mathbf{a}_r$

$+ \dfrac{1}{r} \left[\dfrac{1}{\sin \theta} \dfrac{\partial}{\partial \phi} \dfrac{(\cos \theta)}{r^2} - \dfrac{\partial}{\partial r} (r \cos \theta) \right] \mathbf{a}_\theta$

$$+ \frac{1}{r} \left[\frac{\partial}{\partial r} (r^2 \sin \theta \cos \phi) - \frac{\partial}{\partial \theta} \frac{(\cos \theta)}{r^2} \right] \mathbf{a}_\phi$$

$$= \frac{1}{r \sin \theta} (\cos 2\theta + r \sin \theta \sin \phi) \mathbf{a}_r + \frac{1}{r} (0 - \cos \theta) \mathbf{a}_\theta$$

$$+ \frac{1}{r} \left(2r \sin \theta \cos \phi + \frac{\sin \theta}{r^2} \right) \mathbf{a}_\phi$$

$$= \left(\frac{\cos 2\theta}{r \sin \theta} + \sin \phi \right) \mathbf{a}_r - \frac{\cos \theta}{r} \mathbf{a}_\theta + \left(2 \cos \phi + \frac{1}{r^3} \right) \sin \theta \, \mathbf{a}_\phi$$

PRACTICE EXERCISE 3.8

Determine the curl of each of the vector fields in Practice Exercise 3.6 and evaluate the curls at the specified points.

Answer: (a) $\mathbf{a}_x + y\mathbf{a}_y + (4y - z)\mathbf{a}_z$, $\mathbf{a}_x - 2\mathbf{a}_y - 11\mathbf{a}_z$

(b) $-6\rho z \cos \phi \, \mathbf{a}_\rho + \rho \sin \phi \, \mathbf{a}_\phi + (6z - 1)z \cos \phi \, \mathbf{a}_z$, $5\mathbf{a}_\phi$

(c) $\dfrac{\cot \theta}{r^{1/2}} \mathbf{a}_r - \left(2 \cot \theta \sin \phi + \dfrac{3}{2r^{1/2}} \right) \mathbf{a}_\theta + 2 \sin \theta \cos \phi \, \mathbf{a}_\phi$,

$1.732\mathbf{a}_r - 4.5\mathbf{a}_\theta + 0.5\mathbf{a}_\phi$.

EXAMPLE 3.9

If $\mathbf{A} = \rho \cos \phi \, \mathbf{a}_\rho + \sin \phi \, \mathbf{a}_\phi$, evaluate $\oint \mathbf{A} \cdot d\mathbf{l}$ around the path shown in Figure 3.23. Confirm this by using Stokes's theorem.

Solution:

Let

$$\oint_L \mathbf{A} \cdot d\mathbf{l} = \left[\int_a^b + \int_b^c + \int_c^d + \int_d^a \right] \mathbf{A} \cdot d\mathbf{l}$$

where path L has been divided into segments ab, bc, cd, and da as in Figure 3.23.

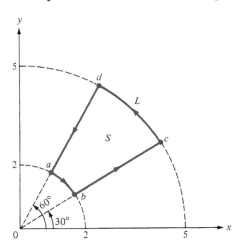

FIGURE 3.23 For Example 3.9.

Along ab, $\rho = 2$ and $d\mathbf{l} = \rho \, d\phi \, \mathbf{a}_\phi$. Hence,

$$\int_a^b \mathbf{A} \cdot d\mathbf{l} = \int_{\phi=60°}^{30°} \rho \sin\phi \, d\phi = 2(-\cos\phi)\Big|_{60°}^{30°} = -(\sqrt{3}-1)$$

Along bc, $\phi = 30°$ and $d\mathbf{l} = d\rho \, \mathbf{a}_\rho$. Hence,

$$\int_b^c \mathbf{A} \cdot d\mathbf{l} = \int_{\rho=2}^5 \rho \cos\phi \, d\rho = \cos 30° \frac{\rho^2}{2}\Big|_2^5 = \frac{21\sqrt{3}}{4}$$

Along cd, $\rho = 5$ and $d\mathbf{l} = \rho \, d\phi \, \mathbf{a}_\phi$. Hence,

$$\int_c^d \mathbf{A} \cdot d\mathbf{l} = \int_{\phi=30°}^{60°} \rho \sin\phi \, d\phi = 5(-\cos\phi)\Big|_{30°}^{60°} = \frac{5}{2}(\sqrt{3}-1)$$

Along da, $\phi = 60°$ and $d\mathbf{l} = d\rho \, \mathbf{a}_\rho$. Hence,

$$\int_d^a \mathbf{A} \cdot d\mathbf{l} = \int_{\rho=5}^2 \rho \cos\phi \, d\rho = \cos 60° \frac{\rho^2}{2}\Big|_5^2 = -\frac{21}{4}$$

Putting all these together results in

$$\oint_L \mathbf{A} \cdot d\mathbf{l} = -\sqrt{3} + 1 + \frac{21\sqrt{3}}{4} + \frac{5\sqrt{3}}{2} - \frac{5}{2} - \frac{21}{4}$$

$$= \frac{27}{4}(\sqrt{3}-1) = 4.941$$

From Stokes's theorem (because L is a closed path),

$$\oint_L \mathbf{A} \cdot d\mathbf{l} = \int_S (\nabla \times \mathbf{A}) \cdot d\mathbf{S}$$

But $d\mathbf{S} = \rho \, d\phi \, d\rho \, \mathbf{a}_z$ and

$$\nabla \times \mathbf{A} = \mathbf{a}_\rho\left[\frac{1}{\rho}\frac{\partial A_z}{\partial \phi} - \frac{\partial A_\phi}{\partial z}\right] + \mathbf{a}_\phi\left[\frac{\partial A_\rho}{\partial z} - \frac{\partial A_z}{\partial \rho}\right] + \mathbf{a}_z\frac{1}{\rho}\left[\frac{\partial}{\partial \rho}(\rho A_\phi) - \frac{\partial A_\rho}{\partial \phi}\right]$$

$$= (0-0)\mathbf{a}_\rho + (0-0)\mathbf{a}_\phi + \frac{1}{\rho}(1+\rho)\sin\phi \, \mathbf{a}_z$$

Hence:

$$\int_S (\nabla \times \mathbf{A}) \cdot d\mathbf{S} = \int_{\phi=30°}^{60°} \int_{\rho=2}^5 \frac{1}{\rho}(1+\rho)\sin\phi \, \rho \, d\rho \, d\phi$$

$$= \int_{30°}^{60°} \sin \phi \, d\phi \int_{2}^{5} (1 + \rho) d\rho$$

$$= -\cos \phi \Big|_{30°}^{60°} \left(\rho + \frac{\rho^2}{2} \right) \Big|_{2}^{5}$$

$$= \frac{27}{4} (\sqrt{3} - 1) = 4.941$$

PRACTICE EXERCISE 3.9

Use Stokes's theorem to confirm your result in Practice Exercise 3.2.

Answer: 1.

EXAMPLE 3.10

For a vector field $\mathbf{A}$, show explicitly that $\nabla \cdot \nabla \times \mathbf{A} = 0$; that is, the divergence of the curl of any vector field is zero.

Solution:

This vector identity, along with the one in Practice Exercise 3.10, is very useful in EM. For simplicity, assume that $\mathbf{A}$ is in Cartesian coordinates.

$$\nabla \cdot \nabla \times \mathbf{A} = \left(\frac{\partial}{\partial x}, \frac{\partial}{\partial y}, \frac{\partial}{\partial z} \right) \cdot \begin{vmatrix} \mathbf{a}_x & \mathbf{a}_y & \mathbf{a}_z \\ \frac{\partial}{\partial x} & \frac{\partial}{\partial y} & \frac{\partial}{\partial z} \\ A_x & A_y & A_z \end{vmatrix}$$

$$= \left(\frac{\partial}{\partial x}, \frac{\partial}{\partial y}, \frac{\partial}{\partial z} \right) \cdot \left[\left(\frac{\partial A_z}{\partial y} - \frac{\partial A_y}{\partial z} \right), -\left(\frac{\partial A_z}{\partial x} - \frac{\partial A_x}{\partial z} \right), \left(\frac{\partial A_y}{\partial x} - \frac{\partial A_x}{\partial y} \right) \right]$$

$$= \frac{\partial}{\partial x} \left(\frac{\partial A_z}{\partial y} - \frac{\partial A_y}{\partial z} \right) - \frac{\partial}{\partial y} \left(\frac{\partial A_z}{\partial x} - \frac{\partial A_x}{\partial z} \right) + \frac{\partial}{\partial z} \left(\frac{\partial A_y}{\partial x} - \frac{\partial A_x}{\partial y} \right)$$

$$= \frac{\partial^2 A_z}{\partial x \, \partial y} - \frac{\partial^2 A_y}{\partial x \, \partial z} - \frac{\partial^2 A_z}{\partial y \, \partial x} + \frac{\partial^2 A_x}{\partial y \, \partial z} + \frac{\partial^2 A_y}{\partial z \, \partial x} - \frac{\partial^2 A_x}{\partial z \, \partial y}$$

$$= 0$$

because $\dfrac{\partial^2 A_z}{\partial x \, \partial y} = \dfrac{\partial^2 A_z}{\partial y \, \partial x}$, and so on.

PRACTICE EXERCISE 3.10

For a scalar field V, show that $\nabla \times \nabla V = 0$; that is, the curl of the gradient of any scalar field vanishes.

Answer: Proof.

3.8 LAPLACIAN OF A SCALAR

For practical reasons, it is expedient to introduce a single operator that is the composite of gradient and divergence operators. This operator is known as the *Laplacian*.

> The **Laplacian** of a scalar field V, written as $\nabla^2 V$, is the divergence of the gradient of V.

Thus, in Cartesian coordinates,

Laplacian $V = \nabla \cdot \nabla V = \nabla^2 V$

$$= \left[\frac{\partial}{\partial x} \mathbf{a}_x + \frac{\partial}{\partial y} \mathbf{a}_y + \frac{\partial}{\partial z} \mathbf{a}_z \right] \cdot \left[\frac{\partial V}{\partial x} \mathbf{a}_x + \frac{\partial V}{\partial y} \mathbf{a}_y + \frac{\partial V}{\partial z} \mathbf{a}_z \right] \tag{3.59}$$

that is,

$$\boxed{\nabla^2 V = \frac{\partial^2 V}{\partial x^2} + \frac{\partial^2 V}{\partial y^2} + \frac{\partial^2 V}{\partial z^2}} \tag{3.60}$$

Notice that the Laplacian of a scalar field is another scalar field.

The Laplacian of V in other coordinate systems can be obtained from eq. (3.60) by transformation. In cylindrical coordinates,

$$\boxed{\nabla^2 V = \frac{1}{\rho} \frac{\partial}{\partial \rho} \left(\rho \frac{\partial V}{\partial \rho} \right) + \frac{1}{\rho^2} \frac{\partial^2 V}{\partial \phi^2} + \frac{\partial^2 V}{\partial z^2}} \tag{3.61}$$

and in spherical coordinates,

$$\boxed{\nabla^2 V = \frac{1}{r^2} \frac{\partial}{\partial r} \left(r^2 \frac{\partial V}{\partial r} \right) + \frac{1}{r^2 \sin \theta} \frac{\partial}{\partial \theta} \left(\sin \theta \frac{\partial V}{\partial \theta} \right) + \frac{1}{r^2 \sin^2 \theta} \frac{\partial^2 V}{\partial \phi^2}} \tag{3.62}$$

A scalar field V is said to be *harmonic* in a given region if its Laplacian vanishes in that region. In other words, if

$$\nabla^2 V = 0 \tag{3.63}$$

is satisfied in the region, the solution for V in eq. (3.63) is harmonic (it is of the form of sine or cosine). Equation (3.63) is called *Laplace's equation*. This equation will be solved in Chapter 6.

We have considered only the Laplacian of a scalar. Since the Laplacian operator ∇^2 is a scalar operator, it is also possible to define the Laplacian of a vector $\mathbf{A}$. In this context, $\nabla^2 \mathbf{A}$ should not be viewed as the divergence of the gradient of $\mathbf{A}$. Rather, $\nabla^2 \mathbf{A}$ is defined as the gradient of the divergence of $\mathbf{A}$ minus the curl of the curl of $\mathbf{A}$. That is,

$$\boxed{\nabla^2 \mathbf{A} = \nabla(\nabla \cdot \mathbf{A}) - \nabla \times \nabla \times \mathbf{A}} \tag{3.64}$$

This equation can be applied in finding $\nabla^2 \mathbf{A}$ in any coordinate system. In the Cartesian system (and only in that system), eq. (3.64) becomes[3]

$$\nabla^2 \mathbf{A} = \nabla^2 A_x \mathbf{a}_x + \nabla^2 A_y \mathbf{a}_y + \nabla^2 A_z \mathbf{a}_z \tag{3.65}$$

EXAMPLE 3.11

Find the Laplacian of the scalar fields of Example 3.3; that is,
(a) $V = e^{-z} \sin 2x \cosh y$
(b) $U = \rho^2 z \cos 2\phi$
(c) $W = 10r \sin^2 \theta \cos \phi$

Solution:

The Laplacian in the Cartesian system can be found by taking the first derivative and later the second derivative.

(a) $\nabla^2 V = \dfrac{\partial^2 V}{\partial x^2} + \dfrac{\partial^2 V}{\partial y^2} + \dfrac{\partial^2 V}{\partial z^2}$

$\qquad = \dfrac{\partial}{\partial x}(2e^{-z} \cos 2x \cosh y) + \dfrac{\partial}{\partial y}(e^{-z} \sin 2x \sinh y)$

$\qquad\quad + \dfrac{\partial}{\partial z}(-e^{-z} \sin 2x \cosh y)$

$\qquad = -4e^{-z} \sin 2x \cosh y + e^{-z} \sin 2x \cosh y + e^{-z} \sin 2x \cosh y$

$\qquad = -2e^{-z} \sin 2x \cosh y$

(b) $\nabla^2 U = \dfrac{1}{\rho} \dfrac{\partial}{\partial \rho}\left(\rho \dfrac{\partial U}{\partial \rho}\right) + \dfrac{1}{\rho^2} \dfrac{\partial^2 U}{\partial \phi^2} + \dfrac{\partial^2 U}{\partial z^2}$

$\qquad = \dfrac{1}{\rho} \dfrac{\partial}{\partial \rho}(2\rho^2 z \cos 2\phi) - \dfrac{1}{\rho^2} 4\rho^2 z \cos 2\phi + 0$

$\qquad = 4z \cos 2\phi - 4z \cos 2\phi$

$\qquad = 0$

(c) $\nabla^2 W = \dfrac{1}{r^2} \dfrac{\partial}{\partial r}\left(r^2 \dfrac{\partial W}{\partial r}\right) + \dfrac{1}{r^2 \sin \theta} \dfrac{\partial}{\partial \theta}\left(\sin \theta \dfrac{\partial W}{\partial \theta}\right) + \dfrac{1}{r^2 \sin^2 \theta} \dfrac{\partial^2 W}{\partial \phi^2}$

$\qquad = \dfrac{1}{r^2} \dfrac{\partial}{\partial r}(10r^2 \sin^2 \theta \cos \phi) + \dfrac{1}{r^2 \sin\theta} \dfrac{\partial}{\partial \theta}(10r \sin 2\theta \sin \theta \cos \phi)$

$\qquad\quad - \dfrac{10r \sin^2 \theta \cos \phi}{r^2 \sin^2 \theta}$

$\qquad = \dfrac{20 \sin^2 \theta \cos \phi}{r} + \dfrac{20r \cos 2\theta \sin \theta \cos \phi}{r^2 \sin \theta}$

$\qquad\quad + \dfrac{10r \sin 2\theta \cos \theta \cos \phi}{r^2 \sin \theta} - \dfrac{10 \cos \phi}{r}$

$\qquad = \dfrac{10 \cos \phi}{r}(2 \sin^2 \theta + 2 \cos 2\theta + 2 \cos^2 \theta - 1)$

$\qquad = \dfrac{10 \cos \phi}{r}(1 + 2 \cos 2\theta)$

---

[3] For explicit formulas for $\nabla^2 \mathbf{A}$ in cylindrical and spherical coordinates, see M. N. O. Sadiku, *Numerical Techniques in Electromagnetics with MATLAB,* 3rd ed. Boca Raton, FL: CRC Press, 2009, p. 647.

PRACTICE EXERCISE 3.11

Determine the Laplacian of the scalar fields of Practice Exercise 3.3, that is,

(a) $U = x^2y + xyz$

(b) $V = \rho z \sin \phi + z^2 \cos^2 \phi + \rho^2$

(c) $f = \cos \theta \sin \phi \ln r + r^2 \phi$

Answer: (a) $2y$, (b) $4 + 2\cos^2 \phi - \dfrac{2z^2}{\rho^2} \cos 2\phi$, (c) $\dfrac{1}{r^2} \cos \theta \sin \phi \, (1 - 2 \ln r$

$\csc^2 \theta \ln r) + 6\phi$.

---

†3.9 CLASSIFICATION OF VECTOR FIELDS

A vector field is uniquely characterized by its divergence and curl. Neither the divergence nor the curl of a vector field is sufficient to completely describe the field. All vector fields can be classified in terms of their vanishing or nonvanishing divergence or curl as follows:

(a) $\nabla \cdot \mathbf{A} = 0, \nabla \times \mathbf{A} = \mathbf{0}$

(b) $\nabla \cdot \mathbf{A} \neq 0, \nabla \times \mathbf{A} = \mathbf{0}$

(c) $\nabla \cdot \mathbf{A} = 0, \nabla \times \mathbf{A} \neq \mathbf{0}$

(d) $\nabla \cdot \mathbf{A} \neq 0, \nabla \times \mathbf{A} \neq \mathbf{0}$

Figure 3.24 illustrates typical fields in these four categories.

A vector field **A** is said to be **solenoidal** (or divergenceless) if $\nabla \cdot \mathbf{A} = 0$

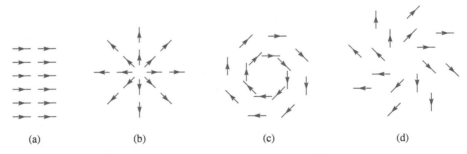

(a)   (b)   (c)   (d)

FIGURE 3.24 Typical fields with vanishing and nonvanishing divergence or curl.

(a) $A = k\mathbf{a}_x, \nabla \cdot \mathbf{A} = 0, \nabla \times \mathbf{A} = \mathbf{0}$,

(b) $A = k\mathbf{r}, \nabla \cdot \mathbf{A} = 3k, \nabla \times \mathbf{A} = \mathbf{0}$,

(c) $A = \mathbf{k} \times \mathbf{r}, \nabla \cdot \mathbf{A} = 0, \nabla \times \mathbf{A} = 2\mathbf{k}$,

(d) $A = \mathbf{k} \times \mathbf{r} + c\mathbf{r}, \nabla \cdot \mathbf{A} = 3c, \nabla \times \mathbf{A} = 2\mathbf{k}$.

Such a field has neither source nor sink of flux. From the divergence theorem,

$$\oint_S \mathbf{A} \cdot d\mathbf{S} = \int_v \nabla \cdot \mathbf{A} \, dv = 0 \tag{3.66}$$

Hence, flux lines of $\mathbf{A}$ entering any closed surface must also leave it. Examples of sole-noidal fields are incompressible fluids, magnetic fields, and conduction current density under steady-state conditions. In general, the field of curl $\mathbf{F}$ (for any $\mathbf{F}$) is purely solenoidal because $\nabla \cdot (\nabla \times \mathbf{F}) = 0$, as shown in Example 3.10. Thus, a solenoidal field $\mathbf{A}$ can always be expressed in terms of another vector $\mathbf{F}$; that is,

if

then
$$\boxed{\begin{array}{l} \nabla \cdot \mathbf{A} = 0 \\ \oint_S \mathbf{A} \cdot d\mathbf{S} = 0 \ \text{ and } \ \mathbf{A} = \nabla \times \mathbf{F} \end{array}} \tag{3.67}$$

A vector field $\mathbf{A}$ is said to be **irrotational** (or potential) if $\nabla \times \mathbf{A} = \mathbf{0}$.

That is, a *curl-free* vector is irrotational.[4] From Stokes's theorem

$$\int_S (\nabla \times \mathbf{A}) \cdot d\mathbf{S} = \oint_L \mathbf{A} \cdot d\mathbf{l} = 0 \tag{3.68}$$

Thus in an irrotational field $\mathbf{A}$, the circulation of $\mathbf{A}$ around a closed path is identically zero. This implies that the line integral of $\mathbf{A}$ is independent of the chosen path. Therefore, an irrotational field is also known as a *conservative field*. Examples of irrotational fields include the electrostatic field and the gravitational field. In general, the field of gradient V (for any scalar V) is purely irrotational, since (see Practice Exercise 3.10)

$$\nabla \times (\nabla V) = \mathbf{0} \tag{3.69}$$

Thus, an irrotational field $\mathbf{A}$ can always be expressed in terms of a scalar field V; that is,

if

then
$$\boxed{\begin{array}{l} \nabla \times \mathbf{A} = \mathbf{0} \\ \oint_L \mathbf{A} \cdot d\mathbf{l} = 0 \ \text{ and } \ \mathbf{A} = -\nabla V \end{array}} \tag{3.70}$$

For this reason, $\mathbf{A}$ may be called a *potential* field and V the scalar potential of $\mathbf{A}$. The negative sign in eq. (3.70) has been inserted for physical reasons that will become evident in Chapter 4.

---

[4] In fact, curl was once known as *rotation*, and curl $\mathbf{A}$ is written as rot $\mathbf{A}$ in some textbooks. This is one reason to use the term *irrotational*.

A vector **A** is uniquely prescribed within a region by its divergence and its curl. If we let

$$\nabla \cdot \mathbf{A} = \rho_v \tag{3.71a}$$

and

$$\nabla \times \mathbf{A} = \boldsymbol{\rho}_S \tag{3.71b}$$

ρ_v can be regarded as the source density of **A** and $\boldsymbol{\rho}_S$ its circulation density. Any vector **A** satisfying eq. (3.71) with both ρ_v and $\boldsymbol{\rho}_S$ vanishing at infinity can be written as the sum of two vectors: one irrotational (zero curl), the other solenoidal (zero divergence). This is called *Helmholtz's theorem.* Thus we may write

$$\mathbf{A} = -\nabla V + \nabla \times \mathbf{B} \tag{3.72}$$

If we let $\mathbf{A}_i = -\nabla V$ and $\mathbf{A}_s = \nabla \times \mathbf{B}$, it is evident from Example 3.10 and Practice Exercise 3.10 that $\nabla \times \mathbf{A}_i = \mathbf{0}$ and $\nabla \cdot \mathbf{A}_s = 0$, showing that $\mathbf{A}_i$ is irrotational and $\mathbf{A}_s$ is solenoidal. Finally, it is evident from eqs. (3.64) and (3.71) that any vector field has a Laplacian that satisfies

$$\nabla^2 \mathbf{A} = \nabla \rho_v - \nabla \times \boldsymbol{\rho}_S \tag{3.73}$$

EXAMPLE 3.12

Show that the vector field **A** is conservative if **A** possesses one of these two properties:

(a) The line integral of the tangential component of **A** along a path extending from a point P to a point Q is independent of the path.

(b) The line integral of the tangential component of **A** around any closed path is zero.

Solution:

(a) If **A** is conservative, $\nabla \times \mathbf{A} = \mathbf{0}$, so there exists a potential V such that

$$\mathbf{A} = -\nabla V = -\left[\frac{\partial V}{\partial x} \mathbf{a}_x + \frac{\partial V}{\partial y} \mathbf{a}_y + \frac{\partial V}{\partial z} \mathbf{a}_z \right]$$

Hence,

$$
\begin{aligned}
\int_P^Q \mathbf{A} \cdot d\mathbf{l} &= -\int_P^Q \left[\frac{\partial V}{\partial x} dx + \frac{\partial V}{\partial y} dy + \frac{\partial V}{\partial z} dz \right] \\
&= -\int_P^Q \left[\frac{\partial V}{\partial x} \frac{dx}{ds} + \frac{\partial V}{\partial y} \frac{dy}{ds} + \frac{\partial V}{\partial z} \frac{dz}{ds} \right] ds \\
&= -\int_P^Q \frac{dV}{ds} ds = -\int_P^Q dV
\end{aligned}
$$

or

$$\int_P^Q \mathbf{A} \cdot d\mathbf{l} = V(P) - V(Q)$$

showing that the line integral depends only on the end points of the curve. Thus, for a conservative field, $\int_P^Q \mathbf{A} \cdot d\mathbf{l}$ is simply the difference in potential at the end points.

(b) If the path is closed, that is, if P and Q coincide, then

$$\oint \mathbf{A} \cdot d\mathbf{l} = V(P) - V(P) = 0$$

PRACTICE EXERCISE 3.12

Show that $\mathbf{B} = (y + z \cos xz)\mathbf{a}_x + x\mathbf{a}_y + x \cos xz \, \mathbf{a}_z$ is conservative, without computing any integrals.

Answer: Proof.

MATLAB 3.1

```
% This script allows the user to compute the integral of
% a function using two different methods:
%    1. the built-in matlab 'quad' function
%    2. user-defined summation
%
%  The user must first create a separate file for the function
%       y = (-1/20)*x^3+(3/5)*x.^2-(21/10)*x+4;
%  The file should be named fun.m and stored in the same
%  directory as this file, and it should contain the following
%  two lines:
%         function y = fun(x)
%         y = (-1/20)*x.^3+(3/5)*x.^2-2.1*x+4;
%
% We will determine the integral of this function from x = 0
% to x = 8
clear

% First we'll plot the function, creating a vector x and y
x=0:0.01:8;
y=fun(x);
figure(1)    % create a figure
plot(x,y, 'LineWidth', 2)      % plot x versus y
axis([0 10 0 4]) % sets the axis appropriately
xlabel('x variable')    % axes labels
ylabel('y variable')    % axes labels

% Next we'll use the built-in Matlab function to find the
% quadrature integral

Q = quad(@fun,0,8);    % The @ is an address operator to
                       % point to fun.m

% Finally we'll create a custom summation to compute the
% integral quadrature integral
```

```
disp('Enter a increment size for the integral, recommended ');
disp(' 0.1 to 1 (the smaller the better, but');
dx=input('smaller requires more computation time)! ... >');

sum=0; % set initial total sum to zero
for x=0:dx:8,
    sum=sum+fun(x)*dx;   % add the partial sums to the total sum
end

disp('')
disp('The computed integrals of the function y(x) between');
disp(' x = 0 and x = 8 are')
% The tab %f outputs the floating point number given in the
% variables Q and sum, similar to C/C++
disp(sprintf(' quad integral ='));
disp(sprintf(' %f\n custom summation integral = %f', Q, sum))

% Now plot the function with the sub-areas used in the
% approximation create rectangular patches for each sub-area
figure(2)   % create another figure number 2
for x=0:dx:8,
    patch([x–dx/2; x–dx/2; x+dx/2; x+dx/2], ...
          [0; fun(x); fun(x); 0], [0.5 0.5 0.5])
end
% now plot original function
hold on
x=0:0.01:8;
y=fun(x);
h=plot(x,y, 'LineWidth', 2)       % plot x versus y
axis([0 10 0 4]) % sets the axis appropriately
xlabel('x variable')    % axes labels
ylabel('y variable')    % axes labels
function y = fun(x)
y = (–1/20)*x.^3+(3/5)*x.^2–2.1*x+4;
```

MATLAB 3.2

```
% This script allows the user to find the divergence and curl
% of a vector field given in symbolic form
% It uses the built-in symbolic derivative function
% called diff() to compute the derivatives
clear
syms x y z   % declare x,y,z to be symbols (variables)

% Prompt the user to enter the symbolic vector
%     For example the user could enter [y*z 4*x*y y]
disp('Enter the symbolic vector (in the format ');
A = input('[ fx(x,y,z) fy(x,y,z) fz(x,y,z)])... \n > ');

% The divergence of A
% e.g. diff(A(2),z) means the derivative of the
```

```
% y-component of vector A with respect to z
divA=diff(A(1),x)+...
    diff(A(2),y)+...
    diff(A(3),z)
% evaluate divergence at point (x,y,z) = (1, -2, 3)
subs(divA,{x,y,z},{1, -2, 3})

% The curl of A
% e.g. diff(A(2),z) means the derivative of the
% y-component of vector A with respect to z
curlA=[diff(A(3),y)-diff(A(2),z),...
    -diff(A(3),x)+diff(A(1),z),...
    diff(A(2),x)-diff(A(1),y)]
% evaluate curl at point (x,y,z) = (1, -2, 3)
subs(curlA,{x,y,z},{1, -2, 3})
```

SUMMARY

1. The differential displacements in the Cartesian, cylindrical, and spherical systems are, respectively,

$$dl = dx\,\mathbf{a}_x + dy\,\mathbf{a}_y + dz\,\mathbf{a}_z$$
$$dl = d\rho\,\mathbf{a}_\rho + \rho\,d\phi\,\mathbf{a}_\phi + dz\,\mathbf{a}_z$$
$$dl = dr\,\mathbf{a}_r + r\,d\theta\,\mathbf{a}_\theta + r\sin\theta\,d\phi\,\mathbf{a}_\phi$$

Note that dl is always taken to be in the positive direction; the direction of the displacement is taken care of by the limits of integration.

2. The differential normal areas in the three systems are, respectively,

$$dS = dy\,dz\,\mathbf{a}_x$$
$$dx\,dz\,\mathbf{a}_y$$
$$dx\,dy\,\mathbf{a}_z$$

$$dS = \rho\,d\phi\,dz\,\mathbf{a}_\rho$$
$$d\rho\,dz\,\mathbf{a}_\phi$$
$$\rho\,d\rho\,d\phi\,\mathbf{a}_z$$

$$dS = r^2\sin\theta\,d\theta\,d\phi\,\mathbf{a}_r$$
$$r\sin\theta\,dr\,d\phi\,\mathbf{a}_\theta$$
$$r\,dr\,d\theta\,\mathbf{a}_\phi$$

Note that dS can be in the positive or negative direction depending on the surface under consideration.

3. The differential volumes in the three systems are

$$dv = dx\,dy\,dz$$
$$dv = \rho\,d\rho\,d\phi\,dz$$
$$dv = r^2\sin\theta\,dr\,d\theta\,d\phi$$

4. The line integral of vector **A** along a path L is given by $\int_L \mathbf{A} \cdot d\mathbf{l}$. If the path is closed, the line integral becomes the circulation of **A** around L, that is, $\oint_L \mathbf{A} \cdot d\mathbf{l}$.

5. The flux or surface integral of a vector **A** across a surface S is defined as $\int_S \mathbf{A} \cdot d\mathbf{S}$. When the surface S is closed, the surface integral becomes the net outward flux of **A** across S, that is, $\oint_S \mathbf{A} \cdot d\mathbf{S}$.

6. The volume integral of a scalar ρ_v over a volume v is defined as $\int_v \rho_v \, dv$.

7. Vector differentiation is performed by using the vector differential operator ∇. The gradient of a scalar field V is denoted by ∇V, the divergence of a vector field **A** by $\nabla \cdot \mathbf{A}$, the curl of **A** by $\nabla \times \mathbf{A}$, and the Laplacian of V by $\nabla^2 V$. All of these are point functions since differentiation is always at a point.

8. The divergence theorem, $\oint_S \mathbf{A} \cdot d\mathbf{S} = \int_v \nabla \cdot \mathbf{A} \, dv$, relates a surface integral over a closed surface to a volume integral.

9. Stokes's theorem, $\oint_L \mathbf{A} \cdot d\mathbf{l} = \int_S (\nabla \times \mathbf{A}) \cdot d\mathbf{S}$, relates a line integral over a closed path to a surface integral.

10. If Laplace's equation, $\nabla^2 V = 0$, is satisfied by a scalar field V in a given region, V is said to be harmonic in that region.

11. A vector field is solenoidal if $\nabla \cdot \mathbf{A} = 0$; it is irrotational or conservative if $\nabla \times \mathbf{A} = \mathbf{0}$.

12. A summary of the vector calculus operations in the three coordinate systems is provided on the inside back cover of the text.

13. The vector identities $\nabla \cdot \nabla \times \mathbf{A} = 0$ and $\nabla \times \nabla V = \mathbf{0}$ are very useful in EM. Other vector identities are in Appendix A.10.

REVIEW QUESTIONS

3.1 Consider the differential volume of Figure 3.25. Match the items in the left-hand column with those on the right.

(a) $d\mathbf{l}$ from A to B (i) $dy\, dz\, \mathbf{a}_x$

(b) $d\mathbf{l}$ from A to D (ii) $-dx\, dz\, \mathbf{a}_y$

(c) $d\mathbf{l}$ from A to E (iii) $dx\, dy\, \mathbf{a}_z$

(d) $d\mathbf{S}$ for face $ABCD$ (iv) $-dx\, dy\, \mathbf{a}_z$

(e) $d\mathbf{S}$ for face $AEHD$ (v) $dx\, \mathbf{a}_x$

(f) $d\mathbf{S}$ for face $DCGH$ (vi) $dy\, \mathbf{a}_y$

(g) $d\mathbf{S}$ for face $ABFE$ (vii) $dz\, \mathbf{a}_z$

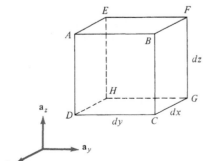

FIGURE 3.25 For Review Question 3.1.

3.2 For the differential volume in Figure 3.26, match the items in the left-hand list with those on the right.

(a) $d\mathbf{l}$ from E to A
(b) $d\mathbf{l}$ from B to A
(c) $d\mathbf{l}$ from D to A
(d) $d\mathbf{S}$ for face $ABCD$
(e) $d\mathbf{S}$ for face $AEHD$
(f) $d\mathbf{S}$ for face $ABFE$
(g) $d\mathbf{S}$ for face $DCGH$

(i) $-\rho \, d\phi \, dz \, \mathbf{a}_\rho$
(ii) $-d\rho \, dz \, \mathbf{a}_\phi$
(iii) $-\rho \, d\rho \, d\phi \, \mathbf{a}_z$
(iv) $\rho \, d\rho \, d\phi \, \mathbf{a}_z$
(v) $d\rho \, \mathbf{a}_\rho$
(vi) $\rho \, d\phi \, \mathbf{a}_\phi$
(vii) $dz \, \mathbf{a}_z$

3.3 Consider the object shown in Figure 3.8. For the volume element, match the items in the left-hand column with those on the right.

(a) $d\mathbf{l}$ from A to D
(b) $d\mathbf{l}$ from E to A
(c) $d\mathbf{l}$ from A to B
(d) $d\mathbf{S}$ for face $EFGH$
(e) $d\mathbf{S}$ for face $AEHD$
(f) $d\mathbf{S}$ for face $ABFE$

(i) $-r^2 \sin\theta \, d\theta \, d\phi \, \mathbf{a}_r$
(ii) $-r \sin\theta \, dr \, d\phi \, \mathbf{a}_\theta$
(iii) $r \, dr \, d\theta \, \mathbf{a}_\phi$
(iv) $dr \, \mathbf{a}_r$
(v) $r \, d\theta \, \mathbf{a}_\theta$
(vi) $r \sin\theta \, d\phi \, \mathbf{a}_\phi$

3.4 If $\mathbf{r} = x\mathbf{a}_x + y\mathbf{a}_y + z\mathbf{a}_z$, the position vector of point (x, y, z) and $r = |\mathbf{r}|$, which of the following is incorrect?

(a) $\nabla r = \mathbf{r}/r$
(b) $\nabla \cdot \mathbf{r} = 1$
(c) $\nabla^2(\mathbf{r} \cdot \mathbf{r}) = 6$
(d) $\nabla \times \mathbf{r} = \mathbf{0}$

3.5 Which of the following is mathematically defined?

(a) $\nabla \times \nabla \cdot \mathbf{A}$
(b) $\nabla \cdot (\nabla \cdot \mathbf{A})$
(c) $\nabla (\nabla V)$
(d) $\nabla (\nabla \cdot \mathbf{A})$

3.6 Which of the following is zero?

(a) grad div
(b) div grad
(c) curl grad
(d) curl curl

3.7 Given field $\mathbf{A} = 3x^2yz\mathbf{a}_x + x^3z\mathbf{a}_y + (x^3y - 2z)\mathbf{a}_z$, it can be said that $\mathbf{A}$ is

(a) Harmonic
(b) Divergenceless
(c) Solenoidal
(d) Rotational
(e) Conservative

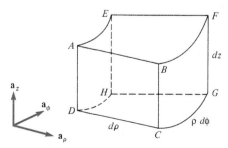

FIGURE 3.26 For Review Question 3.2.

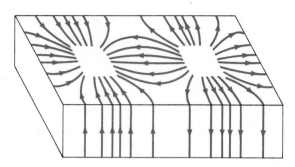

FIGURE 3.27 For Review Question 3.8.

3.8 The surface current density **J** in a rectangular waveguide is plotted in Figure 3.27. It is evident from the figure that **J** diverges at the top wall of the guide, whereas it is divergenceless at the side wall.

(a) True　　　　　　　　　　　　(b) False

3.9 Stokes's theorem is applicable only when a closed path exists and the vector field and its derivatives are continuous within the path.

(a) True　　　　　　　　　　　　(c) Not necessarily

(b) False

3.10 If a vector field **Q** is solenoidal, which of these is true?

(a) $\oint_L \mathbf{Q} \cdot d\mathbf{l} = 0$　　　　　　　　(d) $\nabla \times \mathbf{Q} \neq \mathbf{0}$

(b) $\oint_S \mathbf{Q} \cdot d\mathbf{S} = 0$　　　　　　　　(e) $\nabla^2 \mathbf{Q} = \mathbf{0}$

(c) $\nabla \times \mathbf{Q} = \mathbf{0}$

Answers: 3.1a-(vi), b-(vii), c-(v), d-(i), e-(ii), f-(iv), g-(iii), 3.2a-(vi), b-(v), c-(vii), d-(ii), e-(i), f-(iv), g-(iii), 3.3a-(v), b-(vi), c-(iv), d-(iii), e-(i), f-(ii), 3.4b, 3.5d, 3.6c, 3.7e, 3.8a, 3.9a, 3.10b.

PROBLEMS

Section 3.2—Differential Length, Area, and Volume

3.1 Using the differential length dl, find the length of each of the following curves:

(a) $\rho = 3, \pi/4 < \phi < \pi/2, z = $ constant

(b) $r = 1, \theta = 30°, 0 < \phi < 60°$

(c) $r = 4, 30° < \theta < 90°, \phi = $ constant

3.2 Calculate the areas of the following surfaces using the differential surface area dS:

(a) $\rho = 2, 0 < z < 5, \pi/3 < \phi < \pi/2$

(b) $z = 1, 1 < \rho < 3, 0 < \phi < \pi/4$

(c) $r = 10, \pi/4 < \theta < 2\pi/3, 0 < \phi < 2\pi$

(d) $0 < r < 4, 60° < \theta < 90°, \phi = $ constant

3.3 Use the differential volume dv to determine the volumes of the following regions:

(a) $0 < x < 1, 1 < y < 2, -3 < z < 3$

(b) $2 < \rho < 5, \pi/3 < \phi < \pi, -1 < z < 4$

(c) $1 < r < 3, \pi/2 < \theta < 2\pi/3, \pi/6 < \phi < \pi/2$

3.4 Find the length of a path from $P_1(4, 0°, 0)$ to $P_2(4, 30°, 0)$.

3.5 Calculate the area of the surface defined by $r = 5, 0 < \theta < \pi/4, 0 < \phi < \pi/2$.

3.6 Calculate the volume defined by $2 < \rho < 5, 0 < \phi < 30°, 0 < z < 10$.

Section 3.3—Line, Surface, and Volume Integrals

3.7 Let $H = xy^2\mathbf{a}_x + x^2y\mathbf{a}_y$. Evaluate the line integral along the parabola $x = y^2$ joining point $P(1, 1, 0)$ to point $Q(16, 4, 0)$.

3.8 Evaluate the line integral $\int_L (2x^2 - 4xy)dx + 3xy - 2x^2y)dy$ over the straight path L joining point $P(1, -1, 2)$ to $Q(3, 1, 2)$.

3.9 If the integral $\int_A^B \mathbf{F} \cdot d\mathbf{l}$ is regarded as the work done in moving a particle from A to B, find the work done by the force field

$$\mathbf{F} = 2xy\mathbf{a}_x + (x^2 - z^2)\mathbf{a}_y - 3xz^2\mathbf{a}_z$$

on a particle that travels from $A(0, 0, 0)$ to $B(2, 1, 3)$ along

(a) The segment $(0, 0, 0) \rightarrow (0, 1, 0) \rightarrow (2, 1, 0) \rightarrow (2, 1, 3)$

(b) The straight line $(0, 0, 0)$ to $(2, 1, 3)$

3.10 A vector field is represented by $\mathbf{F} = \rho^2\mathbf{a}_\rho + z\mathbf{a}_\phi + \cos\phi\mathbf{a}_z$ Newtons. Evaluate the work done or $\int_L \mathbf{F} \cdot d\mathbf{l}$, where L is from $P(2, 0°, 0)$ to $Q(2, \pi/4, 3)$. Assume that L consists of the arc $\rho = 2, 0 < \phi < \pi/4, z = 0$, followed by the line $\rho = 2, \phi = \pi/4, 0 < z < 3$.

3.11 If

$$\mathbf{H} = (x - y)\mathbf{a}_x + (x^2 + zy)\mathbf{a}_y + 5yz\mathbf{a}_z$$

evaluate $\int_L \mathbf{H} \cdot d\mathbf{l}$ along the contour of Figure 3.28.

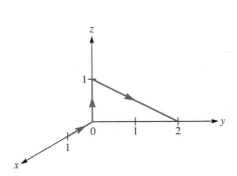

FIGURE 3.28 For Problem 3.11.

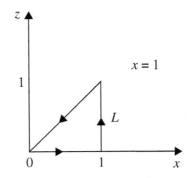

FIGURE 3.29 For Problem 3.12.

3.12 Determine the circulation of $\mathbf{B} = xy\mathbf{a}_x - yz\mathbf{a}_y + xz\mathbf{a}_z$ around the path L on the $x = 1$ plane, shown in Figure 3.29.

3.13 Let $\mathbf{A} = y\mathbf{a}_x + z\mathbf{a}_y + x\mathbf{a}_z$. Find the flux of $\mathbf{A}$ through surface $y = 1, 0 < x < 1, 0 < z < 2$.

3.14 If $\mathbf{D} = x^2z\mathbf{a}_x + y^3\mathbf{a}_y + yz^2\mathbf{a}_z$, calculate the flux of $\mathbf{D}$ passing through the volume bounded by planes $x = -1$, $x = 1$, $y = 0$, $y = 4$, $z = 1$, and $z = 3$.

3.15 A vector field is specified as $\mathbf{A} = r\mathbf{a}_r - 3\mathbf{a}_\theta + 5\phi\mathbf{a}_\phi$. Find the flux of the field out of the closed surface defined by $0 < r < 4, 0 < \theta < \pi/2, 0 < \phi < 3 < \pi/2$.

3.16 (a) Evaluate $\int_v xy\, dv$, where v is defined by $0 < x < 1, 0 < y < 1, 0 < z < 2$.

(b) Determine $\int_v \rho z\, dv$, where v is bounded by $\rho = 1, \rho = 3, \phi = 0, \phi = \pi, z = 0$, and $z = 2$.

Section 3.5—Gradient of a Scalar

3.17 Calculate the gradient of:

(a) $V_1 = 6xy - 2xz + z$

(b) $V_2 = 10\rho \cos \phi - \rho z$

(c) $V_3 = \dfrac{2}{r} \cos \phi$

3.18 Find the gradient of the following scalar fields and evaluate the gradient at the specified point.

(a) $V(x, y, z) = 10xyz - 2x^2z$ at $P(-1, 4, 3)$

(b) $U(\rho, \phi, z) = 2\rho \sin \phi + \rho z$ at $Q(2, 90°, -1)$

(c) $W(r, \theta, \phi) = \dfrac{4}{r} \sin \theta \cos \phi$ at $R(1, \pi/6, \pi/2)$

3.19 If $\mathbf{r} = x\mathbf{a}_x + y\mathbf{a}_y + z\mathbf{a}_z$ is the position vector of point (x, y, z), $r = |\mathbf{r}|$, and n is an integer, show that $\nabla r^n = nr^{n-2}\mathbf{r}$.

3.20 The temperature in an auditorium is given by $T = x^2 + y^2 - z$. A mosquito located at $(1, 1, 2)$ in the auditorium desires to fly in such a direction that it will get warm as soon as possible. In what direction must it fly?

3.21 A family of planes is described by $F = x - 2y + z$. Find a unit normal $\mathbf{a}_n$ to the planes.

3.22 Consider the scalar function $T = r \sin \theta \cos \phi$. Determine the magnitude and direction of the maximum rate of change of T at $P(2, 6°, 30°)$.

3.23 Let $f = x^2y - 2xy^2 + z^3$. Find the directional derivative of f at point $(2, 4, -3)$ in the direction of $\mathbf{a}_x + 2\mathbf{a}_y - \mathbf{a}_z$.

3.24 (a) Using the gradient concept, prove that the angle between two planes

$$ax + by + cz = d$$
$$\alpha x + \beta y + \gamma z = \delta$$

is

$$\theta = \cos^{-1} \frac{a\alpha + b\beta + c\gamma}{\sqrt{(a^2 + b^2 + c^2)(\alpha^2 + \beta^2 + \gamma^2)}}$$

(b) Calculate the angle between two planes $x + 2y + 3z = 5$ and $x + y = 0$.

3.25 Let $V(x, y, z) = 4xye^z$. Find the maximum rate of change of V at $(3, 1, -2)$ and the direction in which it occurs.

3.26 (a) Prove that for scalar fields V and U,

$$\nabla(UV) = U\nabla V + V\nabla U$$

(b) Verify part (a) by assuming that $V = 5x^2y + 2yz$ and $U = 3xyz$.

Section 3.6—Divergence of a Vector and Divergence Theorem

3.27 Evaluate the divergence of the following vector fields:

(a) $\mathbf{A} = xy\mathbf{a}_x + y^2\mathbf{a}_y - xz\mathbf{a}_z$
(b) $\mathbf{B} = \rho z^2\mathbf{a}_\rho + \rho \sin^2\phi\,\mathbf{a}_\phi + 2\rho z \sin^2\phi\,\mathbf{a}_z$
(c) $\mathbf{C} = r\mathbf{a}_r + r\cos^2\theta\,\mathbf{a}_\phi$

3.28 (a) If $\mathbf{A} = x^2y\mathbf{a}_x + xa_y + 2yz\mathbf{a}_z$, find $\nabla \cdot \mathbf{A}$ at point $(-3, 4, 2)$.

(b) Given that $\mathbf{B} = 3\rho \sin \phi\mathbf{a}_\rho - 5\rho^2z\mathbf{a}_\phi + 8z \cos^2\phi\mathbf{a}_z$, find $\nabla \cdot \mathbf{B}$ at point $(5, 30°, 1)$.
(c) Let $\mathbf{C} = r^2 \cos \phi\mathbf{a}_r + 2r\mathbf{a}_\phi$, find $\nabla \cdot \mathbf{C}$ at point $(2, \pi/3, \pi/2)$.

3.29 The heat flow vector $\mathbf{H} = k\nabla T$, where T is the temperature and k is the thermal conductivity. Show that if

$$T = 50 \sin \frac{\pi x}{2} \cosh \frac{\pi y}{2}$$

then $\nabla \cdot \mathbf{H} = 0$.

3.30 (a) Prove that

$$\nabla \cdot (V\mathbf{A}) = V\nabla \cdot \mathbf{A} + \mathbf{A} \cdot \nabla V$$

where V is a scalar field and $\mathbf{A}$ is a vector field.

(b) Evaluate $\nabla \cdot (V\mathbf{A})$ when $\mathbf{A} = 2x\mathbf{a}_x + 3y\mathbf{a}_y - 4z\mathbf{a}_z$ and $V = xyz$.

3.31 If $\mathbf{r} = x\mathbf{a}_x + y\mathbf{a}_y + z\mathbf{a}_z$ and $\mathbf{T} = 2zy\mathbf{a}_x + xy^2\mathbf{a}_y + x^2yz\mathbf{a}_z$, determine

(a) $(\nabla \cdot \mathbf{r})\mathbf{T}$
(b) $(\mathbf{r} \cdot \nabla)\mathbf{T}$
(c) $\nabla \cdot \mathbf{r}(\mathbf{r} \cdot \mathbf{T})$
(d) $(\mathbf{r} \cdot \nabla)r^2$

3.32 If $\mathbf{A} = 2x\mathbf{a}_x - z^2\mathbf{a}_y + 3xy\mathbf{a}_z$, find the flux of $\mathbf{A}$ through a surface defined by $\rho = 2$, $0 < \phi < \pi/2, 0 < z < 1$.

3.33 Let $\mathbf{D} = 2\rho z^2\mathbf{a}_\rho + \rho \cos^2\phi\,\mathbf{a}_z$. Evaluate

(a) $\oint_S \mathbf{D} \cdot d\mathbf{S}$

(b) $\int_v \nabla \cdot \mathbf{D}\,dv$

over the region defined by $2 \le \rho \le 5, -1 \le z \le 1, 0 < \phi < 2\pi$.

3.34 If $\mathbf{H} = 10 \cos \theta \mathbf{a}_r$, evaluate $\int_S \mathbf{H} \cdot d\mathbf{S}$ over a hemisphere defined by $r = 1, 0 < \phi < 2\pi, 0 < \theta < \pi/2$.

3.35 Evaluate both sides of the divergence theorem for the vector field

$$\mathbf{H} = 2xy\mathbf{a}_x + +x^2 + z^2)\mathbf{a}_y + 2yz\mathbf{a}_z$$

and the rectangular region defined by $0 < x < 1, 1 < y < 2, -1 < z < 3$.

3.36 Let $\mathbf{H} = 4\rho^2 \mathbf{a}_\rho - 2z\mathbf{a}_z$. Verify the divergence theorem for the cylindrical region defined by $\rho = 10, 0 < \phi < 2\pi, 0 < z < 3$.

*3.37 Apply the divergence theorem to evaluate $\oint_S \mathbf{A} \cdot d\mathbf{S}$, where $\mathbf{A} = x^2 \mathbf{a}_x + y^2 \mathbf{a}_y + z^2 \mathbf{a}_z$ and S is the surface of the solid bounded by the cylinder $\rho = 1$ and planes $z = 2$ and $z = 4$.

3.38 Verify the divergence theorem for the function $\mathbf{A} = r^2 \mathbf{a}_r + r \sin \theta \cos \phi \, \mathbf{a}_\theta$ over the surface of a quarter of a hemisphere defined by $0 < r < 3, 0 < \phi < \pi/2, 0 < \theta < \pi/2$.

3.39 Calculate the total outward flux of vector

$$\mathbf{F} = \rho^2 \sin \phi \, \mathbf{a}_\rho + z \cos \phi \, \mathbf{a}_\phi + \rho z \mathbf{a}_z$$

through the hollow cylinder defined by $2 \leq \rho \leq 3, 0 \leq z \leq 5$.

Section 3.7—Curl of a Vector and Stokes's Theorem

3.40 Evaluate the curl of the following vector fields:

(a) $\mathbf{A} = xy\mathbf{a}_x + y^2 \mathbf{a}_y - xz\mathbf{a}_z$
(b) $\mathbf{B} = \rho z^2 \mathbf{a}_\rho + \rho \sin^2 \phi \, \mathbf{a}_\phi + 2\rho z \sin^2 \phi \, \mathbf{a}_z$
(c) $\mathbf{C} = r\mathbf{a}_r + r \cos^2 \theta \, \mathbf{a}_\phi$

3.41 Evaluate $\nabla \times \mathbf{A}$ and $\nabla \cdot (\nabla \times \mathbf{A})$ if:

(a) $\mathbf{A} = x^2 y\mathbf{a}_x + y^2 z\mathbf{a}_y - 2xz\mathbf{a}_z$
(b) $\mathbf{A} = \rho^2 z\mathbf{a}_\rho + \rho^3 \mathbf{a}_\phi + 3\rho z^2 \mathbf{a}_z$
(c) $\mathbf{A} = \dfrac{\sin \phi}{r^2} \mathbf{a}_r - \dfrac{\cos \phi}{r^2} \mathbf{a}_\theta$

3.42 Let $\mathbf{H} = \rho \sin \phi \mathbf{a}_\rho + \rho \cos \phi \mathbf{a}_\phi - \rho \mathbf{a}_z$; find $\nabla \times \mathbf{H}$ and $\nabla \times \nabla \times \mathbf{H}$.

3.43 Let $\mathbf{A} = \dfrac{x\mathbf{a}_x + y\mathbf{a}_y + z\mathbf{a}_z}{(x^2 + y^2 + z^2)^{3/2}}$; show that $\nabla \times \mathbf{A} = 0$.

*3.44 Given that $\mathbf{F} = x^2 y\mathbf{a}_x - y\mathbf{a}_y$, find

(a) $\oint_L \mathbf{F} \cdot d\mathbf{l}$, where L is shown in Figure 3.30.

(b) $\int_S (\nabla \times \mathbf{F}) \cdot d\mathbf{S}$, where S is the area bounded by L.

(c) Is Stokes's theorem satisfied?

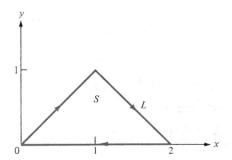

FIGURE 3.30 For Problem 3.44.

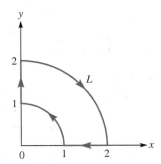

FIGURE 3.31 For Problem 3.45.

3.45 Let $\mathbf{A} = \rho \sin \phi \, \mathbf{a}_\rho + \rho^2 \mathbf{a}_\phi$; evaluate $\oint_L \mathbf{A} \cdot d\mathbf{l}$ if L is the contour of Figure 3.31.

3.46 If $\mathbf{F} = 2\rho z \mathbf{a}_\rho + 3z \sin \phi \, \mathbf{a}_\phi - 4\rho \cos \phi \, \mathbf{a}_z$, verify Stokes's theorem for the open surface defined by $z = 1, 0 < \rho < 2, 0 < \phi < 45°$.

3.47 Let $\mathbf{A} = 4x^2 e^{-y} \mathbf{a}_x - 8xe^{-y} \mathbf{a}_y$. Determine $\nabla \times [\nabla(\nabla \cdot \mathbf{A})]$.

3.48 Let $V = \dfrac{\sin \theta \cos \phi}{r}$. Determine:

  (a) ∇V,   (b) $\nabla \times \nabla V$,   (c) $\nabla \cdot \nabla V$

**3.49 A vector field is given by

$$\mathbf{Q} = \frac{\sqrt{x^2 + y^2 + z^2}}{\sqrt{x^2 + y^2}} \left[(x - y)\mathbf{a}_x + (x + y)\mathbf{a}_y \right]$$

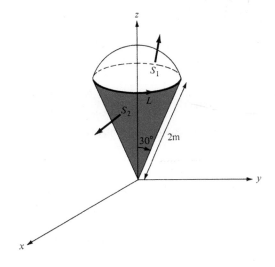

FIGURE 3.32 Volume in form of ice cream cone for Problem 3.49.

****Double asterisks indicate problems of highest difficulty.

Evaluate the following integrals:

(a) $\int_L \mathbf{Q} \cdot d\mathbf{l}$, where L is the circular edge of the volume in the form of an ice cream cone shown in Figure 3.32.

(b) $\int_{S_1} (\nabla \times \mathbf{Q}) \cdot d\mathbf{S}$, where S_1 is the top surface of the volume

(c) $\int_{S_2} (\nabla \times \mathbf{Q}) \cdot d\mathbf{S}$, where S_2 is the slanting surface of the volume

(d) $\int_{S_1} \mathbf{Q} \cdot d\mathbf{S}$

(e) $\int_{S_2} \mathbf{Q} \cdot d\mathbf{S}$

(f) $\int_v \nabla \cdot \mathbf{Q} \, dv$

How do your results in parts (a) to (f) compare?

*3.50 A rigid body spins about a fixed axis through its center with angular velocity ω. If $\mathbf{u}$ is the velocity at any point in the body, show that $\omega = 1/2 \, \nabla \times \mathbf{u}$.

3.51 Given that $H = 2xz\mathbf{a}_x + 5xyz\mathbf{a}_y + 8(y + z)\mathbf{a}_z$, find (a) $\nabla \cdot \mathbf{H}$ (b) $\nabla \times \mathbf{H}$.

3.52 Let $B = r^2\mathbf{a}_r + 4r \cos 2\theta\mathbf{a}_\theta$. Find the divergence and curl of B.

3.53 For a vector field $\mathbf{A}$ and a scalar field V, show in Cartesian coordinates that

(a) $\nabla \cdot (V\nabla V) = V\nabla^2 V + |\nabla V|^2$

(b) $\nabla \times (V\mathbf{A}) = V\nabla \times \mathbf{A} + \nabla V \times \mathbf{A}$

3.54 If $\mathbf{B} = x^2y\mathbf{a}_x + (2x^2 + y)\mathbf{a}_y - (y - z)\mathbf{a}_z$, find

(a) $\nabla \cdot \mathbf{B}$

(b) $\nabla \times \mathbf{B}$

(c) $\nabla (\nabla \cdot \mathbf{B})$

(d) $\nabla \times \nabla \times \mathbf{B}$

Section 3.8—Laplacian of a Scalar

3.55 Find $\nabla^2 V$ for each of the following scalar fields:

(a) $V_1 = x^3 + y^3 + z^3$

(b) $V_2 = \rho z^2 \sin 2\phi$

(c) $V_3 = r^2(1 + \cos \theta \sin \phi)$

3.56 Find the Laplacian of the following scalar fields and compute the value at the specified point.

(a) $U = x^3 y^2 e^{xz}$, $(1, -1, 1)$

(b) $V = \rho^2 z(\cos \phi + \sin \phi)$, $(5, \pi/6, -2)$

(c) $W = e^{-r} \sin \theta \cos \phi$, $(1, \pi/3, \pi/6)$

3.57 If $\mathbf{r} = x\mathbf{a}_x + y\mathbf{a}_y + z\mathbf{a}_z$ is the position vector of point (x, y, z), $r = |\mathbf{r}|$, show that:

(a) $\nabla(\ln r) = \dfrac{\mathbf{r}}{r^2}$

(b) $\nabla^2(\ln r) = \dfrac{1}{r^2}$

3.58

(a) If $U(x, y, z) = xy^2z^3$, find ∇U and $\nabla^2 U$.

(b) If $V(\rho, \phi, z) = \dfrac{\sin \phi}{\rho}$, find ∇V and $\nabla^2 V$.

(c) If $W(r, \theta, \phi,) = r^2 \sin \theta \cos \phi$, find ∇W and $\nabla^2 W$.

3.59 Given that $V = \rho^2 z \cos \phi$, find ∇V and $\nabla^2 V$.

3.60 If $V = \dfrac{5 \cos \phi}{r^2}$, find: (a) ∇V, (b) $\nabla \cdot \nabla V$, (c) $\nabla \times \nabla V$.

3.61 Let $U = 4xyz^2 + 10yz$. Show that $\nabla^2 U = \nabla \cdot \nabla U$.

***3.62** In cylindrical coordinates,

$$\nabla^2 \mathbf{A} = \left(\nabla^2 A_\rho - \frac{2}{\rho^2} \frac{\partial A_\phi}{\partial \phi} - \frac{A_\rho}{\rho^2} \right) \mathbf{a}_\rho + \left(\nabla^2 A_\phi + \frac{2}{\rho^2} \frac{\partial A_\rho}{\partial \phi} - \frac{A_\phi}{\rho^2} \right) \mathbf{a}_\phi + \nabla^2 A_z \mathbf{a}_z$$

If $\mathbf{G} = 2\rho \sin \phi \mathbf{a}_\rho + 4\rho \cos \phi \mathbf{a}_\phi + (z^2 + 1)\rho \mathbf{a}_z$, find $\nabla^2 \mathbf{G}$.

3.63 According to eq. (3.64), $\nabla \times (\nabla \times \mathbf{A}) = \nabla (\nabla \cdot \mathbf{A}) - \nabla^2 \mathbf{A}$. Show that $\mathbf{A} = xz\mathbf{a}_x + z^2\mathbf{a}_y + yz\mathbf{a}_z$ satisfies this vector identity.

Section 3.9—Classification of Vector Fields

3.64 Consider the following vector fields:
$\mathbf{A} = x\mathbf{a}_x + y\mathbf{a}_y + z\mathbf{a}_z$
$\mathbf{B} = 2\rho \cos \phi \mathbf{a}_\rho - 4\rho \sin \phi \mathbf{a}_\phi + 3\mathbf{a}_z$
$\mathbf{C} = \sin \theta \mathbf{a}_r + r \sin \theta \mathbf{a}_\phi$
Which of these fields are (a) solenoidal and (b) irrotational?

3.65 Given the vector field

$$\mathbf{G} = (16xy - z)\mathbf{a}_x + 8x^2\mathbf{a}_y - x\mathbf{a}_z$$

(a) Is $\mathbf{G}$ irrotational (or conservative)?

(b) Find the net flux of $\mathbf{G}$ over the cube $0 < x, y, z < 1$.

(c) Determine the circulation of $\mathbf{G}$ around the edge of the square $z = 0, 0 < x, y < 1$. Assume anticlockwise direction.

3.66 The electric field due to a line charge is given by

$$E = \frac{\lambda}{2\pi\epsilon\rho} \mathbf{a}_\rho$$

where λ is a constant. Show that E is solenoidal. Show that it is also conservative.

3.67 A vector field is given by $H = \frac{10}{r^2}\mathbf{a}_r$. Show that $\oint_L H \cdot d\mathbf{I} = 0$ for any closed path L.

3.68 Show that the vector field $B = (3x^2z + y^2)\mathbf{a}_x + 2xy\mathbf{a}_y + x^3\mathbf{a}_z$ is conservative.

3.69 Show that the vector field $D = (3\rho + 1) \sin \phi\mathbf{a}_z$ is solenoidal.

3.70 The field of an electric dipole is given by

$$E = k \frac{(2\cos\theta\mathbf{a}_r + \sin\theta\mathbf{a}_\theta)}{r^3}$$

where k is a constant. Show that E is conservative.

ELECTROSTATICS

Charles Augustin de Coulomb (1736–1806), a French physicist, was famous for his discoveries in the field of electricity and magnetism. He formulated Coulomb's law, to be discussed in this chapter.

Coulomb was born in Angoulême, France, to a family of wealth and social position. His father's family was well known in the legal profession, and his mother's family was also quite wealthy. Coulomb was educated in Paris and chose the profession of military engineer. Upon his retirement in 1789, Coulomb turned his attention to physics and published seven papers on electricity and magnetism. He was known for his work on electricity, magnetism, and mechanics. He invented a magnetoscope, a magnetometer, and a torsion balance that he employed in establishing Coulomb's law—the law of force between two charged bodies. Coulomb may be said to have extended Newtonian mechanics to a new realm of physics. The unit of electric charge, the coulomb, is named after him.

CHAPTER 4

ELECTROSTATIC FIELDS

Who is wise? He that learns from every one. Who is powerful? He that governs his passions. Who is rich? He that is content. Who is that? Nobody.

—BENJAMIN FRANKLIN

4.1 INTRODUCTION

Having mastered some essential mathematical tools needed for this course, we are now prepared to study the basic concepts of EM. We shall begin with those fundamental concepts that are applicable to static (or time-invariant) electric fields in free space (or vacuum). An electrostatic field is produced by a static charge distribution. A typical example of such a field is found in a cathode-ray tube.

Before we commence our study of electrostatics, it might be helpful to examine briefly the importance of such a study. Electrostatics is a fascinating subject that has grown up in diverse areas of application. Electric power transmission, X-ray machines, and lightning protection are associated with strong electric fields and will require a knowledge of electrostatics to understand and design suitable equipment. The devices used in solid-state electronics are based on electrostatics. These include resistors, capacitors, and active devices such as bipolar and field effect transistors, which are based on control of electron motion by electrostatic fields. Almost all computer peripheral devices, with the exception of magnetic memory, are based on electrostatic fields. Touch pads, capacitance keyboards, cathode-ray tubes, liquid crystal displays, and electrostatic printers are typical examples. In medical work, diagnosis is often carried out with the aid of electrostatics, as incorporated in electrocardiograms, electroencephalograms, and other recordings of the electrical activity of organs including eyes, ears, and the stomach. In industry, electrostatics is applied in a variety of forms such as paint spraying, electrodeposition, electrochemical machining, and separation of fine particles. Electrostatics is used in agriculture to sort seeds, for direct spraying of plants, to measure the moisture content of crops, to spin cotton, and for speed-baking bread and smoking meat.[1,2]

---

[1] For various applications of electrostatics, see J. M. Crowley, *Fundamentals of Applied Electrostatics*. New York: John Wiley & Sons, 1999; A. D. Moore, ed., *Electrostatics and Its Applications*. New York: John Wiley & Sons, 1973; and C. E. Jowett, *Electrostatics in the Electronics Environment*. New York: John Wiley & Sons, 1976.

[2] An interesting story on the magic of electrostatics is found in B. Bolton, *Electromagnetism and Its Applications: An Introduction*. London: Van Nostrand, 1980, p. 2.

We begin our study of electrostatics by investigating the two fundamental laws governing electrostatic fields: (1) Coulomb's law and (2) Gauss's law. Both of these laws are based on experimental studies, and they are interdependent. Although Coulomb's law is applicable in finding the electric field due to any charge configuration, it is easier to use Gauss's law when charge distribution is symmetrical. Based on Coulomb's law, the concept of electric field intensity will be introduced and applied to cases involving point, line, surface, and volume charges. Special problems that can be solved with much effort using Coulomb's law will be solved with ease by applying Gauss's law. Throughout our discussion in this chapter, we will assume that the electric field is in a vacuum or free space. Electric fields in material space will be covered in the next chapter.

4.2 COULOMB'S LAW AND FIELD INTENSITY

Coulomb's law is an experimental law formulated in 1785 by Charles Augustin de Coulomb, then a colonel in the French army. It deals with the force a point charge exerts on another point charge. By a *point charge* we mean a charge that is located on a body whose dimensions are much smaller than other relevant dimensions. For example, a collection of electric charges on a pinhead may be regarded as a point charge. Electrons are regarded as point charges. The polarity of charges may be positive or negative; like charges repel, while unlike charges attract. Charges are generally measured in coulombs (C). One coulomb is approximately equivalent to 6×10^{18} electrons; it is a very large unit of charge because one electron charge $e = -1.6019 \times 10^{-19}$ C.

> **Coulomb's law** states that the force F between two point charges Q_1 and Q_2 is:
>
> 1. Along the line joining them
> 2. Directly proportional to the product Q_1Q_2 of the charges
> 3. Inversely proportional to the square of the distance R between them.[3]

Expressed mathematically,

$$F = \frac{k\,Q_1Q_2}{R^2} \tag{4.1}$$

where k is the proportionality constant whose value depends on the choice of system of units. In SI units, charges Q_1 and Q_2 are in coulombs (C), the distance R is in meters (m), and the force F is in newtons (N) so that $k = 1/4\pi\varepsilon_0$. The constant ε_0 is known as the *permittivity of free space* (in farads per meter) and has the value

$$
\begin{array}{l}
\varepsilon_0 = 8.854 \times 10^{-12} \simeq \dfrac{10^{-9}}{36\pi} \text{ F/m} \\[2mm]
\text{or} \\[1mm]
\quad k = \dfrac{1}{4\pi\varepsilon_0} \simeq 9 \times 10^9 \text{ m/F}
\end{array}
\tag{4.2}
$$

---

[3] Further details of experimental verification of Coulomb's law can be found in W. F. Magie, *A Source Book in Physics*. Cambridge, MA: Harvard Univ. Press, 1963, pp. 408–420.

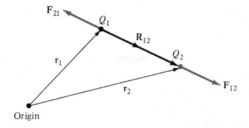

FIGURE 4.1 Coulomb vector force on point charges Q_1 and Q_2.

Thus eq. (4.1) becomes

$$F = \frac{Q_1 Q_2}{4\pi\varepsilon_o R^2} \tag{4.3}$$

If point charges Q_1 and Q_2 are located at points having position vectors $\mathbf{r}_1$ and $\mathbf{r}_2$, then the force $\mathbf{F}_{12}$ on Q_2 due to Q_1, shown in Figure 4.1, is given by

$$\boxed{\mathbf{F}_{12} = \frac{Q_1 Q_2}{4\pi\varepsilon_o R^2} \mathbf{a}_{R_{12}}} \tag{4.4}$$

where

$$\mathbf{R}_{12} = \mathbf{r}_2 - \mathbf{r}_1 \tag{4.5a}$$

$$R = |\mathbf{R}_{12}| \tag{4.5b}$$

$$\mathbf{a}_{R_{12}} = \frac{\mathbf{R}_{12}}{R} \tag{4.5c}$$

By substituting eq. (4.5) into eq. (4.4), we may write eq. (4.4) as

$$\mathbf{F}_{12} = \frac{Q_1 Q_2}{4\pi\varepsilon_o R^3} \mathbf{R}_{12} \tag{4.6a}$$

or

$$\mathbf{F}_{12} = \frac{Q_1 Q_2 (\mathbf{r}_2 - \mathbf{r}_1)}{4\pi\varepsilon_o |\mathbf{r}_2 - \mathbf{r}_1|^3} \tag{4.6b}$$

It is worthwhile to note that

1. As shown in Figure 4.1, the force $\mathbf{F}_{21}$ on Q_1 due to Q_2 is given by

$$\mathbf{F}_{21} = |\mathbf{F}_{12}|\mathbf{a}_{R_{21}} = |\mathbf{F}_{12}|(-\mathbf{a}_{R_{12}})$$

or

$$\mathbf{F}_{21} = -\mathbf{F}_{12} \tag{4.7}$$

since

$$\mathbf{a}_{R_{21}} = -\mathbf{a}_{R_{12}}$$

2. Like charges (charges of the same sign) repel each other, while unlike charges attract. This is illustrated in Figure 4.2.

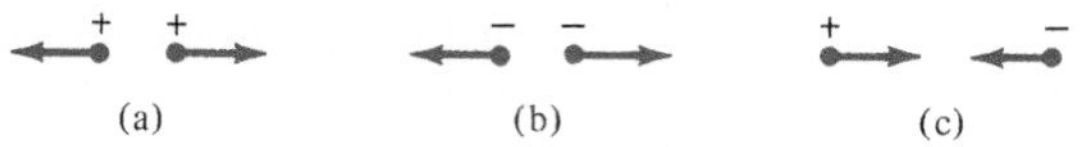

FIGURE 4.2 (a), (b) Like charges repel. (c) Unlike charges attract.

3. The distance R between the charged bodies Q_1 and Q_2 must be large compared with the linear dimensions of the bodies; that is, Q_1 and Q_2 must be point charges.
4. Q_1 and Q_2 must be static (at rest).
5. The signs of Q_1 and Q_2 must be taken into account in eq. (4.4). For like charges, $Q_1 Q_2 > 0$. For unlike charges, $Q_1 Q_2 < 0$.
6. Charges cannot be created or destroyed; the quantity of total charge remains constant.

If we have more than two point charges, we can use the *principle of superposition* to determine the force on a particular charge. The principle states that if there are N charges $Q_1, Q_2, \ldots, Q_N$ located, respectively, at points with position vectors $\mathbf{r}_1, \mathbf{r}_2, \ldots, \mathbf{r}_N$, the resultant force $\mathbf{F}$ on a charge Q located at point $\mathbf{r}$ is the vector sum of the forces exerted on Q by each of the charges $Q_1, Q_2, \ldots, Q_N$. Hence,

$$\mathbf{F} = \mathbf{F}_1 + \mathbf{F}_2 + \mathbf{F}_3 + \cdots + \mathbf{F}_N$$

$$= \frac{QQ_1(\mathbf{r} - \mathbf{r}_1)}{4\pi\varepsilon_o |\mathbf{r} - \mathbf{r}_1|^3} + \frac{QQ_2(\mathbf{r} - \mathbf{r}_2)}{4\pi\varepsilon_o |\mathbf{r} - \mathbf{r}_2|^3} + \cdots + \frac{QQ_N(\mathbf{r} - \mathbf{r}_N)}{4\pi\varepsilon_o |\mathbf{r} - \mathbf{r}_N|^3}$$

or

$$\boxed{\mathbf{F} = \frac{Q}{4\pi\varepsilon_o} \sum_{k=1}^{N} \frac{Q_k(\mathbf{r} - \mathbf{r}_k)}{|\mathbf{r} - \mathbf{r}_k|^3}} \tag{4.8}$$

We can now introduce the concept of *electric field intensity.*

The **electric field intensity** (or **electric field strength**) $\mathbf{E}$ is the force that a unit positive charge experiences when placed in an electric field.

Thus

$$\mathbf{E} = \lim_{Q \to 0} \frac{\mathbf{F}}{Q} \tag{4.9}$$

or simply

$$\boxed{\mathbf{E} = \frac{\mathbf{F}}{Q}} \tag{4.10}$$

For $Q > 0$, the electric field intensity $\mathbf{E}$ is obviously in the direction of the force $\mathbf{F}$ and is measured in newtons per coulomb or volts per meter. The electric field intensity at point $\mathbf{r}$ due to a point charge located at $\mathbf{r}'$ is readily obtained from eqs. (4.6) and (4.10) as

$$\mathbf{E} = \frac{Q}{4\pi\varepsilon_o R^2} \mathbf{a}_R = \frac{Q(\mathbf{r} - \mathbf{r}')}{4\pi\varepsilon_o |\mathbf{r} - \mathbf{r}'|^3} \tag{4.11a}$$

or simply

$$\boxed{\mathbf{E} = \frac{Q}{4\pi\varepsilon_o r^2} \mathbf{a}_r} \tag{4.11b}$$

For N point charges $Q_1, Q_2, \ldots, Q_N$ located at $\mathbf{r}_1, \mathbf{r}_2, \ldots, \mathbf{r}_N$, the electric field intensity at point $\mathbf{r}$ is obtained from eqs. (4.8) and (4.10) as

$$\mathbf{E} = \mathbf{E}_1 + \mathbf{E}_2 + \mathbf{E}_3 + \cdots + \mathbf{E}_N$$

$$= \frac{Q_1(\mathbf{r} - \mathbf{r}_1)}{4\pi\varepsilon_0|\mathbf{r} - \mathbf{r}_1|^3} + \frac{Q_2(\mathbf{r} - \mathbf{r}_2)}{4\pi\varepsilon_0|\mathbf{r} - \mathbf{r}_2|^3} + \cdots + \frac{Q_N(\mathbf{r} - \mathbf{r}_N)}{4\pi\varepsilon_0|\mathbf{r} - \mathbf{r}_N|^3}$$

or

$$\mathbf{E} = \frac{1}{4\pi\varepsilon_0} \sum_{k=1}^{N} \frac{Q_k(\mathbf{r} - \mathbf{r}_k)}{|\mathbf{r} - \mathbf{r}_k|^3} \qquad (4.12)$$

EXAMPLE 4.1

Point charges 1 mC and -2 mC are located at $(3, 2, -1)$ and $(-1, -1, 4)$, respectively. Calculate the electric force on a 10 nC charge located at $(0, 3, 1)$ and the electric field intensity at that point.

Solution:

$$\mathbf{F} = \sum_{k=1,2} \frac{QQ_k}{4\pi\varepsilon_0 R^2} \mathbf{a}_R = \sum_{k=1,2} \frac{QQ_k(\mathbf{r} - \mathbf{r}_k)}{4\pi\varepsilon_0|\mathbf{r} - \mathbf{r}_k|^3}$$

$$= \frac{Q}{4\pi\varepsilon_0} \left\{ \frac{10^{-3}[(0, 3, 1) - (3, 2, -1)]}{|(0, 3, 1) - (3, 2, -1)|^3} - \frac{2 \cdot 10^{-3}[(0, 3, 1) - (-1, -1, 4)]}{|(0, 3, 1) - (-1, -1, 4)|^3} \right\}$$

$$= \frac{10^{-3} \cdot 10 \cdot 10^{-9}}{4\pi \cdot \dfrac{10^{-9}}{36\pi}} \left[\frac{(-3, 1, 2)}{(9 + 1 + 4)^{3/2}} - \frac{2(1, 4, -3)}{(1 + 16 + 9)^{3/2}} \right]$$

$$= 9 \cdot 10^{-2} \left[\frac{(-3, 1, 2)}{14\sqrt{14}} + \frac{(-2, -8, 6)}{26\sqrt{26}} \right]$$

$$\mathbf{F} = -6.512\mathbf{a}_x - 3.713\mathbf{a}_y + 7.509\mathbf{a}_z \text{ mN}$$

At that point,

$$\mathbf{E} = \frac{\mathbf{F}}{Q}$$

$$= (-6.512, -3.713, 7.509) \cdot \frac{10^{-3}}{10 \cdot 10^{-9}}$$

$$\mathbf{E} = -651.2\mathbf{a}_x - 371.3\mathbf{a}_y + 750.9\mathbf{a}_z \text{ kV/m}$$

PRACTICE EXERCISE 4.1

Point charges 5 nC and -2 nC are located at $(2, 0, 4)$ and $(-3, 0, 5)$, respectively.

(a) Determine the force on a 1 nC point charge located at $(1, -3, 7)$.
(b) Find the electric field $\mathbf{E}$ at $(1, -3, 7)$.

Answer: (a) $-1.004\mathbf{a}_x - 1.284\mathbf{a}_y + 1.4\mathbf{a}_z$ nN.
(b) $-1.004\mathbf{a}_x - 1.284\mathbf{a}_y + 1.4\mathbf{a}_z$ V/m.

EXAMPLE 14.2

Two point charges of equal mass m and charge Q are suspended at a common point by two threads of negligible mass and length ℓ. Show that at equilibrium the inclination angle α of each thread to the vertical is given by

$$Q^2 = 16\pi \, \varepsilon_o mg\ell^2 \sin^2 \alpha \tan \alpha$$

If α is very small, show that

$$\alpha = \sqrt[3]{\frac{Q^2}{16\pi\varepsilon_o mg\ell^2}}$$

Solution:

Consider the system of charges as shown in Figure 4.3, where F_e is the electric or Coulomb force, T is the tension in each thread, and mg is the weight of each charge. At A or B

$$T \sin \alpha = F_e$$
$$T \cos \alpha = mg$$

Hence,

$$\frac{\sin \alpha}{\cos \alpha} = \frac{F_e}{mg} = \frac{1}{mg} \cdot \frac{Q^2}{4\pi\varepsilon_o r^2}$$

But $r = AB$ is given by

$$r = 2\ell \sin \alpha$$

Hence,

$$Q^2 \cos \alpha = 16\pi\varepsilon_o mg\ell^2 \sin^3 \alpha$$

or

$$Q^2 = 16\pi\varepsilon_o mg\ell^2 \sin^2 \alpha \tan \alpha$$

as required. When α is very small

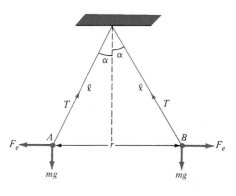

FIGURE 4.3 Suspended charged particles; for Example 4.2.

$$\tan \alpha \simeq \alpha \simeq \sin \alpha$$

and so

$$Q^2 = 16\pi\varepsilon_o mg\ell^2\alpha^3$$

or

$$\alpha = \sqrt[3]{\frac{Q^2}{16\pi\varepsilon_o mg\ell^2}}$$

PRACTICE EXERCISE 4.2

Three identical small spheres of mass m are suspended from a common point by threads of negligible masses and equal length ℓ. A charge Q is divided equally among the spheres, and they come to equilibrium at the corners of a horizontal equilateral triangle whose sides are d. Show that

$$Q^2 = 12\pi\varepsilon_o mgd^3\left[\ell^2 - \frac{d^2}{3}\right]^{-1/2}$$

where g = acceleration due to gravity.

Answer: Proof.

EXAMPLE 4.3   A practical application of electrostatics is in electrostatic separation of solids. For example, Florida phosphate ore, consisting of small particles of quartz and phosphate rock, can be separated into its components by applying a uniform electric field as in Figure 4.4. Assuming zero initial velocity and displacement, determine the separation between the particles after falling 80 cm. Take $E = 500$ kV/m and $Q/m = 9\ \mu$C/kg for both positively and negatively charged particles.

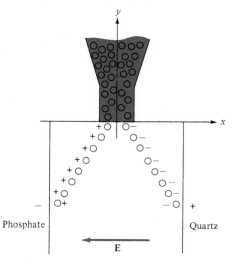

FIGURE 4.4 Electrostatic separation of solids; for Example 4.3.

Solution:

Ignoring the coulombic force between particles, the electrostatic force is acting horizontally while the gravitational force (weight) is acting vertically on the particles. Thus,

$$Q\mathbf{E} = m\frac{d^2x}{dt^2}\mathbf{a}_x$$

or

$$\frac{d^2x}{dt^2} = \frac{Q}{m}E$$

Integrating twice gives

$$x = \frac{Q}{2m}Et^2 + c_1 t + c_2$$

where c_1 and c_2 are integration constants. Similarly,

$$-mg = m\frac{d^2y}{dt^2}$$

or

$$\frac{d^2y}{dt^2} = -g$$

Integrating twice, we get

$$y = -1/2gt^2 + c_3 t + c_4$$

Since the initial displacement is zero,

$$x(t=0) = 0 \rightarrow c_2 = 0$$
$$y(t=0) = 0 \rightarrow c_4 = 0$$

Also, because of zero initial velocity,

$$\frac{dx}{dt}\bigg|_{t=0} = 0 \rightarrow c_1 = 0$$

$$\frac{dy}{dt}\bigg|_{t=0} = 0 \rightarrow c_3 = 0$$

Thus

$$x = \frac{QE}{2m}t^2, \quad y = -\frac{1}{2}gt^2$$

When $y = -80$ cm $= -0.8$ m

$$t^2 = \frac{0.8 \times 2}{9.8} = 0.1633$$

and

$$x = 1/2 \times 9 \times 10^{-6} \times 5 \times 10^5 \times 0.1633 = 0.3673 \text{ m}$$

The separation between the particles is $2x = 73.47$ cm.

PRACTICE EXERCISE 4.3

An ion rocket emits positive cesium ions from a wedge-shaped electrode into the region described by $x > |y|$. The electric field is $\mathbf{E} = -400\mathbf{a}_x + 200\mathbf{a}_y$ kV/m. The ions have single electronic charges $e = -1.6019 \times 10^{-19}$ C and mass $m = 2.22 \times 10^{-25}$ kg, and they travel in a vacuum with zero initial velocity. If the emission is confined to -40 cm $< y < 40$ cm, find the largest value of x that can be reached.

Answer: 0.8 m.

4.3 ELECTRIC FIELDS DUE TO CONTINUOUS CHARGE DISTRIBUTIONS

So far we have considered only forces and electric fields due to point charges, which are essentially charges occupying very small physical space. It is also possible to have continuous charge distribution along a line, on a surface, or in a volume, as illustrated in Figure 4.5.

It is customary to denote the line charge density, surface charge density, and volume charge density by ρ_L (in C/m), ρ_S (in C/m^2), and ρ_v (in C/m^3), respectively. These must not be confused with ρ (without subscript), used for radial distance in cylindrical coordinates.

The charge element dQ and the total charge Q due to these charge distributions are obtained from Figure 4.5 as

$$dQ = \rho_L \, dl \rightarrow Q = \int_L \rho_L \, dl \quad \text{(line charge)} \tag{4.13a}$$

FIGURE 4.5 Various charge distributions and charge elements.

| Point charge | Line charge | Surface charge | Volume charge |

$$dQ = \rho_S \, dS \;\rightarrow\; Q = \int_S \rho_S \, dS \quad \text{(surface charge)} \tag{4.13b}$$

$$dQ = \rho_v \, dv \;\rightarrow\; Q = \int_v \rho_v \, dv \quad \text{(volume charge)} \tag{4.13c}$$

The electric field intensity due to each of the charge distributions ρ_L, ρ_S, and ρ_v may be regarded as the summation of the field contributed by the numerous point charges making up the charge distribution. We treat dQ as a point charge. Thus by replacing Q in eq. (4.11) with charge element $dQ = \rho_L \, dl$, $\rho_S \, dS$, or $\rho_v \, dv$ and integrating, we get

$$\mathbf{E} = \int_L \frac{\rho_L \, dl}{4\pi\varepsilon_o R^2} \, \mathbf{a}_R \quad \text{(line charge)} \tag{4.14}$$

$$\mathbf{E} = \int_S \frac{\rho_S \, dS}{4\pi\varepsilon_o R^2} \, \mathbf{a}_R \quad \text{(surface charge)} \tag{4.15}$$

$$\mathbf{E} = \int_v \frac{\rho_v \, dv}{4\pi\varepsilon_o R^2} \, \mathbf{a}_R \quad \text{(volume charge)} \tag{4.16}$$

It should be noted that R^2 and $\mathbf{a}_R$ vary as the integrals in eqs. (4.14) to (4.16) are evaluated. We shall now apply these formulas to some specific charge distributions.

A. A Line Charge

Consider a line charge with uniform charge density ρ_L extending from A to B along the z-axis as shown in Figure 4.6. The charge element dQ associated with element $dl = dz$ of the line is

$$dQ = \rho_L \, dl = \rho_L \, dz$$

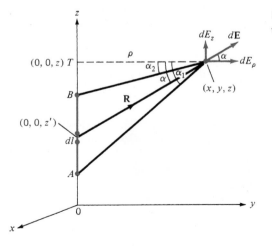

FIGURE 4.6 Evaluation of the **E** field due to a line charge.

and hence the total charge Q is

$$Q = \int_{z_A}^{z_B} \rho_L \, dz \tag{4.17}$$

The electric field intensity $\mathbf{E}$ at an arbitrary point $P(x, y, z)$ can be found by using eq. (4.14). It is important that we learn to derive and substitute each term in eqs. (4.14) to (4.16) for a given charge distribution. It is customary to denote the field point[4] by (x, y, z) and the source point by (x', y', z'). Thus from Figure 4.6,

$$dl = dz'$$
$$\mathbf{R} = (x, y, z) - (0, 0, z') = x\mathbf{a}_x + y\mathbf{a}_y + (z - z')\mathbf{a}_z$$

or

$$\mathbf{R} = \rho\mathbf{a}_\rho + (z - z')\mathbf{a}_z$$
$$R^2 = |\mathbf{R}|^2 = x^2 + y^2 + (z - z')^2 = \rho^2 + (z - z')^2$$
$$\frac{\mathbf{a}_R}{R^2} = \frac{\mathbf{R}}{|\mathbf{R}|^3} = \frac{\rho\mathbf{a}_\rho + (z - z')\mathbf{a}_z}{[\rho^2 + (z - z')^2]^{3/2}}$$

Substituting all this into eq. (4.14), we get

$$\mathbf{E} = \frac{\rho_L}{4\pi\varepsilon_o} \int \frac{\rho\mathbf{a}_\rho + (z - z')\mathbf{a}_z}{[\rho^2 + (z - z')^2]^{3/2}} \, dz' \tag{4.18}$$

To evaluate this, it is convenient that we define α, α_1, and α_2 as in Figure 4.6.

$$R = [\rho^2 + (z - z')^2]^{1/2} = \rho \sec \alpha$$
$$z' = OT - \rho \tan \alpha, \quad dz' = -\rho \sec^2 \alpha \, d\alpha$$

Hence, eq. (4.18) becomes

$$\mathbf{E} = \frac{-\rho_L}{4\pi\varepsilon_o} \int_{\alpha_1}^{\alpha_2} \frac{\rho \sec^2 \alpha \left[\cos \alpha \, \mathbf{a}_\rho + \sin \alpha \, \mathbf{a}_z \right] d\alpha}{\rho^2 \sec^2 \alpha}$$

$$= -\frac{\rho_L}{4\pi\varepsilon_o\rho} \int_{\alpha_1}^{\alpha_2} \left[\cos \alpha \, \mathbf{a}_\rho + \sin \alpha \, \mathbf{a}_z \right] d\alpha \tag{4.19}$$

Thus for a *finite line charge*,

$$\mathbf{E} = \frac{\rho_L}{4\pi\varepsilon_o\rho} \left[-(\sin \alpha_2 - \sin \alpha_1)\mathbf{a}_\rho + (\cos \alpha_2 - \cos \alpha_1)\mathbf{a}_z \right] \tag{4.20}$$

---

[4] The field point is the point at which the field is to be evaluated.

As a special case, for an *infinite line charge*, point B is at $(0, 0, \infty)$ and A at $(0, 0, -\infty)$ so that $\alpha_1 = \pi/2$, $\alpha_2 = -\pi/2$; the z-component vanishes and eq. (4.20) becomes

$$\boxed{\mathbf{E} = \frac{\rho_L}{2\pi\varepsilon_o\rho} \mathbf{a}_\rho} \tag{4.21}$$

Bear in mind that eq. (4.21) is obtained for an infinite line charge along the z-axis so that ρ and $\mathbf{a}_\rho$ have their usual meaning. If the line is not along the z-axis, ρ is the perpendicular distance from the line to the point of interest, and $\mathbf{a}_\rho$ is a unit vector along that distance directed from the line charge to the field point.

B. A Surface Charge

Consider an infinite sheet of charge in the xy-plane with uniform charge density ρ_S. The charge associated with an elemental area dS is

$$dQ = \rho_S \, dS \tag{4.22}$$

From eq. (4.15), the contribution to the $\mathbf{E}$ field at point $P(0, 0, h)$ by the charge dQ on the elemental surface 1 shown in Figure 4.7 is

$$d\mathbf{E} = \frac{dQ}{4\pi\varepsilon_o R^2} \mathbf{a}_R \tag{4.23}$$

From Figure 4.7,

$$\mathbf{R} = \rho(-\mathbf{a}_\rho) + h\mathbf{a}_z, \quad R = |\mathbf{R}| = [\rho^2 + h^2]^{1/2}$$

$$\mathbf{a}_R = \frac{\mathbf{R}}{R}, \quad dQ = \rho_S \, dS = \rho_S \, \rho \, d\phi \, d\rho$$

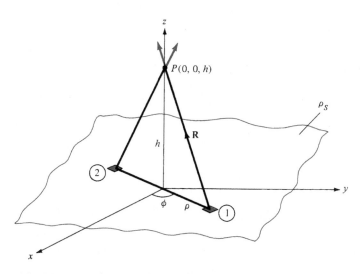

FIGURE 4.7 Evaluation of the $\mathbf{E}$ field due to an infinite sheet of charge.

Substitution of these terms into eq. (4.23) gives

$$dE = \frac{\rho_S \rho \, d\phi \, d\rho \left[-\rho \mathbf{a}_\rho + h\mathbf{a}_z\right]}{4\pi\varepsilon_0[\rho^2 + h^2]^{3/2}} \tag{4.24}$$

Owing to the symmetry of the charge distribution, for every element 1, there is a corresponding element 2 whose contribution along $\mathbf{a}_\rho$ cancels that of element 1, as illustrated in Figure 4.7. Thus the contributions to E_ρ add up to zero so that $\mathbf{E}$ has only z-component. This can also be shown mathematically by replacing $\mathbf{a}_\rho$ with $\cos\phi\,\mathbf{a}_x + \sin\phi\,\mathbf{a}_y$. Integration of $\cos\phi$ or $\sin\phi$ over $0 < \phi < 2\pi$ gives zero. Therefore,

$$\mathbf{E} = \int_S d\mathbf{E}_z = \frac{\rho_S}{4\pi\varepsilon_0} \int_{\phi=0}^{2\pi} \int_{\rho=0}^{\infty} \frac{h\rho \, d\rho \, d\phi}{[\rho^2 + h^2]^{3/2}} \mathbf{a}_z$$

$$= \frac{\rho_S h}{4\pi\varepsilon_0} 2\pi \int_0^\infty [\rho^2 + h^2]^{-3/2} \frac{1}{2} d(\rho^2) \mathbf{a}_z$$

$$= \frac{\rho_S h}{2\varepsilon_0} \left\{ -[\rho^2 + h^2]^{-1/2} \right\}_0^\infty \mathbf{a}_z$$

$$\mathbf{E} = \frac{\rho_S}{2\varepsilon_0} \mathbf{a}_z \tag{4.25}$$

that is, $\mathbf{E}$ has only z-component if the charge is in the xy-plane. Equation (4.25) is valid for $h > 0$; for $h < 0$, we would need to replace $\mathbf{a}_z$ with $-\mathbf{a}_z$. In general, for an *infinite sheet* of charge

$$\boxed{\mathbf{E} = \frac{\rho_S}{2\varepsilon_0} \mathbf{a}_n} \tag{4.26}$$

where $\mathbf{a}_n$ is a unit vector normal to the sheet. From eq. (4.25) or (4.26), we notice that the electric field is normal to the sheet and it is surprisingly independent of the distance between the sheet and the point of observation P. In a parallel-plate capacitor, the electric field existing between the two plates having equal and opposite charges is given by

$$\mathbf{E} = \frac{\rho_S}{2\varepsilon_0} \mathbf{a}_n + \frac{-\rho_S}{2\varepsilon_0} (-\mathbf{a}_n) = \frac{\rho_S}{\varepsilon_0} \mathbf{a}_n \tag{4.27}$$

C. A Volume Charge

Next, let us consider a sphere of radius a centered at the origin. Let the volume of the sphere be filled uniformly with a volume-charge density ρ_v (in C/m³) as shown in Figure 4.8. The charge dQ associated with the elemental volume dv chosen at (ρ', θ', ϕ') is

$$dQ = \rho_v \, dv$$

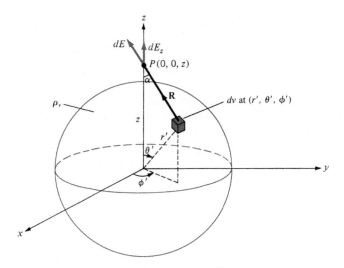

FIGURE 4.8 Evaluation of the **E** field due to a volume charge distribution.

and hence the total charge in a sphere of radius a is

$$Q = \int_v \rho_v \, dv = \rho_v \int_v dv \tag{4.28}$$

$$= \rho_v \frac{4\pi a^3}{3}$$

The electric field $d\mathbf{E}$ outside the sphere at $P(0, 0, z)$ due to the elementary volume charge is

$$d\mathbf{E} = \frac{\rho_v \, dv}{4\pi\varepsilon_o R^2} \mathbf{a}_R$$

where $\mathbf{a}_R = \cos\alpha \, \mathbf{a}_z + \sin\alpha \, \mathbf{a}_\rho$. Owing to the symmetry of the charge distribution, the contributions to E_x or E_y add up to zero. We are left with only E_z, given by

$$E_z = \mathbf{E} \cdot \mathbf{a}_z = \int_v dE \cos\alpha = \frac{\rho_v}{4\pi\varepsilon_o} \int_v \frac{dv \cos\alpha}{R^2} \tag{4.29}$$

Again, we need to derive expressions for dv, R^2, and $\cos\alpha$:

$$dv = r'^2 \sin\theta' \, dr' \, d\theta' \, d\phi' \tag{4.30}$$

Applying the cosine rule to Figure 4.8, we have

$$R^2 = z^2 + r'^2 - 2zr' \cos\theta'$$

$$r'^2 = z^2 + R^2 - 2zR \cos\alpha$$

It is convenient to evaluate the integral in eq. (4.29) in terms of R and r'. Hence we express $\cos\theta'$, $\cos\alpha$, and $\sin\theta'\,d\theta'$ in terms of R and r', that is,

$$\cos\alpha = \frac{z^2 + R^2 - r'^2}{2zR} \tag{4.31a}$$

$$\cos\theta' = \frac{z^2 + r'^2 - R^2}{2zr'} \tag{4.31b}$$

Differentiating eq. (4.31b) with respect to θ' and keeping z and r' fixed, we obtain

$$\sin\theta'\,d\theta' = \frac{R\,dR}{z\,r'} \tag{4.32}$$

As θ' varies from 0 to π, R varies from $(z - r')$ to $(z + r')$ if P is outside the sphere. Substituting eqs. (4.30) to (4.32) into eq. (4.29) yields

$$E_z = \frac{\rho_v}{4\pi\varepsilon_o} \int_{\phi'=0}^{2\pi} d\phi' \int_{r'=0}^{a} \int_{R=z-r'}^{z+r'} r'^2 \frac{R\,dR}{zr'}\,dr'\, \frac{z^2 + R^2 - r'^2}{2zR} \frac{1}{R^2}$$

$$= \frac{\rho_v 2\pi}{8\pi\varepsilon_o z^2} \int_{r'=0}^{a} \int_{R=z-r'}^{z+r'} r'\left[1 + \frac{z^2 - r'^2}{R^2}\right] dR\,dr'$$

$$= \frac{\rho_v \pi}{4\pi\varepsilon_o z^2} \int_{0}^{a} r'\left[R - \frac{(z^2 - r'^2)}{R}\right]_{z-r'}^{z+r'} dr'$$

$$= \frac{\rho_v \pi}{4\pi\varepsilon_o z^2} \int_{0}^{a} 4r'^2\,dr' = \frac{1}{4\pi\varepsilon_o}\frac{1}{z^2}\left(\frac{4}{3}\pi a^3 \rho_v\right)$$

or

$$\mathbf{E} = \frac{Q}{4\pi\varepsilon_o z^2}\,\mathbf{a}_z \tag{4.33}$$

This result is obtained for $\mathbf{E}$ at $P(0, 0, z)$. Owing to the symmetry of the charge distribution, the electric field at $P(r, \theta, \phi)$ is readily obtained from eq. (4.33) as

$$\mathbf{E} = \frac{Q}{4\pi\varepsilon_o r^2}\,\mathbf{a}_r \tag{4.34}$$

which is identical to the electric field at the same point due to a point charge Q located at the origin or the center of the spherical charge distribution. The reason for this will become obvious as we cover Gauss's law in Section 4.5.

EXAMPLE 4.4

A circular ring of radius a carries a uniform charge ρ_L C/m and is placed on the xy-plane with axis the same as the z-axis.

(a) Show that

$$\mathbf{E}(0, 0, h) = \frac{\rho_L a h}{2\varepsilon_o[h^2 + a^2]^{3/2}}\,\mathbf{a}_z$$

(b) What values of h give the maximum value of **E**?

(c) If the total charge on the ring is Q, find **E** as $a \rightarrow 0$.

Solution:

(a) Consider the system as shown in Figure 4.9. Again the trick in finding **E** by using eq. (4.14) is deriving each term in the equation. In this case,

$$dl = a \, d\phi, \quad \mathbf{R} = a(-\mathbf{a}_\rho) + h\mathbf{a}_z$$

$$R = |\mathbf{R}| = [a^2 + h^2]^{1/2}, \quad \mathbf{a}_R = \frac{\mathbf{R}}{R}$$

or

$$\frac{\mathbf{a}_R}{R^2} = \frac{\mathbf{R}}{|\mathbf{R}|^3} = \frac{-a\mathbf{a}_\rho + h\mathbf{a}_z}{[a^2 + h^2]^{3/2}}$$

Hence

$$\mathbf{E} = \frac{\rho_L}{4\pi\varepsilon_o} \int_{\phi=0}^{2\pi} \frac{(-a\mathbf{a}_\rho + h\mathbf{a}_z)}{[a^2 + h^2]^{3/2}} a \, d\phi$$

By symmetry, the contributions along $\mathbf{a}_\rho$ add up to zero. This is evident from the fact that for every element dl there is a corresponding element diametrically opposite that gives an equal but opposite dE_ρ so that the two contributions cancel each other. Thus we are left with the z-component. That is,

$$\mathbf{E} = \frac{\rho_L a h \mathbf{a}_z}{4\pi\varepsilon_o [h^2 + a^2]^{3/2}} \int_0^{2\pi} d\phi = \frac{\rho_L a h \mathbf{a}_z}{2\varepsilon_o [h^2 + a^2]^{3/2}}$$

as required.

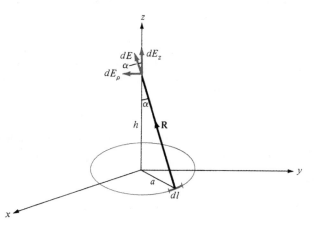

FIGURE 4.9 Charged ring; for Example 4.4.

(b)

$$\frac{d|\mathbf{E}|}{dh} = \frac{\rho_L a}{2\varepsilon_o} \left\{ \frac{[h^2 + a^2]^{3/2}(1) - \dfrac{3}{2}(h)2h[h^2 + a^2]^{1/2}}{[h^2 + a^2]^3} \right\}$$

For maximum **E**, $\dfrac{d|\mathbf{E}|}{dh} = 0$, which implies that

$$[h^2 + a^2]^{1/2}[h^2 + a^2 - 3h^2] = 0$$

$$a^2 - 2h^2 = 0 \quad \text{or} \quad h = \pm\frac{a}{\sqrt{2}}$$

(c) Since the charge is uniformly distributed, the line charge density is

$$\rho_L = \frac{Q}{2\pi a}$$

so that

$$\mathbf{E} = \frac{Qh}{4\pi\varepsilon_o[h^2 + a^2]^{3/2}} \mathbf{a}_z$$

As $a \to 0$

$$\mathbf{E} = \frac{Q}{4\pi\varepsilon_o h^2} \mathbf{a}_z$$

or in general

$$\mathbf{E} = \frac{Q}{4\pi\varepsilon_o r^2} \mathbf{a}_R$$

which is the same as that of a point charge, as one would expect.

PRACTICE EXERCISE 4.4

A circular disk of radius a is uniformly charged with ρ_S C/m². The disk lies on the $z = 0$ plane with its axis along the z-axis.

(a) Show that at point $(0, 0, h)$

$$\mathbf{E} = \frac{\rho_S}{2\varepsilon_o}\left\{1 - \frac{h}{[h^2 + a^2]^{1/2}}\right\}\mathbf{a}_z$$

(b) From this, derive the **E** field due to an infinite sheet of charge on the $z = 0$ plane.

(c) If $a \ll h$, show that **E** is similar to the field due to a point charge.

Answer: (a) Proof, (b) $\dfrac{\rho_S}{2\varepsilon_o}\mathbf{a}_z$, (c) Proof.

EXAMPLE 4.5

The finite sheet $0 \le x \le 1$, $0 \le y \le 1$ on the $z = 0$ plane has a charge density $\rho_S = xy(x^2 + y^2 + 25)^{3/2}$ nC/m^2. Find

(a) The total charge on the sheet
(b) The electric field at $(0, 0, 5)$
(c) The force experienced by a -1 mC charge located at $(0, 0, 5)$

Solution:

(a) $Q = \displaystyle\int_S \rho_S \, dS = \int_0^1 \int_0^1 xy(x^2 + y^2 + 25)^{3/2} \, dx \, dy$ nC

Since $x \, dx = 1/2 \, d(x^2)$, we now integrate with respect to x^2 (or change variables: $x^2 = u$ so that $x \, dx = du/2$).

$$Q = \frac{1}{2} \int_0^1 y \int_0^1 (x^2 + y^2 + 25)^{3/2} \, d(x^2) \, dy \text{ nC}$$

$$= \frac{1}{2} \int_0^1 y \frac{2}{5} (x^2 + y^2 + 25)^{5/2} \Big|_0^1 \, dy$$

$$= \frac{1}{5} \int_0^1 \frac{1}{2} [(y^2 + 26)^{5/2} - (y^2 + 25)^{5/2}] \, d(y^2)$$

$$= \frac{1}{10} \cdot \frac{2}{7} [(y^2 + 26)^{7/2} - (y^2 + 25)^{7/2}] \Big|_0^1$$

$$= \frac{1}{35} [(27)^{7/2} + (25)^{7/2} - 2(26)^{7/2}]$$

$$Q = 33.15 \text{ nC}$$

(b) $\mathbf{E} = \displaystyle\int_S \frac{\rho_S \, dS \, \mathbf{a}_R}{4\pi\varepsilon_0 r^2} = \int_S \frac{\rho_S \, dS \, (\mathbf{r} - \mathbf{r}')}{4\pi\varepsilon_0 |\mathbf{r} - \mathbf{r}'|^3}$

where $\mathbf{r} - \mathbf{r}' = (0, 0, 5) - (x, y, 0) = (-x, -y, 5)$. Hence,

$$\mathbf{E} = \int_0^1 \int_0^1 \frac{10^{-9} xy(x^2 + y^2 + 25)^{3/2}(-x\mathbf{a}_x - y\mathbf{a}_y + 5\mathbf{a}_z) dx \, dy}{4\pi \cdot \dfrac{10^{-9}}{36\pi} (x^2 + y^2 + 25)^{3/2}}$$

$$= 9 \left[-\int_0^1 x^2 \, dx \int_0^1 y \, dy \, \mathbf{a}_x - \int_0^1 x \, dx \int_0^1 y^2 dy \, \mathbf{a}_y + 5 \int_0^1 x \, dx \int_0^1 y \, dy \, \mathbf{a}_z \right]$$

$$= 9 \left(\frac{-1}{6}, \frac{-1}{6}, \frac{5}{4} \right)$$

$$= (-1.5, -1.5, 11.25) \text{ V/m}$$

(c) $\mathbf{F} = q\mathbf{E} = (1.5, 1.5, -11.25)$ mN

PRACTICE EXERCISE 4.5

A square plate described by $-2 \leq x \leq 2$, $-2 \leq y \leq 2$, $z = 0$ carries a charge $12 \left| y \right|$ mC/m². Find the total charge on the plate and the electric field intensity at $(0, 0, 10)$.

Answer: 192 mC, 16.6 $\mathbf{a}_z$ MV/m.

EXAMPLE 4.6

Planes $x = 2$ and $y = -3$, respectively, carry charges 10 nC/m² and 15 nC/m². If the line $x = 0$, $z = 2$ carries charge 10π nC/m, calculate **E** at $(1, 1, -1)$ due to the three charge distributions.

Solution:

Let

$$\mathbf{E} = \mathbf{E}_1 + \mathbf{E}_2 + \mathbf{E}_3$$

where $\mathbf{E}_1$, $\mathbf{E}_2$, and $\mathbf{E}_3$ are, respectively, the contributions to **E** at point $(1, 1, -1)$ due to the infinite sheet 1, infinite sheet 2, and infinite line 3 as shown in Figure 4.10(a). Applying eqs. (4.26) and (4.21) gives

$$\mathbf{E}_1 = \frac{\rho_{S_1}}{2\varepsilon_o}(-\mathbf{a}_x) = -\frac{10 \cdot 10^{-9}}{2 \cdot \dfrac{10^{-9}}{36\pi}}\mathbf{a}_x = -180\pi\mathbf{a}_x$$

$$\mathbf{E}_2 = \frac{\rho_{S_2}}{2\varepsilon_o}\mathbf{a}_y = \frac{15 \cdot 10^{-9}}{2 \cdot \dfrac{10^{-9}}{36\pi}}\mathbf{a}_y = 270\pi\mathbf{a}_y$$

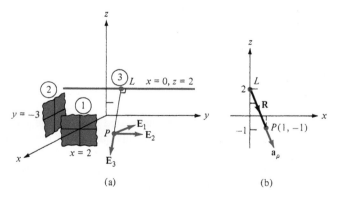

(a)   (b)

FIGURE 4.10 For Example 4.6: (**a**) three charge distributions, (**b**) finding ρ and $\mathbf{a}_\rho$ on plane $y = 1$.

and

$$E_3 = \frac{\rho_L}{2\pi\varepsilon_o\rho} \mathbf{a}_\rho$$

where $\mathbf{a}_\rho$ (not regular $\mathbf{a}_\rho$ but with a similar meaning) is a unit vector along LP perpendicular to the line charge and ρ is the length LP to be determined from Figure 4.10(b). Figure 4.10(b) results from Figure 4.10(a) if we consider plane $y = 1$ on which $\mathbf{E}_3$ lies. From Figure 4.10(b), the distance vector from L to P is

$$\mathbf{R} = -3\mathbf{a}_z + \mathbf{a}_x$$

$$\rho = |\mathbf{R}| = \sqrt{10}, \quad \mathbf{a}_\rho = \frac{\mathbf{R}}{|\mathbf{R}|} = \frac{1}{\sqrt{10}}\mathbf{a}_x - \frac{3}{\sqrt{10}}\mathbf{a}_z$$

Hence,

$$\mathbf{E}_3 = \frac{10\pi \cdot 10^{-9}}{2\pi \cdot \dfrac{10^{-9}}{36\pi}} \cdot \frac{1}{10}(\mathbf{a}_x - 3\mathbf{a}_z)$$

$$= 18\pi(\mathbf{a}_x - 3\mathbf{a}_z)$$

Thus by adding $\mathbf{E}_1$, $\mathbf{E}_2$, and $\mathbf{E}_3$, we obtain the total field as

$$\mathbf{E} = -162\pi\mathbf{a}_x + 270\pi\mathbf{a}_y - 54\pi\mathbf{a}_z \text{ V/m}$$

Note that to obtain $\mathbf{a}_r$, $\mathbf{a}_\rho$, or $\mathbf{a}_n$, which we always need for finding $\mathbf{F}$ or $\mathbf{E}$, we must go from the charge (at position vector $\mathbf{r}'$) to the field point (at position vector $\mathbf{r}$); hence $\mathbf{a}_r$, $\mathbf{a}_\rho$, or $\mathbf{a}_n$ is a unit vector along $\mathbf{r} - \mathbf{r}'$. In addition, $\mathbf{r}$ and $\mathbf{r}'$ are defined locally, not globally. Observe this carefully in Figures 4.6 to 4.10.

PRACTICE EXERCISE 4.6

In Example 4.6 if the line $x = 0$, $z = 2$ is rotated through $90°$ about the point $(0, 2, 2)$ so that it becomes $x = 0$, $y = 2$, find $\mathbf{E}$ at $(1, 1, -1)$.

Answer: $-282.7\mathbf{a}_x + 565.5\mathbf{a}_y$ V/m.

4.4 ELECTRIC FLUX DENSITY

The flux due to the electric field $\mathbf{E}$ can be calculated by using the general definition of flux in eq. (3.13). For practical reasons, however, this quantity is not usually considered to be the most useful flux in electrostatics. Also, eqs. (4.11) to (4.16) show that the electric field intensity is dependent on the medium in which the charge is placed (free space in this chapter). Suppose a new vector field $\mathbf{D}$ is defined by

$$\boxed{\mathbf{D} = \varepsilon_o\mathbf{E}} \tag{4.35}$$

We use eq. (3.13) to define *electric flux* Ψ in terms of **D**, namely,

$$\Psi = \int_S \mathbf{D} \cdot d\mathbf{S} \tag{4.36}$$

In SI units, one line of electric flux emanates from $+1$ C and terminates on -1 C. Therefore, the electric flux is measured in coulombs. Hence, the vector field **D** is called the *electric flux density* and is measured in coulombs per square meter. For historical reasons, the electric flux density is also called *electric displacement*.

From eq. (4.35), it is apparent that all the formulas derived for **E** from Coulomb's law in Sections 4.2 and 4.3 can be used in calculating **D**, except that we have to multiply those formulas by ε_o. For example, for an infinite sheet of charge, eqs. (4.26) and (4.35) give

$$\mathbf{D} = \frac{\rho_S}{2} \mathbf{a}_n \tag{4.37}$$

and for a volume charge distribution, eqs. (4.16) and (4.35) give

$$\mathbf{D} = \int_v \frac{\rho_v \, dv}{4\pi R^2} \mathbf{a}_R \tag{4.38}$$

Note from eqs. (4.37) and (4.38) that **D** is a function of charge and position only; it is independent of the medium.

EXAMPLE 4.7

Determine **D** at $(4, 0, 3)$ if there is a point charge -5π mC at $(4, 0, 0)$ and a line charge 3π mC/m along the y-axis.

Solution:
Let $\mathbf{D} = \mathbf{D}_Q + \mathbf{D}_L$, where $\mathbf{D}_Q$ and $\mathbf{D}_L$ are flux densities due to the point charge and line charge, respectively, as shown in Figure 4.11:

$$\mathbf{D}_Q = \varepsilon_o\mathbf{E} = \frac{Q}{4\pi R^2} \mathbf{a}_R = \frac{Q(\mathbf{r} - \mathbf{r}')}{4\pi |\mathbf{r} - \mathbf{r}'|^3}$$

where $\mathbf{r} - \mathbf{r}' = (4, 0, 3) - (4, 0, 0) = (0, 0, 3)$. Hence,

$$\mathbf{D}_Q = \frac{-5\pi \cdot 10^{-3}(0, 0, 3)}{4\pi |(0, 0, 3)|^3} = -0.139 \, \mathbf{a}_z \text{ mC/m}^2$$

Also

$$\mathbf{D}_L = \frac{\rho_L}{2\pi\rho} \mathbf{a}_\rho$$

In this case

$$\mathbf{a}_\rho = \frac{(4, 0, 3) - (0, 0, 0)}{|(4, 0, 3) - (0, 0, 0)|} = \frac{(4, 0, 3)}{5}$$

$$\rho = |(4, 0, 3) - (0, 0, 0)| = 5$$

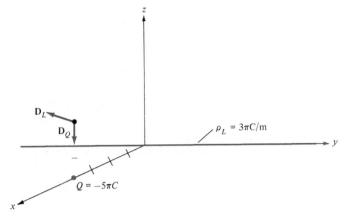

FIGURE 4.11 Flux density **D** due to a point charge and an infinite line charge.

Hence,

$$\mathbf{D}_L = \frac{3\pi}{2\pi(25)}\left(4\mathbf{a}_x + 3\mathbf{a}_z\right) = 0.24\mathbf{a}_x + 0.18\mathbf{a}_z \, \text{mC/m}^2$$

Thus

$$\mathbf{D} = \mathbf{D}_Q + \mathbf{D}_L$$

$$= 240\mathbf{a}_x + 41.1\mathbf{a}_z \, \mu\text{C/m}^2$$

PRACTICE EXERCISE 4.7

A point charge of 30 nC is located at the origin, while plane $y = 3$ carries charge 10 nC/m². Find **D** at $(0, 4, 3)$.

Answer: $5.076\mathbf{a}_y + 0.0573\mathbf{a}_z$ nC/m².

4.5 GAUSS'S LAW—MAXWELL'S EQUATION

Gauss's[5] law constitutes one of the fundamental laws of electromagnetism.

> **Gauss's law** states that the total electric flux ψ through any *closed* surface is equal to the total charge enclosed by that surface.

---

[5] The German mathematician Carl Friedrich Gauss (see Chapter 3 opening) developed the divergence theorem of Section 3.6, popularly known by his name. He was the first physicist to measure electric and magnetic quantities in absolute units. For details on Gauss's measurements, see W. F. Magie, *A Source Book in Physics*. Cambridge, MA: Harvard Univ. Press, 1963, pp. 519–524.

Thus

$$\Psi = Q_{\text{enc}} \tag{4.39}$$

that is,

$$\Psi = \oint_S d\Psi = \oint_S \mathbf{D} \cdot d\mathbf{S}$$

$$= \text{total charge enclosed } Q = \int_v \rho_v \, dv \tag{4.40}$$

or

$$Q = \oint_S \mathbf{D} \cdot d\mathbf{S} = \int_v \rho_v \, dv \tag{4.41}$$

By applying divergence theorem to the middle term in eq. (4.41), we have

$$\oint_S \mathbf{D} \cdot d\mathbf{S} = \int_v \nabla \cdot \mathbf{D} \, dv \tag{4.42}$$

Comparing the two volume integrals in eqs. (4.41) and (4.42) results in

$$\boxed{\rho_v = \nabla \cdot \mathbf{D}} \tag{4.43}$$

which is the first of the four *Maxwell's equations* to be derived. Equation (4.43) states that the volume charge density is the same as the divergence of the electric flux density.[6] It is equivalent to Coulomb's law of force between point charges.

Note that:

1. Equations (4.41) and (4.43) are basically stating Gauss's law in different ways; eq. (4.41) is the integral form, whereas eq. (4.43) is the differential or point form of Gauss's law. Equation (4.43) is sometimes called the *source equation*.
2. Gauss's law is an alternative statement of Coulomb's law; proper application of the divergence theorem to Coulomb's law results in Gauss's law.
3. Gauss's law provides an easy means of finding **E** or **D** for symmetrical charge distributions such as a point charge, an infinite line charge, an infinite cylindrical surface charge, and a spherical distribution of charge. A continuous charge distribution has rectangular symmetry if it depends only on x (or y or z), cylindrical symmetry if it depends only on ρ, or spherical symmetry if it depends only on r (independent of θ and ϕ). It must be stressed that whether the charge distribution is symmetric or not, Gauss's law always holds. For example, consider the charge

---

[6]This should not be surprising to us from the way we defined divergence of a vector in eq. (3.32):

$$\nabla \cdot \mathbf{D} = \lim_{\Delta v \to \phi} \frac{\oint \mathbf{D} \cdot d\mathbf{S}}{\Delta v}, \text{ which reduces to } \frac{\Delta Q}{\Delta v} = \rho_v.$$

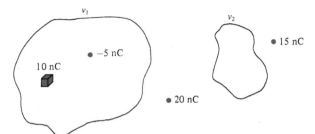

FIGURE 4.12 Illustration of Gauss's law: flux leaving v_1 is 5 nC and that leaving v_2 is 0 C.

distribution in Figure 4.12 where v_1 and v_2 are closed surfaces (or volumes). The total flux leaving v_1 is $10 - 5 = 5$ nC because only 10 nC and -5 nC charges are enclosed by v_1. Although charges 20 nC and 15 nC outside v_1 do contribute to the flux crossing v_1, the net flux crossing v_1, according to Gauss's law, is irrespective of those charges outside v_1. Similarly, the total flux leaving v_2 is zero because no charge is enclosed by v_2. Thus we see that Gauss's law, $\Psi = Q_{enc}$, is still obeyed even though the charge distribution is not symmetric. However, we cannot use the law to determine **E** or **D** when the charge distribution is not symmetric; we must resort to Coulomb's law to determine **E** or **D** in that case.

4.6 APPLICATIONS OF GAUSS'S LAW

The procedure for applying Gauss's law to calculate the electric field involves first knowing whether symmetry exists. Once it has been found that symmetric charge distribution exists, we construct a mathematical closed surface (known as a *Gaussian surface*). The surface is chosen such that **D** is normal or tangential to the Gaussian surface. When **D** is normal to the surface, $\mathbf{D} \cdot d\mathbf{S} = D\,dS$ because **D** is perpendicular to the surface. When **D** is tangential to the surface, $\mathbf{D} \cdot d\mathbf{S} = 0$. Thus we must choose a surface that has some of the symmetry exhibited by the charge distribution. The choice of an appropriate Gaussian surface, where there is symmetry in the charge distribution comes from intuitive reasoning and a slight degree of maturity in the application of Coulomb's law. We shall now apply these basic ideas to the following cases.

A. Point Charge

Suppose a point charge Q is located at the origin. To determine **D** at a point P, it is easy to see that choosing a spherical surface containing P will satisfy symmetry conditions. Thus, a spherical surface centered at the origin is the Gaussian surface in this case and is shown in Figure 4.13.

Since **D** is everywhere normal to the Gaussian surface, that is, $\mathbf{D} = D_r \mathbf{a}_r$, applying Gauss's law ($\Psi = Q_{enc}$) gives

$$Q = \oint_S \mathbf{D} \cdot d\mathbf{S} = D_r \oint_S dS = D_r \, 4\pi r^2 \qquad (4.44)$$

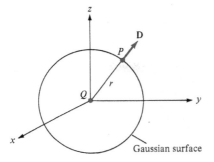

FIGURE 4.13 Gaussian surface about a point charge.

where $\oint dS = \int_{\phi=0}^{2\pi} \int_{\theta=0}^{\pi} r^2 \sin \theta \, d\theta \, d\phi = 4\pi r^2$ is the surface area of the Gaussian surface. Thus

$$\mathbf{D} = \frac{Q}{4\pi r^2} \mathbf{a}_r \tag{4.45}$$

as expected from eqs. (4.11) and (4.35).

B. Infinite Line Charge

Suppose the infinite line of uniform charge ρ_L C/m lies along the z-axis. To determine $\mathbf{D}$ at a point P, we choose a cylindrical surface containing P to satisfy the symmetry condition as shown in Figure 4.14. The electric flux density $\mathbf{D}$ is constant on and normal to the cylindrical Gaussian surface; that is, $\mathbf{D} = D_\rho \mathbf{a}_\rho$. If we apply Gauss's law to an arbitrary length ℓ of the line

$$\rho_L \ell = Q = \int_S \mathbf{D} \cdot d\mathbf{S} = D_\rho \int_S dS = D_\rho \, 2\pi\rho\ell \tag{4.46}$$

where $\int_S dS = 2\pi\rho\ell$ is the surface area of the Gaussian surface. Note that $\int \mathbf{D} \cdot d\mathbf{S}$ evaluated on the top and bottom surfaces of the cylinder is zero, since $\mathbf{D}$ has no z-component; that means that $\mathbf{D}$ is tangential to those surfaces. Thus

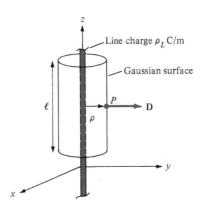

FIGURE 4.14 Gaussian surface about an infinite line charge.

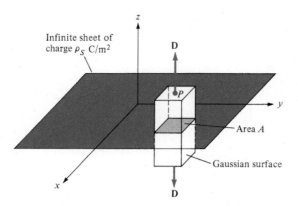

FIGURE 4.15 Gaussian surface about an infinite line sheet of charge.

$$\mathbf{D} = \frac{\rho_L}{2\pi\rho}\mathbf{a}_\rho \tag{4.47}$$

as expected from eqs. (4.21) and (4.35).

C. Infinite Sheet of Charge

Consider an infinite sheet of uniform charge ρ_S C/m^2 lying on the $z = 0$ plane. To determine $\mathbf{D}$ at point P, we choose a rectangular box that is cut symmetrically by the sheet of charge and has two of its faces parallel to the sheet as shown in Figure 4.15. As $\mathbf{D}$ is normal to the sheet, $\mathbf{D} = D_z\mathbf{a}_z$, and applying Gauss's law gives

$$\rho_S \int_S dS = Q = \oint_S \mathbf{D} \cdot d\mathbf{S} = D_z \left[\int_{\text{top}} dS + \int_{\text{bottom}} dS \right] \tag{4.48}$$

Note that $\mathbf{D} \cdot d\mathbf{S}$ evaluated on the sides of the box is zero because $\mathbf{D}$ has no components along $\mathbf{a}_x$ and $\mathbf{a}_y$. If the top and bottom area of the box each has area A, eq. (4.48) becomes

$$\rho_S A = D_z(A + A) \tag{4.49}$$

and thus

$$\mathbf{D} = \frac{\rho_S}{2}\mathbf{a}_z$$

or

$$\mathbf{E} = \frac{\mathbf{D}}{\varepsilon_0} = \frac{\rho_S}{2\varepsilon_0}\mathbf{a}_z \tag{4.50}$$

as expected from eq. (4.25).

D. Uniformly Charged Sphere

Consider a sphere of radius a with a uniform charge ρ_o C/m^3. To determine $\mathbf{D}$ everywhere, we construct Gaussian surfaces for cases $r \leq a$ and $r \geq a$ separately. Since the charge has spherical symmetry, it is obvious that a spherical surface is an appropriate Gaussian surface.

For $r \leq a$, the total charge enclosed by the spherical surface of radius r, as shown in Figure 4.16(a), is

$$Q_{\text{enc}} = \int_v \rho_v dv = \rho_o \int_v dv = \rho_o \int_{\phi=0}^{2\pi} \int_{\theta=0}^{\pi} \int_{r=0}^{r} r^2 \sin\theta \, dr \, d\theta \, d\phi \qquad (4.51)$$

$$= \rho_o \frac{4}{3} \pi r^3$$

and

$$\Psi = \oint_S \mathbf{D} \cdot d\mathbf{S} = D_r \oint_S dS = D_r \int_{\phi=0}^{2\pi} \int_{\theta=0}^{\pi} r^2 \sin\theta \, d\theta \, d\phi$$

$$= D_r \, 4\pi r^2 \qquad (4.52)$$

Hence, $\Psi = Q_{\text{enc}}$ gives

$$D_r \, 4\pi r^2 = \frac{4\pi r^3}{3} \rho_o$$

or

$$\mathbf{D} = \frac{r}{3} \rho_o \, \mathbf{a}_r \quad 0 < r \leq a \qquad (4.53)$$

For $r \geq a$, the Gaussian surface is shown in Figure 4.16(b). The charge enclosed by the surface is the entire charge in this case, that is,

$$Q_{\text{enc}} = \int_v \rho_v \, dv = \rho_o \int_v dv = \rho_o \int_{\phi=0}^{2\pi} \int_{\theta=0}^{\pi} \int_{r=0}^{a} r^2 \sin\theta \, dr \, d\theta \, d\phi$$

$$= \rho_o \frac{4}{3} \pi a^3 \qquad (4.54)$$

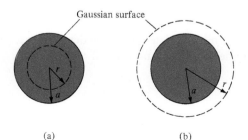

Gaussian surface

(a)   (b)

FIGURE 4.16 Gaussian surface for a uniformly charged sphere when (**a**) $r \leq a$ and (**b**) $r \geq a$.

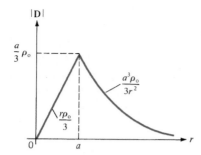

FIGURE 4.17 Sketch of |**D**| against r for a uniformly charged sphere.

while

$$\psi = \oint_S \mathbf{D} \cdot d\mathbf{S} = D_r\, 4\pi r^2 \tag{4.55}$$

just as in eq. (4.52). Hence,

$$D_r\, 4\pi r^2 = \frac{4}{3}\pi a^3 \rho_\circ$$

or

$$\mathbf{D} = \frac{a^3}{3r^2}\rho_\circ \mathbf{a}_r, \qquad r \ge a \tag{4.56}$$

Thus from eqs. (4.53) and (4.56), **D** everywhere is given by

$$\mathbf{D} = \begin{cases} \dfrac{r}{3}\rho_\circ \mathbf{a}_r, & 0 < r \le a \\[2mm] \dfrac{a^3}{3r^2}\rho_\circ \mathbf{a}_r, & r \ge a \end{cases} \tag{4.57}$$

and $|\mathbf{D}|$ is as sketched in Figure 4.17.

Notice from eqs. (4.44), (4.46), (4.48), and (4.52) that the ability to take **D** out of the integral sign is the key to finding **D** using Gauss's law. In other words, **D** must be constant on the Gaussan surface.

EXAMPLE 4.8

Given that $\mathbf{D} = z\rho \cos^2\phi\, \mathbf{a}_z$ C/m², calculate the charge density at $(1, \pi/4, 3)$ and the total charge enclosed by the cylinder of radius 1 m with $-2 \le z \le 2$ m.

Solution:

$$\rho_v = \nabla \cdot \mathbf{D} = \frac{\partial D_z}{\partial z} = \rho \cos^2 \phi$$

At $(1, \pi/4, 3)$, $\rho_v = 1 \cdot \cos^2(\pi/4) = 0.5\ \text{C/m}^3$. The total charge enclosed by the cylinder can be found in two different ways.

Method 1: This method is based directly on the definition of the total volume charge.

$$Q = \int_v \rho_v\, dv = \int_v \rho \cos^2 \phi\, \rho\, d\phi\, d\rho\, dz$$

$$= \int_{z=-2}^{2} dz \int_{\phi=0}^{2\pi} \cos^2 \phi\, d\phi \int_{\rho=0}^{1} \rho^2\, d\rho = 4(\pi)(1/3)$$

$$= \frac{4\pi}{3}\ \text{C}$$

Method 2: Alternatively, we can use Gauss's law

$$Q = \Psi = \oint_S \mathbf{D} \cdot d\mathbf{S} = \left[\int_S + \int_t + \int_b \right] \mathbf{D} \cdot d\mathbf{S}$$

$$= \Psi_s + \Psi_t + \Psi_b$$

where Ψ_s, Ψ_t, and Ψ_b are the flux through the sides (curved surface), the top surface, and the bottom surface of the cylinder, respectively (see Figure 3.18). Since $\mathbf{D}$ does not have component along $\mathbf{a}_\rho$, $\Psi_s = 0$, for Ψ_t, $d\mathbf{S} = \rho\, d\phi\, d\rho\, \mathbf{a}_z$ so

$$\Psi_t = \int_{\rho=0}^{1} \int_{\phi=0}^{2\pi} z\rho \cos^2 \phi\, \rho\, d\phi\, d\rho \bigg|_{z=2} = 2 \int_0^1 \rho^2\, d\rho \int_0^{2\pi} \cos^2 \phi\, d\phi$$

$$= 2 \left(\frac{1}{3} \right) \pi = \frac{2\pi}{3}$$

and for Ψ_b, $d\mathbf{S} = -\rho\, d\phi\, d\rho\, \mathbf{a}_z$, so

$$\Psi_b = - \int_{\rho=0}^{1} \int_{\phi=0}^{2\pi} z\rho \cos^2 \phi\, \rho\, d\phi\, d\rho \bigg|_{z=-2} = 2 \int_0^1 \rho^2\, d\rho \int_0^{2\pi} \cos^2 \phi\, d\phi$$

$$= \frac{2\pi}{3}$$

Thus

$$Q = \Psi = 0 + \frac{2\pi}{3} + \frac{2\pi}{3} = \frac{4\pi}{3}\ \text{C}$$

as obtained earlier.

PRACTICE EXERCISE 4.8

If $\mathbf{D} = (2y^2 + z)\mathbf{a}_x + 4xy\mathbf{a}_y + x\mathbf{a}_z$ C/m^2, find

(a) The volume charge density at $(-1, 0, 3)$
(b) The flux through the cube defined by $0 \le x \le 1, 0 \le y \le 1, 0 \le z \le 1$
(c) The total charge enclosed by the cube

Answer: (a) -4 C/m^3, (b) 2 C, (c) 2 C.

EXAMPLE 4.9

A charge distribution with spherical symmetry has density

$$\rho_v = \begin{cases} \dfrac{\rho_o r}{R}, & 0 \le r \le R \\ 0, & r > R \end{cases}$$

Determine **E** everywhere.

Solution:
The charge distribution is similar to that in Figure 4.16. Since symmetry exists, we can apply Gauss's law to find **E**.

$$\varepsilon_o \oint_S \mathbf{E} \cdot d\mathbf{S} = Q_{enc} = \int_v \rho_v \, dv$$

(a) For $r < R$

$$\varepsilon_o E_r \, 4\pi r^2 = Q_{enc} = \int_0^r \int_0^\pi \int_0^{2\pi} \rho_v r^2 \sin\theta \, d\phi \, d\theta \, dr$$

$$= \int_0^r 4\pi r^2 \frac{\rho_o r}{R} \, dr = \frac{\rho_o \pi r^4}{R}$$

or

$$\mathbf{E} = \frac{\rho_o r^2}{4\varepsilon_o R} \mathbf{a}_r$$

(b) For $r > R$,

$$\varepsilon_o E_r 4\pi r^2 = Q_{enc} = \int_0^r \int_0^\pi \int_0^{2\pi} \rho_v r^2 \sin\theta \, d\phi \, d\theta \, dr$$

$$= \int_0^R \frac{\rho_o r}{R} 4\pi r^2 \, dr + \int_R^r 0 \cdot 4\pi r^2 \, dr$$

$$= \pi \rho_o R^3$$

or

$$E = \frac{\rho_o R^3}{4\varepsilon_o r^2} \mathbf{a}_r$$

PRACTICE EXERCISE 4.9

A charge distribution in free space has $\rho_v = 2r \text{ nC/m}^3$ for $0 \leq r \leq 10$ m and zero otherwise. Determine $\mathbf{E}$ at $r = 2$ m and $r = 12$ m.

Answer: $226\mathbf{a}_r$ V/m, $3.927\mathbf{a}_r$ kV/m.

4.7 ELECTRIC POTENTIAL

From our discussions in the preceding sections, we can obtain the electric field intensity $\mathbf{E}$ due to a charge distribution from Coulomb's law in general or, when the charge distribution is symmetric, from Gauss's law. Another way of obtaining $\mathbf{E}$ is from the electric scalar potential V, to be defined in this section. In a sense, this way of finding $\mathbf{E}$ is easier because it is easier to handle scalars than vectors.

Suppose we wish to move a point charge Q from point A to point B in an electric field $\mathbf{E}$ as shown in Figure 4.18. From Coulomb's law, the force on Q is $\mathbf{F} = Q\mathbf{E}$ so that the *work done* in displacing the charge by $d\mathbf{l}$ is

$$dW = -\mathbf{F} \cdot d\mathbf{l} = -Q\mathbf{E} \cdot d\mathbf{l} \tag{4.58}$$

The negative sign indicates that the work is being done by an external agent. Thus the total work done, or the potential energy required, in moving Q from A to B, is

$$W = -Q \int_A^B \mathbf{E} \cdot d\mathbf{l} \tag{4.59}$$

Dividing W by Q in eq. (4.59) gives the potential energy per unit charge. This quantity, denoted by V_{AB}, is known as the *potential difference* between points A and B. Thus

$$V_{AB} = \frac{W}{Q} = -\int_A^B \mathbf{E} \cdot d\mathbf{l} \tag{4.60}$$

Note that

1. In determining V_{AB}, A is the initial point while B is the final point.
2. If V_{AB} is negative, there is a loss in potential energy in moving Q from A to B; this implies that the work is being done by the field. However, if V_{AB} is positive, there is a gain in potential energy in the movement; an external agent performs the work.
3. V_{AB} is independent of the path taken (to be shown a little later).
4. V_{AB} is measured in joules per coulomb, commonly referred to as volts (V).

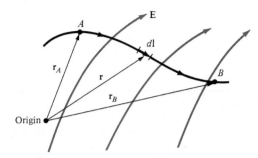

FIGURE 4.18 Displacement of point charge Q in an electrostatic field **E**.

As an example, if the **E** field in Figure 4.18 is due to a point charge Q located at the origin, then

$$\mathbf{E} = \frac{Q}{4\pi\varepsilon_o r^2}\mathbf{a}_r \tag{4.61}$$

so eq. (4.60) becomes

$$V_{AB} = -\int_{r_A}^{r_B} \frac{Q}{4\pi\varepsilon_o r^2}\mathbf{a}_r \cdot dr\,\mathbf{a}_r \tag{4.62a}$$

$$= \frac{Q}{4\pi\varepsilon_o}\left[\frac{1}{r_B} - \frac{1}{r_A}\right]$$

or

$$V_{AB} = V_B - V_A \tag{4.62b}$$

where V_B and V_A are the *potentials* (or *absolute potentials*) at B and A, respectively. Thus the potential difference V_{AB} may be regarded as the potential at B with reference to A. In problems involving point charges, it is customary to choose infinity as reference; that is, we assume the potential at infinity is zero. Thus if $V_A = 0$ as $r_A \rightarrow \infty$ in eq. (4.62), the potential at any point ($r_B \rightarrow r$) due to a point charge Q located at the origin is

$$\boxed{V = \frac{Q}{4\pi\varepsilon_o r}} \tag{4.63}$$

Note from eq. (4.62a) that because **E** points in the radial direction, any contribution from a displacement in the θ or ϕ direction is wiped out by the dot product $\mathbf{E} \cdot d\mathbf{l} = E\cos\alpha\,dl = E\,dr$, where α is the angle between **E** and $d\mathbf{l}$. Hence the potential difference V_{AB} is independent of the path as asserted earlier. In general, vectors whose line integral does not depend on the path of integration are called *conservative*. Thus, **E** is conservative.

The **potential** at any point is the potential difference between that point and a chosen point (or reference point) at which the potential is zero.

In other words, if one assumes zero potential at infinity, the potential at a distance r from the point charge is the work done per unit charge by an external agent in transferring a test charge from infinity to that point. Thus

$$V = -\int_{\infty}^{r} \mathbf{E} \cdot d\mathbf{l} \tag{4.64}$$

If the point charge Q in eq. (4.63) is not located at the origin but at a point whose position vector is $\mathbf{r}'$, the potential $V(x, y, z)$ or simply $V(\mathbf{r})$ at $\mathbf{r}$ becomes

$$V(\mathbf{r}) = \frac{Q}{4\pi\varepsilon_0 |\mathbf{r} - \mathbf{r}'|} \tag{4.65}$$

We have considered the electric potential due to a point charge. The same basic ideas apply to other types of charge distribution because any charge distribution can be regarded as consisting of point charges. The superposition principle, which we applied to electric fields, applies to potentials also. For n point charges $Q_1, Q_2, \ldots, Q_n$ located at points with position vectors $\mathbf{r}_1, \mathbf{r}_2, \ldots, \mathbf{r}_n$, the potential at $\mathbf{r}$ is

$$V(\mathbf{r}) = \frac{Q_1}{4\pi\varepsilon_0 |\mathbf{r} - \mathbf{r}_1|} + \frac{Q_2}{4\pi\varepsilon_0 |\mathbf{r} - \mathbf{r}_2|} + \cdots + \frac{Q_n}{4\pi\varepsilon_0 |\mathbf{r} - \mathbf{r}_n|}$$

or

$$V(\mathbf{r}) = \frac{1}{4\pi\varepsilon_0} \sum_{k=1}^{n} \frac{Q_k}{|\mathbf{r} - \mathbf{r}_k|} \quad \text{(point charges)} \tag{4.66}$$

For continuous charge distributions, we replace Q_k in eq. (4.66) with charge element $\rho_L \, dl$, $\rho_S \, dS$, or $\rho_v \, dv$ and the summation becomes an integration, so the potential at $\mathbf{r}$ becomes

$$V(\mathbf{r}) = \frac{1}{4\pi\varepsilon_0} \int_L \frac{\rho_L(\mathbf{r}')dl'}{|\mathbf{r} - \mathbf{r}'|} \quad \text{(line charge)} \tag{4.67}$$

$$V(\mathbf{r}) = \frac{1}{4\pi\varepsilon_0} \int_S \frac{\rho_S(\mathbf{r}')dS'}{|\mathbf{r} - \mathbf{r}'|} \quad \text{(surface charge)} \tag{4.68}$$

$$V(\mathbf{r}) = \frac{1}{4\pi\varepsilon_0} \int_v \frac{\rho_v(\mathbf{r}')dv'}{|\mathbf{r} - \mathbf{r}'|} \quad \text{(volume charge)} \tag{4.69}$$

where the primed coordinates are used customarily to denote source point location and the unprimed coordinates refer to field point (the point at which V is to be determined).

The following points should be noted:

1. We recall that in obtaining eqs. (4.63) to (4.69), the zero potential (reference) point has been chosen arbitrarily to be at infinity. If any other point is chosen as reference, eq. (4.63), for example, becomes

$$V = \frac{Q}{4\pi\varepsilon_o r} + C \tag{4.70}$$

where C is a constant that is determined at the chosen point of reference. The same idea applies to eqs. (4.65) to (4.69).

2. The potential at a point can be determined in two ways depending on whether the charge distribution or **E** is known. If the charge distribution is known, we use one of eqs. (4.65) to (4.70) depending on the charge distribution. If **E** is known, we simply use

$$V = -\int \mathbf{E} \cdot d\mathbf{l} + C \tag{4.71}$$

The potential difference V_{AB} can be found generally from

$$V_{AB} = V_B - V_A = -\int_A^B \mathbf{E} \cdot d\mathbf{l} = \frac{W}{Q} \tag{4.72}$$

EXAMPLE 4.10

Two point charges $-4\,\mu C$ and $5\,\mu C$ are located at $(2, -1, 3)$ and $(0, 4, -2)$, respectively. Find the potential at $(1, 0, 1)$, assuming zero potential at infinity.

Solution:

Let

$$Q_1 = -4\,\mu C, \quad Q_2 = 5\,\mu C$$

$$V(\mathbf{r}) = \frac{Q_1}{4\pi\varepsilon_o |\mathbf{r} - \mathbf{r}_1|} + \frac{Q_2}{4\pi\varepsilon_o |\mathbf{r} - \mathbf{r}_2|} + C_o$$

If $V(\infty) = 0$, $C_o = 0$,

$$|\mathbf{r} - \mathbf{r}_1| = |(1, 0, 1) - (2, -1, 3)| = |(-1, 1, -2)| = \sqrt{6}$$

$$|\mathbf{r} - \mathbf{r}_2| = |(1, 0, 1) - (0, 4, -2)| = |(1, -4, 3)| = \sqrt{26}$$

Hence

$$V(1, 0, 1) = \frac{10^{-6}}{4\pi \times \dfrac{10^{-9}}{36\pi}} \left[\frac{-4}{\sqrt{6}} + \frac{5}{\sqrt{26}} \right]$$

$$= 9 \times 10^3 \left(-1.633 + 0.9806 \right)$$

$$= -5.872\,\text{kV}$$

PRACTICE EXERCISE 4.10

If point charge $3\,\mu C$ is located at the origin in addition to the two charges of Example 4.10, find the potential at $(-1, 5, 2)$, assuming $V(\infty) = 0$.

Answer: 10.23 kV.

EXAMPLE 4.11

A point charge of 5 nC is located at $(-3, 4, 0)$, while line $y = 1$, $z = 1$ carries uniform charge 2 nC/m.

(a) If $V = 0$ V at $O(0, 0, 0)$, find V at $A(5, 0, 1)$.

(b) If $V = 100$ V at $B(1, 2, 1)$, find V at $C(-2, 5, 3)$.

(c) If $V = -5$ V at O, find V_{BC}.

Solution:

Let the potential at any point be

$$V = V_Q + V_L$$

where V_Q and V_L are the contributions to V at that point due to the point charge and the line charge, respectively. For the point charge,

$$V_Q = -\int \mathbf{E} \cdot d\mathbf{l} = -\int \frac{Q}{4\pi\varepsilon_o r^2} \mathbf{a}_r \cdot dr\, \mathbf{a}_r$$

$$= \frac{Q}{4\pi\varepsilon_o r} + C_1$$

For the infinite line charge,

$$V_L = -\int \mathbf{E} \cdot d\mathbf{l} = -\int \frac{\rho_L}{2\pi\varepsilon_o \rho} \mathbf{a}_\rho \cdot d\rho\, \mathbf{a}_\rho$$

$$= -\frac{\rho_L}{2\pi\varepsilon_o} \ln \rho + C_2$$

Hence,

$$V = -\frac{\rho_L}{2\pi\varepsilon_o} \ln \rho + \frac{Q}{4\pi\varepsilon_o r} + C$$

where $C = C_1 + C_2 =$ constant, ρ is the perpendicular distance from the line $y = 1$, $z = 1$ to the field point, and r is the distance from the point charge to the field point.

(a) If $V = 0$ at $O(0, 0, 0)$, and V at $A(5, 0, 1)$ is to be determined, we must first determine the values of ρ and r at O and A. Finding r is easy; we use eq. (2.31). To find ρ for any point (x, y, z), we utilize the fact that ρ is the perpendicular distance from (x, y, z) to line $y = 1$, $z = 1$, which is parallel to the x-axis. Hence ρ is the distance between (x, y, z) and $(x, 1, 1)$ because the distance vector between the two points is perpendicular to $\mathbf{a}_x$. Thus

$$\rho = |(x, y, z) - (x, 1, 1)| = \sqrt{(y - 1)^2 + (z - 1)^2}$$

Applying this for ρ and eq. (2.31) for r at points O and A, we obtain

$$\rho_O = |(0, 0, 0) - (0, 1, 1)| = \sqrt{2}$$
$$r_O = |(0, 0, 0) - (-3, 4, 0)| = 5$$

$$\rho_A = |(5, 0, 1) - (5, 1, 1)| = 1$$
$$r_A = |(5, 0, 1) - (-3, 4, 0)| = 9$$

Hence,

$$V_O - V_A = -\frac{\rho_L}{2\pi\varepsilon_o}\ln\frac{\rho_O}{\rho_A} + \frac{Q}{4\pi\varepsilon_o}\left[\frac{1}{r_O} - \frac{1}{r_A}\right]$$

$$= \frac{-2 \cdot 10^{-9}}{2\pi \cdot \dfrac{10^{-9}}{36\pi}}\ln\frac{\sqrt{2}}{1} + \frac{5 \cdot 10^{-9}}{4\pi \cdot \dfrac{10^{-9}}{36\pi}}\left[\frac{1}{5} - \frac{1}{9}\right]$$

$$0 - V_A = -36\ln\sqrt{2} + 45\left(\frac{1}{5} - \frac{1}{9}\right)$$

or

$$V_A = 36\ln\sqrt{2} - 4 = 8.477 \text{ V}$$

Notice that we have avoided calculating the constant C by subtracting one potential from another and that it does not matter which one is subtracted from which.

(b) If $V = 100$ at $B(1, 2, 1)$ and V at $C(-2, 5, 3)$ is to be determined, we find

$$\rho_B = |(1, 2, 1) - (1, 1, 1)| = 1$$
$$r_B = |(1, 2, 1) - (-3, 4, 0)| = \sqrt{21}$$
$$\rho_C = |(-2, 5, 3) - (-2, 1, 1)| = \sqrt{20}$$
$$r_C = |(-2, 5, 3) - (-3, 4, 0)| = \sqrt{11}$$
$$V_C - V_B = -\frac{\rho_L}{2\pi\varepsilon_o}\ln\frac{\rho_C}{\rho_B} + \frac{Q}{4\pi\varepsilon_o}\left[\frac{1}{r_C} - \frac{1}{r_B}\right]$$
$$V_C - 100 = -36\ln\frac{\sqrt{20}}{1} + 45 \cdot \left[\frac{1}{\sqrt{11}} - \frac{1}{\sqrt{21}}\right]$$
$$= -50.175 \text{ V}$$

or

$$V_C = 49.825 \text{ V}$$

(c) To find the potential difference between two points, we do not need a potential reference if a common reference is assumed.

$$V_{BC} = V_C - V_B = 49.825 - 100$$
$$= -50.175 \text{ V}$$

> **PRACTICE EXERCISE 4.11**
>
> A point charge of 5 nC is located at the origin. If $V = 2$ V at $(0, 6, -8)$, find
>
> (a) The potential at $A(-3, 2, 6)$
> (b) The potential at $B(1, 5, 7)$
> (c) The potential difference V_{AB}
>
> **Answer:** (a) 3.929 V, (b) 2.696 V, (c) -1.233 V.

4.8 RELATIONSHIP BETWEEN E AND *V*—MAXWELL'S EQUATION

As shown in the preceding section, the potential difference between points A and B is independent of the path taken. Hence,

$$V_{BA} = -V_{AB}$$

that is, $V_{BA} + V_{AB} = \oint_L \mathbf{E} \cdot d\mathbf{l} = 0$
or

$$\oint_L \mathbf{E} \cdot d\mathbf{l} = 0 \tag{4.73}$$

This shows that the line integral of **E** along a closed path as shown in Figure 4.19 must be zero. Physically, this implies that no net work is done in moving a charge along a closed path in an electrostatic field. Applying Stokes's theorem to eq. (4.73) gives

$$\oint_L \mathbf{E} \cdot d\mathbf{l} = \int_S (\nabla \times \mathbf{E}) \cdot d\mathbf{S} = 0$$

or

$$\nabla \times \mathbf{E} = \mathbf{0} \tag{4.74}$$

Any vector field that satisfies eq. (4.73) or (4.74) is said to be *conservative*, or *irrotational*, as discussed in Section 3.9. In other words, vectors whose line integral does not depend on

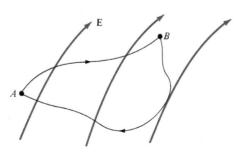

FIGURE 4.19 The conservative nature of an electrostatic field.

the path of integration are called conservative vector fields. Thus an electrostatic field is a conservative field. Equation (4.73) or (4.74) is referred to as *Maxwell's equation* (the second Maxwell's equation to be derived) for static electric fields. Equation (4.73) is the integral form, and eq. (4.74) is the differential form; they both depict the conservative nature of an electrostatic field.

From the way we defined potential, $V = -\int \mathbf{E} \cdot d\mathbf{l}$, it follows that

$$dV = -\mathbf{E} \cdot d\mathbf{l} = -E_x \, dx - E_y \, dy - E_z \, dz$$

But from calculus of multivariables, a total change in $V(x, y, z)$ is the sum of partial changes with respect to x, y, z variables:

$$dV = \frac{\partial V}{\partial x} \, dx + \frac{\partial V}{\partial y} \, dy + \frac{\partial V}{\partial z} \, dz$$

Comparing the two expressions for dV, we obtain

$$E_x = -\frac{\partial V}{\partial x}, \quad E_y = -\frac{\partial V}{\partial y}, \quad E_z = -\frac{\partial V}{\partial z} \tag{4.75}$$

Thus,

$$\boxed{\mathbf{E} = -\nabla V} \tag{4.76}$$

that is, the electric field intensity is the gradient of V. The negative sign shows that the direction of $\mathbf{E}$ is opposite to the direction in which V increases; $\mathbf{E}$ is directed from higher to lower levels of V. Since the curl of the gradient of a scalar function is always zero ($\nabla \times \nabla V = 0$), eq. (4.74) obviously implies that $\mathbf{E}$ must be a gradient of some scalar function. Thus eq. (4.76) could have been obtained from eq. (4.74).

Equation (4.76) shows another way to obtain the $\mathbf{E}$ field apart from using Coulomb's or Gauss's law. That is, if the potential field V is known, the $\mathbf{E}$ can be found by using eq. (4.76). One may wonder how one function V can possibly contain all the information that the three components of $\mathbf{E}$ carry. The three components of $\mathbf{E}$ are not independent of one another: they are explicitly interrelated by the condition $\nabla \times \mathbf{E} = 0$. The potential formulation exploits this feature to maximum advantage, reducing a vector problem to a scalar one.

EXAMPLE 4.12

Given the potential $V = \dfrac{10}{r^2} \sin \theta \cos \phi$,

(a) Find the electric flux density $\mathbf{D}$ at $(2, \pi/2, 0)$.
(b) Calculate the work done in moving a 10 μC charge from point $A(1, 30°, 120°)$ to $B(4, 90°, 60°)$.

Solution:

(a) $\mathbf{D} = \varepsilon_0 \mathbf{E}$

But

$$\mathbf{E} = -\nabla V = -\left[\frac{\partial V}{\partial r} \mathbf{a}_r + \frac{1}{r} \frac{\partial V}{\partial \theta} \mathbf{a}_\theta + \frac{1}{r \sin \theta} \frac{\partial V}{\partial \phi} \mathbf{a}_\phi \right]$$

$$= \frac{20}{r^3} \sin \theta \cos \phi \, \mathbf{a}_r - \frac{10}{r^3} \cos \theta \cos \phi \, \mathbf{a}_\theta + \frac{10}{r^3} \sin \phi \, \mathbf{a}_\phi$$

At $(2, \pi/2, 0)$,

$$\mathbf{D} = \varepsilon_o \mathbf{E} \, (r = 2, \theta = \pi/2, \phi = 0) = \varepsilon_o \left(\frac{20}{8} \mathbf{a}_r - 0\mathbf{a}_\theta + 0\mathbf{a}_\phi \right)$$

$$= 2.5\varepsilon_o \mathbf{a}_r \ \text{C/m}^2 = 22.1 \ \mathbf{a}_r \ \text{pC/m}^2$$

(b) The work done can be found in two ways, using either **E** or *V*.

Method 1:

$$W = -Q \int_L \mathbf{E} \cdot d\mathbf{l} \quad \text{or} \quad -\frac{W}{Q} = \int_L \mathbf{E} \cdot d\mathbf{l}$$

and because the electrostatic field is conservative, the path of integration is immaterial. Hence the work done in moving Q from $A(1, 30°, 120°)$ to $B(4, 90°, 60°)$ is the same as that in moving Q from A to A', from A' to B', and from B' to B, where

$A(1, 30°, 120°)$		$B(4, 90°, 60°)$
$\downarrow d\mathbf{l} = dr \, \mathbf{a}_r$	$d\mathbf{l} = r \, d\theta \, \mathbf{a}_\theta$	$\uparrow d\mathbf{l} = r \sin \theta \, d\phi \, \mathbf{a}_\phi$
$A'(4, 30°, 120°)$	$\rightarrow$	$B'(4, 90°, 120°)$

That is, instead of being moved directly from A to B, Q is moved from $A \rightarrow A', A' \rightarrow B'$, $B' \rightarrow B$, so that only one variable is changed at a time. This makes the line integral much easier to evaluate. Thus

$$\frac{-W}{Q} = -\frac{1}{Q} \left(W_{AA'} + W_{A'B'} + W_{B'B} \right)$$

$$= \left(\int_{AA'} + \int_{A'B'} + \int_{B'B} \right) \mathbf{E} \cdot d\mathbf{l}$$

$$= \int_{r=1}^{4} \frac{20 \sin \theta \cos \phi}{r^3} dr \Bigg|_{\theta = 30°, \phi = 120°}$$

$$+ \int_{\theta=30°}^{90°} \frac{-10 \cos \theta \cos \phi}{r^3} r \, d\theta \Bigg|_{r=4, \phi=120°}$$

$$+ \int_{\phi=120°}^{60°} \frac{10 \sin \phi}{r^3} r \sin \theta \, d\phi \Bigg|_{r=4, \theta=90°}$$

$$= 20 \left(\frac{1}{2} \right) \left(\frac{-1}{2} \right) \left[-\frac{1}{2r^2} \Big|_{r=1}^{4} \right]$$

$$- \frac{10}{16} \frac{(-1)}{2} \sin \theta \Big|_{30°}^{90°} + \frac{10}{16} (1) \left[-\cos \phi \Big|_{120°}^{60°} \right]$$

$$-\frac{W}{Q} = \frac{-75}{32} + \frac{5}{32} - \frac{10}{16}$$

or

$$W = \frac{45}{16} Q = 28.125 \ \mu J$$

Method 2:

Since V is known, this method is much easier.

$$W = -Q \int_A^B \mathbf{E} \cdot d\mathbf{l} = Q V_{AB}$$

$$= Q(V_B - V_A)$$

$$= 10 \left(\frac{10}{16} \sin 90° \cos 60° - \frac{10}{1} \sin 30° \cos 120° \right) \cdot 10^{-6}$$

$$= 10 \left(\frac{10}{32} - \frac{-5}{2} \right) \cdot 10^{-6}$$

$$= 28.125 \ \mu J \text{ as obtained before}$$

PRACTICE EXERCISE 4.12

Given that $\mathbf{E} = (3x^2 + y)\mathbf{a}_x + x\mathbf{a}_y$ kV/m, find the work done in moving a $-2 \ \mu C$ charge from $(0, 5, 0)$ to $(2, -1, 0)$ by taking the straight-line path.

(a) $(0, 5, 0) \rightarrow (2, 5, 0) \rightarrow (2, -1, 0)$

(b) $y = 5 - 3x$

Answer: (a) 12 mJ, (b) 12 mJ.

4.9 AN ELECTRIC DIPOLE AND FLUX LINES

An **electric dipole** is formed when two point charges of equal magnitude but opposite sign are separated by a small distance.

The importance of the field due to a dipole will be evident in the subsequent chapters. Consider the dipole shown in Figure 4.20. The potential at point $P(r, \theta, \phi)$ is given by

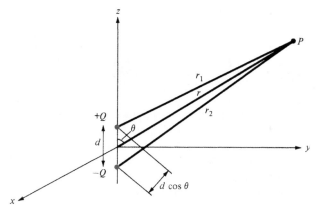

FIGURE 4.20 An electric dipole.

$$V = \frac{Q}{4\pi\varepsilon_o}\left[\frac{1}{r_1} - \frac{1}{r_2}\right] = \frac{Q}{4\pi\varepsilon_o}\left[\frac{r_2 - r_1}{r_1 r_2}\right] \tag{4.77}$$

where r_1 and r_2 are the distances between P and $+Q$ and P and $-Q$, respectively. If $r \gg d$, $r_2 - r_1 \simeq d\cos\theta$, $r_2 r_1 \simeq r^2$, and eq. (4.77) becomes

$$V = \frac{Q}{4\pi\varepsilon_o}\frac{d\cos\theta}{r^2} \tag{4.78}$$

Since $d\cos\theta = \mathbf{d}\cdot\mathbf{a}_r$, where $\mathbf{d} = d\mathbf{a}_z$, if we define

$$\mathbf{p} = Q\mathbf{d} \tag{4.79}$$

as the *dipole moment,* eq. (4.78) may be written as

$$V = \frac{\mathbf{p}\cdot\mathbf{a}_r}{4\pi\varepsilon_o r^2} \tag{4.80}$$

Note that the dipole moment $\mathbf{p}$ is directed from $-Q$ to $+Q$. If the dipole center is not at the origin but at $\mathbf{r}'$, eq. (4.80) becomes

$$\boxed{V(\mathbf{r}) = \frac{\mathbf{p}\cdot(\mathbf{r} - \mathbf{r}')}{4\pi\varepsilon_o|\mathbf{r} - \mathbf{r}'|^3}} \tag{4.81}$$

The electric field due to the dipole with center at the origin, shown in Figure 4.20, can be obtained readily from eqs. (4.76) and (4.78) as

$$\mathbf{E} = -\nabla V = -\left[\frac{\partial V}{\partial r}\mathbf{a}_r + \frac{1}{r}\frac{\partial V}{\partial\theta}\mathbf{a}_\theta\right]$$

$$= \frac{Qd\cos\theta}{2\pi\varepsilon_o r^3}\mathbf{a}_r + \frac{Qd\sin\theta}{4\pi\varepsilon_o r^3}\mathbf{a}_\theta$$

or

$$\mathbf{E} = \frac{p}{4\pi\varepsilon_0 \mathbf{r}^3} (2 \cos\theta \, \mathbf{a}_r + \sin\theta \, \mathbf{a}_\theta) \tag{4.82}$$

where $p = |\mathbf{p}| = Qd$.

Notice that a point charge is a *monopole* and its electric field varies inversely as r^2 while its potential field varies inversely as r [see eqs. (4.61) and (4.63)]. From eqs. (4.80) and (4.82), we notice that the electric field due to a dipole varies inversely as r^3, while its potential varies inversely as r^2. The electric fields due to successive higher-order multipoles (such as a *quadrupole* consisting of two dipoles or an *octupole* consisting of two quadrupoles) vary inversely as $r^4, r^5, r^6, \ldots$, while their corresponding potentials vary inversely as $r^3, r^4, r^5, \ldots$.

The idea of *electric flux* lines (or *electric lines of force* as they are sometimes called) was introduced by Michael Faraday (1791–1867) in his experimental investigation as a way of visualizing the electric field.

> An **electric flux line** is an imaginary path or line drawn in such a way that its direction at any point is the direction of the electric field at that point.

In other words, they are the lines to which the electric flux density **D** is tangential at every point.

Any surface on which the potential is the same throughout is known as an *equipotential surface.* The intersection of an equipotential surface and a plane results in a path or line known as an *equipotential line.* No work is done in moving a charge from one point to another along an equipotential line or surface ($V_A - V_B = 0$) and hence

$$\int_L \mathbf{E} \cdot d\mathbf{l} = 0 \tag{4.83}$$

on the line or surface. From eq. (4.83), we may conclude that the lines of force or flux lines (or the direction of **E**) are always normal to equipotential surfaces. Examples of equipotential surfaces for point charge and a dipole are shown in Figure 4.21. Note from

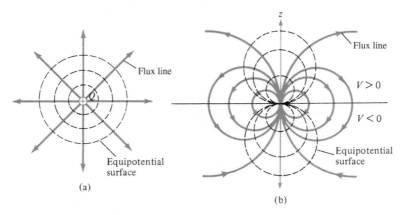

FIGURE 4.21 Equipotential surfaces for (**a**) a point charge and (**b**) an electric dipole.

these examples that the direction of **E** is everywhere normal to the equipotential lines. We shall see the importance of equipotential surfaces when we discuss conducting bodies in electric fields; it will suffice to say at this point that such bodies are equipotential volumes.

A typical application of field mapping (flux lines and equipotential surfaces) is found in the study of the human heart. The human heart beats in response to an electric field potential difference across it. The heart can be characterized as a dipole with the field map similar to that of Figure 4.21(b). Such a field map is useful in detecting abnormal heart position.[7] In Section 14.2, we will discuss a numerical technique for field mapping.

EXAMPLE 4.13

Two dipoles with dipole moments $-5\mathbf{a}_z$ nC · m and $9\mathbf{a}_z$ nC · m are located at points $(0, 0, -2)$ and $(0, 0, 3)$, respectively. Find the potential at the origin.

Solution:

$$V = \sum_{k=1}^{2} \frac{\mathbf{p}_k \cdot \mathbf{r}_k}{4\pi\varepsilon_o r_k^3}$$

$$= \frac{1}{4\pi\varepsilon_o} \left[\frac{\mathbf{p}_1 \cdot \mathbf{r}_1}{r_1^3} + \frac{\mathbf{p}_2 \cdot \mathbf{r}_2}{r_2^3} \right]$$

where

$$\mathbf{p}_1 = -5\mathbf{a}_z, \quad \mathbf{r}_1 = (0, 0, 0) - (0, 0, -2) = 2\mathbf{a}_z, \quad r_1 = |\mathbf{r}_1| = 2$$

$$\mathbf{p}_2 = 9\mathbf{a}_z, \quad \mathbf{r}_2 = (0, 0, 0) - (0, 0, 3) = -3\mathbf{a}_z, \quad r_2 = |\mathbf{r}_2| = 3$$

Hence,

$$V = \frac{1}{4\pi \cdot \dfrac{10^{-9}}{36\pi}} \left[\frac{-10}{2^3} - \frac{27}{3^3} \right] \cdot 10^{-9}$$

$$= -20.25 \text{ V}$$

PRACTICE EXERCISE 4.13

An electric dipole of $100 \, \mathbf{a}_z$ pC · m is located at the origin. Find V and **E** at points

(a) $(0, 0, 10)$

(b) $(1, \pi/3, \pi/2)$

Answer: (a) 9 mV, $1.8\mathbf{a}_r$ mV/m, (b) 0.45 V, $0.9\mathbf{a}_r + 0.7794\mathbf{a}_\theta$ V/m.

---

[7]For more information on this, see R. Plonsey, *Bioelectric Phenomena*, New York: McGraw-Hill, 1969.

4.10 ENERGY DENSITY IN ELECTROSTATIC FIELDS

To determine the energy present in an assembly of charges, we must first determine the amount of work necessary to assemble them. Suppose we wish to position three point charges Q_1, Q_2, and Q_3 in an initially empty space shown shaded in Figure 4.22. No work is required to transfer Q_1 from infinity to P_1 because the space is initially charge free and there is no electric field [from eq. (4.59), $W = 0$]. The work done in transferring Q_2 from infinity to P_2 is equal to the product of Q_2 and the potential V_{21} at P_2 due to Q_1. Similarly, the work done in positioning Q_3 at P_3 is equal to $Q_3(V_{32} + V_{31})$, where V_{32} and V_{31} are the potentials at P_3 due to Q_2 and Q_1, respectively. Hence the total work done in positioning the three charges is

$$W_E = W_1 + W_2 + W_3$$
$$= 0 + Q_2V_{21} + Q_3(V_{31} + V_{32}) \tag{4.84}$$

If the charges were positioned in reverse order,

$$W_E = W_3 + W_2 + W_1$$
$$= 0 + Q_2V_{23} + Q_1(V_{12} + V_{13}) \tag{4.85}$$

where V_{23} is the potential at P_2 due to Q_3, V_{12} and V_{13} are, respectively, the potentials at P_1 due to Q_2 and Q_3. Adding eqs. (4.84) and (4.85) gives

$$2W_E = Q_1(V_{12} + V_{13}) + Q_2(V_{21} + V_{23}) + Q_3(V_{31} + V_{32})$$
$$= Q_1V_1 + Q_2V_2 + Q_3V_3$$

or

$$W_E = \frac{1}{2}(Q_1V_1 + Q_2V_2 + Q_3V_3) \tag{4.86}$$

where V_1, V_2, and V_3 are total potentials at P_1, P_2, and P_3, respectively. In general, if there are n point charges, eq. (4.86) becomes

$$\boxed{W_E = \frac{1}{2}\sum_{k=1}^{n} Q_kV_k} \quad \text{(in joules)} \tag{4.87}$$

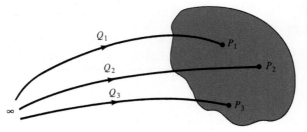

FIGURE 4.22 Assembling of charges.

If, instead of point charges, the region has a continuous charge distribution, the summation in eq. (4.87) becomes integration; that is,

$$W_E = \frac{1}{2} \int_L \rho_L V \, dl \quad \text{(line charge)} \tag{4.88}$$

$$W_E = \frac{1}{2} \int_S \rho_S V \, dS \quad \text{(surface charge)} \tag{4.89}$$

$$W_E = \frac{1}{2} \int_v \rho_v V \, dv \quad \text{(volume charge)} \tag{4.90}$$

Since $\rho_v = \nabla \cdot \mathbf{D}$, eq. (4.90) can be further developed to yield

$$W_E = \frac{1}{2} \int_v (\nabla \cdot \mathbf{D}) \, V \, dv \tag{4.91}$$

But for any vector $\mathbf{A}$ and scalar V, the identity

$$\nabla \cdot V\mathbf{A} = \mathbf{A} \cdot \nabla V + V(\nabla \cdot \mathbf{A})$$

or

$$(\nabla \cdot \mathbf{A})V = \nabla \cdot V\mathbf{A} - \mathbf{A} \cdot \nabla V \tag{4.92}$$

holds. Applying the identity in eqs. (4.92) to (4.91), we get

$$W_E = \frac{1}{2} \int_v (\nabla \cdot V\mathbf{D}) \, dv - \frac{1}{2} \int_v (\mathbf{D} \cdot \nabla V) \, dv \tag{4.93}$$

By applying divergence theorem to the first term on the right-hand side of this equation, we have

$$W_E = \frac{1}{2} \oint_S (V\mathbf{D}) \cdot d\mathbf{S} - \frac{1}{2} \int_v (\mathbf{D} \cdot \nabla V) \, dv \tag{4.94}$$

From Section 4.9, we recall that V varies as $1/r$ and $\mathbf{D}$ as $1/r^2$ for point charges; V varies as $1/r^2$ and $\mathbf{D}$ as $1/r^3$ for dipoles; and so on. Hence, $V\mathbf{D}$ in the first term on the right-hand side of eq. (4.94) must vary at least as $1/r^3$ while dS varies as r^2. Consequently, the first integral in eq. (4.94) must tend to zero as the surface S becomes large. Hence, eq. (4.94) reduces to

$$W_E = -\frac{1}{2} \int_v (\mathbf{D} \cdot \nabla V) \, dv = \frac{1}{2} \int_v (\mathbf{D} \cdot \mathbf{E}) \, dv \tag{4.95}$$

and since $\mathbf{E} = -\nabla V$ and $\mathbf{D} = \varepsilon_o \mathbf{E}$, the electrostatic energy is

$$W_E = \frac{1}{2}\int_v \mathbf{D}\cdot\mathbf{E}\,dv = \frac{1}{2}\int_v \varepsilon_o E^2\,dv \qquad (4.96)$$

From this, we can define electrostatic energy density w_E (in J/m^3) as

$$w_E = \frac{dW_E}{dv} = \frac{1}{2}\mathbf{D}\cdot\mathbf{E} = \frac{1}{2}\varepsilon_o E^2 = \frac{D^2}{2\varepsilon_o} \qquad (4.97)$$

so eq. (4.95) may be written as

$$W_E = \int_v w_E\,dv \qquad (4.98)$$

EXAMPLE 4.14

The point charges -1 nC, 4 nC, and 3 nC are located at $(0,0,0)$, $(0,0,1)$, and $(1,0,0)$, respectively. Find the energy in the system.

Solution:

Method 1:

$$W = W_1 + W_2 + W_3$$
$$= 0 + Q_2 V_{21} + Q_3(V_{31} + V_{32})$$

$$= Q_2 \cdot \frac{Q_1}{4\pi\varepsilon_o\,|(0,0,1) - (0,0,0)|}$$

$$+ \frac{Q_3}{4\pi\varepsilon_o}\left[\frac{Q_1}{|(1,0,0) - (0,0,0)|} + \frac{Q_2}{|(1,0,0) - (0,0,1)|}\right]$$

$$= \frac{1}{4\pi\varepsilon_o}\left(Q_1 Q_2 + Q_1 Q_3 + \frac{Q_2 Q_3}{\sqrt{2}}\right)$$

$$= \frac{1}{4\pi\cdot\dfrac{10^{-9}}{36\pi}}\left(-4 - 3 + \frac{12}{\sqrt{2}}\right)\cdot 10^{-18}$$

$$= 9\left(\frac{12}{\sqrt{2}} - 7\right)\text{nJ} = 13.37\text{ nJ}$$

Method 2:

$$W = \frac{1}{2}\sum_{k=1}^{3} Q_k V_k = \frac{1}{2}(Q_1 V_1 + Q_2 V_2 + Q_3 V_3)$$

$$= \frac{Q_1}{2}\left[\frac{Q_2}{4\pi\varepsilon_o(1)} + \frac{Q_3}{4\pi\varepsilon_o(1)}\right] + \frac{Q_2}{2}\left[\frac{Q_1}{4\pi\varepsilon_o(1)} + \frac{Q_3}{4\pi\varepsilon_o(\sqrt{2})}\right]$$

$$+ \frac{Q_3}{2} \left[\frac{Q_1}{4\pi\varepsilon_o(1)} + \frac{Q_2}{4\pi\varepsilon_o(\sqrt{2})} \right]$$

$$= \frac{1}{4\pi\varepsilon_o} \left(Q_1 Q_2 + Q_1 Q_3 + \frac{Q_2 Q_3}{\sqrt{2}} \right)$$

$$= 9 \left(\frac{12}{\sqrt{2}} - 7 \right) \text{nJ} = 13.37 \text{ nJ}$$

PRACTICE EXERCISE 4.14

Point charges $Q_1 = 1$ nC, $Q_2 = -2$ nC, $Q_3 = 3$ nC, and $Q_4 = -4$ nC are positioned one at a time and in that order at $(0, 0, 0)$, $(1, 0, 0)$, $(0, 0, -1)$, and $(0, 0, 1)$, respectively. Calculate the energy in the system after each charge is positioned.

Answer: $0, -18$ nJ, -29.18 nJ, -68.27 nJ.

EXAMPLE 4.15

A charge distribution with spherical symmetry has density

$$\rho_v = \begin{cases} \rho_o, & 0 \leq r \leq R \\ 0, & r > R \end{cases}$$

Determine V everywhere and the energy stored in region $r < R$.

Solution:
The **D** field has already been found in Section 4.6D using Gauss's law.

(a) For $r \geq R$, $\mathbf{E} = \dfrac{\rho_o R^3}{3\varepsilon_o r^2} \mathbf{a}_r$.

Once **E** is known, V is determined as

$$V = -\int \mathbf{E} \cdot d\mathbf{l} = -\frac{\rho_o R^3}{3\varepsilon_o} \int \frac{1}{r^2} dr$$

$$= \frac{\rho_o R^3}{3\varepsilon_o r} + C_1, \quad r \geq R$$

Since $V(r = \infty) = 0$, $C_1 = 0$.

(b) For $r \leq R$, $\mathbf{E} = \dfrac{\rho_o r}{3\varepsilon_o} \mathbf{a}_r$.

Hence,

$$V = -\int \mathbf{E} \cdot d\mathbf{l} = -\frac{\rho_o}{3\varepsilon_o} \int r \, dr$$

$$= -\frac{\rho_o r^2}{6\varepsilon_o} + C_2$$

From part (a) $V(r = R) = \dfrac{\rho_o R^2}{3\varepsilon_o}$. Hence,

$$\frac{R^2 \rho_o}{3\varepsilon_o} = -\frac{R^2 \rho_o}{6\varepsilon_o} + C_2 \;\rightarrow\; C_2 = \frac{R^2 \rho_o}{2\varepsilon_o}$$

and

$$V = \frac{\rho_o}{6\varepsilon_o} (3R^2 - r^2)$$

Thus from parts (a) and (b)

$$V = \begin{cases} \dfrac{\rho_o R^3}{3\varepsilon_o r}, & r \geq R \\[3mm] \dfrac{\rho_o}{6\varepsilon_o} (3R^2 - r^2), & r \leq R \end{cases}$$

(c)   The energy stored is given by

$$W = \frac{1}{2} \int_v \mathbf{D} \cdot \mathbf{E} \, dv = \frac{1}{2} \varepsilon_o \int_v E^2 \, dv$$

For $r \leq R$,

$$\mathbf{E} = \frac{\rho_o r}{3\varepsilon_o} \mathbf{a}_r$$

Hence,

$$W = \frac{1}{2} \varepsilon_o \frac{\rho_o^2}{9\varepsilon_o^2} \int_{r=0}^{R} \int_{\theta=0}^{\pi} \int_{\phi=0}^{2\pi} r^2 \cdot r^2 \sin\theta \, d\phi \, d\theta \, dr$$

$$= \frac{\rho_o^2}{18\varepsilon_o} 4\pi \cdot \frac{r^5}{5} \Big|_0^R = \frac{2\pi \rho_o^2 R^5}{45\varepsilon_o} \text{ J}$$

PRACTICE EXERCISE 4.15

If $V = x - y + xy + 2z$ V, find $\mathbf{E}$ at $(1, 2, 3)$ and the electrostatic energy stored in a cube of side 2 m centered at the origin.

Answer: $-3\mathbf{a}_x - 2\mathbf{a}_z$ V/m, 0.2358 nJ.

†4.11 APPLICATION NOTE—ELECTROSTATIC DISCHARGE

Electrostatic discharge (ESD) (or static electricity, as it is commonly known) refers to the sudden transfer (discharge) of static charge between objects at different electrostatic potentials. A good example is the "zap" one feels after walking on a synthetic carpet and then touching a metal doorknob.

ESD belongs to a family of electrical problems known as electrical overstress (EOS). Other members of the EOS family include lightning and electromagnetic pulses (EMPs). ESD poses a serious threat to electronic devices and affects the operation of the systems that contain those devices. An ESD can destroy an integrated circuit (IC), shut down a computer system, cause a fuel tank to explode, and so on. ESD is a rapid-discharge event that transfers a finite amount of charge between two bodies at different potentials. ESD costs industry many billions of dollars annually. The damage to an IC depends on the current densities and voltage gradients developed during the event. The harmful effects of ESD are now recognized as a major contributor to poor product yield and long-term unreliability in many electronic assemblies. Most electronics companies now regard all semiconductor devices as ESD sensitive. For this reason, a good understanding of ESD is required in industry. It is considered the responsibility of the design engineer to ensure that electronic systems are designed and protected against damage from ESD.

What causes ESD? Static charge is a result of an unbalanced electrical charge at rest. For example, it is created by insulator surfaces rubbing together or pulling apart. One surface gains electrons, while the other loses electrons. If the charge transfer causes an excess of electrons on an object, the charge is negative. On the other hand, a deficiency of electrons on the object makes the static charge positive. When a static charge moves from one surface to another, it becomes ESD. ESD events occur to balance the charge between two objects. The movement of these charges often occurs rapidly and randomly, leading to high currents.

ESD can occur in one of the following four ways:

- A charged body touches a device such as an IC.
- A charged device touches a grounded surface.
- A charged machine touches a device.
- An electrostatic field induces a voltage across a dielectric that is sufficient to cause breakdown.

There are two sources of ESD-generated events: people and equipment. ESD from a person can vary depending on footwear, posture (standing or sitting), and what the person has in his or her hand (metal or dielectric). The capacitance of a person could double if the individual were sitting instead of standing. The generated voltage is the driving force behind the ESD event. For example, walking across a synthetic carpet on a dry day may generate a potential of 20 kV on the person's body.

An ESD event takes place in the following four stages.

1. *Charge generation*: This could be triboelectricity, induction, or conduction. Triboelectricity requires physical contact between two different materials or the rubbing together of two materials. For example, a person who walks across a synthetic carpet becomes charged by the process of triboelectrification. Fundamental electrostatics tells us that some materials tend to charge positively, while others tend to charge negatively. The triboelectric series (see Table 4.1) summarizes this propensity. A material near the top of

TABLE 4.1 The Triboelectric Series

Material	Polarity of Charge
Air	+
Human hands	
Rabbit fur	
Glass	
Mica	
Human hair	
Fur	
Lead	
Silk	
Aluminum	
Paper	
Cotton	
Steel	
Wood	
Amber	
Wax	
Hard rubber	
Nickel, copper	
Gold	
Polyester	
Polyethylene	
PVC (vinyl)	
Silicon	
Teflon	−

Table 4.1 is charged positively when rubbed by a material below it. For example, combing your hair with a hard rubber comb leaves your hair positively charged and the comb negatively charged. Inductive charging takes place when a conducting object comes close to a charged object and is then removed. Conductive charging, which involves the physical contact and balancing of voltage between two objects at different potentials, often occurs during automated testing.

2. *Charge transfer:* This is the second stage in an ESD event. Charge transfers from the higher potential body to the lower potential body until the potentials between them are equal. Charge transfer is characterized by the capacitance of the two bodies involved and the impedance between them.

3. *Device response:* At this stage, we analyze how a circuit responds to a pulse and how it withstands the redistribution of charge. When an ESD event begins, charge starts to redistribute, and this movement of charge generates currents and induces voltages.

4. *Device failure:* The last stage involves assessing the kind of failure, if any. This is when we determine whether the device survived. There are three kinds of failure: hard failure (i.e., physical destruction), soft failure, and latent failure.

The importance of ESD has led standards organizations to develop guidelines for control and prevention of ESD. The ESD Association has developed a standard known as ANSI/ESD S20.20 (2007) to establish and maintain ESD control. The standard identifies and describes key measurement processes to qualify a company's ESD control program. Here is a short list of dos and don'ts.

- Treat everything as static sensitive.
- Touch something grounded before handling electronic assemblies or components.
- Wear a grounded wrist strap whenever possible (see Figure 4.23).

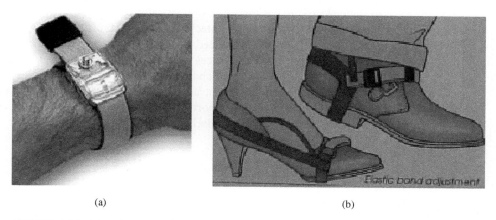

(a)   (b)

FIGURE 4.23 (**a**) Wrist strap. (**b**) Foot grounders.

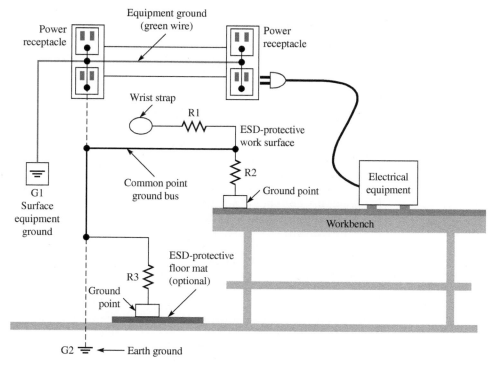

FIGURE 4.24 A typical ESD-protected workstation.

- Keep the relative humidity at 40% or greater.
- Don't touch any leads, pins, or traces when handling charged devices.
- Don't move around a lot.
- Don't touch electronic devices if you are getting static discharges.

A static-control wrist strap is an inexpensive way to minimize the risk of ESD. Use of garments designed to protect electronic component assembly operations from damage due to static electricity has increased dramatically.

Today, design for ESD protection is critical but difficult owing to the reduction in device sizes, high operating speeds, factory automation, and uncontrolled user environments. ESD protection of devices involves minimizing the environmental exposure by providing means to minimize charge generation and charge transfer. It is applied during production, transportation, and handling of most electronic products. A typical ESD-protected workstation is portrayed in Figure 4.24. Electronic devices are protected against ESD by a strategy to discharge ESD events that may occur when the device is exposed to ESD.

MATLAB 4.1

```
% This script allows the user to input a number of charges
% and compute the electric field at a particular coordinate
% observation point due to these charges
clear

n = input('Enter number of charges in the system... \n >  ');
if isempty(n); n = 1; end
Q=zeros(n,4);      % create a matrix of zeros, with n rows
                          % and 4 columns

r=input('Enter observation location [x y z]... \n >  ');
if isempty(r); r = [0 0 0]; end

% loop through all the charges that the user input
% and collect the observation points for each charge
for index=1:n,
    disp(sprintf('Enter position of charge number '));
    disp(sprintf('%d in the format [x y z]...',index));
    Q(index,1:3) = input('>  ');
    disp(sprintf('Enter the charge value of charge '));
    disp(sprintf('number %d...',index));
    Q(index,4) = input('>  ');
end

% add the E-field at the observation point due to all charges
Etotal=0;   % set the initial total field sum to zero
for index=1:n,
  rtemp=r-Q(index,1:3);
  rtemp_unitvector = rtemp/norm(rtemp);
```

```
% electric field due to a charge
  Etemp=Q(index,4)/(4*pi*8.86e-12*(norm(rtemp))^2)*rtemp_unit-
vector;
  Etotal=Etemp+Etotal; % add the partial sums to the total
end
% Display output
disp(sprintf('The total electric field at point [x y z] '))
disp(sprintf(' = [%d %d %d] is ', r(1), r(2), r(3)))
Etotal
```

MATLAB 4.2

```
% This script computes the results of example 4.5 by
% numerical integration
clear
format short    % prints only 6 decimal places

% Part (a) Computation of total charge
% prompt user for the integration increment used by the
% summation loop
dx = input('Enter the integration increment dx... \n > ');
dy = input('Enter the integration increment dy... \n > ');

% perform the double integral
total_charge=0;
for x=0:dx:1,  % loop through 0 <= x <= 1
    for y=0:dy:1,   % loop through 0 <= x <= 1
        total_charge=rho_s_fun(x,y)*dx*dy+total_charge;
    end
end
% Display results
disp(sprintf('The total charge computed by manual double'));
disp(sprintf(' integration is %d nC',total_charge))
% Double integral evaluation using the built-in
% function dblquad
% The user must write the integrand function as a
% separate file, in this case rho_s_fun
total_charge=dblquad(@rho_s_fun, 0, 1, 0, 1);
disp(sprintf('The total charge computed by the functional'))
disp(sprintf(' double integration is %d nC',total_charge))
% Part (b) Evaluation of electric field
% perform the double integral
Etotal=0;   % initial sum
for x=0:dx:1,
    for y=0:dy:1,
        rminusrprime=[-x -y 5];   % | r - rprime |
        numerator=rho_s_fun(x,y)*1e-9*dx*dy*rminusrprime;
            % the 1e-9 is because the charge is in nC
```

```
          denominator=4*pi*1e-9/(36*pi)*norm(rminusrprime)^3;
              %  |  r - rprime  |  ^ 3
          Etotal=Etotal+numerator/denominator;
              % add partial sum to initial sum
      end
end
% Display results
disp(sprintf('The electric field computed by manual double'));
disp(sprintf(' integration is (%d, %d, %d) V/m',...
    Etotal(1), Etotal(2), Etotal(3)))
```

SUMMARY

1. The two fundamental laws for electrostatic fields (Coulomb's and Gauss's) are presented in this chapter. Coulomb's law of force states that

$$\mathbf{F} = \frac{Q_1 Q_2}{4\pi\varepsilon_o R^2}\,\mathbf{a}_R$$

2. Based on Coulomb's law, we define the electric field intensity $\mathbf{E}$ as the force per unit charge; that is,

$$\mathbf{E} = \frac{Q}{4\pi\varepsilon_o R^2}\,\mathbf{a}_R = \frac{Q\,\mathbf{R}}{4\pi\varepsilon_o R^3} \quad \text{(point charge only)}$$

3. For a continuous charge distribution, the total charge is given by

$$Q = \int_L \rho_L \, dl \qquad \text{for line charge}$$

$$Q = \int_S \rho_S \, dS \qquad \text{for surface charge}$$

$$Q = \int_v \rho_v \, dv \qquad \text{for volume charge}$$

The $\mathbf{E}$ field due to a continuous charge distribution is obtained from the formula for point charge by replacing Q with $dQ = \rho_L \, dl$, $dQ = \rho_S \, dS$ or $dQ = \rho_v \, dv$ and integrating over the line, surface, or volume, respectively.

4. For an infinite line charge,

$$\mathbf{E} = \frac{\rho_L}{2\pi\varepsilon_o \rho}\,\mathbf{a}_\rho$$

and for an infinite sheet of charge,

$$\mathbf{E} = \frac{\rho_S}{2\varepsilon_o} \mathbf{a}_n$$

5. The electric flux density $\mathbf{D}$ is related to the electric field intensity (in free space) as

$$\mathbf{D} = \varepsilon_o \mathbf{E}$$

The electric flux through a surface S is

$$\Psi = \int_S \mathbf{D} \cdot d\mathbf{S}$$

6. Gauss's law states that the net electric flux penetrating a closed surface is equal to the total charge enclosed, that is, $\Psi = Q_{enc}$. Hence,

$$\Psi = \oint_S \mathbf{D} \cdot d\mathbf{S} = Q_{enc} = \int_v \rho_v \, dv$$

or

$$\rho_v = \nabla \cdot \mathbf{D} \quad \text{(first Maxwell equation to be derived)}$$

When charge distribution is symmetric, so that a Gaussian surface (where $\mathbf{D} = D_n \mathbf{a}_n$ is constant) can be found, Gauss's law is useful in determining $\mathbf{D}$; that is,

$$D_n \oint_S dS = Q_{enc} \quad \text{or} \quad D_n = \frac{Q_{enc}}{S}$$

7. The total work done, or the electric potential energy, to move a point charge Q from point A to B in an electric field $\mathbf{E}$ is

$$W = -Q \int_A^B \mathbf{E} \cdot d\mathbf{l}$$

8. The potential at $\mathbf{r}$ due to a point charge Q at $\mathbf{r}'$ is

$$V(\mathbf{r}) = \frac{Q}{4\pi\varepsilon_o |\mathbf{r} - \mathbf{r}'|} + C$$

where C is evaluated at a given reference potential point; for example, $C = 0$ if $V(\mathbf{r} \to \infty) = 0$. To determine the potential due to a continuous charge distribution, we replace Q in the formula for point charge by $dQ = \rho_L \, dl$, $dQ = \rho_S \, dS$, or $dQ = \rho_v \, dv$ and integrate over the line, surface, or volume, respectively.

9. If the charge distribution is not known, but the field intensity **E** is given, we find the potential by using

$$V = -\int_L \mathbf{E} \cdot d\mathbf{l} + C$$

10. The potential difference V_{AB}, the potential at B with reference to A, is

$$V_{AB} = -\int_A^B \mathbf{E} \cdot d\mathbf{l} = \frac{W}{Q} = V_B - V_A$$

11. Since an electrostatic field is conservative (the net work done along a closed path in a static **E** field is zero),

$$\oint_L \mathbf{E} \cdot d\mathbf{l} = 0$$

or

$$\nabla \times \mathbf{E} = \mathbf{0} \quad (\text{second Maxwell equation to be derived})$$

12. Given the potential field, the corresponding electric field is found by using

$$\mathbf{E} = -\nabla V$$

13. For an electric dipole centered at $\mathbf{r}'$ with dipole moment $\mathbf{p}$, the potential at $\mathbf{r}$ is given by

$$V(\mathbf{r}) = \frac{\mathbf{p} \cdot (\mathbf{r} - \mathbf{r}')}{4\pi\varepsilon_0 |\mathbf{r} - \mathbf{r}'|^3}$$

14. The flux density **D** is tangential to the electric flux lines at every point. An equipotential surface (or line) is one on which $V = \text{constant}$. At every point, the equipotential line is orthogonal to the electric flux line.

15. The electrostatic energy due to n point charges is

$$W_E = \frac{1}{2} \sum_{k=1}^{n} Q_k V_k$$

For a continuous volume charge distribution,

$$W_E = \frac{1}{2} \int_v \mathbf{D} \cdot \mathbf{E} \, dv = \frac{1}{2} \int_v \varepsilon_0 |\mathbf{E}|^2 \, dv$$

16. Electrostatic discharge (ESD) refers to the sudden transfer of static charge between objects at different electrostatic potentials. Since all semiconductor devices are regarded as ESD sensitive, a good understandng of ESD is required in industry.

4.1 Point charges $Q_1 = 1$ nC and $Q_2 = 2$ nC are at a distance apart. Which of the following statements are incorrect?

(a) The force on Q_1 is repulsive.

(b) The force on Q_2 is the same in magnitude as that on Q_1.

(c) As the distance between them decreases, the force on Q_1 increases linearly.

(d) The force on Q_2 is along the line joining them.

(e) A point charge $Q_3 = -3$ nC located at the midpoint between Q_1 and Q_2 experiences no net force.

4.2 Plane $z = 10$ m carries charge 20 nC/m^2. The electric field intensity at the origin is

(a) $-10 \, \mathbf{a}_z$ V/m

(b) $-18\pi \, \mathbf{a}_z$ V/m

(c) $-72\pi \, \mathbf{a}_z$ V/m

(d) $-360\pi \, \mathbf{a}_z$ V/m

4.3 Point charges 30 nC, -20 nC, and 10 nC are located at $(-1, 0, 2)$, $(0, 0, 0)$, and $(1, 5, -1)$, respectively. The total flux leaving a cube of side 6 m centered at the origin is

(a) -20 nC

(b) 10 nC

(c) 20 nC

(d) 30 nC

(e) 60 nC

4.4 The electric flux density on a spherical surface $r = b$ is the same for a point charge Q located at the origin and for charge Q uniformly distributed on surface $r = a(a < b)$.

(a) Yes

(b) No

(c) Not necessarily

4.5 The work done by the force $\mathbf{F} = 4\mathbf{a}_x - 3\mathbf{a}_y + 2\mathbf{a}_z$ N in giving a 1 nC charge a displacement of $10\mathbf{a}_x + 2\mathbf{a}_y - 7\mathbf{a}_z$ m is

(a) 103 nJ

(b) 60 nJ

(c) 64 nJ

(d) 20 nJ

4.6 By saying that the electrostatic field is conservative, we do *not* mean that

(a) It is the gradient of a scalar potential.

(b) Its circulation is identically zero.

(c) Its curl is identically zero.

(d) The work done in a closed path inside the field is zero.

(e) The potential difference between any two points is zero.

4.7 Suppose a uniform electric field exists in the room in which you are working, such that the lines of force are horizontal and at right angles to one wall. As you walk toward the wall from which the lines of force emerge into the room, are you walking toward

(a) points of higher potential?

(b) points of lower potential?

(c) points of the same potential (equipotential line)?

4.8 A charge Q is uniformly distributed throughout a sphere of radius a. Taking the potential at infinity as zero, the potential at $r = b < a$ is

(a) $-\int_\infty^b \dfrac{Qr}{4\pi\varepsilon_o a^3}\, dr$

(b) $-\int_\infty^b \dfrac{Q}{4\pi\varepsilon_o r^2}\, dr$

(c) $-\int_\infty^a \dfrac{Q}{4\pi\varepsilon_o r^2}\, dr - \int_a^b \dfrac{Qr}{4\pi\varepsilon_o a^3}\, dr$

(d) $-\int_\infty^a \dfrac{Q}{4\pi\varepsilon_o r^3}\, dr$

4.9 A potential field is given by $V = 3x^2 y - yz$. Which of the following is not true?

(a) At point $(1, 0, -1)$, V and $\mathbf{E}$ vanish.

(b) $x^2 y = 1$ is an equipotential line on the xy-plane.

(c) The equipotential surface $V = -8$ passes through point $P+2, -1, 4)$.

(d) The electric field at P is $12\mathbf{a}_x - 8\mathbf{a}_y - \mathbf{a}_z$ V/m.

(e) A unit normal to the equipotential surface $V = -8$ at P is $-0.83\mathbf{a}_x + 0.55\mathbf{a}_y + 0.07\mathbf{a}_z$.

4.10 An electric potential field is produced by point charges 1 μC and 4 μC located at $(-2, 1, 5)$ and $(1, 3, -1)$, respectively. The energy stored in the field is

(a) 2.57 mJ

(b) 5.14 mJ

(c) 10.28 mJ

(d) None of the above

Answers: 4.1c,e, 4.2d, 4.3b, 4.4a, 4.5d, 4.6e, 4.7a, 4.8c, 4.9a, 4.10b.

PROBLEMS

Section 4.2—Coulomb's Law and Field Intensity

4.1 Point charges $Q_1 = 5\,\mu$C and $Q_2 = -4\,\mu$C are placed at $(3, 2, 1)$ and $(-4, 0, 6)$, respectively. Determine the force on Q_1.

4.2 Point charges Q_1 and Q_2 are, respectively, located at $(4, 0, -3)$ and $(2, 0, 1)$. If $Q_2 = 4$ nC, find Q_1 such that

(a) The $\mathbf{E}$ at $(5, 0, 6)$ has no z-component.

(b) The force on a test charge at $(5, 0, 6)$ has no x-component.

4.3 A point Q is located at (a, 0, 0), while another charge $-Q$ is at (−a, 0, 0). Find E at:
(a) (0, 0, 0) (b) (0, a, 0), (c) (a, 0, a).

4.4 Determine the electric field intensity required to levitate a body 2 kg in mass and charged with −4 mC.

Section 4.3—Electric Fields due to Continuous Charge Distributions

4.5 Determine the total charge

(a) On line $0 < x < 5$ m if $\rho_L = 12x^2$ mC/m

(b) On the cylinder $\rho = 3, 0 < z < 4$ m if $\rho_S = \rho z^2$ nC/m²

(c) Within the sphere $r = 4$ m if $\rho_v = \dfrac{10}{r \sin \theta}$ C/m³

4.6 A cube is defined by $0 < x < a, 0 < y < a$, and $0 < z < a$. If it is charged with $\rho_v = \dfrac{\rho_o x}{a}$, where ρ_o is a constant, calculate the total charge in the cube.

4.7 A volume charge with density $\rho_v = 5\rho^2 z$ mC/m³ exists in a region defined by $0 < \rho < 2, 0 < z < 1, 30° < \phi < 90°$. Calculate the total charge in the region.

4.8 Given that $\rho_s = 6xy$ C/m², calculate the total charge on the triangular region in Figure 4.25.

4.9 A wedge-shaped surface has its corners located at (0, 0, 4), (2, 0, 4), and (2, 3, 4). If the surface has charge distribution with $\rho_s = 10x^2 yz$ mC/m², find the total charge on the surface.

4.10 Given that $\rho_v = 4\rho^2 z \cos \phi$ nC/m³, find the total charge contained in a wedge defined by $0 < \rho < 2, 0 < \phi < \pi/4, 0 < z < 1$.

4.11 Line $0 < x < 1$ m is charged with density $12x^2$ nC/m. (a) Find the total charge. (b) Determine the electric field intensity at (0, 0, 1000 m).

4.12 A uniform charge 12 μC/m is formed on a loop described by $x^2 + y^2 = 9$ on the $z = 0$ plane. Determine the force exerted on a 4 mC point charge at (0, 0, 4).

4.13 An annular disk of inner radius a and outer radius b is placed on the xy-plane and centered at the origin. If the disk carries uniform charge with density ρ_s, find E at (0, 0, h).

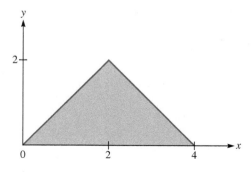

FIGURE 4.25 For Problem 4.8.

4.14 (a) An infinite sheet at $z = 0$ has a uniform charge density 12 nC/m². Find E on both sides of the planar sheet.

(b) A second sheet at $z = 4$ has a uniform charge density of -12 nC/m². Show that E exists only between the planar sheets. Find E.

4.15 Plane $x + 2y = 5$ carries charge $\rho_S = 6$ nC/m². Determine E at $(-1, 0, 1)$.

4.16 Plane $x = 0$ has a uniform charge density ρ_s, while plane $x = a$ has $-\rho_s$. Determine the electric field intensity in regions (a) $x < 0$, (b) $0 < x < a$, (c) $x > a$.

4.17 Three surface charge distributions are located in free space as follows: 10 μC/m² at $x = 2$, -20 μC/m² at $y = -3$, and 30 μC/m² at $z = 5$. Determine E at (a) $P(5, -1, 4)$, (b) $R(0, -2, 1)$, (c) $Q(3, -4, 10)$.

4.18 The gravitation force between two bodies of masses m_1 and m_2 is

$$\mathbf{F}_g = \frac{Gm_1m_2}{r^2}\mathbf{a}_r$$

where $G = 6.67 \times 10^{-11}$ N(m/kg)². Find the ratio of the electrostatic and gravitational forces between two electrons.

Section 4.4—Electric Flux Density

*4.19 State Gauss's law. Deduce Coulomb's law from Gauss's law, thereby affirming that Gauss's law is an alternative statement of Coulomb's law and that Coulomb's law is implicit in Maxwell's equation $\nabla \cdot \mathbf{D} = \rho_v$.

4.20 Three point charges are located in the $z = 0$ plane: a charge $+Q$ at point $(-1, 0)$, a charge $+Q$ at point $(1, 0)$, and a charge $-2Q$ at point $(0, 1)$. Determine the electric flux density at $(0, 0)$.

4.21 A ring placed along $y^2 + z^2 = 4$, $x = 0$ carries a uniform charge of 5 μC/m.

(a) Find $\mathbf{D}$ at $P(3, 0, 0)$.

(b) If two identical point charges Q are placed at $(0, -3, 0)$ and $(0, 3, 0)$ in addition to the ring, find the value of Q such that $\mathbf{D} = 0$ at P.

4.22 The electric flux density in free space is given by $\mathbf{D} = y^2\mathbf{a}_x + 2xy\mathbf{a}_y - 4za_z$ nC/m²(a). Find the volume charge density. (b) Determine the flux through surface $x = 3$, $0 < y < 6$, $0 < z < 5$.

4.23 In free space, $\mathbf{E} = 12\rho z\cos\phi\mathbf{a}_\rho - 6\rho z\sin\phi\mathbf{a}_\phi + 6\rho^2\cos\phi\mathbf{a}_z$ V/m. Find the electric flux through surface $\phi = 90°$ $0 < \rho < 2$, $0 < z < 5$.

4.24 If $\mathbf{D} = \sin\theta\sin\phi\mathbf{a}_r + \cos\theta\sin\phi\mathbf{a}_\theta + \cos\phi\mathbf{a}_\phi$ nC/m², find: (a) the charge density at $(2, 30°, 60°)$, (b) the flux through $r = 2$, $0 < \theta < 30°$, $0 < \phi < 60°$.

Sections 4.5 and 4.6—Gauss's Law and Applications

4.25   Determine the charge density due to each of the following electric flux densities:

(a) $\mathbf{D} = 8xy\mathbf{a}_x + 4x^2\mathbf{a}_y$ C/m^2

(b) $\mathbf{D} = 4\rho \sin \phi\, \mathbf{a}_\rho + 2\rho \cos \phi\, \mathbf{a}_\phi + 2z^2\mathbf{a}_z$ C/m^2

(c) $\mathbf{D} = \dfrac{2 \cos \theta}{r^3}\mathbf{a}_r + \dfrac{\sin \theta}{r^3}\mathbf{a}_\theta$ C/m^2

4.26   A cube with 2 m sides ($0 < x, y, z < 2$ m) carries a charge with density $\rho_v = 12xyz$ mC/m^3. (a) Calculate the total charge. (b) Find the total outward flux from the cube.

4.27   If spherical surfaces $r = 1$ m and $r = 2$ m, respectively, carry uniform surface charge densities 8 nC/m^2 and -6 mC/m^2, find $\mathbf{D}$ at $r = 3$ m.

4.28   A sphere of radius a is centered at the origin. If $\rho_v = \begin{cases} 5r^{1/2}, & 0 < r < a \\ 0, & \text{otherwise} \end{cases}$. Determine E everywhere.

4.29   Let $\mathbf{D} = 2xy\mathbf{a}_x + x^2\mathbf{a}_y$ C/m^2 and find

(a) The volume charge density ρ_v.

(b) The flux through surface $0 < x < 1, 0 < z < 1, y = 1$.

(c) The total charge contained in the region $0 < x < 1, 0 < y < 1, 0 < z < 1$.

4.30   In a certain region, the electric field is given by

$$\mathbf{D} = 2\rho(z + 1)\cos \phi\, \mathbf{a}_\rho - \rho(z + 1)\sin \phi\, \mathbf{a}_\phi + \rho^2 \cos \phi\, \mathbf{a}_z\ \mu\text{C/m}^2$$

(a) Find the charge density.

(b) Calculate the total charge enclosed by the volume $0 < \rho < 2$, $0 < \phi < \pi/2$, $0 < z < 4$.

(c) Confirm Gauss's law by finding the net flux through the surface of the volume in (b).

*4.31   The Thomson model of a hydrogen atom is a sphere of positive charge with an electron (a point charge) at its center. The total positive charge equals the electronic charge e. Prove that when the electron is at a distance r from the center of the sphere of positive charge, it is attracted with a force

$$F = \frac{e^2 r}{4\pi\varepsilon_o R^3}$$

where R is the radius of the sphere.

4.32   A long coaxial cable has an inner conductor with radius a and outer conductor with radius b. If the inner conductor has $\rho_s = \rho_o/\rho$, where ρ_o is a constant, determine E everywhere.

4.33   Let

$$\rho_v = \begin{cases} \dfrac{10}{r^2}\,\text{mC/m}^3, & 1 < r < 4 \\ 0, & r > 4 \end{cases}$$

(a) Find the net flux crossing surface $r = 2$ m and $r = 6$ m.

(b) Determine **D** at $r = 1$ m and $r = 5$ m.

4.34   A spherical region of radius a has total charge Q. If the charge is uniformly distributed, apply Gauss's law to find **D** both inside and outside the sphere.

Sections 4.7 and 4.8—Electric Potential and Relationship with E

4.35   Two point charges $Q = 2$ nC and $Q = -4$ nC are located at $(1, 0, 3)$ and $(-2, 1, 5)$, respectively. Determine the potential at $P(1, -2, 3)$.

4.36   A charge of 8 nC is placed at each of the four corners of a square of sides 4 cm long. Calculate the electrical potential at the point 3 cm above the center of the square.

4.37   (a) A total charge $Q = 60\ \mu$C is split into two equal charges located at 180° intervals around a circular loop of radius 4 m. Find the potential at the center of the loop.

(b) If Q is split into three equal charges spaced at 120° intervals around the loop, find the potential at the center.

(c) If in the limit $\rho_L = \dfrac{Q}{8\pi}$, find the potential at the center.

4.38   Three point charges $Q_1 = 1$ mC, $Q_2 = -2$ mC, and $Q_3 = 3$ mC are, respectively, located at $(0, 0, 4)$, $(-2, 5, 1)$, and $(3, -4, 6)$.

(a) Find the potential V_P at $P(-1, 1, 2)$.

(b) Calculate the potential difference V_{PQ} if Q is $(1, 2, 3)$.

4.39   The potential distribution in free space is given by

$$V = \rho^2 e^{-z} \sin \phi \text{ V}$$

Calculate the electric field strength at $(4, \pi/4, -1)$.

4.40 $V = x^2 y(z + 3)$ V. Find

(a) **E** at $(3, 4, -6)$

(b) the charge within the cube $0 < x < 1, 0 < y < 1, 0 < z < 1$.

4.41   The volume charge density inside an atomic nucleus of radius a is $\rho_v = \rho_o\left(\dfrac{1 - r^2}{a^2}\right)$ where ρ_o is a constant.

(a) Calculate the total charge.

(b) Determine **E** and V outside the nucleus.

(c) Determine **E** and V inside the nucleus.

(d) Prove that **E** is maximum at $r = 0.745a$.

4.42 Let charge Q be uniformly distributed on a circular ring defined by $a < \rho < b$ and shown in Figure 4.26. Find $\mathbf{D}$ at $(0, 0, h)$.

4.43 If $\mathbf{D} = 4x\mathbf{a}_x - 10y^2\mathbf{a}_y + z^2\mathbf{a}_z$ C/m^2, find the charge density at $P(1, 2, 3)$.

4.44 A 10 nC charge is uniformly distributed over a spherical shell $r = 3$ cm, and a -5 nC charge is uniformly distributed over another spherical shell $r = 5$ cm. Find $\mathbf{D}$ for regions $r < 3$ cm, 3 cm $< r <$ 5 cm, $r > 5$ cm.

4.45 In free space, an electric field is given by

$$\mathbf{E} = \begin{cases} E_0(\rho/a)\mathbf{a}_\rho, & 0 < \rho < a \\ 0, & \text{otherwise} \end{cases}$$

Calculate the volume charge density.

4.46 The electric field intensity in free space is given by

$$\mathbf{E} = 2\,xyz\mathbf{a}_x + x^2z\mathbf{a}_y + x^2y\mathbf{a}_z \text{ V/m}$$

Calculate the amount of work necessary to move a 2 μC charge from $(2, 1, -1)$ to $(5, 1, 2)$.

4.47 Given that $\mathbf{E} = 12\rho z\cos\phi\mathbf{a}_\rho - 6\rho z\sin\phi\mathbf{a}_\phi + 6\rho^2\cos\phi\mathbf{a}_z$ (a) find the volume charge density at $A(2, 180°, -1)$, (b) calculate the work done in transferring a 10 μC charge from A to $B(2, 0°, -1)$.

4.48 In an electric field $\mathbf{E} = 20r\sin\theta\,\mathbf{a}_r + 10r\cos\theta\,\mathbf{a}_\theta$ V/m, calculate the energy expended in transferring a 10 nC charge

(a) From $A(5, 30°, 0°)$ to $B(5, 90°, 0°)$
(b) From A to $C(10, 30°, 0°)$
(c) From A to $D(5, 30°, 60°)$
(d) From A to $E(10, 90°, 60°)$

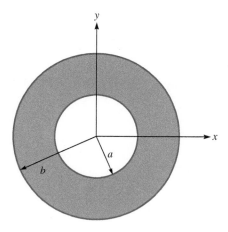

FIGURE 4.26 For Problem 4.42.

4.49 Let $\mathbf{E} = \dfrac{10}{r^2}\mathbf{a}_z$ V/m. Find V_{AB}, where A is $(1, \pi/4, \pi/2)$ and B is $(5, \pi, 0)$.

4.50 In free space, $\mathbf{E} = 20x\mathbf{a}_x + 40y\mathbf{a}_y - 10z\mathbf{a}_z$ V/m. Calculate the work done in transferring a 2 mC charge along the arc $\rho = 2, 0 < \phi < \pi/2$ in the $z = 0$ plane.

4.51 A sheet of charge with density $\rho_s = 40$ nC/m² occupies the $x = 0$ plane. Determine the work done in moving a 10 μC charge from point $A(3, 4, -1)$ to point $B(1, 2, 6)$.

4.52 For each of the following potential distributions, find the electric field intensity and the volume charge distribution:

(a) $V = 2x^2 + 4y^2$

(b) $V = 10\rho^2 \sin \phi + 6\rho z$

(c) $V = 5r^2 \cos \theta \sin \phi$

4.53 Let $V = \rho e^{-z}\sin \phi$. (a) Find $\mathbf{E}$. (b) Show that $\mathbf{E}$ is conservative.

4.54 In free space, $V = \dfrac{1}{r^3} \sin \theta \cos \phi$. Find $\mathbf{D}$ at $(1, 30°, 60°)$.

4.55 Each of two concentric spherical shells has inner radius a and outer radius b. If the inner shell carries charge Q, while the outer shell carries charge $-Q$, determine the potential difference V_{ab} between the shells.

4.56 A uniform surface charge with density ρ_s exists on a hemispherical surface with $r = a$ and $\theta \le \pi/2$. Calculate the electric potential at the center.

4.57 If $\mathbf{D} = 2\rho \sin \phi \mathbf{a}_\rho - \dfrac{\cos \phi}{2\rho}\mathbf{a}_\phi$ C/m², determine whether $\mathbf{D}$ is a genuine electric flux density. Determine the flux crossing $\rho = 1, 0 \le \phi \le \pi/4, 0 < z < 1$.

*4.58 (a) Prove that when a particle of constant mass and charge is accelerated from rest in an electric field, its final velocity is proportional to the square root of the potential difference through which it is accelerated.

(b) Find the magnitude of the proportionality constant if the particle is an electron.

(c) Through what voltage must an electron be accelerated, assuming no change in its mass, to require a velocity one-tenth that of light? (At such velocities, the mass of a body becomes appreciably larger than its "rest mass" and cannot be considered constant.)

*4.59 An electron is projected with an initial velocity $u_o = 10^7$ m/s into the uniform field between the parallel plates of Figure 4.27. It enters the field midway between the plates. If the electron just misses the upper plate as it emerges from the field,

(a) find the electric field intensity.

(b) calculate the electron's velocity as it emerges from the field. Neglect edge effects.

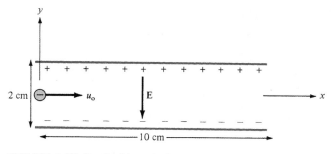

FIGURE 4.27 For Problem 4.59.

Section 4.9—Electric Dipole and Flux Lines

4.60   An electric dipole with $\mathbf{p} = pa_z\ \text{C}\cdot\text{m}$ is placed at $(x, z) = (0, 0)$. If the potential at $(0, 1\ \text{nm})$ is 9 V, find the potential at $(1\ \text{nm}, 1\ \text{nm})$.

4.61   Point charges Q and $-Q$ are located at $(0, d/2, 0)$ and $(0, -d/2, 0)$. Show that at point (r, θ, ϕ), where $r \gg d$,

$$V = \frac{Qd \sin\theta \sin\phi}{4\pi\varepsilon_o r^2}$$

Find the corresponding **E** field.

4.62   A dipole has dipole moment $p = 2\mathbf{a}_x + 6\mathbf{a}_y - 4\mathbf{a}_z\ \mu\text{C}\cdot\text{m}$. If the dipole is located in free space at $(2, 3, -1)$, find the potential at $(4, 0, 1)$.

4.63   A z-directed dipole has E in eq. (4.82). Determine the values of θ that will make **E** have no z-component.

Section 4.10—Energy Density

4.64   Determine the amount of work needed to transfer two charges of 40 nC and –50 nC from infinity to locations (0, 0, 1) and (2, 0, 0), respectively.

4.65   If $V = 2x^2 + 6y^2$ V in free space, find the energy stored in a volume defined by $-1 \le x \le 1$, $-1 \le y \le 1$, and $-1 \le z \le 1$.

4.66   Find the energy stored in the hemispherical region $r \le 2$ m, $0 < \theta < \pi, 0 < \phi < \pi$ where

$$\mathbf{E} = 2r \sin\theta \cos\phi\ \mathbf{a}_r + r\cos\theta \cos\phi\ \mathbf{a}_\theta - r\sin\phi\ \mathbf{a}_\phi\ \text{V/m}$$

exists.

4.67   A spherical conductor of radius a carries a surface charge with density ρ_o. Determine the potential energy in terms of a.

4.68   In free space, $\mathbf{E} = y^2\mathbf{a}_x + 2xy\mathbf{a}_y - 4z\mathbf{a}_z$ V/m. Determine the energy stored in the region defined by $0 < x < 2$, $-1 < y < 1$, $0 < z < 4$.

4.69   In free space, $V = \rho e^{-z} \sin\phi$. (a) Find **E**. (b) Determine the energy stored in the region $0 < \rho < 1, 0 < \phi < 2\pi, 0 < z < 2$.

CAREERS IN ELECTROMAGNETICS

Electromagnetics is the branch of electrical engineering (or physics) that deals with the analysis and application of electric and magnetic fields. It is necessary for the understanding of all forms of light. Without an understanding of electromagnetics, there would be no radios, televisions, telephones, computers, or CD players.

The principles of electromagnetics (EM) are applied in various allied disciplines, such as electric machines, electromechanical energy conversion, radar meteorology, remote sensing, satellite communications, bioelectromagnetics, electromagnetic interference and compatibility, plasmas, and fiber optics. EM devices include electric motors and generators, transformers, electromagnets, magnetic levitation systems, antennas, radars, microwave ovens, microwave dishes, superconductors, and electrocardiograms. The design of these devices requires a thorough knowledge of the laws and principles of EM.

EM is regarded as one of the more difficult disciplines in electrical engineering. One reason is that EM phenomena are rather abstract. But those who enjoy working with mathematics and are able to visualize the invisible should consider careers in EM, since few electrical engineers specialize in this area. To specialize in EM, one should consider taking courses such as Antennas, Microwaves, Wave Propagation, Electromagnetic Compatibility, and Computational Electromagnetics. Electrical engineers who specialize in EM are needed in the microwave industry, radio/TV broadcasting stations, electromagnetic research laboratories, and the communications industry.

CHAPTER 5

ELECTRIC FIELDS IN MATERIAL SPACE

Knowledge will forever govern ignorance: and a people who mean to be their own Governors, must arm themselves with the power which knowledge gives.

—JAMES MADISON

5.1 INTRODUCTION

In the last chapter, we considered electrostatic fields in free space or a space that has no materials in it. Thus what we have developed so far under electrostatics may be regarded as the "vacuum" field theory. By the same token, what we shall develop in this chapter may be regarded as the theory of electric phenomena in material space. As will soon be evident, most of the formulas derived in Chapter 4 are still applicable, though some may require modification.

Just as electric fields can exist in free space, they can exist in material media. Materials are broadly classified in terms of their electrical properties as conductors and nonconductors. Nonconducting materials are usually referred to as *insulators* or *dielectrics*. A brief discussion of the electrical properties of materials in general will be given to provide a basis for understanding the concepts of conduction, electric current, and polarization. Further discussion will be on some properties of dielectric materials such as susceptibility, permittivity, linearity, isotropy, homogeneity, dielectric strength, and relaxation time. The concept of boundary conditions for electric fields existing in two different media will be introduced.

5.2 PROPERTIES OF MATERIALS

A discussion of the electrical properties of materials may seem out of place in a text of this kind. But questions such as why an electron does not leave a conductor surface, why a current-carrying wire remains uncharged, why materials behave differently in an electric field, and why waves travel with less speed in conductors than in dielectrics are easily answered by considering the electrical properties of materials. A thorough discussion of this subject is usually found in texts on physical electronics. Here, a brief discussion will suffice to help us understand the mechanism by which materials influence an electric field.

In a broad sense, materials may be classified in terms of their *conductivity* σ, in mhos per meter ($\mho$/m) or, more usually siemens per meter (S/m), as conductors and nonconductors, or technically as metals and insulators (or dielectrics). The conductivity of a material usually depends on temperature and frequency. A material with *high conductivity* ($\sigma \gg 1$) is referred to as a *metal*, whereas one with *low conductivity* ($\sigma \ll 1$) is referred to as an *insulator*. A material whose conductivity lies somewhere between those of metals and insulators is called a *semiconductor*. The values of conductivity of some common materials are shown in Table B.1 in Appendix B. From this table, it is clear that materials such as copper and aluminum are metals, silicon and germanium are semiconductors, and glass and rubber are insulators.

The conductivity of metals generally increases with decrease in temperature. At temperatures near absolute zero ($T = 0$ K), some conductors exhibit infinite conductivity and are called *superconductors*. Lead and aluminum are typical examples of such metals. The conductivity of lead at 4 K is of the order of 10^{20} S/m. The interested reader is referred to the literature on superconductivity.[1]

We shall be concerned only with metals and insulators in this text. Microscopically, the major difference between a metal and an insulator lies in the number of electrons available for conduction of current. Dielectric materials have few electrons available for conduction of current, whereas metals have an abundance of free electrons. Further discussion on the behavior of conductors and dielectrics in an electric field will be given in subsequent sections.

5.3 CONVECTION AND CONDUCTION CURRENTS

Electric voltage (or potential difference) and current are two fundamental quantities in electrical engineering. We considered potential in the last chapter. Before examining how the electric field behaves in a conductor or dielectric, it is appropriate to consider electric current. Electric current is generally caused by the motion of electric charges.

> The **current** (in amperes) through a given area is the electric charge passing through the area per unit time.

That is,

$$I = \frac{dQ}{dt} \tag{5.1}$$

Thus in a current of one ampere, charge is being transferred at a rate of one coulomb per second.

We now introduce the concept of *current density* **J**. If current ΔI flows through a planar surface ΔS, the current density is

$$\mathbf{J} = \frac{\Delta I}{\Delta S}$$

---

[1] The August 1989 issue of the *Proceedings of IEEE* was devoted to "Applications of Superconductivity."

or

$$\Delta I = J \Delta S \tag{5.2}$$

assuming that the current density is perpendicular to the surface. If the current density is not normal to the surface,

$$\Delta I = \mathbf{J} \cdot \Delta \mathbf{S} \tag{5.3}$$

Thus, the total current flowing through a surface S is

$$\boxed{I = \int_S \mathbf{J} \cdot d\mathbf{S}} \tag{5.4}$$

Depending on how I is produced, there are different kinds of current density: convection current density, conduction current density, and displacement current density. We will consider convection and conduction current densities here; displacement current density will be considered in Chapter 9. What we need to keep in mind is that eq. (5.4) applies to any kind of current density. Compared with the general definition of flux in eq. (3.13), eq. (5.4) shows that the current I through S is merely the flux of the current density $\mathbf{J}$.

CASE A: CONVECTION CURRENT

Convection current, as distinct from conduction current, does not involve conductors and consequently does not satisfy Ohm's law. It occurs when current flows through an insulating medium such as liquid, rarefied gas, or a vacuum. A beam of electrons in a vacuum tube, for example, is a convection current.

Consider a filament of Figure 5.1. If there is a flow of charge, of density ρ_v, at velocity $\mathbf{u} = u_y \mathbf{a}_y$, from eq. (5.1), the current through the filament is

$$\Delta I = \frac{\Delta Q}{\Delta t} = \rho_v \Delta S \frac{\Delta y}{\Delta t} = \rho_v \Delta S\, u_y \tag{5.5}$$

The **current density** at a given point is the current through a unit normal area at that point.

The y-directed current density J_y is given by

$$J_y = \frac{\Delta I}{\Delta S} = \rho_v u_y \tag{5.6}$$

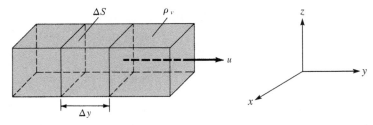

FIGURE 5.1 Current in a filament.

Hence, in general

$$\boxed{\mathbf{J} = \rho_v \mathbf{u}}$$

(5.7)

The current I is the *convection current* and J is the *convection current density* in amperes per square meter (A/m^2).

CASE B: CONDUCTION CURRENT

Conduction current requires a conductor. A conductor is characterized by a large number of free electrons that provide conduction current due to an impressed electric field. When an electric field $\mathbf{E}$ is applied, the force on an electron with charge $-e$ is

$$\mathbf{F} = -e\mathbf{E}$$

(5.8)

Since the electron is not in free space, it will not experience an average acceleration under the influence of the electric field. Rather, it suffers constant collisions with the atomic lattice and drifts from one atom to another. If an electron with mass m is moving in an electric field $\mathbf{E}$ with an average drift velocity $\mathbf{u}$, according to Newton's law, the average change in momentum of the free electron must match the applied force. Thus,

$$\frac{m\mathbf{u}}{\tau} = -e\mathbf{E}$$

(5.9a)

or

$$\mathbf{u} = -\frac{e\tau}{m}\mathbf{E}$$

(5.9b)

where τ is the average time interval between collisions. This indicates that the drift velocity of the electron is directly proportional to the applied field. If there are n electrons per unit volume, the electronic charge density is given by

$$\rho_v = -ne$$

(5.10)

Thus the *conduction current density* is

$$\mathbf{J} = \rho_v\mathbf{u} = \frac{ne^2\tau}{m}\mathbf{E} = \sigma\mathbf{E}$$

or

$$\boxed{\mathbf{J} = \sigma\mathbf{E}}$$

(5.11)

where $\sigma = ne^2\tau/m$ is the conductivity of the conductor. As mentioned earlier, the values of σ for common materials are provided in Table B.1 in Appendix B. The relationship in eq. (5.11) is known as the point form of *Ohm's law*.

5.4 CONDUCTORS

A conductor has an abundance of charge that is free to move. We will consider two cases involving a conductor.

CASE A: ISOLATED CONDUCTOR

Consider an isolated conductor, such as shown in Figure 5.2(a). When an external electric field $\mathbf{E}_e$ is applied, the positive free charges are pushed along the same direction as the applied field, while the negative free charges move in the opposite direction. This charge migration takes place very quickly. The free charges do two things. First, they accumulate on the surface of the conductor and form an *induced surface charge*. Second, the induced charges set up an internal induced field $\mathbf{E}_i$, which cancels the externally applied field $\mathbf{E}_e$. The result is illustrated in Figure 5.2(b). This leads to an important property of a conductor:

> A **perfect conductor** ($\sigma = \infty$) cannot contain an electrostatic field within it.

A conductor is called an *equipotential* body, implying that the potential is the same everywhere in the conductor. This is based on the fact that $\mathbf{E} = -\nabla V = \mathbf{0}$.

Another way of looking at this is to consider Ohm's law, $\mathbf{J} = \sigma \mathbf{E}$. To maintain a finite current density $\mathbf{J}$, in a perfect conductor ($\sigma \rightarrow \infty$), requires that the electric field inside the conductor $\sigma = \infty$ vanish. In other words, $\mathbf{E} \rightarrow \mathbf{0}$ because $\sigma \rightarrow \infty$ in a perfect conductor. If some charges are introduced in the interior of such a conductor, the charges will move to the surface and redistribute themselves quickly in such a manner that the field inside the conductor vanishes. According to Gauss's law, if $\mathbf{E} = \mathbf{0}$, the charge density ρ_v must be zero. We conclude again that a perfect conductor cannot contain an electrostatic field within it. Under static conditions,

$$\boxed{\mathbf{E} = \mathbf{0}, \ \rho_v = 0, \ V_{ab} = 0 \text{ inside a conductor}} \tag{5.12}$$

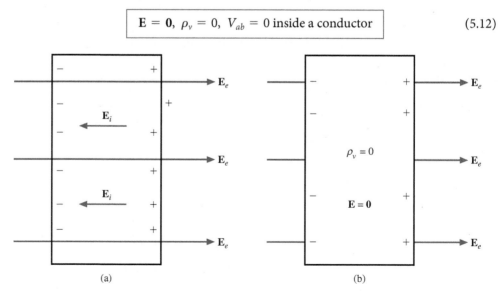

(a)   (b)

FIGURE 5.2 (**a**) An isolated conductor under the influence of an applied field. (**b**) A conductor has zero electric field under static conditions.

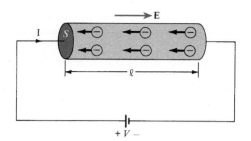

FIGURE 5.3 A conductor of uniform cross section under an applied **E** field.

where V_{ab} is the potential difference between points a and b in the conductor. This implies that a conductor is equipotential medium since the electric potential is the same at every point.

CASE B: CONDUCTOR MAINTAINED AT A POTENTIAL

We now consider a conductor whose ends are maintained at a potential difference V, as shown in Figure 5.3. Note that in this case, $\mathbf{E} \neq 0$ inside the conductor, as in Figure 5.2. What is the difference? There is no static equilibrium in Figure 5.3, since the conductor is not isolated but is wired to a source of electromotive force, which compels the free charges to move and prevents the eventual establishment of electrostatic equilibrium. Thus in the case of Figure 5.3, an electric field must exist inside the conductor to sustain the flow of current. As the electrons move, they encounter some damping forces called *resistance*. Based on Ohm's law in eq. (5.11), we will derive the resistance of the conducting material. Suppose the conductor has a *uniform* cross section of area S and is of length ℓ. The direction of the electric field $\mathbf{E}$ produced is the same as the direction of the flow of positive charges or current I. This direction is opposite to the direction of the flow of electrons. The electric field applied is uniform, and its magnitude is given by

$$E = \frac{V}{\ell} \tag{5.13}$$

Since the conductor has a uniform cross section,

$$J = \frac{I}{S} \tag{5.14}$$

Substituting eqs. (5.11) and (5.13) into eq. (5.14) gives

$$\frac{I}{S} = \sigma E = \frac{\sigma V}{\ell} \tag{5.15}$$

Hence,

$$\boxed{R = \frac{V}{I} = \frac{\ell}{\sigma S}} \tag{5.16}$$

or

$$R = \frac{\rho_c \ell}{S}$$

where $\rho_c = 1/\sigma$ is the *resistivity* of the material. Equation (5.16) is useful in determining the resistance of any conductor of uniform cross section. If the cross section of the conductor is not uniform, eq. (5.16) is not applicable. However, the basic definition of resistance R as the ratio of the potential difference V between the two ends of the conductor to the current I through the conductor still applies. Therefore, applying eqs. (4.60) and (5.4) gives the resistance of a conductor of nonuniform cross section; that is,

$$\boxed{R = \frac{V}{I} = \frac{\int_L \mathbf{E} \cdot d\mathbf{l}}{\int_S \sigma \mathbf{E} \cdot d\mathbf{S}}} \tag{5.17}$$

Note that the negative sign before $V = -\int \mathbf{E} \cdot d\mathbf{l}$ is dropped in eq. (5.17) because $\int \mathbf{E} \cdot d\mathbf{l} < 0$ if $I > 0$. Equation (5.17) will not be utilized until we get to Section 6.5.

Power P (in watts) is defined as the rate of change of energy W (in joules) or force times velocity. Hence,

$$P = \int_v \rho_v \, dv \, \mathbf{E} \cdot \mathbf{u} = \int_v \mathbf{E} \cdot \rho_v \mathbf{u} \, dv$$

or

$$P = \int_v \mathbf{E} \cdot \mathbf{J} \, dv \tag{5.18}$$

which is known as *Joule's law.* The power density w_P (in W/m^3) is given by the integrand in eq. (5.18); that is,

$$w_P = \frac{dP}{dv} = \mathbf{E} \cdot \mathbf{J} = \sigma \, |\mathbf{E}|^2 \tag{5.19}$$

For a conductor with uniform cross section, $dv = dS \, dl$, so eq. (5.18) becomes

$$P = \int_L E \, dl \int_S J \, dS = VI$$

or

$$P = I^2 R \tag{5.20}$$

which is the more common form of Joule's law in electric circuit theory.

EXAMPLE 5.1

If $\mathbf{J} = \dfrac{1}{r^3} (2 \cos \theta \, \mathbf{a}_r + \sin \theta \, \mathbf{a}_\theta)$ A/m^2, calculate the current passing through

(a) A hemispherical shell of radius 20 cm, $0 < \theta < \pi/2, 0 < \phi < 2\pi$
(b) A spherical shell of radius 10 cm

Solution:

$I = \int_S \mathbf{J} \cdot d\mathbf{S}$,   where $d\mathbf{S} = r^2 \sin \theta \, d\phi \, d\theta \, \mathbf{a}_r$ in this case.

(a) $I = \displaystyle\int_{\theta=0}^{\pi/2} \int_{\phi=0}^{2\pi} \frac{1}{r^3} 2 \cos\theta \, r^2 \sin\theta \, d\phi \, d\theta \Bigg|_{r=0.2}$

$= \dfrac{2}{r} 2\pi \displaystyle\int_{\theta=0}^{\pi/2} \sin\theta \, d(\sin\theta)\Bigg|_{r=0.2}$

$= \dfrac{4\pi}{0.2} \dfrac{\sin^2\theta}{2}\Bigg|_0^{\pi/2} = 10\pi = 31.4 \text{ A}$

(b) The only difference here is that we have $0 \le \theta \le \pi$ instead of $0 \le \theta \le \pi/2$ and $r = 0.1$ m. Hence,

$$I = \frac{4\pi}{0.1} \frac{\sin^2\theta}{2}\Bigg|_0^{\pi} = 0$$

Alternatively, for this case

$$I = \oint_S \mathbf{J} \cdot d\mathbf{S} = \int_v \nabla \cdot \mathbf{J} \, dv = 0$$

since $\nabla \cdot \mathbf{J} = 0$. We can show this:

$$\nabla \cdot \mathbf{J} = \frac{1}{r^2} \frac{\partial}{\partial r}\left[r^2 \frac{2}{r} \cos\theta \right] + \frac{1}{r\sin\theta} \frac{\partial}{\partial\theta}\left[\frac{1}{r^3} \sin^2\theta \right] = \frac{-2}{r^4} \cos\theta + \frac{2}{r^4} \cos\theta = 0$$

PRACTICE EXERCISE 5.1

For the current density $\mathbf{J} = 10z \sin^2\phi \, \mathbf{a}_\rho$ A/m^2, find the current through the cylindrical surface $\rho = 2$, $1 \le z \le 5$ m.

Answer: 754 A.

EXAMPLE 5.2

A typical example of convective charge transport is found in the Van de Graaff generator, where charge is transported on a moving belt from the base to the dome as shown in Figure 5.4. If a surface charge density 10^{-7} C/m^2 is transported by the belt at a velocity of 2 m/s, calculate the charge collected in 5 s. Take the width of the belt as 10 cm.

Solution:

If ρ_S = surface charge density, u = speed of the belt, and w = width of the belt, the current on the dome is

$$I = \rho_S u w$$

The total charge collected in $t = 5$ s is

$$Q = It = \rho_S u w t = 10^{-7} \times 2 \times 0.1 \times 5$$

$$= 100 \text{ nC}$$

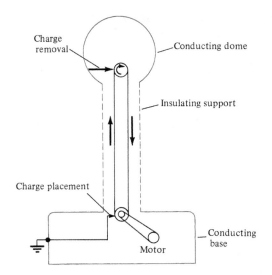

FIGURE 5.4 Van de Graaff generator; for Example 5.2.

PRACTICE EXERCISE 5.2

In a Van de Graaff generator, $w = 0.1$ m, $u = 10$ m/s, and from the dome to the ground there are leakage paths having a total resistance of 10^{14} Ω. If the belt carries charge 0.5 $\mu C/m^2$, find the potential difference between the dome and the base. *Note:* In the steady state, the current through the leakage path is equal to the charge transported per unit time by the belt.

Answer: 50 mV.

EXAMPLE 5.3

A wire of diameter 1 mm and conductivity 5×10^7 S/m has 10^{29} free electrons per cubic meter when an electric field of 10 mV/m is applied. Determine

(a) The charge density of free electrons
(b) The current density
(c) The current in the wire
(d) The drift velocity of the electrons (take the electronic charge as $e = -1.6 \times 10^{-19}$ C)

Solution:

(In this particular problem, convection and conduction currents are the same.)

(a) $\rho_v = ne = (10^{29})(-1.6 \times 10^{-19}) = -1.6 \times 10^{10}$ C/m^3

(b) $J = \sigma E = (5 \times 10^7)(10 \times 10^{-3}) = 500 \, \text{kA/m}^2$

(c) $I = JS = (5 \times 10^5)\left(\dfrac{\pi d^2}{4}\right) = \dfrac{5\pi}{4} \times 10^{-6} \times 10^5 = 0.393$ A

(d) Since $J = \rho_v u$, $u = \dfrac{J}{\rho_v} = \dfrac{5 \times 10^5}{1.6 \times 10^{10}} = 3.125 \times 10^{-5}$ m/s

EXAMPLE 5.4

A lead ($\sigma = 5 \times 10^6$ S/m) bar of square cross section has a hole bored along its length of 4 m so that its cross section becomes that of Figure 5.5. Find the resistance between the square ends.

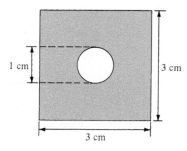

FIGURE 5.5 Cross section of the lead bar of Example 5.4.

Solution:

Since the cross section of the bar is uniform, we may apply eq. (5.16); that is,

$$R = \frac{\ell}{\sigma S}$$

where $S = d^2 - \pi r^2 = 3^2 - \pi \left(\frac{1}{2}\right)^2 = \left(9 - \frac{\pi}{4}\right)$ cm^2.

Hence,

$$R = \frac{4}{5 \times 10^6 (9 - \pi/4) \times 10^{-4}} = 974 \, \mu\Omega$$

5.5 POLARIZATION IN DIELECTRICS

In Section 5.2, we noticed that the main difference between a conductor and a dielectric lies in the availability of free electrons in the outermost atomic shells to conduct current. Although the charges in a dielectric are not able to move about freely, they are bound by finite forces, and we may certainly expect a displacement when an external force is applied.

To understand the macroscopic effect of an electric field on a dielectric, consider an atom of the dielectric as consisting of a negative charge $-Q$ (electron cloud) and a positive charge $+Q$ (nucleus) as in Figure 5.6(a). A similar picture can be adopted for a dielectric molecule; we can treat the nuclei in molecules as point charges and the electronic structure as a single cloud of negative charge. Since we have equal amounts of positive and negative charge, the whole atom or molecule is electrically neutral. When an electric field **E** is applied, the positive charge is displaced from its equilibrium position in the direction of **E** by the force $\mathbf{F}_+ = Q\mathbf{E}$, while the negative charge is displaced in the opposite direction by the force $\mathbf{F}_- = Q\mathbf{E}$. A dipole results from the displacement of the charges, and the dielectric is said to be *polarized*. In the polarized state, the electron cloud is distorted by the applied electric field **E.** This distorted charge distribution is equivalent, by the principle of superposition, to the original distribution plus a dipole whose moment is

$$\mathbf{p} = Q\mathbf{d} \tag{5.21}$$

where **d** is the distance vector from $-Q$ to $+Q$ of the dipole as in Figure 5.6(b). If there are N dipoles in a volume Δv of the dielectric, the total dipole moment due to the electric field is

$$Q_1\mathbf{d}_1 + Q_2\mathbf{d}_2 + \cdots + Q_N\mathbf{d}_N = \sum_{k=1}^{N} Q_k\mathbf{d}_k \tag{5.22}$$

As a measure of intensity of the polarization, we define *polarization* **P** (in coulombs per meter squared) as the dipole moment per unit volume of the dielectric; that is,

$$\mathbf{P} = \lim_{\Delta v \to 0} \frac{\sum_{k=1}^{N} Q_k\mathbf{d}_k}{\Delta v} \tag{5.23}$$

Thus we conclude that the major effect of the electric field **E** on a dielectric is the creation of dipole moments that align themselves in the direction of **E**. This type of dielectric

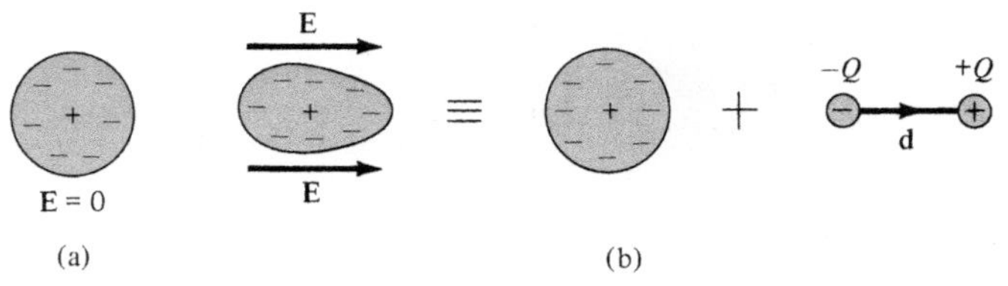

FIGURE 5.6 Polarization of a nonpolar atom or molecule.

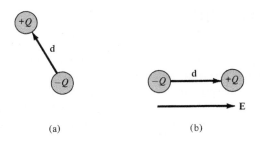

FIGURE 5.7 Polarization of a polar molecule: (**a**) permanent dipole (**E** = **0**), (**b**) alignment of permanent dipole (**E** ≠ **0**).

is said to be *nonpolar*. Examples of such dielectrics are hydrogen, oxygen, nitrogen, and the rare gases. Nonpolar dielectric molecules do not possess dipoles until the application of the electric field as we have noticed. Other types of molecule such as water, sulfur dioxide, hydrochloric acid, and polystyrene have built-in permanent dipoles that are randomly oriented as shown in Figure 5.7(a) and are said to be *polar*. When an electric field **E** is applied to a polar molecule, the permanent dipole experiences a torque tending to align its dipole moment parallel with **E** as in Figure 5.7(b).

Let us now calculate the field due to a polarized dielectric. Consider the dielectric material shown in Figure 5.8 as consisting of dipoles with dipole moment **P** per unit volume. According to eq. (4.80), the potential dV at an exterior point O due to the dipole moment **P** dv' is

$$dV = \frac{\mathbf{P} \cdot \mathbf{a}_R \, dv'}{4\pi\varepsilon_o R^2} \tag{5.24}$$

where $R^2 = (x - x')^2 + (y - y')^2 + (z - z')^2$ and R is the distance between the volume element dv' at (x', y', z') and the field point $O\,(x, y, z)$. We can transform eq. (5.24) into a form that facilitates physical interpretation. It is readily shown (see Section 7.7) that the gradient of $1/R$ with respect to the primed coordinates is

$$\nabla'\left(\frac{1}{R}\right) = \frac{\mathbf{a}_R}{R^2}$$

where ∇' is the del operator with respect to (x', y', z'). Thus,

$$\frac{\mathbf{P} \cdot \mathbf{a}_R}{R^2} = \mathbf{P} \cdot \nabla'\left(\frac{1}{R}\right)$$

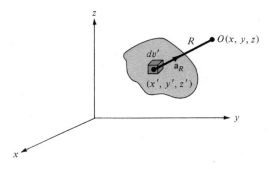

FIGURE 5.8 A block of dielectric material with dipole moment **P** per unit volume.

Applying the vector identity $\nabla' \cdot f\mathbf{A} = f\nabla' \cdot \mathbf{A} + \mathbf{A} \cdot \nabla'f$,

$$\frac{\mathbf{P} \cdot \mathbf{a}_R}{R^2} = \nabla' \cdot \left(\frac{\mathbf{P}}{R}\right) - \frac{\nabla' \cdot \mathbf{P}}{R} \tag{5.25}$$

Substituting this into eq. (5.24) and integrating over the entire volume v' of the dielectric, we obtain

$$V = \int_{v'} \frac{1}{4\pi\varepsilon_0} \left[\nabla' \cdot \frac{\mathbf{P}}{R} - \frac{1}{R}\nabla' \cdot \mathbf{P}\right] dv'$$

Applying divergence theorem to the first term leads finally to

$$V = \oint_{S'} \frac{\mathbf{P} \cdot \mathbf{a}'_n}{4\pi\varepsilon_0 R} dS' + \int_{v'} \frac{-\nabla' \cdot \mathbf{P}}{4\pi\varepsilon_0 R} dv' \tag{5.26}$$

where $\mathbf{a}'_n$ is the outward unit normal to surface S' of the dielectric. Comparing the two terms on the right side of eq. (5.26) with eqs. (4.68) and (4.69) shows that the two terms denote the potential due to surface and volume charge distributions with densities (upon dropping the primes):

$$\boxed{\begin{aligned} \rho_{ps} &= \mathbf{P} \cdot \mathbf{a}_n \\ \rho_{pv} &= -\nabla \cdot \mathbf{P} \end{aligned}} \tag{5.27a}\tag{5.27b}$$

In other words, eq. (5.26) reveals that where polarization occurs, an equivalent volume charge density ρ_{pv} is formed throughout the dielectric, while an equivalent surface charge density ρ_{ps} is formed over the surface of the dielectric. We refer to ρ_{ps} and ρ_{pv} as *bound* (or *polarization*) *surface* and *volume charge densities*, respectively, as distinct from *free* surface and volume charge densities ρ_S and ρ_v. Bound charges are those that are not free to move within the dielectric material; they are caused by the displacement that occurs on a molecular scale during polarization. Free charges are those that are capable of moving over macroscopic distance, as do electrons in a conductor; they are the stuff we control. The total positive bound charge on surface S bounding the dielectric is

$$Q_b = \oint \mathbf{P} \cdot d\mathbf{S} = \oint \rho_{ps} \, dS \tag{5.28a}$$

while the charge that remains inside S is

$$-Q_b = \int_v \rho_{pv} \, dv = -\int_v \nabla \cdot \mathbf{P} \, dv \tag{5.28b}$$

If the entire dielectric were electrically neutral prior to application of the electric field and if we have not added any free charge, the dielectric will remain electrically neutral. Thus the total charge of the dielectric material remains zero, that is,

$$\text{total charge} = \oint_S \rho_{ps}\, dS + \int_v \rho_{pv}\, dv = Q_b - Q_b = 0$$

We now consider the case in which the dielectric region contains free charge. If ρ_v is the volume density of free charge, the total volume charge density ρ_t is given by

$$\rho_t = \rho_v + \rho_{pv} = \nabla \cdot \varepsilon_o \mathbf{E} \tag{5.29}$$

Hence,

$$\begin{aligned}
\rho_v &= \nabla \cdot \varepsilon_o \mathbf{E} - \rho_{pv} \\
&= \nabla \cdot (\varepsilon_o \mathbf{E} + \mathbf{P}) \\
&= \nabla \cdot \mathbf{D}
\end{aligned} \tag{5.30}$$

where

$$\boxed{\mathbf{D} = \varepsilon_o \mathbf{E} + \mathbf{P}} \tag{5.31}$$

We conclude that the net effect of the dielectric on the electric field $\mathbf{E}$ is to increase $\mathbf{D}$ inside it by the amount $\mathbf{P}$. In other words, the application of $\mathbf{E}$ to the dielectric material causes the flux density to be greater than it would be in free space. It should be noted that the definition of $\mathbf{D}$ in eq. (4.35) for free space is a special case of that in eq. (5.31) because $\mathbf{P} = 0$ in free space.

For some dielectrics, $\mathbf{P}$ is proportional to the applied electric field $\mathbf{E}$, and we have

$$\boxed{\mathbf{P} = \chi_e \varepsilon_o \mathbf{E}} \tag{5.32}$$

where χ_e, known as the *electric susceptibility* of the material, is more or less a measure of how susceptible (or sensitive) a given dielectric is to electric fields.

5.6 DIELECTRIC CONSTANT AND STRENGTH

By substituting eq. (5.32) into eq. (5.31), we obtain

$$\mathbf{D} = \varepsilon_o (1 + \chi_e) \mathbf{E} = \varepsilon_o \varepsilon_r \mathbf{E} \tag{5.33}$$

or

$$\boxed{\mathbf{D} = \varepsilon \mathbf{E}} \tag{5.34}$$

where

$$\boxed{\varepsilon = \varepsilon_o \varepsilon_r} \tag{5.35}$$

and

$$\boxed{\varepsilon_r = 1 + \chi_e = \frac{\varepsilon}{\varepsilon_o}}$$

(5.36)

In eqs. (5.33) to (5.36), ε is called the *permittivity* of the dielectric, ε_o is the permittivity of free space, defined in eq. (4.2) as approximately $10^{-9}/36\pi$ F/m, and ε_r is called the *dielectric constant* or *relatve permittivity*.

> The **dielectric constant** (or **relative permittivity**) ε_r is the ratio of the permittivity of the dielectric to that of free space.

It should also be noticed that ε_r and χ_e are dimensionless, whereas ε and ε_o are in farads per meter. The approximate values of the dielectric constants of some common materials are given in Table B.2 in Appendix B. The values given in Table B.2 are for static or low-frequency (<1000 Hz) fields; the values may change at high frequencies. Note from the table that ε_r is always greater than or equal to unity. For free space $\varepsilon_r = 1$.

The theory of dielectrics we have discussed so far assumes ideal dielectrics. Practically speaking, no dielectric is ideal. When the electric field in a dielectric is sufficiently large, it begins to pull electrons completely out of the molecules, and the dielectric becomes conducting. *Dielectric breakdown* is said to have occurred when a dielectric becomes conducting. Dielectric breakdown occurs in all kinds of dielectric materials (gases, liquids, or solids) and depends on the nature of the material, temperature, humidity, and the amount of time that the field is applied. The minimum value of the electric field at which dielectric breakdown occurs is called the *dielectric strength* of the dielectric material.

> The **dielectric strength** is the maximum electric field that a dielectric can tolerate or withstand without electrical breakdown.

Table B.2 also lists the dielectric strength of some common dielectrics. Since our theory of dielectrics does not apply after dielectric breakdown has taken place, we shall always assume ideal dielectric and avoid dielectric breakdown.

†5.7 LINEAR, ISOTROPIC, AND HOMOGENEOUS DIELECTRICS

A material is said to be *linear* if **D** varies linearly with **E** and *nonlinear* otherwise. Materials for which ε (or σ) does not vary in the region being considered and is therefore the same at all points (i.e., independent of x, y, z) are said to be *homogeneous*. They are said to be *inhomogeneous* (or nonhomogeneous) when ε is dependent on the space coordinates. The atmosphere is a typical example of an inhomogeneous medium; its permittivity varies with altitude. Materials for which **D** and **E** are in the same direction are said to be *isotropic*. That is, is, isotropic dielectrics are those that have the same properties in all directions. In *anisotropic* (or nonisotropic) materials, **D**, **E**, and **P** are not parallel; ε or χ_e has nine components that are collectively referred to as a *tensor*. For example, instead of eq. (5.34), we have

$$
\begin{bmatrix} D_x \\ D_y \\ D_z \end{bmatrix} = \begin{bmatrix} \varepsilon_{xx} & \varepsilon_{xy} & \varepsilon_{xz} \\ \varepsilon_{yz} & \varepsilon_{yy} & \varepsilon_{yz} \\ \varepsilon_{zx} & \varepsilon_{zy} & \varepsilon_{zz} \end{bmatrix} \begin{bmatrix} E_x \\ E_y \\ E_z \end{bmatrix} \tag{5.37}
$$

for anisotropic materials. Crystalline materials and magnetized plasma are anisotropic.

> A **dielectric material** (in which **D** = ε**E** applies) is linear if ε does not change with the applied **E** field, homogeneous if ε does not change from point to point, and isotropic if ε does not change with direction. Although eqs. (5.24) to (5.31) are for dielectric materials in general, eqs. (5.32) to (5.34) are only for linear, isotropic materials.

The same idea holds for a conducting material in which **J** = σ**E** applies. The material is linear if σ does not vary with **E**, homogeneous if σ is the same at all points, and isotropic if σ does not vary with direction.

For most of the time, we will be concerned only with linear, isotropic, and homogeneous media. Such media are called *simple materials*. For such media, all formulas derived in Chapter 4 for free space can be applied by merely replacing ε_o with $\varepsilon_o\varepsilon_r$. Thus Coulomb's law of eq. (4.4), for example, becomes

$$
\mathbf{F} = \frac{Q_1 Q_2}{4\pi\varepsilon_o\varepsilon_r R^2}\,\mathbf{a}_R \tag{5.38}
$$

and eq. (4.96) becomes

$$
W = \frac{1}{2}\int_v \varepsilon_o\varepsilon_r E^2\,dv \tag{5.39}
$$

when applied to a dielectric medium.

EXAMPLE 5.5

A dielectric cube of side L and center at the origin has a radial polarization given by **P** = $a\mathbf{r}$, where a is a constant and $\mathbf{r} = x\mathbf{a}_x + y\mathbf{a}_y + z\mathbf{a}_z$. Find all bound charge densities and show explicitly that the total bound charge vanishes.

Solution:

For each of the six faces of the cube, there is a surface charge density ρ_{ps}. For the face located at $x = L/2$, for example,

$$
\rho_{ps} = \mathbf{P} \cdot \mathbf{a}_x \Big|_{x=L/2} = ax \Big|_{x=L/2} = \frac{aL}{2}
$$

The total bound surface charge is

$$Q_s = \int_S \rho_{ps} \, dS = 6 \int_{-L/2}^{L/2} \int_{-L/2}^{L/2} \rho_{ps} \, dy \, dz = \frac{6aL}{2} L^2$$

$$= 3aL^3$$

The bound volume charge density is given by

$$\rho_{pv} = -\nabla \cdot \mathbf{P} = -(a + a + a) = -3a$$

and the total bound volume charge is

$$Q_v = \int_v \rho_{pv} \, dv = -3a \int_v dv = -3aL^3$$

Hence, the total charge is

$$Q_t = Q_s + Q_v = 3aL^3 - 3aL^3 = 0$$

PRACTICE EXERCISE 5.5

A thin rod of cross-sectional area A extends along the x-axis from $x = 0$ to $x = L$. The polarization of the rod is along its length and is given by $P_x = ax^2 + b$. Calculate ρ_{pv} and ρ_{ps} at each end. Show explicitly that the total bound charge vanishes in this case.

Answer: $0, -2aL, -b, aL^2 + b$, proof.

EXAMPLE 5.6

The electric field intensity in polystyrene ($\varepsilon_r = 2.55$) filling the space between the plates of a parallel-plate capacitor is 10 kV/m. The distance between the plates is 1.5 mm. Calculate:

(a) D
(b) P
(c) The surface charge density of free charge on the plates
(d) The surface density of polarization charge
(e) The potential difference between the plates

Solution:

(a) $D = \varepsilon_o \varepsilon_r E = \dfrac{10^{-9}}{36\pi} \times (2.55) \times 10^4 = 225.4 \text{ nC/m}^2$

(b) $P = \chi_e \varepsilon_o E = (1.55) \times \dfrac{10^{-9}}{36\pi} \times 10^4 = 137 \text{ nC/m}^2$

(c) $\rho_S = \mathbf{D} \cdot \mathbf{a}_n = \pm D_n = \pm 225.4 \text{ nC/m}^2$

(d) $\rho_{ps} = \mathbf{P} \cdot \mathbf{a}_n = \pm P_n = \pm 137 \text{ nC/m}^2$

(e) $V = Ed = 10^4(1.5 \times 10^{-3}) = 15 \text{ V}$

EXAMPLE 5.7

A dielectric sphere ($\varepsilon_r = 5.7$) of radius 10 cm has a point charge of 2 pC placed at its center. Calculate:

(a) The surface density of polarization charge on the surface of the sphere
(b) The force exerted by the charge on a-4 pC point charge placed on the sphere

Solution:

(a) Assuming that the point charge is located at the origin, we apply Coulomb's or Gauss's law to obtain

$$\mathbf{E} = \frac{Q}{4\pi\varepsilon_o\varepsilon_r r^2}\,\mathbf{a}_r$$

$$\mathbf{P} = \chi_e\varepsilon_o\mathbf{E} = \frac{\chi_e Q}{4\pi\varepsilon_r r^2}\,\mathbf{a}_r$$

$$\rho_{ps} = \mathbf{P}\cdot\mathbf{a}_r = \frac{(\varepsilon_r - 1)Q}{4\pi\varepsilon_r r^2} = \frac{(4.7)\,2 \times 10^{-12}}{4\pi(5.7)\,100 \times 10^{-4}}$$

$$= 13.12 \text{ pC/m}^2$$

(b) From Coulomb's law, we have

$$\mathbf{F} = \frac{Q_1 Q_2}{4\pi\varepsilon_o\varepsilon_r r^2}\,\mathbf{a}_r = \frac{(-4)(2) \times 10^{-24}}{4\pi \times \dfrac{10^{-9}}{36\pi}\,(5.7)\,100 \times 10^{-4}}\,\mathbf{a}_r$$

$$= -1.263\mathbf{a}_r \text{ pN}$$

EXAMPLE 5.8

Find the force with which the plates of a parallel-plate capacitor attract each other. Also determine the pressure on the surface of the plate due to the field.

Solution:

From eq. (4.26), the electric field intensity on the surface of each plate is

$$\mathbf{E} = \frac{\rho_S}{2\varepsilon}\mathbf{a}_n$$

where $\mathbf{a}_n$ is a unit normal to the plate and ρ_S is the surface charge density. The total force on each plate is

$$\mathbf{F} = Q\mathbf{E} = \rho_S S \cdot \frac{\rho_S}{2\varepsilon}\mathbf{a}_n = \frac{\rho_S^2 S}{2\varepsilon_o \varepsilon_r}\mathbf{a}_n$$

or

$$F = \frac{\rho_S^2 S}{2\varepsilon} = \frac{Q^2}{2\varepsilon S}$$

The pressure of force per area is $\dfrac{\rho_S^2}{2\varepsilon_o \varepsilon_r}$. Notice that the dielectric affects the force or pressure.

PRACTICE EXERCISE 5.8

Shown in Figure 5.9 is a potential-measuring device known as an *electrometer*. It is basically a parallel-plate capacitor with the guarded plate being suspended from a balance arm so that the force F on it is measurable in terms of weight. If S is the area of each plate, show that

$$V_1 - V_2 = \left[\frac{2\,Fd^2}{\varepsilon_o S}\right]^{1/2}$$

Answer: Proof.

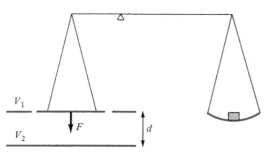

FIGURE 5.9 An electrometer; for Practice Exercise 5.8.

5.8 CONTINUITY EQUATION AND RELAXATION TIME

From the principle of charge conservation, the time rate of decrease of charge within a given volume must be equal to the net outward current flow through the surface of the volume. Thus current I_{out} coming out of the closed surface is

$$I_{out} = \oint_S \mathbf{J} \cdot d\mathbf{S} = \frac{-dQ_{in}}{dt} \tag{5.40}$$

where Q_{in} is the total charge enclosed by the closed surface. Invoking the divergence theorem, we write

$$\oint_S \mathbf{J} \cdot d\mathbf{S} = \int_v \nabla \cdot \mathbf{J} \, dv \tag{5.41}$$

But

$$\frac{-dQ_{in}}{dt} = -\frac{d}{dt} \int_v \rho_v \, dv = -\int_v \frac{\partial \rho_v}{\partial t} \, dv \tag{5.42}$$

Substituting eqs. (5.41) and (5.42) into eq. (5.40) gives

$$\int_v \nabla \cdot \mathbf{J} \, dv = -\int_v \frac{\partial \rho_v}{\partial t} \, dv$$

or

$$\boxed{\nabla \cdot \mathbf{J} = -\frac{\partial \rho_v}{\partial t}} \tag{5.43}$$

which is called the *continuity of current equation* or just *continuity equation*. It must be kept in mind that the continuity equation is derived from the principle of conservation of charge and essentially states that there can be no accumulation of charge at any point. For steady currents, $\partial \rho_v / \partial t = 0$, and hence $\nabla \cdot \mathbf{J} = 0$, showing that the total charge leaving a volume is the same as the total charge entering it. Kirchhoff's current law follows from this.

Having discussed the continuity equation and the properties σ and ε of materials, it is appropriate to consider the effect of introducing charge at some *interior* point of a given material (conductor or dielectric). We make use of eq. (5.43) in conjunction with Ohm's law

$$\mathbf{J} = \sigma \, \mathbf{E} \tag{5.44}$$

and Gauss's law

$$\nabla \cdot \mathbf{E} = \frac{\rho_v}{\varepsilon} \tag{5.45}$$

Substituting eqs. (5.44) and (5.45) into eq. (5.43) yields

$$\nabla \cdot \sigma \mathbf{E} = \frac{\sigma \rho_v}{\varepsilon} = -\frac{\partial \rho_v}{\partial t}$$

or

$$\frac{\partial \rho_v}{\partial t} + \frac{\sigma}{\varepsilon} \rho_v = 0 \tag{5.46}$$

This is a homogeneous linear ordinary differential equation. By separating variables in eq. (5.46), we get

$$\frac{\partial \rho_v}{\rho_v} = -\frac{\sigma}{\varepsilon} \partial t \tag{5.47}$$

and integrating both sides gives

$$\ln \rho_v = -\frac{\sigma t}{\varepsilon} + \ln \rho_{vo}$$

where $\ln \rho_{vo}$ is a constant of integration. Thus

$$\boxed{\rho_v = \rho_{vo} e^{-t/T_r}} \tag{5.48}$$

where

$$\boxed{T_r = \frac{\varepsilon}{\sigma}} \tag{5.49}$$

and T_r is the time constant in seconds.

In eq. (5.48), ρ_{vo} is the initial charge density (i.e., ρ_v at $t = 0$). The equation shows that the introduction of charge at some interior point of the material results in a decay of volume charge density ρ_v. Associated with the decay is charge movement from the interior point at which it was introduced to the surface of the material. The time constant T_r is known as the *relaxation time* or *rearrangement time*.

> **Relaxation time** is the time it takes a charge placed in the interior of a material to drop to e^{-1} ($= 36.8\%$) of its initial value.

Relaxation time is short for good conductors and long for good dielectrics. For example, for copper $\sigma = 5.8 \times 10^7$ S/m, $\varepsilon_r = 1$, and

$$T_r = \frac{\varepsilon_r \varepsilon_o}{\sigma} = 1 \times \frac{10^{-9}}{36\pi} \times \frac{1}{5.8 \times 10^7}$$

$$= 1.53 \times 10^{-19} \, s \tag{5.50}$$

showing a rapid decay of charge placed inside copper. This implies that for good conductors, the relaxation time is so short that most of the charge will vanish from any interior point and appear at the surface (as surface charge) almost instantaneously. On the other hand, for fused quartz, for instance, $\sigma = 10^{-17}$ S/m, $\varepsilon_r = 5.0$,

$$T_r = 5 \times \frac{10^{-9}}{36\pi} \times \frac{1}{10^{-17}}$$

$$= 51.2 \, days \tag{5.51}$$

showing a very large relaxation time. Thus for good dielectrics, one may consider the introduced charge to remain wherever placed for times up to days.

5.9 BOUNDARY CONDITIONS

So far, we have considered the existence of the electric field in a homogeneous medium. If the field exists in a region consisting of two different media, the conditions that the field must satisfy at the interface separating the media are called *boundary conditions*. These conditions are helpful in determining the field on one side of the boundary if the field on the other side is known. Obviously, the conditions will be dictated by the types of material the media are made of. We shall consider the boundary conditions at an interface separating

- Dielectric (ε_{r1}) and dielectric (ε_{r2})
- Conductor and dielectric
- Conductor and free space

To determine the boundary conditions, we need to use Maxwell's equations:

$$\oint_L \mathbf{E} \cdot d\mathbf{l} = 0 \tag{5.52}$$

and

$$\oint_S \mathbf{D} \cdot d\mathbf{S} = Q_{enc} \tag{5.53}$$

where Q_{enc} is the free charge enclosed by the surface S. Also we need to decompose the electric field intensity $\mathbf{E}$ into two orthogonal components:

$$\mathbf{E} = \mathbf{E}_t + \mathbf{E}_n \tag{5.54}$$

where $\mathbf{E}_t$ and $\mathbf{E}_n$ are, respectively, the tangential and normal components of $\mathbf{E}$ to the interface of interest. A similar decomposition can be done for the electric flux density $\mathbf{D}$.

A. Dielectric–Dielectric Boundary Conditions

Consider the **E** field existing in a region that consists of two different dielectrics characterized by $\varepsilon_1 = \varepsilon_o \varepsilon_{r1}$ and $\varepsilon_2 = \varepsilon_o \varepsilon_{r2}$ as shown in Figure 5.10(a). The fields $\mathbf{E}_1$ and $\mathbf{E}_2$ in media 1 and 2, respectively, can be decomposed as

$$\mathbf{E}_1 = \mathbf{E}_{1t} + \mathbf{E}_{1n} \tag{5.55a}$$

$$\mathbf{E}_2 = \mathbf{E}_{2t} + \mathbf{E}_{2n} \tag{5.55b}$$

We apply eq. (5.52) to the closed path *abcda* of Figure 5.10(a), assuming that the path is very small with respect to the spatial variation of **E**. We obtain

$$0 = E_{1t}\,\Delta w - E_{1n}\frac{\Delta h}{2} - E_{2n}\frac{\Delta h}{2} - E_{2t}\,\Delta w + E_{2n}\frac{\Delta h}{2} + E_{1n}\frac{\Delta h}{2} \tag{5.56}$$

where $E_t = |\mathbf{E}_t|$ and $E_n = |\mathbf{E}_n|$. The $\dfrac{\Delta h}{2}$ terms cancel, and eq. (5.56) becomes

$$0 = (E_{1t} - E_{2t})\,\Delta w$$

or

$$\boxed{E_{1t} = E_{2t}} \tag{5.57}$$

Thus the tangential components of **E** are the same on the two sides of the boundary. In other words, E_t undergoes no change on the boundary and it is said to be *continuous* across the boundary. Since $\mathbf{D} = \varepsilon\mathbf{E} = \mathbf{D}_t + \mathbf{D}_n$, eq. (5.57) can be written as

$$\frac{D_{1t}}{\varepsilon_1} = E_{1t} = E_{2t} = \frac{D_{2t}}{\varepsilon_2}$$

or

$$\frac{D_{1t}}{\varepsilon_1} = \frac{D_{2t}}{\varepsilon_2} \tag{5.58}$$

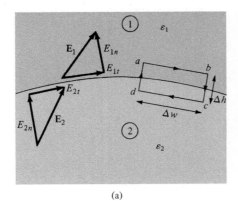

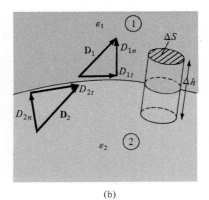

(a)   (b)

FIGURE 5.10 Dielectric–dielectric boundary: (**a**) determining $E_{1t} = E_{2t}$, (**b**) determining $D_{1n} = D_{2n}$.

that is, D_t undergoes some change across the interface. Hence D_t is said to be *discontinuous* across the interface.

Similarly, we apply eq. (5.53) to the pillbox (cylindrical Gaussian surface) of Figure 5.10(b). The contribution due to the sides vanishes. Allowing $\Delta h \to 0$ gives

$$\Delta Q = \rho_S \Delta S = D_{1n} \Delta S - D_{2n} \Delta S$$

or

$$\boxed{D_{1n} - D_{2n} = \rho_S} \tag{5.59}$$

where ρ_S is the free charge density placed deliberately at the boundary. It should be borne in mind that eq. (5.59) is based on the assumption that **D** is directed from region 2 to region 1 and eq. (5.59) must be applied accordingly. If no free charges exist at the interface (i.e., charges are not deliberately placed there), $\rho_S = 0$ and eq. (5.59) becomes

$$\boxed{D_{1n} = D_{2n}} \tag{5.60}$$

Thus the normal component of **D** is continuous across the interface; that is, D_n undergoes no change at the boundary. Since $\mathbf{D} = \varepsilon \mathbf{E}$, eq. (5.60) can be written as

$$\varepsilon_1 E_{1n} = \varepsilon_2 E_{2n} \tag{5.61}$$

showing that the normal component of **E** is discontinuous at the boundary. Equations (5.57) and (5.59) or (5.60) are collectively referred to as *boundary conditions;* they must be satisfied by an electric field at the boundary separating two different dielectrics.

As mentioned earlier, the boundary conditions are usually applied in finding the electric field on one side of the boundary given the field on the other side. Besides this, we can use the boundary conditions to determine the "refraction" of the electric field across the interface. Consider $\mathbf{D}_1$ or $\mathbf{E}_1$ and $\mathbf{D}_2$ or $\mathbf{E}_2$ making angles θ_1 and θ_2 with the *normal* to the interface as illustrated in Figure 5.11. Using eq. (5.57), we have

$$E_1 \sin \theta_1 = E_{1t} = E_{2t} = E_2 \sin \theta_2$$

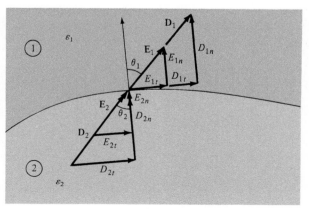

FIGURE 5.11 Refraction of **D** or **E** at a dielectric–dielectric boundary.

or

$$E_1 \sin \theta_1 = E_2 \sin \theta_2 \tag{5.62}$$

Similarly, by applying eq. (5.60) or (5.61), we get

$$\varepsilon_1 E_1 \cos \theta_1 = D_{1n} = D_{2n} = \varepsilon_2 E_2 \cos \theta_2$$

or

$$\varepsilon_1 E_1 \cos \theta_1 = \varepsilon_2 E_2 \cos \theta_2 \tag{5.63}$$

Dividing eq. (5.62) by eq. (5.63) gives

$$\frac{\tan \theta_1}{\varepsilon_1} = \frac{\tan \theta_2}{\varepsilon_2} \tag{5.64}$$

Since $\varepsilon_1 = \varepsilon_0 \varepsilon_{r1}$ and $\varepsilon_2 = \varepsilon_0 \varepsilon_{r2}$, eq. (5.64) becomes

$$\boxed{\frac{\tan \theta_1}{\tan \theta_2} = \frac{\varepsilon_{r1}}{\varepsilon_{r2}}} \tag{5.65}$$

This is the *law of refraction* of the electric field at a boundary free of charge (since $\rho_S = 0$ is assumed at the interface). Thus, in general, an interface between two dielectrics produces bending of the flux lines as a result of unequal polarization charges that accumulate on the opposite sides of the interface.

B. Conductor–Dielectric Boundary Conditions

Figure 5.12 shows the case of conductor–dielectric boundary conditions. The conductor is assumed to be perfect (i.e., $\sigma \rightarrow \infty$ or $\rho_c \rightarrow 0$). Although such a conductor is not realizable for most practical purposes, we may regard conductors such as copper and silver as though they were perfect conductors.

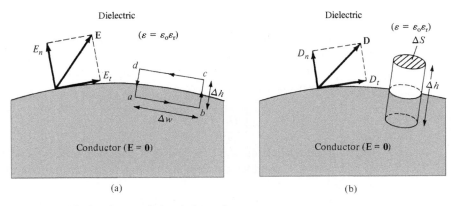

FIGURE 5.12 Conductor–dielectric boundary.

To determine the boundary conditions for a conductor–dielectric interface, we follow the same procedure used for the dielectric–dielectric interface except that we incorporate the fact that $\mathbf{E} = \mathbf{0}$ inside the conductor. Applying eq. (5.52) to the closed path *abcda* of Figure 5.12(a) gives

$$0 = 0 \cdot \Delta w + 0 \cdot \frac{\Delta h}{2} + E_n \cdot \frac{\Delta h}{2} - E_t \cdot \Delta w - E_n \cdot \frac{\Delta h}{2} - 0 \cdot \frac{\Delta h}{2} \qquad (5.66)$$

As $\Delta h \rightarrow 0$,

$$E_t = 0 \qquad (5.67)$$

Similarly, by applying eq. (5.53) to the cylindrical pillbox of Figure 5.12(b) and letting $\Delta h \rightarrow 0$, we get

$$\Delta Q = D_n \cdot \Delta S - 0 \cdot \Delta S \qquad (5.68)$$

because $\mathbf{D} = \varepsilon\mathbf{E} = \mathbf{0}$ inside the conductor. Equation (5.68) may be written as

$$D_n = \frac{\Delta Q}{\Delta S} = \rho_S$$

or

$$D_n = \rho_S \qquad (5.69)$$

Thus under static conditions, the following conclusions can be made about a perfect conductor:

1. No electric field may exist *within* a conductor; that is, considering our conclusion in Section 5.4,

$$\boxed{\rho_v = 0, \quad \mathbf{E} = \mathbf{0}} \qquad (5.70)$$

2. Since $\mathbf{E} = -\nabla V = \mathbf{0}$, there can be no potential difference between any two points in the conductor; that is, a conductor is an equipotential body.
3. An electric field $\mathbf{E}$ must be external to the conductor and must be *normal* to its surface; that is,

$$\boxed{D_t = \varepsilon_o\varepsilon_r E_t = 0, \quad D_n = \varepsilon_o\varepsilon_r E_n = \rho_S} \qquad (5.71)$$

An important application of the fact that $\mathbf{E} = \mathbf{0}$ inside a conductor is in *electrostatic screening* or *shielding*. If conductor A kept at zero potential surrounds conductor B as shown in Figure 5.13, B is said to be electrically screened by A from other electric circuits, such as conductor C, outside A. Similarly, conductor C outside A is screened by A from B. Thus

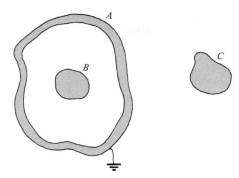

FIGURE 5.13 Electrostatic screening.

conductor *A* acts like a screen or shield, and the electrical conditions inside and outside the screen are completely independent of each other.

C. Conductor–Free Space Boundary Conditions

The conductor–free space boundary conditions, illustrated in Figure 5.14, comprise a special case of conductor–dielectric conditions. The boundary conditions at the interface between a conductor and free space can be obtained from eq. (5.71) by replacing ε_r by 1 (because free space may be regarded as a special dielectric for which $\varepsilon_r = 1$). The electric field **E** must be external to the conductor and normal to its surface. Thus the boundary conditions are

$$\boxed{D_t = \varepsilon_o E_t = 0, \quad D_n = \varepsilon_o E_n = \rho_S}$$

(5.72)

It should be noted again that eq. (5.72) implies that the **E** field must approach the conducting surface normally.

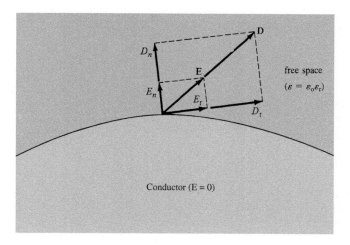

FIGURE 5.14 Conductor–free space boundary.

EXAMPLE 5.9

Two extensive homogeneous isotropic dielectrics meet on plane $z = 0$. For $z > 0$, $\varepsilon_{r1} = 4$ and for $z < 0$, $\varepsilon_{r2} = 3$. A uniform electric field $\mathbf{E}_1 = 5\mathbf{a}_x - 2\mathbf{a}_y + 3\mathbf{a}_z$ kV/m exists for $z \geq 0$. Find

(a) $\mathbf{E}_2$ for $z \leq 0$
(b) The angles $\mathbf{E}_1$ and $\mathbf{E}_2$ make with the interface
(c) The energy densities (in J/m^3) in both dielectrics
(d) The energy within a cube of side 2 m centered at $(3, 4, -5)$

Solution:

Let the problem be as illustrated in Figure 5.15.

(a) Since $\mathbf{a}_z$ is normal to the boundary plane, we obtain the normal components as

$$E_{1n} = \mathbf{E}_1 \cdot \mathbf{a}_n = \mathbf{E}_1 \cdot \mathbf{a}_z = 3$$
$$\mathbf{E}_{1n} = 3\mathbf{a}_z$$
$$\mathbf{E}_{2n} = (\mathbf{E}_2 \cdot \mathbf{a}_z)\mathbf{a}_z$$

Also

$$\mathbf{E} = \mathbf{E}_n + \mathbf{E}_t$$

Hence,

$$\mathbf{E}_{1t} = \mathbf{E}_1 - \mathbf{E}_{1n} = 5\mathbf{a}_x - 2\mathbf{a}_y$$

Thus

$$\mathbf{E}_{2t} = \mathbf{E}_{1t} = 5\mathbf{a}_x - 2\mathbf{a}_y$$

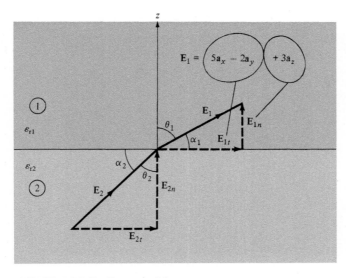

FIGURE 5.15 For Example 5.9.

Similarly,

$$\mathbf{D}_{2n} = \mathbf{D}_{1n} \;\rightarrow\; \varepsilon_{r2}\mathbf{E}_{2n} = \varepsilon_{r1}\mathbf{E}_{1n}$$

or

$$\mathbf{E}_{2n} = \frac{\varepsilon_{r1}}{\varepsilon_{r2}}\,\mathbf{E}_{1n} = \frac{4}{3}(3\mathbf{a}_z) = 4\mathbf{a}_z$$

Thus

$$\mathbf{E}_2 = \mathbf{E}_{2t} + \mathbf{E}_{2n}$$
$$= 5\mathbf{a}_x - 2\mathbf{a}_y + 4\mathbf{a}_z \text{ kV/m}$$

(b) Let α_1 and α_2 be the angles $\mathbf{E}_1$ and $\mathbf{E}_2$ they make with the interface while θ_1 and θ_2 are the angles they make with the normal to the interface as shown in Figure 5.15; that is,

$$\alpha_1 = 90 - \theta_1$$
$$\alpha_2 = 90 - \theta_2$$

Since $E_{1n} = 3$ and $E_{1t} = \sqrt{25 + 4} = \sqrt{29}$

$$\tan\theta_1 = \frac{E_{1t}}{E_{1n}} = \frac{\sqrt{29}}{3} = 1.795 \;\rightarrow\; \theta_1 = 60.9°$$

Hence,

$$\alpha_1 = 29.1°$$

Alternatively,

$$\mathbf{E}_1 \cdot \mathbf{a}_n = |\mathbf{E}_1| \cdot 1 \cdot \cos\theta_1$$

or

$$\cos\theta_1 = \frac{3}{\sqrt{38}} = 0.4867 \;\rightarrow\; \theta_1 = 60.9°$$

Similarly,

$$E_{2n} = 4, \quad E_{2t} = E_{1t} = \sqrt{29}$$

$$\tan\theta_2 = \frac{E_{2t}}{E_{2n}} = \frac{\sqrt{29}}{4} = 1.346 \;\rightarrow\; \theta_2 = 53.4°$$

Hence,

$$\alpha_2 = 36.6°$$

Note that $\dfrac{\tan \theta_1}{\tan \theta_2} = \dfrac{\varepsilon_{r1}}{\varepsilon_{r2}}$ is satisfied.

(c) The energy densities are given by

$$w_{E1} = \frac{1}{2}\varepsilon_1|\mathbf{E}_1|^2 = \frac{1}{2} \times 4 \times \frac{10^{-9}}{36\pi} \times (25 + 4 + 9) \times 10^6$$
$$= 672 \ \mu J/m^3$$

$$w_{E2} = \frac{1}{2}\varepsilon_2|\mathbf{E}_2|^2 = \frac{1}{2} \times 3 \times \frac{10^{-9}}{36\pi}(25 + 4 + 16) \times 10^6$$
$$= 597 \ \mu J/m^3$$

(d) At the center $(3, 4, -5)$ of the cube of side 2 m, $z = -5 < 0$; that is, the cube is in region 2 with $2 \le x \le 4, 3 \le y \le 5, -6 \le z \le -4$. Hence

$$W_E = \int w_{E2} \, dv = \int_{x=2}^{4} \int_{y=3}^{5} \int_{z=-6}^{-4} w_{E2} \, dz \, dy \, dz = w_{E2}(2)(2)(2)$$

$$= 597 \times 8 \ \mu J = 4.776 \ mJ$$

PRACTICE EXERCISE 5.9

A homogeneous dielectric $(\varepsilon_r = 2.5)$ fills region 1 $(x < 0)$ while region 2 $(x > 0)$ is free space.

(a) If $\mathbf{D}_1 = 12\mathbf{a}_x - 10\mathbf{a}_y + 4\mathbf{a}_z$ nC/m², find $\mathbf{D}_2$ and θ_2.

(b) If $E_2 = 12$ V/m and $\theta_2 = 60°$, find E_1 and θ_1. Take θ_1 and θ_2 as defined in Example 5.9.

Answer: (a) $12\mathbf{a}_x - 4\mathbf{a}_y + 1.6\mathbf{a}_z$ nC/m², 19.75°, (b) 10.67 V/m, 77°.

EXAMPLE 5.10

Region $y < 0$ consists of a perfect conductor while region $y > 0$ is a dielectric medium $(\varepsilon_{1r} = 2)$ as in Figure 5.16. If there is a surface charge of 2 nC/m² on the conductor, determine $\mathbf{E}$ and $\mathbf{D}$ at

(a) $A(3, -2, 2)$
(b) $B(-4, 1, 5)$

Solution:

(a) Point $A(3, -2, 2)$ is in the conductor since $y = -2 < 0$ at A. Hence,

$$\mathbf{E} = 0 = \mathbf{D}$$

(b) Point $B(-4, 1, 5)$ is in the dielectric medium since $y = 1 > 0$ at B.

$$D_n = \rho_S = 2 \ nC/m^2$$

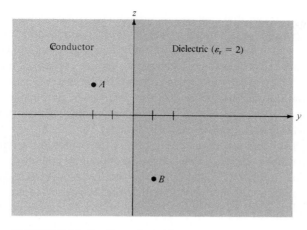

FIGURE 5.16 For Example 5.10.

Hence,

$$\mathbf{D} = 2\mathbf{a}_y \text{ nC/m}^2$$

and

$$\mathbf{E} = \frac{\mathbf{D}}{\varepsilon_0 \varepsilon_r} = 2 \times 10^{-9} \times \frac{36\pi}{2} \times 10^9 \mathbf{a}_y = 36\pi\, \mathbf{a}_y$$
$$= 113.1\, \mathbf{a}_y \text{ V/m}$$

PRACTICE EXERCISE 5.10

It is found that $\mathbf{E} = 60\mathbf{a}_x + 20\mathbf{a}_y - 30\mathbf{a}_z$ mV/m at a particular point on the interface between air and a conducting surface. Find $\mathbf{D}$ and ρ_S at that point.

Answer: $0.531\mathbf{a}_x + 0.177\mathbf{a}_y - 0.265\mathbf{a}_z$ pC/m^2, 0.619 pC/m^2.

---

†5.10 APPLICATION NOTE—MATERIALS WITH HIGH DIELECTRIC CONSTANT

This section is included in recognition of the growing importance of high dielectric constant materials to the semiconductor industry. As we noticed earlier in this chapter, the dielectric constant of a material is a property that determines its ability to become electrically polarized. The higher the dielectric constant, the more charge you can store, and the smaller you can make electronic circuits. High dielectric constant materials are increasingly important for pushing the state of the art in semiconductor integrated circuits. These materials, which find numerous technological applications, are necessary when high capacitance values are required. For example, to reduce the size of a dielectric resonator it is necessary to increase the dielectric constant of the material used. This is because at a fixed frequency, the diameter of the resonator is inversely proportional to the square root of the dielectric constant. Unfortunately, the higher the dielectric constant of a material, the higher its dielectric loss, as will be shown in Chapter 10.

High dielectric constants have been discovered in oxides of the type $ACu_3Ti_4O_{12}$. The most exceptional behavior is exhibited by a perovskite-related oxide containing calcium (Ca), copper (Cu), titanium (Ti), and oxygen (O) in the formula $CaCu_3Ti_4O_{12}$. This material is unusual in that it has an extremely high dielectric constant—about 11,000 (measured at 100 kHz). In addition, unlike most dielectric materials, this one retains its enormously high dielectric constant over a wide range of temperatures, from 100 to 600 degrees kelvin (K) (or -173 to $327°C$), making it ideal for a wide range of applications.

High dielectric constant materials are of great interest for other high-performance electric devices as well. One technology currently under development uses barium strontium titanate (BST), planned for use in dynamic random access memories (DRAMs). Although the dielectric constants are considerable, one disadvantage is the need for platinum electrodes. Another example occurs in radio frequency identification (RFID) chips, which require high capacitance to store charge. Frequently these use separate discrete devices, which are undesirably high in cost and low in yield.[2]

5.11 APPLICATION NOTE—GRAPHENE

All solid materials are supposed to have three dimensions. But graphene is a two-dimensional material made up of a single planar array of carbon atoms densely packed in a honeycomb or chicken-wire fashion, as shown in Figure 5.17. It has the smallest thickness and yet is one of the strongest of solids. Both the electrical conductivity and the thermal conductivity of graphene are very, very high. Graphene is almost completely transparent, yet so dense that even the smallest atom, helium, cannot pass through it. Graphene has drawn enormous curiosity on account of its unusual properties, which have many potential applications.

Every pencil lead has graphite, and a line drawn by a pencil is a primitive form of graphene. Around 1947 Philip Wallace first studied the theoretical aspects of graphite as its thickness was reduced. The name *graphene* was first coined in 1987 by S. Mouras and coworkers to describe the graphite layers that had various compounds inserted between them. In a sense, carbon nanotubes are rolled-up graphene sheets.

Originally, graphene was thought to be unstable in its free form; but in 2003, Andre Geim and Kostya Novoselov at the University of Manchester succeeded in producing the

---

[2] For more information about high dielectric constant materials, see H. S. Nalwa, *Handbook of Low and High Dielectric Constant Materials and Their Applications*. San Diego, CA: Academic Press, 1999, vols. 1 and 2.

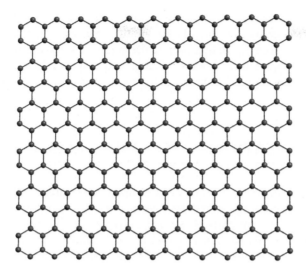

FIGURE 5.17 The structure of graphene.

first isolated graphene flakes, and their work was published in 2004. Their groundbreaking experiments, for which they received the 2010 Nobel Prize for Physics, showed how isolated graphene can be put to use in real-life applications. After the 2004 publication by Geim and Novoselov, other researchers began studying the properties of graphene. The Manchester group further showed that graphene at room temperature exhibits the quantum Hall effect, which had not been seen in other materials. The carrier mobility of graphene is very high, and this property can be exploited in making fast electronic devices. Also, graphene could be used as a chemical sensor to detect molecules of adsorption. With its carbon–carbon bond length of about 0.142 nm, graphene can also be considered as an indefinitely large aromatic molecule, the limiting case of the family of polycyclic aromatic hydrocarbons.

Electrodes with a very high surface area and very low electrical resistance can be made from graphene. Adding graphene to epoxy composites may result in stronger/stiffer components than can be made from epoxy composites containing a similar weight of carbon nanotubes. Graphene appears to bond better to the polymers in the epoxy, allowing more effective coupling of the graphene into the structure of the composite. This property could result in the manufacture of components with high strength-to-weight ratio for such uses as windmill blades or aircraft components.

Today, materials used in making solar cells are expensive, and the required manufacturing techniques are complicated. But if graphene is used as an electrode, while buckyballs and carbon nanotubes are employed to absorb light and generate electrons, it is possible to make solar cells more efficiently and at lower cost. Other potential applications of graphene include making high-speed electronic transistors, integrated circuits, and low-cost display screens for mobile devices. Lithography techniques can be used to fabricate integrated circuits based on graphene. It is also forecast that graphene can replace indium-based electrodes in organic light-emitting diodes (OLEDs), which are used in electronic device display screens that require low power consumption. The use of graphene instead of indium not only reduces the cost but eliminates the use of metals in the OLED, which may make devices easier to recycle. In yet another application, graphene layers are used to increase the binding energy of hydrogen to the graphene surface in a fuel tank, resulting in more hydrogen storage and therefore a lighter weight fuel tank. Such a component could be useful in the development of practical hydrogen-fueled cars.

†5.12 APPLICATION NOTE—PIEZOELECTRICS

The surfaces of many crystals acquire charge upon deformation. While the total charge in the material cannot change merely by deformation, charges of opposite sign whose total is zero appear at different parts of the surface as a result of mechanical stimulus applied to them. Such materials are called piezoelectrics. Examples include quartz, Rochelle salt, tourmaline, and many other crystalline materials. Experiments conducted with mechanical inputs in various directions reveal that only in certain fixed directions called polar axes the effect is pronounced to a maximal degree, with charges being induced on the parts of the surface perpendicular to these polar axes. On the surfaces opposing the polar axes, charges of opposite polarity are found under uniform strain. Further, it is found that if the deformation is reversed, that is, if the deformation is compression instead of expansion, or vice versa, the polarities of the induced charges also get reversed. The direction of the force need not be in the direction of the polar axes, but if the resulting stress has a component along the polar axes, one finds the accumulation of the charges on the surfaces.

Since different directions along a polar axis are not equivalent, if a crystal is rotated through 180° around an axis perpendicular to the polar axis, the latter coincides with itself but the crystal will not. As a result, crystals having a center of symmetry cannot be piezoelectrics. The necessary condition for the piezoelectric effect to manifest upon application of uniform deformation is therefore the absence of a center of symmetry in the given crystal. The symmetry properties of the crystal lattice are determined by the polar axes. Generally, a crystal has multiple polar axes. The piezoelectric effect was discovered by Pierre and Jacques Curie in 1880.

Piezoelectric properties depend on temperature. If at a certain temperature the crystal lattice is rearranged so that a center of symmetry is formed, piezoelectric properties of the crystal vanish at this temperature. If a material were to exhibit strong electromechanical coupling, the polarized atoms and molecules must be aligned well. The dipoles are oriented with respect to one another through a process called poling. Poling is usually brought about by heating the piezoelectric material up above its Curie temperature and then placing it in a strong electric field (typically, 2000 V/mm). The combination of heating and electric field produces motion of the electronic dipoles. Since the material is softer at higher temperatures, heating permits the dipoles to rotate freely. The electric field produces an alignment of the dipoles along the direction of the electric field. What resembles annealing to some extent, a quick reduction in the temperature and removal of the electric field produces a material whose electric dipoles are oriented in the same direction. This direction is referred to as the poling direction of the material. Ionic crystals are found to possess piezoelectric properties. There exists some difference in the deformation of sublattice of positive ions compared to that of negative ions causing crystal polarization and consequential surface charge distribution. To a first approximation, the polarization is directly proportional to the strain and in turn to the external force. The electric potential difference between the oppositely charged faces is therefore proportional to the applied force, which is exploited in numerous transducer applications such as pressure transducers, microphones, automation, and remote event detection.

Just as mechanical force applied to the piezoelectric crystal causes charges and hence potential to appear across the faces as per the direct piezoelectric effect, application of external electric field can bring about deformation of the crystal, and this is the inverse piezoelectric effect. When a piezoelectric body is deformed, work is expended to raise the energy of elastic deformation and also the energy of the electric field appearing as a result of the piezoelectric effect. In this event, it is necessary to overcome an additional force besides the elastic force of the crystal, which impedes the deformation. This is responsible for the inverse piezoelectric effect. As a compensatory measure, we should apply an external electric field opposite to that arising from the direct piezoelectric effect. This establishes that, to deform the piezoelectric in a given dimension by an external field, this field must be equal and opposite to the field that would appear under the given deformation due to the direct piezoelectric effect. If a certain potential difference appears between the faces of a piezoelectric, which are perpendicular to its polar axis, upon a deformation along this axis, a potential difference of the same magnitude but of opposite sign must be applied to these faces to attain the same deformation without applying mechanical forces. The mechanism of the inverse piezoelectric effect is similar to that of the direct effect: under the action of an external field, the crystal sublattices of positive and negative ions are deformed differently, which causes physical deformation of the crystal. The inverse piezoelectric effect also has numerous practical applications. For instance, quartz ultrasonic vibrators are widely used.

Two of the most popular piezoelectric materials are lead-zirconate-titanate (PZT) which is a ceramic, and polyvinylidene fluoride (PVDF), which is a polymer. In addition to the piezoelectric effect, piezoelectric materials exhibit a pyroelectric effect, according to which electric charges begin to appear when the material is subjected to temperature. This effect is used as the underlying principle of several thermal sensors. A sublattice of positive ions in some piezoelectrics turns out to be displaced relative to the sublattice of negative ions in the state of thermodynamic equilibrium. As a result, such crystals are polarized in the absence of an external electric field. Thus, these crystals possess a spontaneous electric polarization. Usually, the presence of such a spontaneous polarization is masked by free surface charges induced on the surface of the crystal from the surrounding medium by the electric field due to spontaneous polarization. This process occurs until the electric field is completely neutralized, that is, until the presence of spontaneous polarization is totally masked. However, as the temperature of the sample changes, for example, as a result of heating, the ionic sublattices become displaced relative to one another, which causes a change in spontaneous polarization, and electric charges appear on the surface of the crystal. The appearance of these charges is called the direct pyroelectric effect, and the corresponding crystals are called pyroelectrics. Every pyroelectric is a piezoelectric, but the converse is not true. This is due to the fact that a pyroelectric has a preferred direction along which spontaneous polarization takes place, while a piezoelectric generally does not have such a direction. The inverse pyroelectric effect is also known to exist: a variation of the electric field in an adiabatically isolated pyroelectric is accompanied by a change in its temperature. The existence of the inverse effect can be proved on the basis of a thermodynamic analysis of the process and be demonstrated experimentally. When conditions are suitable for spontaneous polarlzation, a dielectric tends to go over to such a state in which, on the one hand, spontaneous polarization exists and, on the other, the field energy is minimum. Under these conditions, domains are formed. The factors that weaken the interaction of dipole moments of molecules cause the disappearance of spontaneous polarization and the transition from the ferroelectric state to the state of a polar dielectric. Piezoelectric materials are used widely in transducers such as ultrasonic transmitters and receivers, in sonar for underwater applications, and as actuators for precision positioning devices.

MATLAB 5.1

```
% This script computes parts (a) and (b) for Example 5.1
% using discrete summation approximation for the integration
clear
% the parameters of the shell
r = 0.2;
% Part (a)
sum=0;            % set initial total sum to zero
theta_inc=1/10;   % choose a suitably small increment
                              % for the integral
phi_inc=1/10;     % choose a suitably small increment
                              % for the integral
dtheta=theta_inc*pi/2;
dphi=phi_inc*2*pi;
for theta=0:dtheta:pi/2, % outer integral loop
  for phi=0:dphi:2*pi,  % inner integral loop
      % add the partial sums to the total sum
    sum=sum + 1/r^3*2*cos(theta)*r^2*sin(theta)*dtheta*dphi;
  end
end
% display the output
disp('')
disp(sprintf('The total current through the '))
disp(sprintf(' hemispherical shell is %f A', sum))
% Part (b)
sum=0;            % set initial total sum to zero
r = 0.1;
dtheta=theta_inc*pi;
dphi=phi_inc*2*pi;
for theta=0:dtheta:pi, % outer integral loop
  for phi=0:dphi:2*pi,  % inner integral loop
      % add the partial sums to the total sum
    sum=sum + 1/r^3*2*cos(theta)*r^2*sin(theta)*dtheta*dphi;
  end
end
% display the output
disp('')
disp(sprintf('The total current through the'))
disp(sprintf(' spherical shell is %f A', sum))
```

MATLAB 5.2

```
% This script allows the user to enter an electric field
% on either side of a dielectric boundary and compute the
% electric field on the other side of the boundary
%
% The boundary is assumed to be the plane z=0, with E1 the
% field in
% the region z >=0 and E2 the field in the region z <= 0
%
% inputs: E1 or E2, er1 and er2 (the relative permittivities
% of both media outputs: E1 or E2, the field not input by
% the user
clear
% prompt user for input materials
disp('Enter the relative permittivity in the region ');
er1 = input(' z > 0... \n > ');
if isempty(er1); er1 = 1; elseif er1 < 1; er1 = 1; end
                            % check if dielectric is physical
disp('Enter the relative permittivity in the region ');
er2 = input(' z < 0... \n > ');
if isempty(er2); er2 = 1; elseif er2 < 1; er2 = 1; end
                            % check if dielectric is physical
% prompt the user for the region
disp('Enter the side of the interface where the electric');
side = input('field is known (given)... \n > ');
% if user entered something other than "r" "c" or "s"
% set default as "r"
if isempty(side); side = 1; elseif side > 2; side = 2; end
                            % check if dielectric is physical
if side == 1;
    % prompt the user for the field
    disp('Enter the electric field in side 1 in the ');
    E1 = input(' form [Ex Ey Ez]... \n >');
    E1n = E1(3)*[0 0 1];   % normal direction is +z
    E2n = E1n*er1/er2;  % e-field boundary condition
                                    % for normal component
    E1t = E1 - E1n;   % tangential component of E1
    E2t = E1t;          % e-field boundary condition for
                                    % tangential component
    E2 = E2t + E2n;
elseif side == 2;
    % prompt the user for the field
    disp('Enter the electric field in side 2 in the ');
    E2 = input(' form [Ex Ey Ez]... \n >');
    E2n = E2(3)*[0 0 1];    % normal direction is +z
    E1n = E2n*er2/er1;  % e-field boundary condition
                                    % for normal component
    E2t = E2 - E2n;   % tangential component of E2
```

```
        E1t = E2t;         % e-field boundary condition for
                                      %tangential component

        E1 = E1t + E1n;
    else
        disp('Invalid specification, please re-try \n');
    end
    % Display results
    disp(sprintf('The electric fields are '));
    disp(sprintf('\n E1 = (%d, %d, %d) V/m',E1(1), E1(2), E1(3)));
    disp(sprintf('\n E2 = (%d, %d, %d) V/m',E2(1), E2(2), E2(3)));
```

SUMMARY

1. Materials can be classified roughly as conductors ($\sigma \gg 1$, $\varepsilon_r = 1$) and dielectrics ($\sigma \ll 1$, $\varepsilon_r \geq 1$) in terms of their electrical properties σ and ε_r, where σ is the conductivity and ε_r is the dielectric constant or relative permittivity.

2. Electric current is the flux of electric current density through a surface; that is,

$$I = \int \mathbf{J} \cdot d\mathbf{S}$$

3. The resistance of a conductor of uniform cross section is

$$R = \frac{\ell}{\sigma S}$$

4. The macroscopic effect of polarization on a given volume of a dielectric material is to "paint" its surface with a bound charge $Q_b = \oint_S \rho_{ps}\, dS$ and leave within it an accumulation of bound charge $Q_b = \int_v \rho_{pv}\, dv$, where $\rho_{ps} = \mathbf{P} \cdot \mathbf{a}_n$ and $\rho_{pv} = -\nabla \cdot \mathbf{P}$.

5. In a dielectric medium, the $\mathbf{D}$ and $\mathbf{E}$ fields are related as $\mathbf{D} = \varepsilon\mathbf{E}$, where $\varepsilon = \varepsilon_o \varepsilon_r$ is the permittivity of the medium while $\mathbf{E}$ and $\mathbf{P}$ are related as $\mathbf{P} = \chi_e \varepsilon_o \mathbf{E}$.

6. The electric susceptibility $\chi_e (= \varepsilon_r - 1)$ of a dielectric measures the sensitivity of the material to an electric field.

7. A dielectric material is linear if $\mathbf{D} = \varepsilon\mathbf{E}$ holds, that is, if ε is independent of E. It is homogeneous if ε is independent of position. It is isotropic if ε is a scalar.

8. The principle of charge conservation, the basis of Kirchhoff's current law, is stated in the continuity equation

$$\nabla \cdot \mathbf{J} + \frac{\partial \rho_v}{\partial t} = 0$$

9. The relaxation time, $T_r = \varepsilon/\sigma$, of a material is the time taken by a charge placed in its interior to decrease by a factor of ε^{-1} or to $\approx 37\%$ of its original magnitude.

10. Boundary conditions must be satisfied by an electric field existing in two different media separated by an interface. For a dielectric–dielectric interface

$$E_{1t} = E_{2t}$$

$$D_{1n} - D_{2n} = \rho_S \quad \text{or} \quad D_{1n} = D_{2n} \quad \text{if} \quad \rho_S = 0$$

For a dielectric–conductor interface,

$$E_t = 0, \quad D_n = \varepsilon E_n = \rho_S$$

because $\mathbf{E} = \mathbf{0}$ inside the conductor.

11. Materials of high dielectric constant are of great interest for high-performance electronic devices.

REVIEW QUESTIONS

5.1 Which is *not* an example of convection current?

(a) A moving charged belt

(b) Electronic movement in a vacuum tube

(c) An electron beam in a television tube

(d) Electric current flowing in a copper wire

5.2 What happens when a steady potential difference is applied across the ends of a conducting wire?

(a) All electrons move with a constant velocity.

(b) All electrons move with a constant acceleration.

(c) The random electronic motion will, on the average, be equivalent to a constant velocity of each electron.

(d) The random electronic motion will, on the average, be equivalent to a nonzero constant acceleration of each electron.

5.3 The formula $R = \ell/(\sigma S)$ is for thin wires.

(a) True   (c) Not necessarily

(b) False

5.4 Seawater has $\varepsilon_r = 80$. Its permittivity is

(a) 81   (c) 5.162×10^{-10} F/m

(b) 79   (d) 7.074×10^{-10} F/m

5.5 Both ε_o and χ_e are dimensionless.

(a) True   (b) False

5.6 If $\nabla \cdot \mathbf{D} = \varepsilon \nabla \cdot \mathbf{E}$ and $\nabla \cdot \mathbf{J} = \sigma \nabla \cdot \mathbf{E}$ in a given material, the material is said to be

(a) Linear   (d) Linear and homogeneous

(b) Homogeneous   (e) Linear and isotropic

(c) Isotropic   (f) Isotropic and homogeneous

5.7 The relaxation time of mica $(\sigma = 10^{-15}\text{ S/m}, \varepsilon_r = 6)$ is

(a) $5 \times 10^{-10}\text{ s}$

(d) 10 hr

(b) 10^{-6} s

(e) 15 hr

(c) 5 hr

5.8 The uniform fields shown in Figure 5.18 are near a dielectric–dielectric boundary but on opposite sides of it. Which configurations are correct? Assume that the boundary is charge free and that $\varepsilon_2 > \varepsilon_1$.

5.9 Which of the following statements are incorrect?

(a) The conductivities of conductors and insulators vary with temperature and frequency.

(b) A conductor is an equipotential body in steady state, and **E** is always tangential to the conductor.

(c) Nonpolar molecules have no permanent dipoles.

(d) In a linear dielectric, P varies linearly with E.

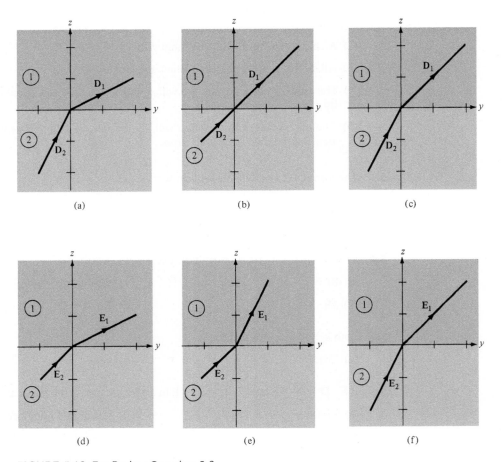

FIGURE 5.18 For Review Question 5.8.

5.10 The electric conditions (charge and potential) inside and outside an electric screening are completely independent of one another.

(a) True   (b) False

Answers: 5.1d, 5.2c, 5.3c, 5.4d, 5.5b, 5.6d, 5.7e, 5.8e, 5.9b, 5.10a.

PROBLEMS

Section 5.3—Convection and Conduction Currents

5.1 Let the current density be $\mathbf{J} = e^{-x}\cos 4\, y\mathbf{a}_x + e^{-x}\sin 4\, y\mathbf{a}_y$ A/m². Determine the current crossing the surface $x = 2,\ 0 < y < \pi/3,\ 0 < z < 4$.

5.2 In a certain region, $\mathbf{J} = \dfrac{10}{r}\, e^{-10^3 t}\mathbf{a}_r$ A/m² . Determine how much current is crossing surface $r = 4$ m at $t = 2$ ms.

5.3 Given that $\mathbf{J} = \dfrac{10}{\rho}\sin\phi\, \mathbf{a}_\rho$ A/m², determine the current flowing through the surface $\rho = 2, 0 < \phi < \pi, 0 < z < 5$ m.

5.4 In a cylindrical conductor of radius 4 mm, the current density is $\mathbf{J} = 5e^{-10\rho}\mathbf{a}_z$ A/m². Find the current through the conductor.

5.5 The current density is

$$\mathbf{J} = \frac{20\, \cos\theta}{r + 3}\mathbf{a}_r\ \text{A/m}^2$$

Determine the current through the surface $r = 3,\ \pi/4 < \theta < \pi/2, 0 < \phi < 2\pi$.

Section 5.4—Conductors

5.6 A 1 MΩ resistor is formed by a cylinder of graphite–clay mixture having a length of 2 cm and a radius of 4 mm. Determine the conductivity of the resistor.

5.7 If the ends of a cylindrical bar of carbon ($\sigma = 3 \times 10^4$ S/m) of radius 5 mm and length 8 cm are maintained at a potential difference of 9 V, find (a) the resistance of the bar, (b) the current through the bar, (c) the power dissipated in the bar.

5.8 A conducting wire is 2 mm in radius and 100 m in length. When a dc voltage of 9 V is applied to the wire, it results in a current of 0.3 A. Find: (a) the E-field in the wire, (b) the conductivity of the wire.

5.9 Two wires have the same diameter and same resistance. If one is made of copper, and the other is of silver, which wire is longer?

5.10 A long wire with circular cross section has a diameter of 4 mm. The wire is 5 m long and it carries 2 A when a 12 V voltage is applied across its ends. Determine the conductivity of the wire.

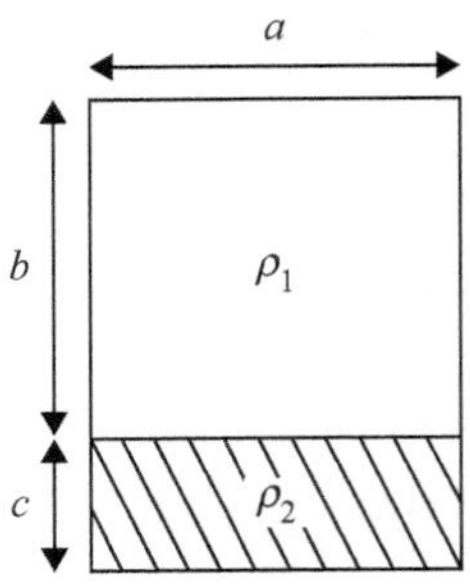

FIGURE 5.19 For Problem 5.12.

5.11 A composite conductor 10 m long consists of an inner core of steel of radius 1.5 cm and an outer sheath of copper whose thickness is 0.5 cm. Take the resistivities of copper and steel as 1.77×10^{-8} and $11.8 \times 10^{-8}\ \Omega \cdot$ m, respectively.

(a) Determine the resistance of the conductor.

(b) If the total current in the conductor is 60 A, what current flows in each metal?

(c) Find the resistance of a solid copper conductor of the same length and cross-sectional areas as the sheath.

5.12 The cross section of a conductor made with two materials with resistivities ρ_1 and ρ_2 is shown in Figure 5.19. Find the resistance of length ℓ of the conductor.

5.13 A 12 V voltage is applied across the ends of a silver wire of length 12.4 m and radius 0.84 mm. Determine the current through the wire.

Sections 5.5–5.7—Polarization and Dielectric Constant

5.14 At a particular temperature and pressure, a helium gas contains 5×10^{25} atoms/m³. If a 10 kV/m field applied to the gas causes an average electron cloud shift of 10^{-18} m, find the dielectric constant of helium.

5.15 A dielectric material contains 2×10^{19} polar molecules/m³, each of dipole moment 1.8×10^{-27} C $\cdot$ m. Assuming that all the dipoles are aligned in the direction of the electric field $\mathbf{E} = 10^5 \mathbf{a}_x$ V/m, find $\mathbf{P}$ and ε_r.

5.16 A 10 mC point charge is embedded in wood, which has $\varepsilon = 4.0$. Assuming that the charge is located at the origin, find $\mathbf{P}$ at $r = 1$ m.

5.17 In a certain dielectric for which $\varepsilon_r = 3.5$, given that $\mathbf{P} = \dfrac{100}{\rho}\mathbf{a}_\rho$ nC/m², find $\mathbf{E}$ and $\mathbf{D}$ at $\rho = 2$ m.

5.18 A cylindrical slab has a polarization given by $\mathbf{P} = p_o\,\rho\mathbf{a}_\rho$. Find the polarization charge density ρ_{pv} inside the slab and its surface charge density ρ_{ps}.

5.19 A spherical shell has $r = 1.2$ cm and $r = 2.6$ cm as inner and outer radii, respectively. If $\mathbf{P} = 4r\mathbf{a}_r$ pC/m², determine (a) the total bound surface charge on the inner surface, (b) the total bound surface charge on the outer surface, (c) the total bound volume charge.

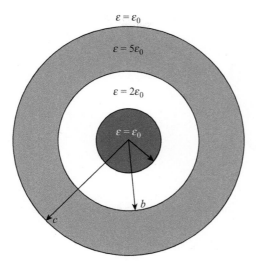

$\varepsilon = \varepsilon_0$

$\varepsilon = 5\varepsilon_0$

$\varepsilon = 2\varepsilon_0$

$\varepsilon = \varepsilon_0$

b

c

FIGURE 5.20 For Problem 5.23.

5.20 In a slab of Teflon ($\varepsilon = 2.1\,\varepsilon_0$), $\mathbf{E} = 6\mathbf{a}_x + 12\mathbf{a}_y - 20\mathbf{a}_z$ V/m, find $\mathbf{D}$ and $\mathbf{P}$.

5.21 The potential distribution in a dielectric material ($\varepsilon = 8\varepsilon_0$) is $V = 4x^2 yz^3$ V. Find V, $\mathbf{E}$, and $\mathbf{P}$ at point $(-2, 5, 3)$.

5.22 In a dielectric material ($\varepsilon = 5\varepsilon_0$), the potential field $V = 10x^2 yz - 5z^2$ V, determine (a) $\mathbf{E}$, (b) $\mathbf{D}$, (c) $\mathbf{P}$, (d) ρ_v.

5.23 Concentric spheres $r = a$, $r = b$, and $r = c$ have charges 4 C, -6 C, and 10 C, respectively, placed on them. If the regions separating them are filled with different dielectrics as shown in Figure 5.20, find $\mathbf{E}$, $\mathbf{D}$, and $\mathbf{P}$ everywhere.

5.24 Consider Figure 5.21 as a spherical dielectric shell so that $\varepsilon = \varepsilon_0\varepsilon_r$ for $a < r < b$ and $\varepsilon = \varepsilon_0$ for $0 < r < a$. If a charge Q is placed at the center of the shell, find

(a) $\mathbf{P}$ for $a < r < b$

(b) ρ_{pv} for $a < r < b$

(c) ρ_{ps} at $r = a$ and $r = b$

5.25 Two point charges in free space are separated by distance d and exert a force 2.6 nN on each other. The force becomes 1.5 nN when the free space is replaced by a homogeneous dielectric material. Calculate the dielectric constant of the material.

*5.26 A conducting sphere of radius a has a total charge Q uniformly distributed on its surface.

(a) If the sphere is embedded in a medium with permittivity ε, find the energy stored.

(b) Repeat part (a) if the permittivity varies as $\varepsilon = \varepsilon_0\left(1 + \dfrac{a}{r}\right)^2$.

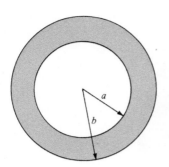

FIGURE 5.21 For Problem 5.24.

5.27 A solid sphere of radius a and dielectric constant ε_r has a uniform volume charge density of ρ_o.

(a) At the center of the sphere, show that

$$V = \frac{\rho_o a^2}{6\varepsilon_o \varepsilon_r}(2\varepsilon_r + 1)$$

(b) Find the potential at the surface of the sphere.

5.28 In an anisotropic medium, **D** is related to **E** as

$$\begin{bmatrix} D_x \\ D_y \\ D_z \end{bmatrix} = \varepsilon_o \begin{bmatrix} 4 & 1 & 1 \\ 1 & 3 & 1 \\ 1 & 1 & 2 \end{bmatrix} \begin{bmatrix} E_x \\ E_y \\ E_z \end{bmatrix}$$

Find **D** due to $\mathbf{E} = E_o(\mathbf{a}_x + \mathbf{a}_y - \mathbf{a}_z)$ V/m.

Section 5.8—Continuity Equation and Relaxation Time

5.29 For static (time-independent) fields, which of the following current densities are possible?

(a) $\mathbf{J} = 2x^3 y \mathbf{a}_x + 4x^2 z^2 \mathbf{a}_y - 6x^2 yz \mathbf{a}_z$ (b) $\mathbf{J} = xy \mathbf{a}_x + y(z + 1)\mathbf{a}_y + 2yz \mathbf{a}_z$

(c) $\mathbf{J} = \dfrac{z^2}{\rho}\mathbf{a}_\rho + z \cos \phi \, \mathbf{a}_z$ (d) $\mathbf{J} = \dfrac{\sin \theta}{r^2}\mathbf{a}_r$

5.30 If $\mathbf{J} = e^{-2y} \sin 2x \mathbf{a}_x + e^{-2y} \cos 2x \mathbf{a}_y + z \mathbf{a}_z$ A/m², find the rate of change of the electric charge density.

5.31 If $\mathbf{J} = \dfrac{100}{\rho^2}\mathbf{a}_\rho$ A/m², find (a) the time rate of increase in the volume charge density, (b) the total current passing through surface defined by $\rho = 2, 0 < z < 1, 0 < \phi < 2\pi$.

5.32 An excess charge placed within a conducting medium becomes one-half of its initial value in 80 μs. Calculate the conductivity of the medium and the relaxation time. Assume that its dielectric constant is 7.5.

5.33   Let ρ_v be the volume charge density of charges in motion. If $\boldsymbol{u}$ is their velocity, show that

$$(\boldsymbol{u} \cdot \nabla)\rho_v + \rho_v\nabla \cdot \boldsymbol{u} + \frac{\partial \rho_v}{\partial t} = 0.$$

5.34   The current density is given by $\mathbf{J} = 0.5 \sin \pi x \mathbf{a}_x$ A/m². Determine the time rate of increase of the charge density (i.e., $\delta\rho_v/\delta t$) at point $(2, 4, -3)$.

5.35   Determine the relaxation time for each of the following media:

(a) Hard rubber $(\sigma = 10^{-15}$ S/m, $\varepsilon = 3.1\varepsilon_o)$
(b) Mica $(\sigma = 10^{-15}$ S/m, $\varepsilon = 6\varepsilon_o)$
(c) Distilled water $(\sigma = 10^{-4}$ S/m, $\varepsilon = 80\varepsilon_o)$

5.36   Lightning strikes a dielectric sphere of radius 20 mm for which $\varepsilon_r = 2.5$, $\sigma = 5 \times 10^{-6}$ S/m and deposits uniformly a charge of 1 C. Determine the initial volume charge density and the volume charge density 2 μs later.

Section 5.9—Boundary Conditions

5.37   Show that the normal and tangential components of the current density $\mathbf{J}$ at the interface between two media with conductivities s_1 and s_2 satisfy

$$J_{1n} = J_{2n}, \quad \frac{J_{1t}}{J_{2t}} = \frac{\sigma_1}{\sigma_2}$$

5.38   Let $z < 0$ be region 1 with dielectric constant $\varepsilon_{r1} = 4$, while $z > 0$ is region 2 with $\varepsilon_{r2} = 7.5$. Given that $\mathbf{E}_1 = 60\mathbf{a}_x - 100\mathbf{a}_y + 40\mathbf{a}_z$ V/m, (a) find $\mathbf{P}_1$, (b) calculate $\mathbf{D}_2$.

5.39   Region 1 is $x < 0$ with, $\boldsymbol{\varepsilon}_1 = 4\varepsilon_o$, while region 2 is $x > 0$ with $\boldsymbol{\varepsilon} = 2\varepsilon_o$. If $\mathbf{E}_2 = 6\mathbf{a}_x - 10\mathbf{a}_y + 8\mathbf{a}_z$ V/m, (a) find $\mathbf{P}_1$, and $\mathbf{P}_2$, (b) calculate the energy densities in both regions.

5.40   A dielectric interface is defined by $4x + 3y = 10$ m. The region including the origin is free space, where $\mathbf{D}_1 = 2\mathbf{a}_x - 4\mathbf{a}_y + 6.5\mathbf{a}_z$ nC/m². In the other region, $\varepsilon_{r2} = 2.5$. Find $\mathbf{D}_2$ and the angle θ_2 that $\mathbf{D}_2$ makes with the normal.

5.41   Regions 1 and 2 have permittivities $\boldsymbol{\varepsilon}_1 = 2\varepsilon_o$ and $\boldsymbol{\varepsilon}_2 = 5\varepsilon_o$. The regions are separated by a plane whose equation is $x + 2y + z = 1$ such that $x + 2y + z > 1$ is region 1. If $\mathbf{E}_1 = 20\mathbf{a}_x - 10\mathbf{a}_y + 40\mathbf{a}_z$ V/m, find: (a) the normal and tangential components of $\mathbf{E}_1$, (b) $\mathbf{E}_2$.

5.42   Given that $\mathbf{E}_1 = 10\mathbf{a}_x - 6\mathbf{a}_y + 12\mathbf{a}_z$ V/m in Figure 5.22, find (a) $\mathbf{P}_1$, (b) $\mathbf{E}_2$ and the angle $\mathbf{E}_2$ makes with the y-axis, (c) the energy density in each region.

5.43   Two homogeneous dielectric regions 1 $(\rho \leq 4$ cm) and 2 $(\rho \geq 4$ cm) have dielectric constants 3.5 and 1.5, respectively. If $\mathbf{D}_2 = 12\mathbf{a}_\rho - 6\mathbf{a}_\phi + 9\mathbf{a}_z$ nC/m², calculate (a) $\mathbf{E}_1$ and $\mathbf{D}_1$, (b) $\mathbf{P}_2$ and ρ_{pv2}, (c) the energy density for each region.

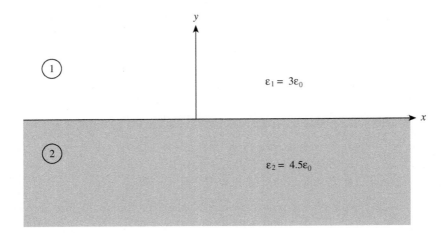

FIGURE 5.22 For Problem 5.42.

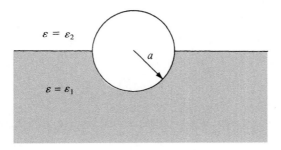

FIGURE 5.23 For Problem 5.44.

5.44 A conducting sphere of radius a is half-embedded in a liquid dielectric medium of permittivity ε_1 as in Figure 5.23. The region above the liquid is a gas of permittivity ε_2. If the total free charge on the sphere is Q, determine the electric field intensity everywhere.

5.45 A dielectric sphere $\boldsymbol{\varepsilon}_1 = 2\boldsymbol{\varepsilon}_o$ is buried in a medium with $\boldsymbol{\varepsilon}_2 = 6\boldsymbol{\varepsilon}_o$. Given that $\mathbf{E}_2 = 10\sin\theta\mathbf{a}_r + 5\cos\theta\mathbf{a}_\theta$ in the medium, calculate E1 and D1 in the dielectric sphere.

*5.46 Two parallel sheets of glass $(\varepsilon_r = 8.5)$ mounted vertically are separated by a uniform air gap between their inner surface. The sheets, properly sealed, are immersed in oil $(\varepsilon_r = 3.0)$ as shown in Figure 5.24. A uniform electric field of strength 2 kV/m in the horizontal direction exists in the oil. Calculate the magnitude and direction of the electric field in the glass and in the enclosed air gap when (a) the field is normal to the glass surfaces and (b) the field in the oil makes an angle of 75° with a normal to the glass surfaces. Ignore edge effects.

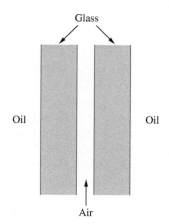

Glass

Oil

Oil

Air

FIGURE 5.24 For Problem 5.46.

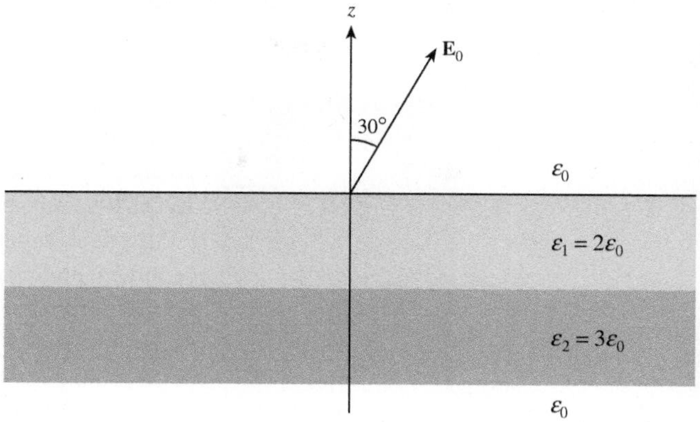

z

$\mathbf{E}_0$

30°

ε_0

$\varepsilon_1 = 2\varepsilon_0$

$\varepsilon_2 = 3\varepsilon_0$

ε_0

FIGURE 5.25 For Problem 5.49.

5.47 At a point on a conducting surface, $\mathbf{E} = 30\mathbf{a}_x - 40\mathbf{a}_y + 20\mathbf{a}_z$ mV/m. Calculate the surface charge density at that point.

5.48 (a) Given that $\mathbf{E} = 15\mathbf{a}_x - 8\mathbf{a}_z$ V/m at a point on a conductor surface, what is the surface charge density at that point? Assume $\varepsilon = \varepsilon_0$.

(b) Region $y \geq 2$ is occupied by a conductor. If the surface charge on the conductor is -20 nC/m², find $\mathbf{D}$ just outside the conductor.

5.49 Two planar slabs of equal thickness but with different dielectric constants are shown in Figure 5.25. $\mathbf{E}_0$ in air makes an angle of 30° with the z-axis. Calculate the angle that $\mathbf{E}$ makes with the z-axis in each of the two dielectric layers.

Pierre-Simon de Laplace (1749–1827), a French astronomer and mathematician, discovered the Laplace transform and Laplace's equation, to be discussed in this chapter. He believed the world was entirely deterministic. To Laplace, the universe was nothing but a giant problem in calculus.

Born of humble origins in Beaumont-en-Auge, Normandy, Laplace became a professor of mathematics at the age of 20. His mathematical abilities inspired the famous mathematician Siméon Poisson, who called Laplace the Isaac Newton of France. Laplace made important contributions in potential theory, probability theory, astronomy, and celestial mechanics. He was widely known for his work *Traité de Mécanique Céleste (Celestial Mechanics)*, which supplemented the work of Newton on astronomy. Laplace is one of the few giants in the history of probability and statistics. He was born and died a Catholic.

Siméon-Denis Poisson (1781–1840), a French mathematical physicist whose name is attached to a wide area of ideas: Poisson's integral, Poisson's equation in potential theory (to be discussed in this chapter), Poisson brackets in differential equations, Poisson's ratio in elasticity, the Poisson distribution in probability theory, and Poisson's constant in electricity.

Born at Pithviers, south of Paris, the son of a retired soldier, Siméon Poisson was originally forced to study medicine by his family, but he began to study mathematics in 1798 at the École Polytechnique at the age of 17. His abilities excited the interest of his teachers Lagrange and Laplace, whose friendship he retained to the end of their lives. A paper on finite differences, written when Poisson was 18, attracted the attention of Legendre. Poisson's chief interest lay in the application of mathematics to physics, especially in electrostatics and magnetism. Poisson made important contributions to mechanics, theory of elasticity, optics, calculus, differential geometry, and probability theory. He published between 300 and 400 mathematical works.

CHAPTER 6

ELECTROSTATIC BOUNDARY-VALUE PROBLEMS

Wise men profit more from fools than fools from wise men; for the wise men shun the mistakes of the fools, but fools do not imitate the successes of the wise.

—MARCUS P. CATO

6.1 INTRODUCTION

The procedure for determining the electric field $\mathbf{E}$ in the preceding chapters has generally been to use either Coulomb's law or Gauss's law when the charge distribution is known, or $\mathbf{E} = -\nabla V$ when the potential V is known throughout the region. In most practical situations, however, neither the charge distribution nor the potential distribution is known.

In this chapter, we shall consider practical electrostatic problems where only electrostatic conditions (charge and potential) at some boundaries are known and it is desired to find $\mathbf{E}$ and V throughout the region. Such problems are usually tackled using Poisson's or Laplace's equation or the method of images, and they are usually referred to as *boundary-value* problems. The concepts of resistance and capacitance will be covered. We shall use Laplace's equation in deriving the resistance of an object and the capacitance of a capacitor. Example 6.5 should be given special attention because we will refer to it often in the remaining part of the text.

6.2 POISSON'S AND LAPLACE'S EQUATIONS

Poisson's and Laplace's equations are easily derived from Gauss's law (for a linear, isotropic material medium):

$$\nabla \cdot \mathbf{D} = \nabla \cdot \varepsilon \mathbf{E} = \rho_v \tag{6.1}$$

and

$$\mathbf{E} = -\nabla V \tag{6.2}$$

Substituting eq. (6.2) into eq. (6.1) gives

$$\nabla \cdot (-\varepsilon \nabla V) = \rho_v \tag{6.3}$$

for an inhomogeneous medium. For a homogeneous medium, eq. (6.3) becomes

$$\boxed{\nabla^2 V = -\frac{\rho_v}{\varepsilon}} \tag{6.4}$$

This is known as *Poisson's equation*. A special case of this equation occurs when $\rho_v = 0$ (i.e., for a charge-free region). Equation (6.4) then becomes

$$\boxed{\nabla^2 V = 0} \tag{6.5}$$

which is known as *Laplace's equation*. Note that in taking ε out of the left-hand side of eq. (6.3) to obtain eq. (6.4), we have assumed that ε is constant throughout the region in which V is defined; for an inhomogeneous region, ε is not constant and eq. (6.4) does not follow eq. (6.3). Equation (6.3) is Poisson's equation for an inhomogeneous medium; it becomes Laplace's equation for an inhomogeneous medium when $\rho_v = 0$.

Recall that the Laplacian operator ∇^2 was derived in Section 3.8. Thus Laplace's equation in Cartesian, cylindrical, or spherical coordinates, respectively, is given by

$$\boxed{\frac{\partial^2 V}{\partial x^2} + \frac{\partial^2 V}{\partial y^2} + \frac{\partial^2 V}{\partial z^2} = 0} \tag{6.6}$$

$$\boxed{\frac{1}{\rho}\frac{\partial}{\partial \rho}\left(\rho\frac{\partial V}{\partial \rho}\right) + \frac{1}{\rho^2}\frac{\partial^2 V}{\partial \phi^2} + \frac{\partial^2 V}{\partial z^2} = 0} \tag{6.7}$$

$$\boxed{\frac{1}{r^2}\frac{\partial}{\partial r}\left(r^2\frac{\partial V}{\partial r}\right) + \frac{1}{r^2\sin\theta}\frac{\partial}{\partial \theta}\left(\sin\theta\frac{\partial V}{\partial \theta}\right) + \frac{1}{r^2\sin^2\theta}\frac{\partial^2 V}{\partial \phi^2} = 0} \tag{6.8}$$

depending on the coordinate variables used to express V, that is, $V(x, y, z)$, $V(\rho, \phi, z)$, or $V(r, \theta, \phi)$. Poisson's equation in those coordinate systems may be obtained by simply replacing zero on the right-hand side of eqs. (6.6), (6.7), and (6.8) with $-\rho_v/\varepsilon$.

Laplace's equation is of primary importance in solving electrostatic problems involving a set of conductors maintained at different potentials. Examples of such problems include capacitors and vacuum tube diodes. Laplace's and Poisson's equations are not only useful in solving electrostatic field problem; they are used in various other field problems. For example, V would be interpreted as magnetic potential in magnetostatics, as temperature in heat conduction, as stress function in fluid flow, and as pressure head in seepage.

†6.3 UNIQUENESS THEOREM

Since there are several methods (analytical, graphical, numerical, experimental, etc.) of solving a given problem, we may wonder whether solving Laplace's equation in different ways gives different solutions. Therefore, before we begin to solve Laplace's equation, we should answer this question: if a solution of Laplace's equation satisfies a given set of boundary conditions, is this the only possible solution? The answer is yes: there is only one solution. We say that the solution is unique. Thus any solution of Laplace's equation that satisfies the same boundary conditions must be the only solution regardless of the method used. This is known as the *uniqueness theorem*. The theorem applies to any solution of Poisson's or Laplace's equation in a given region or closed surface.

The theorem is proved by contradiction. We assume that there are two solutions V_1 and V_2 of Laplace's equation, both of which satisfy the prescribed boundary conditions. Thus

$$\nabla^2 V_1 = 0, \quad \nabla^2 V_2 = 0 \tag{6.9a}$$

$$V_1 = V_2 \quad \text{on the boundary} \tag{6.9b}$$

We consider their difference

$$V_d = V_2 - V_1 \tag{6.10}$$

which obeys

$$\nabla^2 V_d = \nabla^2 V_2 - \nabla^2 V_1 = 0 \tag{6.11a}$$

$$V_d = 0 \quad \text{on the boundary} \tag{6.11b}$$

according to eq. (6.9). From the divergence theorem

$$\int_v \nabla \cdot \mathbf{A} \, dv = \oint_S \mathbf{A} \cdot d\mathbf{S} \tag{6.12}$$

where S is the surface surrounding volume v and is the boundary of the original problem. We let $\mathbf{A} = V_d \nabla V_d$ and use a vector identity

$$\nabla \cdot \mathbf{A} = \nabla \cdot (V_d \nabla V_d) = V_d \nabla^2 V_d + \nabla V_d \cdot \nabla V_d$$

But $\nabla^2 V_d = 0$ according to eq. (6.11a), so

$$\nabla \cdot \mathbf{A} = \nabla V_d \cdot \nabla V_d \tag{6.13}$$

Substituting eq. (6.13) into eq. (6.12) gives

$$\int_v \nabla V_d \cdot \nabla V_d \, dv = \oint_S V_d \nabla V_d \cdot d\mathbf{S} \tag{6.14}$$

From eqs. (6.9) and (6.11), it is evident that the right-hand side of eq. (6.14) vanishes.

Hence,

$$\int_v |\nabla V_d|^2 \, dv = 0$$

Since the integrand is everywhere positive,

$$|\nabla V|_d = 0 \tag{6.15a}$$

or

$$V_d = V_2 - V_1 = \text{constant everywhere in } v \tag{6.15b}$$

But eq. (6.15) must be consistent with eq. (6.9b). Hence, $V_d = 0$ or $V_1 = V_2$ everywhere, showing that V_1 and V_2 cannot be different solutions of the same problem.

> This is the **uniqueness theorem:** If a solution to Laplace's equation can be found that satisfies the boundary conditions, then the solution is unique.

Similar steps can be taken to show that the theorem applies to Poisson's equation and to prove the theorem for the case where the electric field (potential gradient) is specified on the boundary.

Before we begin to solve boundary-value problems, we should bear in mind the three things that uniquely describe a problem:

1. The appropriate differential equation (Laplace's or Poisson's equation in this chapter)
2. The solution region
3. The prescribed boundary conditions

A problem does not have a unique solution and cannot be solved completely if any of the three items is missing.

6.4 GENERAL PROCEDURES FOR SOLVING POISSON'S OR LAPLACE'S EQUATION

The following general procedure may be taken in solving a given boundary-value problem involving Poisson's or Laplace's equation:

1. Solve Laplace's (if $\rho_v = 0$) or Poisson's (if $\rho_v \neq 0$) equation using either (a) direct integration when V is a function of one variable or (b) separation of variables if V is a function of more than one variable. The solution at this point is not unique but is expressed in terms of unknown integration constants to be determined.
2. Apply the boundary conditions to determine a unique solution for V. Imposing the given boundary conditions makes the solution unique.
3. Having obtained V, find $\mathbf{E}$ using $\mathbf{E} = -\nabla V$, $\mathbf{D}$ from $\mathbf{D} = \varepsilon \mathbf{E}$, and $\mathbf{J}$ from $\mathbf{J} = \sigma \mathbf{E}$.

4. If required, find the charge Q induced on a conductor using $Q = \int_s \rho_S \, dS$, where $\rho_S = D_n$ and D_n is the component of **D** normal to the conductor. If necessary, the capacitance of two conductors can be found using $C = Q/V$ or the resistance of an object can be found by using $R = V/I$, where $I = \int_s \mathbf{J} \cdot d\mathbf{S}$.

Solving Laplace's (or Poisson's) equation, as in step 1, is not always as complicated as it may seem. In some cases, the solution may be obtained by mere inspection of the problem. Also a solution may be checked by going backward and finding out if it satisfies both Laplace's (or Poisson's) equation and the prescribed boundary condition.

EXAMPLE 6.1

Current-carrying components in high-voltage power equipment can be cooled to carry away the heat caused by ohmic losses. A means of pumping is based on the force transmitted to the cooling fluid by charges in an electric field. Electrohydrodynamic (EHD) pumping is modeled in Figure 6.1. The region between the electrodes contains a uniform charge ρ_o, which is generated at the left electrode and collected at the right electrode. Calculate the pressure of the pump if $\rho_o = 25 \text{ mC/m}^3$ and $V_o = 22 \text{ kV}$.

Solution:

Since $\rho_v \neq 0$, we apply Poisson's equation

$$\nabla^2 V = -\frac{\rho_v}{\varepsilon}$$

The boundary conditions $V(z = 0) = V_o$ and $V(z = d) = 0$ show that V depends only on z (there is no ρ or ϕ dependence). Hence,

$$\frac{d^2 V}{dz^2} = \frac{-\rho_o}{\varepsilon}$$

Integrating once gives

$$\frac{dV}{dz} = \frac{-\rho_o z}{\varepsilon} + A$$

Integrating again yields

$$V = -\frac{\rho_o z^2}{2\varepsilon} + Az + B$$

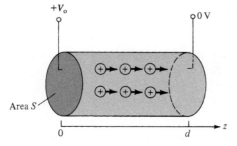

FIGURE 6.1 An electrohydrodynamic pump; for Example 6.1.

where A and B are integration constants to be determined by applying the boundary conditions. When $z = 0$, $V = V_o$,

$$V_o = -0 + 0 + B \rightarrow B = V_o$$

When $z = d$, $V = 0$,

$$0 = -\frac{\rho_o d^2}{2\varepsilon} + Ad + V_o$$

or

$$A = \frac{\rho_o d}{2\varepsilon} - \frac{V_o}{d}$$

The electric field is given by

$$\mathbf{E} = -\nabla V = -\frac{dV}{dz}\mathbf{a}_z = \left(\frac{\rho_o z}{\varepsilon} - A\right)\mathbf{a}_z$$

$$= \left[\frac{V_o}{d} + \frac{\rho_o}{\varepsilon}\left(z - \frac{d}{2}\right)\right]\mathbf{a}_z$$

The net force is

$$\mathbf{F} = \int_v \rho_v \mathbf{E}\, dv = \rho_o \int dS \int_{z=0}^d \mathbf{E}\, dz$$

$$= \rho_o S\left[\frac{V_o z}{d} + \frac{\rho_o}{2\varepsilon}(z^2 - dz)\right]\Big|_0^d \mathbf{a}_z$$

$$\mathbf{F} = \rho_o S V_o \mathbf{a}_z$$

The force per unit area or pressure is

$$\rho = \frac{F}{S} = \rho_o V_o = 25 \times 10^{-3} \times 22 \times 10^3 = 550 \text{ N/m}^2$$

PRACTICE EXERCISE 6.1

In a one-dimensional device, the charge density is given by $\rho_v = \rho_o x/a$. If $\mathbf{E} = \mathbf{0}$ at $x = 0$ and $V = 0$ at $x = a$, find V and $\mathbf{E}$.

Answer: $\dfrac{\rho_o}{6\varepsilon a}(a^3 - x^3)$, $\dfrac{\rho_o x^2}{2a\varepsilon}\mathbf{a}_x$.

EXAMPLE 6.2

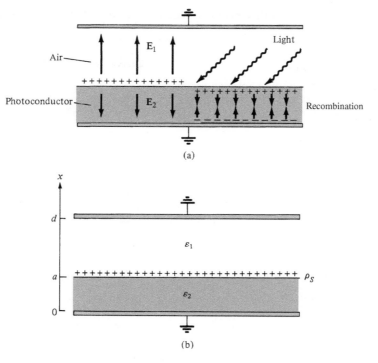

FIGURE 6.2 For Example 6.2.

The xerographic copying machine is an important application of electrostatics. The surface of the photoconductor is initially charged uniformly as in Figure 6.2(a). When light from the document to be copied is focused on the photoconductor, the charges on the lower surface combine with those on the upper surface to neutralize each other. The image is developed by pouring a charged black powder over the surface of the photoconductor. The electric field attracts the charged powder, which is later transferred to paper and melted to form a permanent image. We want to determine the electric field below and above the surface of the photoconductor.

Solution:

Consider the modeled version of Figure 6.2(a) shown in Figure 6.2(b). Since $\rho_v = 0$ in this case, we apply Laplace's equation. Also the potential depends only on x. Thus

$$\nabla^2 V = \frac{d^2V}{dx^2} = 0$$

Integrating twice gives

$$V = Ax + B$$

Let the potentials above and below $x = a$ be V_1 and V_2, respectively:

$$V_1 = A_1x + B_1, \quad x > a \tag{6.2.1a}$$

$$V_2 = A_2x + B_2, \quad x < a \tag{6.2.1b}$$

The boundary conditions at the grounded electrodes are

$$V_1(x = d) = 0 \tag{6.2.2a}$$

$$V_2(x = 0) = 0 \tag{6.2.2b}$$

At the surface of the photoconductor,

$$V_1(x = a) = V_2(x = a) \tag{6.2.3a}$$

$$D_{1n} - D_{2n} = \rho_S \Big|_{x=a} \tag{6.2.3b}$$

We use the four conditions in eqs. (6.2.2) and (6.2.3) to determine the four unknown constants A_1, A_2, B_1, and B_2. From eqs. (6.2.1) and (6.2.2),

$$0 = A_1 d + B_1 \;\rightarrow\; B_1 = -A_1 d \tag{6.2.4a}$$

$$0 = 0 + B_2 \;\rightarrow\; B_2 \;\; = 0 \tag{6.2.4b}$$

From eqs. (6.2.1) and (6.2.3a),

$$A_1 a + B_1 = A_2 a \tag{6.2.5}$$

To apply eq. (6.2.3b), recall that $\mathbf{D} = \varepsilon \mathbf{E} = -\varepsilon \nabla V$ so that

$$\rho_S = D_{1n} - D_{2n} = \varepsilon_1 E_{1n} - \varepsilon_2 E_{2n} = -\varepsilon_1 \frac{dV_1}{dx} + \varepsilon_2 \frac{dV_2}{dx}$$

or

$$\rho_S = -\varepsilon_1 A_1 + \varepsilon_2 A_2 \tag{6.2.6}$$

Solving for A_1 and A_2 in eqs. (6.2.4) to (6.2.6), we obtain

$$\mathbf{E}_1 = -A_1 \mathbf{a}_x = \frac{\rho_S \mathbf{a}_x}{\varepsilon_1 \left[1 + \dfrac{\varepsilon_2}{\varepsilon_1} \dfrac{d}{a} - \dfrac{\varepsilon_2}{\varepsilon_1} \right]}, \quad a \le x \le d$$

$$\mathbf{E}_2 = -A_2 \mathbf{a}_x = \frac{-\rho_S \left(\dfrac{d}{a} - 1 \right) \mathbf{a}_x}{\varepsilon_1 \left[1 + \dfrac{\varepsilon_2}{\varepsilon_1} \dfrac{d}{a} - \dfrac{\varepsilon_2}{\varepsilon_1} \right]}, \quad 0 \le x \le a$$

PRACTICE EXERCISE 6.2

For the model of Figure 6.2(b), if $\rho_S = 0$ and the upper electrode is maintained at V_o while the lower electrode is grounded, show that

$$\mathbf{E}_1 = \frac{-V_o \mathbf{a}_x}{d - a + \dfrac{\varepsilon_1}{\varepsilon_2} a}, \quad \mathbf{E}_2 = \frac{-V_o \mathbf{a}_x}{a + \dfrac{\varepsilon_2}{\varepsilon_1} d - \dfrac{\varepsilon_2}{\varepsilon_1} a}$$

Answer: Proof.

EXAMPLE 6.3

Semi-infinite conducting planes at $\phi = 0$ and $\phi = \pi/6$ are separated by an infinitesimal insulating gap as shown in Figure 6.3. If $V(\phi = 0) = 0$ and $V(\phi = \pi/6) = 100$ V, calculate V and $\mathbf{E}$ in the region between the planes.

Solution:

Since V depends only on ϕ, Laplace's equation in cylindrical coordinates becomes

$$\nabla^2 V = \frac{1}{\rho^2}\frac{d^2V}{d\phi^2} = 0$$

Since $\rho = 0$ is excluded owing to the insulating gap, we can multiply by ρ^2 to obtain

$$\frac{d^2V}{d\phi^2} = 0$$

which is integrated twice to give

$$V = A\phi + B$$

We apply the boundary conditions to determine constants A and B. When $\phi = 0$, $V = 0$,

$$0 = 0 + B \rightarrow B = 0$$

When $\phi = \phi_o$, $V = V_o$,

$$V_o = A\phi_o \rightarrow A = \frac{V_o}{\phi_o}$$

Hence,

$$V = \frac{V_o}{\phi_o}\phi$$

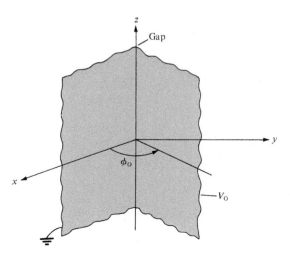

FIGURE 6.3 Potential $V(\phi)$ due to semi-infinite conducting planes.

and

$$\mathbf{E} = -\nabla V = -\frac{1}{\rho}\frac{dV}{d\phi}\mathbf{a}_\phi = -\frac{V_o}{\rho\phi_o}\mathbf{a}_\phi$$

Substituting $V_o = 100$ and $\phi_o = \pi/6$ gives

$$V = \frac{600}{\pi}\phi \quad \text{and} \quad \mathbf{E} = \frac{-600}{\pi\rho}\mathbf{a}_\phi$$

Check: $\nabla^2 V = 0$, $V(\phi = 0) = 0$, $V(\phi = \pi/6) = 100$.

PRACTICE EXERCISE 6.3

Two conducting plates of size 1×5 m are inclined at $45°$ to each other with a gap of width 4 mm separating them as shown in Figure 6.4. Determine an approximate value of the charge per plate if the plates are maintained at a potential difference of 50 V. Assume that the medium between them has $\varepsilon_r = 1.5$.

Answer: 22.2 nC.

EXAMPLE 6.4

Two conducting cones ($\theta = \pi/10$ and $\theta = \pi/6$) of infinite extent are separated by an infinitesimal gap at $r = 0$. If $V(\theta = \pi/10) = 0$ and $V(\theta = \pi/6) = 50$ V, find V and $\mathbf{E}$ between the cones.

Solution:

Consider the coaxial cone of Figure 6.5, where the gap serves as an insulator between the two conducting cones. Here V depends only on θ, so Laplace's equation in spherical coordinates becomes

$$\nabla^2 V = \frac{1}{r^2 \sin\theta}\frac{d}{d\theta}\left[\sin\theta\frac{dV}{d\theta}\right] = 0$$

FIGURE 6.4 For Practice Exercise 6.3.

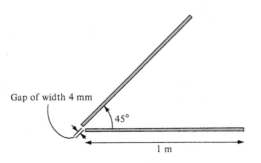

Gap of width 4 mm

$45°$

1 m

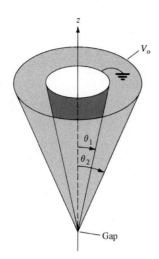

Since $r = 0$ and $\theta = 0, \pi$ are excluded, we can multiply by $r^2 \sin \theta$ to get

$$\frac{d}{d\theta}\left[\sin \theta \frac{dV}{d\theta}\right] = 0$$

Integrating once gives

$$\sin \theta \frac{dV}{d\theta} = A$$

or

$$\frac{dV}{d\theta} = \frac{A}{\sin \theta}$$

Integrating this results in

$$V = A \int \frac{d\theta}{\sin \theta} = A \int \frac{d\theta}{2 \cos \theta/2 \sin \theta/2}$$

$$= A \int \frac{1/2 \sec^2 \theta/2 \; d\theta}{\tan \theta/2}$$

$$= A \int \frac{d(\tan \theta/2)}{\tan \theta/2}$$

$$= A \ln(\tan \theta/2) + B$$

We now apply the boundary conditions to determine the integration constants A and B.

$$V(\theta = \theta_1) = 0 \;\rightarrow\; 0 = A \ln(\tan \theta_1/2) + B$$

or

$$B = -A \ln(\tan \theta_1/2)$$

Hence,

$$V = A \ln\left[\frac{\tan \theta/2}{\tan \theta_1/2}\right]$$

Also

$$V(\theta = \theta_2) = V_o \rightarrow V_o = A \ln\left[\frac{\tan \theta_2/2}{\tan \theta_1/2}\right]$$

or

$$A = \frac{V_o}{\ln\left[\dfrac{\tan \theta_2/2}{\tan \theta_1/2}\right]}$$

Thus

$$V = \frac{V_o \ln\left[\dfrac{\tan \theta/2}{\tan \theta_1/2}\right]}{\ln\left[\dfrac{\tan \theta_2/2}{\tan \theta_1/2}\right]}$$

$$\mathbf{E} = -\nabla V = -\frac{1}{r}\frac{dV}{d\theta}\mathbf{a}_\theta = -\frac{A}{r \sin \theta}\mathbf{a}_\theta$$

$$= -\frac{V_o}{r \sin \theta \ln\left[\dfrac{\tan \theta_2/2}{\tan \theta_1/2}\right]}\mathbf{a}_\theta$$

Taking $\theta_1 = \pi/10$, $\theta_2 = \pi/6$, and $V_o = 50$ gives

$$V = \frac{50 \ln\left[\dfrac{\tan \theta/2}{\tan \pi/20}\right]}{\ln\left[\dfrac{\tan \pi/12}{\tan \pi/20}\right]} = 95.1 \ln\left[\frac{\tan \theta/2}{0.1584}\right] V$$

and

$$\mathbf{E} = -\frac{95.1}{r \sin \theta}\mathbf{a}_\theta \; V/m$$

Check: $\nabla^2 V = 0$, $V(\theta = \pi/10) = 0$, $V(\theta = \pi/6) = V_o$.

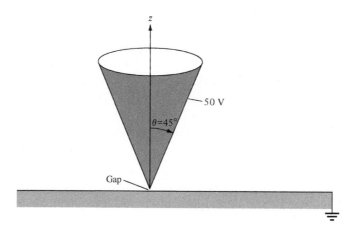

FIGURE 6.6 For Practice Exercise 6.4.

PRACTICE EXERCISE 6.4

A large conducting cone $(\theta = 45°)$ is placed on a conducting plane with a tiny gap separating it from the plane as shown in Figure 6.6. If the cone is connected to a 50 V source, find V and $\mathbf{E}$ at $(-3, 4, 2)$.

Answer: 27.87 V, $-11.35\mathbf{a}_\theta$ V/m.

EXAMPLE 6.5

(a) Determine the potential function for the region inside the rectangular trough of infinite length whose cross section is shown in Figure 6.7.
(b) For $V_\text{o} = 100$ V and $b = 2a$, find the potential at $x = a/2$, $y = 3a/4$.

Solution:

(a) The potential V in this case depends on x and y. Laplace's equation becomes

$$\nabla^2 V = \frac{\partial^2 V}{\partial x^2} + \frac{\partial^2 V}{\partial y^2} = 0 \tag{6.5.1}$$

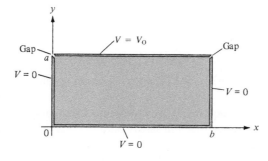

FIGURE 6.7 Potential $V(x, y)$ due to a conducting rectangular trough; for Example 6.5.

We have to solve this equation subject to the following boundary conditions:

$$V(x = 0, 0 \leq y < a) = 0 \tag{6.5.2a}$$

$$V(x = b, 0 \leq y < a) = 0 \tag{6.5.2b}$$

$$V(0 \leq x \leq b, y = 0) = 0 \tag{6.5.2c}$$

$$V(0 < x < b, y = a) = V_o \tag{6.5.2d}$$

We solve eq. (6.5.1) by the method of *separation of variables;* that is, we seek a product solution of V. Let

$$V(x, y) = X(x) \, Y(y) \tag{6.5.3}$$

where X is a function of x only and Y is a function of y only. Substituting eq. (6.5.3) into eq. (6.5.1) yields

$$X''Y + Y''X = 0$$

Dividing through by XY and separating X from Y gives

$$-\frac{X''}{X} = \frac{Y''}{Y} \tag{6.5.4a}$$

Since the left-hand side of this equation is a function of x only and the right-hand side is a function of y only, for the equality to hold, both sides must be equal to a constant λ; that is,

$$-\frac{X''}{X} = \frac{Y''}{Y} = \lambda \tag{6.5.4b}$$

The constant λ is known as the *separation constant.* From eq. (6.5.4b), we obtain

$$X'' + \lambda X = 0 \tag{6.5.5a}$$

and

$$Y'' - \lambda Y = 0 \tag{6.5.5b}$$

Thus the variables have been separated at this point and we refer to eq. (6.5.5) as *separated equations.* We can solve for $X(x)$ and $Y(y)$ separately and then substitute our solutions into eq. (6.5.3). To do this requires that the boundary conditions in eq. (6.5.2) be separated, if possible. We separate them as follows:

$$V(0, y) = X(0)Y(y) = 0 \rightarrow X(0) = 0 \tag{6.5.6a}$$

$$V(b, y) = X(b)Y(y) = 0 \rightarrow X(b) = 0 \tag{6.5.6b}$$

$$V(x, 0) = X(x)Y(0) = 0 \rightarrow Y(0) = 0 \tag{6.5.6c}$$

$$V(x, a) = X(x)Y(a) = V_o \, (\text{inseparable}) \tag{6.5.6d}$$

To solve for $X(x)$ and $Y(y)$ in eq. (6.5.5), we impose the boundary conditions in eq. (6.5.6). We consider possible values of λ that will satisfy both the separated equations in eq. (6.5.5) and the conditions in eq. (6.5.6).

CASE 1.

If $\lambda = 0$, then eq. (6.5.5a) becomes

$$X'' = 0 \quad \text{or} \quad \frac{d^2X}{dx^2} = 0$$

which, upon integrating twice, yields

$$X = Ax + B \tag{6.5.7}$$

The boundary conditions in eqs. (6.5.6a) and (6.5.6b) imply that

$$X(x = 0) = 0 \rightarrow 0 = 0 + B \quad \text{or} \quad B = 0$$

and

$$X(x = b) = 0 \rightarrow 0 = A \cdot b + 0 \quad \text{or} \quad A = 0$$

because $b \neq 0$. Hence our solution for X in eq. (6.5.7) becomes

$$X(x) = 0$$

which makes $V = 0$ in eq. (6.5.3). Thus we regard $X(x) = 0$ as a trivial solution and we conclude that $\lambda \neq 0$.

CASE 2.

If $\lambda < 0$, say $\lambda = -\alpha^2$, then eq. (6.5.5a) becomes

$$X'' - \alpha^2 X = 0 \quad \text{or} \quad (D^2 - \alpha^2)X = 0$$

where $D = \dfrac{d}{dx}$, that is,

$$DX = \pm \alpha X \tag{6.5.8}$$

showing that we have two possible solutions corresponding to the plus and minus signs. For the plus sign, eq. (6.5.8) becomes

$$\frac{dX}{dx} = \alpha X \quad \text{or} \quad \frac{dX}{X} = \alpha \, dx$$

Hence,

$$\int \frac{dX}{X} = \int \alpha \, dx \quad \text{or} \quad \ln X = \alpha x + \ln A_1$$

where $\ln A_1$ is a constant of integration. Thus

$$X = A_1 e^{\alpha x} \tag{6.5.9a}$$

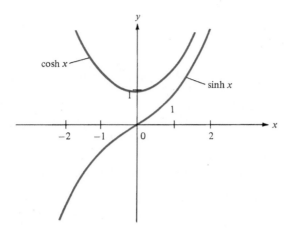

FIGURE 6.8 Sketch of cosh x and sinh x showing that sinh $x = 0$ if and only if $x = 0$; for Case 2 of Example 6.5.

Similarly, for the minus sign, solving eq. (6.5.8) gives

$$X = A_2 e^{-\alpha x} \tag{6.5.9b}$$

The total solution consists of what we have in eqs. (6.5.9a) and (6.5.9b); that is,

$$X(x) = A_1 e^{\alpha x} + A_2 e^{-\alpha x} \tag{6.5.10}$$

Since $\cosh \alpha x = (e^{\alpha x} + e^{-\alpha x})/2$ and $\sinh \alpha x = (e^{\alpha x} - e^{-\alpha x})/2$ or $e^{\alpha x} = \cosh \alpha x + \sinh \alpha x$ and $e^{-\alpha x} = \cosh \alpha x - \sinh \alpha x$, eq. (6.5.10) can be written as

$$X(x) = B_1 \cosh \alpha x + B_2 \sinh \alpha x \tag{6.5.11}$$

where $B_1 = A_1 + A_2$ and $B_2 = A_1 - A_2$. In view of the given boundary conditions, we prefer eq. (6.5.11) to eq. (6.5.10) as the solution. Again, eqs. (6.5.6a) and (6.5.6b) require that

$$X(x = 0) = 0 \rightarrow 0 = B_1 \cdot (1) + B_2 \cdot (0) \quad \text{or} \quad B_1 = 0$$

and

$$X(x = b) = 0 \rightarrow 0 = 0 + B_2 \sinh \alpha b$$

Since $\alpha \neq 0$ and $b \neq 0$, $\sinh \alpha b$ cannot be zero. This is due to the fact that $\sinh x = 0$ if and only if $x = 0$ as shown in Figure 6.8. Hence $B_2 = 0$ and

$$X(x) = 0$$

This is also a trivial solution and we conclude that λ cannot be less than zero.

CASE 3.

If $\lambda > 0$, say $\lambda = \beta^2$, then eq. (6.5.5a) becomes

$$X'' + \beta^2 X = 0$$

that is,

$$(D^2 + \beta^2)X = 0 \quad \text{or} \quad DX = \pm j\beta X \tag{6.5.12}$$

where $j = \sqrt{-1}$. From eqs. (6.5.8) and (6.5.12), we notice that the difference between Cases 2 and 3 is the replacement of α by $j\beta$. By taking the same procedure as in Case 2, we obtain the solution as

$$X(x) = C_o e^{j\beta x} + C_1 e^{-j\beta x} \tag{6.5.13a}$$

Since $e^{j\beta x} = \cos \beta x + j \sin \beta x$ and $e^{-j\beta x} = \cos \beta x - j \sin \beta x$, eq. (6.5.13a) can be written as

$$X(x) = g_o \cos \beta x + g_1 \sin \beta x \tag{6.5.13b}$$

where $g_o = C_o + C_1$ and $g_1 = j(C_o - C_1)$.

In view of the given boundary conditions, we prefer to use eq. (6.5.13b). Imposing the conditions in eqs. (6.5.6a) and (6.5.6b) yields

$$X(x = 0) = 0 \rightarrow 0 = g_o \cdot (1) + 0 \quad \text{or} \quad g_o = 0$$

and

$$X(x = b) = 0 \rightarrow 0 = 0 + g_1 \sin \beta b$$

Suppose $g_1 \neq 0$ (otherwise we get a trivial solution), then

$$\sin \beta b = 0 = \sin n\pi \rightarrow \beta b = n\pi$$
$$\beta = \frac{n\pi}{b}, \quad n = 1, 2, 3, 4, \ldots \tag{6.5.14}$$

Note that, unlike sinh x, which is zero only when $x = 0$, sin x is zero at an infinite number of points as shown in Figure 6.9. It should also be noted that $n \neq 0$ because $\beta \neq 0$; we have already considered the possibility $\beta = 0$ in Case 1, where we ended up with a trivial solution. Also we do not need to consider $n = -1, -2, -3, -4, \ldots$ because $\lambda = \beta^2$

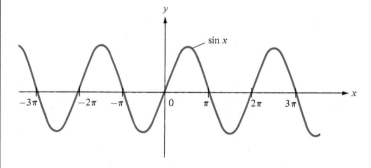

FIGURE 6.9 Sketch of sin x showing that sin $x = 0$ at infinite number of points; for Case 3 of Example 6.5.

would remain the same for positive and negative integer values of n. Thus for a given n, eq. (6.5.13b) becomes

$$X_n(x) = g_n \sin \frac{n\pi x}{b} \tag{6.5.15}$$

Having found $X(x)$ and

$$\lambda = \beta^2 = \frac{n^2\pi^2}{b^2} \tag{6.5.16}$$

we solve eq. (6.5.5b), which is now

$$Y'' - \beta^2 Y = 0$$

The solution to this is similar to eq. (6.5.11) obtained in Case 2; that is,

$$Y(y) = h_o \cosh \beta y + h_1 \sinh \beta y$$

The boundary condition in eq. (6.5.6c) implies that

$$Y(y = 0) = 0 \rightarrow 0 = h_o \cdot (1) + 0 \quad \text{or} \quad h_o = 0$$

Hence our solution for $Y(y)$ becomes

$$Y_n(y) = h_n \sinh \frac{n\pi y}{b} \tag{6.5.17}$$

Substituting eqs. (6.5.15) and (6.5.17), which are the solutions to the separated equations in eq. (6.5.5), into the product solution in eq. (6.5.3) gives

$$V_n(x, y) = g_n h_n \sin \frac{n\pi x}{b} \sinh \frac{n\pi y}{b}$$

This shows that there are many possible solutions V_1, V_2, V_3, V_4, and so on, for $n = 1, 2, 3, 4$, and so on.

By the *superposition theorem*, if V_1, V_2, V_3, ..., V_n are solutions of Laplace's equation, the linear combination

$$V = c_1 V_1 + c_2 V_2 + c_3 V_3 + \cdots + c_n V_n$$

(where c_1, c_2, c_3, ..., c_n are constants) is also a solution of Laplace's equation. Thus the solution to eq. (6.5.1) is

$$V(x, y) = \sum_{n=1}^{\infty} c_n \sin \frac{n\pi x}{b} \sinh \frac{n\pi y}{b} \tag{6.5.18}$$

where $c_n = g_n h_n$ are the coefficients to be determined from the boundary condition in eq. (6.5.6d). Imposing this condition gives

$$V(x, y = a) = V_o = \sum_{n=1}^{\infty} c_n \sin \frac{n\pi x}{b} \sinh \frac{n\pi a}{b} \tag{6.5.19}$$

which is a Fourier series expansion of V_o. Multiplying both sides of eq. (6.5.19) by $\sin m\pi x/b$ and integrating over $0 < x < b$ gives

$$\int_0^b V_o \sin \frac{m\pi x}{b} \, dx = \sum_{n=1}^{\infty} c_n \sinh \frac{n\pi a}{b} \int_0^b \sin \frac{m\pi x}{b} \sin \frac{n\pi x}{b} \, dx \qquad (6.5.20)$$

By the orthogonality property of the sine or cosine function (see Appendix A.9).

$$\int_0^{\pi} \sin mx \sin nx \, dx = \begin{bmatrix} 0, & m \neq n \\ \pi/2, & m = n \end{bmatrix}$$

Incorporating this property in eq. (6.5.20) means that all terms on the right-hand side of eq. (6.5.20) will vanish except one term in which $m = n$. Thus eq. (6.5.20) reduces to

$$\int_0^b V_o \sin \frac{n\pi x}{b} \, dx = c_n \sinh \frac{n\pi a}{b} \int_0^b \sin^2 \frac{n\pi x}{b} \, dx$$

$$-V_o \frac{b}{n\pi} \cos \frac{n\pi x}{b}\bigg|_0^b = c_n \sinh \frac{n\pi a}{b} \frac{1}{2} \int_0^b \left(1 - \cos \frac{2n\pi x}{b}\right) dx$$

$$\frac{V_o b}{n\pi} (1 - \cos n\pi) = c_n \sinh \frac{n\pi a}{b} \cdot \frac{b}{2}$$

or

$$c_n \sinh \frac{n\pi a}{b} = \frac{2V_o}{n\pi} (1 - \cos n\pi)$$

$$= \begin{cases} \dfrac{4V_o}{n\pi}, & n = 1, 3, 5, \dots \\ 0, & n = 2, 4, 6, \dots \end{cases}$$

that is,

$$c_n = \begin{cases} \dfrac{4V_o}{n\pi \sinh \dfrac{n\pi a}{b}}, & n = \text{odd} \\ 0, & n = \text{even} \end{cases} \qquad (6.5.21)$$

Substituting this into eq. (6.5.18) gives the complete solution as

$$\boxed{V(x, y) = \frac{4V_o}{\pi} \sum_{n=1,3,5,\dots}^{\infty} \frac{\sin \dfrac{n\pi x}{b} \sinh \dfrac{n\pi y}{b}}{n \sinh \dfrac{n\pi a}{b}}} \qquad (6.5.22)$$

Check: $\nabla^2 V = 0$, $V(x = 0, y) = 0 = V(x = b, y) = V(x, y = 0)$, $V(x, y = a) = V_o$. The solution in eq. (6.5.22) should not be a surprise; it can be guessed by mere observation of the potential system in Figure 6.7. From this figure, we notice that along x, V varies from 0 (at $x = 0$) to 0 (at $x = b$) and only a sine function can satisfy this requirement. Similarly, along y, V varies from 0 (at $y = 0$) to V_o (at $y = a$) and only a hyperbolic sine function can satisfy this. Thus we should expect the solution as in eq. (6.5.22).

To determine the potential for each point (x, y) in the trough, we take the first few terms of the convergent infinite series in eq. (6.5.22). Taking four or five terms may be sufficient.

(b) For $x = a/2$ and $y = 3a/4$, where $b = 2a$, we have

$$V\left(\frac{a}{2}, \frac{3a}{4}\right) = \frac{4V_o}{\pi} \sum_{n=1,3,5,\ldots}^{\infty} \frac{\sin n\pi/4 \, \sinh 3n\pi/8}{n \, \sinh n\pi/2}$$

$$= \frac{4V_o}{\pi} \left[\frac{\sin \pi/4 \, \sinh 3\pi/8}{\sinh \pi/2} + \frac{\sin 3\pi/4 \, \sinh 9\pi/8}{3 \, \sinh 3\pi/2} \right.$$

$$\left. + \frac{\sin 5\pi/4 \, \sinh 15\pi/8}{5 \, \sinh 5\pi/2} + \cdots \right]$$

$$= \frac{4V_o}{\pi} (0.4517 + 0.0725 - 0.01985 - 0.00645 + 0.00229 + \cdots)$$

$$= 0.6374V_o$$

It is instructive to consider a special case of $a = b = 1$ m and $V_o = 100$ V. The potentials at some specific points are calculated by using eq. (6.5.22), and the result is displayed in Figure 6.10(a). The corresponding flux lines and equipotential lines are shown in Figure 6.10(b). A simple MATLAB program based on eq. (6.5.22) is displayed in Figure 6.11. This self-explanatory program can be used to calculate $V(x, y)$ at any point within the trough. In Figure 6.11, $V(x = b/4, y = 3a/4)$ is typically calculated and found to be 43.2 V.

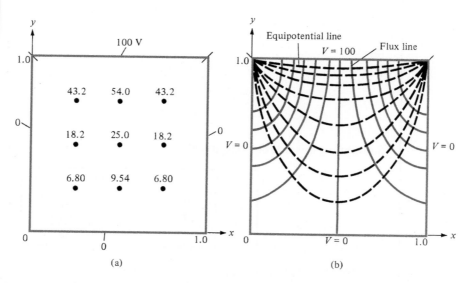

FIGURE 6.10 For Example 6.5: (**a**) $V(x, y)$ calculated at some points, (**b**) sketch of flux lines and equipotential lines.

(a)
```
h=0.1;
vo=100;
a=1.0;
b=a;
c=4*vo/pi;
IMAX = a/h;
JMAX = b/h;
NMAX = 13;
for I=1:IMAX
     x=h*I;
     for J=1:JMAX
          y=h*J;
          sum=0.0;
          for n =1:2:NMAX
               a1=sin(n*pi*x/b);
               a2=sinh(n*pi*y/b);
               a3=n*sinh(n*pi*a/b);
               sum= sum + c*a1*a2/a3;
          end
          V(I,J)=sum;
     end
end
mesh(V);
```

(b)

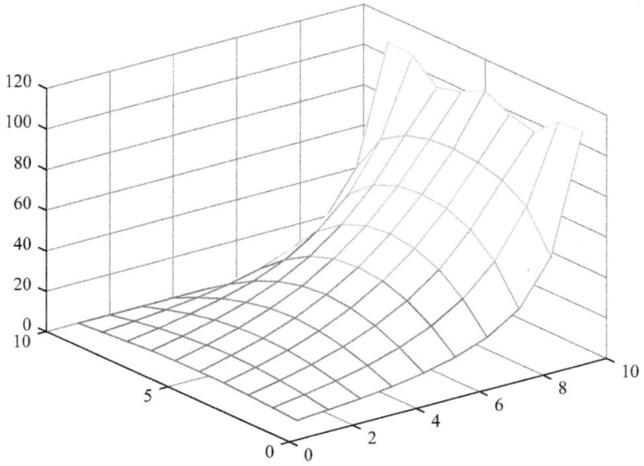

FIGURE 6.11 (a) MATLAB program for Example 6.5, (b) the output of the MATLAB program.

PRACTICE EXERCISE 6.5

For the problem in Example 6.5, take $V_o = 100$ V, $b = 2a = 2$ m, and find V and **E** at

(a) $(x, y) = (a, a/2)$
(b) $(x, y) = (3a/2, a/4)$

Answer: (a) 44.51 V, $-99.25\mathbf{a}_y$ V/m, (b) 16.5 V, $20.6\mathbf{a}_x - 70.34\mathbf{a}_y$ V/m.

EXAMPLE 6.6

Find the potential distribution in Example 6.5 if V_o is not constant but

(a) $V_o = 10 \sin 3\pi x/b$, $y = a$, $0 \le x \le b$

(b) $V_o = 2 \sin \dfrac{\pi x}{b} + \dfrac{1}{10} \sin \dfrac{5\pi x}{b}$, $y = a$, $0 \le x \le b$

Solution:

(a) In Example 6.5, every step before eq. (6.5.19) remains the same; that is, the solution is of the form

$$V(x, y) = \sum_{n=1}^{\infty} c_n \sin \frac{n\pi x}{b} \sinh \frac{n\pi y}{b} \tag{6.6.1}$$

in accordance with eq. (6.5.18). But instead of eq. (6.5.19), we now have

$$V(y = a) = V_o = 10 \sin \frac{3\pi x}{b} = \sum_{n=1}^{\infty} c_n \sin \frac{n\pi x}{b} \sinh \frac{n\pi a}{b}$$

By equating the coefficients of the sine terms on both sides, we obtain

$$c_n = 0, \quad n \ne 3$$

For $n = 3$,

$$10 = c_3 \sinh \frac{3\pi a}{b}$$

or

$$c_3 = \frac{10}{\sinh \dfrac{3\pi a}{b}}$$

Thus the solution in eq. (6.6.1) becomes

$$V(x, y) = 10 \sin \frac{3\pi x}{b} \frac{\sinh \dfrac{3\pi y}{b}}{\sinh \dfrac{3\pi a}{b}}$$

(b) Similarly, instead of eq. (6.5.19), we have

$$V_o = V(y = a)$$

or

$$2 \sin \frac{\pi x}{b} + \frac{1}{10} \sin \frac{5\pi x}{b} = \sum_{n=1}^{\infty} c_n \sin \frac{n\pi x}{b} \sinh \frac{n\pi a}{b}$$

Equating the coefficient of the sine terms:

$$c_n = 0, \quad n \ne 1, 5$$

For $n = 1$,

$$2 = c_1 \sinh \frac{\pi a}{b} \quad \text{or} \quad c_1 = \frac{2}{\sinh \dfrac{\pi a}{b}}$$

For $n = 5$,

$$\frac{1}{10} = c_5 \sinh \frac{5\pi a}{b} \quad \text{or} \quad c_5 = \frac{1}{10 \sinh \dfrac{5\pi a}{b}}$$

Hence,

$$V(x, y) = \frac{2 \sin \dfrac{\pi x}{b} \sinh \dfrac{\pi y}{b}}{\sinh \dfrac{\pi a}{b}} + \frac{\sin \dfrac{5\pi x}{b} \sinh \dfrac{5\pi y}{b}}{10 \sinh \dfrac{5\pi a}{b}}$$

PRACTICE EXERCISE 6.6

In Example 6.5, suppose everything remains the same except that V_o is replaced by $V_o \sin \dfrac{7\pi x}{b}, 0 \le x \le b, y = a$. Find $V(x, y)$.

Answer: $\dfrac{V_o \sin \dfrac{7\pi x}{b} \sinh \dfrac{7\pi y}{b}}{\sinh \dfrac{7\pi a}{b}}$.

EXAMPLE 6.7

Obtain the separated differential equations for potential distribution $V(\rho, \phi, z)$ in a charge-free region.

Solution:

This example, like Example 6.5, further illustrates the method of separation of variables. Since the region is free of charge, we need to solve Laplace's equation in cylindrical coordinates; that is,

$$\nabla^2 V = \frac{1}{\rho} \frac{\partial}{\partial \rho} \left(\rho \frac{\partial V}{\partial \rho} \right) + \frac{1}{\rho^2} \frac{\partial^2 V}{\partial \phi^2} + \frac{\partial^2 V}{\partial z^2} = 0 \tag{6.7.1}$$

We let

$$V(\rho, \phi, z) = R(\rho) \, \Phi(\phi) \, Z(z) \tag{6.7.2}$$

where R, Φ, and Z are, respectively, functions of ρ, ϕ, and z. Substituting eq. (6.7.2) into eq. (6.7.1) gives

$$\frac{\Phi Z}{\rho} \frac{d}{d\rho}\left(\rho \frac{dR}{d\rho}\right) + \frac{RZ}{\rho^2} \frac{d^2\Phi}{d\phi^2} + R\Phi \frac{d^2Z}{dz^2} = 0 \qquad (6.7.3)$$

We divide through by $R\Phi Z$ to obtain

$$\frac{1}{\rho R} \frac{d}{d\rho}\left(\rho \frac{dR}{d\rho}\right) + \frac{1}{\rho^2\Phi} \frac{d^2\Phi}{d\phi^2} = -\frac{1}{Z} \frac{d^2Z}{dz^2} \qquad (6.7.4)$$

The right-hand side of this equation is solely a function of z, whereas the left-hand side does not depend on z. For the two sides to be equal, they must be constant; that is,

$$\frac{1}{\rho R} \frac{d}{d\rho}\left(\rho \frac{dR}{d\rho}\right) + \frac{1}{\rho^2\Phi} \frac{d^2\Phi}{d\phi^2} = -\frac{1}{Z} \frac{d^2Z}{dz^2} = -\lambda^2 \qquad (6.7.5)$$

where $-\lambda^2$ is a separation constant. Equation (6.7.5) can be separated into two parts:

$$\frac{1}{Z} \frac{d^2Z}{dz^2} = \lambda^2 \qquad (6.7.6)$$

or

$$Z'' - \lambda^2 Z = 0 \qquad (6.7.7)$$

and

$$\frac{\rho}{R} \frac{d}{d\rho}\left(\rho \frac{dR}{d\rho}\right) + \lambda^2\rho^2 + \frac{1}{\Phi} \frac{d^2\Phi}{d\phi^2} = 0 \qquad (6.7.8)$$

Equation (6.7.8) can be written as

$$\frac{\rho^2}{R} \frac{d^2R}{d\rho^2} + \frac{\rho}{R} \frac{dR}{d\rho} + \lambda^2\rho^2 = -\frac{1}{\Phi} \frac{d^2\Phi}{d\phi^2} = \mu^2 \qquad (6.7.9)$$

where μ^2 is another separation constant. Equation (6.7.9) is separated as

$$\Phi'' + \mu^2\Phi = 0 \qquad (6.7.10)$$

and

$$\rho^2 R'' + \rho R' + (\rho^2\lambda^2 - \mu^2)R = 0 \qquad (6.7.11)$$

Equations (6.7.7), (6.7.10), and (6.7.11) are the required separated differential equations. Equation (6.7.7) has a solution similar to the solution obtained in Case 2 of Example 6.5; that is,

$$Z(z) = c_1 \cosh \lambda z + c_2 \sinh \lambda z \qquad (6.7.12)$$

The solution to eq. (6.7.10) is similar to the solution obtained in Case 3 of Example 6.5; that is,

$$\Phi(\phi) = c_3 \cos \mu\phi + c_4 \sin \mu\phi \qquad (6.7.13)$$

Equation (6.7.11) is known as the *Bessel differential equation* and its solution is beyond the scope of this text.[1]

PRACTICE EXERCISE 6.7

Repeat Example 6.7 for $V(r, \theta, \phi)$.

Answer: If $V(r, \theta, \phi) = R(r) F(\theta) \Phi(\phi)$, $\Phi'' + \lambda^2 \Phi = 0$, $R'' + \dfrac{2}{r} R' - \dfrac{\mu^2}{r^2} R = 0$,
$F'' + \cos\theta\, F' + (\mu^2 \sin\theta - \lambda^2 \csc\theta) F = 0$.

6.5 RESISTANCE AND CAPACITANCE

In Section 5.4 the concept of resistance was covered and we derived eq. (5.16) for finding the resistance of a conductor of uniform cross section. If the cross section of the conductor is not uniform, eq. (5.16) becomes invalid and the resistance is obtained from eq. (5.17):

$$\boxed{R = \frac{V}{I} = \frac{\int_L \mathbf{E} \cdot d\mathbf{l}}{\int_S \sigma \mathbf{E} \cdot d\mathbf{S}}} \qquad (6.16)$$

The problem of finding the resistance of a conductor of nonuniform cross section can be treated as a boundary-value problem. Using eq. (6.16), the resistance R (or conductance $G = 1/R$) of a given conducting material can be found by following these steps:

1. Choose a suitable coordinate system.
2. Assume V_o as the potential difference between conductor terminals.
3. Solve Laplace's equation $\nabla^2 V = 0$ to obtain V. Then determine $\mathbf{E}$ from $\mathbf{E} = -\nabla V$ and find I from $I = \int_S \sigma \mathbf{E} \cdot d\mathbf{S}$.
4. Finally, obtain R as V_o/I.

In essence, we assume V_o, find I, and determine $R = V_o/I$. Alternatively, it is possible to assume current I_o, find the corresponding potential difference V, and determine R from

---

[1] For a complete solution of Laplace's equation in cylindrical or spherical coordinates, see, for example, D. T. Paris and F. K. Hurd, *Basic Electromagnetic Theory*. New York: McGraw-Hill, 1969, pp. 150–159.

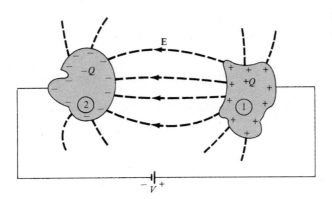

FIGURE 6.12 A two-conductor capacitor.

$R = V/I_o$. As will be discussed shortly, the capacitance of a capacitor is obtained using a similar technique.

Generally speaking, to have a capacitor we must have two (or more) conductors carrying equal but opposite charges. This implies that all the flux lines leaving one conductor must necessarily terminate at the surface of the other conductor. The conductors are sometimes referred to as the *plates* of the capacitor. The plates may be separated by free space or a dielectric.

Consider the two-conductor capacitor of Figure 6.12. The conductors are maintained at a potential difference V given by

$$V = V_1 - V_2 = -\int_2^1 \mathbf{E} \cdot d\mathbf{l} \tag{6.17}$$

where $\mathbf{E}$ is the electric field existing between the conductors and conductor 1 is assumed to carry a positive charge. (Note that the $\mathbf{E}$ field is always normal to the conducting surfaces.)

We define the *capacitance C* of the capacitor as the ratio of the magnitude of the charge on one of the plates to the potential difference between them; that is,

$$\boxed{C = \frac{Q}{V} = \frac{\varepsilon \int_S \mathbf{E} \cdot d\mathbf{S}}{\int_L \mathbf{E} \cdot d\mathbf{l}}} \tag{6.18}$$

The negative sign before $V = -\int_L \mathbf{E} \cdot d\mathbf{l}$ has been dropped because we are interested in the absolute value of V. The capacitance C is a physical property of the capacitor and is measured in farads (F). Most capacitances are practically much smaller than a farad and are specified in microfarads (μF) or picofarads (pF). We can use eq. (6.18) to obtain C for any given two-conductor capacitance by following either of these methods:

1. Assuming Q and determining V in terms of Q (involving Gauss's law)

$$C = \frac{Q \text{ (assume)}}{V \text{ (find)}}$$

2. Assuming V and determining Q in terms of V (involving solving Laplace's equation)

$$C = \frac{Q \quad \text{(find)}}{V \quad \text{(assume)}}$$

We shall use the former method here, and the latter method will be illustrated in Examples 6.10 and 6.11. The former method involves taking the following steps:

1. Choose a suitable coordinate system.
2. Let the two conducting plates carry charges $+Q$ and $-Q$.
3. Determine **E** by using Coulomb's or Gauss's law and find V from $V = -\int_L \mathbf{E} \cdot d\mathbf{l}$. The negative sign may be ignored in this case because we are interested in the absolute value of V.
4. Finally, obtain C from $C = Q/V$.

We will now apply this mathematically attractive procedure to determine the capacitance of some important two-conductor configurations.

A. Parallel-Plate Capacitor

Consider the parallel-plate capacitor of Figure 6.13(a). Suppose that each of the plates has an area S and they are separated by a distance d. We assume that plates 1 and 2, respectively, carry charges $+Q$ and $-Q$ uniformly distributed on them so that

$$\rho_S = \frac{Q}{S} \tag{6.19}$$

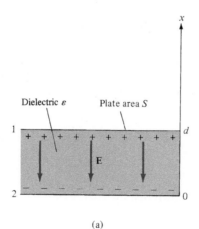

(a)

FIGURE 6.13 (**a**) Parallel-plate capacitor. (**b**) Fringing effect due to a parallel-plate capacitor.

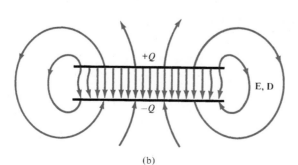

(b)

An ideal parallel-plate capacitor is one in which the plate separation d is very small compared with the dimensions of the plate. Assuming such an ideal case, the fringing field at the edge of the plates, as illustrated in Figure 6.13(b), can be ignored so that the field between them is considered uniform. If the space between the plates is filled with a homogeneous dielectric with permittivity ε and we ignore flux fringing at the edges of the plates, from eq. (4.27), $\mathbf{D} = -\rho_S \mathbf{a}_x$ or

$$\mathbf{E} = \frac{\rho_S}{\varepsilon}(-\mathbf{a}_x)$$

$$= -\frac{Q}{\varepsilon S}\mathbf{a}_x \tag{6.20}$$

Hence,

$$V = -\int_2^1 \mathbf{E} \cdot d\mathbf{l} = -\int_0^d \left[-\frac{Q}{\varepsilon S}\mathbf{a}_x \right] \cdot dx\, \mathbf{a}_x = \frac{Qd}{\varepsilon S} \tag{6.21}$$

and thus for a parallel-plate capacitor

$$\boxed{C = \frac{Q}{V} = \frac{\varepsilon S}{d}} \tag{6.22}$$

This formula offers a means of measuring the dielectric constant ε_r of a given dielectric. By measuring the capacitance C of a parallel-plate capacitor with the space between the plates filled with the dielectric and the capacitance C_o with air between the plates, we find ε_r from

$$\varepsilon_r = \frac{C}{C_o} \tag{6.23}$$

Using eq. (4.96), it can be shown that the energy stored in a capacitor is given by

$$\boxed{W_E = \frac{1}{2}CV^2 = \frac{1}{2}QV = \frac{Q^2}{2C}} \tag{6.24}$$

To verify this for a parallel-plate capacitor, we substitute eq. (6.20) into eq. (4.96) and obtain

$$W_E = \frac{1}{2}\int_v \varepsilon \frac{Q^2}{\varepsilon^2 S^2}\, dv = \frac{\varepsilon Q^2 S d}{2\varepsilon^2 S^2}$$

$$= \frac{Q^2}{2}\left(\frac{d}{\varepsilon S}\right) = \frac{Q^2}{2C} = \frac{1}{2}QV$$

as expected.

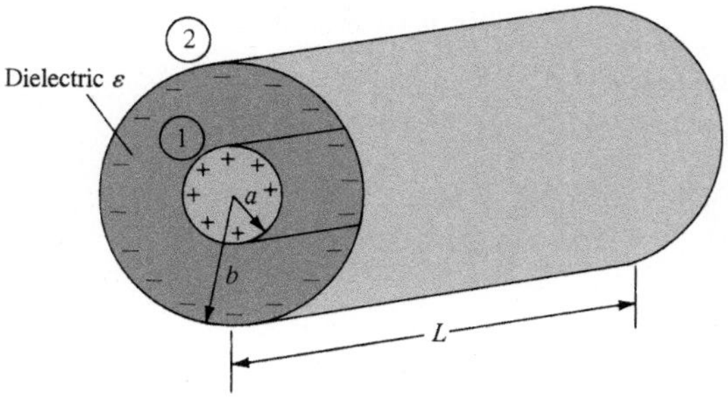

FIGURE 6.14 A coaxial capacitor.

B. Coaxial Capacitor

A coaxial capacitor is essentially a coaxial cable or coaxial cylindrical capacitor. Consider length L of two coaxial conductors of inner radius a and outer radius b $(b > a)$ as shown in Figure 6.14. Let the space between the conductors be filled with a homogeneous dielectric with permittivity ε. We assume that conductors 1 and 2, respectively, carry $+Q$ and $-Q$ uniformly distributed on them. By applying Gauss's law to an arbitrary Gaussian cylindrical surface of radius ρ $(a < \rho < b)$, we obtain

$$Q = \varepsilon \oint_S \mathbf{E} \cdot d\mathbf{S} = \varepsilon E_\rho 2\pi\rho L \tag{6.25}$$

Hence,

$$\mathbf{E} = \frac{Q}{2\pi\varepsilon\rho L} \mathbf{a}_\rho \tag{6.26}$$

Neglecting flux fringing at the cylinder ends,

$$V = -\int_2^1 \mathbf{E} \cdot d\mathbf{l} = -\int_b^a \left[\frac{Q}{2\pi\varepsilon\rho L} \mathbf{a}_\rho \right] \cdot d\rho\, \mathbf{a}_\rho \tag{6.27a}$$

$$= \frac{Q}{2\pi\varepsilon L} \ln \frac{b}{a} \tag{6.27b}$$

Thus the capacitance of a coaxial cylinder is given by

$$\boxed{C = \frac{Q}{V} = \frac{2\pi\varepsilon L}{\ln \dfrac{b}{a}}} \tag{6.28}$$

C. Spherical Capacitor

A spherical capacitor is the case of two concentric spherical conductors. Consider the inner sphere of radius a and outer sphere of radius b $(b > a)$ separated by a dielectric medium with permittivity ε as shown in Figure 6.15. We assume charges $+Q$ and $-Q$ on the inner

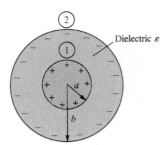

FIGURE 6.15 A spherical capacitor.

Dielectric ε

and outer spheres, respectively. By applying Gauss's law to an arbitrary Gaussian spherical surface of radius r $(a < r < b)$, we have

$$Q = \varepsilon \oint_S \mathbf{E} \cdot d\mathbf{S} = \varepsilon E_r 4\pi r^2 \tag{6.29}$$

that is,

$$\mathbf{E} = \frac{Q}{4\pi\varepsilon r^2} \mathbf{a}_r \tag{6.30}$$

The potential difference between the conductors is

$$V = -\int_2^1 \mathbf{E} \cdot d\mathbf{l} = -\int_b^a \left[\frac{Q}{4\pi\varepsilon r^2} \mathbf{a}_r \right] \cdot dr\, \mathbf{a}_r$$

$$= \frac{Q}{4\pi\varepsilon} \left[\frac{1}{a} - \frac{1}{b} \right] \tag{6.31}$$

Thus the capacitance of the spherical capacitor is

$$\boxed{C = \frac{Q}{V} = \frac{4\pi\varepsilon}{\dfrac{1}{a} - \dfrac{1}{b}}} \tag{6.32}$$

By letting $b \to \infty$, $C = 4\pi\varepsilon a$, which is the capacitance of a spherical capacitor whose outer plate is infinitely large. Such is the case of a spherical conductor at a large distance from other conducting bodies—the *isolated sphere*. Even an irregularly shaped object of about the same size as the sphere will have nearly the same capacitance. This fact is useful in estimating the stray capacitance of an isolated body or piece of equipment.

Recall from network theory that if two capacitors with capacitance C_1 and C_2 are in series (i.e., they have the same charge on them) as shown in Figure 6.16(a), the total capacitance is

$$\frac{1}{C} = \frac{1}{C_1} + \frac{1}{C_2}$$

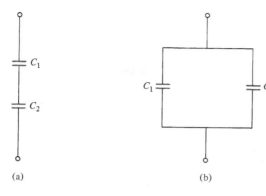

FIGURE 6.16 Capacitors (**a**) in series and (**b**) in parallel.

(a)   (b)

or

$$C = \frac{C_1 C_2}{C_1 + C_2} \tag{6.33}$$

If the capacitors are in parallel (i.e., if they have the same voltage across their plates) as shown in Figure 6.16(b), the total capacitance is

$$C = C_1 + C_2 \tag{6.34}$$

Let us reconsider the expressions for finding the resistance R and the capacitance C of an electrical system. The expressions were given in eqs. (6.16) and (6.18):

$$R = \frac{V}{I} = \frac{\int_L \mathbf{E} \cdot d\mathbf{l}}{\int_S \sigma \mathbf{E} \cdot d\mathbf{S}} \tag{6.16}$$

$$C = \frac{Q}{V} = \frac{\varepsilon \oint_S \mathbf{E} \cdot d\mathbf{S}}{\int_L \mathbf{E} \cdot d\mathbf{l}} \tag{6.18}$$

The product of these expressions yields

$$\boxed{RC = \frac{\varepsilon}{\sigma}} \tag{6.35}$$

which is the relaxation time T_r of the medium separating the conductors. It should be remarked that eq. (6.35) is valid only when the medium is homogeneous; this is easily inferred from eqs. (6.16) and (6.18). Assuming homogeneous media, the resistance of various capacitors mentioned earlier can be readily obtained using eq. (6.35). The following examples are provided to illustrate this idea.

For a parallel-plate capacitor,

$$C = \frac{\varepsilon S}{d}, \quad R = \frac{d}{\sigma S} \tag{6.36}$$

For a cylindrical capacitor,

$$C = \frac{2\pi\varepsilon L}{\ln\dfrac{b}{a}}, \quad R = \frac{\ln\dfrac{b}{a}}{2\pi\sigma L} \tag{6.37}$$

For a spherical capacitor,

$$C = \frac{4\pi\varepsilon}{\dfrac{1}{a} - \dfrac{1}{b}}, \quad R = \frac{\dfrac{1}{a} - \dfrac{1}{b}}{4\pi\sigma} \tag{6.38}$$

And finally for an isolated spherical conductor,

$$C = 4\pi\varepsilon a, \quad R = \frac{1}{4\pi\sigma a} \tag{6.39}$$

It should be noted that the resistance R in each of eqs. (6.35) to (6.39) is not the resistance of the capacitor plate but the leakage resistance between the plates; therefore, σ in those equations is the conductivity of the dielectric medium separating the plates.

EXAMPLE 6.8

A metal bar of conductivity σ is bent to form a flat 90° sector of inner radius a, outer radius b, and thickness t as shown in Figure 6.17. Show that (a) the resistance of the bar between the vertical curved surfaces at $\rho = a$ and $\rho = b$ is

$$R = \frac{2\ln\dfrac{b}{a}}{\sigma\pi t}$$

and (b) the resistance between the two horizontal surfaces at $z = 0$ and $z = t$ is

$$R' = \frac{4t}{\sigma\pi(b^2 - a^2)}$$

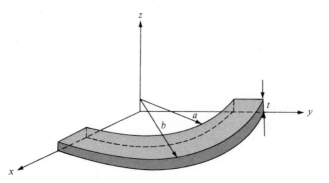

FIGURE 6.17 Bent metal bar for Example 6.8.

Solution:

(a)   Between the vertical curved ends located at $\rho = a$ and $\rho = b$, the bar has a nonuniform cross section and hence eq. (5.16) does not apply. We have to use eq. (6.16). Let a potential difference V_o be maintained between the curved surfaces at $\rho = a$ and $\rho = b$ so that $V(\rho = a) = 0$ and $V(\rho = b) = V_o$. We solve for V in Laplace's equation $\nabla^2 V = 0$ in cylindrical coordinates. Since $V = V(\rho)$,

$$\nabla^2 V = \frac{1}{\rho}\frac{d}{d\rho}\left(\rho\frac{dV}{d\rho}\right) = 0$$

As $\rho = 0$ is excluded, upon multiplying by ρ and integrating once, this becomes

$$\rho\frac{dV}{d\rho} = A$$

or

$$\frac{dV}{d\rho} = \frac{A}{\rho}$$

Integrating once again yields

$$V = A\ln\rho + B$$

where A and B are constants of integration to be determined from the boundary conditions.

$$V(\rho = a) = 0 \rightarrow 0 = A\ln a + B \quad \text{or} \quad B = -A\ln a$$

$$V(\rho = b) = V_o \rightarrow V_o = A\ln b + B = A\ln b - A\ln a = A\ln\frac{b}{a} \quad \text{or} \quad A = \frac{V_o}{\ln\frac{b}{a}}$$

Hence,

$$V = A\ln\rho - A\ln a = A\ln\frac{\rho}{a} = \frac{V_o}{\ln\frac{b}{a}}\ln\frac{\rho}{a}$$

$$\mathbf{E} = -\nabla V = -\frac{dV}{d\rho}\mathbf{a}_\rho = -\frac{A}{\rho}\mathbf{a}_\rho = -\frac{V_o}{\rho\ln\frac{b}{a}}\mathbf{a}_\rho$$

$$\mathbf{J} = \sigma\mathbf{E}, \quad d\mathbf{S} = -\rho\, d\phi\, dz\, \mathbf{a}_\rho$$

$$I = \int_S \mathbf{J}\cdot d\mathbf{S} = \int_{\phi=0}^{\pi/2}\int_{z=0}^{t}\frac{V_o\sigma}{\rho\ln\frac{b}{a}}\,dz\,\rho\,d\phi = \frac{\pi}{2}\frac{tV_o\sigma}{\ln\frac{b}{a}}$$

Thus

$$R = \frac{V_o}{I} = \frac{2 \ln \dfrac{b}{a}}{\sigma \pi t}$$

as required.

(b) Let V_o be the potential difference between the two horizontal surfaces so that $V(z = 0) = 0$ and $V(z = t) = V_o$. $V = V(z)$, so Laplace's equation $\nabla^2 V = 0$ becomes

$$\frac{d^2V}{dz^2} = 0$$

Integrating twice gives

$$V = Az + B$$

We apply the boundary conditions to determine A and B:

$$V(\ z = 0) \ = 0 \ \rightarrow 0 \ = 0 \ + B \quad \text{or} \quad B = 0$$

$$V(\ z = t) \ = V_o \ \rightarrow V_o \ = At \quad \text{or} \quad A = \frac{V_o}{t}$$

Hence,

$$V = \frac{V_o}{t} z$$

$$\mathbf{E} = -\nabla V = -\frac{dV}{dz}\mathbf{a}_z = -\frac{V_o}{t}\mathbf{a}_z$$

$$\mathbf{J} = \sigma \mathbf{E} = -\frac{\sigma V_o}{t}\mathbf{a}_z, \quad d\mathbf{S} = -\rho \, d\phi \, d\rho \, \mathbf{a}_z$$

$$I = \int_S \mathbf{J} \cdot d\mathbf{S} = \int_{\rho=0}^{b} \int_{\phi=0}^{\pi/2} \frac{V_0 \sigma}{t} \rho \, d\phi \, d\rho$$

$$= \frac{V_o \sigma}{t} \cdot \frac{\pi}{2} \frac{\rho^2}{2}\Big|_a^b = \frac{V_o \sigma \pi (b^2 - a^2)}{4t}$$

Thus

$$R' = \frac{V_o}{I} = \frac{4t}{\sigma \pi (b^2 - a^2)}$$

Alternatively, for this case, the cross section of the bar is uniform between the horizontal surfaces at $z = 0$ and $z = t$ and eq. (5.16) holds. Hence,

$$R' = \frac{\ell}{\sigma S} = \frac{t}{\sigma \frac{\pi}{4}(b^2 - a^2)}$$

$$= \frac{4t}{\sigma \pi (b^2 - a^2)}$$

as required.

PRACTICE EXERCISE 6.8

A disk of thickness t has radius b and a central hole of radius a. Taking the conductivity of the disk as σ, find the resistance between

(a) The hole and the rim of the disk
(b) The two flat sides of the disk

Answer: (a) $\dfrac{\ln \dfrac{b}{a}}{2\pi t \sigma}$, (b) $\dfrac{t}{\sigma \pi (b^2 - a^2)}$.

EXAMPLE 6.9

A coaxial cable contains an insulating material of conductivity σ. If the radius of the central wire is a and that of the sheath is b, show that the conductance of the cable per unit length is [see eq. (6.37)]

$$G = \frac{2\pi \sigma}{\ln \dfrac{b}{a}}$$

Solution:

Consider length L of the coaxial cable as shown in Figure 6.14. Let V_o be the potential difference between the inner and outer conductors so that $V(\rho = a) = 0$ and $V(\rho = b) = V_o$, which allows V and E to be found just as in part (a) of Example 6.8. Hence,

$$\mathbf{J} = \sigma \mathbf{E} = \frac{-\sigma V_o}{\rho \ln \dfrac{b}{a}} \mathbf{a}_\rho, \quad d\mathbf{S} = -\rho \, d\phi \, dz \, \mathbf{a}_\rho$$

$$I = \int_S \mathbf{J} \cdot d\mathbf{S} = \int_{\phi=0}^{2\pi} \int_{z=0}^{L} \frac{V_o \sigma}{\rho \ln \dfrac{b}{a}} \rho \, dz \, d\phi$$

$$= \frac{2\pi L \sigma V_o}{\ln \dfrac{b}{a}}$$

The resistance of the cable of length is given by

$$R = \frac{V_o}{I} \cdot \frac{+}{L} = \frac{\ln\frac{b}{a}}{-\pi\sigma}$$

and the conductance per unit length is

$$G = \frac{1}{R} = \frac{2\pi\sigma}{\ln\left(\frac{b}{a}\right)}$$

as required.

PRACTICE EXERCISE 6.9

A coaxial cable contains an insulating material of conductivity σ_1 in its upper half and another material of conductivity σ_2 in its lower half (similar to the situation shown later in Figure 6.19b). If the radius of the central wire is a and that of the sheath is b, show that the leakage resistance of length ℓ of the cable is

$$R = \frac{1}{\pi\ell(\sigma_1 + \sigma_2)} \ln\frac{b}{a}$$

Answer: Proof.

EXAMPLE 6.10

Conducting spherical shells with radii $a = 10$ cm and $b = 30$ cm are maintained at a potential difference of 100 V such that $V(r = b) = 0$ and $V(r = a) = 100$ V. Determine V and E in the region between the shells. If $\varepsilon_r = 2.5$ in the region, determine the total charge induced on the shells and the capacitance of the capacitor.

Solution:

Consider the spherical shells shown in Figure 6.18 and assume that V depends only on r. Hence Laplace's equation becomes

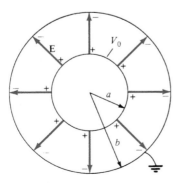

FIGURE 6.18 Potential $V(r)$ due to conducting spherical shells.

$$\nabla^2 V = \frac{1}{r^2}\frac{d}{dr}\left[r^2\frac{dV}{dr}\right] = 0$$

Since $r \neq 0$ in the region of interest, we multiply through by r^2 to obtain

$$\frac{d}{dr}\left[r^2\frac{dV}{dr}\right] = 0$$

Integrating once gives

$$r^2\frac{dV}{dr} = A$$

or

$$\frac{dV}{dr} = \frac{A}{r^2}$$

Integrating again gives

$$V = -\frac{A}{r} + B$$

As usual, constants A and B are determined from the boundary conditions.

When $r = b$, $V = 0 \rightarrow 0 = -\frac{A}{b} + B$ or $B = \frac{A}{b}$

Hence,

$$V = A\left[\frac{1}{b} - \frac{1}{r}\right]$$

Also when $r = a$, $V = V_o \rightarrow V_o = A\left[\frac{1}{b} - \frac{1}{a}\right]$

or

$$A = \frac{V_o}{\dfrac{1}{b} - \dfrac{1}{a}}$$

Thus

$$V = V_o\frac{\left[\dfrac{1}{r} - \dfrac{1}{b}\right]}{\dfrac{1}{a} - \dfrac{1}{b}}$$

$$E = -\nabla V = -\frac{dV}{dr}\,a_r = -\frac{A}{r^2}\,a_r$$

$$= \frac{V_o}{r^2\left[\frac{1}{a} - \frac{1}{b}\right]}\,a_r$$

$$Q = \int_S \varepsilon E \cdot dS = \int_{\theta=0}^{\pi}\int_{\phi=0}^{2\pi} \frac{\varepsilon_o \varepsilon_r V_o}{r^2\left[\frac{1}{a} - \frac{1}{b}\right]}\,r^2 \sin\theta\,d\phi\,d\theta$$

$$= \frac{4\pi\varepsilon_o\varepsilon_r V_o}{\frac{1}{a} - \frac{1}{b}}$$

Alternatively,

$$\rho_s = D_n = \varepsilon E_r, \quad Q = \int_s \rho_s dS$$

The capacitance is easily determined as

$$C = \frac{Q}{V_o} = \frac{4\pi\varepsilon}{\frac{1}{a} - \frac{1}{b}}$$

which is the same as we obtained in eq. (6.32); there in Section 6.5, we assumed Q and found the corresponding V_o, but here we assumed V_o and found the corresponding Q to determine C. Substituting $a = 0.1$ m, $b = 0.3$ m, $V_o = 100$ V yields

$$V = 100\,\frac{\left[\frac{1}{r} - \frac{10}{3}\right]}{10 - 10/3} = 15\left[\frac{1}{r} - \frac{10}{3}\right]\,V$$

Check: $\nabla^2 V = 0$, $V(r = 0.3 \text{ m}) = 0$, $V(r = 0.1 \text{ m}) = 100$.

$$E = \frac{100}{r^2\left[10 - 10/3\right]}\,a_r = \frac{15}{r^2}\,a_r \text{ V/m}$$

$$Q = \pm 4\pi \cdot \frac{10^{-9}}{36\pi} \cdot \frac{(2.5) \cdot (100)}{10 - 10/3}$$

$$= \pm 4.167 \text{ nC}$$

The positive charge is induced on the inner shell; the negative charge is induced on the outer shell. Also

$$C = \frac{|Q|}{V_o} = \frac{4.167 \times 10^{-9}}{100} = 41.67 \text{ pF}$$

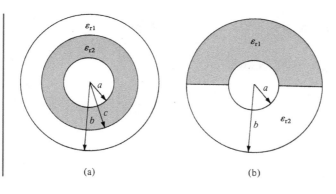

FIGURE 6.19 For Practice
Exercises 6.9, 6.10, and 6.12.

(a)   (b)

PRACTICE EXERCISE 6.10

If Figure 6.19 represents the cross sections of two spherical capacitors, determine their capacitances. Let $a = 1$ mm, $b = 3$ mm, $c = 2$ mm, $\varepsilon_{r1} = 2.5$, and $\varepsilon_{r2} = 3.5$.

Answer: (a) 0.53 pF, (b) 0.5 pF.

EXAMPLE 6.11

In Section 6.5, it was mentioned that the capacitance $C = Q/V$ of a capacitor can be found by either assuming Q and finding V, as in Section 6.5, or by assuming V and finding Q, as in Example 6.10. Use the latter method to derive eq. (6.22).

Solution:

Assume that the parallel plates in Figure 6.13 are maintained at a potential difference V_o so that $V(x = 0)$ and $V(x = d) = V_o$. This necessitates solving a one-dimensional boundary-value problem; that is, we solve Laplace's equation

$$\nabla^2 V = \frac{d^2 V}{dx^2} = 0$$

Integrating twice gives

$$V = Ax + B$$

where A and B are integration constants to be determined from the boundary conditions. At $x = 0$, $V = 0 \rightarrow 0 = 0 + B$, or $B = 0$, and at $x = d$, $V = V_o \rightarrow V_o = Ad + 0$ or $A = V_o/d$.

Hence,

$$V = \frac{V_o}{d} x$$

Notice that this solution satisfies Laplace's equation and the boundary conditions.

We have assumed the potential difference between the plates to be V_o. Our goal is to find the charge Q on either plate so that we can eventually find the capacitance $C = Q/V_o$. The charge on either plate is

$$Q = \int_S \rho_S \, dS$$

But $\rho_S = \mathbf{D} \cdot \mathbf{a}_n = \varepsilon \mathbf{E} \cdot \mathbf{a}_n$, where

$$\mathbf{E} = -\nabla V = -\frac{dV}{dx}\mathbf{a}_x = -A\mathbf{a}_x = -\frac{V_o}{d}\mathbf{a}_x$$

On the lower plate, $\mathbf{a}_n = \mathbf{a}_x$, so

$$\rho_S = -\frac{\varepsilon V_o}{d} \quad \text{and} \quad Q = -\frac{\varepsilon V_o S}{d}$$

On the upper plate, $\mathbf{a}_n = -\mathbf{a}_x$, so

$$\rho_S = \frac{\varepsilon V_o}{d} \quad \text{and} \quad Q = \frac{\varepsilon V_o S}{d}$$

As expected, Q is equal but opposite on each plate. Thus

$$C = \frac{|Q|}{V_o} = \frac{\varepsilon S}{d}$$

which is in agreement with eq. (6.22).

PRACTICE EXERCISE 6.11

Derive the formula for the capacitance $C = Q/V_o$ of a cylindrical capacitor in eq. (6.28) by assuming V_o and finding Q.

EXAMPLE 6.12

Determine the capacitance of each of the capacitors in Figure 6.20. Take $\varepsilon_{r1} = 4$, $\varepsilon_{r2} = 6$, $d = 5$ mm, $S = 30$ cm^2.

Solution:

(a) Since **D** and **E** are normal to the dielectric interface, the capacitor in Figure 6.20(a) can be treated as consisting of two capacitors C_1 and C_2 in series as in Figure 6.16(a).

$$C_1 = \frac{\varepsilon_o \varepsilon_{r1} S}{d/2} = \frac{2\varepsilon_o \varepsilon_{r1} S}{d}, \quad C_2 = \frac{2\varepsilon_o \varepsilon_{r2} S}{d}$$

The total capacitor C is given by

$$C = \frac{C_1 C_2}{C_1 + C_2} = \frac{2\varepsilon_o S}{d} \frac{(\varepsilon_{r1}\varepsilon_{r2})}{\varepsilon_{r1} + \varepsilon_{r2}}$$

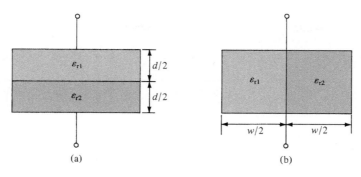

FIGURE 6.20 For Example 6.12.

$$= 2 \cdot \frac{10^{-9}}{36\pi} \cdot \frac{30 \times 10^{-4}}{5 \times 10^{-3}} \cdot \frac{4 \times 6}{10} \tag{6.12.1}$$

$$C = 25.46 \text{ pF}$$

(b) In this case, $\mathbf{D}$ and $\mathbf{E}$ are parallel to the dielectric interface. We may treat the capacitor as consisting of two capacitors C_1 and C_2 in parallel (the same voltage across C_1 and C_2) as in Figure 6.16(b).

$$C_1 = \frac{\varepsilon_o \varepsilon_{r1} S/2}{d} = \frac{\varepsilon_o \varepsilon_{r1} S}{2d}, \quad C_2 = \frac{\varepsilon_o \varepsilon_{r2} S}{2d}$$

The total capacitance is

$$C = C_1 + C_2 = \frac{\varepsilon_o S}{2d} (\varepsilon_{r1} + \varepsilon_{r2})$$

$$= \frac{10^{-9}}{36\pi} \cdot \frac{30 \times 10^{-4}}{2 \cdot (5 \times 10^{-3})} \cdot 10 \tag{6.12.2}$$

$$C = 26.53 \text{ pF}$$

Notice that when $\varepsilon_{r1} = \varepsilon_{r2} = \varepsilon_r$, eqs. (6.12.1) and (6.12.2) agree with eq. (6.22) as expected.

PRACTICE EXERCISE 6.12

Determine the capacitance of 10 m length of the cylindrical capacitors shown in Figure 6.19. Take $a = 1$ mm, $b = 3$ mm, $c = 2$ mm, $\varepsilon_{r1} = 2.5$, and $\varepsilon_{r2} = 3.5$.

Answer: (a) 1.54 F, (b) 1.52 nF.

EXAMPLE 6.13

A cylindrical capacitor has radii $a = 1$ cm and $b = 2.5$ cm. If the space between the plates is filled with an inhomogeneous dielectric with $\varepsilon_r = (10 + \rho)/\rho$, where ρ is in centimeters, find the capacitance per meter of the capacitor.

Solution:

The procedure is the same as that taken in Section 6.5 except that eq. (6.27a) now becomes

$$V = -\int_b^a \frac{Q}{2\pi\varepsilon_o\varepsilon_r\rho L}\, d\rho = -\frac{Q}{2\pi\varepsilon_o L}\int_b^a \frac{d\rho}{\rho\left(\dfrac{10 + \rho}{\rho}\right)}$$

$$= \frac{-Q}{2\pi\varepsilon_o L}\int_b^a \frac{d\rho}{10 + \rho} = \frac{-Q}{2\pi\varepsilon_o L} \ln\left(10 + \rho\right)\Big|_b^a$$

$$= \frac{Q}{2\pi\varepsilon_o L} \ln \frac{10 + b}{10 + a}$$

Thus the capacitance per meter is $(L = 1\,\text{m})$

$$C = \frac{Q}{V} = \frac{2\pi\varepsilon_o}{\ln \dfrac{10 + b}{10 + a}} = 2\pi \cdot \frac{10^{-9}}{36\pi} \cdot \frac{1}{\ln \dfrac{12.5}{11.0}}$$

$$C = 434.6\,\text{pF/m}$$

PRACTICE EXERCISE 6.13

A spherical capacitor with $a = 1.5$ cm, $b = 4$ cm has an inhomogeneous dielectric of $\varepsilon = 10\varepsilon_o/r$. Calculate the capacitance of the capacitor.

Answer: 1.13 nF.

6.6 METHOD OF IMAGES

The method of images, introduced by Lord Kelvin in 1848, is commonly used to determine V, **E**, **D**, and ρ_S due to charges in the presence of conductors. By this method, we avoid solving Poisson's or Laplace's equation but rather utilize the fact that a conducting surface is an equipotential. Although the method does not apply to all electrostatic problems, it can reduce a formidable problem to a simple one.

> The **image theory** states that a given charge configuration above an infinite grounded perfect conducting plane may be replaced by the charge configuration itself, its image, and an equipotential surface in place of the conducting plane.

Typical examples of point, line, and volume charge configurations are portrayed in Figure 6.21(a), and their corresponding image configurations are in Figure 6.21(b).

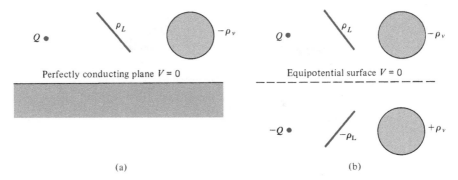

FIGURE 6.21 Image system: (**a**) charge configurations above a perfectly conducting plane, (**b**) image configuration with the conducting plane replaced by equipotential surface.

In applying the image method, two conditions must always be satisfied:

1. The image charge(s) must be located in the conducting region.
2. The image charge(s) must be located such that on the conducting surface(s) the potential is zero or constant.

The first condition is necessary to satisfy Poisson's equation, and the second condition ensures that the boundary conditions are satisfied. Let us now apply the image theory to two specific problems.

A. A Point Charge above a Grounded Conducting Plane

Consider a point charge Q placed at a distance h from a perfect conducting plane of infinite extent as in Figure 6.22(a). The image configuration is in Figure 6.22(b). The electric field in the region above the plane at point $P(x, y, z)$ is given by

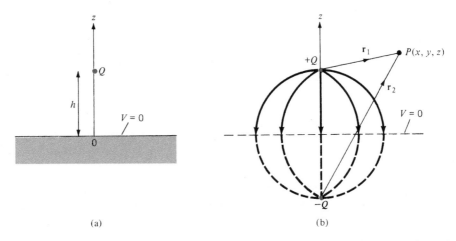

FIGURE 6.22 (**a**) Point charge and grounded conducting plane. (**b**) Image configuration and field lines.

$$\mathbf{E} = \mathbf{E}_+ + \mathbf{E}_- \tag{6.40}$$

$$= \frac{Q\,\mathbf{r}_1}{4\pi\varepsilon_o r_1^3} + \frac{-Q\,\mathbf{r}_2}{4\pi\varepsilon_o r_2^3} \tag{6.41}$$

The distance vectors $\mathbf{r}_1$ and $\mathbf{r}_2$ are given by

$$\mathbf{r}_1 = (x, y, z) - (0, 0, h) = (x, y, z - h) \tag{6.42}$$
$$\mathbf{r}_2 = (x, y, z) - (0, 0, -h) = (x, y, z + h) \tag{6.43}$$

so eq. (6.41) becomes

$$\mathbf{E} = \frac{Q}{4\pi\varepsilon_o}\left[\frac{x\mathbf{a}_x + y\mathbf{a}_y + (z - h)\mathbf{a}_z}{[x^2 + y^2 + (z - h)^2]^{3/2}} - \frac{x\mathbf{a}_x + y\mathbf{a}_y + (z + h)\mathbf{a}_z}{[x^2 + y^2 + (z + h)^2]^{3/2}}\right] \tag{6.44}$$

It should be noted that when $z = 0$, $\mathbf{E}$ has only the z-component, confirming that $\mathbf{E}$ is normal to the conducting surface.

The potential at P is easily obtained from eq. (6.41) or (6.44) using $V = -\int_L \mathbf{E} \cdot d\mathbf{l}$. Thus

$$V = V_+ + V_-$$

$$= \frac{Q}{4\pi\varepsilon_o r_1} + \frac{-Q}{4\pi\varepsilon_o r_2} \tag{6.45}$$

$$V = \frac{Q}{4\pi\varepsilon_o}\left\{\frac{1}{[x^2 + y^2 + (z - h)^2]^{1/2}} - \frac{1}{[x^2 + y^2 + (z + h)^2]^{1/2}}\right\}$$

for $z \geq 0$ and $V = 0$ for $z \leq 0$. Note that $V(z = 0) = 0$.

The surface charge density of the induced charge can also be obtained from eq. (6.44) as

$$\rho_S = D_n = \varepsilon_o E_n\Big|_{z=0}$$

$$= \frac{-Qh}{2\pi[x^2 + y^2 + h^2]^{3/2}} \tag{6.46}$$

The total induced charge on the conducting plane is

$$Q_i = \int \rho_S \, dS = \int_{-\infty}^{\infty}\int_{-\infty}^{\infty} \frac{-Qh \, dx \, dy}{2\pi[x^2 + y^2 + h^2]^{3/2}} \tag{6.47}$$

By changing variables, $\rho^2 = x^2 + y^2$, $dx \, dy = \rho \, d\rho \, d\phi$, and we have

$$Q_i = -\frac{Qh}{2\pi}\int_0^{2\pi}\int_0^{\infty} \frac{\rho \, d\rho \, d\phi}{[\rho^2 + h^2]^{3/2}} \tag{6.48}$$

Integrating over ϕ gives 2π, and letting $\rho \, d\rho = \frac{1}{2}d\,(\rho^2)$, we obtain

$$Q_i = -\frac{Qh}{2\pi} \, 2\pi \int_0^\infty [\rho^2 + h^2]^{-3/2} \frac{1}{2} d(\rho^2)$$

$$= \frac{Qh}{[\rho^2 + h^2]^{1/2}} \bigg|_0^\infty \tag{6.49}$$

$$= -Q$$

as expected, because all flux lines terminating on the conductor would have terminated on the image charge if the conductor were absent.

B. A Line Charge above a Grounded Conducting Plane

Consider an infinite line charge with density ρ_L C/m located at a distance h from the grounded conducting plane at $z = 0$. This may be regarded as a problem of a long conductor over the earth. The image system of Figure 6.22(b) applies to the line charge except that Q is replaced by ρ_L. The infinite line charge ρ_L may be assumed to be at $x = 0$, $z = h$, and the image $-\rho_L$ at $x = 0$, $z = -h$ so that the two are parallel to the y-axis. The electric field at point P is given (from eq. 4.21) by

$$\mathbf{E} = \mathbf{E}_+ + \mathbf{E}_- \tag{6.50}$$

$$= \frac{\rho_L}{2\pi\varepsilon_o\rho_1}\mathbf{a}_{\rho 1} + \frac{-\rho_L}{2\pi\varepsilon_o\rho_2}\mathbf{a}_{\rho 2} \tag{6.51}$$

The distance vectors $\boldsymbol{\rho}_1$ and $\boldsymbol{\rho}_2$ are given by

$$\boldsymbol{\rho}_1 = (x, y, z) - (0, y, h) = (x, 0, z - h) \tag{6.52}$$

$$\boldsymbol{\rho}_2 = (x, y, z) - (0, y, -h) = (x, 0, z + h) \tag{6.53}$$

so eq. (6.51) becomes

$$\mathbf{E} = \frac{\rho_L}{2\pi\varepsilon_o}\left[\frac{x\mathbf{a}_x + (z - h)\mathbf{a}_z}{x^2 + (z - h)^2} - \frac{x\mathbf{a}_x + (z + h)\mathbf{a}_z}{x^2 + (z + h)^2}\right] \tag{6.54}$$

Again, notice that when $z = 0$, $\mathbf{E}$ has only the z-component, confirming that $\mathbf{E}$ is normal to the conducting surface.

The potential at P is obtained from eq. (6.51) or (6.54) using $V = -\int_L \mathbf{E} \cdot d\mathbf{l}$. Thus

$$V = V_+ + V_-$$

$$= -\frac{\rho_L}{2\pi\varepsilon_o}\ln\rho_1 - \frac{-\rho_L}{2\pi\varepsilon_o}\ln\rho_2$$

$$= -\frac{\rho_L}{2\pi\varepsilon_o}\ln\frac{\rho_1}{\rho_2} \tag{6.55}$$

Substituting $\rho_1 = |\boldsymbol{\rho}_1|$ and $\rho_2 = |\boldsymbol{\rho}_2|$ in eqs. (6.52) and (6.53) into eq. (6.55) gives

$$V = -\frac{\rho_L}{2\pi\varepsilon_o}\ln\left[\frac{x^2 + (z-h)^2}{x^2 + (z+h)^2}\right]^{1/2} \tag{6.56}$$

for $z \ge 0$ and $V = 0$ for $z \le 0$. Note that $V(z = 0) = 0$.

The surface charge induced on the conducting plane is given by

$$\rho_S = D_n = \varepsilon_o E_z\bigg|_{z=0} = \frac{-\rho_L h}{\pi(x^2 + h^2)} \tag{6.57}$$

The induced charge per length on the conducting plane is

$$\rho_i = \int_L \rho_S \, dx = -\frac{\rho_L h}{\pi}\int_{-\infty}^{\infty}\frac{dx}{x^2 + h^2} \tag{6.58}$$

By letting $x = h\tan\alpha$, eq. (6.58) becomes

$$\rho_i = -\frac{\rho_L h}{\pi}\int_{-\pi/2}^{\pi/2}\frac{d\alpha}{h} \tag{6.59}$$

$$= -\rho_L$$

as expected.

EXAMPLE 6.14 A point charge Q is located at point $(a, 0, b)$ between two semi-infinite conducting planes intersecting at right angles as in Figure 6.23. Determine the potential at point $P(x, y, z)$ in region $z \ge 0$ and $x \ge 0$ and the force on Q.

Solution:

The image configuration is shown in Figure 6.24. Three image charges are necessary to satisfy the two conditions listed at the beginning of this section. From Figure 6.24(a), the potential at point $P(x, y, z)$ is the superposition of the potentials at P due to the four point charges; that is,

$$V = \frac{Q}{4\pi\varepsilon_o}\left[\frac{1}{r_1} - \frac{1}{r_2} + \frac{1}{r_3} - \frac{1}{r_4}\right]$$

where

$$r_1 = [(x-a)^2 + y^2 + (z-b)^2]^{1/2}$$
$$r_2 = [(x+a)^2 + y^2 + (z-b)^2]^{1/2}$$
$$r_3 = [(x+a)^2 + y^2 + (z+b)^2]^{1/2}$$
$$r_4 = [(x-a)^2 + y^2 + (z+b)^2]^{1/2}$$

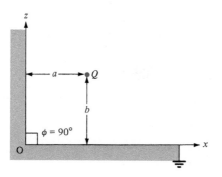

FIGURE 6.23 Point charge between two semi-infinite conducting planes.

From Figure 6.24(b), the net force on Q is

$$\mathbf{F} = \mathbf{F}_1 + \mathbf{F}_2 + \mathbf{F}_3$$

$$= -\frac{Q^2}{4\pi\varepsilon_o(2b)^2}\,\mathbf{a}_z - \frac{Q^2}{4\pi\varepsilon_o(2a)^2}\,\mathbf{a}_x + \frac{Q^2(2a\mathbf{a}_x + 2b\mathbf{a}_z)}{4\pi\varepsilon_o[(2a)^2 + (2b)^2]^{3/2}}$$

$$= \frac{Q^2}{16\pi\varepsilon_o}\left\{\left[\frac{a}{(a^2 + b^2)^{3/2}} - \frac{1}{a^2}\right]\mathbf{a}_x + \left[\frac{b}{(a^2 + b^2)^{3/2}} - \frac{1}{b^2}\right]\mathbf{a}_z\right\}$$

The electric field due to this system can be determined similarly, and the charge induced on the planes can also be found.

In general, when the method of images is used for a system consisting of a point charge between two semi-infinite conducting planes inclined at an angle ϕ (in degrees), the number of images is given by

$$\boxed{\mathbf{N} = \left(\frac{360°}{\phi} - + \right)}$$

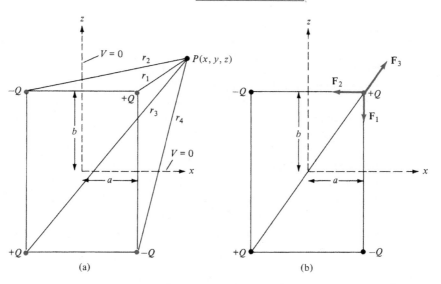

(a)   (b)

FIGURE 6.24 Determining **(a)** the potential at P and **(b)** the force on charge Q.

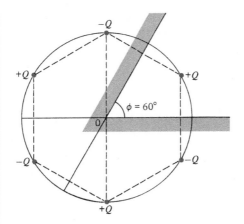

FIGURE 6.25 Point charge between two semi-infinite conducting walls inclined at $\phi = 60°$ to each other.

because the charge and its images all lie on a circle. For example, when $\phi = 180°$, $N = 1$ as in the case of Figure 6.22; for $\phi = 90°$, $N = 3$ as in the case of Figure 6.23; and for $\phi = 60°$, we expect $N = 5$ as shown in Figure 6.25.

PRACTICE EXERCISE 6.14

If the point charge $Q = 10$ nC in Figure 6.25 is 10 cm away from point O and along the line bisecting $\phi = 60°$, find the magnitude of the force on Q due to the charge induced on the conducting walls.

Answer: 60.54 μN.

---

†6.7 APPLICATION NOTE—CAPACITANCE OF MICROSTRIP LINES

The increasing application of integrated circuits at microwave frequencies has generated interest in the use of rectangular and circular microstrip disk capacitors as lumped-element circuits. The fringing field effects of such capacitors were first observed in 1877 by Kirchhoff, who used conformal mapping to account for the fringing. But his analysis was limited by the assumption that the capacitor is air filled. In microstrip applications, the capacitor plates are separated by a dielectric material instead of free space. Lately, others have come up with better approximate closed-form solutions to the problem taking into account the presence of the dielectric material and fringing. We consider only the circular disk capacitor.

The geometry of the circular microstrip capacitor, with radius r and separation distance d, is shown in Figure 6.26. Again, if disk area $S(S = \pi r^2)$ is very large compared with the separation distance (i.e., $\sqrt{S} \gg d$), then fringing is minimal and the capacitance is given by

$$C = \frac{\varepsilon_o \varepsilon_r \pi r^2}{d} \tag{6.60}$$

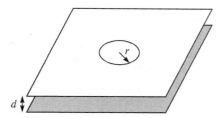

FIGURE 6.26 Circular microstrip capacitor.

Several researchers have attempted to account for the effect of fringing and to obtain a closed-form solution. We consider the following cases.

CASE 1.

According to Kirchhoff,[2] the fringing capacitance is

$$\Delta C = \varepsilon_o \varepsilon_r r \left(\log \frac{16\pi r}{d} - 1 \right) \tag{6.61}$$

so that the total capacitance is

$$C_T = \frac{\varepsilon_o \varepsilon_r \pi r^2}{d} + \varepsilon_o \varepsilon_r r \left(\log \frac{16\pi r}{d} - 1 \right) \tag{6.62}$$

It should be noted that Kirchhoff's approximation is valid only for $\varepsilon_r = 1$.

CASE 2.

According to Chew and Kong,[3] the total capacitance including fringing is

$$C_T = \frac{\varepsilon_o \varepsilon_r \pi r^2}{d} \left\{ 1 + \frac{2d}{\pi \varepsilon_r r} \left[\ln\left(\frac{r}{2d}\right) + (1.41\varepsilon_r + 1.77) + \frac{d}{r}(0.268\varepsilon_r + 1.65) \right] \right\} \tag{6.63}$$

CASE 3.

Wheeler used interpolation to match the three cases of small, medium, and large disk sizes. According to Wheeler,[4] we first define the following

$$C_{ks} = \varepsilon_o r \left[4(1 + \varepsilon_r) + \frac{\varepsilon_r \pi r}{d} \right] \tag{6.64}$$

[2] L. D. Landau and E. M. Lifshitz, *Electrodynamics of Continuous Media.* Oxford: Pergamon Press, 1960, p. 20.

[3] W. C. Chew and J. A. Kong, "Effects of fringing fields on the capacitance of circular microstrip disk," *IEEE Transactions on Microwave Theory and Techniques*, vol. 28, no. 2, Feb. 1980, pp. 98–103.

[4] H. A. Wheeler, "A simple formula for the capacitance of a disc on dielectric on a plane," *IEEE Transactions on Microwave Theory and Techniques,* vol. 30, no. 11, Nov. 1982, pp. 2050–2054.

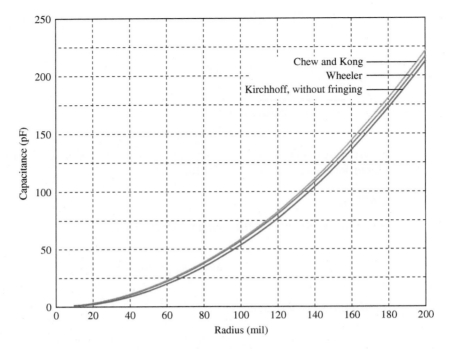

FIGURE 6.27 Capacitance of the circular microstrip capacitor.

where $k = \varepsilon_r$. When $k = 1$, eq. (6.64) becomes

$$C_{1s} = \varepsilon_o r\left[8 + \frac{\pi r}{d}\right] \tag{6.65}$$

The total capacitance is

$$C_T = \frac{C_{ks}}{k_c C_{1s}} C_1 + \left(1 - \frac{1}{k_c}\right)C_2 C_{ks} \tag{6.66}$$

where

$$C_1 = \varepsilon_o r\left[8 + \frac{\pi r}{d} + \frac{2}{3}\ln\left(\frac{1 + 0.8(r/d)^2 + (0.31r/d)^4}{1 + 0.9(r/d)}\right)\right] \tag{6.67}$$

$$C_2 = 1 - \frac{1}{4 + 2.6\dfrac{r}{d} + 2.9\dfrac{d}{r}} \tag{6.68}$$

$$k_c = 0.37 + 0.63\varepsilon_r \tag{6.69}$$

A MATLAB program was developed by using eqs. (6.62) to (6.69). With specific values of $d = 10$ mil and $\varepsilon_r = 74.04$, the values of C and C_T for $10 < r < 200$ mil are plotted in Figure 6.27 for the three cases. The curve for Kirchhoff's approximation coincides with the case without fringing.

6.8 APPLICATION NOTE—RF MEMS

Radio-frequency microelectromechanical systems (RF MEMS) are electronic devices with a submillimeter-sized moving part in the form of a beam, comb, disk, or ring, capable of providing RF functionality. Figure 6.28 shows one example. Another class consists of bulk or surface micromachined devices, such as thin-film bulk acoustic resonators (FBARs), which rely on energy transduction from the electrical energy domain to the acoustic energy domain and vice versa to provide RF functionality. A myriad of devices can be made using RF MEMS. Examples include RF MEMS switches, switched capacitors, varactors, and vibrating RF MEMS resonators. Several national laboratories and universities are actively engaged in developing these devices. Most notably, IBM Research Laboratory, Hughes Research Laboratories, Northeastern University in cooperation with Analog Devices, Raytheon, Rockwell Science, Westinghouse Research Laboratories, and the University of Michigan, Ann Arbor, and a few others are known to pioneer in this area of research.

Modeling RF MEMS devices is a challenge because they exhibit nonlinear behavior and hysteresis. The electrostatic force in MEMS, which depends on the square of the drive voltage, causes electrostatic force strengthening. This results in pull-in of the electrodes and limits the linearity of these devices. Also, residual charge at the interfaces influences the switching times. The Verilog-A hardware description language (HDL) is well suited for RF MEMS device modeling by means of nonlinear-physics-based equations. In addition, simplified empirical equations or model-order-reduced equations can be used to trade accuracy for feasibility. Device models that use empirical equations in addition to physics-based equations are referred to as compact models.

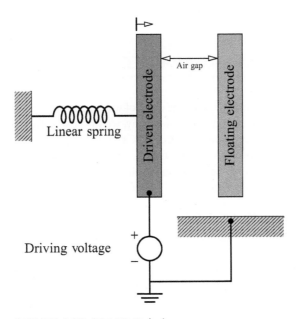

FIGURE 6.28 RF MEMS device.

RF MEMS have a deflecting cantilever or fixed-fixed beam and can be classified further by actuation method (electrostatic, electrothermal, magnetostatic, or piezoelectric), axis of deflection (lateral or vertical), circuit configuration (series or shunt), suspension (cantilever or fixed-fixed), or contact interface (capacitive or ohmic). Electrostatically actuated RF MEMS switches offer low insertion loss and high isolation, good linearity, greater power handling, and a higher electrical quality factor Q_e. A capacitive fixed-fixed beam RF MEMS switch is a micromachined capacitor with a moving top electrode called the beam. The beam, which is sometimes perforated, is suspended by springs over a lower electrode, separated by an air gap and a dielectric. It is generally connected in shunt with the transmission line. This type of switch is generally used for the X- to W-band, which lies in the range of 8 to 110 GHz. An ohmic cantilever RF MEMS switch is capacitive in the up-state but makes an ohmic contact in the down-state. It is an asymmetrical device with the clamp designated as source, the bias electrode designated as gate, and the contact electrode designated as drain, analogous to the field effect transistor. An ohmic cantilever switch is generally connected in series with the transmission line and is used from dc to about 40 GHz into the Ka-band.

A vibrating RF MEMS resonator has a vibrating beam, comb, disk, or ring (wine glass), which is sufficiently isolated from the surroundings to obtain a high mechanical quality factor Q_m. Vacuum encapsulation with an ambient pressure $P < 10^{-5}$ mbar results in negligible air damping. The vibrating RF MEMS resonator is often driven into a weakly nonlinear regime to increase the energy storage, taking advantage of the reduction in stored energy that occurs as the size is reduced. Vibrating RF MEMS resonators have a precise self-referenced oscillation frequency. Reference oscillators are used in local oscillators, which are implemented as voltage-controlled oscillators (VCOs). RF MEMS are extensively used in broadband wireless radio communications. Also, RF MEMS switches, switched capacitors, and varactors are applied in electronically scanned arrays, software-defined radios, reconfigurable antennas, and tunable bandpass filters.

The opening vignette of Chapter 12 provides an additional discussion of RF MEMS.

†6.9 APPLICATION NOTE—SUPERCAPACITORS

Historically, capacitors came in the range of picofarads and microfarads. Electrolytic capacitors had higher values of capacitance. But today, we find capacitors even of few kilofarads capacitance. They are called supercapacitors (SCs), ultracapacitors, or electric double-layer capacitors and are becoming more popular due to their growing use in electric vehicles. The internal structure of a typical SC is shown in Figure 6.29. SCs have a lower voltage rating. They are supposed to bridge the gap between electrolytic capacitors and batteries. As they store more energy per unit volume or mass than electrolytic capacitors, they seem to gain more prominence than batteries.

The charging and discharging times of SCs are very, very short compared to rechargeable batteries. For applications requiring long-term compact energy storage and regenerative braking in electric vehicles, SCs prove to be a good choice. Even in cranes and elevators, with short-term energy storage or burst-mode power delivery, these units can be used. The higher value of capacitance of SCs is contributed from one or a combination of more than one source such as faradic electron charge-transfer or formation of electrostatic double layer. The inner surface of each electrode in these devices is not smooth, but rather padded with activated porous carbon resulting in a surface area that is about 100,000 times as large as the surface area of an ordinary capacitor. The large surface area is not the only key feature. Charges in the form of ions stick to the inner surfaces of the electrodes. Thereby, the effective distance between the positive and the negative charges at each electrode is of the order of nanometers. Large surface area and small distance of separation are among the prime reasons for high nominal values of capacitance of SCs.

A third source of increasing the value of capacitance is by choosing high value of permittivity for the film that separates the electrodes. Barium titanate and its composites, such as barium strontium titanate, have recently been fabricated in nanometer-sized crystals and thin films. If such nanometer-sized crystals are deposited on the internal surface of the activated carbon electrode, this will result in a very substantial increase in the overall dielectric constant of the capacitor because the spacing between the electrolyte and the surface of the electrode will essentially remain a few nanometers thick, while the relative permittivity is increased substantially. The dielectric constant of powdered $BaSrTiO_3$, for example, is typically 12,000 to 15,000 at room temperature, whereas, many electrolytes have it in the range of 37–65.

Although research in the field of the electrochemical double-layer capacitors (EDLCs) began in early 1950s, it was only after the late 1970s that the technology of SCs appeared on the market. Compared to the rechargeable batteries, EDLCs have the additional advantage that they have a relatively low internal resistance and can store and deliver energy at higher power rating. As for their constructional details, the double-layer capacitors usually are made of two carbon electrodes isolated from one another through a porous membrane. The entire assembly is immersed in an electrolyte, which allows the ionic flow between the electrodes. The membrane under ideal conditions precludes the possibility of any electrical short circuit between the electrodes. A current-collecting plate is connected with each of the electrodes with the intent of minimizing the internal resistance. Most of the energy is stored in a polarized liquid, which is formed when a voltage is applied to the capacitor terminals. Thus, a double layer of negative and positive charges is generated on the contact surface between the electrode and the electrolyte, whence comes the name, EDLC. The separation of charge is of the order of a few angstrom units, much smaller than in a conventional capacitor.

Pseudocapacitance is established by faradic electron charge transfer with redox reactions, intercalation, or electrosorption. Electrochemical pseudocapacitors use metal oxide or conducting polymer electrodes with a high amount of electrochemical pseudocapacitance in addition to the double-layer capacitance. Another category is the hybrid capacitor, such as the lithium-ion capacitor, that uses electrodes with differing characteristics: one exhibiting mostly electrostatic capacitance and the other mostly electrochemical capacitance. Between the two electrodes, the electrolyte such as manganese dioxide or a conducting polymer serves as an ionic conductive pathway. This is part of the second electrode and is a feature that distinguishes them from the conventional electrolytic capacitors.

Supercapacitors are polarized by design with asymmetric electrodes or, for symmetric electrodes, by a potential applied during manufacture.

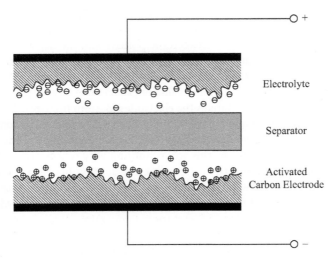

FIGURE 6.29 Cross-sectional view of a supercapacitor.

MATLAB 6.1   Consider a trace of thickness $t = 15\ \mu$m, length $l = 8$ mm, and with exponentially increasing width from $w_1 = 0.2$ mm to $w_2 = 2.5$ mm connected to a 10 V source by perfect electric conductor (PEC) slabs at the ends. Such a shape is commonly encountered around via pads or near impedance transformers. The trace has conductivity $\sigma = 6.5 \times 10^5$ S/m. Determine the total resistance of the trace by integrating the incremental cross-sectional resistance.

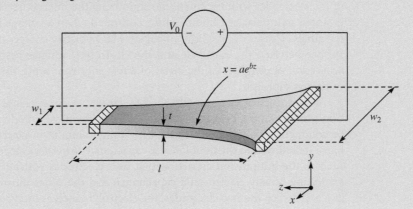

To simplify, we can place the origin between the PEC–trace junction on the wide side of the trace, at the center of the bottom side.

The incremental resistance is a slab of length dz in the xy-plane given by

$$dR = \frac{dz}{\sigma t w(z)}$$

We must determine the function $w(z)$. Since the trace tapers according to a known expression, we can determine the coefficients a, b:

$$\frac{w(z)}{2} = x(z) = ae^{bz}$$

We know that

$$x(z = 0) = \frac{w_2}{2}, x(z = l) = \frac{w_1}{2}$$

So,

$$x(z = 0) = a = \frac{w_2}{2}$$

And

$$x(z = l) = ae^{bl} = \frac{w_2}{2}e^{bl} = \frac{w_1}{2}$$

$$w_2 e^{bl} = w_1$$

$$b = \frac{1}{\ell} \ln \frac{w_1}{w_2}$$

This integral can be solved as

$$R = \int_0^l dR = \int_0^l \frac{dz}{\sigma t w(z)} = \int_0^l \frac{dz}{\sigma t 2x(z)} = \frac{1}{2\sigma t}\int_0^l \frac{dz}{x(z)}$$

$$= \frac{1}{w_2 \sigma t}\int_0^l e^{-\frac{1}{l}\ln \frac{w_1}{w_2} z} dz$$

Plugging in numbers for the geometric parameters, $R = 1.49\ \Omega$.

```
% This script computes the integral of an exponential trace
% a function using two different methods:
%    1. the built-in matlab 'quad' function
%    2. user-defined summation
%
% The user must first create a separate file for the function
```

```
%           y = (-1/20)*x^3+(3/5)*x.^2-(21/10)*x+4;
% The file should be named fun.m and stored in the same directory
% as this file, and it should contain the following two lines:
%           function y = fun(x)
%           y = (-1/20)*x.^3+(3/5)*x.^2-2.1*x+4;
%
% We will determine the integral of this function from
% x = 0 to x = 8
clear

% the parameters of the trace
w1 = 0.2e-3;
w2 = 2.5e-3;
cond =  6.5e5; % conductivity
t = 15e-6; % thickness
l = 8e-3;  % length

sum=0; % set initial total sum to zero
dz=1/1000;
for z=0:dz:1,
    sum=sum+1/(w2*cond*t)*exp(-1/l*log(w1/w2)*z)*dz;
                % add the partial sums to the total sum
end

disp('')
disp(sprintf('The total resistance of the trace is %f', sum))
```

SUMMARY

1. Boundary-value problems are those in which the potentials or their derivatives at the boundaries of a region are specified and we are to determine the potential field within the region. They are solved by using Poisson's equation if $\rho_v \neq 0$ or Laplace's equation if $\rho_v = 0$.

2. In a nonhomogeneous region, Poisson's equation is

$$\nabla \cdot \varepsilon \nabla V = -\rho_v$$

For a homogeneous region, ε is independent of space variables. Poisson's equation becomes

$$\nabla^2 V = -\frac{\rho_v}{\varepsilon}$$

In a charge-free region ($\rho_v = 0$), Poisson's equation becomes Laplace's equation; that is,

$$\nabla^2 V = 0$$

3. We solve the differential equation resulting from Poisson's or Laplace's equation by integrating twice if V depends on one variable or by the method of separation of variables if V is a function of more than one variable. We then apply the prescribed boundary conditions to obtain a unique solution.

4. The uniqueness theorem states that if V satisfies Poisson's or Laplace's equation and the prescribed boundary condition, V is the only possible solution for that problem. This enables us to find the solution to a given problem via any expedient means because we are assured of one, and only one, solution.

5. The problem of finding the resistance R of an object or the capacitance C of a capacitor may be treated as a boundary-value problem. To determine R, we assume a potential difference V_o between the ends of the object, solve Laplace's equation, find $I = \int_S \sigma \mathbf{E} \cdot d\mathbf{S}$, and obtain $R = V_o/I$. Similarly, to determine C, we assume a potential difference of V_o between the plates of the capacitor, solve Laplace's equation, find $Q = \int_S \varepsilon \mathbf{E} \cdot d\mathbf{S}$, and obtain $C = Q/V_o$.

6. A boundary-value problem involving an infinite conducting plane or wedge may be solved by using the method of images. This basically entails replacing the charge configuration by itself, its image, and an equipotential surface in place of the conducting plane. Thus the original problem is replaced by "an image problem," which is solved by using techniques covered in Chapters 4 and 5.

7. The computation of the capacitance of microstrip lines has become important because such lines are used in microwave devices. Three formulas for finding the capacitance of a circular microstrip line have been presented.

REVIEW QUESTIONS

6.1 Equation $\nabla \cdot (-\varepsilon \nabla V) = \rho_v$ may be regarded as Poisson's equation for an inhomogeneous medium.

(a) True   (b) False

6.2 In cylindrical coordinates, the equation

$$\frac{\partial^2 \psi}{\partial \rho^2} + \frac{1}{\rho}\frac{\partial \psi}{\partial \rho} + \frac{\partial^2 \psi}{\partial z^2} + 10 = 0$$

is called

(a) Maxwell's equation   (d) Helmholtz's equation
(b) Laplace's equation   (e) Lorentz's equation
(c) Poisson's equation

6.3 Two potential functions V_1 and V_2 satisfy Laplace's equation within a closed region and assume the same values on its surface. V_1 must be equal to V_2.

(a) True   (c) Not necessarily
(b) False

6.4 Which of the following potentials does not satisfy Laplace's equation?

(a) $V = 2x + 5$ (d) $V = \dfrac{10}{r}$

(b) $V = 10\,xy$ (e) $V = \rho \cos \phi + 10$

(c) $V = r \cos \phi$

6.5 Which of the following is not true?

(a) $-5 \cos 3x$ is a solution to $\phi''(x) + 9\phi(x) = 0$

(b) $10 \sin 2x$ is a solution to $\phi''(x) - 4\phi(x) = 0$

(c) $-4 \cosh 3y$ is a solution to $R''(y) - 9R(y) = 0$

(d) $\sinh 2y$ is a solution to $R''(y) - 4R(y) = 0$

(e) $\dfrac{g''(x)}{g(x)} = -\dfrac{h''(y)}{h(y)} = f(z) = -1$ where $g(x) = \sin x$ and $h(y) = \sinh y$

6.6 If $V_1 = X_1 Y_1$ is a product solution of Laplace's equation, which of these are not solutions of Laplace's equation?

(a) $-10X_1 Y_1$

(b) $X_1 Y_1 + 2xy$

(c) $X_1 Y_1 - x + y$

(d) $X_1 + Y_1$

(e) $(X_1 - 2)(Y_1 + 3)$

6.7 The capacitance of a capacitor filled by a linear dielectric is independent of the charge on the plates and the potential difference between the plates.

(a) True

(b) False

6.8 A parallel-plate capacitor connected to a battery stores twice as much charge with a given dielectric as it does with air as dielectric. The susceptibility of the dielectric is

(a) 0

(b) 1

(c) 2

(d) 3

(e) 4

6.9 A potential difference V_o is applied to a mercury column in a cylindrical container. The mercury is now poured into another cylindrical container of half the radius and the same potential difference V_o applied across the ends. As a result of this change of space, the resistance will be increased

(a) 2 times

(b) 4 times

(c) 8 times

(d) 16 times

6.10 Two conducting plates are inclined at an angle $30°$ to each other with a point charge between them. The number of image charges is

(a) 12

(b) 11

(c) 6

(d) 5

(e) 3

Answers: 6.1a, 6.2c, 6.3a, 6.4c, 6.5b, 6.6d,e, 6.7a, 6.8b, 6.9d, 6.10b.

PROBLEMS

Section 6.2—Poisson's and Laplace's Equations

6.1 Given $V = 5x^3 y^2 z$ and $\varepsilon = 2.25\varepsilon_o$, find (a) **E** at point $P(-3, 1, 2)$, (b) ρ_v at P.

6.2 Let $V = \dfrac{10 \cos \theta \sin \phi}{r^2}$ and $\varepsilon = \varepsilon_o$. (a) Find **E** at point $P(1, 60°, 30°)$. (b) Determine ρ_v at P.

6.3 Conducting sheets are located at $y = 1$ and $y = 3$ planes. The space between them is filled with a nonuniform charge distribution $\rho_v = \dfrac{y}{4\pi}$ nC/m^3 and $\varepsilon = 4\varepsilon_o$. Assuming that $V(y = 1) = 0$ and $V(y = 3) = 50$ V, find $V(y = 2)$.

6.4 In free space, $V = 10\rho^{0.8}$ V. Find **E** and the volume charge density at $\rho = 0.6$ m.

6.5 A certain material occupies the space between two conducting slabs located at $y = \pm 2$ cm. When heated, the material emits electrons such that $\rho_v = 50(1 - y^2) \, \mu$C/m^3. If the slabs are both held at 30 kV, find the potential distribution within the slabs. Take $\varepsilon = 3\varepsilon_o$.

6.6 Two large flat metal sheets are located at $z = 0$ and $z = d$ and are maintained at 0 and V_o, respectively. The charge density between the sheets is $\rho_v(z) = \rho_o z/d$, where ρ_o is a constant. Determine the potential at all points between the plates.

6.7 In cylindrical coordinates, $V = 0$ at $\rho = 2$ m and $V = 60$ V at $\rho = 5$ m due to charge distribution $\rho_v = \dfrac{10}{\rho}$ pC/m^3. If $\varepsilon_r = 3.6$, find **E**.

6.8 The region between two cylinders $\rho = a$ and $\rho = b$ has charge density ρ_o. Determine the potential distribution V.

6.9 The dielectric region ($\varepsilon = 6\varepsilon_o$) between a pair of concentric spheres $r = 1$ and $r = 4$ has charge distribution $\rho_v = \dfrac{10}{r}$ nC/m^3. If $V(r = 1) = 0$ and $V(r = 4) = 50$ V, determine $V(r = 2)$.

6.10 Determine whether each of the following potentials satisfies Laplace's equation.

(a) $V_1 = 3xyz + y - z^2$

(b) $V_2 = \dfrac{10\sin\phi}{\rho}$

(c) $V_3 = \dfrac{5\sin\phi}{r}$

6.11 Given $V = x^3y + yz + cz^2$, find c such that V satifies Laplace's equation.

6.12 The potential field $V = 2x^2yz - y^3z$ exists in a dielectric medium having $\varepsilon = 2\varepsilon_o$. (a) Does V satisfy Laplace's equation? (b) Calculate the total charge within the unit cube $0 < x < 1$ m, $0 < y < 1$ m, $0 < z < 1$ m.

6.13 Two conducting coaxial cylinders are located at $\rho = 1$ cm and $\rho = 1.5$ cm. The inner conductor is maintained at 50 V while the outer one is grounded. If the cylinders are separated by a dielectric material with $\varepsilon = 4\varepsilon_o$, find the surface charge density on the inner conductor.

6.14 Consider the conducting plates shown in Figure 6.30. If $V(z = 0) = 0$ and $V(z = 2$ mm$) = 50$ V, determine V, **E**, and **D** in the dielectric region ($\varepsilon_r = 1.5$) between the plates and ρ_S on the plates.

6.15 The cylindrical structure whose cross section is in Figure 6.31 has inner and outer radii of 5 mm and 15 mm, respectively. If $V(\rho = 5$ mm$) = 100$ V and $V(\rho = 15$ mm$) = 0$ V, calculate V, **E**, and **D** at $\rho = 10$ mm and ρ_S on each plate. Take $\varepsilon_r = 2.0$.

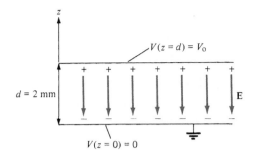

FIGURE 6.30 For Problem 6.14.

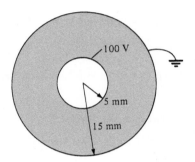

FIGURE 6.31 Cylindrical capacitor of Problem 6.15.

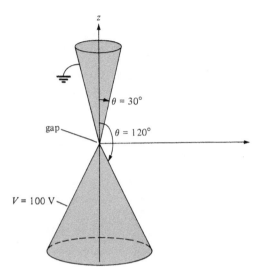

FIGURE 6.32 Conducting cones of Problem 6.20.

6.16 In cylindrical coordinates, $V = 50$ V on plane $\phi = \pi/2$ and $V = 0$ on plane $\phi = 0$. Assuming that the planes are insulated along the z-axis, determine $\mathbf{E}$ between the planes.

*6.17 (a) Show that $V = V_o(1 - a^2/\rho^2)\,\rho \sin \phi$ (where V_o is constant) satisfies Laplace's equation. (b) Determine $\mathbf{E}$ for $\rho^2 \gg a^2$.

6.18 Two conducting planes are located at $x = 0$ and $x = 50$ mm. The zero voltage reference is at $x = 20$ mm. Given that $\mathbf{E} = -110\mathbf{a}_x$ V/m, calculate the conductor voltages.

6.19 The region between concentric spherical conducting shells $r = 0.5$ m and $r = 1$ m is charge free. If $V(r = 0.5) = -50$ V and $V(r = 1) = 50$ V, determine the potential distribution and the electric field strength in the region between the shells.

6.20 Find V and $\mathbf{E}$ at $(3, 0, 4)$ due to the two conducting cones of infinite extent shown in Figure 6.32.

*6.21 The inner and outer electrodes of a diode are coaxial cylinders of radii $a = 0.6$ mm and $b = 30$ mm, respectively. The inner electrode is maintained at 70 V, while the outer electrode is grounded. (a) Assuming that the length of the electrodes $\ell \gg a, b$ and ignoring the effects of space charge, calculate the potential at $\rho = 15$ mm. (b) If an electron is injected radially through a small hole in the inner electrode with velocity 10^7 m/s, find its velocity at $\rho = 15$ mm.

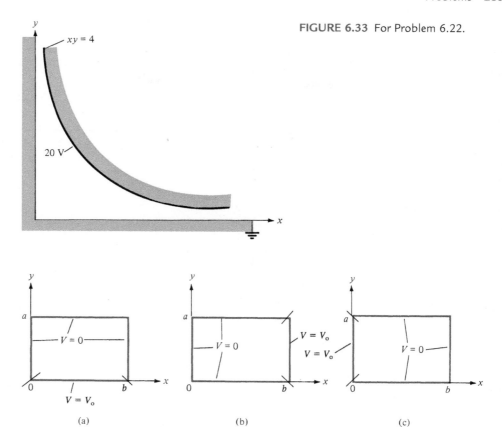

FIGURE 6.33 For Problem 6.22.

FIGURE 6.34 For Problem 6.23.

6.22 An electrode with a hyperbolic shape $(xy = 4)$ is placed above a grounded right-angle corner as in Figure 6.33. Calculate V and $\mathbf{E}$ at point $(1, 2, 0)$ when the electrode is connected to a 20 V source.

*6.23 Solve Laplace's equation for the two-dimensional electrostatic systems of Figure 6.34 and find the potential $V(x, y)$.

*6.24 Find the potential $V(x, y)$ due to the two-dimensional systems of Figure 6.35.

6.25 A conducting strip is defined as shown in Figure 6.35(b). The potential distribution is

$$V(x,y) = \frac{4V_{o}}{\pi} \sum_{n=\text{odd}}^{\infty} \frac{\sin\left(\dfrac{n\pi y}{a}\right)}{n} \exp(-n\pi x/a)$$

Find the electric field $\mathbf{E}$.

6.26 Figure 6.36 shows the cross-sectional view of an infinitely long rectangular slot. Find the potential distribution in the slot.

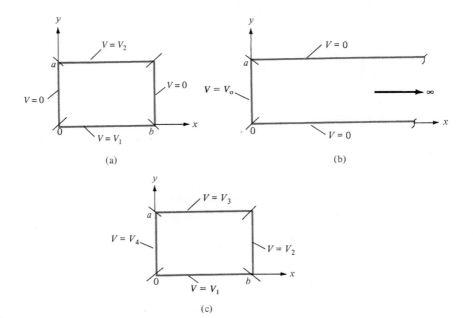

(a)   (b)

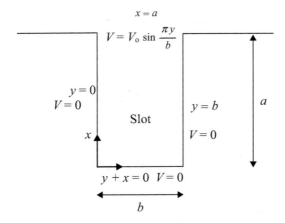

(c)

FIGURE 6.35 For Problems 6.24 and 6.25.

$$x = a$$

$$V = V_0 \sin \frac{\pi y}{b}$$

$y = 0$
$V = 0$

Slot

$y = b$
$V = 0$

a

x

$y + x = 0 \;\; V = 0$

b

FIGURE 6.36 For Problem 6.26.

6.27 By letting $V(\rho, \phi) = R(\rho)\Phi(\phi)$ be the solution of Laplace's equation in a region where $\rho \neq 0$, show that the separated differential equations for R and Φ are

$$R'' + \frac{R'}{\rho} - \frac{\lambda}{\rho^2} R = 0$$

and

$$\Phi'' + \lambda \Phi = 0$$

where λ is the separation constant.

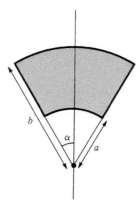

FIGURE 6.37 For Problem 6.30.

6.28 A potential in spherical coordinates is a function of r and θ but not ϕ. Assuming that $V(r, \theta) = R(r)F(\theta)$, obtain the separated differential equations for R and F in a region for which $\rho_v = 0$.

Section 6.5—Resistance and Capacitance

6.29 Show that the resistance of the bar of Figure 6.17 between the vertical ends located at $\phi = 0$ and $\phi = \pi/2$ is

$$R = \frac{\pi}{2\sigma t \ln \dfrac{b}{a}}$$

*6.30 Show that the resistance of the sector of a spherical shell of conductivity σ, with cross section shown in Figure 6.37 (where $0 \leq \phi < 2\pi$), between its base (i.e., from $r = a$ to $r = b$) is

$$R = \frac{1}{2\pi\sigma(1 - \cos\alpha)}\left[\frac{1}{a} - \frac{1}{b}\right]$$

6.31 A spherical shell has inner and outer radii a and b, respectively. Assume that the shell has a uniform conductivity σ and that it has copper electrodes plated on the inner and outer surfaces. Show that

$$R = \frac{1}{4\pi\sigma}\left(\frac{1}{a} - \frac{1}{b}\right)$$

*6.32 A hollow conducting hemisphere of radius a is buried with its flat face lying flush with the earth's surface, thereby serving as an earthing electrode. If the conductivity of earth is σ, show that the leakage conductance between the electrode and earth is $2\pi a\sigma$.

6.33 Another method of finding the capacitance of a capacitor is by using energy considerations, that is,

$$C = \frac{2W_E}{V_o^2} = \frac{1}{V_o^2}\int \varepsilon|\mathbf{E}|^2\, dv$$

Using this approach, derive eqs. (6.22), (6.28), and (6.32).

6.34 A cylindrical capacitor has inner radius a and outer radius b. The region between the cylinders has conductivity σ. Determine the conductance per unit length of the capacitor.

6.35 A coaxial cable with inner radius a and outer radius b has a steady-state voltage V across it. Determine the power loss per unit length. Assume that the conductivity of the region between the cylinders is σ.

6.36 In an integrated circuit, a capacitor is formed by growing a silicon dioxide layer ($\varepsilon_r = 4$) of thickness 1 μm over the conducting silicon substrate and covering it with a metal electrode of area S. Determine S if a capacitance of 2 nF is desired.

6.37 Calculate the capacitance of the parallel-plate capacitor shown in Figure 6.38.

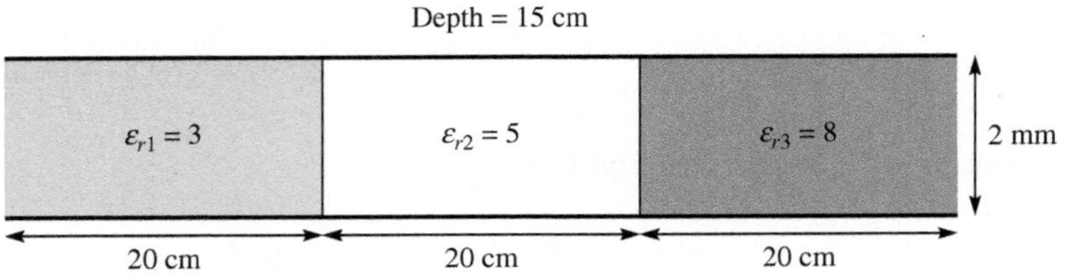

Depth = 15 cm
εr1 = 3
εr2 = 5
εr3 = 8
2 mm
20 cm
20 cm
20 cm

FIGURE 6.38 For Problem 6.37.

ε0
ε0εr
ε0
0
a
2a
3a
x

FIGURE 6.39 For Problem 6.38.

10 cm2
2 mm

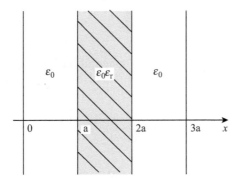

FIGURE 6.40 For Problem 6.39.

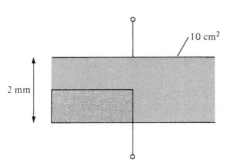

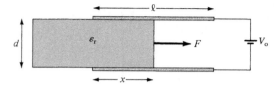

FIGURE 6.41 For Problem 6.41.

6.38 A capacitor consists of two infinitely large plates of area A and $3a$ apart as shown in Figure 6.39. If a dielectric slab of thickness a is located midway between the plates, determine the capacitance.

6.39 The parallel-plate capacitor of Figure 6.40 is quarter-filled with mica ($\varepsilon_r = 6$). Find the capacitance of the capacitor.

6.40 To appreciate the physical size of 1 F capacitor, consider a parallel-plate capacitor filled with air and with separation distance of 1 mm. Find the area of the plates to provide a capacitance of 1 F.

*6.41 An air-filled parallel plate capacitor of length L, width a, and plate separation d has its plates maintained at constant potential difference V_0. If a dielectric slab of dielectric constant ε_r is slid between the plates and is withdrawn until only a length x remains between the plates as in Figure 6.41, show that the force tending to restore the slab to its original position is

$$F = \frac{\varepsilon_0 (\varepsilon_r - 1)\, a\, V_0^2}{2d}$$

6.42 A parallel-plate capacitor has plate area 200 cm^2 and plate separation of 3 mm. The charge density is 1 μC/m^2 and air is the dielectric. Find

(a) The capacitance of the capacitor

(b) The voltage between the plates

(c) The force with which the plates attract each other

6.43 The capacitance of a parallel-plate capacitor is 56 μF when the dielectric material is in place. The capacitance drops to 32 μF when the dielectric material is removed. Calculate the dielectric constant ε_r of the material.

6.44 A parallel-plate capacitor has a 4 mm plate separation, 0.5 m^2 surface area per plate, and a dielectric with $\varepsilon_r = 6.8$. If the plates are maintained at 9 V potential difference, calculate (a) the capacitance, (b) the charge density on each plate.

6.45 A parallel-plate capacitor remains connected to a voltage source while the separation between the plates changes from d to $3d$. Express new values of C, Q, E, and W in terms of the old values C_0, Q_0, E_0, and W_0.

6.46 A parallel-plate capacitor has plate area 40 cm^2. The dielectric has two layers with permittivity $\varepsilon_1 = 4\varepsilon_o$ and $\varepsilon_2 = 6\varepsilon_o$, and each layer is 2 mm thick. If the capacitor is connected to a voltage 12 V, calculate: (a) the capacitance of the capacitor, (b) the total charge on each plate, (c) the values of E, D, and P.

6.47 The space between spherical conducting shells $r = 5$ cm and $r = 10$ cm is filled with a dielectric material for which $\varepsilon = 2.25\varepsilon_o$. The two shells are maintained at a potential difference of 80 V. (a) Find the capacitance of the system. (b) Calculate the charge density on shell $r = 5$ cm.

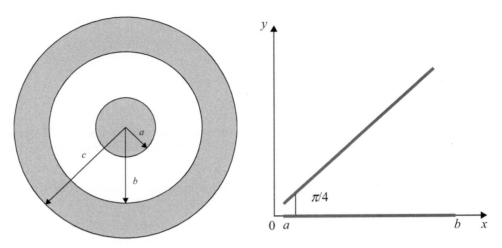

FIGURE 6.42 For Problem 6.49.   **FIGURE 6.43** For Problem 6.52.

6.48 A spherical capacitor has inner radius d and outer radius a. Concentric with the spherical conductors and lying between them is a spherical shell of outer radius c and inner radius b. If the regions $d < r < c$, $c < r < b$, and $b < r < a$ are filled with materials with permittivities ε_1, ε_2, and ε_3, respectively, determine the capacitance of the system.

6.49 Determine the capacitance of a conducting sphere surrounded by a thick spherical shell as shown in Figure 6.42.

6.50 The coaxial cable in Figure 6.14 has two dielectrics with ε_{r1} for $a < \rho < c$ and ε_{r2} for $c < \rho < b$ where $a < c < b$. Determine the capacitance of the system.

6.51 A coaxial cable has inner radius of 5 mm and outer radius of 8 mm. If the cable is 3 km long, calculate its capacitance. Assume $\varepsilon = 2.5\varepsilon_o$.

6.52 A capacitor consists of two plates with equal width $(b - a)$, and a length L in the z-direction. The plates are separated by $\phi = \pi/4$, as shown in Figure 6.43. Assume that the plates are separated by a dielectric material $(\varepsilon = \varepsilon_o\varepsilon_r)$ and ignore fringing. Determine the capacitance.

6.53 A segment of the cylindrical capacitor is defined by $\rho_1 < \rho < \rho_2$, $0 < \phi < \alpha$. If $V(\phi = 0) = 0$ and $V(\phi = \alpha) = V_o$, show that the capacitance of the segment is

$$C = \frac{\varepsilon L}{\alpha}\ln\left(\frac{\rho_2}{\rho_1}\right),$$ where L is the length and ε is the permittivity of the dielectric.

*6.54 In an ink-jet printer the drops are charged by surrounding the jet of radius 20 μm with a concentric cylinder of radius 600 μm as in Figure 6.44. Calculate the minimum voltage required to generate a charge 50 fC on the drop if the length of the jet inside the cylinder is 100 μm. Take $\varepsilon = \varepsilon_o$.

6.55 The cross section of a cable is shown in Figure 6.45. Determine the capacitance per unit length.

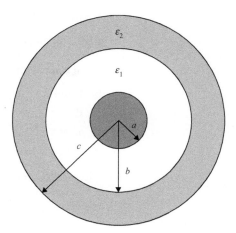

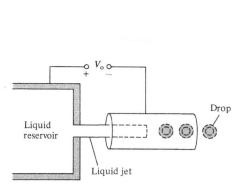

FIGURE 6.44 Simplified geometry of an ink-jet printer; for Problem 6.54.

FIGURE 6.45 For Problem 6.55.

FIGURE 6.46 For Problem 6.57.

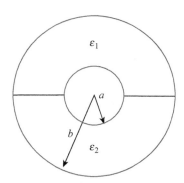

6.56 A spherical capacitor has inner radius of $a = 2$ cm and outer radius of $b = 4$ cm. The interior is a dielectric material with $\varepsilon_o \varepsilon_r$. The outer conductor is grounded while the inner one is maintained at 100 V. (a) Determine ε_r if the surface charge density on the inner conductor is 400 nC/m². (b) Find the capacitance of the structure.

6.57 One half of the dielectric region of a spherical capacitor has permittivity ε_1 while the other half has ε_2 as shown in Figure 6.46. Show that the capacitance of the system is given by

$$C = \frac{2\pi(\varepsilon_1 + \varepsilon_2)ab}{b - a}$$

*6.58 A spherical capacitor has an inner conductor of radius a carrying charge Q and is maintained at zero potential. If the outer conductor contracts from a radius b to c under internal forces, prove that the work performed by the electric field as a result of the contraction is

$$W = \frac{Q^2(b - c)}{8\pi\varepsilon bc}$$

*6.59 A parallel-plate capacitor has its plates at $x = 0, d$ and the space between the plates is filled with an inhomogeneous material with permittivity $\varepsilon = \varepsilon_0\left(1 + \dfrac{x}{d}\right)$. If the plate at $x = d$ is maintained at V_0 while the plate at $x = 0$ is grounded, find:

(a) V and **E**

(b) **P**

(c) ρ_{P_s} at $x = 0, d$

(d) the capacitance, assuming that each plate has area S

6.60 Two parallel conducting plates are located at $x = d$ and $x = -d$. The plate at $x = d$ is held at V_0, while the plate at $x = -d$ is grounded. If the space between the plates is filled with an inhomogeneous dielectric medium with

$$\varepsilon = \frac{2\varepsilon_o}{1 + \left(\dfrac{x}{d}\right)^2}$$

find the capacitance. Assume that each plate has an area S.

6.61 A spherical capacitor has inner radius a and outer radius b and is filled with an inhomogeneous dielectric with $\varepsilon = \varepsilon_o k/r^2$. Show that the capacitance of the capacitor is

$$C = \frac{4\pi\varepsilon_o k}{b - a}$$

6.62 If the earth is regarded as a spherical capacitor, what is its capacitance? Assume the radius of the earth to be approximately 6370 km.

6.63 A capacitor is formed by two coaxial metal cylinders of radii $a = 1$ mm and $b = 5$ mm. If the space between the cylinders is filled with a dielectric having $\varepsilon_r = 3(1 + \rho)$, $a < \rho < b$, and ρ is in millimeters, determine the capacitance per meter.

6.64 A two-wire transmission line is formed with two identical wires which are widely separated. If the radius of each wire is a and the center-to-center spacing is **D**, the approximate formula for the capacitance per unit length is

$$C = \frac{\pi\varepsilon}{\ln\left(\dfrac{D - a}{a}\right)}$$

where $D \gg a$. For $\varepsilon = 4\varepsilon_o$ and $D/a = 12$, calculate C.

Section 6.6—Method of Images

6.65 A 10-nC point charge is located at point (0, 0, 10 m) above a grounded conducting plane $z = 0$. (a) Find the surface charge density at point $(2, -4, 0)$. (b) Calculate the total charge on the plate.

6.66 Two point charges of 3 nC and -4 nC are placed, respectively, at $(0, 0, 1\,\text{m})$ and $(0, 0, 2\,\text{m})$ while an infinite conducting plane is at $z = 0$. Determine

(a) The total charge induced on the plane

(b) The magnitude of the force of attraction between the charges and the plane

*6.67 A point charge of 10 μC is located at $(1, 1, 1)$ and the positive portions of the coordinate planes are occupied by three mutually perpendicular plane conductors maintained at zero potential. Find the force on the charge due to the conductors.

6.68 A point charge Q is placed between two earthed intersecting conducting planes that are inclined at $45°$ to each other. Determine the number of image charges and their locations.

6.69 Infinite line $x = 3$, $z = 4$ carries 16 nC/m and is located in free space above the conducting plane $z = 0$. (a) Find $\mathbf{E}$ at $(2, -2, 3)$. (b) Calculate the induced surface charge density on the conducting plane at $(5, -6, 0)$.

6.70 In free space, infinite planes $y = 4$ and $y = 8$ carry charges 20 nC/m^2 and 30 nC/m^2, respectively. If plane $y = 2$ is grounded, calculate $\mathbf{E}$ at $P(0, 0, 0)$ and $Q(-4, 6, 2)$.

MAGNETOSTATICS

Jean-Baptiste Biot (1774–1862), a French physicist and mathematician, made advances in geometry, astronomy, elasticity, electricity, magnetism, heat, and optics.

Born in Paris, he studied at the École Polytechnique in Paris, where he realized his potential. Biot studied a wide range of mathematical topics, mostly on the applied mathematics. Biot, together with the French physicist Felix Savart, discovered that the magnetic field intensity of a current flowing through a wire varies inversely with the distance from the wire. This relation, now known as Biot–Savart's law, will be covered in this chapter. Biot discovered that when light passes through some substances, including sugar solutions, the plane of polarization of the light is rotated by an amount that depends on the color of the light. In addition to his scientific pursuits, Biot was a prolific writer. He completed over 250 works of various types, the most renowned of which is his *Elementary Treatise on Physical Astronomy* (1805).

André-Marie Ampère (1775–1836), a French physicist, mathematician, and natural philosopher, is best known for defining a way to measure the flow of current. He has been called the Newton of electricity.

Born at Polemieux, near Lyons, Ampère took a passionate delight in the pursuit of knowledge from his very infancy. Although André never attended school, he received an excellent education. His father taught him Latin, which enabled him to master the works of Euler and Bernouilli. André poured over his studies of mathematics and soon began to create his own theories and ideas. His reading embraced a wide range of knowledge—history, travels, poetry, philosophy, metaphysics, and the natural sciences. As an adult, Ampère was notoriously absent-minded. He became a professor of mathematics at the École Polytechnique, and later at the Collège de France. He developed Oersted's discovery of the link between electric and magnetic fields and introduced the concepts of current element and the force between current elements. Ampère made several contributions to electromagnetism, including the formulation of the law that bears his name, which will be discussed in this chapter. The ampere unit of electric current is named after him.

CHAPTER 7

MAGNETOSTATIC FIELDS

The highest happiness on earth is in marriage. Every man who is happily married is a successful man even if he has failed in everything else.

—WILLIAM L. PHELPS

7.1 INTRODUCTION

In Chapters 4 to 6, we limited our discussions to static electric fields characterized by **E** or **D**. We now focus our attention on static magnetic fields, which are characterized by **H** or **B**. There are similarities and dissimilarities between electric and magnetic fields. As **E** and **D** are related according to $\mathbf{D} = \varepsilon\mathbf{E}$ for linear, isotropic material space, **H** and **B** are related according to $\mathbf{B} = \mu\mathbf{H}$. Table 7.1 further shows the analogy between electric and magnetic field quantities. Some of the magnetic field quantities will be introduced later in this chapter, and others will be presented in the next. The analogy is presented here to show that most of the equations we have derived for the electric fields may be readily used to obtain corresponding equations for magnetic fields if the equivalent analogous quantities are substituted. This way it does not appear as if we are learning new concepts.

A definite link between electric and magnetic fields was established by Oersted[1] in 1820. As we have noticed, an electrostatic field is produced by static or stationary charges. If the charges are moving with constant velocity, a static magnetic (or magnetostatic) field is produced. A magnetostatic field is produced by a constant current flow (or direct current). This current flow may be due to magnetization currents as in permanent magnets, electron-beam currents as in vacuum tubes, or conduction currents as in current-carrying wires. In this chapter, we consider magnetic fields in free space due to direct current. Magnetostatic fields in material space are covered in Chapter 8.

Our study of magnetostatics is not a dispensable luxury but an indispensable necessity. Motors, transformers, microphones, compasses, telephone bell ringers, television focusing controls, advertising displays, magnetically levitated high-speed vehicles, memory stores, magnetic separators, and so on, which play an important role in our everyday life,[2] could not have been developed without an understanding of magnetic phenomena.

---

[1]Hans Christian Oersted (1777–1851), a Danish professor of physics, after 13 years of frustrating efforts discovered that electricity could produce magnetism.

[2]Various applications of magnetism can be found in J. K. Watson, *Applications of Magnetism*. New York: John Wiley & Sons, 1980.

TABLE 7.1 Analogy between Electric and Magnetic Fields*

Term	Electric	Magnetic
Basic laws	$\mathbf{F} = \dfrac{Q_1 Q_2}{4\pi\varepsilon R^2}\mathbf{a}_R$	$d\mathbf{B} = \dfrac{\mu_o I\, d\mathbf{l} \times \mathbf{a}_R}{4\pi R^2}$
	$\oint \mathbf{D}\cdot d\mathbf{S} = Q_{\text{enc}}$	$\oint \mathbf{H}\cdot d\mathbf{l} = I_{\text{enc}}$
Force law	$\mathbf{F} = Q\mathbf{E}$	$\mathbf{F} = Q\mathbf{u}\times\mathbf{B}$
Source element	dQ	$d Q\mathbf{u} = I d\mathbf{l}$
Field intensity	$E = \dfrac{V}{\ell}\ (\text{V/m})$	$H = \dfrac{I}{\ell}\ (\text{A/m})$
Flux density	$D = \dfrac{\psi}{S}\ (\text{C/m}^2)$	$B = \dfrac{\psi}{S}\ (\text{Wb/m}^2)$
Relationship between fields	$\mathbf{D} = \varepsilon\mathbf{E}$	$\mathbf{B} = \mu\mathbf{H}$
Potentials	$\mathbf{E} = -\nabla V$	$\mathbf{H} = -\nabla V_m\ (\mathbf{J}=0)$
	$V = \displaystyle\int_L \dfrac{\rho_L dl}{4\pi\varepsilon R}$	$\mathbf{A} = \displaystyle\int_L \dfrac{\mu I\, d\ell}{4\pi R}$
Flux	$\psi = \int \mathbf{D}\cdot d\mathbf{S}$	$\psi = \int_s \mathbf{B}\cdot d\mathbf{S}$
	$\psi = Q = CV$	$\psi = LI$
	$I = C\dfrac{dV}{dt}$	$V = L\dfrac{dI}{dt}$
Energy density	$w_E = \dfrac{1}{2}\mathbf{D}\cdot\mathbf{E}$	$w_m = \dfrac{1}{2}\mathbf{B}\cdot\mathbf{H}$
Poisson's equation	$\nabla^2 V = -\dfrac{\rho_v}{\varepsilon}$	$\nabla^2\mathbf{A} = -\mu\mathbf{J}$

*A similar analogy can be found in R. S. Elliot, "Electromagnetic theory: a simplified representation," *IEEE Transactions on Education,* vol. E-24, no. 4, Nov. 1981, pp. 294–296.

There are two major laws governing magnetostatic fields: (1) Biot–Savart's law,[3] and (2) Ampère's circuit law. Like Coulomb's law, Biot–Savart's law is the general law of magnetostatics. Just as Gauss's law is a special case of Coulomb's law, Ampère's law is a special case of Biot–Savart's law and is easily applied in problems involving symmetrical current distribution. The two laws of magnetostatics are stated and applied first, with their derivations provided later in the chapter.

7.2 BIOT–SAVART'S LAW

Biot–Savart's law states that the differential magnetic field intensity dH produced at a point P, as shown in Figure 7.1, by the differential current element $I\,dl$ is proportional to the product $I\,dl$ and the sine of the angle α between the element and the line joining P to the element and is inversely proportional to the square of the distance R between P and the element.

[3]The experiments and analyses of the effect of a current element were carried out by Ampère and by Jean-Baptiste Biot and Felix Savart around 1820.

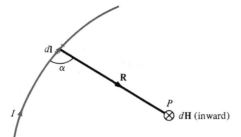

FIGURE 7.1 Magnetic field $d\mathbf{H}$ at P due to current element $I\, d\mathbf{l}$.

That is,

$$dH \propto \frac{I\, dl \sin \alpha}{R^2} \tag{7.1}$$

or

$$dH = \frac{kI\, dl \sin \alpha}{R^2} \tag{7.2}$$

where k is the constant of proportionality. In SI units, $k = 1/4\pi$, so eq. (7.2) becomes

$$dH = \frac{I\, dl \sin \alpha}{4\pi R^2} \tag{7.3}$$

From the definition of cross product in eq. (1.21), it is easy to notice that eq. (7.3) is better put in vector form as

$$\boxed{d\mathbf{H} = \frac{I\, d\mathbf{l} \times \mathbf{a}_R}{4\pi R^2} = \frac{I\, d\mathbf{l} \times \mathbf{R}}{4\pi R^3}} \tag{7.4}$$

where $R = |\mathbf{R}|$ and $\mathbf{a}_R = \mathbf{R}/R$; $\mathbf{R}$ and $d\mathbf{l}$ are illustrated in Figure 7.1. Thus the direction of $d\mathbf{H}$ can be determined by the right-hand rule with the right-hand thumb pointing in the direction of the current and the right-hand fingers encircling the wire in the direction of $d\mathbf{H}$ as shown in Figure 7.2(a). Alternatively, we can use the right-handed-screw rule to determine the direction of $d\mathbf{H}$: with the screw placed along the wire and pointed in the direction of current flow, the direction of rotation of the screw is the direction of $d\mathbf{H}$ as in Figure 7.2(b).

It is customary to represent the direction of the magnetic field intensity $\mathbf{H}$ (or current I) by a small circle with a dot or cross sign depending on whether $\mathbf{H}$ (or I) is out of the page, or into it respectively, as illustrated in Figure 7.3.

Just as we can have different charge configurations (see Figure 4.5), we can have different current distributions: line current, surface current, and volume current as shown in Figure 7.4. If we define $\mathbf{K}$ as the surface current density in amperes per meter and $\mathbf{J}$ as the volume current density in amperes per meter squared, the source elements are related as

$$I\ d\mathbf{l} \equiv \mathbf{K}\, dS \equiv \mathbf{J}\, dv \tag{7.5}$$

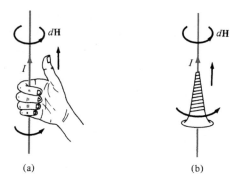

FIGURE 7.2 Determining the direction of $d\mathbf{H}$ using (**a**) the right-hand rule or (**b**) the right-handed-screw rule.

FIGURE 7.3 Conventional representation of $\mathbf{H}$ (or I) (**a**) out of the page and (**b**) into the page.

Thus in terms of the distributed current sources, the Biot–Savart's law as in eq. (7.4) becomes

$$\mathbf{H} = \int_L \frac{I\,d\mathbf{l} \times \mathbf{a}_R}{4\pi R^2} \quad \text{(line current)} \tag{7.6}$$

$$\mathbf{H} = \int_S \frac{\mathbf{K}\,dS \times \mathbf{a}_R}{4\pi R^2} \quad \text{(surface current)} \tag{7.7}$$

$$\mathbf{H} = \int_v \frac{\mathbf{J}\,dv \times \mathbf{a}_R}{4\pi R^2} \quad \text{(volume current)} \tag{7.8}$$

where $\mathbf{a}_R$ is a unit vector pointing from the differential element of current to the point of interest.

As an example, let us apply eq. (7.6) to determine the field due to a *straight* current-carrying filamentary conductor of finite length AB as in Figure 7.5. We assume that the

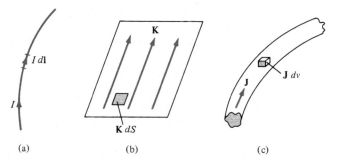

FIGURE 7.4 Current distributions: (**a**) line current, (**b**) surface current, (**c**) volume current.

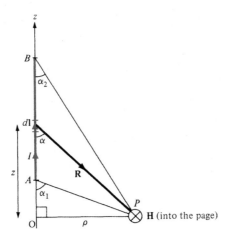

FIGURE 7.5 Field at point *P* due to a straight filamentary conductor.

conductor is along the *z*-axis with its upper and lower ends, respectively, subtending angles α_2 and α_1 at *P*, the point at which **H** is to be determined. Particular note should be taken of this assumption, as the formula to be derived will have to be applied accordingly. Note that current flows from point *A*, where $\alpha = \alpha_1$, to point *B*, where $\alpha = \alpha_2$. If we consider the contribution $d\mathbf{H}$ at *P* due to an element $d\mathbf{l}$ at $(0, 0, z)$,

$$d\mathbf{H} = \frac{I\,d\mathbf{l} \times \mathbf{R}}{4\pi R^3} \tag{7.9}$$

But $d\mathbf{l} = dz\,\mathbf{a}_z$ and $\mathbf{R} = \rho\mathbf{a}_\rho - z\mathbf{a}_z$, so

$$d\mathbf{l} \times \mathbf{R} = \rho\,dz\,\mathbf{a}_\phi \tag{7.10}$$

Hence,

$$\mathbf{H} = \int \frac{I\rho\,dz}{4\pi[\rho^2 + z^2]^{3/2}} \mathbf{a}_\phi \tag{7.11}$$

Letting $z = \rho \cot\alpha$, $dz = -\rho \csc^2\alpha\,d\alpha$, $[\rho^2 + z^2]^{3/2} = \rho^3 \csc^3\alpha$, and eq. (7.11) becomes

$$\mathbf{H} = -\frac{1}{4\pi}\int_{\alpha_1}^{\alpha_2} \frac{\rho^2 \csc^2\alpha\,d\alpha}{\rho^3 \csc^3\alpha} \mathbf{a}_\phi$$

$$= -\frac{I}{4\pi\rho}\mathbf{a}_\phi \int_{\alpha_1}^{\alpha_2} \sin\alpha\,d\alpha$$

or

$$\boxed{\mathbf{H} = \frac{I}{4\pi\rho}(\cos\alpha_2 - \cos\alpha_1)\mathbf{a}_\phi} \tag{7.12}$$

This expression is generally applicable for any straight filamentary conductor. The conductor need not lie on the z-axis, but it must be straight. Notice from eq. (7.12) that **H** is always along the unit vector $\mathbf{a}_\phi$ (i.e., along concentric circular paths) irrespective of the length of the wire or the point of interest P. As a special case, when the conductor is *semi-infinite* (with respect to P) so that point A is now at $O(0, 0, 0)$ while B is at $(0, 0, \infty)$, $\alpha_1 = 90°$, $\alpha_2 = 0°$, and eq. (7.12) becomes

$$\mathbf{H} = \frac{I}{4\pi\rho}\,\mathbf{a}_\phi \tag{7.13}$$

Another special case is found when the conductor is *infinite* in length. For this case, point A is at $(0, 0, -\infty)$ while B is at $(0, 0, \infty)$; $\alpha_1 = 180°$, $\alpha_2 = 0°$, and eq. (7.12) reduces to

$$\mathbf{H} = \frac{I}{2\pi\rho}\,\mathbf{a}_\phi \tag{7.14}$$

To find unit vector $\mathbf{a}_\phi$ in eqs. (7.12) to (7.14) is not always easy. A simple approach is to determine $\mathbf{a}_\phi$ from

$$\mathbf{a}_\phi = \mathbf{a}_\ell \times \mathbf{a}_\rho \tag{7.15}$$

where $\mathbf{a}_\ell$ is a unit vector along the line current and $\mathbf{a}_\rho$ is a unit vector along the perpendicular line from the line current to the field point.

EXAMPLE 7.1

The conducting triangular loop in Figure 7.6(a) carries a current of 10 A. Find **H** at $(0, 0, 5)$ due to side 1 of the loop.

Solution:

This example illustrates how eq. (7.12) is applied to any straight, thin, current-carrying conductor. The key point to keep in mind in applying eq. (7.12) is figuring out α_1, α_2, ρ, and $\mathbf{a}_\phi$. To find **H** at $(0, 0, 5)$ due to side 1 of the loop in Figure 7.6(a), consider Figure 7.6(b), where side 1 is treated as a straight conductor. Notice that we join the point of interest $(0, 0, 5)$ to the beginning and end of the line current. Observe that α_1, α_2, and ρ are assigned in the same manner as in Figure 7.5 on which eq. (7.12) is based:

$$\cos \alpha_1 = \cos 90° = 0, \quad \cos \alpha_2 = \frac{2}{\sqrt{29}}, \quad \rho = 5$$

To determine $\mathbf{a}_\phi$ is often the hardest part of applying eq. (7.12). According to eq. (7.15), $\mathbf{a}_\ell = \mathbf{a}_x$ and $\mathbf{a}_\rho = \mathbf{a}_z$, so

$$\mathbf{a}_\phi = \mathbf{a}_x \times \mathbf{a}_z = -\mathbf{a}_y$$

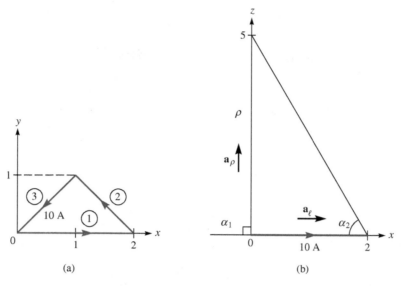

FIGURE 7.6 For Example 7.1: (**a**) conducting triangular loop, (**b**) side 1 of the loop.

Hence,

$$\mathbf{H}_1 = \frac{I}{4\pi\rho}(\cos\alpha_2 - \cos\alpha_1)\mathbf{a}_\phi = \frac{10}{4\pi(5)}\left(\frac{2}{\sqrt{29}} - 0\right)(-\mathbf{a}_y)$$

$$= -59.1\mathbf{a}_y \, \text{mA/m}$$

PRACTICE EXERCISE 7.1

Find **H** at $(0, 0, 5)$ due to side 3 of the triangular loop in Figure 7.6(a).

Answer: $-30.63\mathbf{a}_x + 30.63\mathbf{a}_y \, \text{mA/m}.$

EXAMPLE 7.2

Find **H** at $(-3, 4, 0)$ due to the current filament shown in Figure 7.7(a).

Solution:

Let $\mathbf{H} = \mathbf{H}_1 + \mathbf{H}_2$, where $\mathbf{H}_1$ and $\mathbf{H}_2$ are the contributions to the magnetic field intensity at $P(-3, 4, 0)$ due to the portions of the filament along x and z, respectively.

$$\mathbf{H}_2 = \frac{I}{4\pi\rho}(\cos\alpha_2 - \cos\alpha_1)\mathbf{a}_\phi$$

At $P(-3, 4, 0)$, $\rho = (9 + 16)^{1/2} = 5$, $\alpha_1 = 90°$, $\alpha_2 = 0°$, and $\mathbf{a}_\phi$ is obtained as a unit vector along the circular path through P on plane $z = 0$ as in Figure 7.7(b). The direction of $\mathbf{a}_\phi$

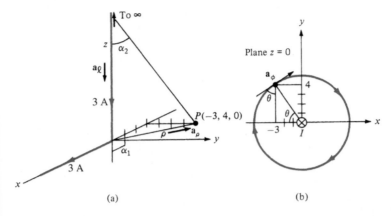

FIGURE 7.7 For Example 7.2: (**a**) current filament along semi-infinite x- and z-axes, $\mathbf{a}_\ell$ and $\mathbf{a}_p$ for $\mathbf{H}_2$ only; (**b**) determining $\mathbf{a}_p$ for $\mathbf{H}_2$.

is determined using the right-handed-screw rule or the right-hand rule. From the geometry in Figure 7.7(b),

$$\mathbf{a}_\phi = \sin\theta\,\mathbf{a}_x + \cos\theta\,\mathbf{a}_y = \frac{4}{5}\mathbf{a}_x + \frac{3}{5}\mathbf{a}_y$$

Alternatively, we can determine $\mathbf{a}_\phi$ from eq. (7.15). At point P, $\mathbf{a}_\ell$ and $\mathbf{a}_p$ are as illustrated in Figure 7.7(a) for $\mathbf{H}_2$. Hence,

$$\mathbf{a}_\phi = -\mathbf{a}_z \times \left(-\frac{3}{5}\mathbf{a}_x + \frac{4}{5}\mathbf{a}_y \right) = \frac{4}{5}\mathbf{a}_x + \frac{3}{5}\mathbf{a}_y$$

as obtained before. Thus

$$\mathbf{H}_2 = \frac{3}{4\pi(5)}(1 - 0)\frac{(4\mathbf{a}_x + 3\mathbf{a}_y)}{5}$$

$$= 38.2\mathbf{a}_x + 28.65\mathbf{a}_y \text{ mA/m}$$

It should be noted that in this case $\mathbf{a}_\phi$ happens to be the negative of the regular $\mathbf{a}_\phi$ of cylindrical coordinates. $\mathbf{H}_2$ could have also been obtained in cylindrical coordinates as

$$\mathbf{H}_2 = \frac{3}{4\pi(5)}(1 - 0)(-\mathbf{a}_\phi)$$

$$= -47.75\mathbf{a}_\phi \text{ mA/m}$$

Similarly, for $\mathbf{H}_1$ at P, $\rho = 4$, $\alpha_2 = 0°$, $\cos\alpha_1 = 3/5$, and $\mathbf{a}_\phi = \mathbf{a}_z$ or $\mathbf{a}_\phi = \mathbf{a}_\ell \times \mathbf{a}_p = \mathbf{a}_x \times \mathbf{a}_y = \mathbf{a}_z$. Hence,

$$\mathbf{H}_1 = \frac{3}{4\pi(4)}\left(1 - \frac{3}{5} \right)\mathbf{a}_z$$

$$= 23.87\mathbf{a}_z \text{ mA/m}$$

Thus

$$\mathbf{H} = \mathbf{H}_1 + \mathbf{H}_2 = 38.2\mathbf{a}_x + 28.65\mathbf{a}_y + 23.87\mathbf{a}_z \text{ mA/m}$$

or

$$\mathbf{H} = -47.75\mathbf{a}_\phi + 23.87\mathbf{a}_z \text{ mA/m}$$

Notice that although the current filaments appear to be semi-infinite (they occupy the positive z- and x-axes), it is only the filament along the z-axis that is semi-infinite with respect to point P. Thus $\mathbf{H}_2$ could have been found by using eq. (7.13), but the equation could not have been used to find $\mathbf{H}_1$ because the filament along the x-axis is not semi-infinite with respect to P.

PRACTICE EXERCISE 7.2

The positive y-axis (semi-infinite line with respect to the origin) carries a filamentary current of 2 A in the $-\mathbf{a}_y$ direction. Assume it is part of a large circuit. Find $\mathbf{H}$ at

(a) $A(2, 3, 0)$
(b) $B(3, 12, -4)$

Answer: (a) $145.8\mathbf{a}_z$ mA/m, (b) $48.97\mathbf{a}_x + 36.73\mathbf{a}_z$ mA/m.

EXAMPLE 7.3

A circular loop located on $x^2 + y^2 = 9, z = 0$ carries a direct current of 10 A along $\mathbf{a}_\phi$. Determine $\mathbf{H}$ at $(0, 0, 4)$ and $(0, 0, -4)$.

Solution:

Consider the circular loop shown in Figure 7.8(a). The magnetic field intensity $d\mathbf{H}$ at point $P(0, 0, h)$ contributed by current element $I\, d\mathbf{l}$ is given by Biot–Savart's law:

$$d\mathbf{H} = \frac{I\, d\mathbf{l} \times \mathbf{R}}{4\pi R^3}$$

where $d\mathbf{l} = \rho\, d\phi\, \mathbf{a}_\phi$, $\mathbf{R} = (0, 0, h) - (x, y, 0) = -\rho\mathbf{a}_\rho + h\mathbf{a}_z$, and

$$d\mathbf{l} \times \mathbf{R} = \begin{vmatrix} \mathbf{a}_\rho & \mathbf{a}_\phi & \mathbf{a}_z \\ 0 & \rho\, d\phi & 0 \\ -\rho & 0 & h \end{vmatrix} = \rho h\, d\phi\, \mathbf{a}_\rho + \rho^2\, d\phi\, \mathbf{a}_z$$

Hence,

$$d\mathbf{H} = \frac{I}{4\pi[\rho^2 + h^2]^{3/2}}(\rho h\, d\phi\, \mathbf{a}_\rho + \rho^2\, d\phi\, \mathbf{a}_z) = dH_\rho\mathbf{a}_\rho + dH_z\mathbf{a}_z$$

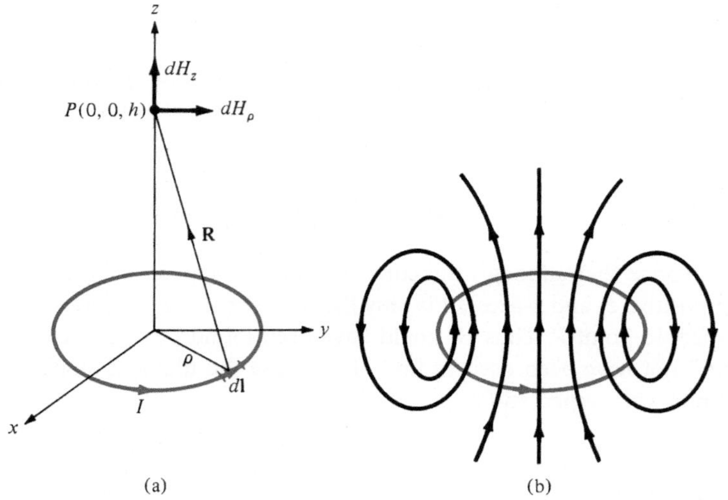

FIGURE 7.8 For Example 7.3: (**a**) circular current loop, (**b**) flux lines due to the current loop.

By symmetry, the contributions along $\mathbf{a}_\rho$ add up to zero because the radial components produced by current element pairs 180° apart cancel. This may also be shown mathematically by writing $\mathbf{a}_\rho$ in rectangular coordinate systems (i.e., $\mathbf{a}_\rho = \cos\phi\,\mathbf{a}_x + \sin\phi\,\mathbf{a}_y$). Integrating $\cos\phi$ or $\sin\phi$ over $0 \le \phi \le 2\pi$ gives zero, thereby showing that $H_\rho = 0$. Thus

$$\mathbf{H} = \int dH_z\,\mathbf{a}_z = \int_0^{2\pi} \frac{I\rho^2\,d\phi\,\mathbf{a}_z}{4\pi[\rho^2 + h^2]^{3/2}} = \frac{I\rho^2 2\pi\mathbf{a}_z}{4\pi[\rho^2 + h^2]^{3/2}}$$

or

$$\mathbf{H} = \frac{I\rho^2\mathbf{a}_z}{2[\rho^2 + h^2]^{3/2}}$$

(a) Substituting $I = 10\ A, \rho = 3, h = 4$ gives

$$\mathbf{H}(0, 0, 4) = \frac{10(3)^2\mathbf{a}_z}{2[9 + 16]^{3/2}} = 0.36\mathbf{a}_z\ \text{A/m}$$

(b) Notice from $d\mathbf{l} \times \mathbf{R}$ in the Biot–Savart law that if h is replaced by $-h$, the z-component of $d\mathbf{H}$ remains the same while the ρ-component still adds up to zero due to the axial symmetry of the loop. Hence

$$\mathbf{H}(0, 0, -4) = \mathbf{H}(0, 0, 4) = 0.36\mathbf{a}_z\ \text{A/m}$$

The flux lines due to the circular current loop are sketched in Figure 7.8(b).

PRACTICE EXERCISE 7.3

A thin ring of radius 5 cm is placed on plane $z = 1$ cm so that its center is at $(0, 0, 1$ cm$)$. If the ring carries 50 mA along $\mathbf{a}_\phi$, find $\mathbf{H}$ at

(a) $(0, 0, -1$ cm$)$
(b) $(0, 0, 10$ cm$)$

Answer: (a) $400\mathbf{a}_z$ mA/m, (b) $57.3\mathbf{a}_z$ mA/m.

EXAMPLE 7.4

A solenoid of length ℓ and radius a consists of N turns of wire carrying current I. Show that at point P along its axis,

$$\mathbf{H} = \frac{nI}{2}(\cos \theta_2 - \cos \theta_1)\mathbf{a}_z$$

where $n = N/\ell$, θ_1 and θ_2 are the angles subtended at P by the end turns as illustrated in Figure 7.9. Also show that if $\ell \gg a$, at the center of the solenoid,

$$\mathbf{H} = nI\mathbf{a}_z$$

Solution:

Consider the cross section of the solenoid as shown in Figure 7.9. Since the solenoid consists of circular loops, we apply the result of Example 7.3. The contribution to the magnetic field H at P by an element of the solenoid of length dz is

$$dH_z = \frac{I \, dl \, a^2}{2[a^2 + z^2]^{3/2}} = \frac{Ia^2 n \, dz}{2[a^2 + z^2]^{3/2}}$$

where $dl = n \, dz = (N/\ell) \, dz$. From Figure 7.9, $\tan \theta = a/z$; that is,

$$dz = -a \csc^2 \theta \, d\theta = -\frac{[z^2 + a^2]^{3/2}}{a^2} \sin \theta \, d\theta$$

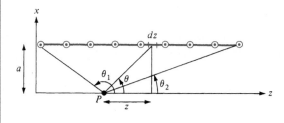

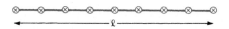

FIGURE 7.9 For Example 7.4; cross section of a solenoid.

Hence,

$$dH_z = -\frac{nI}{2} \sin \theta \, d\theta$$

or

$$H_z = -\frac{nI}{2} \int_{\theta_1}^{\theta_2} \sin \theta \, d\theta$$

Thus

$$\mathbf{H} = \frac{nI}{2} (\cos \theta_2 - \cos \theta_1) \mathbf{a}_z$$

as required. Substituting $n = N/\ell$ gives

$$\mathbf{H} = \frac{NI}{2\ell} (\cos \theta_2 - \cos \theta_1) \mathbf{a}_z$$

At the center of the solenoid,

$$\cos \theta_2 = \frac{\ell/2}{[a^2 + \ell^2/4]^{1/2}} = -\cos \theta_1$$

and

$$\mathbf{H} = \frac{In\ell}{2[a^2 + \ell^2/4]^{1/2}} \mathbf{a}_z$$

If $\ell \gg a$ or $\theta_2 \simeq 0°, \theta_1 \simeq 180°$,

$$\mathbf{H} = nI \mathbf{a}_z = \frac{NI}{\ell} \mathbf{a}_z$$

PRACTICE EXERCISE 7.4

The solenoid of Figure 7.9 has 2000 turns, a length of 75 cm, and a radius of 5 cm. If it carries a current of 50 mA along $\mathbf{a}_\phi$, find $\mathbf{H}$ at

(a) $(0, 0, 0)$
(b) $(0, 0, 75 \text{ cm})$
(c) $(0, 0, 50 \text{ cm})$

Answer: (a) $66.52 \mathbf{a}_z$ A/m, (b) $66.52 \mathbf{a}_z$ A/m, (c) $131.7 \mathbf{a}_z$ A/m.

7.3 AMPÈRE'S CIRCUIT LAW—MAXWELL'S EQUATION

> **Ampère's circuit law** states that the line integral of **H** around a *closed* path is the same as the net current I_{enc} enclosed by the path.

In other words, the circulation of **H** equals I_{enc}; that is,

$$\oint_L \mathbf{H} \cdot d\mathbf{l} = I_{enc} \qquad (7.16)$$

Ampère's law is similar to Gauss's law, since Ampère's law is easily applied to determine **H** when the current distribution is symmetrical. It should be noted that eq. (7.16) always holds regardless of whether the current distribution is symmetrical or not, but we can use the equation to determine **H** only when a symmetrical current distribution exists. Ampère's law is a special case of Biot–Savart's law; the former may be derived from the latter.

By applying Stokes's theorem to the left-hand side of eq. (7.16), we obtain

$$I_{enc} = \oint_L \mathbf{H} \cdot d\mathbf{l} = \int_S (\nabla \times \mathbf{H}) \cdot d\mathbf{S} \qquad (7.17)$$

But

$$I_{enc} = \int_S \mathbf{J} \cdot d\mathbf{S} \qquad (7.18)$$

Comparing the surface integrals in eqs. (7.17) and (7.18) clearly reveals that

$$\nabla \times \mathbf{H} = \mathbf{J} \qquad (7.19)$$

This is the third Maxwell's equation to be derived; it is essentially Ampère's law in differential (or point) form, whereas eq. (7.16) is the integral form. From eq. (7.19), we should observe that $\nabla \times \mathbf{H} = \mathbf{J} \neq \mathbf{0}$; that is, a magnetostatic field is not conservative.

7.4 APPLICATIONS OF AMPÈRE'S LAW

We now apply Ampère's circuit law to determine **H** for some symmetrical current distributions as we did for Gauss's law. We will consider an infinite line current, an infinite sheet of current, and an infinitely long coaxial transmission line. In each case, we apply $\oint_L \mathbf{H} \cdot d\mathbf{l} = I_{enc}$. For symmetrical current distribution, **H** is either parallel or perpendicular to $d\mathbf{l}$. When **H** is parallel to $d\mathbf{l}$, $|\mathbf{H}| =$ constant.

A. Infinite Line Current

Consider an infinitely long filamentary current I along the z-axis as in Figure 7.10. To determine **H** at an observation point P, we allow a closed path to pass through P. This path, on which Ampère's law is to be applied, is known as an *Amperian path* (analogous to the term "Gaussian surface"). We choose a concentric circle as the Amperian path in view of eq. (7.14), which shows that **H** is constant provided ρ is constant. Since this path encloses the whole current I, according to Ampère's law,

$$I = \int_L H_\phi \mathbf{a}_\phi \cdot \rho \, d\phi \, \mathbf{a}_\phi = H_\phi \int_L \rho \, d\phi = H_\phi \cdot 2\pi\rho$$

or

$$\mathbf{H} = \frac{I}{2\pi\rho} \mathbf{a}_\phi \tag{7.20}$$

as expected from eq. (7.14).

B. Infinite Sheet of Current

Consider an infinite current sheet in the $z = 0$ plane. If the sheet has a uniform current density $\mathbf{K} = K_y \mathbf{a}_y$ A/m as shown in Figure 7.11, applying Ampère's law to the rectangular closed path 1-2-3-4-1 (Amperian path) gives

$$\oint \mathbf{H} \cdot d\mathbf{l} = I_{\text{enc}} = K_y b \tag{7.21a}$$

To evaluate the integral, we first need to have an idea of what **H** is like. To achieve this, we regard the infinite sheet as comprising filaments; $d\mathbf{H}$ above or below the sheet due to a pair of filamentary currents can be found by using eqs. (7.14) and (7.15). As evident in Figure 7.11(b), the resultant $d\mathbf{H}$ has only an x-component. Also, **H** on one side of the sheet is the negative of that on the other side. Owing to the infinite extent of the sheet, the sheet can be

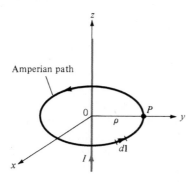

FIGURE 7.10 Ampère's law applied to an infinite filamentary line current.

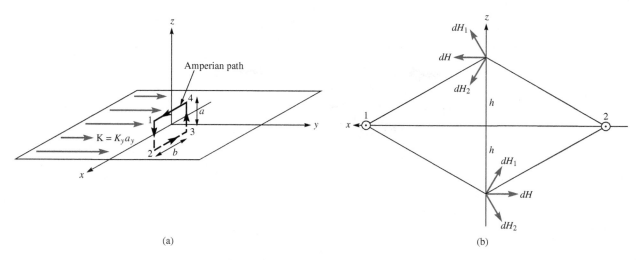

FIGURE 7.11 Application of Ampère's law to an infinite sheet: (**a**) closed path 1-2-3-4-1, (**b**) symmetrical pair of current filaments with current along $\mathbf{a}_y$.

regarded as consisting of such filamentary pairs so that the characteristics of **H** for a pair are the same for the infinite current sheet, that is,

$$\mathbf{H} = \begin{cases} H_o\mathbf{a}_x & z > 0 \\ -H_o\mathbf{a}_x & z < 0 \end{cases} \tag{7.21b}$$

where H_o is yet to be determined. Evaluating the line integral of **H** in eq. (7.21a) along the closed path in Figure 7.11(a) gives

$$\oint \mathbf{H} \cdot d\mathbf{l} = \left(\int_1^2 + \int_2^3 + \int_3^4 + \int_4^1 \right) \mathbf{H} \cdot d\mathbf{l}$$
$$= 0(-a) + (-H_o)(-b) + 0(a) + H_o(b) \tag{7.21c}$$
$$= 2H_o b$$

From eqs. (7.21a) and (7.21c), we obtain $H_o = \dfrac{1}{2} K_y$. Substituting H_o in eq. (7.21b) gives

$$\mathbf{H} = \begin{cases} \dfrac{1}{2} K_y \mathbf{a}_x, & z > 0 \\ -\dfrac{1}{2} K_y \mathbf{a}_x, & z < 0 \end{cases} \tag{7.22}$$

In general, for an infinite sheet of current density **K** A/m,

$$\boxed{\mathbf{H} = \frac{1}{2} \mathbf{K} \times \mathbf{a}_n} \tag{7.23}$$

where $\mathbf{a}_n$ is a unit normal vector directed from the current sheet to the point of interest.

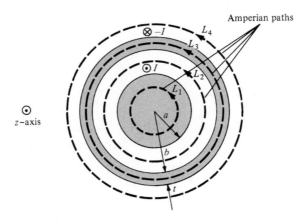

FIGURE 7.12 Cross section of the transmission line; the positive z-direction is out of the page.

C. Infinitely Long Coaxial Transmission Line

Consider an infinitely long transmission line consisting of two concentric cylinders having their axes along the z-axis. The cross section of the line is shown in Figure 7.12, where the z-axis is out of the page. The inner conductor has radius a and carries current I, while the outer conductor has inner radius b and thickness t and carries return current $-I$. We want to determine **H** everywhere, assuming that current is uniformly distributed in both conductors. Since the current distribution is symmetrical, we apply Ampère's law along the Amperian path for each of the four possible regions: $0 \leq \rho \leq a$, $a \leq \rho \leq b$, $b \leq \rho \leq b + t$, and $\rho \geq b + t$.

For region $0 \leq \rho \leq a$, we apply Ampère's law to path L_1, giving

$$\oint_{L_1} \mathbf{H} \cdot d\mathbf{l} = I_{\text{enc}} = \int_S \mathbf{J} \cdot d\mathbf{S} \tag{7.24}$$

Since the current is uniformly distributed over the cross section,

$$\mathbf{J} = \frac{I}{\pi a^2} \mathbf{a}_z, \quad d\mathbf{S} = \rho \, d\phi \, d\rho \, \mathbf{a}_z$$

$$I_{\text{enc}} = \int_S \mathbf{J} \cdot d\mathbf{S} = \frac{I}{\pi a^2} \int_{\phi=0}^{2\pi} \int_{\rho=0}^{\rho} \rho \, d\phi \, d\rho = \frac{I}{\pi a^2} \pi \rho^2 = \frac{I\rho^2}{a^2}$$

Hence eq. (7.24) becomes

$$H_\phi \int_{L_1} dl = H_\phi \, 2\pi\rho = \frac{I\rho^2}{a^2}$$

or

$$H_\phi = \frac{I\rho}{2\pi a^2} \tag{7.25}$$

For region $a \leq \rho \leq b$, we use path L_2 as the Amperian path,

$$\oint_{L_2} \mathbf{H} \cdot d\mathbf{l} = I_{enc} = I$$

$$H_\phi 2\pi\rho = I$$

or

$$H_\phi = \frac{I}{2\pi\rho} \tag{7.26}$$

since the whole current I is enclosed by L_2. Notice that eq. (7.26) is the same as eq. (7.14), and it is independent of a. For region $b \leq \rho \leq b + t$, we use path L_3, getting

$$\oint_{L_3} \mathbf{H} \cdot d\mathbf{l} = H_\phi \cdot 2\pi\rho = I_{enc} \tag{7.27a}$$

where

$$I_{enc} = I + \int \mathbf{J} \cdot d\mathbf{S}$$

and $\mathbf{J}$ in this case is the current density (current per unit area) of the outer conductor and is along $-\mathbf{a}_z$, that is,

$$\mathbf{J} = -\frac{I}{\pi[(b+t)^2 - b^2]} \mathbf{a}_z$$

Thus

$$I_{enc} = I - \frac{I}{\pi[(b+t)^2 - b^2]} \int_{\phi=0}^{2\pi} \int_{\rho=b}^{\rho} \rho \, d\rho \, d\phi$$

$$= I\left[1 - \frac{\rho^2 - b^2}{t^2 + 2bt}\right]$$

Substituting this in eq. (7.27a), we have

$$H_\phi = \frac{I}{2\pi\rho}\left[1 - \frac{\rho^2 - b^2}{t^2 + 2bt}\right] \tag{7.27b}$$

For region $\rho \geq b + t$, we use path L_4, getting

$$\oint_{L_4} \mathbf{H} \cdot d\mathbf{l} = I - I = 0$$

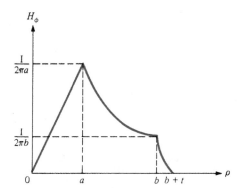

FIGURE 7.13 Plot of H_ϕ against ρ.

or

$$H_\phi = 0 \tag{7.28}$$

Putting eqs. (7.25) to (7.28) together gives

$$\mathbf{H} = \begin{cases} \dfrac{I\rho}{2\pi a^2}\, \mathbf{a}_\phi, & 0 \le \rho \le a \\[2mm] \dfrac{I}{2\pi\rho}\, \mathbf{a}_\phi, & a \le \rho \le b \\[2mm] \dfrac{I}{2\pi\rho}\left[1 - \dfrac{\rho^2 - b^2}{t^2 + 2bt}\right] \mathbf{a}_\phi, & b \le \rho \le b + t \\[2mm] 0, & \rho \ge b + t \end{cases} \tag{7.29}$$

The magnitude of **H** is sketched in Figure 7.13.

From these examples, it can be observed that the ability to take **H** from under the integral sign is the key to using Ampère's law to determine **H**. In other words, Ampère's law can be used to find **H** only due to symmetric current distributions for which it is possible to find a closed path over which **H** is constant in magnitude.

EXAMPLE 7.5

Planes $z = 0$ and $z = 4$ carry current $\mathbf{K} = -10\mathbf{a}_x$ A/m and $\mathbf{K} = 10\mathbf{a}_x$ A/m, respectively. Determine **H** at

(a) $(1, 1, 1)$
(b) $(0, -3, 10)$

Solution:

The parallel current sheets are shown in Figure 7.14. Let

$$\mathbf{H} = \mathbf{H}_o + \mathbf{H}_4$$

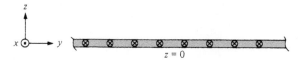

FIGURE 7.14 For Example 7.5: parallel infinite current sheets.

where $\mathbf{H}_o$ and $\mathbf{H}_4$ are the contributions due to the current sheets $z = 0$ and $z = 4$, respectively. We make use of eq. (7.23).

(a) At $(1, 1, 1)$, which is between the plates $(0 < z = 1 < 4)$,

$$\mathbf{H}_o = 1/2 \, \mathbf{K} \times \mathbf{a}_n = 1/2(-10\mathbf{a}_x) \times \mathbf{a}_z = 5\mathbf{a}_y \, \text{A/m}$$

$$\mathbf{H}_4 = 1/2 \, \mathbf{K} \times \mathbf{a}_n = 1/2(10\mathbf{a}_x) \times (-\mathbf{a}_z) = 5\mathbf{a}_y \, \text{A/m}$$

Hence,

$$\mathbf{H} = 10\mathbf{a}_y \, \text{A/m}$$

(b) At $(0, -3, 10)$, which is above the two sheets $(z = 10 > 4 > 0)$,

$$\mathbf{H}_o = 1/2(-10\mathbf{a}_x) \times \mathbf{a}_z = 5\mathbf{a}_y \, \text{A/m}$$

$$\mathbf{H}_4 = 1/2(10\mathbf{a}_x) \times \mathbf{a}_z = -5\mathbf{a}_y \, \text{A/m}$$

Hence,

$$\mathbf{H} = 0 \, \text{A/m}$$

PRACTICE EXERCISE 7.5

Plane $y = 1$ carries current $\mathbf{K} = 50\mathbf{a}_z$ mA/m. Find $\mathbf{H}$ at
(a) $(0, 0, 0)$
(b) $(1, 5, -3)$

Answer: (a) $25\mathbf{a}_x$ mA/m, (b) $-25\mathbf{a}_x$ mA/m.

EXAMPLE 7.6

A toroid whose dimensions are shown in Figure 7.15 has N turns and carries current I. Determine H inside and outside the toroid.

Solution:

We apply Ampère's circuit law to the Amperian path, which is a circle of radius ρ shown dashed in Figure 7.15. Since N wires cut through this path each carrying current I, the net current enclosed by the Amperian path is NI. Hence,

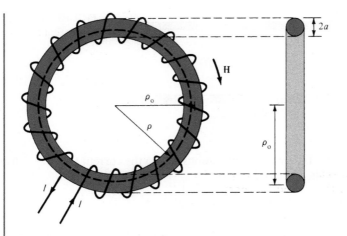

FIGURE 7.15 For Example 7.6: a toroid with a circular cross section.

$$\oint \mathbf{H} \cdot d\mathbf{l} = I_{\text{enc}} \rightarrow H \cdot 2\pi\rho = NI$$

or

$$H = \frac{NI}{2\pi\rho}, \quad \text{for } \rho_{\text{o}} - a < \rho < \rho_{\text{o}} + a$$

where ρ_{o} is the mean radius of the toroid as shown in Figure 7.15. An approximate value of H is

$$H_{\text{approx}} = \frac{NI}{2\pi\rho_{\text{o}}} = \frac{NI}{\ell}$$

Notice that this is the same as the formula obtained for H for points well inside a very long solenoid $(\ell \gg a)$. Thus a straight solenoid may be regarded as a special toroidal coil for which $\rho_{\text{o}} \rightarrow \infty$. Outside the toroid, the current enclosed by an Amperian path is $NI - NI = 0$ and hence $H = 0$.

PRACTICE EXERCISE 7.6

A toroid of circular cross section whose center is at the origin and axis the same as the z-axis has 1000 turns with $\rho_{\text{o}} = 10$ cm, $a = 1$ cm. If the toroid carries a 100 mA current, find $|H|$ at

(a) $(3 \text{ cm}, -4 \text{ cm}, 0)$
(b) $(6 \text{ cm}, 9 \text{ cm}, 0)$

Answer: (a) 0, (b) 147.1 A/m.

7.5 MAGNETIC FLUX DENSITY—MAXWELL'S EQUATION

The magnetic flux density **B** is similar to the electric flux density **D**. As **D** $= \varepsilon_o$**E** in free space, the magnetic flux density **B** is related to the magnetic field intensity **H** according to

$$\mathbf{B} = \mu_o \mathbf{H} \tag{7.30}$$

where μ_o is a constant known as the *permeability of free space*. The constant is in Henrys per meter (H/m) and has the value of

$$\mu_o = 4\pi \times 10^{-7} \text{ H/m} \tag{7.31}$$

The precise definition of the magnetic flux density **B**, in terms of the magnetic force, will be given in the next chapter.

The magnetic flux through a surface S is given by

$$\Psi = \int_S \mathbf{B} \cdot d\mathbf{S} \tag{7.32}$$

where the magnetic flux Ψ is in webers (Wb) and the magnetic flux density is in webers per square meter (Wb/m^2) or teslas (T).

A magnetic flux line is a path to which **B** is tangential at every point on the line. It is a line along which the needle of a magnetic compass will orient itself if placed in the presence of a magnetic field. For example, the magnetic flux lines due to a straight long wire are shown in Figure 7.16. The flux lines are determined by using the same principle followed in Section 4.10 for the electric flux lines. The direction of **B** is taken as that indicated as "north" by the needle of the magnetic compass. Notice that each flux line is closed and has no beginning or end. Though Figure 7.16 is for a straight, current-carrying conductor, it is generally true that magnetic flux lines are closed and do not cross each other regardless of the current distribution.

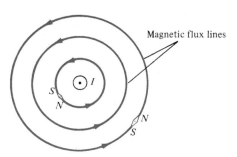

Magnetic flux lines

FIGURE 7.16 Magnetic flux lines due to a straight wire with current coming out of the page.

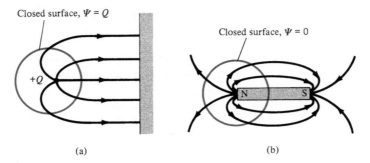

FIGURE 7.17 Flux leaving a closed surface due to (**a**) isolated electric charge $\psi = \oint_S \mathbf{D} \cdot d\mathbf{S} = Q$, (**b**) magnetic charge, $\psi = \oint_S \mathbf{B} \cdot d\mathbf{S} = 0$.

In an electrostatic field, the flux passing through a closed surface is the same as the charge enclosed; that is, $\psi = \oint_S \mathbf{D} \cdot d\mathbf{S} = Q$. Thus it is possible to have an isolated electric charge as shown in Figure 7.17(a), which also reveals that electric flux lines are not necessarily closed. Unlike electric flux lines, magnetic flux lines always close upon themselves as in Figure 7.17(b). This is because *it is not possible to have isolated magnetic poles (or magnetic charges)*. For example, if we desire to have an isolated magnetic pole by dividing a magnetic bar successively into two, we end up with pieces each having north and south poles as illustrated in Figure 7.18. We find it impossible to separate the north pole from the south pole.

An **isolated magnetic** charge does not exist.

Thus the total flux through a closed surface in a magnetic field must be zero; that is,

$$\oint_S \mathbf{B} \cdot d\mathbf{S} = 0 \tag{7.33}$$

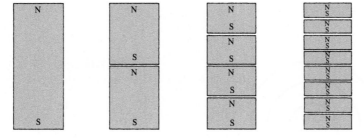

FIGURE 7.18 Successive division of a bar magnet results in pieces with north and south poles, showing that magnetic poles cannot be isolated.

This equation is referred to as the *law of conservation of magnetic flux* or *Gauss's law for magnetostatic fields*, just as $\oint_S \mathbf{D} \cdot d\mathbf{S} = Q$ is Gauss's law for electrostatic fields. Although the magnetostatic field is not conservative, magnetic flux is conserved.

By applying the divergence theorem to eq. (7.33), we obtain

$$\oint_S \mathbf{B} \cdot d\mathbf{S} = \int_v \nabla \cdot \mathbf{B} \, dv = 0$$

or

$$\boxed{\nabla \cdot \mathbf{B} = 0} \tag{7.34}$$

This equation is the fourth Maxwell's equation to be derived. Equation (7.33) or (7.34) shows that magnetostatic fields have no sources or sinks. Equation (7.34) suggests that magnetic field lines are always continuous.

7.6 MAXWELL'S EQUATIONS FOR STATIC FIELDS

Having derived Maxwell's four equations for static fields, we may take a moment to put them together as in Table 7.2. From the table, we notice that the order in which the equations are presented differs from the order in which they were derived. This was done for the sake of clarity.

The choice between differential and integral forms of the equations depends on a given problem. It is evident from Table 7.2 that a vector field is defined completely by specifying its curl and its divergence. A field can be electric or magnetic only if it satisfies the corresponding Maxwell equations (see Problems 7.40 and 7.41). It should be noted that Maxwell's equations as in Table 7.2 are only for static electric and magnetic fields. As will be discussed in Chapter 9, the divergence equations will remain the same for time-varying EM fields, but the curl equations will have to be modified.

TABLE 7.2 Maxwell's Equations for Static Electric and Magnetic Fields

Differential (or Point) Form	Integral Form	Remarks
$\nabla \cdot \mathbf{D} = \rho_v$	$\oint_S \mathbf{D} \cdot d\mathbf{S} = \int_v \rho_v \, dv$	Gauss's law
$\nabla \cdot \mathbf{B} = 0$	$\oint_S \mathbf{B} \cdot d\mathbf{S} = 0$	Nonexistence of magnetic monopole
$\nabla \times \mathbf{E} = \mathbf{0}$	$\oint_L \mathbf{E} \cdot d\mathbf{l} = 0$	Conservative nature of electrostatic field
$\nabla \times \mathbf{H} = \mathbf{J}$	$\oint_L \mathbf{H} \cdot d\mathbf{l} = \int_S \mathbf{J} \cdot d\mathbf{S}$	Ampère's law

7.7 MAGNETIC SCALAR AND VECTOR POTENTIALS

We recall that some electrostatic field problems were simplified by relating the electric potential V to the electric field intensity $\mathbf{E}(\mathbf{E} = -\nabla V)$. Similarly, we can define a potential associated with magnetostatic field $\mathbf{B}$. In fact, the magnetic potential could be scalar V_m or vector $\mathbf{A}$. To define V_m and $\mathbf{A}$ involves recalling two important identities (see Example 3.10 and Practice Exercise 3.10):

$$\nabla \times (\nabla V) = \mathbf{0} \tag{7.35a}$$

$$\nabla \cdot (\nabla \times \mathbf{A}) = 0 \tag{7.35b}$$

which must always hold for any scalar field V and vector field $\mathbf{A}$.

Just as $\mathbf{E} = -\nabla V$, we define the *magnetic scalar potential* V_m (in amperes) as related to $\mathbf{H}$ according to

$$\boxed{\mathbf{H} = -\nabla V_m} \qquad \text{if } \mathbf{J} = \mathbf{0} \tag{7.36}$$

The condition attached to this equation is important and will be explained. Combining eq. (7.36) and eq. (7.19) gives

$$\mathbf{J} = \nabla \times \mathbf{H} = \nabla \times (-\nabla V_m) = \mathbf{0} \tag{7.37}$$

since V_m must satisfy the condition in eq. (7.35a). Thus the magnetic scalar potential V_m is only defined in a region where $\mathbf{J} = \mathbf{0}$ as in eq. (7.36). We should also note that V_m satisfies Laplace's equation just as V does for electrostatic fields; hence,

$$\nabla^2 V_m = 0, \quad (\mathbf{J} = \mathbf{0}) \tag{7.38}$$

We know that for a magnetostatic field, $\nabla \cdot \mathbf{B} = 0$ as stated in eq. (7.34). To satisfy eqs. (7.34) and (7.35b) simultaneously, we can define the *magnetic vector potential* $\mathbf{A}$ (in Wb/m) such that

$$\boxed{\mathbf{B} = \nabla \times \mathbf{A}} \tag{7.39}$$

Just as we defined

$$V = \int \frac{dQ}{4\pi\varepsilon_o R} \tag{7.40}$$

we can define

$$\boxed{\mathbf{A} = \int_L \frac{\mu_o I \, d\mathbf{l}}{4\pi R}} \qquad \text{for line current} \tag{7.41}$$

$$\boxed{\mathbf{A} = \int_S \frac{\mu_o \mathbf{K}\, dS}{4\pi R}} \qquad \text{for surface current} \tag{7.42}$$

$$\boxed{\mathbf{A} = \int_v \frac{\mu_o \mathbf{J}\, dv}{4\pi R}} \qquad \text{for volume current} \tag{7.43}$$

Rather than obtaining eqs. (7.41) to (7.43) from eq. (7.40), an alternative approach would be to obtain eqs. (7.41) to (7.43) from eqs. (7.6) to (7.8). For example, we can derive eq. (7.41) from eq. (7.6) in conjunction with eq. (7.39). To do this, we write eq. (7.6) as

$$\mathbf{B} = \frac{\mu_o}{4\pi} \int_L \frac{I\, d\mathbf{l}' \times \mathbf{R}}{R^3} \tag{7.44}$$

where $\mathbf{R}$ is the distance vector from the line element $d\mathbf{l}'$ at the source point (x', y', z') to the field point (x, y, z) as shown in Figure 7.19 and $R = |\mathbf{R}|$, that is,

$$R = |\mathbf{r} - \mathbf{r}'| = [(x - x')^2 + (y - y')^2 + (z - z')^2]^{1/2} \tag{7.45}$$

Hence,

$$\nabla\!\left(\frac{1}{R}\right) = -\frac{(x - x')\mathbf{a}_x + (y - y')\mathbf{a}_y + (z - z')\mathbf{a}_z}{[(x - x')^2 + (y - y')^2 + (z - z')^2]^{3/2}} = -\frac{\mathbf{R}}{R^3}$$

or

$$\frac{\mathbf{R}}{R^3} = -\nabla\!\left(\frac{1}{R}\right) \quad \left(= \frac{\mathbf{a}_R}{R^2}\right) \tag{7.46}$$

where the differentiation is with respect to x, y, and z. Substituting this into eq. (7.44), we obtain

$$\mathbf{B} = -\frac{\mu_o}{4\pi} \int_L I\, d\mathbf{l}' \times \nabla\!\left(\frac{1}{R}\right) \tag{7.47}$$

FIGURE 7.19 Illustration of the source point (x', y', z') and the field point (x, y, z).

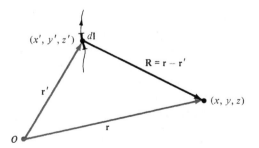

We apply the vector identity

$$\nabla \times (f\mathbf{F}) = f\nabla \times \mathbf{F} + (\nabla f) \times \mathbf{F} \tag{7.48}$$

where f is a scalar field and $\mathbf{F}$ is a vector field. Taking $f = 1/R$ and $\mathbf{F} = d\mathbf{l}'$, we have

$$d\mathbf{l}' \times \nabla\left(\frac{1}{R}\right) = \frac{1}{R}\nabla \times d\mathbf{l}' - \nabla \times \left(\frac{d\mathbf{l}'}{R}\right)$$

Since ∇ operates with respect to (x, y, z) while $d\mathbf{l}'$ is a function of (x', y', z'), $\nabla \times d\mathbf{l}' = \mathbf{0}$. Hence,

$$d\mathbf{l}' \times \nabla\left(\frac{1}{R}\right) = -\nabla \times \frac{d\mathbf{l}'}{R} \tag{7.49}$$

With this equation, eq. (7.47) reduces to

$$\mathbf{B} = \nabla \times \int_L \frac{\mu_o I \, d\mathbf{l}'}{4\pi R} \tag{7.50}$$

Comparing eq. (7.50) with eq. (7.39) shows that

$$\mathbf{A} = \int_L \frac{\mu_o I \, d\mathbf{l}'}{4\pi R}$$

verifying eq. (7.41).

By substituting eq. (7.39) into eq. (7.32) and applying Stokes's theorem, we obtain

$$\Psi = \int_S \mathbf{B} \cdot d\mathbf{S} = \int_S (\nabla \times \mathbf{A}) \cdot d\mathbf{S} = \oint_L \mathbf{A} \cdot d\mathbf{l}$$

or

$$\boxed{\Psi = \oint_L \mathbf{A} \cdot d\mathbf{l}} \tag{7.51}$$

Thus the magnetic flux through a given area can be found by using either eq. (7.32) or (7.51). Also, the magnetic field can be determined by using either V_m or $\mathbf{A}$; the choice is dictated by the nature of the given problem except that V_m can be used only in a source-free region. The use of the magnetic vector potential provides a powerful, elegant approach to solving EM problems, particularly those relating to antennas. As we shall notice in Chapter 13, it is more convenient to find $\mathbf{B}$ by first finding $\mathbf{A}$ in antenna problems.

EXAMPLE 7.7

Given the magnetic vector potential $\mathbf{A} = -\rho^2/4\, \mathbf{a}_z$ Wb/m, calculate the total magnetic flux crossing the surface $\phi = \pi/2$, $1 \le \rho \le 2$ m, $0 \le z \le 5$ m.

Solution:

We can solve this problem in two different ways: using eq. (7.32) or eq. (7.51).

Method 1:

$$\mathbf{B} = \nabla \times \mathbf{A} = -\frac{\partial A_z}{\partial \rho}\mathbf{a}_\phi = \frac{\rho}{2}\mathbf{a}_\phi, \quad d\mathbf{S} = d\rho\, dz\, \mathbf{a}_\phi$$

Hence,

$$\Psi = \int_S \mathbf{B} \cdot d\mathbf{S} = \frac{1}{2}\int_{z=0}^{5}\int_{\rho=1}^{2} \rho\, d\rho\, dz = \frac{1}{4}\rho^2\Big|_{2}^{1}(5) = \frac{15}{4}$$

$$\Psi = 3.75 \text{ Wb}$$

Method 2:

We use

$$\Psi = \oint_L \mathbf{A} \cdot d\mathbf{l} = \Psi_1 + \Psi_2 + \Psi_3 + \Psi_4$$

where L is the path bounding surface S; Ψ_1, Ψ_2, Ψ_3, and Ψ_4 are, respectively, the evaluations of $\int_L \mathbf{A} \cdot d\mathbf{l}$ along the segments of L labeled 1 to 4 in Figure 7.20. Since $\mathbf{A}$ has only a z-component,

$$\Psi_1 = 0 = \Psi_3$$

FIGURE 7.20 For Example 7.7.

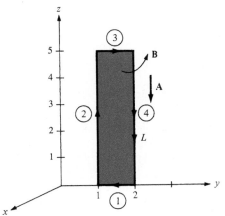

That is,

$$\Psi = \Psi_2 + \Psi_4 = -\frac{1}{4}\left[(1)^2 \int_0^5 dz + (2)^2 \int_5^0 dz\right]$$

$$= -\frac{1}{4}(1-4)(5) = \frac{15}{4}$$

$$= 3.75 \text{ Wb}$$

as obtained by Method 1. Note that the direction of the path L must agree with that of $d\mathbf{S}$.

PRACTICE EXERCISE 7.7

A current distribution gives rise to the vector magnetic potential $\mathbf{A} = x^2 y \mathbf{a}_x + y^2 x \mathbf{a}_y - 4xyz \mathbf{a}_z$ Wb/m. Calculate the following:

(a) $\mathbf{B}$ at $(-1, 2, 5)$
(b) The flux through the surface defined by $z = 1, 0 \le x \le 1, -1 \le y \le 4$

Answer: (a) $20\mathbf{a}_x + 40\mathbf{a}_y + 3\mathbf{a}_z$ Wb/m^2, (b) 20 Wb.

EXAMPLE 7.8

If plane $z = 0$ carries uniform current $\mathbf{K} = K_y \mathbf{a}_y$,

$$\mathbf{H} = \begin{cases} 1/2 \, K_y \mathbf{a}_x, & z > 0 \\ -1/2 \, K_y \mathbf{a}_x, & z < 0 \end{cases}$$

This was obtained in Section 7.4 by using Ampère's law. Obtain this by using the concept of vector magnetic potential.

Solution:

Consider the current sheet as in Figure 7.21. From eq. (7.42),

$$d\mathbf{A} = \frac{\mu_o \mathbf{K} \, dS}{4\pi R}$$

In this problem, $\mathbf{K} = K_y \mathbf{a}_y$, $dS = dx' \, dy'$, and for $z > 0$,

$$R = |\mathbf{R}| = |(0, 0, z) - (x', y', 0)|$$

$$= [(x')^2 + (y')^2 + z^2]^{1/2} \tag{7.8.1}$$

where the primed coordinates are for the source point while the unprimed coordinates are for the field point. It is necessary (and customary) to distinguish between the two points to avoid confusion (see Figure 7.19). Hence

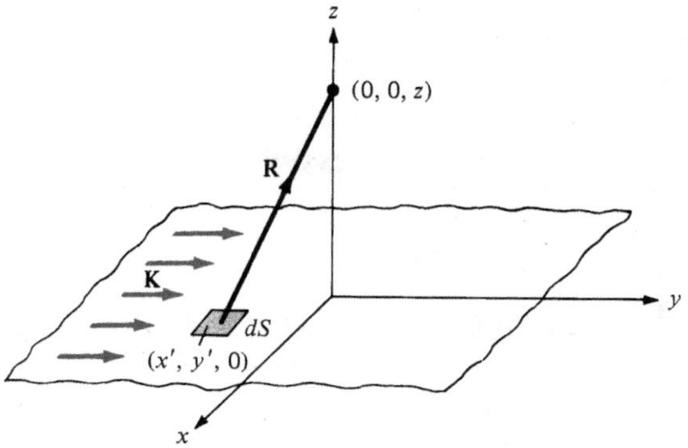

FIGURE 7.21 For Example 7.8: infinite current sheet.

$$dA = \frac{\mu_o K_y \, dx' \, dy' \, \mathbf{a}_y}{4\pi[(x')^2 + (y')^2 + z^2]^{1/2}}$$

$$dB = \nabla \times dA = -\frac{\partial}{\partial z} d A_y \, \mathbf{a}_x$$

$$= \frac{\mu_o K_y z \, dx' \, dy' \, \mathbf{a}_x}{4\pi[(x')^2 + (y')^2 + z^2]^{3/2}}$$

$$\mathbf{B} = \frac{\mu_o K_y z \mathbf{a}_x}{4\pi} \int_{-\infty}^{\infty} \int_{-\infty}^{\infty} \frac{dx' \, dy'}{[(x')^2 + (y')^2 + z^2]^{3/2}} \qquad (7.8.2)$$

In the integrand, we may change coordinates from Cartesian to cylindrical for convenience so that

$$\mathbf{B} = \frac{\mu_o K_y z \mathbf{a}_x}{4\pi} \int_{\rho'=0}^{\infty} \int_{\phi'=0}^{2\pi} \frac{\rho' \, d\phi' \, d\rho'}{[(\rho')^2 + z^2]^{3/2}}$$

$$= \frac{\mu_o K_y z \mathbf{a}_x}{4\pi} 2\pi \int_0^{\infty} [(\rho')^2 + z^2]^{-3/2} \, 1/2 \, d[(\rho')^2]$$

$$= \frac{\mu_o K_y z \mathbf{a}_x}{2} \frac{-1}{[(\rho')^2 + z^2]^{1/2}} \bigg|_{\rho'=0}^{\infty}$$

$$= \frac{\mu_o K_y \mathbf{a}_x}{2}$$

Hence

$$\mathbf{H} = \frac{\mathbf{B}}{\mu_o} = \frac{K_y}{2} \mathbf{a}_x, \quad \text{for } z > 0$$

By simply replacing z by $-z$ in eq. (7.8.2) and following the same procedure, we obtain

$$\mathbf{H} = -\frac{K_y}{2} \mathbf{a}_x, \quad \text{for } z < 0$$

PRACTICE EXERCISE 7.8

Repeat Example 7.8 by using Biot–Savart's law to determine **H** at points $(0, 0, h)$ and $(0, 0, -h)$.

†7.8 DERIVATION OF BIOT–SAVART'S LAW AND AMPÈRE'S LAW

Both Biot–Savart's law and Ampère's law may be derived by using the concept of magnetic vector potential. The derivation will involve the use of the vector identities in eq. (7.48) and

$$\nabla \times \nabla \times \mathbf{A} = \nabla(\nabla \cdot \mathbf{A}) - \nabla^2 \mathbf{A} \tag{7.52}$$

Since Biot–Savart's law as given in eq. (7.4) is defined in terms of line current, we begin our derivation with eqs. (7.39) and (7.41); that is,

$$\mathbf{B} = \nabla \times \oint_L \frac{\mu_o I \, d\mathbf{l}'}{4\pi R} = \frac{\mu_o I}{4\pi} \oint_L \nabla \times \frac{1}{R} \, d\mathbf{l}' \tag{7.53}$$

where R is as defined in eq. (7.45). If the vector identity in eq. (7.48) is applied by letting $\mathbf{F} = d\mathbf{l}'$ and $f = 1/R$, eq. (7.53) becomes

$$\mathbf{B} = \frac{\mu_o I}{4\pi} \oint_L \left[\frac{1}{R} \nabla \times d\mathbf{l}' + \left(\nabla \frac{1}{R} \right) \times d\mathbf{l}' \right] \tag{7.54}$$

Since ∇ operates with respect to (x, y, z) and $d\mathbf{l}'$ is a function of (x', y', z'), $\nabla \times d\mathbf{l}' = \mathbf{0}$. Also

$$\frac{1}{R} = [(x - x')^2 + (y - y')^2 + (z - z')^2]^{-1/2} \tag{7.55}$$

$$\nabla \left[\frac{1}{R} \right] = -\frac{(x - x')\mathbf{a}_x + (y - y')\mathbf{a}_y + (z - z')\mathbf{a}_z}{[(x - x')^2 + (y - y')^2 + (z - z')^2]^{3/2}} = -\frac{\mathbf{a}_R}{R^2} \tag{7.56}$$

where $\mathbf{a}_R$ is a unit vector from the source point to the field point. Thus eq. (7.54) (upon dropping the prime in $d\mathbf{l}'$) becomes

$$\mathbf{B} = \frac{\mu_o I}{4\pi} \oint_L \frac{d\mathbf{l} \times \mathbf{a}_R}{R^2} \tag{7.57}$$

which is Biot–Savart's law.

Using the identity in eq. (7.52) with eq. (7.39), we obtain

$$\nabla \times \mathbf{B} = \nabla(\nabla \cdot \mathbf{A}) - \nabla^2 \mathbf{A} \qquad (7.58)$$

For reasons that will be obvious in Chapter 9, we choose

$$\boxed{\nabla \cdot \mathbf{A} = 0} \qquad (7.59)$$

which is called Coulomb's gauge. Upon replacing $\mathbf{B}$ with $\mu_o \mathbf{H}$ and using eq. (7.19), eq. (7.58) becomes

$$\nabla^2 \mathbf{A} = -\mu_o \nabla \times \mathbf{H}$$

or

$$\boxed{\nabla^2 \mathbf{A} = -\mu_o \mathbf{J}} \qquad (7.60)$$

which is called the *vector Poisson equation*. It is similar to Poisson's equation $(\nabla^2 V = -\rho_v/\varepsilon)$ in electrostatics. In Cartesian coordinates, eq. (7.60) may be decomposed into three scalar equations:

$$\begin{aligned} \nabla^2 A_x &= -\mu_o J_x \\ \nabla^2 A_y &= -\mu_o J_y \\ \nabla^2 A_z &= -\mu_o J_z \end{aligned} \qquad (7.61)$$

which may be regarded as the *scalar Poisson equations.*

It can also be shown that Ampère's circuit law is consistent with our definition of the magnetic vector potential. From Stokes's theorem and eq. (7.39),

$$\begin{aligned} \oint_L \mathbf{H} \cdot d\mathbf{l} &= \int_S \nabla \times \mathbf{H} \cdot d\mathbf{S} \\ &= \frac{1}{\mu_o} \int_S \nabla \times (\nabla \times \mathbf{A}) \cdot d\mathbf{S} \end{aligned} \qquad (7.62)$$

From eqs. (7.52), (7.59), and (7.60),

$$\nabla \times \nabla \times \mathbf{A} = -\nabla^2 \mathbf{A} = \mu_o \mathbf{J}$$

Substituting this into eq. (7.62) yields

$$\oint_L \mathbf{H} \cdot d\mathbf{l} = \int_S \mathbf{J} \cdot d\mathbf{S} = I$$

which is Ampère's circuit law.

†7.9 APPLICATION NOTE—LIGHTNING

Lightning is the discharge of static electricity generated in clouds by natural processes.

Lightning may also be regarded as a transient, high-current electric discharge. It is a major natural source of electromagnetic radiation that interferes with modern electronics and communication systems. Lightning strikes somewhere on the surface of the earth about 100 times every second. Lightning, the thunderbolt from mythology, has long been feared as an atmospheric flash of supernatural origins: the great weapon of the gods. Today, scientific rather than mystical techniques are used to explain lightning, with experimental procedures replacing intuitive concepts. Yet, we remain in awe of lightning, which still shines with its mystery, and rightly so. Deaths and injuries to livestock and other animals, thousands of forest and brush fires, and millions of dollars in damage to buildings, communications systems, power lines, and electrical systems are among the results of lightning.

Since lightning can reach from clouds to the ground or to other clouds, lightning may be classified into two types: (1) cloud-to-cloud and (2) cloud-to-ground. A typical cloud-to-ground lighting is shown in Figure 7.22. The cloud-to-cloud discharge is more common and is important for aircraft in flight. However, cloud-to-ground lightning has been studied more extensively because of its practical interest (e.g., as the cause of injuries and death or disturbances in power and communication systems). A typical cloud-to-ground lightning carries about 10 C to 20 C at an average height of 5 km above the ground. The portion of the cloud-to-ground discharge that produces physical damage at ground level by virtue of its high current is called the return stroke. The current in a return stroke is typically 10 kA but can be as high as 200 kA.

Under good weather conditions, an electric field of the order 100 V/m exists near the earth's surface. Movements inside a cloud cause the cloud to become an electric dipole, with negative charges in the lower part and positive charges in the upper part.

FIGURE 7.22 A cloud-to-ground lightning.

The approach of the negatively charged particles to the ground induces more positive charges, especially on tall, sharp structures. A lightning bolt follows the path of least resistance at the moment of initiation; this is rarely a straight line, and it is unique for each strike. However, if we assume that lightning strokes arrive in the vertical direction, we can estimate the striking distance as a function of the amplitude of the current of the return stroke. The base striking distance D in meters, and the current I, in kiloamperes, are related as

$$D = 10I^{0.65} \tag{7.63}$$

Humans and animals within the striking distance may be hurt.

A common way to protect people, buildings, and other structures from lightning is to use lightning rods. Originally developed by Benjamin Franklin, a lightning rod is a pointed metal rod attached to the roof of a building. It is connected to a copper or aluminum wire, and the wire is connected to a conductive grid buried in the ground nearby. Lightning rods provide a low-resistance path to ground that can be used to conduct the enormous electrical currents when lightning strikes occur. When lightning strikes, the system attempts to carry the harmful electrical current away from the structure and safely to ground.

7.10 APPLICATION NOTE—POLYWELLS

A polywell is a polyhedral group of metal rings; inside each ring is a coil, which produces a magnetic field. As schematically illustrated in Figure 7.23, the position of each of the rings and the direction of current flow in each coil are set to create a null magnetic field at

FIGURE 7.23 Coils in a polywell.

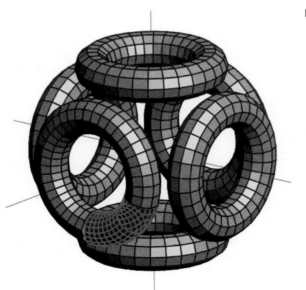

the center of the cube. On two opposing sides of the cube a stream of electrons is injected through the rings. These injected electrons are pushed by the magnetic field toward the (magnetically null) center of the cube, forming a cloud of electrons. When this cloud of electrons is large enough, it will create what is known as an electrostatic potential well. The polywell's name comes from its polyhedral shape and the electrostatic potential well produces.

In a nuclear fusion reactor two lighter atomic particles fuse together to form a heavier particle, releasing large amounts of energy. The normal activity inside a nuclear fusion reactor is as follows: Two spherically concentric, gridded electrodes create a radial electric field that acts as an electrostatic potential well. Then the radial electric field accelerates ions to fusion-revelant energies and confines them in the central grid region.

The fusion reactor system, however, suffers from substantial energy loss due to collisions between the grid itself and the ions. The polywell overcomes this problem by replacing the physical cathode with a virtual cathode, the electron cloud. In the polywell the ion streams are injected into the polyhedron through the remaining four rings. These ions are attracted to the electron cloud and are accelerated to the energy at which fusion can occur.

All these parts—the polywell and the electron and ion guns—are encapsulated in a collection sphere, with all of this inside a vacuum chamber. This collection sphere captures the energy released from the fusion process in the form of alpha particles, which come from the fusion, inside the electron cloud, of boron and hydrogen ions. The use of boron and hydrogen in nuclear fusion is becoming more popular than the use of deuterium and tritium as fuel. Unlike the fusion of deuterium and tritium, the fusion of boron and hydrogen produces little to no radiation and, since the only by-product is helium, there is no radioactive waste.

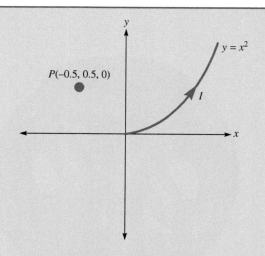

Suppose a 0.5 mA segment of current travels along the parabola $y = x^2$ between $a = (0, 0, 0)$ and $b = (1, 1, 0)$ cm. Using the Biot–Savart law, determine the magnetic field at point $P(-0.5, 0.5, 0)$ due to the segment.

We will find the general solution and evaluate at the observation point. We use the Biot–Savart law to obtain the magnetic field at point P:

$$\mathbf{H} = \int_L \frac{Id\mathbf{l} \times \mathbf{a}_R}{4\pi R^2}$$

The unit vector from the incremental current filament at $(x', y', 0)$ to the observation point P (x, y, z) is

$$\mathbf{a}_R = \frac{(x - x')\mathbf{a}_x + (y - y')\mathbf{a}_y + z\mathbf{a}_z}{R}$$

$$R = \sqrt{(x - x')^2 + (y - y')^2 + z^2}$$

The incremental current element is given by

$$Id\mathbf{L} = I(dx'\mathbf{a}_x + dy'\mathbf{a}_y)$$

The cross product is

$$Id\mathbf{L} \times \mathbf{a}_R = I(dx'\mathbf{a}_y + dy'\mathbf{a}_y) \times \frac{(x - x')\mathbf{a}_x + (y - y')\mathbf{a}_y + z\mathbf{a}_z}{\sqrt{(x - x')^2 + (y - y')^2 + (z - z')^2}}$$

Thus

$$\mathbf{H} = \int_L \frac{Id\mathbf{L} \times \mathbf{a}_R}{4\pi R^2} = \frac{I}{4\pi} \int_a^b \frac{[(y - y')dx' - (x - x')dy']\mathbf{a}_z - z\,dx'\mathbf{a}_y + z\,dy'\mathbf{a}_x}{[(x - x')^2 + (y - y')^2 + z^2]^{1.5}}$$

This integral is numerically evaluated as $0.85I$; thus the magnetic field at P is given by:

$$\mathbf{H} = 1.9437\mathbf{a}_z \frac{\text{mA}}{\text{m}}$$

```
clear
I=0.5e-3;    % the current value

% prompt for observation point
disp('Enter the observation point (in the ');
p0 = input('format [x y z])... \n > ');

if isempty(p0); p0 = [0 0 0]; end

xpstart = 0; xpend = 1e-2;    % start and end points for
                             %integration variable x prime
dxp=1e-7;    % integration variable increment dx

H = [0, 0, 0]; % initial field values before integration sum
zp = 0;          % current lies only in the xy-plane

for xp=xpstart:dxp:xpend,    % begin integration loop
    yp=xp^2*1e2;    % make substitution for y prime in terms
                    % of x prime
```

```
                          % the 1e2 is to offset the 1e-2 squared
                          % term which relates x prime and y prime
                          % in space
        dyp=2*X*dxp;      % make substitution for dy prime in terms
                                % of dx prime

        num = [(p0(3)-zp)*dyp,-(p0(3)-zp)*dxp,((p0(2)-yp)*dxp-...
        (p0(1)-xp)*dyp)];    % numerator
        den = ((p0(1)-xp)^2+(p0(2)-yp)^2)^(3/2);    % denominator

        H = H + num/den; % total field including all three coordinates

end

H=H*I/(4*pi);

% display the output
disp('')
disp('The magnetic field at');
disp(sprintf(' (%f, %f, %f) cm \nis (%f %f %f) A/m', ...
    p0(1), p0(2), p0(2), H(1), H(2), H(3)))
```

MATLAB 7.2

```
% This script allows the user to specify a current
% directed out of the page (+z direction) that lies on the origin,
% is assumed infinite, and points in the z direction
% and plot the vector magnetic field in the xy-plane
%
%
% inputs: I (value of the current), x and y limits of the plot
% outputs: the magnetic field vector plot
clear

% prompt user for input materials
disp('Enter the graph limits ');
plotlim = input(' [xmin xmax ymin ymax]... \n > ');
if isempty(plotlim); plotlim = [-1 1 -1 1]; end
                                            % check if entered
correctly
I = input('Enter the current in Amperes... \n > ');
if isempty(I); I = 1; end    % check if current is entered

dx=(plotlim(2)-plotlim(1))/10;
dy=(plotlim(4)-plotlim(3))/10;
xrange=plotlim(1):dx:plotlim(2);
yrange=plotlim(3):dy:plotlim(4);

[X,Y]=meshgrid(xrange,yrange);
U=zeros(length(xrange), length(yrange));
V=zeros(length(xrange), length(yrange));
for x=1:length(xrange)
```

```
        for y=1:length(yrange)
            r=sqrt(xrange(x)^2+yrange(y)^2);
                        % the distance from the current
            phiuvector=[-yrange(y),xrange(x)]/r;
                        % the unit vector in the phi direction
            H=I/(2*pi*r)*phiuvector;
                        % Ampere's law for an infinite current
                        % fill matrices which contain the vector
                        % components in x and y direction
            U(y, x)=H(1);    % vector x corresponds to columns
            V(y, x)=H(2);    % vector x corresponds to columns
        end
    end

    % Display results
    figure
    quiver(xrange,yrange,U,V)
    axis square
    axis(plotlim)
    xlabel('X location (m)')
    ylabel('Y location (m)')
    disp('Value of first vector to the right of');
    disp(sprintf(' origin = %f A/m',I/(2*pi*dx)))
```

SUMMARY

1. The basic laws (Biot–Savart's and Ampère's) that govern magnetostatic fields are discussed. Biot–Savart's law, which is similar to Coulomb's law, states that the magnetic field intensity $d\mathbf{H}$ at $\mathbf{r}$ due to current element $I\,d\mathbf{l}$ at $\mathbf{r}'$ is

$$d\mathbf{H} = \frac{I\,d\mathbf{l} \times \mathbf{R}}{4\pi R^3} \quad (\text{in A/m})$$

where $\mathbf{R} = \mathbf{r} - \mathbf{r}'$ and $R = |\mathbf{R}|$. For surface or volume current distribution, we replace $I\,d\mathbf{l}$ with $\mathbf{K}\,dS$ or $\mathbf{J}\,dv$, respectively; that is,

$$I\,d\mathbf{l} \equiv \mathbf{K}\,dS \equiv \mathbf{J}\,dv$$

2. Ampère's circuit law, which is similar to Gauss's law, states that the circulation of $\mathbf{H}$ around a closed path is equal to the current enclosed by the path; that is,

$$\oint_L \mathbf{H} \cdot d\mathbf{l} = I_{\text{enc}} = \int_S \mathbf{J} \cdot d\mathbf{S}$$

or

$$\nabla \times \mathbf{H} = \mathbf{J} \quad \text{(third Maxwell equation to be derived)}$$

When current distribution is symmetric so that an Amperian path (on which $\mathbf{H} = H_\phi \mathbf{a}_\phi$ is constant) can be found, Ampère's law is useful in determining $\mathbf{H}$; that is,

$$H_\phi \oint_L dl = I_{\text{enc}} \quad \text{or} \quad H_\phi = \frac{I_{\text{enc}}}{\ell}$$

3. The magnetic flux through a surface S is given by

$$\Psi = \int_S \mathbf{B} \cdot d\mathbf{S} \quad \text{(in Wb)}$$

where $\mathbf{B}$ is the magnetic flux density (in Wb/m^2). In free space,

$$\mathbf{B} = \mu_o \mathbf{H}$$

where $\mu_o = 4\pi \times 10^{-7}$ H/m = permeability of free space.
4. Since an isolated or free magnetic monopole does not exist, the net magnetic flux through a closed surface is zero:

$$\Psi = \oint_S \mathbf{B} \cdot d\mathbf{S} = 0$$

or

$$\nabla \cdot \mathbf{B} = 0 \quad \text{(fourth Maxwell equation to be derived)}$$

5. At this point, all four Maxwell equations for static EM fields have been derived, namely:

$$\nabla \cdot \mathbf{D} = \rho_v$$
$$\nabla \cdot \mathbf{B} = 0$$
$$\nabla \times \mathbf{E} = \mathbf{0}$$
$$\nabla \times \mathbf{H} = \mathbf{J}$$

6. The magnetic scalar potential V_m is defined as

$$\mathbf{H} = -\nabla V_m, \quad \text{if } \mathbf{J} = \mathbf{0}$$

and the magnetic vector potential $\mathbf{A}$ as

$$\mathbf{B} = \nabla \times \mathbf{A}$$

where $\nabla \cdot \mathbf{A} = 0$. With the definition of $\mathbf{A}$, the magnetic flux through a surface S can be found from

$$\Psi = \oint_L \mathbf{A} \cdot d\mathbf{l}$$

where L is the closed path defining surface S (see Figure 3.21). Rather than using Biot–Savart's law, the magnetic field due to a current distribution may be found by using $\mathbf{A}$, a powerful approach that is particularly useful in antenna theory. For a current element $I\,d\mathbf{l}$ at $\mathbf{r}'$, the magnetic vector potential at $\mathbf{r}$ is

$$\mathbf{A} = \int \frac{\mu_o I\,d\mathbf{l}}{4\pi R}, \quad R = |\mathbf{r} - \mathbf{r}'|$$

7. Elements of similarity between electric and magnetic fields exist. Some of these are listed in Table 7.1. Corresponding to Poisson's equation $\nabla^2 V = -\rho_v/\varepsilon$, for example, is

$$\nabla^2 \mathbf{A} = -\mu_o \mathbf{J}$$

8. Lightning may be regarded as a transient, high-current electric discharge. A common way to protect people, buildings, and other structures from lightning is to use lightning rods.

REVIEW QUESTIONS

7.1 One of the following is not a source of magnetostatic fields:

(a) A dc current in a wire
(b) A permanent magnet
(c) An accelerated charge
(d) An electric field linearly changing with time
(e) A charged disk rotating at uniform speed

7.2 Identify the configuration in Figure 7.24 that is not a correct representation of I and $\mathbf{H}$.

7.3 Consider points A, B, C, D, and E on a circle of radius 2 as shown in Figure 7.25. The items in the right-hand list are the values of $\mathbf{a}_\phi$ at different points on the circle. Match these items with the points in the list on the left.

(a) A (i) $\mathbf{a}_x$
(b) B (ii) $-\mathbf{a}_x$
(c) C (iii) $\mathbf{a}_y$
(d) D (iv) $-\mathbf{a}_y$

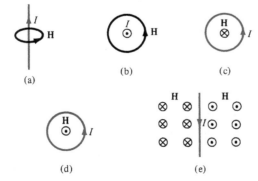

FIGURE 7.24 For Review Question 7.2.

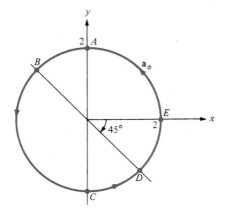

FIGURE 7.25 For Review Question 7.3.

(e) E (v) $\dfrac{\mathbf{a}_x + \mathbf{a}_y}{\sqrt{2}}$

(vi) $\dfrac{-\mathbf{a}_x - \mathbf{a}_y}{\sqrt{2}}$

(vii) $\dfrac{-\mathbf{a}_x + \mathbf{a}_y}{\sqrt{2}}$

(viii) $\dfrac{\mathbf{a}_x - \mathbf{a}_y}{\sqrt{2}}$

7.4 The z-axis carries filamentary current of 10π A along $\mathbf{a}_z$. Which of these is incorrect?

(a) $\mathbf{H} = -\mathbf{a}_x$ A/m at $(0, 5, 0)$

(b) $\mathbf{H} = \mathbf{a}_\phi$ A/m at $(5, \pi/4, 0)$

(c) $\mathbf{H} = -0.8\mathbf{a}_x - 0.6\mathbf{a}_y$ at $(-3, 4, 0)$

(d) $\mathbf{H} = -\mathbf{a}_\phi$ at $(5, 3\pi/2, 0)$

7.5 Plane $y = 0$ carries a uniform current of $30\mathbf{a}_z$ mA/m. At $(1, 10, -2)$, the magnetic field intensity is

(a) $-15\mathbf{a}_x$ mA/m

(b) $15\mathbf{a}_x$ mA/m

(c) $477.5\mathbf{a}_y$ μA/m

(d) $18.85\mathbf{a}_y$ nA/m

(e) None of the above

7.6 For the currents and closed paths of Figure 7.26, calculate the value of $\oint_L \mathbf{H} \cdot d\mathbf{l}$.

7.7 Which of these statements is not characteristic of a static magnetic field?

(a) It is solenoidal.

(b) It is conservative.

(c) It has no sinks or sources.

(d) Magnetic flux lines are always closed.

(e) The total number of flux lines entering a given region is equal to the total number of flux lines leaving the region.

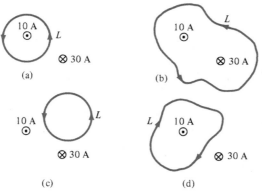

FIGURE 7.26 For Review Question 7.6.

7.8 Two identical coaxial circular coils carry the same current I but in opposite directions. The magnitude of the magnetic field **B** at a point on the axis midway between the coils is

(a) Zero

(b) The same as that produced by one coil

(c) Twice that produced by one coil

(d) Half that produced by one coil.

7.9 Which one of these equations is not Maxwell's equation for a static electromagnetic field in a linear homogeneous medium?

(a) $\nabla \cdot \mathbf{B} = 0$

(b) $\nabla \times \mathbf{D} = 0$

(c) $\oint_L \mathbf{B} \cdot d\mathbf{l} = \mu_o I$

(d) $\oint_S \mathbf{D} \cdot d\mathbf{S} = Q$

(e) $\nabla^2 \mathbf{A} = \mu_o \mathbf{J}$

7.10 Two bar magnets with their north poles having strength $Q_{m1} = 10\,\text{A} \cdot \text{m}$ and $Q_{m2} = 10\,\text{A} \cdot \text{m}$ (magnetic charges) are placed inside a volume as shown in Figure 7.27. The magnetic flux leaving the volume is

(a) 200 Wb

(b) 30 Wb

(c) 10 Wb

(d) 0 Wb

(e) −10 Wb

Answers: 7.1c, 7.2c, 7.3 (a)-(ii), (b)-(vi), (c)-(i), (d)-(v), (e)-(iii), 7.4d, 7.5a, 7.6 (a) 10 A, (b) −20 A, (c) 0, (d) −10 A, 7.7b, 7.8a, 7.9e, 7.10d.

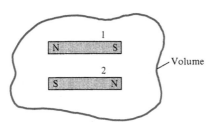

FIGURE 7.27 For Review Question 7.10.

PROBLEMS

Section 7.2—Biot–Savart's Law

7.1 (a) State Biot–Savart's law.

(b) The y- and z-axes, respectively, carry filamentary currents 10 A along $\mathbf{a}_y$ and 20 A along $-\mathbf{a}_z$. Find $\mathbf{H}$ at $(-3, 4, 5)$.

7.2 A long, straight wire carries current 2A. Calculate the distance from the wire when the magnetic field strength is 10 mA/m.

7.3 Two infinitely long wires, placed parallel to the z-axis, carry currents 10 A in opposite directions as shown in Figure 7.28. Find $\mathbf{H}$ at point P.

7.4 Two current elements $I_1 dl_1 = 4 \times 10^{-5}\, \mathbf{a}_x$ A.m at $(0, 0, 0)$ and $I_2 dl_2 = 6 \times 10^{-5}\, \mathbf{a}_y$ A.m at $(0, 0, 1)$ are in free space. Find $\mathbf{H}$ at $(3, 1, -2)$.

7.5 A conducting filament carries current I from point $A(0, 0, a)$ to point $B(0, 0, b)$. Show that at point $P(x, y, 0)$,

$$\mathbf{H} = \frac{I}{4\pi\sqrt{x^2 + y^3}}\left[\frac{b}{\sqrt{x^2 + y^2 + b^2}} - \frac{a}{\sqrt{x^2 + y^2 + a^2}}\right]\mathbf{a}_\phi$$

7.6 Consider AB in Figure 7.29 as part of an electric circuit. Find $\mathbf{H}$ at the origin due to AB.

7.7 Line $x = 0, y = 0, 0 \le z \le 10$ m carries current 2 A along $\mathbf{a}_z$. Calculate $\mathbf{H}$ at points

(a) $(5, 0, 0)$ (c) $(5, 15, 0)$

(b) $(5, 5, 0)$ (d) $(5, -15, 0)$

*__7.8__ (a) Find $\mathbf{H}$ at $(0, 0, 5)$ due to side 2 of the triangular loop in Figure 7.6(a).

(b) Find $\mathbf{H}$ at $(0, 0, 5)$ due to the entire loop.

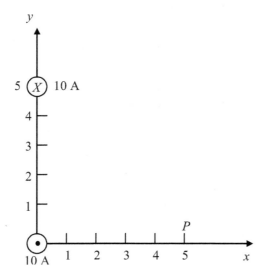

FIGURE 7.28 For Problem 7.3.

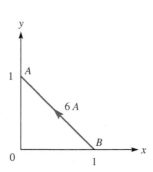

FIGURE 7.29 For Problem 7.6.

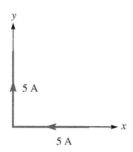

FIGURE 7.30 Current filament for Problem 7.9.

7.9 An infinitely long conductor is bent into an L shape as shown in Figure 7.30. If a direct current of 5 A flows in the conductor, find the magnetic field intensity at (a) $(2, 2, 0)$, (b) $(0, -2, 0)$, and (c) $(0, 0, 2)$.

7.10 Find **H** at the center C of an equilateral triangular loop of side 4 m carrying 5 A of current as in Figure 7.31.

7.11 A rectangular loop carrying 10 A of current is placed on $z = 0$ plane as shown in Figure 7.32. Evaluate **H** at

(a) $(2, 2, 0)$ (b) $(4, 2, 0)$

(c) $(4, 8, 0)$ (d) $(0, 0, 2)$

7.12 A square conducting loop of side 4 cm lies on the $z = 0$ plane and is centered at the origin. If it carries a current 5 mA in the counterclockwise direction, find **H** at the center of the loop.

*7.13 (a) A filamentary loop carrying current I is bent to assume the shape of a regular polygon of n sides. Show that at the center of the polygon

$$H = \frac{nI}{2\pi r} \sin \frac{\pi}{n}$$

where r is the radius of the circle circumscribed by the polygon.

(b) Apply this for the cases of $n = 3$ and $n = 4$ and see if your results agree with those for the triangular loop of Problem 7.10.

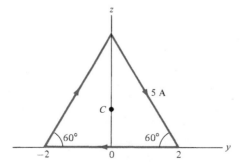

FIGURE 7.31 Equilateral triangular loop for Problem 7.10.

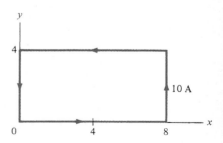

FIGURE 7.32 Rectangular loop of Problem 7.11.

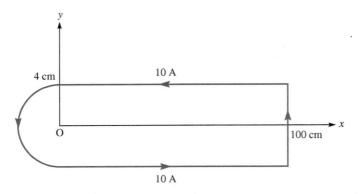

FIGURE 7.33 Filamentary loop of Problem 7.14 (not drawn to scale).

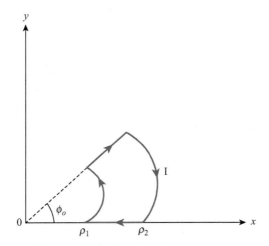

FIGURE 7.34 Problem 7.15.

(c) As n becomes large, show that the result of part (a) becomes that of the circular loop of Example 7.3.

7.14 For the filamentary loop shown in Figure 7.33, find the magnetic field strength at O.

7.15 Figure 7.34 shows a portion of a circular loop. Find **H** at the origin.

7.16 Two identical loops are parallel and separated by distance d as shown in Figure 7.35. Calculate **H** at $(0, 0, d)$ assuming that $a = 3$ cm, $d = 4$ cm, and $I = 10$ A.

7.17 A solenoid of radius 4 mm and length 2 cm has 150 turns/m and carries a current of 500 mA. Find (a) $|\mathbf{H}|$ at the center, (b) $|\mathbf{H}|$ at the ends of the solenoid.

7.18 Plane $x = 10$ carries a current of 100 mA/m along $\mathbf{a}_z$, while line $x = 1$, $y = -2$ carries a filamentary current of 20π mA along $\mathbf{a}_z$. Determine **H** at $(4, 3, 2)$.

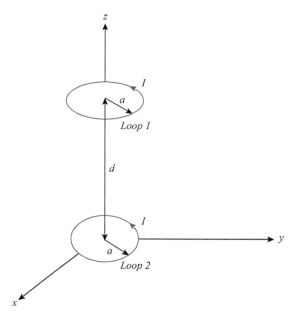

FIGURE 7.35 Problem 7.16.

Section 7.3—Ampère's Circuit Law

7.19 (a) State Ampère's circuit law.

(b) A hollow conducting cylinder has inner radius a and outer radius b and carries current I along the positive z-direction. Find **H** everywhere.

7.20 Current sheets of $20\mathbf{a}_x$ A/m and $-20\mathbf{a}_x$ A/m are located at $y = 1$ and $y = -1$, respectively. Find **H** in region $-1 < y < 1$.

7.21 The $z = 0$ plane carries current $\mathbf{K} = 10\mathbf{a}_x$ A/m, while current filament situated at $y = 0, z = 6$ carries current I along $\mathbf{a}_x$. Find I such that $\mathbf{H}(0, 0, 3) = \mathbf{0}$.

7.22 A conducting cylinder of radius a carries current I along $+\mathbf{a}_z$. (a) Use Ampère's law to find **H** for $\rho < a$ and $\rho > a$. (b) Find **J**.

7.23 An infinitely long cylindrical conductor of radius a is placed along the z-axis. If the current density is $\mathbf{J} = \dfrac{J_o}{\rho}\mathbf{a}_z$, where J_o is constant, find **H** everywhere.

7.24 Let $\mathbf{H} = y^2\mathbf{a}_x + x^2\mathbf{a}_y$ A/m. (a) Find **J**. (b) Determine the current through the strip $z = 1, 0 < x < 2, 1 < y < 5$.

7.25 Let $\mathbf{H} = k_o\left(\dfrac{\rho}{a}\right)\mathbf{a}_\phi$, $\rho < a$, where k_o is a constant. (a) Find **J** for $\rho < a$. (b) Find **H** for $\rho > a$.

7.26 Let $\mathbf{H} = y^2\mathbf{a}_x + x^2\mathbf{a}_y$ A/m. Find **J** at $(1, -4, 7)$.

7.27 Assume a conductor, $\mathbf{H} = 10^3\rho^2\mathbf{a}_\phi$ A/m. (a) Find **J**. (b) Calculate the current through the surface $0 < \rho < 2, 0 < \phi < 2\pi, z = 0$.

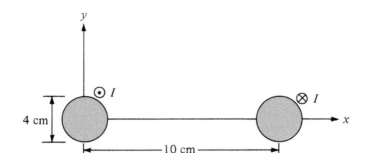

FIGURE 7.36 Two-wire line of Problem 7.30.

7.28 A cylindrical conductor of radius $a = 1$ cm carries current $\mathbf{I}$ which produces $\mathbf{H} = 4\rho a_\phi$ A/m. Find $\mathbf{I}$.

7.29 An infinitely long filamentary wire carries a current of 2 A along the z-axis in the $+z$-direction. Calculate the following:

(a) $\mathbf{B}$ at $(-3, 4, 7)$

(b) The flux through the square loop described by $2 \le \rho \le 6, 0 \le z \le 4, \phi = 90°$.

7.30 Consider the two-wire transmission line whose cross section is illustrated in Figure 7.36. Each wire is of radius 2 cm, and the wires are separated 10 cm. The wire centered at $(0, 0)$ carries a current of 5 A, while the other centered at $(10\text{ cm}, 0)$ carries the return current. Find $\mathbf{H}$ at

(a) $(5\text{ cm}, 0)$

(b) $(10\text{ cm}, 5\text{ cm})$

7.31 An electron beam forms a current of density

$$\mathbf{J} = \begin{cases} J_o(1-\rho^2/a^2)\mathbf{a}_z, & \rho < a \\ 0, & \rho > a \end{cases}$$

(a) Determine the total current.

(b) Find the magnetic field intensity everywhere.

Section 7.5—Magnetic Flux Density

7.32 Determine the magnetic flux through a rectangular loop $(a \times b)$ due to an infinitely long conductor carrying current I as shown in Figure 7.37. The loop and the straight conductors are separated by distance d.

7.33 A semicircular loop of radius a in free space carries a current I. Determine the magnetic flux density at the center of the loop.

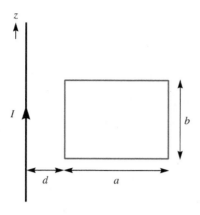

FIGURE 7.37 For Problem 7.32.

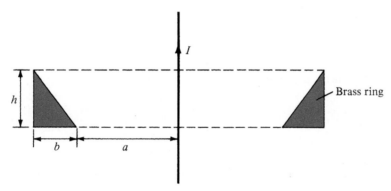

FIGURE 7.38 Cross section of a brass ring enclosing a long straight wire; for Problem 7.35.

7.34 In free space, the magnetic flux density is

$$\mathbf{B} = y^2\mathbf{a}_x + z^2\mathbf{a}_y + x^2\mathbf{a}_z \ \text{Wb/m}^2$$

(a) Show that **B** is a magnetic field

(b) Find the magnetic flux through $x = 1, 0 < y < 1, 1 < z < 4$.

(c) Calculate **J**.

(d) Determine the total magnetic flux through the surface of a cube defined by $0 < x < 2, 0 < y < 2, 0 < z < 2$.

*7.35 A brass ring with triangular cross section encircles a very long straight wire concentrically as in Figure 7.38. If the wire carries a current I, show that the total number of magnetic flux lines in the ring is

$$\Psi = \frac{\mu_0 I h}{2\pi b}\left[b - a \ln\frac{a+b}{b} \right]$$

Calculate Ψ if $a = 30$ cm, $b = 10$ cm, $h = 5$ cm, and $I = 10$ A.

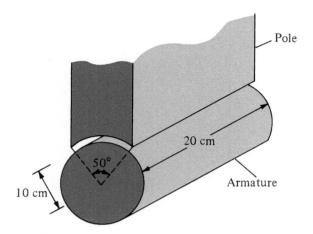

FIGURE 7.39 Electric motor pole of Problem 7.36.

7.36 The electric motor shown in Figure 7.39 has field

$$\mathbf{H} = \frac{10^6}{\rho} \sin 2\phi \, \mathbf{a}_\rho \text{ A/m}$$

Calculate the flux per pole passing through the air gap if the axial length of the pole is 20 cm.

7.37 In free space, $\mathbf{B} = \dfrac{20}{\rho} \sin^2 \phi \mathbf{a}_z$ Wb/m². Determine the magnetic flux crossing the strip $z = 0, 1 < \rho < 2$ m, $0 < \phi < \pi/4$.

7.38 If $\mathbf{B} = \dfrac{2}{r^3} \cos \theta \mathbf{a}_r + \dfrac{1}{r^3} \sin \theta \mathbf{a}_\theta$ Wb/m², find the magnetic flux through the spherical cap $r = 1, \theta < \pi/3$.

7.39 In a hydrogen atom, an electron revolves at velocity 2.2×10^6 m/s. Calculate the magnetic flux density at the center of the electron's orbit. Assume that the radius of the orbit is $R = 5.3 \times 10^{-11}$ m.

Section 7.6—Maxwell's Equations

7.40 Consider the following arbitrary fields. Find out which of them can possibly represent an electrostatic or magnetostatic field in free space.

(a) $\mathbf{A} = y \cos ax\mathbf{a}_x + (y + e^{-x})\mathbf{a}_z$

(b) $\mathbf{B} = \dfrac{20}{\rho} \mathbf{a}_\rho$

(c) $\mathbf{C} = r^2 \sin \theta \, \mathbf{a}_\phi$

7.41 Reconsider Problem 7.40 for the following fields.

(a) $\mathbf{D} = y^2z\mathbf{a}_x + 2(x + 1)yz\mathbf{a}_y - (x + 1)z^2\mathbf{a}_z$

(b) $\mathbf{E} = \dfrac{(z + 1)}{\rho}\cos\phi\,\mathbf{a}_\rho + \dfrac{\sin\phi}{\rho}\mathbf{a}_z$

(c) $\mathbf{F} = \dfrac{1}{r^2}(2\cos\theta\,\mathbf{a}_r + \sin\theta\,\mathbf{a}_\theta)$

Section 7.7—Magnetic Scalar and Vector Potentials

7.42 A current element of length L carries current I in the z direction. Show that at a very distant point,

$$\mathbf{A} = \frac{\mu_o IL}{4\pi r}\mathbf{a}_z$$

Find $\mathbf{B}$.

7.43 In free space, $\mathbf{A} = 10\sin\pi\, y\mathbf{a}_x + (4 + \cos\pi\, x)\mathbf{a}_z$ Wb/m. Find $\mathbf{H}$ and $\mathbf{J}$.

7.44 Given that $\mathbf{A} = \dfrac{2\cos\theta}{r^3}\mathbf{a}_r + \dfrac{\sin\theta}{r^3}\mathbf{a}_\theta$ Wb/m exists in free space.

(a) show that $\nabla \cdot \mathbf{A} = 0$

(b) Find $\mathbf{B}$ at point $T(1, 30°, 60°)$

7.45 For a current distribution in free space,

$$\mathbf{A} = (2x^2y + yz)\mathbf{a}_x + (xy^2 - xz^3)\mathbf{a}_y - (6xyz - 2x^2y^2)\mathbf{a}_z \text{ Wb/m}$$

(a) Calculate $\mathbf{B}$.

(b) Find the magnetic flux through a loop described by $x = 1, 0 < y < 2, 0 < z < 2$.

(c) Show that $\nabla \cdot \mathbf{A} = 0$ and $\nabla \cdot \mathbf{B} = 0$.

7.46 In free space, a small circular loop of current produces

$$\mathbf{A} = \frac{k}{r^2}\sin\theta\,\mathbf{a}_\phi$$

where k is a constant. Find $\mathbf{B}$.

7.47 The magnetic vector potential of a current distribution in free space is given by

$$\mathbf{A} = 15e^{-\rho}\sin\phi\,\mathbf{a}_z \text{ Wb/m}$$

Find $\mathbf{H}$ at $(3, \pi/4, -10)$. Calculate the flux through $\rho = 5, 0 \le \phi \le \pi/2, 0 \le z \le 10$.

7.48 Given that $\mathbf{A} = \dfrac{10}{r} \sin \theta \mathbf{a}_\phi$ Wb/m, find $\mathbf{H}$ at point $(4, 60°, 30°)$.

7.49 An infinitely long conductor of radius a carries a uniform current with $\mathbf{J} = J_o \mathbf{a}_z$. Show that the magnetic vector potential for $\rho < a$ is

$$\mathbf{A} = -\frac{1}{4} \mu_o J_o \rho^2 \mathbf{a}_z$$

7.50 Find the $\mathbf{B}$ field corresponding to the magnetic vector potential

$$\mathbf{A} = \sin \frac{\pi x}{2} \cos \frac{\pi y}{2} \mathbf{a}_z$$

7.51 The magnetic vector potential at a distant point from a small circular loop is given by

$$\mathbf{A} = \frac{A_o}{r^2} \sin \theta \, \mathbf{a}_\phi \text{ Wb/m}$$

where A_o is a constant. Determine the magnetic flux density $\mathbf{B}$.

7.52 The magnetic field intensity in a certain conducting medium is

$$\mathbf{H} = xy^2 \mathbf{a}_x + x^2 z \mathbf{a}_y - y^2 z \mathbf{a}_z \text{ A/m}$$

(a) Calculate the current density at point $P(2, -1, 3)$.

(b) What is $\dfrac{\partial \rho_v}{\partial t}$ at P?

7.53 Let $\mathbf{A} = 10\rho^2 \mathbf{a}_z \, \mu\text{Wb/m}$.

(a) Find $\mathbf{H}$ and $\mathbf{J}$.

(b) Determine the total current crossing the surface $z = 1, 0 \le \rho \le 2, 0 \le \phi \le 2\pi$.

7.54 Prove that the magnetic scalar potential at $(0, 0, z)$ due to a circular loop of radius a shown in Figure 7.8(a) is

$$V_m = \frac{I}{2} \left[1 - \frac{z}{[z^2 + a^2]^{1/2}} \right]$$

7.55 The z-axis carries a filamentary current 12 A along $\mathbf{a}_z$. Calculate V_m at $(4, 30°, -2)$ if $V_m = 0$ at $(10, 60°, 7)$.

7.56 Plane $z = -2$ carries a current of $50\mathbf{a}_y$ A/m. If $V_m = 0$ at the origin, find V_m at

(a) $(-2, 0, 5)$

(b) $(10, 3, 1)$

7.57 Prove in cylindrical coordinates that

(a) $\nabla \times (\nabla V) = \mathbf{0}$

(b) $\nabla \cdot (\nabla \times \mathbf{A}) = 0$

*__7.58__ If $\mathbf{R} = \mathbf{r} - \mathbf{r}'$ and $R = |\mathbf{R}|$, show that

$$\nabla \frac{1}{R} = -\nabla' \frac{1}{R} = -\frac{\mathbf{R}}{R^3}$$

where ∇ and ∇' are del operators with respect to (x, y, z) and (x', y', z'), respectively.

MAGNETIC RESONANCE IMAGING (MRI)

Magnetic resonance imaging (MRI), an exciting technique for probing the human body, was introduced into clinical practice by the early 1980s. In less than 10 years, it became a primary diagnostic tool in several clinical areas such as neurology and orthopedics. Improvements in MRI technology and in computer technology have led to continued growth in the clinical capabilities of the technique. No other technique has proven to be so uniquely flexible and dynamic.

What is MRI? When placed in a static magnetic field, certain atomic nuclei assume one of two states: one has a higher energy level and the other has a lower energy level. The energy difference between the two states is linearly proportional to the strength of the applied magnetic field. (This is called the Zeeman effect.) Thus, the MRI signals received by a probe can be analyzed to study the properties of the nuclei and their environment. MRI stems from the application of nuclear magnetic resonance (NMR) to radiological imaging. Unlike other imaging techniques, such as X-ray computed tomography, MRI does not require exposure of the subject to ionizing radiation and hence is considered safe. It provides more information than other imaging techniques because MRI signals are sensitive to several tissue parameters.

An MRI machine consists of a magnet and a giant cube 7 feet tall by 7 feet wide by 10 feet long ($2\text{ m} \times 2\text{ m} \times 3\text{ m}$), although new models are rapidly shrinking. There is a horizontal tube running through the magnet from front to back. The magnets in use today in MRI machines are in the range of 0.5 T to 2 T. (There is no scientific evidence that fields in that range produce harmful effects in humans.) The patient, lying on his or her back, slides into the tube on a special table. Once the body part to be scanned is in the exact center of the magnetic field, the scan can begin.

MRI has changed from a curiosity to the technique of choice for a wide variety of diseases in various regions of the human body. It has been lauded as a technique that represents a breakthrough in medical diagnosis. Today, an estimated 60 million MRI scans are performed annually to visualize patients' internal structures and diagnose a number of conditions including tumors, stroke damage, heart and brain diseases, and back problems.

CHAPTER 8

MAGNETIC FORCES, MATERIALS, AND DEVICES

Always be kind to your A and B students. Someday one of them will return to your campus as a good professor. And also be kind to your C students. Someday one of them will return and build you a two-million dollar science laboratory.

—YALE UNIVERSITY PRESIDENT

8.1 INTRODUCTION

Having considered the basic laws and techniques commonly used in calculating magnetic field **B** due to current-carrying elements, we are prepared to study the force a magnetic field exerts on charged particles, current elements, and loops. Such a study is important to problems on electrical devices such as ammeters, voltmeters, galvanometers, cyclotrons, plasmas, motors, and magnetohydrodynamic generators. The precise definition of the magnetic field, deliberately sidestepped in the preceding chapter, will be given here. The concepts of magnetic moments and dipole will also be considered.

Furthermore, we will consider magnetic fields in material media, as opposed to the magnetic fields in vacuum or free space examined in the preceding chapter. The results of Chapter 7 need only some modification to account for the presence of materials in a magnetic field. Further discussions will cover inductors, inductances, magnetic energy, and magnetic circuits.

8.2 FORCES DUE TO MAGNETIC FIELDS

There are at least three ways in which force due to magnetic fields can be experienced. The force can be (a) due to a moving charged particle in a **B** field, (b) on a current element in an external **B** field, or (c) between two current elements.

A. Force on a Charged Particle

According to our discussion in Chapter 4, the electric force $\mathbf{F}_e$ on a stationary or moving electric charge Q in an electric field is given by Coulomb's experimental law and is related to the electric field intensity **E** as

$$\mathbf{F}_e = Q\mathbf{E} \tag{8.1}$$

This shows that if Q is positive, $\mathbf{F}_e$ and **E** have the same direction.

A magnetic field can exert force only on a moving charge. From experiments, it is found that the magnetic force $\mathbf{F}_m$ experienced by a charge Q moving with a velocity $\mathbf{u}$ in a magnetic field $\mathbf{B}$ is

$$\mathbf{F}_m = Q\mathbf{u} \times \mathbf{B} \qquad (8.2)$$

This clearly shows that $\mathbf{F}_m$ is perpendicular to both $\mathbf{u}$ and $\mathbf{B}$.

From eqs. (8.1) and (8.2), a comparison between the electric force $\mathbf{F}_e$ and the magnetic force $\mathbf{F}_m$ can be made. We see that $\mathbf{F}_e$ is independent of the velocity of the charge and can perform work on the charge and change its kinetic energy. Unlike $\mathbf{F}_e$, $\mathbf{F}_m$ depends on the charge velocity and is normal to it. However, $\mathbf{F}_m$ cannot perform work because it is at right angles to the direction of motion of the charge $(\mathbf{F}_m \cdot d\mathbf{l} = 0)$; it does not cause an increase in kinetic energy of the charge. The magnitude of $\mathbf{F}_m$ is generally small in comparison to $\mathbf{F}_e$ except at high velocities.

For a moving charge Q in the presence of both electric and magnetic fields, the total force on the charge is given by,

$$\mathbf{F} = \mathbf{F}_e + \mathbf{F}_m$$

or

$$\boxed{\mathbf{F} = Q(\mathbf{E} + \mathbf{u} \times \mathbf{B})} \qquad (8.3)$$

This is known as the *Lorentz force equation.*[1] It relates mechanical force to electrical force. If the mass of the charged particle moving in $\mathbf{E}$ and $\mathbf{B}$ fields is m, by Newton's second law of motion.

$$\mathbf{F} = m\frac{d\mathbf{u}}{dt} = Q(\mathbf{E} + \mathbf{u} \times \mathbf{B}) \qquad (8.4)$$

The solution to this equation is important in determining the motion of charged particles in $\mathbf{E}$ and $\mathbf{B}$ fields. We should bear in mind that in such fields, energy can be transferred only by means of the electric field. A summary on the force exerted on a charged particle is given in Table 8.1.

Since eq. (8.2) is closely parallel to eq. (8.1), which defines the electric field, some authors and instructors prefer to begin their discussions on magnetostatics from eq. (8.2), just as discussions on electrostatics usually begin with Coulomb's force law.

B. Force on a Current Element

To determine the force on a current element $I\,d\mathbf{l}$ of a current-carrying conductor due to the magnetic field $\mathbf{B}$, we modify eq. (8.2) using the fact that for convection current [see eq. (5.7)]:

$$\mathbf{J} = \rho_v \mathbf{u} \qquad (8.5)$$

---

[1] After Hendrik Lorentz (1853–1928), who first applied the equation of motion in electric fields.

TABLE 8.1 Force on a Charged Particle

State of Particle	E Field	B Field	Combined E and B Fields
Stationary	QE	—	QE
Moving	QE	$Q\mathbf{u} \times \mathbf{B}$	$Q(\mathbf{E} + \mathbf{u} \times \mathbf{B})$

From eq. (7.5), we recall the relationship between current elements:

$$I\,d\mathbf{l} = \mathbf{K}\,dS = \mathbf{J}\,dv \tag{8.6}$$

Combining eqs. (8.5) and (8.6) yields

$$I\,d\mathbf{l} = \rho_v \mathbf{u}\,dv = dQ\,\mathbf{u}$$

Alternatively, $I\,d\mathbf{l} = \dfrac{dQ}{dt}d\mathbf{l} = dQ\dfrac{d\mathbf{l}}{dt} = dQ\,\mathbf{u}$

Hence,

$$\boxed{I\,d\mathbf{l} = dQ\,\mathbf{u}} \tag{8.7}$$

This shows that an elemental charge dQ moving with velocity $\mathbf{u}$ (thereby producing convection current element $dQ\,\mathbf{u}$) is equivalent to a conduction current element $I\,d\mathbf{l}$. Thus the force on a current element $I\,d\mathbf{l}$ in a magnetic field $\mathbf{B}$ is found from eq. (8.2) by merely replacing $Q\mathbf{u}$ by $I\,d\mathbf{l}$; that is,

$$d\mathbf{F} = I\,d\mathbf{l} \times \mathbf{B} \tag{8.8}$$

If the current I is through a closed path L or circuit, the force on the circuit is given by

$$\boxed{\mathbf{F} = \oint_L I\,d\mathbf{l} \times \mathbf{B}} \tag{8.9}$$

In using eq. (8.8) or (8.9), we should keep in mind that the magnetic field produced by the current element $I\,d\mathbf{l}$ does not exert force on the element itself, just as a point charge does not exert force on itself. The $\mathbf{B}$ field that exerts force on $I\,d\mathbf{l}$ must be due to another element. In other words, the $\mathbf{B}$ field in eq. (8.8) or (8.9) is external to the current element $I\,d\mathbf{l}$. If instead of the line current element $I\,d\mathbf{l}$, we have surface current elements $\mathbf{K}\,dS$ or a volume current element $\mathbf{J}\,dv$, we simply make use of eq. (8.6) so that eq. (8.8) becomes

$$d\mathbf{F} = \mathbf{K}\,dS \times \mathbf{B} \quad \text{or} \quad d\mathbf{F} = \mathbf{J}\,dv \times \mathbf{B} \tag{8.8'}$$

while eq. (8.9) becomes

$$\mathbf{F} = \int_S \mathbf{K}\,dS \times \mathbf{B} \quad \text{or} \quad \mathbf{F} = \int_v \mathbf{J}\,dv \times \mathbf{B} \tag{8.9'}$$

From eq. (8.8)

> The **magnetic field B** is defined as the force per unit current element.

Alternatively, **B** may be defined from eq. (8.2) as the vector that satisfies $\mathbf{F}_m/q = \mathbf{u} \times \mathbf{B}$, just as we defined electric field **E** as the force per unit charge, $\mathbf{F}_e/q$. Both these definitions of **B** show that **B** describes the force properties of a magnetic field.

C. Force between Two Current Elements

Let us now consider the force between two elements $I_1\, d\mathbf{l}_1$ and $I_2\, d\mathbf{l}_2$. According to Biot–Savart's law, both current elements produce magnetic fields. So we may find the force $d(d\mathbf{F}_1)$ on element $I_1\, d\mathbf{l}_1$ due to the field $d\mathbf{B}_2$ produced by element $I_2\, d\mathbf{l}_2$ as shown in Figure 8.1. From eq. (8.8),

$$d(d\mathbf{F}_1) = I_1\, d\mathbf{l}_1 \times d\mathbf{B}_2 \tag{8.10}$$

But from Biot–Savart's law,

$$d\mathbf{B}_2 = \frac{\mu_o I_2\, d\mathbf{l}_2 \times \mathbf{a}_{R_{21}}}{4\pi R_{21}^2} \tag{8.11}$$

Hence,

$$d(d\mathbf{F}_1) = \frac{\mu_o I_1\, d\mathbf{l}_1 \times (I_2\, d\mathbf{l}_2 \times \mathbf{a}_{R_{21}})}{4\pi R_{21}^2} \tag{8.12}$$

This equation is essentially the law of force between two current elements and is analogous to Coulomb's law, which expresses the force between two stationary charges. From eq. (8.12), we obtain the total force $\mathbf{F}_1$ on current loop 1 due to current loop 2 shown in Figure 8.1 as

$$\mathbf{F}_1 = \frac{\mu_o I_1 I_2}{4\pi} \oint_{L_1} \oint_{L_2} \frac{d\mathbf{l}_1 \times (d\mathbf{l}_2 \times \mathbf{a}_{R_{21}})}{R_{21}^2} \tag{8.13}$$

FIGURE 8.1 Force between two current loops.

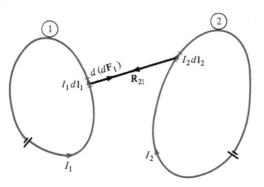

Although this equation appears complicated, we should remember that it is based on eq. (8.10). It is eq. (8.9) or (8.10) that is of fundamental importance.

The force $\mathbf{F}_2$ on loop 2 due to the magnetic field $\mathbf{B}_1$ from loop 1 is obtained from eq. (8.13) by interchanging subscripts 1 and 2. It can be shown that $\mathbf{F}_2 = -\mathbf{F}_1$; thus $\mathbf{F}_1$ and $\mathbf{F}_2$ obey Newton's third law that action and reaction are equal and opposite. It is worthwhile to mention that eq. (8.13) was experimentally established by Oersted and Ampère; Biot and Savart (Ampère's colleagues) actually based their law on it.

EXAMPLE 8.1

A charged particle of mass 2 kg and charge 3 C starts at point $(1, -2, 0)$ with velocity $4\mathbf{a}_x + 3\mathbf{a}_z$ m/s in an electric field $12\mathbf{a}_x + 10\mathbf{a}_y$ V/m. At time $t = 1$ s, determine

(a) The acceleration of the particle
(b) Its velocity
(c) Its kinetic energy
(d) Its position

Solution:

(a) This is an initial-value problem because initial values are given. According to Newton's second law of motion,

$$\mathbf{F} = m\mathbf{a} = Q\mathbf{E}$$

where $\mathbf{a}$ is the acceleration of the particle. Hence,

$$\mathbf{a} = \frac{Q\mathbf{E}}{m} = \frac{3}{2}(12\mathbf{a}_x + 10\mathbf{a}_y) = 18\mathbf{a}_x + 15\mathbf{a}_y \text{ m/s}^2$$

$$\mathbf{a} = \frac{d\mathbf{u}}{dt} = \frac{d}{dt}(u_x, u_y, u_z) = 18\mathbf{a}_x + 15\mathbf{a}_y$$

(b) Equating components and then integrating, we obtain

$$\frac{du_x}{dt} = 18 \rightarrow u_x = 18t + A \tag{8.1.1}$$

$$\frac{du_y}{dt} = 15 \rightarrow u_y = 15t + B \tag{8.1.2}$$

$$\frac{du_z}{dt} = 0 \rightarrow u_z = C \tag{8.1.3}$$

where A, B, and C are integration constants. But at $t = 0$, $\mathbf{u} = 4\mathbf{a}_x + 3\mathbf{a}_z$. Hence,

$$u_x(t = 0) = 4 \rightarrow 4 = 0 + A \quad \text{or} \quad A = 4$$

$$u_y(t = 0) = 0 \rightarrow 0 = 0 + B \quad \text{or} \quad B = 0$$

$$u_z(t = 0) = 3 \rightarrow 3 = C$$

Substituting the values of A, B, and C into eqs. (8.1.1) to (8.1.3) gives

$$\mathbf{u}(t) = (u_x, u_y, u_z) = (18t + 4, 15t, 3)$$

Hence

$$\mathbf{u}(t = 1\text{ s}) = 22\mathbf{a}_x + 15\mathbf{a}_y + 3\mathbf{a}_z \text{ m/s}$$

(c) Kinetic energy (K.E.) $= \dfrac{1}{2}m\,|\mathbf{u}|^2 = \dfrac{1}{2}(2)(22^2 + 15^2 + 3^2)$

$$= 718 \text{ J}$$

(d) $\mathbf{u} = \dfrac{d\mathbf{l}}{dt} = \dfrac{d}{dt}(x, y, z) = (18t + 4, 15t, 3)$

Equating components yields

$$\frac{dx}{dt} = u_x = 18t + 4 \;\rightarrow\; x = 9t^2 + 4t + A_1 \tag{8.1.4}$$

$$\frac{dy}{dt} = u_y = 15t \;\rightarrow\; y = 7.5t^2 + B_1 \tag{8.1.5}$$

$$\frac{dz}{dt} = u_z = 3 \;\rightarrow\; z = 3t + C_1 \tag{8.1.6}$$

At $t = 0$, $(x, y, z) = (1, -2, 0)$; hence,

$$x(t = 0) = 1 \rightarrow 1 \quad\;\; = 0 + A_1 \quad \text{or} \quad A_1 = 1$$
$$y(t = 0) = -2 \rightarrow -2 = 0 + B_1 \quad \text{or} \quad B_1 = -2$$
$$z(t = 0) = 0 \rightarrow 0 \quad\;\; = 0 + C_1 \quad \text{or} \quad C_1 = 0$$

Substituting the values of A_1, B_1, and C_1 into eqs. (8.1.4) to (8.1.6), we obtain

$$(x, y, z) = (9t^2 + 4t + 1, 7.5t^2 - 2, 3t) \tag{8.1.7}$$

Hence, at $t = 1$, $(x, y, z) = (14, 5.5, 3)$.

By eliminating t in eq. (8.1.7), the motion of the particle may be described in terms of x, y, and z.

PRACTICE EXERCISE 8.1

A charged particle of mass 1 kg and charge 2 C starts at the origin with zero initial velocity in a region where $\mathbf{E} = 3\mathbf{a}_z$ V/m. Find the following:

(a) The force on the particle

(b) The time it takes to reach point $P(0, 0, 12\text{ m})$

(c) Its velocity and acceleration at P

(d) Its K.E. at P

Answer: (a) $6a_z$ N, (b) 2 s, (c) $12a_z$ m/s, $6a_z$ m/s², (d) 72 J.

A charged particle of mass 2 kg and 1 C starts at the origin with velocity $3a_y$ m/s and travels in a region of uniform magnetic field $\mathbf{B} = 10a_z$ Wb/m². At $t = 4$ s, do the following.

(a) Calculate the velocity and acceleration of the particle.

(b) Calculate the magnetic force on it.

(c) Determine its K.E. and location.

(d) Find the particle's trajectory by eliminating t.

(e) Show that its K.E. remains constant.

Solution:

(a) $\mathbf{F} = m\dfrac{d\mathbf{u}}{dt} = Q\mathbf{u} \times \mathbf{B}$

$$\mathbf{a} = \frac{d\mathbf{u}}{dt} = \frac{Q}{m}\mathbf{u} \times \mathbf{B}$$

Hence

$$\frac{d}{dt}(u_x\mathbf{a}_x + u_y\mathbf{a}_y + u_z\mathbf{a}_z) = \frac{1}{2}\begin{vmatrix} \mathbf{a}_x & \mathbf{a}_y & \mathbf{a}_z \\ u_x & u_y & u_z \\ 0 & 0 & 10 \end{vmatrix} = 5(u_y\mathbf{a}_x - u_x\mathbf{a}_y)$$

By equating components, we get

$$\frac{du_x}{dt} = 5u_y \tag{8.2.1}$$

$$\frac{du_y}{dt} = -5u_x \tag{8.2.2}$$

$$\frac{du_z}{dt} = 0 \rightarrow u_z = C_o \tag{8.2.3}$$

We can eliminate u_x or u_y in eqs. (8.2.1) and (8.2.2) by taking second derivatives of one equation and making use of the other. Thus

$$\frac{d^2u_x}{dt^2} = 5\frac{du_y}{dt} = -25u_x$$

or

$$\frac{d^2u_x}{dt^2} + 25u_x = 0$$

which is a linear differential equation with solution (see Case 3 of Example 6.5)

$$u_x = C_1 \cos 5t + C_2 \sin 5t \qquad (8.2.4)$$

From eqs. (8.2.1) and (8.2.4),

$$5u_y = \frac{du_x}{dt} = -5C_1 \sin 5t + 5C_2 \cos 5t \qquad (8.2.5)$$

or

$$u_y = -C_1 \sin 5t + C_2 \cos 5t$$

We now determine constants C_0, C_1, and C_2 using the initial conditions. At $t = 0$, $\mathbf{u} = 3\mathbf{a}_y$. Hence,

$$u_x = 0 \rightarrow 0 = C_1 \cdot 1 + C_2 \cdot 0 \rightarrow C_1 = 0$$
$$u_y = 3 \rightarrow 3 = -C_1 \cdot 0 + C_2 \cdot 1 \rightarrow C_2 = 3$$
$$u_z = 0 \rightarrow 0 = C_0$$

Substituting the values of C_0, C_1, and C_2 into eqs. (8.2.3) to (8.2.5) gives

$$\mathbf{u} = (u_x, u_y, u_z) = (3 \sin 5t, 3 \cos 5t, 0) \qquad (8.2.6)$$

Hence,

$$\mathbf{u}(t = 4) = (3 \sin 20, 3 \cos 20, 0)$$
$$= 2.739\mathbf{a}_x + 1.224\mathbf{a}_y \text{ m/s}$$
$$\mathbf{a} = \frac{d\mathbf{u}}{dt} = (15 \cos 5t, -15 \sin 5t, 0)$$

and

$$\mathbf{a}(t = 4) = 6.121\mathbf{a}_x - 13.694\mathbf{a}_y \text{ m/s}^2$$

(b) $$\mathbf{F} = m\mathbf{a} = 12.2\mathbf{a}_x - 27.4\mathbf{a}_y \text{ N}$$

or

$$\mathbf{F} = Q\mathbf{u} \times \mathbf{B} = (1)(2.739\mathbf{a}_x + 1.224\mathbf{a}_y) \times 10\mathbf{a}_z$$
$$= 12.2\mathbf{a}_x - 27.4\mathbf{a}_y \text{ N}$$

(c) K.E. $= \frac{1}{2}m|\mathbf{u}|^2 = \frac{1}{2}(2)(2.739^2 + 1.224^2) = 9 \text{ J}$

$$u_x = \frac{dx}{dt} = 3 \sin 5t \rightarrow x = -\frac{3}{5} \cos 5t + b_1 \qquad (8.2.7)$$

$$u_y = \frac{dy}{dt} = 3 \cos 5t \rightarrow y = \frac{3}{5} \sin 5t + b_2 \qquad (8.2.8)$$

$$u_z = \frac{dz}{dt} = 0 \rightarrow z = b_3 \qquad (8.2.9)$$

where b_1, b_2, and b_3 are integration constants. At $t = 0$, $(x, y, z) = (0, 0, 0)$ and hence,

$$x(t = 0) = 0 \rightarrow 0 = -\frac{3}{5} \cdot 1 + b_1 \rightarrow b_1 = 0.6$$

$$y(t = 0) = 0 \rightarrow 0 = \frac{3}{5} \cdot 0 + b_2 \rightarrow b_2 = 0$$

$$z(t = 0) = 0 \rightarrow 0 = b_3$$

Substituting the values of b_1, b_2, and b_3 into eqs. (8.2.7) to (8.2.9), we obtain

$$(x, y, z) = (0.6 - 0.6 \cos 5t, 0.6 \sin 5t, 0) \qquad (8.2.10)$$

At $t = 4$ s,

$$(x, y, z) = (0.3552, 0.5478, 0)$$

(d) From eq. (8.2.10), we eliminate t by noting that

$$(x - 0.6)^2 + y^2 = (0.6)^2 (\cos^2 5t + \sin^2 5t), \quad z = 0$$

or

$$(x - 0.6)^2 + y^2 = (0.6)^2, \quad z = 0$$

which is a circle on plane $z = 0$, centered at $(0.6, 0, 0)$ and of radius 0.6 m. Thus the particle gyrates in an orbit about a magnetic field line.

(e) $$\text{K.E.} = \frac{1}{2} m |\mathbf{u}|^2 = \frac{1}{2}(2)(9 \cos^2 5t + 9 \sin^2 5t) = 9 \text{ J}$$

which is the same as the K.E. at $t = 0$ and $t = 4$ s. Thus the uniform magnetic field has no effect on the K.E. of the particle.

Note that the angular velocity $\omega = QB/m$ and the radius of the orbit $r = u_o/\omega$, where u_o is the initial speed. An interesting application of the idea in this example is found in a common method of focusing a beam of electrons. The method employs a uniform magnetic field directed parallel to the desired beam as shown in Figure 8.2. Each electron emerging from the electron gun follows a helical path and returns to the axis at the same focal point with other electrons. If the screen of a cathode-ray tube were at this point, a single spot would appear on the screen.

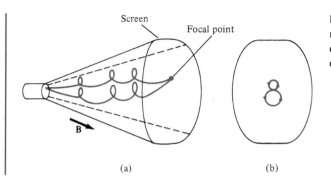

FIGURE 8.2 For Example 8.2: magnetic focusing of a beam of electrons: (**a**) helical paths of electrons, (**b**) end view of paths.

(a)   (b)

PRACTICE EXERCISE 8.2

A proton of mass m is projected into a uniform field $\mathbf{B} = B_o\mathbf{a}_z$ with an initial velocity $\alpha\mathbf{a}_x + \beta\mathbf{a}_z$. (a) Find the differential equations that the position vector $\mathbf{r} = x\mathbf{a}_x + y\mathbf{a}_y + z\mathbf{a}_z$ must satisfy. (b) Show that a solution to these equations is

$$x = \frac{\alpha}{\omega}\sin \omega t, \quad y = \frac{\alpha}{\omega}\cos \omega t, \quad z = \beta t$$

where $\omega = eB_o/m$ and e is the charge on the proton. (c) Show that this solution describes a circular helix in space.

Answer: (a) $\dfrac{dx}{dt} = \alpha \cos \omega t, \dfrac{dy}{dt} = -\alpha \sin \omega t, \dfrac{dz}{dt} = \beta$, (b) and (c) Proof.

EXAMPLE 8.3

A charged particle moves with a uniform velocity $4\mathbf{a}_x$ m/s in a region where $\mathbf{E} = 20\,\mathbf{a}_y$ V/m and $\mathbf{B} = B_o\mathbf{a}_z$ Wb/m^2. Determine B_o such that the velocity of the particle remains constant.

Solution:

If the particle moves with a constant velocity, it is implied that its acceleration is zero. In other words, the particle experiences no net force. Hence,

$$\mathbf{0} = \mathbf{F} = m\mathbf{a} = Q(\mathbf{E} + \mathbf{u} \times \mathbf{B})$$

$$\mathbf{0} = Q(20\mathbf{a}_y + 4\mathbf{a}_x \times B_o\mathbf{a}_z)$$

or

$$-20\mathbf{a}_y = -4B_o\mathbf{a}_y$$

Thus $B_o = 5$.

This example illustrates an important principle employed in a velocity filter shown in Figure 8.3. In this application, $\mathbf{E}$, $\mathbf{B}$, and $\mathbf{u}$ are mutually perpendicular so that $Q\mathbf{u} \times \mathbf{B}$ is

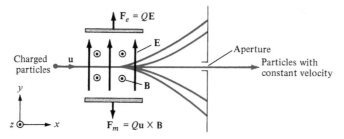

FIGURE 8.3 For Example 8.3: a velocity filter for charged particles.

directed opposite to QE, regardless of the sign of the charge. When the magnitudes of the two vectors are equal,

$$QuB = QE$$

or

$$u = \frac{E}{B}$$

This is the required (critical) speed to balance out the two parts of the Lorentz force. Particles with this speed are undeflected by the fields; they are "filtered" through the aperture. Particles with other speeds are deflected down or up, depending on whether their speeds are greater or less than this critical speed.

PRACTICE EXERCISE 8.3

Uniform **E** and **B** fields are oriented at right angles to each other. An electron moves with a speed of 8×10^6 m/s at right angles to both fields and passes undeflected through the field.

(a) If the magnitude of **B** is 0.5 mWb/m², find the value of **E**.
(b) Will this filter work for positive and negative charges and any value of mass?

Answer: (a) 4 kV/m, (b) yes.

EXAMPLE 8.4

A rectangular loop carrying current I_2 is placed parallel to an infinitely long filamentary wire carrying current I_1 as shown in Figure 8.4(a). Show that the force experienced by the loop is given by

$$\mathbf{F}_\ell = -\frac{\mu_0 I_1 I_2 b}{2\pi}\left[\frac{1}{\rho_0} - \frac{1}{\rho_0 + a}\right]\mathbf{a}_\rho \text{ N}$$

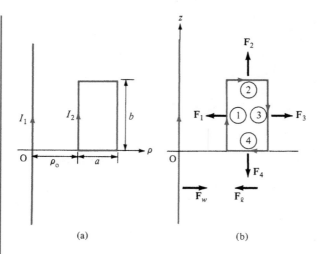

FIGURE 8.4 For Example 8.4: (**a**) rectangular loop inside the field produced by an infinitely long wire, (**b**) forces acting on the loop and wire.

(a)

(b)

Solution:

Let the force on the loop be

$$\mathbf{F}_\ell = \mathbf{F}_1 + \mathbf{F}_2 + \mathbf{F}_3 + \mathbf{F}_4 = I_2 \oint_L d\mathbf{l}_2 \times \mathbf{B}_1$$

where $\mathbf{F}_1$, $\mathbf{F}_2$, $\mathbf{F}_3$, and $\mathbf{F}_4$ are, respectively, the forces exerted on sides of the loop labeled 1, 2, 3, and 4 in Figure 8.4(b). Owing to the infinitely long wire

$$\mathbf{B}_1 = \frac{\mu_o I_1}{2\pi\rho_o} \mathbf{a}_\phi$$

Hence,

$$\mathbf{F}_1 = I_2 \int_L d\mathbf{l}_2 \times \mathbf{B}_1 = I_2 \int_{z=0}^{b} dz\, \mathbf{a}_z \times \frac{\mu_o I_1}{2\pi\rho_o} \mathbf{a}_\phi$$

$$= -\frac{\mu_o I_1 I_2 b}{2\pi\rho_o} \mathbf{a}_\rho \quad (\text{attractive})$$

$\mathbf{F}_1$ is attractive because it is directed toward the long wire; that is, $\mathbf{F}_1$ is along $-\mathbf{a}_\rho$ because loop side 1 and the long wire carry currents along the same direction. Similarly,

$$\mathbf{F}_3 = I_2 \int_L d\mathbf{l}_2 \times \mathbf{B}_1 = I_2 \int_{z=b}^{0} dz\, \mathbf{a}_z \times \frac{\mu_o I_1}{2\pi(\rho_o + a)} \mathbf{a}_\phi$$

$$= \frac{\mu_o I_1 I_2 b}{2\pi(\rho_o + a)} \mathbf{a}_\rho \quad (\text{repulsive})$$

$$\mathbf{F}_2 = I_2 \int_{\rho=\rho_o}^{\rho_o+a} d\rho\, \mathbf{a}_\rho \times \frac{\mu_o I_1\, \mathbf{a}_\phi}{2\pi\rho}$$

$$= \frac{\mu_o I_1 I_2}{2\pi} \ln \frac{\rho_o + a}{\rho_o} \mathbf{a}_z \quad \text{(parallel)}$$

$$\mathbf{F}_4 = I_2 \int_{\rho = \rho_o + a}^{\rho_o} d\rho \, \mathbf{a}_\rho \times \frac{\mu_o I_1 \mathbf{a}_\phi}{2\pi\rho}$$

$$= -\frac{\mu_o I_1 I_2}{2\pi} \ln \frac{\rho_o + a}{\rho_o} \mathbf{a}_z \quad \text{(parallel)}$$

The total force $\mathbf{F}_\ell$ on the loop is the sum of $\mathbf{F}_1$, $\mathbf{F}_2$, $\mathbf{F}_3$, and $\mathbf{F}_4$; that is,

$$\mathbf{F}_\ell = \frac{\mu_o I_1 I_2 b}{2\pi} \left[\frac{1}{\rho_o} - \frac{1}{\rho_o + a} \right] (-\mathbf{a}_\rho)$$

which is an attractive force trying to draw the loop toward the wire. The force $\mathbf{F}_w$ on the wire, by Newton's third law, is $-\mathbf{F}_\ell$; see Figure 8.4(b).

PRACTICE EXERCISE 8.4

In Example 8.4, find the force experienced by the infinitely long wire if $I_1 = 10$ A, $I_2 = 5$ A, $\rho_o = 20$ cm, $a = 10$ cm, $b = 30$ cm.

Answer: $5\mathbf{a}_\rho \, \mu$N.

8.3 MAGNETIC TORQUE AND MOMENT

Now that we have considered the force on a current loop in a magnetic field, we can determine the torque on it. The concept of a current loop experiencing a torque in a magnetic field is of paramount importance in understanding the behavior of orbiting charged particles, dc motors, and generators. If the loop is placed parallel to a magnetic field, it experiences a force that tends to rotate it.

The **torque T** (or mechanical moment of force) on the loop is the vector product of the moment arm **r** and the force **F**.

That is,

$$\mathbf{T} = \mathbf{r} \times \mathbf{F} \tag{8.14}$$

and its units are newton-meters $(\text{N} \cdot \text{m})$.

Let us apply this to a rectangular loop of length ℓ and width w placed in a uniform magnetic field **B** as shown in Figure 8.5(a). From Figure 8.5(a), we notice that $d\mathbf{l}$ is parallel to **B** along sides AB and CD of the loop and no force is exerted on those sides. Thus

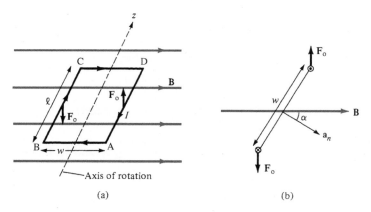

FIGURE 8.5 (a) Rectangular planar loop in a uniform magnetic field. **(b)** Cross-sectional view of part (a).

$$\mathbf{F} = I \int_B^C d\mathbf{l} \times \mathbf{B} + I \int_D^A d\mathbf{l} \times \mathbf{B}$$

$$= I \int_0^\ell dz \, \mathbf{a}_z \times \mathbf{B} + I \int_\ell^0 dz \, \mathbf{a}_z \times \mathbf{B}$$

or

$$\mathbf{F} = \mathbf{F}_o - \mathbf{F}_o = 0 \tag{8.15}$$

where $|\mathbf{F}_o| = IB\ell$ because **B** is uniform. Thus, no force is exerted on the loop as a whole. However, $\mathbf{F}_o$ and $-\mathbf{F}_o$ act at different points on the loop, thereby creating a couple. If the normal to the plane of the loop makes an angle α with **B**, as shown in the cross-sectional view of Figure 8.5(b), the torque on the loop is

$$|\mathbf{T}| = |\mathbf{F}_o| \, w \sin \alpha$$

or

$$T = BI\ell w \sin \alpha \tag{8.16}$$

But $\ell w = S$, the area of the loop. Hence,

$$T = BIS \sin \alpha \tag{8.17}$$

We define the quantity

$$\boxed{\mathbf{m} = IS\mathbf{a}_n} \tag{8.18}$$

as the *magnetic dipole moment* (in A · m²) of the loop. In eq. (8.18), $\mathbf{a}_n$ is a unit normal vector to the plane of the loop and its direction is determined by the right-hand rule: fingers in the direction of current and thumb along $\mathbf{a}_n$.

The **magnetic dipole moment** is the product of current and area of the loop; its direction is normal to the loop.

Introducing eq. (8.18) in eq. (8.17), we obtain

$$\boxed{\mathbf{T} = \mathbf{m} \times \mathbf{B}} \tag{8.19}$$

Although this expression was obtained by using a rectangular loop, it is generally applicable in determining the torque on a planar loop of any arbitrary shape. The only limitation is that the magnetic field must be uniform. It should be noted that the torque is in the direction of the axis of rotation (the z-axis in the case of Figure 8.5(a)). It is directed with the aim of reducing α so that $\mathbf{m}$ and $\mathbf{B}$ are in the same direction. In an equilibrium position (when $\mathbf{m}$ and $\mathbf{B}$ are in the same direction), the loop is perpendicular to the magnetic field and the torque will be zero as well as the sum of the forces on the loop.

8.4 A MAGNETIC DIPOLE

A **magnetic dipole** consists of a bar magnet or small current-carrying loop. The reason for this and what we mean by "small" will soon be evident. Let us determine the magnetic field $\mathbf{B}$ at an observation point $P(r, \theta, \phi)$ due to a circular loop carrying current I as in Figure 8.6. The magnetic vector potential at P is

$$\mathbf{A} = \frac{\mu_o I}{4\pi} \oint \frac{d\mathbf{l}}{r} \tag{8.20}$$

It can be shown that in the far field $r \gg a$, so that the loop appears small at the observation point, $\mathbf{A}$ has only ϕ-component and it is given by

$$\mathbf{A} = \frac{\mu_o I \pi a^2 \sin \theta \, \mathbf{a}_\phi}{4\pi r^2} \tag{8.21a}$$

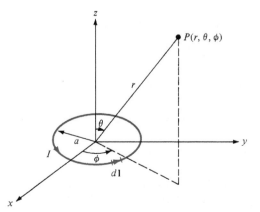

FIGURE 8.6 Magnetic field at P due to a current loop.

or

$$A = \frac{\mu_o\, \mathbf{m} \times \mathbf{a}_r}{4\pi r^2}$$

(8.21b)

where $\mathbf{m} = I\pi a^2 \mathbf{a}_z$, the magnetic moment of the loop, and $\mathbf{a}_z \times \mathbf{a}_r = \sin\theta\, \mathbf{a}_\phi$. We determine the magnetic flux density **B** from $\mathbf{B} = \nabla \times \mathbf{A}$ as

$$\mathbf{B} = \frac{\mu_o m}{4\pi r^3}(2\cos\theta\, \mathbf{a}_r + \sin\theta\, \mathbf{a}_\theta)$$

(8.22)

TABLE 8.2 Comparison between Electric and Magnetic Monopoles and Dipoles

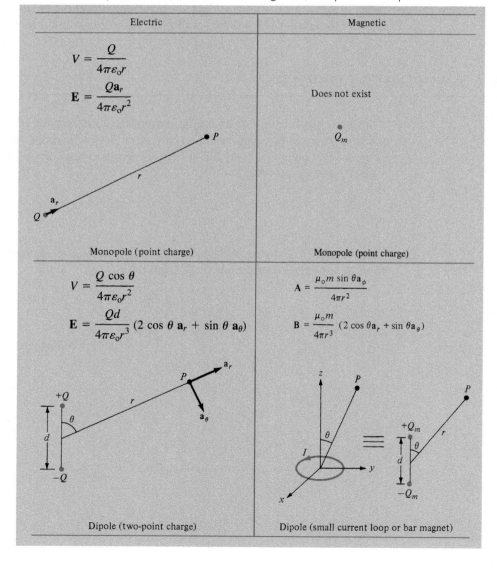

Electric	Magnetic
$V = \dfrac{Q}{4\pi\varepsilon_0 r}$ $\mathbf{E} = \dfrac{Q\mathbf{a}_r}{4\pi\varepsilon_0 r^2}$	Does not exist
Monopole (point charge)	Monopole (point charge)
$V = \dfrac{Q\cos\theta}{4\pi\varepsilon_0 r^2}$ $\mathbf{E} = \dfrac{Qd}{4\pi\varepsilon_0 r^3}(2\cos\theta\, \mathbf{a}_r + \sin\theta\, \mathbf{a}_\theta)$	$A = \dfrac{\mu_o m \sin\theta\, \mathbf{a}_\phi}{4\pi r^2}$ $\mathbf{B} = \dfrac{\mu_o m}{4\pi r^3}(2\cos\theta\, \mathbf{a}_r + \sin\theta\, \mathbf{a}_\theta)$
Dipole (two-point charge)	Dipole (small current loop or bar magnet)

It is interesting to compare eqs. (8.21) and (8.22) with similar expressions in eqs. (4.80) and (4.82) for electrical potential V and electric field intensity $\mathbf{E}$ due to an electric dipole. This comparison is done in Table 8.2, in which we notice the striking similarities between $\mathbf{B}$ in the far field due to a small current loop and $\mathbf{E}$ in the far field due to an electric dipole. It is therefore reasonable to regard a small current loop as a magnetic dipole. The $\mathbf{B}$ lines due to a magnetic dipole are similar to the $\mathbf{E}$ lines due to an electric dipole. Figure 8.7(a) illustrates the $\mathbf{B}$ lines around the magnetic dipole $\mathbf{m} = I\mathbf{S}$.

A short permanent magnetic bar, shown in Figure 8.7(b), may also be regarded as a magnetic dipole. Observe that the $\mathbf{B}$ lines due to the bar are similar to those due to a small current loop in Figure 8.7(a).

Consider the bar magnet of Figure 8.8. If Q_m is an isolated magnetic charge (pole strength) and ℓ is the length of the bar, the bar has a dipole moment $Q_m\ell$. (Notice that Q_m does exist; however, it does not exist without an associated $-Q_m$. See Table 8.2.) When the bar is in a uniform magnetic field $\mathbf{B}$, it experiences a torque

$$\mathbf{T} = \mathbf{m} \times \mathbf{B} = Q_m\boldsymbol{\ell} \times \mathbf{B} \tag{8.23}$$

where $\boldsymbol{\ell}$ points south to north. The torque tends to align the bar with the external magnetic field. The force acting on the magnetic charge is given by

$$\mathbf{F} = Q_m\mathbf{B} \tag{8.24}$$

Since both a small current loop and a bar magnet produce magnetic dipoles, they are equivalent if they produce the same torque in a given $\mathbf{B}$ field, that is, when

$$T = Q_m\ell B = ISB \tag{8.25}$$

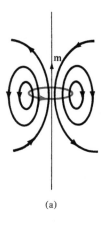

(a)

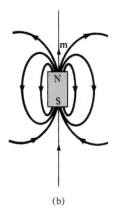

(b)

FIGURE 8.7 The $\mathbf{B}$ lines due to magnetic dipoles: (**a**) a small current loop with $\mathbf{m} = I\mathbf{S}$, (**b**) a bar magnet with $\mathbf{m} = Q_m\boldsymbol{\ell}$.

FIGURE 8.8 A bar magnet in an external magnetic field.

Hence,

$$Q_m \ell = IS \tag{8.26}$$

showing that they must have the same dipole moment.

EXAMPLE 8.5

Determine the magnetic moment of an electric circuit formed by the triangular loop of Figure 8.9.

Solution:

If a plane intercepts the coordinate axes at $(a, 0, 0)$, $(0, b, 0)$, and $(0, 0, c)$, its equation is given by

$$\frac{x}{a} + \frac{y}{b} + \frac{z}{c} = 1 \longrightarrow bcx + cay + abz = abc$$

For the present problem, $a = b = c = 2$. Hence

$$x + y + z = 2$$

Thus, we can use

$$\mathbf{m} = IS\mathbf{a}_n$$

where

$$S = \text{loop area} = \frac{1}{2} \times \text{base} \times \text{height} = \frac{1}{2}(2\sqrt{2})(2\sqrt{2})\sin 60°$$

$$= 4 \sin 60°$$

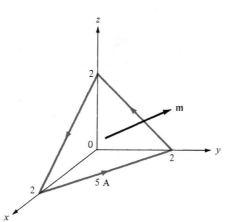

FIGURE 8.9 Triangular loop of Example 8.5.

If we define the plane surface by a function

$$f(x, y, z) = x + y + z - 2 = 0$$

$$\mathbf{a}_n = \pm \frac{\nabla f}{|\nabla f|} = \pm \frac{(\mathbf{a}_x + \mathbf{a}_y + \mathbf{a}_z)}{\sqrt{3}}$$

We choose the plus sign in view of the direction of the current in the loop (using the right-hand rule, **m** is directed as in Figure 8.9). Hence

$$\mathbf{m} = 5(4 \sin 60°)\frac{(\mathbf{a}_x + \mathbf{a}_y + \mathbf{a}_z)}{\sqrt{3}}$$

$$= 10(\mathbf{a}_x + \mathbf{a}_y + \mathbf{a}_z) \, \text{A} \cdot \text{m}^2$$

PRACTICE EXERCISE 8.5

A rectangular coil of area 10 cm^2 carrying current of 50 A lies on plane $2x + 6y - 3z = 7$ such that the magnetic moment of the coil is directed away from the origin. Calculate its magnetic moment.

Answer: $(1.429\mathbf{a}_x + 4.286\mathbf{a}_y - 2.143\mathbf{a}_z) \times 10^{-2} \, \text{A} \cdot \text{m}^2.$

EXAMPLE 8.6

A small current loop L_1 with magnetic moment $5\mathbf{a}_z \, \text{A} \cdot \text{m}^2$ is located at the origin while another small loop current L_2 with magnetic moment $3\mathbf{a}_y \, \text{A} \cdot \text{m}^2$ is located at $(4, -3, 10)$. Determine the torque on L_2.

Solution:

The torque $\mathbf{T}_2$ on the loop L_2 is due to the field $\mathbf{B}_1$ produced by loop L_1. Hence,

$$\mathbf{T}_2 = \mathbf{m}_2 \times \mathbf{B}_1$$

Since $\mathbf{m}_1$ for loop L_1 is along $\mathbf{a}_z$, we find $\mathbf{B}_1$ using eq. (8.22):

$$\mathbf{B}_1 = \frac{\mu_o m_1}{4\pi r^3}(2 \cos \theta \, \mathbf{a}_r + \sin \theta \, \mathbf{a}_\theta)$$

Using eq. (2.23), we transform $\mathbf{m}_2$ from Cartesian to spherical coordinates:

$$\mathbf{m}_2 = 3\mathbf{a}_y = 3(\sin \theta \sin \phi \, \mathbf{a}_r + \cos \theta \sin \phi \, \mathbf{a}_\theta + \cos \phi \, \mathbf{a}_\phi)$$

At $(4, -3, 10)$,

$$r = \sqrt{4^2 + (-3)^2 + 10^2} = 5\sqrt{5}$$

$$\tan \theta = \frac{\rho}{z} = \frac{5}{10} = \frac{1}{2} \rightarrow \sin \theta = \frac{1}{\sqrt{5}}, \quad \cos \theta = \frac{2}{\sqrt{5}}$$

$$\tan \phi = \frac{y}{x} = \frac{-3}{4} \rightarrow \sin \phi = \frac{-3}{5}, \quad \cos \phi = \frac{4}{5}$$

Hence,

$$\mathbf{B}_1 = \frac{4\pi \times 10^{-7} \times 5}{4\pi\, 625 \sqrt{5}} \left(\frac{4}{\sqrt{5}} \mathbf{a}_r + \frac{1}{\sqrt{5}} \mathbf{a}_\theta \right)$$

$$= \frac{10^{-7}}{625} (4\mathbf{a}_r + \mathbf{a}_\theta)$$

$$\mathbf{m}_2 = 3 \left[-\frac{3\mathbf{a}_r}{5\sqrt{5}} - \frac{6\mathbf{a}_\theta}{5\sqrt{5}} + \frac{4\mathbf{a}_\phi}{5} \right]$$

and

$$\mathbf{T} = \frac{10^{-7}(3)}{625(5\sqrt{5})} (-3\mathbf{a}_r - 6\mathbf{a}_\theta + 4\sqrt{5}\mathbf{a}_\phi) \times (4\mathbf{a}_r + \mathbf{a}_\phi)$$

$$= 4.293 \times 10^{-11} (-8.944\mathbf{a}_r + 35.777\mathbf{a}_\theta + 21\mathbf{a}_\phi)$$

$$= -0.384\mathbf{a}_r + 1.536\mathbf{a}_\theta + 0.9015\mathbf{a}_\phi \text{ nN} \cdot \text{m}$$

PRACTICE EXERCISE 8.6

The coil of Practice Exercise 8.5 is surrounded by a uniform field $0.6\mathbf{a}_x + 0.4\mathbf{a}_y + 0.5\mathbf{a}_z$ Wb/m².

(a) Find the torque on the coil.
(b) Show that the torque on the coil is maximum if placed on plane $2x - 8y + 4z = \sqrt{84}$. Calculate the magnitude of the maximum torque.

Answer: (a) $0.03\mathbf{a}_x - 0.02\mathbf{a}_y - 0.02\mathbf{a}_z$ N · m, (b) 0.0439 N · m.

8.5 MAGNETIZATION IN MATERIALS

Our discussion here will parallel that on polarization of materials in an electric field. We shall assume that our atomic model is that of an electron orbiting about a positive nucleus.

We know that a given material is composed of atoms. Each atom may be regarded as consisting of electrons orbiting about a central positive nucleus; the electrons also rotate (or spin) about their own axes. Thus an internal magnetic field is produced by electrons orbiting around the nucleus as in Figure 8.10(a) or electrons spinning as in Figure 8.10(b). Both these electronic motions produce internal magnetic fields $\mathbf{B}_i$ that are similar to the magnetic field produced by a current loop of Figure 8.11. The equivalent current loop has a magnetic moment of $\mathbf{m} = I_b S \mathbf{a}_n$, where S is the area of the loop and I_b is the bound current (bound to the atom).

Without an external $\mathbf{B}$ field applied to the material, the sum of $\mathbf{m}$'s is zero due to random orientation as in Figure 8.12(a). When an external $\mathbf{B}$ field is applied, the magnetic moments of the electrons more or less align themselves with $\mathbf{B}$ so that the net magnetic moment is not zero, as illustrated in Figure 8.12(b).

> The **magnetization M**, in amperes per meter, is the magnetic dipole moment per unit volume.

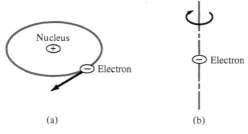

(a) (b)

FIGURE 8.10 (**a**) Electron orbiting around the nucleus. (**b**) Electron spin.

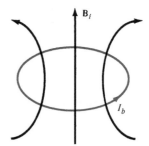

FIGURE 8.11 Circular current loop equivalent to electronic motion of Figure 8.10.

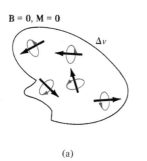

(a)

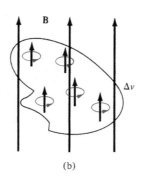

(b)

FIGURE 8.12 Magnetic dipole moment in a volume Δv: (**a**) before $\mathbf{B}$ is applied, (**b**) after $\mathbf{B}$ is applied.

If there are N atoms in a given volume Δv and the kth atom has a magnetic moment $\mathbf{m}_k$,

$$\mathbf{M} = \lim_{\Delta v \to 0} \frac{\sum_{k=1}^{N} \mathbf{m}_k}{\Delta v} \tag{8.27}$$

A medium for which $\mathbf{M}$ is not zero everywhere is said to be magnetized. For a differential volume dv', the magnetic moment is $d\mathbf{m} = \mathbf{M}\, dv'$. Recall that we denote the field point by the unprimed coordinates (x, y, z) and the source point by the primed coordinates (x', y', z'). From eq. (8.21b), the vector magnetic potential due to $d\mathbf{m}$ is

$$d\mathbf{A} = \frac{\mu_o \mathbf{M} \times \mathbf{a}_R}{4\pi R^2}\, dv' = \frac{\mu_o \mathbf{M} \times \mathbf{R}}{4\pi R^3}\, dv'$$

From eq. (7.46) we can write

$$\frac{\mathbf{R}}{R^3} = \nabla'\frac{1}{R}$$

Hence,

$$\mathbf{A} = \frac{\mu_o}{4\pi} \int_{v'} \mathbf{M} \times \nabla'\frac{1}{R}\, dv' \tag{8.28}$$

Using eq. (7.48) gives

$$\mathbf{M} \times \nabla'\frac{1}{R} = \frac{1}{R}\nabla' \times \mathbf{M} - \nabla' \times \frac{\mathbf{M}}{R}$$

Substituting this into eq. (8.28) yields

$$\mathbf{A} = \frac{\mu_o}{4\pi} \int_{v'} \frac{\nabla' \times \mathbf{M}}{R}\, dv' - \frac{\mu_o}{4\pi} \int_{v'} \nabla' \times \frac{\mathbf{M}}{R}\, dv'$$

Applying the vector identity

$$\int_{v'} \nabla' \times \mathbf{F}\, dv' = -\oint_{S'} \mathbf{F} \times d\mathbf{S}$$

to the second integral, we obtain

$$\mathbf{A} = \frac{\mu_o}{4\pi} \int_{v'} \frac{\nabla' \times \mathbf{M}}{R}\, dv' + \frac{\mu_o}{4\pi} \oint_{S'} \frac{\mathbf{M} \times \mathbf{a}_n}{R}\, dS'$$

$$= \frac{\mu_o}{4\pi} \int_{v'} \frac{\mathbf{J}_b\, dv'}{R} + \frac{\mu_o}{4\pi} \oint_{S'} \frac{\mathbf{K}_b\, dS'}{R} \tag{8.29}$$

Comparing eq. (8.29) with eqs. (7.42) and (7.43) (upon dropping the primes) gives

$$\boxed{\mathbf{J}_b = \nabla \times \mathbf{M}} \tag{8.30}$$

and

$$\boxed{\mathbf{K}_b = \mathbf{M} \times \mathbf{a}_n}$$ (8.31)

where $\mathbf{J}_b$ is the *bound volume current density* or *magnetization volume current density*, in amperes per meter squared, $\mathbf{K}_b$ is the *bound surface current density*, in amperes per meter, and $\mathbf{a}_n$ is a unit vector normal to the surface. Equation (8.29) shows that the potential of a magnetic body is due to a volume current density $\mathbf{J}_b$ throughout the body and a surface current $\mathbf{K}_b$ on the surface of the body. The vector $\mathbf{M}$ is analogous to the polarization $\mathbf{P}$ in dielectrics and is sometimes called the *magnetic polarization density* of the medium. In another sense, $\mathbf{M}$ is analogous to $\mathbf{H}$ and they both have the same units. In this respect, as $\mathbf{J} = \nabla \times \mathbf{H}$, so $\mathbf{J}_b = \nabla \times \mathbf{M}$. Also, $\mathbf{J}_b$ and $\mathbf{K}_b$ for a magnetized body are similar to ρ_{pv} and ρ_{ps} for a polarized body. As is evident in eqs. (8.29) to (8.31), $\mathbf{J}_b$ and $\mathbf{K}_b$ can be derived from $\mathbf{M}$; therefore, $\mathbf{J}_b$ and $\mathbf{K}_b$ are not commonly used.

In free space, $\mathbf{M} = \mathbf{0}$ and we have

$$\nabla \times \mathbf{H} = \mathbf{J}_f \quad \text{or} \quad \nabla \times \left(\frac{\mathbf{B}}{\mu_o} \right) = \mathbf{J}_f$$ (8.32)

where $\mathbf{J}_f$ is the free current volume density. In a material medium $\mathbf{M} \neq \mathbf{0}$, and as a result, $\mathbf{B}$ changes so that

$$\nabla \times \left(\frac{\mathbf{B}}{\mu_o} \right) = \mathbf{J}_f + \mathbf{J}_b = \mathbf{J}$$

$$= \nabla \times \mathbf{H} + \nabla \times \mathbf{M}$$

or

$$\boxed{\mathbf{B} = \mu_o(\mathbf{H} + \mathbf{M})}$$ (8.33)

The relationship in eq. (8.33) holds for all materials whether they are linear or not. The concepts of linearity, isotropy, and homogeneity introduced in Section 5.7 for dielectric media equally apply here for magnetic media. For linear materials, $\mathbf{M}$ (in A/m) depends linearly on $\mathbf{H}$ such that

$$\boxed{\mathbf{M} = \chi_m \mathbf{H}}$$ (8.34)

where χ_m is a dimensionless quantity (ratio of M to H) called *magnetic susceptibility* of the medium. It is more or less a measure of how susceptible (or sensitive) the material is to a magnetic field. Substituting eq. (8.34) into eq. (8.33) yields

$$\mathbf{B} = \mu_o(1 + \chi_m)\mathbf{H} = \mu\mathbf{H}$$ (8.35)

or

$$\boxed{\mathbf{B} = \mu_o\mu_r\mathbf{H}}$$ (8.36)

where

$$\boxed{\mu_r = 1 + \chi_m = \frac{\mu}{\mu_o}} \tag{8.37}$$

The quantity $\mu = \mu_o\mu_r$ is called the *permeability* of the material and is measured in henrys per meter; the henry is the unit of inductance and will be defined a little later. The dimensionless quantity μ_r is the ratio of the permeability of a given material to that of free space and is known as the *relative permeability* of the material.

It should be borne in mind that the relationships in eqs. (8.34) to (8.37) hold only for linear and isotropic materials. If the materials are anisotropic (e.g., crystals), eq. (8.33) still holds but eqs. (8.34) to (8.37) do not apply. In this case, μ has nine terms [similar to ε in eq. (5.37)] and, consequently, the fields **B**, **H**, and **M** are no longer parallel.

†8.6 CLASSIFICATION OF MATERIALS

In general, we may use the magnetic susceptibility χ_m or the relative permeability μ_r to classify materials in terms of their magnetic property or behavior. A material is said to be *nonmagnetic* if $\chi_m = 0$ (or $\mu_r = 1$); it is magnetic otherwise. Free space, air, and materials with $\chi_m = 0$ (or $\mu_r \approx 1$) are regarded as nonmagnetic.

Roughly speaking, materials may be grouped into three major classes: diamagnetic, paramagnetic, and ferromagnetic. This rough classification is depicted in Figure 8.13. A material is said to be *diamagnetic* if it has $\mu_r \lesssim 1$ (i.e., very small negative χ_m). It is *paramagnetic* if $\mu_r \gtrsim 1$ (i.e., very small positive χ_m). If $\mu_r \gg 1$ (i.e., very large positive χ_m), the material is *ferromagnetic*. Table B.3 in Appendix B presents the values μ_r for some materials. From Table B.3, it is apparent that for most practical purposes we may assume that $\mu_r \simeq 1$ for diamagnetic and paramagnetic materials. Thus, we may regard diamagnetic and paramagnetic materials as linear and nonmagnetic. Ferromagnetic materials are always nonlinear and magnetic except when their temperatures are above curie temperature (to be explained later). The reason for this will become evident as we more closely examine each of these three types of magnetic material.

Diamagnetism occurs when the magnetic fields in a material that are due to electronic motions of orbiting and spinning completely cancel each other. Thus, the permanent (or intrinsic) magnetic moment of each atom is zero and such materials are weakly affected by a magnetic field. For most diamagnetic materials (e.g., bismuth, lead, copper, silicon, diamond, sodium chloride), χ_m is of the order of -10^{-5}. In certain materials, called *superconductors*, "perfect diamagnetism" occurs at temperatures near absolute zero: $\chi_m = -1$ or $\mu_r = 0$ and $B = 0$. Thus superconductors cannot contain magnetic fields.[2] Except for superconductors, the diamagnetic properties of materials are seldom used in practice. Although the diamagnetic effect is overshadowed by other stronger effects in some materials, all materials exhibit diamagnetism.

---

[2] An excellent treatment of superconductors is found in M. A. Plonus, *Applied Electromagnetics.* New York: McGraw-Hill, 1978, pp. 375–388. Also, the August 1989 issue of the *Proceedings of IEEE* is devoted to superconductivity.

Materials whose atoms have nonzero permanent magnetic moment may be paramagnetic or ferromagnetic. *Paramagnetism* occurs when the magnetic fields produced in a material by orbital and spinning electrons do not cancel completely. Unlike diamagnetism, paramagnetism is temperature dependent. For most paramagnetic materials (e.g., air, platinum, tungsten, potassium), X_m is of the order $+10^{-5}$ to $+10^{-3}$ and is temperature dependent. Such materials find application in masers.

Ferromagnetism occurs in materials whose atoms have relatively large permanent magnetic moment. They are called ferromagnetic materials because the best-known member is iron. Other members are cobalt, nickel, and their alloys. Ferromagnetic materials are very useful in practice. As distinct from diamagnetic and paramagnetic materials, ferromagnetic materials have the following properties:

1. They are capable of being magnetized very strongly by a magnetic field.
2. They retain a considerable amount of their magnetization when removed from the field.
3. They lose their ferromagnetic properties and become linear paramagnetic materials when the temperature is raised above a certain temperature known as the *curie temperature*. Thus if a permanent magnet is heated above its curie temperature (770°C for iron), it loses its magnetization completely.
4. They are nonlinear; that is, the constitutive relation $\mathbf{B} = \mu_o\mu_r\mathbf{H}$ does not hold for ferromagnetic materials because μ_r depends on $\mathbf{B}$ and cannot be represented by a single value.

Thus, the values of μ_r cited in Table B.3 for ferromagnetics are only typical. For example, for nickel $\mu_r = 50$ under some conditions and 600 under other conditions.

As mentioned in Section 5.9 for conductors, ferromagnetic materials, such as iron and steel, are used for screening (or shielding) to protect sensitive electrical devices from disturbances from strong magnetic fields. In the example of a typical iron shield shown in Figure 8.14(a), the compass is protected. Without the iron shield, as in Figure 8.14(b), the compass gives an erroneous reading owing to the effect of the external magnetic field. For perfect screening, it is required that the shield have infinite permeability.

Even though $\mathbf{B} = \mu_o(\mathbf{H} + \mathbf{M})$ holds for all materials including ferromagnetics, the relationship between $\mathbf{B}$ and $\mathbf{H}$ depends on previous magnetization of a ferromagnetic material—its "magnetic history." Instead of having a linear relationship between $\mathbf{B}$ and $\mathbf{H}$ (i.e., $\mathbf{B} = \mu\mathbf{H}$), it is only possible to represent the relationship by a *magnetization curve* or *B–H curve*.

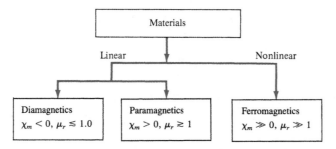

FIGURE 8.13 Classification of materials.

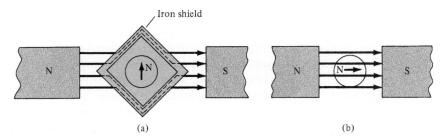

Iron shield

(a)   (b)

FIGURE 8.14 Magnetic screening: (**a**) iron shield protecting a small compass, (**b**) compass gives erroneous reading without the shield.

A typical B–H curve is shown in Figure 8.15. First, note the nonlinear relationship between B and H. Second, at any point on the curve, μ is given by the ratio B/H and not by dB/dH, the slope of the curve.

If we assume that the ferromagnetic material whose B–H curve in Figure 8.15 is initially unmagnetized, as H increases (owing to increase in current) from O to maximum applied field intensity H_{max}, curve OP is produced. This curve is referred to as the *virgin* or *initial magnetization curve*. After reaching saturation at P, if H is decreased, B does not follow the initial curve but lags behind H. This phenomenon of B lagging behind H is called *hysteresis* (which means "to lag" in Greek).

If H is reduced to zero, B is not reduced to zero but to B_r, which is referred to as the *permanent flux density*. The value of B_r depends on H_{max}, the maximum applied field intensity. The existence of B_r is the cause of having permanent magnets. If H increases negatively (by reversing the direction of current), B becomes zero when H becomes H_c, which is known as the *coercive field intensity*. Materials for which H_c is small are said to be magnetically hard. The value of H_c also depends on H_{max}.

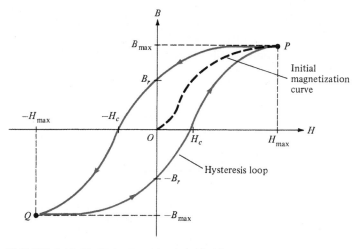

FIGURE 8.15 Typical magnetization (B–H) curve.

Further increase in H in the negative direction to reach Q and a reverse in its direction to reach P gives a closed curve called a *hysteresis loop.* Hysteresis loops vary in shape from one material to another. Some ferrites, for example, have an almost rectangular hysteresis loop and are used in digital computers as magnetic information storage devices. The area of a hysteresis loop gives the energy loss (hysteresis loss) per unit volume during one cycle of the periodic magnetization of the ferromagnetic material. This energy loss is in the form of heat. It is therefore desirable that materials used in electric generators, motors, and transformers have tall but narrow hysteresis loops so that hysteresis losses are minimal.

EXAMPLE 8.7

Region $0 \leq z \leq 2$ m is occupied by an infinite slab of permeable material ($\mu_r = 2.5$). If $\mathbf{B} = 10y\mathbf{a}_x - 5x\mathbf{a}_y$ mWb/m^2 within the slab, determine: (a) $\mathbf{J}$, (b) $\mathbf{J}_b$, (c) $\mathbf{M}$, (d) $\mathbf{K}_b$ on $z = 0$.

Solution:

(a) By definition,

$$\mathbf{J} = \nabla \times \mathbf{H} = \nabla \times \frac{\mathbf{B}}{\mu_o \mu_r} = \frac{1}{4\pi \times 10^{-7}(2.5)}\left(\frac{\partial B_y}{\partial x} - \frac{\partial B_x}{\partial y}\right)\mathbf{a}_z$$

$$= \frac{10^6}{\pi}(-5 - 10)10^{-3}\mathbf{a}_z = -4.775\mathbf{a}_z \text{ kA/m}^2$$

(b) $\mathbf{J}_b = \chi_m \mathbf{J} = (\mu_r - 1)\mathbf{J} = 1.5(-4.775\mathbf{a}_z) \cdot 10^3$
$\qquad = -7.163\mathbf{a}_z \text{ kA/m}^2$

(c) $\mathbf{M} = \chi_m \mathbf{H} = \chi_m \dfrac{\mathbf{B}}{\mu_o \mu_r} = \dfrac{1.5(10y\mathbf{a}_x - 5x\mathbf{a}_y) \cdot 10^{-3}}{4\pi \times 10^{-7}(2.5)}$
$\qquad = 4.775y\mathbf{a}_x - 2.387x\mathbf{a}_y \text{ kA/m}$

(d) $\mathbf{K}_b = \mathbf{M} \times \mathbf{a}_n$. Since $z = 0$ is the lower side of the slab occupying $0 \leq z \leq 2$, $\mathbf{a}_n = -\mathbf{a}_z$. Hence,

$$\mathbf{K}_b = (4.775y\mathbf{a}_x - 2.387x\mathbf{a}_y) \times (-\mathbf{a}_z)$$

$$= 2.387x\mathbf{a}_x + 4.775y\mathbf{a}_y \text{ kA/m}$$

PRACTICE EXERCISE 8.7

In a certain region ($\mu = 4.6\mu_o$),

$$\mathbf{B} = 10e^{-y}\mathbf{a}_z \text{ mWb/m}^2$$

find: (a) χ_m, (b) $\mathbf{H}$, (c) $\mathbf{M}$.

Answer: (a) 3.6, (b) $1730e^{-y}\mathbf{a}_z$ A/m, (c) $6228e^{-y}\mathbf{a}_z$ A/m.

8.7 MAGNETIC BOUNDARY CONDITIONS

We define magnetic boundary conditions as the conditions that $\mathbf{H}$ (or $\mathbf{B}$) field must satisfy at the boundary between two different media. Our derivations here are similar to those in Section 5.9. We make use of Gauss's law for magnetic fields

$$\oint_S \mathbf{B} \cdot d\mathbf{S} = 0 \tag{8.38}$$

and Ampère's circuit law

$$\oint_L \mathbf{H} \cdot d\mathbf{l} = I \tag{8.39}$$

Consider the boundary between two magnetic media 1 and 2, characterized, respectively, by μ_1 and μ_2 as in Figure 8.16. Applying eq. (8.38) to the pillbox (Gaussian surface) of Figure 8.16(a) and allowing $\Delta h \rightarrow 0$, we obtain

$$B_{1n}\,\Delta S - B_{2n}\,\Delta S = 0 \tag{8.40}$$

Thus

$$\boxed{\mathbf{B}_{1n} = \mathbf{B}_{2n}} \quad \text{or} \quad \mu_1 \mathbf{H}_{1n} = \mu_2 \mathbf{H}_{2n} \tag{8.41}$$

since $\mathbf{B} = \mu\mathbf{H}$. Equation (8.41) shows that the normal component of $\mathbf{B}$ is continuous at the boundary. It also shows that the normal component of $\mathbf{H}$ is discontinuous at the boundary; $\mathbf{H}$ undergoes some change at the interface.

Similarly, we apply eq. (8.39) to the closed path $abcda$ of Figure 8.16(b), where surface current $\mathbf{K}$ on the boundary is assumed normal to the path. We obtain

$$K \cdot \Delta w = H_{1t} \cdot \Delta w + H_{1n} \cdot \frac{\Delta h}{2} + H_{2n} \cdot \frac{\Delta h}{2}$$

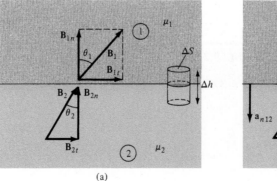

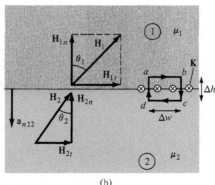

(a)   (b)

FIGURE 8.16 Boundary conditions between two magnetic media: (**a**) for $\mathbf{B}$, (**b**) for $\mathbf{H}$.

$$-H_{2t} \cdot \Delta w - H_{2n} \cdot \frac{\Delta h}{2} - H_{1n} \cdot \frac{\Delta h}{2} \tag{8.42}$$

As $\Delta h \rightarrow 0$, eq. (8.42) leads to

$$H_{1t} - H_{2t} = K \tag{8.43}$$

This shows that the tangential component of H is also discontinuous. Equation (8.43) may be written in terms of B as

$$\frac{B_{1t}}{\mu_1} - \frac{B_{2t}}{\mu_2} = K \tag{8.44}$$

In the general case, eq. (8.43) becomes

$$\boxed{(\mathbf{H}_1 - \mathbf{H}_2) \times \mathbf{a}_{n12} = \mathbf{K}} \tag{8.45}$$

where $\mathbf{a}_{n12}$ is a unit vector normal to the interface and is directed from medium 1 to medium 2. If the boundary is free of current or the media are not conductors (for $\mathbf{K}$ is free current density), $\mathbf{K} = 0$ and eq. (8.43) becomes

$$\boxed{H_{1t} = H_{2t}} \qquad \text{or} \qquad \frac{\mathbf{B}_{1t}}{\mu_1} = \frac{\mathbf{B}_{2t}}{\mu_2} \tag{8.46}$$

Thus the tangential component of $\mathbf{H}$ is continuous while that of $\mathbf{B}$ is discontinuous at the boundary.

If the fields make an angle θ with the normal to the interface, eq. (8.41) results in

$$B_1 \cos \theta_1 = B_{1n} = B_{2n} = B_2 \cos \theta_2 \tag{8.47}$$

while eq. (8.46) produces

$$\frac{B_1}{\mu_1} \sin \theta_1 = H_{1t} = H_{2t} = \frac{B_2}{\mu_2} \sin \theta_2 \tag{8.48}$$

Dividing eq. (8.48) by eq. (8.47) gives

$$\boxed{\frac{\tan \theta_1}{\tan \theta_2} = \frac{\mu_1}{\mu_2}} \tag{8.49}$$

which is [similar to eq. (5.65)] the law of refraction for magnetic flux lines at a boundary with no surface current.

EXAMPLE 8.8

Given that $\mathbf{H}_1 = -2\mathbf{a}_x + 6\mathbf{a}_y + 4\mathbf{a}_z$ A/m in region $y - x - 2 \leq 0$, where $\mu_1 = 5\mu_o$, calculate

(a) $\mathbf{M}_1$ and $\mathbf{B}_1$
(b) $\mathbf{H}_2$ and $\mathbf{B}_2$ in region $y - x - 2 \geq 0$, where $\mu_2 = 2\mu_o$

Solution:

Since $y - x - 2 = 0$ is a plane, $y - x \leq 2$ or $y \leq x + 2$ is region 1 in Figure 8.17. A point in this region may be used to confirm this. For example, the origin $(0, 0)$ is in this region because $0 - 0 - 2 < 0$. If we let the surface of the plane be described by $f(x, y) = y - x - 2$, a unit vector normal to the plane is given by

$$\mathbf{a}_n = \frac{\nabla f}{|\nabla f|} = \frac{\mathbf{a}_y - \mathbf{a}_x}{\sqrt{2}}$$

(a)
$$\mathbf{M}_1 = \chi_{m1}\mathbf{H}_1 = (\mu_{r1} - 1)\,\mathbf{H}_1 = (5 - 1)(-2, 6, 4)$$
$$= -8\mathbf{a}_x + 24\mathbf{a}_y + 16\mathbf{a}_z \text{ A/m}$$
$$\mathbf{B}_1 = \mu_1\mathbf{H}_1 = \mu_0\mu_{r1}\mathbf{H}_1 = 4\pi \times 10^{-7}(5)(-2, 6, 4)$$
$$= -12.57\mathbf{a}_x + 37.7\mathbf{a}_y + 25.13\mathbf{a}_z \,\mu \text{ Wb/m}^2$$

(b) $\mathbf{H}_{1n} = (\mathbf{H}_1 \cdot \mathbf{a}_n)\mathbf{a}_n = \left[(-2, 6, 4) \cdot \dfrac{(-1, 1, 0)}{\sqrt{2}} \right] \dfrac{(-1, 1, 0)}{\sqrt{2}}$

$$= -4\mathbf{a}_x + 4\mathbf{a}_y$$

But

$$\mathbf{H}_1 = \mathbf{H}_{1n} + \mathbf{H}_{1t}$$

Hence,

$$\mathbf{H}_{1t} = \mathbf{H}_1 - \mathbf{H}_{1n} = (-2, 6, 4) - (-4, 4, 0)$$
$$= 2\mathbf{a}_x + 2\mathbf{a}_y + 4\mathbf{a}_z$$

Using the boundary conditions, we have

$$\mathbf{H}_{2t} = \mathbf{H}_{1t} = 2\mathbf{a}_x + 2\mathbf{a}_y + 4\mathbf{a}_z$$
$$\mathbf{B}_{2n} = \mathbf{B}_{1n} \;\rightarrow\; \mu_2\mathbf{H}_{2n} = \mu_1\mathbf{H}_{1n}$$

FIGURE 8.17 For Example 8.8.

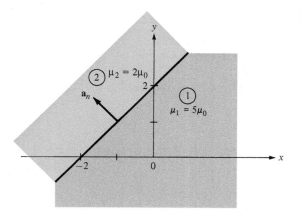

or

$$\mathbf{H}_{2n} = \frac{\mu_1}{\mu_2} \mathbf{H}_{1n} = \frac{5}{2}(-4\mathbf{a}_x + 4\mathbf{a}_y) = -10\mathbf{a}_x + 10\mathbf{a}_y$$

Thus

$$\mathbf{H}_2 = \mathbf{H}_{2n} + \mathbf{H}_{2t} = -8\mathbf{a}_x + 12\mathbf{a}_y + 4\mathbf{a}_z \text{ A/m}$$

and

$$\mathbf{B}_2 = \mu_2\mathbf{H}_2 = \mu_o\mu_{r2}\mathbf{H}_2 = (4\pi \times 10^{-7})(2)(-8, 12, 4)$$
$$= -20.11\mathbf{a}_x + 30.16\mathbf{a}_y + 10.05\mathbf{a}_z \,\mu \text{ Wb/m}^2$$

PRACTICE EXERCISE   8.8

Region 1, described by $3x + 4y \geq 10$, is free space, whereas region 2, described by $3x + 4y \leq 10$, is a magnetic material for which $\mu \simeq 10\mu_o$. Assuming that the boundary between the material and free space is current free, find $\mathbf{B}_2$ if $\mathbf{B}_1 = 0.1\mathbf{a}_x + 0.4\mathbf{a}_y + 0.2\mathbf{a}_z \text{ Wb/m}^2$.

Answer: $-1.052\mathbf{a}_x + 1.264\mathbf{a}_y + 2\mathbf{a}_z \text{ Wb/m}^2$.

EXAMPLE 8.9

The xy-plane serves as the interface between two different media. Medium 1 ($z < 0$) is filled with a material whose $\mu_r = 6$, and medium 2 ($z > 0$) is filled with a material whose $\mu_r = 4$. If the interface carries current $(1/\mu_o)\,\mathbf{a}_y$ mA/m, and $\mathbf{B}_2 = 5\mathbf{a}_x + 8\mathbf{a}_z$ mWb/m^2, find $\mathbf{H}_1$ and $\mathbf{B}_1$.

Solution:

In Example 8.8, $\mathbf{K} = \mathbf{0}$, so eq. (8.46) was appropriate. In this example, however, $\mathbf{K} \neq \mathbf{0}$, and we must resort to eq. (8.45) in addition to eq. (8.41). Consider the problem as illustrated in Figure 8.18. Let $\mathbf{B}_1 = (B_x, B_y, B_z)$ in mWb/m^2.

$$\mathbf{B}_{1n} = \mathbf{B}_{2n} = 8\mathbf{a}_z \rightarrow B_z = 8 \tag{8.9.1}$$

But

$$\mathbf{H}_2 = \frac{\mathbf{B}_2}{\mu_2} = \frac{1}{4\mu_o}(5\mathbf{a}_x + 8\mathbf{a}_z)\text{mA/m} \tag{8.9.2}$$

and

$$\mathbf{H}_1 = \frac{\mathbf{B}_1}{\mu_1} = \frac{1}{6\mu_o}(B_x\mathbf{a}_x + B_y\mathbf{a}_y + B_z\mathbf{a}_z) \text{ mA/m} \tag{8.9.3}$$

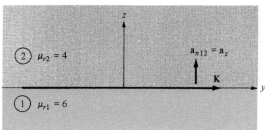

FIGURE 8.18 For Example 8.9.

Having found the normal components, we can find the tangential components by using

$$(\mathbf{H}_1 - \mathbf{H}_2) \times \mathbf{a}_{n12} = \mathbf{K}$$

or

$$\mathbf{H}_1 \times \mathbf{a}_{n12} = \mathbf{H}_2 \times \mathbf{a}_{n12} + \mathbf{K} \qquad (8.9.4)$$

Substituting eqs. (8.9.2) and (8.9.3) into eq. (8.9.4) gives

$$\frac{1}{6\mu_\text{o}}(B_x\mathbf{a}_x + B_y\mathbf{a}_y + B_z\mathbf{a}_z) \times \mathbf{a}_z = \frac{1}{4\mu_\text{o}}(5\mathbf{a}_x + 8\mathbf{a}_z) \times \mathbf{a}_z + \frac{1}{\mu_\text{o}}\mathbf{a}_y$$

Equating components yields

$$B_y = 0, \quad \frac{-B_x}{6} = \frac{-5}{4} + 1, \quad \text{or} \quad B_x = \frac{6}{4} = 1.5 \qquad (8.9.5)$$

From eqs. (8.9.1) and (8.9.5), we have

$$\mathbf{B}_1 = 1.5\mathbf{a}_x + 8\mathbf{a}_z \, \text{mWb/m}^2$$

$$\mathbf{H}_1 = \frac{\mathbf{B}_1}{\mu_1} = \frac{1}{\mu_\text{o}}(0.25\mathbf{a}_x + 1.33\mathbf{a}_z) \, \text{mA/m}$$

and

$$\mathbf{H}_2 = \frac{1}{\mu_\text{o}}(1.25\mathbf{a}_x + 2\mathbf{a}_z) \, \text{mA/m}$$

Note that H_{1x} is $1/\mu_\text{o}$ mA/m less than H_{2x} because of the current sheet and also that $B_{1n} = B_{2n}$.

PRACTICE EXERCISE 8.9

A unit normal vector from region 2 $(\mu = 2\mu_\text{o})$ to region 1 $(\mu = \mu_\text{o})$ is $\mathbf{a}_{n21} = (6\mathbf{a}_x + 2\mathbf{a}_y - 3\mathbf{a}_z)/7$. If $\mathbf{H}_1 = 10\mathbf{a}_x + \mathbf{a}_y + 12\mathbf{a}_z$ A/m and $\mathbf{H}_2 = H_{2x}\mathbf{a}_x - 5\mathbf{a}_y + 4\mathbf{a}_z$ A/m, determine

(a) $\mathbf{H}_{2x}$
(b) The surface current density **K** on the interface
(c) The angles $\mathbf{B}_1$ and $\mathbf{B}_2$ make with the normal to the interface

Answer: (a) 5.833, (b) $4.86\mathbf{a}_x - 8.64\mathbf{a}_y + 3.95\mathbf{a}_z$ A/m, (c) 76.27°, 77.62°.

8.8 INDUCTORS AND INDUCTANCES

A circuit (or closed conducting path) carrying current I produces a magnetic field **B** that causes a flux $\Psi = \int \mathbf{B} \cdot d\mathbf{S}$ to pass through each turn of the circuit as shown in Figure 8.19. If the circuit has N identical turns, we define the *flux linkage* λ as

$$\lambda = N\Psi \tag{8.50}$$

Also, if the medium surrounding the circuit is linear, the flux linkage λ is proportional to the current I producing it; that is,

$$\lambda \propto I$$

or

$$\lambda = LI \tag{8.51}$$

where L is a constant of proportionality called the *inductance* of the circuit. The inductance L is a property of the physical arrangement of the circuit. It is the ability of the physical arrangement to store magnetic energy. A circuit or part of a circuit that has inductance is called an *inductor*. The **inductance** L of an inductor is the ratio of the magnetic flux linkage λ to the current I through the inductor.

$$\boxed{L = \frac{\lambda}{I} = \frac{N\Psi}{I}} \tag{8.52}$$

The unit of inductance is the henry (H), which is the same as webers per ampere. Since the henry is a fairly large unit, inductances are usually expressed in millihenrys (mH).

FIGURE 8.19 Magnetic field **B** produced by a circuit.

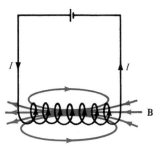

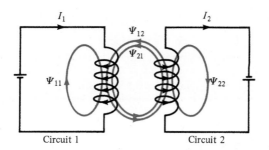

FIGURE 8.20 Magnetic interaction between two circuits.

The inductance defined by eq. (8.52) is commonly referred to as *self-inductance,* since the linkages are produced by the inductor itself. Like capacitance, inductance may be regarded as a measure of how much magnetic energy is stored in an inductor. The magnetic energy (in joules) stored in an inductor is expressed in circuit theory as

$$W_m = \frac{1}{2}LI^2 \tag{8.53}$$

or

$$\boxed{L = \frac{2W_m}{I^2}} \tag{8.54}$$

Thus the self-inductance of a circuit may be defined or calculated from energy considerations.

If instead of having a single circuit, we have two circuits carrying current I_1 and I_2 as shown in Figure 8.20, a magnetic interaction exists between the circuits. Four component fluxes Ψ_{11}, Ψ_{12}, Ψ_{21}, and Ψ_{22} are produced. The flux Ψ_{12}, for example, is the flux passing through circuit 1 due to current I_2 in circuit 2. If $\mathbf{B}_2$ is the magnetic flux density due to I_2 and S_1 is the area of circuit 1, then

$$\Psi_{12} = \int_{S_1} \mathbf{B}_2 \cdot d\mathbf{S} \tag{8.55}$$

The *mutual inductance* M_{12} is the ratio of the flux linkage $\lambda_{12} = N_1\Psi_{12}$ on circuit 1 to current I_2.

$$\boxed{M_{12} = \frac{\lambda_{12}}{I_2} = \frac{N_1\Psi_{12}}{I_2}} \tag{8.56}$$

Similarly, the mutual inductance M_{21} is defined as the flux linkages of circuit 2 per unit current I_1; that is,

$$M_{21} = \frac{\lambda_{21}}{I_1} = \frac{N_2\Psi_{21}}{I_1} \tag{8.57a}$$

It can be shown by using energy concepts that if the medium surrounding the circuits is linear (i.e., in the absence of ferromagnetic material),

$$M_{12} = M_{21} \tag{8.57b}$$

The mutual inductance M_{12} or M_{21} is expressed in henrys and should not be confused with the magnetization vector $\mathbf{M}$ expressed in amperes per meter. Mutual inductance is fundamental to the operation of transformers.

We define the self-inductance of circuits 1 and 2, respectively, as

$$L_1 = \frac{\lambda_{11}}{I_1} = \frac{N_1 \Psi_1}{I_1} \tag{8.58}$$

and

$$L_2 = \frac{\lambda_{22}}{I_2} = \frac{N_2 \Psi_2}{I_2} \tag{8.59}$$

where $\Psi_1 = \Psi_{11} + \Psi_{12}$ and $\Psi_2 = \Psi_{21} + \Psi_{22}$. The total energy in the magnetic field is the sum of the energies due to L_1, L_2, and M_{12} (or M_{21}); that is,

$$W_m = W_1 + W_2 + W_{12}$$
$$= \frac{1}{2} L_1 I_1^2 + \frac{1}{2} L_2 I_2^2 \pm M_{12} I_1 I_2 \tag{8.60}$$

The positive sign is taken if currents I_1 and I_2 flow such that the magnetic fields of the two circuits strengthen each other. If the currents flow such that their magnetic fields oppose each other, the negative sign is taken.

As mentioned earlier, an inductor is a conductor arranged in a shape appropriate to store magnetic energy. Typical examples of inductors are toroids, solenoids, coaxial transmission lines, and parallel-wire transmission lines. The inductance of each of these inductors can be determined by following a procedure similar to that taken in determining the capacitance of a capacitor. For a given inductor, we find the self-inductance L by taking these steps:

1. Choose a suitable coordinate system.
2. Let the inductor carry current I.
3. Determine $\mathbf{B}$ from Biot–Savart's law (or from Ampère's law if symmetry exists) and calculate Ψ from $\Psi = \int_s \mathbf{B} \cdot d\mathbf{S}$.
4. Finally find L from $L = \dfrac{\lambda}{I} = \dfrac{N\Psi}{I}$.

The mutual inductance between two circuits may be calculated by taking a similar procedure.

In an inductor such as a coaxial or a parallel-wire transmission line, the inductance produced by the flux internal to the conductor is called the *internal inductance* L_{in} while that produced by the flux external to it is called *external inductance* L_{ext}. The total inductance L is

$$L = L_{in} + L_{ext} \tag{8.61}$$

Just as it was shown that for capacitors

$$RC = \frac{\varepsilon}{\sigma} \tag{6.35}$$

it can be shown that

$$\boxed{L_{ext}C = \mu\varepsilon} \tag{8.62}$$

Thus L_{ext} may be calculated using eq. (8.62) if C is known.

A collection of formulas for some fundamental circuit elements is presented in Table 8.3. All formulas can be derived by taking the steps just outlined.[3]

8.9 MAGNETIC ENERGY

Just as the potential energy in an electrostatic field was derived as

$$W_E = \frac{1}{2}\int \mathbf{D} \cdot \mathbf{E}\, dv = \frac{1}{2}\int \varepsilon E^2\, dv \tag{4.96}$$

we would like to derive a similar expression for the energy in a magnetostatic field. A simple approach is using the magnetic energy in the field of an inductor. From eq. (8.53),

$$W_m = \frac{1}{2}LI^2 \tag{8.53}$$

The energy is stored in the magnetic field **B** of the inductor. We would like to express eq. (8.53) in terms of **B** or **H**.

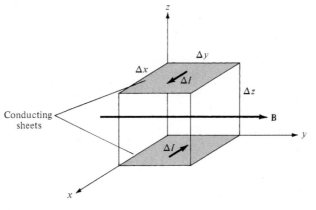

FIGURE 8.21 A differential volume in a magnetic field.

---

[3] Additional formulas can be found in standard electrical handbooks or in H. Knoepfel, *Pulsed High Magnetic Fields.* Amsterdam: North-Holland, 1970, pp. 312–324.

TABLE 8.3 A Collection of Formulas for Inductance of Common Elements

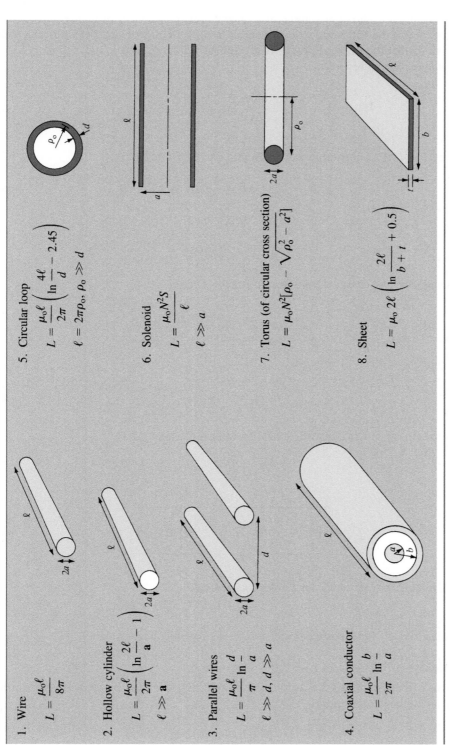

1. Wire

$$L = \frac{\mu_o \ell}{8\pi}$$

2. Hollow cylinder

$$L = \frac{\mu_o \ell}{2\pi} \left(\ln \frac{2\ell}{\mathbf{a}} - 1 \right)$$

$$\ell \gg \mathbf{a}$$

3. Parallel wires

$$L = \frac{\mu_o \ell}{\pi} \ln \frac{d}{a}$$

$$\ell \gg d, d \gg a$$

4. Coaxial conductor

$$L = \frac{\mu_o \ell}{2\pi} \ln \frac{b}{a}$$

5. Circular loop

$$L = \frac{\mu_o \ell}{2\pi} \left(\ln \frac{4\ell}{d} - 2.45 \right)$$

$$\ell = 2\pi \rho_o, \ \rho_o \gg d$$

6. Solenoid

$$L = \frac{\mu_o N^2 S}{\ell}$$

$$\ell \gg a$$

7. Torus (of circular cross section)

$$L = \mu_o N^2 \left[\rho_o - \sqrt{\rho_o^2 - a^2} \right]$$

8. Sheet

$$L = \mu_o 2\ell \left(\ln \frac{2\ell}{b + t} + 0.5 \right)$$

Consider a differential volume in a magnetic field as shown in Figure 8.21. Let the volume be covered with conducting sheets at the top and bottom surfaces with current ΔI. We assume that the whole region is filled with such differential volumes. From eq. (8.52), each volume has an inductance

$$\Delta L = \frac{\Delta \Psi}{\Delta I} = \frac{\mu H \, \Delta x \, \Delta z}{\Delta I} \tag{8.63}$$

where $\Delta I = H \, \Delta y$. Substituting eq. (8.63) into eq. (8.53), we have

$$\Delta W_m = \frac{1}{2}\Delta L \, \Delta I^2 = \frac{1}{2} \mu H^2 \, \Delta x \, \Delta y \, \Delta z \tag{8.64}$$

or

$$\Delta W_m = \frac{1}{2} \mu H^2 \, \Delta v$$

The magnetostatic energy density w_m (in J/m^3) is defined as

$$w_m = \lim_{\Delta v \to 0} \frac{\Delta W_m}{\Delta v} = \frac{1}{2} \mu H^2$$

Hence,

$$w_m = \frac{1}{2}\mu H^2 = \frac{1}{2} \mathbf{B} \cdot \mathbf{H} = \frac{B^2}{2\mu} \tag{8.65}$$

Thus the energy in a magnetostatic field in a linear medium is

$$W_m = \int_v w_m \, dv$$

or

$$\boxed{W_m = \frac{1}{2}\int_v \mathbf{B} \cdot \mathbf{H} \, dv = \frac{1}{2}\int_v \mu H^2 \, dv} \tag{8.66}$$

which is similar to eq. (4.96) for an elctrostatic field.

EXAMPLE 8.10

Calculate the self-inductance per unit length of an infinitely long solenoid.

Solution:

We recall from Example 7.4 that for an infinitely long solenoid, the magnetic flux inside the solenoid per unit length is

$$B = \mu H = \mu In$$

where $n = N/\ell$ = number of turns per unit length. If S is the cross-sectional area of the solenoid, the total flux through the cross section is

$$\Psi = BS = \mu InS$$

Since this flux is only for a unit length of the solenoid, the linkage per unit length is

$$\lambda' = \frac{\lambda}{\ell} = n\Psi = \mu n^2 IS$$

and thus the inductance per unit length is

$$L' = \frac{L}{\ell} = \frac{\lambda'}{I} = \mu n^2 S$$

$$\boxed{L' = \mu n^2 S} \quad \text{H/m}$$

PRACTICE EXERCISE 8.10

A very long solenoid with 2×2 cm cross section has an iron core $(\mu_r = 1000)$ and 4000 turns per meter. It carries a current of 500 mA. Find the following:

(a) Its self-inductance per meter
(b) The energy per meter stored in its field

Answer: (a) 8.042 H/m, (b) 1.005 J/m.

EXAMPLE 8.11

Determine the self-inductance of a coaxial cable of inner radius a and outer radius b.

Solution:

The self-inductance of the inductor can be found in two different ways: by taking the four steps given in Section 8.8 or by using eqs. (8.54) and (8.66).

Method 1: Consider the cross section of the cable as shown in Figure 8.22. We recall from eq. (7.29) that by applying Ampère's circuit law, we obtained for region 1 $(0 \le \rho \le a)$,

$$\mathbf{B}_1 = \frac{\mu I \rho}{2\pi a^2} \mathbf{a}_\phi$$

and for region 2 $(a \le \rho \le b)$,

$$\mathbf{B}_2 = \frac{\mu I}{2\pi \rho} \mathbf{a}_\phi$$

We first find the internal inductance L_{in} by considering the flux linkages due to the inner conductor. From Figure 8.22(a), the flux leaving a differential shell of thickness $d\rho$ is

$$dΨ_1 = B_1 \, dρ \, dz = \frac{μIρ}{2πa^2} \, dρ \, dz$$

The flux linkage is $dΨ_1$ multiplied by the ratio of the area within the path enclosing the flux to the total area, that is,

$$dλ_1 = dΨ_1 \cdot \frac{I_{enc}}{I} = dΨ_1 \cdot \frac{πρ^2}{πa^2}$$

because I is uniformly distributed over the cross section for dc excitation. Thus, the total flux linkages within the differential flux element are

$$dλ_1 = \frac{μIρ \, dρ \, dz}{2πa^2} \cdot \frac{ρ^2}{a^2}$$

For length $ℓ$ of the cable,

$$λ_1 = \int_{ρ=0}^{a} \int_{z=0}^{ℓ} \frac{μIρ^3 \, dρ \, dz}{2πa^4} = \frac{μIℓ}{8π}$$

$$L_{in} = \frac{λ_1}{I} = \frac{μℓ}{8π} \tag{8.11.1}$$

The internal inductance per unit length, given by

$$\boxed{L'_{in} = \frac{L_{in}}{ℓ} = \frac{μ}{8π} \quad \text{H/m}} \tag{8.11.2}$$

is independent of the radius of the conductor or wire. Since the inductance does not depend on a, we can make the wire as thin as possible. Thus eqs. (8.11.1) and (8.11.2) are also applicable to finding the inductance of any infinitely long straight conductor of finite radius.

We now determine the external inductance L_{ext} by considering the flux linkages between the inner and the outer conductor as in Figure 8.22(b). For a differential shell of thickness $dρ$,

$$dΨ_2 = B_2 \, dρ \, dz = \frac{μI}{2πρ} \, dρ \, dz$$

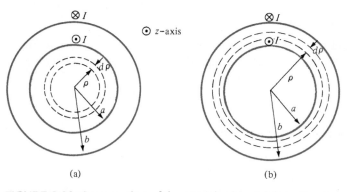

(a)   (b)

FIGURE 8.22 Cross section of the coaxial cable: (**a**) for region 1, $0 < ρ < a$, (**b**) for region 2, $a < ρ < b$; for Example 8.11.

In this case, the total current I is enclosed within the path enclosing the flux. Hence,

$$\lambda_2 = \Psi_2 = \int_{\rho=a}^{b} \int_{z=0}^{\ell} \frac{\mu I \, d\rho \, dz}{2\pi\rho} = \frac{\mu I \ell}{2\pi} \ln\frac{b}{a}$$

$$L_{ext} = \frac{\lambda_2}{I} = \frac{\mu\ell}{2\pi} \ln\frac{b}{a}$$

Thus

$$L = L_{in} + L_{ext} = \frac{\mu\ell}{2\pi} \left[\frac{1}{4} + \ln\frac{b}{a}\right]$$

or the inductance per length is

$$\boxed{L' = \frac{L}{\ell} = \frac{\mu}{2\pi} \left[\frac{1}{4} + \ln\frac{b}{a}\right]} \quad \text{H/m}$$

Method 2: It is easier to use eqs. (8.54) and (8.66) to determine L, that is,

$$W_m = \frac{1}{2} L I^2 \quad \text{or} \quad L = \frac{2W_m}{I^2}$$

where

$$W_m = \frac{1}{2} \int \mathbf{B} \cdot \mathbf{H} \, dv = \int \frac{B^2}{2\mu} \, dv$$

Hence

$$L_{in} = \frac{2}{I^2} \int_v \frac{B_1^2}{2\mu} \, dv = \frac{1}{I^2\mu} \iiint \frac{\mu^2 I^2 \rho^2}{4\pi^2 a^4} \rho \, d\rho \, d\phi \, dz$$

$$= \frac{\mu}{4\pi^2 a^4} \int_0^{\ell} dz \int_0^{2\pi} d\phi \int_0^{a} \rho^3 \, d\rho = \frac{\mu\ell}{8\pi}$$

$$L_{ext} = \frac{2}{I^2} \int_v \frac{B_2^2}{2\mu} \, dv = \frac{1}{I^2\mu} \iiint \frac{\mu^2 I^2}{4\pi^2 \rho^2} \rho \, d\rho \, d\phi \, dz$$

$$= \frac{\mu}{4\pi^2} \int_0^{\ell} dz \int_0^{2\pi} d\phi \int_a^{b} \frac{d\rho}{\rho} = \frac{\mu\ell}{2\pi} \ln\frac{b}{a}$$

and

$$L = L_{in} + L_{ext} = \frac{\mu\ell}{2\pi} \left[\frac{1}{4} + \ln\frac{b}{a}\right]$$

as obtained previously.

EXAMPLE 8.12

Determine the inductance per unit length of a two-wire transmission line with separation distance d. Each wire has radius a as shown in Figure 11.2 (b).

Solution:

We use the two methods of Example 8.11.

Method 1: We determine L_{in} just as we did in Example 8.11. Thus for region $0 \le \rho \le a$, we obtain

$$\lambda_1 = \frac{\mu I\ell}{8\pi}$$

as before. For region $a \le \rho \le d - a$, the flux linkages between the wires are

$$\lambda_2 = \Psi_2 = \int_{\rho=a}^{d-a}\int_{z=0}^{\ell} \frac{\mu I}{2\pi\rho}\,d\rho\,dz = \frac{\mu I\ell}{2\pi}\ln\frac{d-a}{a}$$

The flux linkages produced by wire 1 are

$$\lambda_1 + \lambda_2 = \frac{\mu I\ell}{8\pi} + \frac{\mu I\ell}{2\pi}\ln\frac{d-a}{a}$$

By symmetry, the same amount of flux is produced by current $-I$ in wire 2. Hence the total linkages are

$$\lambda = 2(\lambda_1 + \lambda_2) = \frac{\mu I\ell}{\pi}\left[\frac{1}{4} + \ln\frac{d-a}{a}\right] = LI$$

If $d \gg a$, the self-inductance per unit length is

$$\boxed{L' = \frac{L}{\ell} = \frac{\mu}{\pi}\left[\frac{1}{4} + \ln\frac{d}{a}\right]} \quad \text{H/m}$$

Method 2: From Example 8.11, we have

$$L_{in} = \frac{\mu\ell}{8\pi}$$

Now

$$L_{ext} = \frac{2}{I^2} \int \frac{B^2 \, dv}{2\mu} = \frac{1}{I^2\mu} \iiint \frac{\mu^2 I^2}{4\pi^2\rho^2} \rho \, d\rho \, d\phi \, dz$$

$$= \frac{\mu}{4\pi^2} \int_0^\ell dz \int_0^{2\pi} d\phi \int_a^{d-a} \frac{d\rho}{\rho}$$

$$= \frac{\mu\ell}{2\pi} \ln \frac{d-a}{a}$$

Since the two wires are symmetrical,

$$L = 2 \left(L_{in} + L_{ext} \right)$$

$$= \frac{\mu\ell}{\pi} \left[\frac{1}{4} + \ln \frac{d-a}{a} \right] \text{H}$$

as obtained earlier.

PRACTICE EXERCISE   8.12

Two #10 copper wires (2.588 mm in diameter) are placed parallel in air with a separation distance d between them. If the inductance of each wire is 1.2 μH/m, calculate

(a) L_{in} and L_{ext} per meter for each wire
(b) The separation distance d

Answer: (a) 0.05 and 115 μH/m, (b) 40.79 cm.

EXAMPLE 8.13

Two coaxial circular wires of radii a and b $(b > a)$ are separated by distance h $(h \gg a, b)$ as shown in Figure 8.23. Find the mutual inductance between the wires.

Solution:

Let current I_1 flow in wire 1. At an arbitrary point P on wire 2, the magnetic vector potential due to wire 1 is given by eq. (8.21a), namely

$$\mathbf{A}_1 = \frac{\mu I_1 a^2 \sin\theta}{4r^2} \mathbf{a}_\phi = \frac{\mu I_1 a^2 b \mathbf{a}_\phi}{4\left[h^2 + b^2\right]^{3/2}}$$

If $h \gg b$

$$\mathbf{A}_1 \simeq \frac{\mu I_1 a^2 b}{4h^3} \mathbf{a}_\phi$$

Hence,

$$\Psi_{12} = \oint \mathbf{A}_1 \cdot d\mathbf{l}_2 = \frac{\mu I_1 a^2 b}{4h^3} 2\pi b = \frac{\mu \pi I_1 a^2 b^2}{2h^3}$$

and

$$M_{12} = \frac{\Psi_{12}}{I_1} = \frac{\mu \pi a^2 b^2}{2h^3}$$

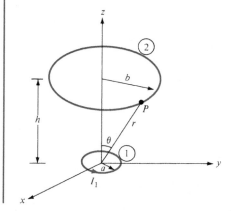

FIGURE 8.23 Two coaxial circular wires; for Example 8.13.

PRACTICE EXERCISE   8.13

Find the mutual inductance of two coplanar concentric circular loops of radii 2 m and 3 m.

Answer: 2.632 μH.

---

†8.10 MAGNETIC CIRCUITS

The concept of magnetic circuits is based on solving some magnetic field problems by using a circuit approach. Magnetic devices such as toroids, transformers, motors, generators, and relays may be considered as magnetic circuits. The analysis of such circuits is made simple if an analogy between magnetic circuits and electric circuits is exploited. Once this has been done, we can directly apply concepts in electric circuits to solve their analogous magnetic circuits.

The analogy between magnetic and electric circuits is summarized in Table 8.4 and portrayed in Figure 8.24. The reader is advised to pause and study Table 8.4 and Figure 8.24. First, we notice from Table 8.4 that two terms are new. We define the *magnetomotive force* (mmf) $\mathscr{F}$, in ampere-turns ($A \cdot t$), as

$$\mathscr{F} = NI = \oint \mathbf{H} \cdot d\mathbf{l} \qquad (8.67)$$

The source of mmf in magnetic circuits is usually a coil-carrying current as in Figure 8.24. We also define *reluctance* $\mathscr{R}$, in ampere-turns per weber, as

$$\mathscr{R} = \frac{\ell}{\mu S} \qquad (8.68)$$

where ℓ and S are, respectively, the mean length and the cross-sectional area of the magnetic core. The reciprocal of reluctance is *permeance* $\mathscr{P}$. The basic relationship for circuit elements is Ohm's law $(V = IR)$:

$$\mathscr{F} = \Psi \mathscr{R} \qquad (8.69)$$

Based on this, Kirchhoff's current and voltage laws can be applied to nodes and loops of a given magnetic circuit just as in an electric circuit. The rules of adding voltages and for

TABLE 8.4 Analogy between Electric and Magnetic Circuits

Electric	Magnetic
Conductivity σ	Permeability μ
Field intensity E	Field intensity H
Current $I = \int \mathbf{J} \cdot d\mathbf{S}$	Magnetic flux $\Psi = \int \mathbf{B} \cdot d\mathbf{S}$
Current density $J = \dfrac{I}{S} = \sigma E$	Flux density $B = \dfrac{\Psi}{S} = \mu H$
Electromotive force (emf) V	Magnetomotive force (mmf) $\mathscr{F}$
Resistance R	Reluctance $\mathscr{R}$
Conductance $G = \dfrac{1}{R}$	Permeance $\mathscr{P} = \dfrac{1}{\mathscr{R}}$
Ohm's law $R = \dfrac{V}{I} = \dfrac{\ell}{\sigma S}$	Ohm's law $\mathscr{R} = \dfrac{\mathscr{F}}{\Psi} = \dfrac{\ell}{\mu S}$
$\qquad$ or $\quad V = E\ell = IR$	$\qquad$ or $\quad \mathscr{F} = H\ell = \Psi \mathscr{R} = NI$
Kirchhoff's laws:	Kirchhoff's laws:
$\qquad \Sigma I = 0$	$\qquad \Sigma \Psi = 0$
$\Sigma V - \Sigma RI = 0$	$\Sigma \mathscr{F} - \Sigma \mathscr{R} \Psi = 0$

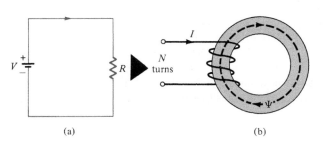

(a)   (b)

FIGURE 8.24 Analogy between (**a**) an electric circuit and (**b**) a magnetic circuit.

combining series and parallel resistances also hold for mmfs and reluctances. Thus for n magnetic circuit elements in series

$$\Psi_1 = \Psi_2 = \Psi_3 = \cdots = \Psi_n \tag{8.70}$$

and

$$\mathcal{F} = \mathcal{F}_1 + \mathcal{F}_2 + \cdots + \mathcal{F}_n \tag{8.71}$$

For n magnetic circuit elements in parallel,

$$\Psi = \Psi_1 + \Psi_2 + \Psi_3 + \cdots + \Psi_n \tag{8.72}$$

and

$$\mathcal{F}_1 = \mathcal{F}_2 = \mathcal{F}_3 = \cdots = \mathcal{F}_n \tag{8.73}$$

Some differences between electric and magnetic circuits should be pointed out. Unlike an electric circuit, where the current I flows, magnetic flux does not flow. Also, conductivity σ is independent of current density J in an electric circuit, whereas permeability μ varies with flux density B in a magnetic circuit. This is because ferromagnetic (nonlinear) materials are normally used in most practical magnetic devices. These differences notwithstanding, the magnetic circuit concept serves in the approximate analysis of practical magnetic devices.

†8.11 FORCE ON MAGNETIC MATERIALS

It is of practical interest to determine the force that a magnetic field exerts on a piece of magnetic material in the field. This is useful in electromechanical systems such as electromagnets, relays, and rotating machines and in magnetic levitation (see Section 8.12). Consider, for example, an electromagnet made of iron of constant relative permeability as shown in Figure 8.25. The coil has N turns and carries a current I. If we ignore fringing, the magnetic field in the air gap is the same as that in iron $(B_{1n} = B_{2n})$. To find the force between the two pieces of iron, we calculate the change in the total energy that would result were the two pieces of the magnetic circuit separated by a differential displacement $d\mathbf{l}$. The work required to effect the displacement is equal to the change in stored energy in the air gap (assuming constant current), that is,

$$-F \, dl = dW_m = 2\left[\frac{1}{2}\frac{B^2}{\mu_o} S \, dl\right] \tag{8.74}$$

where S is the cross-sectional area of the gap, the factor 2 accounts for the two air gaps, and the negative sign indicates that the force acts to reduce the air gap (or that the force is attractive). Thus

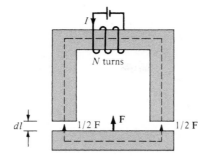

FIGURE 8.25 An electromagnet.

$$F = -2\left(\frac{B^2 S}{2\mu_o}\right) \tag{8.75}$$

Note that the force is exerted on the lower piece and not on the current-carrying upper piece giving rise to the field. The tractive force across a *single* gap can be obtained from eq. (8.75) as

$$F = -\frac{B^2 S}{2\mu_o} \tag{8.76}$$

Notice the similarity between eq. (8.76) and that derived in Example 5.8 for electrostatic case. Equation (8.76) can be used to calculate the forces in many types of devices including relays, rotating machines, and magnetic levitation. The tractive pressure (in N/m^2) in a magnetized surface is

$$p = \frac{F}{S} = \frac{B^2}{2\mu_o} = \frac{1}{2}BH \tag{8.77}$$

which is the same as the enery density w_m in the air gap.

EXAMPLE 8.14

The toroidal core of Figure 8.26(a) has $\rho_o = 10$ cm and a circular cross section with $a = 1$ cm. If the core is made of steel ($\mu = 1000\ \mu_o$) and has a coil with 200 turns, calculate the amount of current that will produce a flux of 0.5 mWb in the core.

Solution:

This problem can be solved in two different ways: by using the magnetic field approach (direct) or by using the electric circuit analog (indirect).

Method 1: Since ρ_o is large compared with a, from Example 7.6,

$$B = \frac{\mu NI}{\ell} = \frac{\mu_o \mu_r NI}{2\pi\rho_o}$$

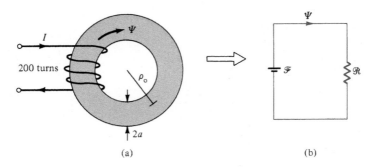

FIGURE 8.26 (**a**) Toroidal core of Example 8.14. (**b**) Its equivalent electric circuit analog.

Hence,

$$\Psi = BS = \frac{\mu_o \mu_r NI \, \pi a^2}{2\pi \rho_o}$$

or

$$I = \frac{2\rho_o \Psi}{\mu_o \mu_r Na^2} = \frac{2(10 \times 10^{-2})(0.5 \times 10^{-3})}{4\pi \times 10^{-7}(1000)(200)(1 \times 10^{-4})}$$

$$= \frac{100}{8\pi} \quad = 3.979 \text{ A}$$

Method 2: The toroidal core in Figure 8.26(a) is analogous to the electric circuit of Figure 8.26(b). From the circuit and Table 8.4,

$$\mathscr{F} = NI = \Psi \mathscr{R} = \Psi \frac{\ell}{\mu S} = \Psi \frac{2\pi \rho_o}{\mu_o \mu_r \pi a^2}$$

or

$$I = \frac{2\rho_o \Psi}{\mu_o \mu_r Na^2} = 3.979 \text{ A}$$

as obtained with Method 1.

PRACTICE EXERCISE 8.14

A conductor of radius a is bent into a circular loop of mean radius ρ_o (see Figure 8.26(a)). If $\rho_o = 10$ cm and $2a = 1$ cm, calculate the internal inductance of the loop.

Answer: 31.42 nH.

EXAMPLE 8.15

In the magnetic circuit of Figure 8.27, calculate the current in the coil that will produce a magnetic flux density of 1.5 Wb/m² in the air gap, assuming that $\mu = 50\mu_o$ and that all branches have the same cross-sectional area of 10 cm².

Solution:

The magnetic circuit of Figure 8.27 is analogous to the electric circuit of Figure 8.28. In Figure 8.27, $\mathcal{R}_1$, $\mathcal{R}_2$, $\mathcal{R}_3$, and $\mathcal{R}_a$ are the reluctances in paths 143, 123, 35 and 16, and 56 (air gap), respectively. Thus

$$\mathcal{R}_1 = \mathcal{R}_2 = \frac{\ell}{\mu_o\mu_r S} = \frac{30 \times 10^{-2}}{(4\pi \times 10^{-7})(50)(10 \times 10^{-4})}$$

$$= \frac{3 \times 10^8}{20\pi}$$

$$\mathcal{R}_3 = \frac{9 \times 10^{-2}}{(4\pi \times 10^{-7})(50)(10 \times 10^{-4})} = \frac{0.9 \times 10^8}{20\pi}$$

$$\mathcal{R}_a = \frac{1 \times 10^{-2}}{(4\pi \times 10^{-7})(1)(10 \times 10^{-4})} = \frac{5 \times 10^8}{20\pi}$$

We combine $\mathcal{R}_1$ and $\mathcal{R}_2$ as resistors in parallel. Hence,

$$\mathcal{R}_1 \| \mathcal{R}_2 = \frac{\mathcal{R}_1\mathcal{R}_2}{\mathcal{R}_1 + \mathcal{R}_2} = \frac{\mathcal{R}_1}{2} = \frac{1.5 \times 10^8}{20\pi}$$

The total reluctance is

$$\mathcal{R}_T = \mathcal{R}_a + \mathcal{R}_3 + \mathcal{R}_1 \| \mathcal{R}_2 = \frac{7.4 \times 10^8}{20\pi}$$

The mmf is

$$\mathcal{F} = NI = \Psi_a \mathcal{R}_T$$

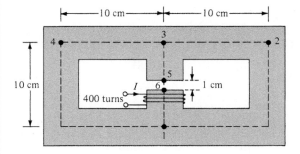

FIGURE 8.27 Magnetic circuit for Example 8.15.

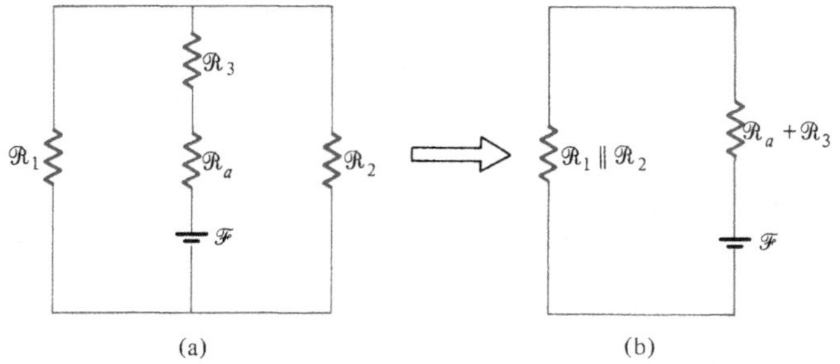

FIGURE 8.28 Electric circuit analog of the magnetic circuit in Figure 8.27.

But $\Psi_a = \Psi = B_a S$. Hence

$$I = \frac{B_a S \mathcal{R}_T}{N} = \frac{1.5 \times 10 \times 10^{-4} \times 7.4 \times 10^8}{400 \times 20\pi}$$

$$= 44.16 \text{ A}$$

PRACTICE EXERCISE 8.15

The toroid of Figure 8.26(a) has a coil of 1000 turns wound on its core. If $\rho_o = 10$ cm and $a = 1$ cm, find the current required to establish a magnetic flux of 0.5 mWb

(a) If the core is nonmagnetic
(b) If the core has $\mu_r = 500$

Answer: (a) 795.8 A, (b) 1.592 A.

EXAMPLE 8.16

A U-shaped electromagnet shown in Figure 8.29 is designed to lift a 400 kg mass (which includes the mass of the keeper). The iron yoke ($\mu_r = 3000$) has a cross section of 40 cm^2 and mean length of 50 cm, and the air gaps are each 0.1 mm long. Neglecting the reluctance of the keeper, calculate the number of turns in the coil when the excitation current is 1 A.

Solution:

The tractive force across the two air gaps must balance the weight. Hence

$$F = 2\frac{(B_a^2 S)}{2\mu_o} = mg$$

or

$$B_a^2 = \frac{mg\mu_o}{S} = \frac{400 \times 9.8 \times 4\pi \times 10^{-7}}{40 \times 10^{-4}}$$

$$B_a = 1.11 \text{ Wb/m}^2$$

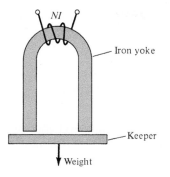

FIGURE 8.29 U-shaped electromagnet; for Example 8.16.

Iron yoke

Keeper

Weight

But

$$\mathcal{F} = NI = \Psi(\mathcal{R}_a + \mathcal{R}_i)$$

$$\mathcal{R}_a = \frac{\ell_a}{\mu S} = \frac{2 \times 0.1 \times 10^{-3}}{4\pi \times 10^{-7} \times 40 \times 10^{-4}} = \frac{6 \times 10^6}{48\pi}$$

$$\mathcal{R}_i = \frac{\ell_i}{\mu_o \mu_r S} = \frac{50 \times 10^{-2}}{4\pi \times 10^{-7} \times 3000 \times 40 \times 10^{-4}} = \frac{5 \times 10^6}{48\pi}$$

$$\mathcal{F}_a = \frac{\mathcal{R}_a}{\mathcal{R}_a + \mathcal{R}_i} \mathcal{F} = \frac{6}{6 + 5} NI = \frac{6}{11} NI$$

Since

$$\mathcal{F}_a = H_a \ell_a = \frac{B_a \ell_a}{\mu_o}$$

$$N = \frac{11}{6} \frac{B_a \ell_a}{\mu_o I} = \frac{11 \times 1.11 \times 0.1 \times 10^{-3}}{6 \times 4\pi \times 10^{-7} \times 1}$$

$$N = 162$$

PRACTICE EXERCISE   8.16

Find the force across the air gap of the magnetic circuit of Example 8.15.

Answer: 895.2 N.

---

†8.12 APPLICATION NOTE—MAGNETIC LEVITATION

Overcoming the grip of earth's gravity has been a major challenge for years. However, scientists and engineers have found many ways to achieve levitation. For example, a helicopter may be regarded as a levitation device, one that uses a stream of air to keep the aircraft floating.

> **Magnetic levitation** (maglev) is a way of using electromagnetic fields to levitate things without any noise or the need for liquid fuel or air.

Thus, maglev is the means of floating one magnet over another. According to a theorem due to Earnshaw, it is impossible to achieve static levitation by means of any combination of fixed magnets and electric charges. Static levitation means stable suspension of an object against gravity. There are, however, ways to levitate by getting round the assumptions of the theorem. Magnetic levitation employs diamagnetism, an intrinsic property of many materials referring to their ability to expel temporarily a portion of an external magnetic field. As a result, diamagnetic materials repel and are repelled by strong magnetic fields.

Superconductors are ideal diamagnetics and completely expel magnetic fields at low temperatures. It is possible to levitate superconductors and other diamagnetic materials. This property is also used in maglev trains. It has become commonplace to see the new high-temperature superconducting materials levitated in this way. A superconductor is perfectly diamagnetic, which means that it expels a magnetic field. Other diamagnetic materials are commonplace and can also be levitated in a magnetic field if it is strong enough.

There are two types of maglev: electromagnetic levitation (EML), which uses the attractive force between electromagnets on the levitated object and the circuit on the ground, and electrodynamic levitation (EDL), which makes use of the repulsive force between magnets (superconductive magnets) on the levitated object and induced current in the secondary circuit on the ground. Any type of maglev system consists of three subsystems: a magnetic suspension, a propulsion motor, and a power system. The magnetic suspension is supposed to ensure a stable suspension of a vehicle in its own magnetic field. The propulsion motor should produce a propulsion force sufficient for a continuous flight of the vehicle along an assigned track with a given speed. The power system provides uninterrupted power supply.

As discussed in Section 8.5, materials in a magnetic field will become magnetized. Most materials such as water, wood, and plastic are diamagnetic, which means that they are repelled by magnetic fields. This repulsive force is, however, very weak compared with the attractive force a ferromagnetic material such as iron will experience in a magnetic field. As shown in Figure 8.30, if the repulsive force due to a magnetic field on a diamagnetic object is exactly equal to the weight of the object, then the object may be levitated in air. The magnetic fields required for this type of levitation are very large, typically 17 T. To produce such large fields requires using superconductive magnets. Thus, maglev relies on superconductors in practical applications.

Levitating trains and levitating displays are but two examples of electromagnetic levitation. The need for fast and reliable transportation is increasing throughout the world. High-speed rail has been the solution for many countries. A maglev train is a train-like vehicle that is suspended in the air above the track and propelled forward using the repulsive and attractive forces of magnetism, as shown in Figure 8.31. Trains are fast, comfortable, and energy efficient. Conventional railroads operate at speeds below 300 km/hr, while maglev vehicles are designed for operating speeds of up to 500 km/hr. A major

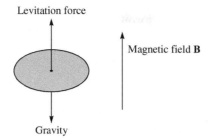

Levitation force

Gravity

Magnetic field **B**

FIGURE 8.30 A levitated object.

FIGURE 8.31 Maglev train.

advantage of maglev systems is their ability to operate in almost all weather conditions; they are prepared for icy conditions because they do not require overhead power lines—parts that are subject to freezing on conventional railroads.

Maglev technology is a reality. Japan and Germany have invested billions of dollars into the research and development of their maglev systems. In the United States, communities from Florida to California are considering building maglev systems.

8.13 APPLICATION NOTE—SQUIDs

SQUIDs are superconducting quantum interference devices. They are the most sensitive detectors of magnetic flux known. They are quite versatile, capable of measuring any physical quantity that can be eventually converted to magnetic flux. It is a highly sensitive magnetometer capable of measuring magnetic fields as feeble as even 5×10^{-18} T. The need for accurate measurement of magnetic field arises in various applications like geomagnetism, biomagnetism, magnetic microscopy, space magnetometry, magneto-cardiography, nuclear magnetic resonance (NMR) or low magnetic field magnetic resonance imaging (MRI), metrological application, magnetic microscopy, and nondestructive testing or evaluation and magnetic anomaly detection. It is hypothesized that certain animals generate tiny levels of magnetic flux from their brain in order to navigate.

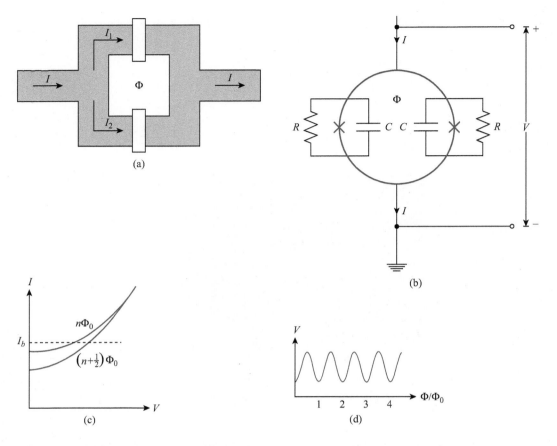

FIGURE 8.32 (a) The structure of a DC SQUID, (b) schematic and the shunt resistive load, (c) V-I characteristic for $\Phi = n\Phi_0$ and for $\Phi = (n + \frac{1}{2})\Phi_0$ cases, (d) voltage-flux response characteristics.

SQUIDs greatly help in such investigations. Environmental magnetic noise is likely to corrupt the magnetic field measurements unless the measuring device has a high figure of merit; hence, SQUIDs are proposed as a viable option.

There are two central ideas in the development of SQUIDs: (i) magnetic flux quantization and (ii) Josephson tunneling. Just as fundamental charge of an electron is $e = 1.6022 \times 10^{-19}$ C, magnetic flux is quantized in steps of $\Phi_0 = \dfrac{h}{2e} = 2.0678$ femto webers, where h is Planck's constant. The Josephson tunneling effect is the phenomenon of supercurrent, a current that flows without any voltage applied across a junction, which consists of two superconductors coupled by a weak link. The weak link may be composed of a thin insulating barrier, known as a superconductor–insulator–superconductor junction, or SIS, a short section of nonsuperconducting metal or SNS, or a physical constriction that weakens the superconductivity at the point of contact. SQUIDs are made of a superconducting loop with Josephson junctions. There are two types of SQUIDs: (i) direct current SQUIDs or DC SQUIDs and (ii) radio frequency SQUIDs or RF SQUIDs. Comparatively, RF SQUIDs are less sensitive than DC counterparts, as they employ only one Josephson junction, and for this reason they are less expensive.

Essentially, a DC SQUID, as depicted in Figure 8.32(a), consists of two junctions that are connected in parallel on a flux-to-voltage transducer. a superconducting loop of inductance L. Each junction is resistively shunted to eliminate hysteresis on the volt-ampere characteristic. Figure 8.32(b) shows the schematic with shunted resistive loading and Figure 8.32(c) shows a typical characteristic for flux of the form $\Phi = n\Phi_0$ and $\Phi = (n + 1/2) \Phi_0$, where Φ is the applied magnetic flux and n is an integer. If we bias the SQUID with a constant current above a threshold, the voltage across the SQUID oscillates with period Φ_0, as we steadily increase Φ, as indicated in Figure 8.32(d). The SQUID is generally operated on the steep part of the V–Φ curve where the slope is maximum. Thus, the SQUID produces an output voltage in response to even a small input flux and is effectively a flux-to-voltage transducer.

A DC SQUID employing two resistively shunted Josephson tunnel junctions as shown in Figure 8.32(a) is typically constructed from thin films of superconductors of low transition-temperature T_c. The SQUID is coupled to an integrated superconducting coil carrying a signal source. A bias current entering the SQUID splits into the two parallel branches containing a Josephson junction each. In the absence of any external magnetic field, the input current splits into the two branches equally. If a small external magnetic field is applied to the superconducting loop, a screening current begins to circulate in the loop that generates a magnetic field canceling the applied external flux. The induced current is in the same direction as the incoming bias-current in one of the branches of the superconducting loop and is opposite to it in the other branch. As soon as the current in either branch exceeds the critical current of the Josephson junction, a voltage appears across the junction.

For a given constant biasing current into the SQUID device, the measured voltage oscillates with the changes in phase at the two junctions, which depends upon the change in the magnetic flux. Thus one might estimate the change in magnitude of the incident flux in terms of the voltage alterations.

The RF SQUID involves a single Josephson junction interrupting the current flow around a superconducting loop and is operated with a radio frequency flux bias. In both cases, the output from the SQUID is periodic with period in the magnetic flux applied to the loop. One generally is able to detect an output signal corresponding to a flux change of extremely small magnitude. Instruments based on low critical temperature SQUIDs include magnetometers, magnetic gradiometers, voltmeters, susceptometers, amplifiers, and displacement sensors; their applications vary from neuromagnetism and magnetotelluric sounding to the detection of gravity waves and magnetic resonance. The applications of SQUIDs are wide ranging, from the detection of tiny magnetic fields produced by the human brain and the measurement of fluctuating geomagnetic fields in remote areas to the detection of gravity waves and the observation of spin noise in an ensemble of magnetic nuclei.

MATLAB 8.1

```
% This script computes the results for Example 8.1
%
% It uses the 'dsolve' function which solves symbolic
% differential equations
% the arguments to the function are
%    1. the differential equation, with D and D2 in
%       front of the variable denoting 1st and 2nd derivative
%    2. the 1st order initial value
%    3. the 2nd order initial value
%    4. the independent variable
clear
syms at ax ay az t % this statement makes symbolic variables
                        % for acceleration & position conponents
at=[ax, ay, az];   % group the acceleration conponents

% part a
a=[12, 10, 0]*3/2;
% display function is similar to printf in c/c++
disp(sprintf('Part a\nThe acceleration '))
disp(sprintf('is (%f, %f, %f) m/s', a(1), a(2), a(3)))

% part b
% Solve for the velocity (each line solves for one component
v=[dsolve('Dvx=ax','vx(0)=4','t'), ...  % x
    dsolve('Dvy=ay','vy(0)=0','t'),...  % y
    dsolve('Dvz=az','vz(0)=3','t')];    % z

v=subs(v,{ax,ay,az},{a(1), a(2), a(3)}); % Replace velocity
                                        % variable with numbers
disp(sprintf('\n\nPart b\nThe general velocity is given by'))
                                        % display
pretty(v)            % make variable expression look algebraic
v=subs(v,{t},1);     % determine numerical v at time t
disp('The velocity at (1, -2, 0) is ');
disp(sprintf('(%f, %f, %f) m/s', v(1), v(2), v(3)))

% part c
disp('\n\nPart c\nThe kinetic energy ') % display
disp(sprintf(' is %f J',0.5*2*norm(v)^2))
% part d
% Solve for the position (each line solves for one component
p=[dsolve('D2px=ax','px(0)=1','Dpx(0)=4','t'), ...   % x
    dsolve('D2py=ay','py(0)=-2','Dpy(0)=0','t'),...  % y
    dsolve('D2pz=az','pz(0)=0','Dpz(0)=3','t')];      %z
% Find the acceleration, and replace the acceleration variable
% components with the actual numbers
p=subs(p,{ax,ay,az},{a(1), a(2), a(3)});
disp(sprintf('\n\nPart d\nThe general position is given by'))
```

```
% display
pretty(p)               % make variable expression look algebraic
p=subs(p,{t},1);        % determine numerical v at time t
disp('The position at time t = 1 s ');
disp(sprintf(' is (%f, %f, %f) m/s', p(1), p(2), p(3)))
```

SUMMARY

1. The Lorentz force equation

$$\mathbf{F} = Q(\mathbf{E} + \mathbf{u} \times \mathbf{B}) = m\frac{d\mathbf{u}}{dt}$$

relates the force acting on a particle with charge Q in the presence of EM fields. It expresses the fundamental law relating EM to mechanics.

2. Based on the Lorentz force law, the force experienced by a current element $I\,d\mathbf{l}$ in a magnetic field $\mathbf{B}$ is

$$d\mathbf{F} = I\,d\mathbf{l} \times \mathbf{B}$$

From this, the magnetic field $\mathbf{B}$ is defined as the force per unit current element.

3. The torque on a current loop with magnetic moment $\mathbf{m}$ in a uniform magnetic field $\mathbf{B}$ is

$$\mathbf{T} = \mathbf{m} \times \mathbf{B} = IS\mathbf{a}_n \times \mathbf{B}$$

4. A magnetic dipole is a bar magnet or a small filamental current loop; it is so called because its $\mathbf{B}$ field lines are similar to the $\mathbf{E}$ field lines of an electric dipole.

5. When a material is subjected to a magnetic field, it becomes magnetized. The magnetization $\mathbf{M}$ is the magnetic dipole moment per unit volume of the material. For linear material,

$$\mathbf{M} = \chi_m\mathbf{H}$$

where χ_m is the magnetic susceptibility of the material.

6. In terms of their magnetic properties, materials are either linear (diamagnetic or paramagnetic) or nonlinear (ferromagnetic). For linear materials,

$$\mathbf{B} = \mu\mathbf{H} = \mu_o\mu_r\mathbf{H} = \mu_o(1 + \chi_m)\mathbf{H} = \mu_o(\mathbf{H} + \mathbf{M})$$

where μ = permeability and $\mu_r = \mu/\mu_o$ = relative permeability of the material. For nonlinear material, $B = \mu(H)\,H$, that is, μ does not have a fixed value; the relationship between B and H is usually represented by a magnetization curve.

7. The boundary conditions that **H** or **B** must satisfy at the interface between two different media are

$$\mathbf{B}_{1n} = \mathbf{B}_{2n}$$

$$(\mathbf{H}_1 - \mathbf{H}_2) \times \mathbf{a}_{n12} = \mathbf{K} \ \ or \ \ \mathbf{H}_{1t} = \mathbf{H}_{2t}, \ \ if \ \mathbf{K} = \mathbf{0},$$

where $\mathbf{a}_{n12}$ is a unit vector directed from medium 1 to medium 2.

8. Energy in a magnetostatic field is given by

$$W_m = \frac{1}{2} \int_v \mathbf{B} \cdot \mathbf{H} \, dv$$

For an inductor carrying current I

$$W_m = \frac{1}{2} L I^2$$

Thus the inductance L can be found by using

$$L = \frac{\int_v \mathbf{B} \cdot \mathbf{H} \, dv}{I^2}$$

9. The inductance L of an inductor can also be determined from its basic definition: the ratio of the magnetic flux linkage to the current through the inductor, that is,

$$L = \frac{\lambda}{I} = \frac{N\Psi}{I}$$

Thus by assuming current I, we determine **B** and $\Psi = \int_s \mathbf{B} \cdot d\mathbf{S}$, and finally find $L = N\Psi/I$.

10. A magnetic circuit can be analyzed in the same way as an electric circuit. We simply keep in mind the similarity between

$$\mathcal{F} = NI = \oint \mathbf{H} \cdot d\mathbf{l} = \Psi \mathcal{R} \quad and \quad V = IR$$

that is,

$$\mathcal{F} \leftrightarrow V, \ \Psi \leftrightarrow I, \ \mathcal{R} \leftrightarrow R$$

Thus we can apply Ohm's and Kirchhoff's laws to magnetic circuits just as we apply them to electric circuits.

11. The magnetic pressure (or force per unit surface area) on a piece of magnetic material is

$$P = \frac{F}{S} = \frac{1}{2} BH = \frac{B^2}{2\mu_o}$$

where B is the magnetic field at the surface of the material.

12. Magnetic levitation (maglev) is a way of using EM fields to levitate objects. One important area of application of maglev is transportation. Conventional railroads operate at speeds below 300 km/hr, while maglev vehicles are designed for operating speeds of up to 500 km/hr.

REVIEW QUESTIONS

8.1 Which of the following statements are not true about electric force $\mathbf{F}_e$ and magnetic force $\mathbf{F}_m$ on a charged particle?

(a) $\mathbf{E}$ and $\mathbf{F}_e$ are parallel to each other, whereas $\mathbf{B}$ and $\mathbf{F}_m$ are perpendicular to each other.
(b) Both $\mathbf{F}_e$ and $\mathbf{F}_m$ depend on the velocity of the charged particle.
(c) Both $\mathbf{F}_e$ and $\mathbf{F}_m$ can perform work.
(d) Both $\mathbf{F}_e$ and $\mathbf{F}_m$ are produced when a charged particle moves at a constant velocity.
(e) $\mathbf{F}_m$ is generally small in magnitude in comparison to $\mathbf{F}_e$.
(f) $\mathbf{F}_e$ is an accelerating force, whereas $\mathbf{F}_m$ is a purely deflecting force.

8.2 Two thin parallel wires carry currents along the same direction. The force experienced by one due to the other is

(a) Parallel to the lines
(b) Perpendicular to the lines and attractive
(c) Perpendicular to the lines and repulsive
(d) Zero

8.3 The force on differential length $d\mathbf{l}$ at point P in the conducting circular loop in Figure 8.33 is

(a) Outward along OP
(b) Inward along OP
(c) In the direction of the magnetic field
(d) Tangential to the loop at P

8.4 The resultant force on the circular loop in Figure 8.33 has the magnitude of

(a) $2\pi\rho_o IB$ (c) $2\rho_o IB$
(b) $\pi\rho_o^2 IB$ (d) Zero

8.5 What is the unit of magnetic charge?

(a) Ampere-meter squared   (c) Ampere
(b) Coulomb   (d) Ampere-meter

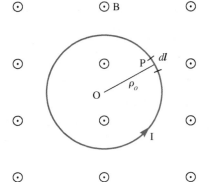

FIGURE 8.33 For Review Questions 8.3 and 8.4.

8.6 Which of these materials requires the lowest value of magnetic field strength to magnetize it?

(a) Nickel

(b) Silver

(c) Tungsten

(d) Sodium chloride

8.7 Identify the statement that is not true of ferromagnetic materials.

(a) They have a large χ_m.

(b) They have a fixed value of μ_r.

(c) Energy loss is proportional to the area of the hysteresis loop.

(d) They lose their nonlinearity property above the curie temperature.

8.8 Which of these formulas is wrong?

(a) $B_{1n} = B_{2n}$

(b) $B_2 = \sqrt{B_{2n}^2 + B_{2t}^2}$

(c) $H_1 = H_{1n} + H_{1t}$

(d) $\mathbf{a}_{n21} \times (\mathbf{H}_1 - \mathbf{H}_2) = \mathbf{K}$, where $\mathbf{a}_{n21}$ is a unit vector normal to the interface and directed from region 2 to region 1.

8.9 Each of the following pairs consists of an electric circuit term and the corresponding magnetic circuit term. Which pairs are not corresponding?

(a) V and $\mathscr{F}$

(b) G and $\mathscr{P}$

(c) ε and μ

(d) IR and $H\mathscr{R}$

(e) $\Sigma I = 0$ and $\Sigma \Psi = 0$

8.10 A multilayer coil of 2000 turns of fine wire is 20 mm long and has a thickness 5 mm of winding. If the coil carries a current of 5 mA, the mmf generated is

(a) 10 A $\cdot$ t

(b) 500 A $\cdot$ t

(c) 2000 A $\cdot$ t

(d) None of the above

Answers: 8.1 b,c, 8.2b, 8.3a, 8.4d, 8.5c, 8.6a, 8.7b, 8.8c, 8.9c,d, 8.10a.

Section 8.2—Forces due to Magnetic Fields

8.1 A 4 mC charge has velocity $\mathbf{u} = 1.4\mathbf{a}_x - 3.2\mathbf{a}_y - \mathbf{a}_z$ m/s at point $P(2, 5, -3)$ in the presence of $\mathbf{E} = 2xyz\mathbf{a}_x + x^2z\mathbf{a}_y + x^2y\mathbf{a}_z$ V/m and $\mathbf{B} = y^2\mathbf{a}_x + z^2\mathbf{a}_y + x^2\mathbf{a}_z$ Wb/m². Find the force on the charge at P.

8.2 An electron ($m = 9.11 \times 10^{-31}$ kg) moves in a circular orbit of radius 0.4×10^{-10} m with an angular velocity of 2×10^{16} rad/s. Find the centripetal force required to hold the electron.

8.3 A 1 mC charge with velocity $10\mathbf{a}_x - 2\mathbf{a}_y + 6\mathbf{a}_z$ m/s enters a region where the magnetic flux density is $25\mathbf{a}_z$ Wb/m². (a) Calculate the force on the charge. (b) Determine the electric field intensity necessary to make the velocity of the charge constant.

8.4 Assume an electric field intensity of 20 kV/m and a magnetic flux density of $5\ \mu$Wb/m² exist in a region. Find the ratio of the magnitudes of electric and magnetic forces on an electron that has attained a velocity of 0.5×10^8 m/s.

8.5 A -2 mC charge starts at point $(0, 1, 2)$ with a velocity of $5\mathbf{a}_x$ m/s in a magnetic field $\mathbf{B} = 6\mathbf{a}_y$ Wb/m². Determine the position and velocity of the particle after 10 s, assuming that the mass of the charge is 1 gram. Describe the motion of the charge.

*8.6 By injecting an electron beam normally to the plane edge of a uniform field $B_o\mathbf{a}_z$, electrons can be dispersed according to their velocity as in Figure 8.34.

   (a) Show that the electrons would be ejected out of the field in paths parallel to the input beam as shown.

   (b) Derive an expression for the exit distance d above the entry point.

8.7 Two large conducting plates are 8 cm apart and have a potential difference 12 kV. A drop of oil with mass 0.4 g is suspended in space between the plates. Find the charge on the drop.

8.8 A straight conductor 0.2 m long carries a current 4.5 A along $\mathbf{a}_x$. If the conductor lies in the magnetic field $\mathbf{B} = 2.5(\mathbf{a}_y + \mathbf{a}_z)$ mWb/m², calculate the force on the conductor.

8.9 Determine $|\mathbf{B}|$ that will produce the same force on a charged particle moving at 140 m/s that an electric field of 12 kV/m produces.

*8.10 Three infinite lines L_1, L_2, and L_3 defined by $x = 0$, $y = 0$; $x = 0$, $y = 4$; $x = 3$, $y = 4$, respectively, carry filamentary currents -100 A, 200 A, and 300 A along $\mathbf{a}_z$. Find the force per unit length on

   (a) L_2 due to L_1
   (b) L_1 due to L_2
   (c) L_3 due to L_1
   (d) L_3 due to L_1 and L_2.

   State whether each force is repulsive or attractive.

8.11 Two infinitely long parallel wires are separated by a distance of 20 cm. If the wires carry current of 10 A in opposite directions, calculate the force on the wires.

8.12 A conductor 2 m long carrying a current of 3 A is placed parallel to the z-axis at distance $\rho_o = 10$ cm as shown in Figure 8.35. If the field in the region is $\cos(\phi/3)\ \mathbf{a}_\rho$ Wb/m², how much work is required to rotate the conductor one revolution about the z-axis?

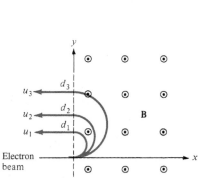

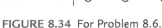

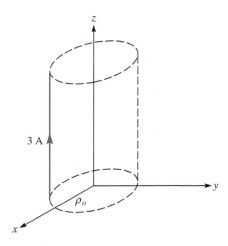

FIGURE 8.34 For Problem 8.6.

FIGURE 8.35 For Problem 8.12.

*8.13 A conducting triangular loop carrying a current of 2 A is located close to an infinitely long, straight conductor with a current of 5 A, as shown in Figure 8.36. Calculate (a) the force on side 1 of the triangular loop and (b) the total force on the loop.

*8.14 A three-phase transmission line consists of three conductors that are supported at points A, B, and C to form an equilateral triangle as shown in Figure 8.37. At one instant, conductors A and B both carry a current of 75 A while conductor C carries a return current of 150 A. Find the force per meter on conductor C at that instant.

8.15 A current sheet with $\mathbf{K} = 10\mathbf{a}_x$ A/m lies in free space in the $z = 2$ m plane. A filamentary conductor on the x-axis carries a current of 2.5 A in the $\mathbf{a}_x$-direction. Determine the force per unit length on the conductor.

8.16 The magnetic field in a certain region is $\mathbf{B} = 40\,\mathbf{a}_x$ mWb/m^2. A conductor that is 2 m in length lies in the z-axis and carries a current of 5 A in the $\mathbf{a}_z$-direction. Calculate the force on the conductor.

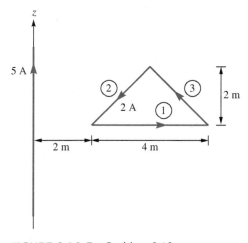

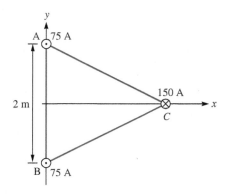

FIGURE 8.36 For Problem 8.13.

FIGURE 8.37 For Problem 8.14.

Sections 8.3 and 8.4—Magnetic Torque, Moments, and Dipole

*8.17 A rectangular loop shown in Figure 8.38 carries current $I = 10$ A and is situated in the field $\mathbf{B} = 4.5(\mathbf{a}_y - \mathbf{a}_z)$ Wb/m². Find the torque on the loop.

8.18 A 60-turn coil carries a current of 2 A and lies in the plane $x + 2y - 5z = 12$ such that the magnetic moment $\mathbf{m}$ of the coil is directed away from the origin. Calculate $\mathbf{m}$, assuming that the area of the coil is 8 cm².

8.19 The earth has a magnetic moment of about 8×10^{22} A · m² and its radius is 6370 km. Imagine that there is a loop around the equator and determine how much current in the loop would result in the same magnetic moment.

8.20 A triangular loop is placed in the x-z plane, as shown in Figure 8.39. Assume that a dc current $I = 2$ A flows in the loop and that $\mathbf{B} = 30\mathbf{a}_z$ m Wb/m exists in the region. Find the forces and torque on the loop.

8.21 A loop with 50 turns and surface area of 12 cm² carries a current of 4 A. If the loop rotates in a uniform magnetic field of 100 mWb/m², find the torque exerted on the loop.

8.22 High-current circuit breakers typically consist of coils that generate a magnetic field to blow out the arc formed when the contacts open. An arc 30 mm long carries a current of 520 A in a direction perpendicular to a magnetic flux density of 0.4 mWb/m². Determine the magnetic force on the arc.

Section 8.5—Magnetization in Materials

8.23 For a linear, isotropic, and homogeneous magnetic medium, show that $M = \dfrac{\chi_m}{\mu_o(1 + \chi_m)}B$.

8.24 A block of iron ($\mu = 5000\mu_o$) is placed in a uniform magnetic field with 1.5 Wb/m². If iron consists of 8.5×10^{28} atoms/m³, calculate (a) the magnetization M, (b) the average magnetic moment.

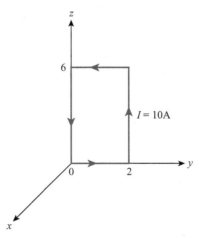

FIGURE 8.38 For Problem 8.16.

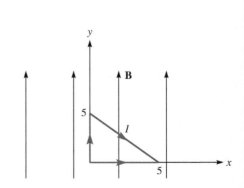

FIGURE 8.39 For Problem 8.20.

8.25 In a magnetic material, with $\chi_m = 6.5$, the magnetization is $\mathbf{M} = 24y^2\mathbf{a}_z$ A/m. Find μ_r, $\mathbf{H}$, and $\mathbf{J}$ at $y = 2$ cm.

8.26 In a ferromagnetic material ($\mu = 80\mu_o$), $\mathbf{B} = 20x\mathbf{a}_y$ mWb/m². Determine: (a) μ_r, (b) χ_m, (c) $\mathbf{H}$, (d) $\mathbf{M}$, (e) $\mathbf{J}_b$.

8.27 An electromagnet is made of a ferromagnetic material whose magnetization curve can be approximated by

$$B(H) = B_oH/(H_o + H) \text{ mWb/m}^2$$

where $B_o = 2$ Wb/m² and $H_o = 100$ A/m
Find μ_r when $H = 250$ A/m.

8.28 An infinitely long cylindrical conductor of radius a and permeability $\mu_o\mu_r$ is placed along the z-axis. If the conductor carries a uniformly distributed current I along $\mathbf{a}_z$, find $\mathbf{M}$ and $\mathbf{J}_b$ for $0 < \rho < a$.

Section 8.7—Magnetic Boundary Conditions

*8.29 (a) For the boundary between two magnetic media such as is shown in Figure 8.16, show that the boundary conditions on the magnetization vector are

$$\frac{M_{1t}}{\chi_{m1}} - \frac{M_{2t}}{\chi_{m2}} = K \quad \text{and} \quad \frac{\mu_1}{\chi_{m1}}M_{1n} = \frac{\mu_2}{\chi_{m2}}M_{2n}$$

(b) If the boundary is not current free, show that instead of eq. (8.49), we obtain

$$\frac{\tan\theta_1}{\tan\theta_2} = \frac{\mu_1}{\mu_2}\left[1 + \frac{K\mu_2}{B_2\sin\theta_2}\right]$$

8.30 Region 1, for which $\mu_1 = 2.5\mu_o$, is defined by $z < 0$, while region 2, for which $\mu_2 = 4\mu_o$, is defined by $z > 0$. If $\mathbf{B}_1 = 6\mathbf{a}_x - 4.2\mathbf{a}_y + 1.8\mathbf{a}_z$ mWb/m², find $\mathbf{H}_2$ and the angle $\mathbf{H}_2$ makes with the interface.

8.31 In medium 1 ($z < 0$) $\mu_1 = 5\mu_o$, while in medium 2 ($z > 0$) $\mu_2 = 2\mu_o$. If $\mathbf{B}_1 = 4\mathbf{a}_x - 10\mathbf{a}_y + 12\mathbf{a}_z$ mWb/m², find $\mathbf{B}_2$ and the energy density in medium 2.

8.32 In region $x < 0$, $\mu = \mu_o$, a uniform magnetic field makes angle 42° with the normal to the interface. Calculate the angle the field makes with the normal in region $x > 0$, $\mu = 6.5\mu_o$.

8.33 A current sheet with $\mathbf{K} = 12\mathbf{a}_y$ A/m is placed at $x = 0$, which separates region 1, $x < 0$, $\mu = 2\mu_o$ and region 2, $x > 0$, $\mu = 4\mu_o$. If $\mathbf{H}_1 = 10\mathbf{a}_x + 6\mathbf{a}_z$ A/m, find $\mathbf{H}_2$.

8.34 Suppose space is divided into region 1 ($y < 0$, $\mu_1 = \mu_o\mu_{r1}$) and region 2 ($y < 0$, $\mu_2 = \mu_o\mu_{r2}$). If $\mathbf{H}_1 = \alpha\mathbf{a}_x + \beta\mathbf{a}_y + \delta\mathbf{a}_z$ A/m, find $\mathbf{H}_2$.

8.35 If $\mu_1 = 2\mu_o$ for region 1 ($0 < \phi < \pi$) and $\mu_2 = 5\mu_o$ for region 2 ($\pi < \phi < 2\pi$) and $\mathbf{B}_2 = 10\mathbf{a}_\rho + 15\mathbf{a}_\phi - 20\mathbf{a}_z$ mWb/m². Calculate (a) $\mathbf{B}_1$, (b) the energy densities in the two media.

*8.36 Region 1 is defined by $x - y + 2z > 5$ with $\mu_1 = 2\mu_o$, while region 2 is defined by $x - y + 2z < 5$ with $\mu_2 = 5\mu_o$. If $\mathbf{H}_1 = 40\mathbf{a}_x + 20\mathbf{a}_y - 30\mathbf{a}_z$ A/m, find (a) $\mathbf{H}_{1n}$, (b) $\mathbf{H}_{2t}$, (c) $\mathbf{B}_2$.

8.37 Inside a right circular cylinder, $\mu_1 = 800\,\mu_o$, while the exterior is free space. Given that $\mathbf{B}_1 = \mu_o(22\mathbf{a}_\rho + 45\mathbf{a}_\phi)$ Wb/m², determine $\mathbf{B}_2$ just outside the cylinder.

8.38 The plane $z = 0$ separates air $(z \ge 0, \mu = \mu_o)$ from iron $(z \le 0, \mu = 200\mu_o)$. Given that

$$\mathbf{H} = 10\mathbf{a}_x + 15\mathbf{a}_y - 3\mathbf{a}_z \text{ A/m}$$

in air, find $\mathbf{B}$ in iron and the angle it makes with the interface.

8.39 Region $0 \le z \le 2$ m is filled with an infinite slab of magnetic material $(\mu = 2.5\mu_o)$. If the surfaces of the slab at $z = 0$ and $z = 2$, respectively, carry surface currents $30\mathbf{a}_x$ A/m and $-40\mathbf{a}_x$ A/m as in Figure 8.40, calculate $\mathbf{H}$ and $\mathbf{B}$ for
(a) $z < 0$
(b) $0 < z < 2$
(c) $z > 2$

8.40 Medium 1 is free space and is defined by $r < a$, while medium 2 is a magnetic material with permeability μ_2 and defined by $r > a$. The magnetic flux densities in the media are:

$$\mathbf{B}_1 = B_{o1}\left[\left(1 + \frac{1.6a^3}{r^3}\right)\cos\theta\,\mathbf{a}_r - \left(1 - \frac{0.8a^3}{r^3}\right)\sin\theta\,\mathbf{a}_\theta\right]$$

$$\mathbf{B}_2 = B_{o2}(\cos\theta\,\mathbf{a}_r - \sin\theta\,\mathbf{a}_\theta)$$

Find μ_2.

Section 8.8—Inductors and Inductance

*8.41 (a) If the cross section of the toroid of Figure 7.15 is a square of side a, show that the self-inductance of the toroid is

$$L = \frac{\mu_o N^2 a}{2\pi}\ln\left[\frac{2\rho_o + a}{2\rho_o - a}\right]$$

  (b) If the toroid has a circular cross section as in Figure 7.15, show that

$$L = \frac{\mu_o N^2 a^2}{2\rho_o}$$

where $\rho_o \gg a$.

FIGURE 8.40 For Problem 8.39.

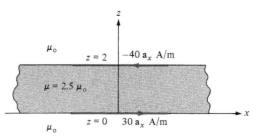

8.42 An air-filled toroid of square cross section has inner radius 3 cm, outer radius 5 cm, and height 2 cm. How many turns are required to produce an inductance of 45 μH?

8.43 A wire of radius 2 mm is 40 m long. Calculate its inductance. Assume $\mu = \mu_o$.

8.44 A coaxial cable has an internal inductance that is twice the external inductance. If the inner radius is 6.5 mm, calculate the outer radius.

8.45 A hollow cylinder of radius $a = 2$ cm is 10 m long. Find the inductance of the cylinder. (See Table 8.3.)

8.46 Show that the mutual inductance between the rectangular loop and the infinite line current of Figure 8.4 is

$$M_{12} = \frac{\mu b}{2\pi} \ln\left[\frac{a + \rho_o}{\rho_o}\right]$$

Calculate M_{12} when $a = b = \rho_o = 1$ m.

***8.47** Prove that the mutual inductance between the close-wound coaxial solenoids of length ℓ_1 and ℓ_2 ($\ell_1 \gg \ell_2$), turns N_1 and N_2, and radii r_1 and r_2 with $r_1 \simeq r_2$ is

$$M_{12} = \frac{\mu N_1 N_2}{\ell_1} \pi r_1^2$$

***8.48** A loop resides outside the region between two parallel long wires carrying currents in opposite directions as shown in Figure 8.41. Find the total flux linking the loop.

Section 8.9—Magnetic Energy

8.49 A coaxial cable consists of an inner conductor of radius 1.2 cm and an outer conductor of radius 1.8 cm. The two conductors are separated by an insulating medium ($\mu = 4\mu_o$). If the cable is 3 m long and carries 25 mA current, calculate the energy stored in the medium.

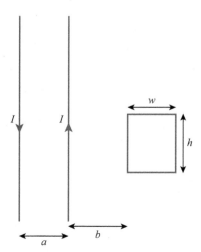

FIGURE 8.41 For Problem 8.48.

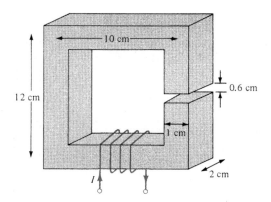

FIGURE 8.42 For Problem 8.54.   FIGURE 8.43 For Problem 8.55.

8.50 In a certain region for which $\chi_m = 19$,

$$\mathbf{H} = 5x^2yz\mathbf{a}_x + 10xy^2z\mathbf{a}_y - 15xyz^2\mathbf{a}_z \text{ A/m}$$

How much energy is stored in $0 < x < 1, 0 < y < 2, -1 < z < 2$?

8.51 The magnetic field in a material space $(\mu = 15\mu_o)$ is given by

$$\mathbf{B} = 4\mathbf{a}_x + 12\mathbf{a}_y \text{ mWb/m}^2$$

Calculate the energy stored in region $0 < x < 2, 0 < y < 3, 0 < z < 4$.

Section 8.10—Magnetic Circuits

8.52 A cobalt ring $(\mu_r = 600)$ has a mean radius of 30 cm. If a coil wound on the ring carries 12 A, calculate the number of turns required to establish an average magnetic flux density of 1.5 Wb/m² in the ring.

8.53 Refer to Figure 8.27. If the current in the coil is 0.5 A, find the mmf and the magnetic field intensity in the air gap. Assume that $\mu = 500\mu_o$ and that all branches have the same cross-sectional area of 10 cm².

8.54 The magnetic circuit of Figure 8.42 has a current of 10 A in the coil of 2000 turns. Assume that all branches have the same cross section of 2 cm² and that the material of the core is iron with $\mu_r = 1500$. Calculate R, $\mathcal{F}$, and Ψ for

(a) The core

(b) The air gap

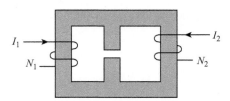

FIGURE 8.44 For Problem 8.56.

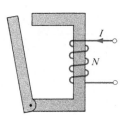

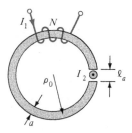

FIGURE 8.45 For Problem 8.58.

FIGURE 8.46 For Problem 8.59.

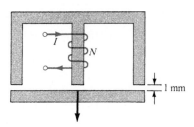

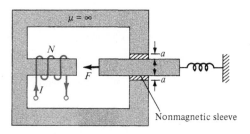

FIGURE 8.47 For Problem 8.60.

FIGURE 8.48 For Problem 8.61.

8.55 Consider the magnetic circuit in Figure 8.42. Assuming that the core ($\mu = 1000\mu_o$) has a uniform cross section of 4 cm², determine the flux density in the air gap.

8.56 For the magnetic circuit shown in Figure 8.44, draw the equivalent electric circuit. Assume that all the sections have constant cross-sectional areas.

8.57 An air gap in an electric machine has length 4.4 mm and area 4.82×10^{-2} m². Find the reluctance of the gap.

Section 8.11—Force on Magnetic Materials

8.58 An electromagnetic relay is modeled as shown in Figure 8.45. What force is on the armature (moving part) of the relay if the flux in the air gap is 2 mWb? The area of the gap is 0.3 cm², and its length 1.5 mm.

8.59 A toroid with air gap, shown in Figure 8.46, has a square cross section. A long conductor carrying current I_2 is inserted in the air gap. If $I_1 = 200$ mA, $N = 750$, $\rho_o = 10$ cm, $a = 5$ mm, and $\ell_a = 1$ mm, calculate

(a) The force across the gap when $I_2 = 0$ and the relative permeability of the toroid is 300.

(b) The force on the conductor when $I_2 = 2$ mA and the permeability of the toroid is infinite. Neglect fringing in the gap in both cases.

8.60 A section of an electromagnet with a plate below it carrying a load as shown in Figure 8.47. The electromagnet has a contact area of 200 cm² per pole, and the middle pole has a winding of 1000 turns with $I = 3$ A. Calculate the maximum mass that can be lifted. Assume that the reluctance of the electromagnet and the plate is negligible.

8.61 Figure 8.48 shows the cross section of an electromechanical system in which the plunger moves freely between two nonmagnetic sleeves. Assuming that all legs have the same cross-sectional area S, show that

$$\mathbf{F} = -\frac{2\,N^2 I^2 \mu_{o} S}{(a + 2x)^2}\,\mathbf{a}_x$$

WAVES AND APPLICATIONS

Michael Faraday (1791–1867), an English chemist and physicist, is known for his pioneering experiments in electricity and magnetism. Many consider him the greatest experimentalist who ever lived.

Born at Newington, near London, to a poor family, he received little more than an elementary education. During a seven-year apprenticeship as a bookbinder, Faraday developed his interest in science and in particular chemistry. As a result, Faraday started a second apprenticeship in chemistry. Following in the footsteps of Benjamin Franklin and other early scientists, Michael Faraday studied the nature of electricity. Later in life, Faraday became professor of chemistry at the Royal Institution. He discovered benzene and formulated the second law of electrolysis. Faraday's greatest contribution to science was in the field of electricity. Faraday's introduction of the concept of lines of force was initially rejected by most of the mathematical physicists of Europe. He discovered electromagnetic induction (to be covered in this chapter), the battery, the electric arc (plasmas), and the Faraday cage (electrostatics). His biggest breakthrough was his invention of the electric motor and dynamo (or generator). Despite his achievements, Faraday remained a modest and humble person. In his day, Faraday was deeply religious. The unit of capacitance, the farad, is named after him.

James Clerk Maxwell (1831–1879), Scottish mathematician and physicist, published physical and mathematical theories of the electromagnetic field.

Born at Edinburgh, Scotland, Maxwell showed an early understanding and love for the field of mathematics. Dissatisfied with the toys he was given, he made his own scientific toys at the age of 8! Maxwell was a true genius who made several contributions to the scientific community, but his most important achievement was his development of the equations of electromagnetic waves, which we now call Maxwell's equations. In 1931, on the centennial anniversary of Maxwell's birth, Einstein described

Maxwell's work as the *"most profound and the most fruitful that physics has experienced since the time of Newton."* Without Maxwell's work, radio and television could not exist. The 1888 announcement by the German physics professor Heinrich Rudolf Hertz (see Chapter 10) that he had transmitted and received electromagnetic waves was almost universally received as a glorious confirmation of Maxwell's equations. The maxwell (Mx), the unit of measurement of magnetic flux in the centimeter-gram-second (cgs) system of units, was named in his honor.

CHAPTER 9

MAXWELL'S EQUATIONS

Some people make enemies instead of friends because it is less trouble.

—E. C. MCKENZIE

9.1 INTRODUCTION

In Part 2 (Chapters 4–6) of this text, we mainly concentrated our efforts on electrostatic fields denoted by $\mathbf{E}(x, y, z)$; Part 3 (Chapters 7 and 8) was devoted to magnetostatic fields represented by $\mathbf{H}(x, y, z)$. We have therefore restricted our discussions to static, or time-invariant, EM fields. Henceforth, we shall examine situations in which electric and magnetic fields are dynamic, or time varying. It should be mentioned first that in static EM fields, electric and magnetic fields are independent of each other, whereas in dynamic EM fields, the two fields are interdependent. In other words, a time-varying electric field necessarily involves a corresponding time-varying magnetic field. Second, time-varying EM fields, represented by $\mathbf{E}(x, y, z, t)$ and $\mathbf{H}(x, y, z, t)$, are of more practical value than static EM fields. However, familiarity with static fields provides a good background for understanding dynamic fields. Third, recall that electrostatic fields are usually produced by static electric charges, whereas magnetostatic fields are due to motion of electric charges with uniform velocity (direct current) or static magnetic charges (magnetic poles); time-varying fields or waves are usually due to accelerated charges or time-varying currents such as shown in Figure 9.1. Any pulsating current will produce radiation (time-varying fields). It is worth noting that pulsating current of the type shown in Figure 9.1(b) is the cause of radiated emission in digital logic boards. In summary:

stationary charges	→ electrostatic fields
steady currents	→ magnetostatic fields
time-varying currents	→ electromagnetic fields (or waves)

Our aim in this chapter is to lay a firm foundation for our subsequent studies. This will involve introducing two major concepts: (1) electromotive force based on Faraday's experiments and (2) displacement current, which resulted from Maxwell's hypothesis. As a result of these concepts, Maxwell's equations as presented in Section 7.6 and the boundary conditions for static EM fields will be modified to account for the time variation of the fields. Maxwell's equations, which summarize the laws of electromagnetism, shall be

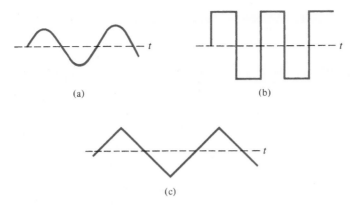

FIGURE 9.1 Examples of time-varying current: (**a**) sinusoidal, (**b**) rectangular, (**c**) triangular.

the basis of our discussions in the remaining part of the text. For this reason, Section 9.5 should be regarded as the heart of this text.

9.2 FARADAY'S LAW

After Oersted's experimental discovery (upon which Biot–Savart and Ampère based their laws) that a steady current produces a magnetic field, it seemed logical to find out whether magnetism would produce electricity. In 1831, about 11 years after Oersted's discovery, Michael Faraday in London and Joseph Henry in New York discovered that a time-varying magnetic field would produce an electric current.[1]

According to Faraday's experiments, a static magnetic field produces no current flow; but in a closed circuit, a time-varying field produces an induced voltage (called *electromotive force* or simply emf) that causes a flow of current.

> Faraday discovered that the **induced emf,** V_{emf} (in volts) in any closed circuit is equal to the time rate of change of the magnetic flux linkage by the circuit.

This is called *Faraday's law*, and it can be expressed as

$$V_{\text{emf}} = -\frac{d\lambda}{dt} = -N\frac{d\Psi}{dt} \tag{9.1}$$

where $\lambda = N\Psi$ is the flux linkage, N is the number of turns in the circuit, and Ψ is the flux through each turn. The negative sign shows that the induced voltage acts in such a way as

---

[1] For details on the experiments of Michael Faraday (1791–1867) and Joseph Henry (1797–1878), see W. F. Magie, *A Source Book in Physics.* Cambridge, MA: Harvard Univ. Press, 1963, pp. 472–519.

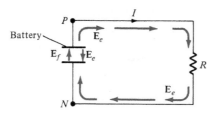

FIGURE 9.2 A circuit showing emf-producing field $\mathbf{E}_f$ and electrostatic field $\mathbf{E}_e$.

to oppose the flux producing it. This behavior is described as Lenz's law[2]. Lenz's law states the direction of current flow in the circuit is such that the induced magnetic field produced by the induced current opposes change in the original magnetic field.

Recall that we described an electric field as one in which electric charges experience force. The electric fields considered so far are caused by electric charges; in such fields, the flux lines begin and end on the charges. However, electric fields of other kinds are not directly caused by electric charges. These are emf-produced fields. Sources of emf include electric generators, batteries, thermocouples, fuel cells, and photovoltaic cells, which all convert nonelectrical energy into electrical energy.

Consider the electric circuit of Figure 9.2, where the battery is a source of emf. The electrochemical action of the battery results in an emf-produced field $\mathbf{E}_f$. Due to the accumulation of charge at the battery terminals, an electrostatic field $\mathbf{E}_e$ ($= -\nabla V$) also exists. The total electric field at any point is

$$\mathbf{E} = \mathbf{E}_f + \mathbf{E}_e \tag{9.2}$$

Note that $\mathbf{E}_f$ is zero outside the battery, $\mathbf{E}_f$ and $\mathbf{E}_e$ have opposite directions in the battery, and the direction of $\mathbf{E}_e$ inside the battery is opposite to that outside it. If we integrate eq. (9.2) over the closed circuit, we have

$$\oint_L \mathbf{E} \cdot d\mathbf{l} = \oint_L \mathbf{E}_f \cdot d\mathbf{l} + 0 = \int_N^P \mathbf{E}_f \cdot d\mathbf{l} \quad \text{(through battery)} \tag{9.3a}$$

where $\oint \mathbf{E}_e \cdot d\mathbf{l} = 0$ because $\mathbf{E}_e$ is conservative. The emf of the battery is the line integral of the emf-produced field, that is,

$$V_{\text{emf}} = \int_N^P \mathbf{E}_f \cdot d\mathbf{l} = -\int_N^P \mathbf{E}_e \cdot d\mathbf{l} = IR \tag{9.3b}$$

since $\mathbf{E}_f$ and $\mathbf{E}_e$ are equal but opposite within the battery (see Figure 9.2). It may also be regarded as the potential difference ($V_P - V_N$) between the battery's open-circuit terminals. It is important to note the following facts.

1. An electrostatic field $\mathbf{E}_e$ cannot maintain a steady current in a closed circuit, since $\oint_L \mathbf{E}_e \cdot d\mathbf{l} = 0 = IR$.
2. An emf-produced field $\mathbf{E}_f$ is nonconservative.
3. Except in electrostatics, voltage and potential difference are usually not equivalent.

---

[2]After Heinrich Friedrich Emil Lenz (1804–1865), a Russian professor of physics.

9.3 TRANSFORMER AND MOTIONAL ELECTROMOTIVE FORCES

Having considered the connection between emf and electric field, we may examine how Faraday's law links electric and magnetic fields. For a circuit with a single turn ($N = 1$), eq. (9.1) becomes

$$V_{emf} = -\frac{d\Psi}{dt}$$

(9.4)

In terms of **E** and **B**, eq. (9.4) can be written as

$$V_{emf} = \oint_L \mathbf{E} \cdot d\mathbf{l} = -\frac{d}{dt} \int_S \mathbf{B} \cdot d\mathbf{S}$$

(9.5)

where Ψ has been replaced by $\int_S \mathbf{B} \cdot d\mathbf{S}$ and S is the surface area of the circuit bounded by the closed path L. It is clear from eq. (9.5) that in a time-varying situation, both electric and magnetic fields are present and are interrelated. Note that $d\mathbf{l}$ and $d\mathbf{S}$ in eq. (9.5) are in accordance with the right-hand rule as well as Stokes's theorem. This should be observed in Figure 9.3. The variation of flux with time as in eq. (9.1) or eq. (9.5) may be caused in three ways:

1. By having a stationary loop in a time-varying **B** field
2. By having a time-varying loop area in a static **B** field
3. By having a time-varying loop area in a time-varying **B** field

Each of these will be considered separately.

A. Stationary Loop in Time-Varying B Field (Transformer emf)

In Figure 9.3 a stationary conducting loop is in a time-varying magnetic **B** field. Equation (9.5) becomes

$$V_{emf} = \oint_L \mathbf{E} \cdot d\mathbf{l} = -\int_S \frac{\partial \mathbf{B}}{\partial t} \cdot d\mathbf{S}$$

(9.6)

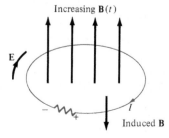

Increasing **B**(t)

E

Induced **B**

FIGURE 9.3 Induced emf due to a stationary loop in a time-varying **B** field.

This emf induced by the time-varying current (producing the time-varying **B** field) in a stationary loop is often referred to as *transformer emf* in power analysis, since it is due to transformer action. By applying Stokes's theorem to the middle term in eq. (9.6), we obtain

$$\int_S (\nabla \times \mathbf{E}) \cdot d\mathbf{S} = -\int_S \frac{\partial \mathbf{B}}{\partial t} \cdot d\mathbf{S} \tag{9.7}$$

For the two integrals to be equal, their integrands must be equal; that is,

$$\boxed{\nabla \times \mathbf{E} = -\frac{\partial \mathbf{B}}{\partial t}} \tag{9.8}$$

This is one of the Maxwell's equations for time-varying fields. It shows that the time-varying **E** field is not conservative ($\nabla \times \mathbf{E} \neq \mathbf{0}$). This does not imply that the principles of energy conservation are violated. The work done in taking a charge about a closed path in a time-varying electric field, for example, is due to the energy from the time-varying magnetic field. Observe that Figure 9.3 obeys Lenz's law: the induced current I flows such as to produce a magnetic field that opposes the change in $\mathbf{B}(t)$.

B. Moving Loop in Static B Field (Motional emf)

When a conducting loop is moving in a static **B** field, an emf is induced in the loop. We recall from eq. (8.2) that the force on a charge moving with uniform velocity **u** in a magnetic field **B** is

$$\mathbf{F}_m = Q\mathbf{u} \times \mathbf{B} \tag{8.2}$$

We define the *motional electric field* $\mathbf{E}_m$ as

$$\mathbf{E}_m = \frac{\mathbf{F}_m}{Q} = \mathbf{u} \times \mathbf{B} \tag{9.9}$$

If we consider a conducting loop, moving with uniform velocity **u** as consisting of a large number of free electrons, the emf induced in the loop is

$$\boxed{V_{\text{emf}} = \oint_L \mathbf{E}_m \cdot d\mathbf{l} = \oint_L (\mathbf{u} \times \mathbf{B}) \cdot d\mathbf{l}} \tag{9.10}$$

This type of emf is called *motional emf* or *flux-cutting emf* because it is due to motional action. It is the kind of emf found in electrical machines such as motors, generators, and alternators. Figure 9.4 illustrates a two-pole dc machine with one armature coil and a two-bar commutator. Although the analysis of the dc machine is beyond the scope of this text, we can see that voltage is generated as the coil rotates within the magnetic field. Another

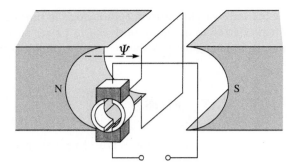

FIGURE 9.4 A direct-current machine.

example of motional emf is illustrated in Figure 9.5, where a rod is moving between a pair of rails. In this example, **B** and **u** are perpendicular, so eq. (9.9) in conjunction with eq. (8.2) becomes

$$\mathbf{F}_m = I\boldsymbol{\ell} \times \mathbf{B} \tag{9.11}$$

or

$$F_m = I\ell B \tag{9.12}$$

and eq. (9.10) becomes

$$V_{\text{emf}} = uB\ell \tag{9.13}$$

By applying Stokes's theorem to eq. (9.10), we have

$$\int_S (\nabla \times \mathbf{E}_m) \cdot d\mathbf{S} = \int_S \nabla \times (\mathbf{u} \times \mathbf{B}) \cdot d\mathbf{S}$$

or

$$\boxed{\nabla \times \mathbf{E}_m = \nabla \times (\mathbf{u} \times \mathbf{B})} \tag{9.14}$$

Notice that unlike eq. (9.6), there is no need for a minus sign in eq. (9.10) because Lenz's law is already accounted for.

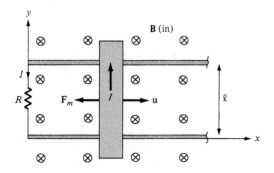

FIGURE 9.5 Induced emf due to a moving loop in a static **B** field.

To apply eq. (9.10) is not always easy; some care must be exercised. The following points should be noted.

1. The integral in eq. (9.10) is zero along the portion of the loop where $\mathbf{u} = \mathbf{0}$. Thus $d\mathbf{l}$ is taken along the portion of the loop that is cutting the field (along the rod in Figure 9.5), where $\mathbf{u}$ has nonzero value.
2. The direction of the induced current is the same as that of $\mathbf{E}_m$ or $\mathbf{u} \times \mathbf{B}$. The limits of the integral in eq. (9.10) are selected in the direction opposite to the induced current, thereby satisfying Lenz's law. In Figure 9.5, for example, the integration over L is along $-\mathbf{a}_y$, whereas induced current flows in the rod along $\mathbf{a}_y$.

C. Moving Loop in Time-Varying Field

In the general case, a moving conducting loop is in a time-varying magnetic field. Both transformer emf and motional emf are present. Combining eqs. (9.6) and (9.10) gives the total emf as

$$V_{\text{emf}} = \oint_L \mathbf{E} \cdot d\mathbf{l} = -\int_S \frac{\partial \mathbf{B}}{\partial t} \cdot d\mathbf{S} + \oint_L (\mathbf{u} \times \mathbf{B}) \cdot d\mathbf{l} \qquad (9.15)$$

or from eqs. (9.8) and (9.14),

$$\nabla \times \mathbf{E} = -\frac{\partial \mathbf{B}}{\partial t} + \nabla \times (\mathbf{u} \times \mathbf{B}) \qquad (9.16)$$

Note that eq. (9.15) is equivalent to eq. (9.4), so V_{emf} can be found using either eq. (9.15) or (9.4). In fact, eq. (9.4) can always be applied in place of eqs. (9.6), (9.10), and (915).

EXAMPLE 9.1

A conducting bar can slide freely over two conducting rails as shown in Figure 9.6. Calculate the induced voltage in the bar

(a) If the bar is stationed at $y = 8$ cm and $\mathbf{B} = 4 \cos 10^6 t \mathbf{a}_z$ mWb/m^2
(b) If the bar slides at a velocity $\mathbf{u} = 20\mathbf{a}_y$ m/s and $\mathbf{B} = 4\mathbf{a}_z$ mWb/m^2
(c) If the bar slides at a velocity $\mathbf{u} = 20\mathbf{a}_y$ m/s and $\mathbf{B} = 4 \cos (10^6 t - y) \mathbf{a}_z$ mWb/m^2

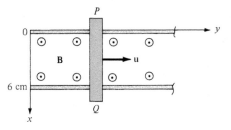

FIGURE 9.6 For Example 9.1.

Solution:

(a) In this case, we have transformer emf given by

$$V_{emf} = -\int_S \frac{\partial \mathbf{B}}{\partial t} \cdot d\mathbf{S} = \int_{y=0}^{0.08} \int_{x=0}^{0.06} 4(10^{-3})(10^6) \sin 10^6 t \, dx \, dy$$

$$= 4(10^3)(0.08)(0.06) \sin 10^6 t$$

$$= 19.2 \sin 10^6 t \text{ V}$$

The polarity of the induced voltage (according to Lenz's law) is such that point P on the bar is at lower potential than Q when $\mathbf{B}$ is increasing.

(b) This is the case of motional emf:

$$V_{emf} = \int_L (\mathbf{u} \times \mathbf{B}) \cdot d\mathbf{l} = \int_{x=\ell}^0 (u\mathbf{a}_y \times B\mathbf{a}_z) \cdot dx \, \mathbf{a}_x$$

$$= -uB\ell = -20(4 \times 10^{-3})(0.06)$$

$$= -4.8 \text{ mV}$$

(c) Both transformer emf and motional emf are present in this case. This problem can be solved in two ways.

Method 1: Using eq. (9.15), we write

$$V_{emf} = -\int_S \frac{\partial \mathbf{B}}{\partial t} \cdot d\mathbf{S} + \int_L (\mathbf{u} \times \mathbf{B}) \cdot d\mathbf{l} \tag{9.1.1}$$

$$= \int_{x=0}^{0.06} \int_0^y 4.10^{-3}(10^6) \sin(10^6 t - y') dy' \, dx$$

$$+ \int_{0.06}^0 [20\mathbf{a}_y \times 4.10^{-3} \cos(10^6 t - y)\mathbf{a}_z] \cdot dx \, \mathbf{a}_x$$

$$= 240 \cos(10^6 t - y')\Big|_0^y - 80(10^{-3})(0.06) \cos(10^6 t - y)$$

$$= 240 \cos(10^6 t - y) - 240 \cos 10^6 t - 4.8(10^{-3}) \cos(10^6 t - y)$$

$$\simeq 240 \cos(10^6 t - y) - 240 \cos 10^6 t \tag{9.1.2}$$

because the motional emf is negligible compared with the transformer emf. Using trigonometric identity, we write

$$\cos A - \cos B = -2 \sin \frac{A + B}{2} \sin \frac{A - B}{2}$$

$$V_{emf} = -480 \sin\left(10^6 t - \frac{y}{2}\right) \sin \frac{y}{2} \text{ V} \tag{9.1.3}$$

Method 2: Alternatively, we can apply eq. (9.4), namely,

$$V_{emf} = -\frac{\partial \Psi}{\partial t} \qquad (9.1.4)$$

where

$$\Psi = \int \mathbf{B} \cdot d\mathbf{S}$$

$$= \int_{y=0}^{y} \int_{x=0}^{0.06} 4 \cos(10^6 t - y) \, dx \, dy$$

$$= -4(0.06) \sin(10^6 t - y) \Big|_{y=0}^{y}$$

$$= -0.24 \sin(10^6 t - y) + 0.24 \sin 10^6 t \, \text{mWb}$$

But

$$\frac{dy}{dt} = u \rightarrow y = ut = 20t$$

Hence,

$$\Psi = -0.24 \sin(10^6 t - 20t) + 0.24 \sin 10^6 t \, \text{mWb}$$

$$V_{emf} = -\frac{\partial \Psi}{\partial t} = 0.24(10^6 - 20) \cos(10^6 t - 20t) - 0.24(10^6) \cos 10^6 t \, \text{mV}$$

$$\approx 240 \cos(10^6 t - y) - 240 \cos 10^6 t \, \text{V} \qquad (9.1.5)$$

which is the same result in (9.1.2). Notice that in eq. (9.1.1), the dependence of y on time is taken care of in $\int (\mathbf{u} \times \mathbf{B}) \cdot d\mathbf{l}$, and we should not be bothered by it in $\partial \mathbf{B}/\partial t$. Why? Because in computing the transformer emf, the loop is assumed stationary. This is a subtle point one must keep in mind in applying eq. (9.1.1). For the same reason, the second method is always easier.

PRACTICE EXERCISE 9.1

Consider the loop of Figure 9.5. If $\mathbf{B} = 0.5\mathbf{a}_z$ Wb/m², $R = 20 \, \Omega$, $\ell = 10$ cm, and the rod is moving with a constant velocity of $8\mathbf{a}_x$ m/s, find

(a) The induced emf in the rod

(b) The current through the resistor

(c) The motional force on the rod

(d) The power dissipated by the resistor.

Answer: (a) 0.4 V, (b) 20 mA, (c) $-\mathbf{a}_x$ mN, (d) 8 mW.

EXAMPLE 9.2

The loop shown in Figure 9.7 is inside a uniform magnetic field $\mathbf{B} = 50\mathbf{a}_x \, \text{mWb/m}^2$. If side DC of the loop cuts the flux lines at the frequency of 50 Hz and the loop lies in the yz-plane at time $t = 0$, find

(a) The induced emf at $t = 1$ ms

(b) The induced current at $t = 3$ ms

Solution:

(a) Since the **B** field is time invariant, the induced emf is motional, that is,

$$V_{\text{emf}} = \int_L (\mathbf{u} \times \mathbf{B}) \cdot d\mathbf{l}$$

where

$$d\mathbf{l} = d\mathbf{l}_{DC} = dz \, \mathbf{a}_z, \quad \mathbf{u} = \frac{d\mathbf{l}'}{dt} = \frac{\rho \, d\phi}{dt} \mathbf{a}_\phi = \rho\omega\mathbf{a}_\phi$$

$$\rho = AD = 4 \, \text{cm}, \quad \omega = 2\pi f = 100\pi$$

Because **u** and $d\mathbf{l}$ are in cylindrical coordinates, we transform **B** into cylindrical coordinates by using eq. (2.9):

$$\mathbf{B} = B_o\mathbf{a}_x = B_o(\cos\phi \, \mathbf{a}_\rho - \sin\phi \, \mathbf{a}_\phi)$$

where $B_o = 0.05$. Hence,

$$\mathbf{u} \times \mathbf{B} = \begin{vmatrix} \mathbf{a}_\rho & \mathbf{a}_\phi & \mathbf{a}_z \\ 0 & \rho\omega & 0 \\ B_o\cos\phi & -B_o\sin\phi & 0 \end{vmatrix} = -\rho\omega B_o \cos\phi \, \mathbf{a}_z$$

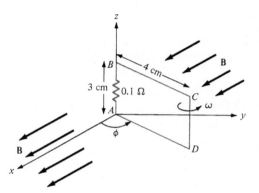

FIGURE 9.7 For Example 9.2; polarity is for increasing emf.

and

$$(\mathbf{u} \times \mathbf{B}) \cdot d\mathbf{l} = -\rho\omega B_o \cos\phi \, dz = -0.04(100\pi)(0.05) \cos\phi \, dz$$
$$= -0.2\pi \cos\phi \, dz$$

$$V_{emf} = \int_{z=0}^{0.03} -0.2\pi \cos\phi \, dz = -6\pi \cos\phi \, \text{mV}$$

To determine ϕ, recall that

$$\omega = \frac{d\phi}{dt} \rightarrow \phi = \omega t + C_o$$

where C_o is an integration constant. At $t = 0$, $\phi = \pi/2$ because the loop is in the yz-plane at that time, $C_o = \pi/2$. Hence,

$$\phi = \omega t + \frac{\pi}{2}$$

and

$$V_{emf} = -6\pi \cos\left(\omega t + \frac{\pi}{2}\right) = 6\pi \sin(100\pi t) \, \text{mV}$$

$$\text{At } t = 1 \text{ ms}, V_{emf} = 6\pi \sin(0.1\pi) = 5.825 \text{ mV}$$

(b) The current induced is

$$i = \frac{V_{emf}}{R} = 60\pi \sin(100\pi t) \, \text{mA}$$

At $t = 3$ ms,

$$i = 60\pi \sin(0.3\pi) \, \text{mA} = 0.1525 \text{ A}$$

PRACTICE EXERCISE 9.2

Rework Example 9.2 with everything the same except that the **B** field is changed to:

(a) $\mathbf{B} = 50\mathbf{a}_y$ mWb/m²—that is, the magnetic field is oriented along the y-direction
(b) $\mathbf{B} = 0.02t\mathbf{a}_x$ Wb/m²—that is, the magnetic field is time varying.

Answer: (a) -17.93 mV, -0.1108 A, (b) 20.5 μV, -41.92 mA.

EXAMPLE 9.3

The magnetic circuit of Figure 9.8 has a uniform cross section of 10^{-3} m². If the circuit is energized by a current $i_1(t) = 3 \sin 100\pi t$ A in the coil of $N_1 = 200$ turns, find the emf induced in the coil of $N_2 = 100$ turns. Assume that $\mu = 500 \, \mu_o$.

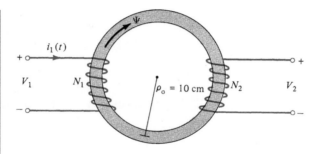

FIGURE 9.8 Magnetic circuit of Example 9.3.

Solution:

The flux in the circuit is

$$\Psi = \frac{\mathcal{F}}{\mathcal{R}} = \frac{N_1 i_1}{\ell/\mu S} = \frac{N_1 i_1 \mu S}{2\pi \rho_o}$$

According to Faraday's law, the emf induced in the second coil is

$$V_2 = -N_2 \frac{d\Psi}{dt} = -\frac{N_1 N_2 \mu S}{2\pi \rho_o} \frac{di_1}{dt}$$

$$= -\frac{100 \cdot (200) \cdot (500) \cdot (4\pi \times 10^{-7}) \cdot (10^{-3}) \cdot 300\pi \cos 100\pi t}{2\pi(10 \times 10^{-2})}$$

$$= -6\pi \cos 100\pi t \text{ V}$$

PRACTICE EXERCISE 9.3

A magnetic core of uniform cross section 4 cm^2 is connected to a 120 V, 60 Hz generator as shown in Figure 9.9. Calculate the induced emf V_2 in the secondary coil.

Answer: 72 V.

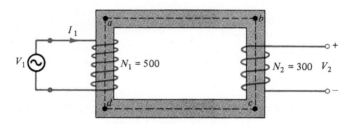

FIGURE 9.9 For Practice Exercise 9.3.

9.4 DISPLACEMENT CURRENT

In Section 9.3 we have essentially reconsidered Maxwell's curl equation for electrostatic fields and modified it for time-varying situations to satisfy Faraday's law. We shall now reconsider Maxwell's curl equation for magnetic fields (Ampère's circuit law) for time-varying conditions.

For static EM fields, we recall that

$$\nabla \times \mathbf{H} = \mathbf{J} \tag{9.17}$$

But the divergence of the curl of any vector field is identically zero (see Example 3.10). Hence,

$$\nabla \cdot (\nabla \times \mathbf{H}) = 0 = \nabla \cdot \mathbf{J} \tag{9.18}$$

The continuity of current in eq. (5.43), however, requires that

$$\nabla \cdot \mathbf{J} = -\frac{\partial \rho_v}{\partial t} \neq 0 \tag{9.19}$$

Thus eqs. (9.18) and (9.19) are obviously incompatible for time-varying conditions. We must modify eq. (9.17) to agree with eq. (9.19). To do this, we add a term to eq. (9.17) so that it becomes

$$\nabla \times \mathbf{H} = \mathbf{J} + \mathbf{J}_d \tag{9.20}$$

where $\mathbf{J}_d$ is to be determined and defined. Again, the divergence of the curl of any vector is zero. Hence:

$$\nabla \cdot (\nabla \times \mathbf{H}) = 0 = \nabla \cdot \mathbf{J} + \nabla \cdot \mathbf{J}_d \tag{9.21}$$

In order for eq. (9.21) to agree with eq. (9.19),

$$\nabla \cdot \mathbf{J}_d = -\nabla \cdot \mathbf{J} = \frac{\partial \rho_v}{\partial t} = \frac{\partial}{\partial t} (\nabla \cdot \mathbf{D}) = \nabla \cdot \frac{\partial \mathbf{D}}{\partial t} \tag{9.22a}$$

or

$$\boxed{\mathbf{J}_d = \frac{\partial \mathbf{D}}{dt}} \tag{9.22b}$$

Substituting eq. (9.22b) into eq. (9.20) results in

$$\boxed{\nabla \times \mathbf{H} = \mathbf{J} + \frac{\partial \mathbf{D}}{\partial t}} \tag{9.23}$$

This is Maxwell's equation (based on Ampère's circuit law) for a time-varying field. The term $\mathbf{J}_d = \partial \mathbf{D}/\partial t$ is known as *displacement current density* and $\mathbf{J}$ is the conduction current density $(\mathbf{J} = \sigma\mathbf{E})$.[3] The insertion of $\mathbf{J}_d$ into eq. (9.17) was one of the major contributions of Maxwell. Without the term $\mathbf{J}_d$, the propagation of electromagnetic waves (e.g., radio or TV waves) would be impossible. At low frequencies, $\mathbf{J}_d$ is usually neglected compared with $\mathbf{J}$. However, at radio frequencies, the two terms are comparable. At the time of Maxwell, high-frequency sources were not available and eq. (9.23) could not be verified experimentally. It was years later that Hertz succeeded in generating and detecting radio waves, thereby verifying eq. (9.23). This is one of the rare cases of a mathematical argument paving the way for experimental investigation.

Based on the displacement current density, we define the *displacement current* as

$$I_d = \int_S \mathbf{J}_d \cdot d\mathbf{S} = \int_S \frac{\partial \mathbf{D}}{\partial t} \cdot d\mathbf{S} \tag{9.24}$$

We must bear in mind that displacement current is a result of time-varying electric field. A typical example of such current is the current through a capacitor when an alternating voltage source is applied to its plates. This example, shown in Figure 9.10, serves to illustrate the need for the displacement current. Applying an unmodified form of Ampère's circuit law to a closed path L shown in Figure 9.10(a) gives

$$\oint_L \mathbf{H} \cdot d\mathbf{l} = \int_{S_1} \mathbf{J} \cdot d\mathbf{S} = I_{\text{enc}} = I \tag{9.25}$$

where I is the current through the conductor and S_1 is the flat surface bounded by L. If we use the balloon-shaped surface S_2 that passes between the capacitor plates, as in Figure 9.10(b),

$$\oint_L \mathbf{H} \cdot d\mathbf{l} = \int_{S_2} \mathbf{J} \cdot d\mathbf{S} = I_{\text{enc}} = 0 \tag{9.26}$$

because no conduction current $(\mathbf{J} = 0)$ flows through S_2. This is contradictory in view of the fact that the same closed path L is used. To resolve the conflict, we need to include

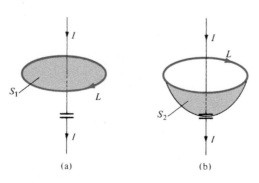

FIGURE 9.10 Two surfaces of integration showing the need for $\mathbf{J}_d$ in Ampère's circuit law.

(a)

(b)

---

[3] Recall that we also have $\mathbf{J} = \rho_v \mathbf{u}$ as the convection current density.

the displacement current in Ampère's circuit law. The total current density is $\mathbf{J} + \mathbf{J}_d$. In eq. (9.25), $\mathbf{J}_d = \mathbf{0}$, so that the equation remains valid. In eq. (9.26), $\mathbf{J} = \mathbf{0}$, so that

$$\oint_L \mathbf{H} \cdot d\mathbf{l} = \int_{S_2} \mathbf{J}_d \cdot d\mathbf{S} = \frac{d}{dt} \int_{S_2} \mathbf{D} \cdot d\mathbf{S} = \frac{dQ}{dt} = I \qquad (9.27)$$

So we obtain the same current for either surface, although it is conduction current in S_1 and displacement current in S_2.

EXAMPLE 9.4

A parallel-plate capacitor with plate area of 5 cm^2 and plate separation of 3 mm has a voltage 50 sin $10^3 t$ V applied to its plates. Calculate the displacement current assuming $\varepsilon = 2\varepsilon_0$.

Solution:

$$D = \varepsilon E = \varepsilon \frac{V}{d}$$

$$J_d = \frac{\partial D}{\partial t} = \frac{\varepsilon}{d} \frac{dV}{dt}$$

Hence,

$$I_d = J_d \cdot S = \frac{\varepsilon S}{d} \frac{dV}{dt} = C \frac{dV}{dt}$$

which is the same as the conduction current, given by

$$I_c = \frac{dQ}{dt} = S \frac{d\rho_s}{dt} = S \frac{dD}{dt} = \varepsilon S \frac{dE}{dt} = \frac{\varepsilon S}{d} \frac{dV}{dt} = C \frac{dV}{dt}$$

$$I_d = 2 \cdot \frac{10^{-9}}{36\pi} \cdot \frac{5 \times 10^{-4}}{3 \times 10^{-3}} \cdot 10^3 \times 50 \cos 10^3 t$$

$$= 147.4 \cos 10^3 t \text{ nA}$$

PRACTICE EXERCISE 9.4

In free space, $\mathbf{E} = 20 \cos(\omega t - 50x) \, \mathbf{a}_y$ V/m. Calculate

(a) $\mathbf{J}_d$
(b) $\mathbf{H}$
(c) ω

Answer: (a) $-20\omega\varepsilon_0 \sin(\omega t - 50x) \, \mathbf{a}_y$ A/m^2, (b) $0.4 \, \omega\varepsilon_0 \cos(\omega t - 50x) \, \mathbf{a}_z$ A/m, (c) 1.5×10^{10} rad/s.

9.5 MAXWELL'S EQUATIONS IN FINAL FORMS

The Scottish physicist James Clerk Maxwell (1831–1879) is regarded as the founder of electromagnetic theory in its present form. Maxwell's celebrated work led to the discovery of electromagnetic waves.[4] Through his theoretical efforts when he was between 35 and 40 years old, Maxwell published the first unified theory of electricity and magnetism. The theory comprised all previously known results, both experimental and theoretical, on electricity and magnetism. It further introduced displacement current and predicted the existence of electromagnetic waves. Maxwell's equations were not fully accepted by many scientists until 1888, when they were confirmed by Heinrich Rudolf Hertz (1857–1894). The German physicist was successful in generating and detecting radio waves.

The laws of electromagnetism that Maxwell put together in the form of four equations were presented in Table 7.2 in Section 7.6 for static conditions. The more generalized forms of these equations are those for time-varying conditions shown in Table 9.1. We notice from the table that the divergence equations remain the same, while the curl equations have been modified. The integral form of Maxwell's equations depicts the underlying physical laws, whereas the differential form is used more frequently in solving problems. For a field to "qualify" as an electromagnetic field, it must satisfy all four Maxwell's equations. The importance of Maxwell's equations cannot be overemphasized because they summarize all known laws of electromagnetism. We shall often refer to them in the remainder of this text.

Since this section is meant to be a compendium of our discussion in this text, it is worthwhile to mention other equations that go hand in hand with Maxwell's equations. The Lorentz force equation

$$\mathbf{F} = Q(\mathbf{E} + \mathbf{u} \times \mathbf{B}) \tag{9.28}$$

TABLE 9.1 Generalized Forms of Maxwell's Equations

Differential Form	Integral Form	Remarks
$\nabla \cdot \mathbf{D} = \rho_v$	$\oint_S \mathbf{D} \cdot d\mathbf{S} = \int_v \rho_v \, dv$	Gauss's law
$\nabla \cdot \mathbf{B} = 0$	$\oint_S \mathbf{B} \cdot d\mathbf{S} = 0$	Nonexistence of isolated magnetic charge*
$\nabla \times \mathbf{E} = -\dfrac{\partial \mathbf{B}}{\partial t}$	$\oint_L \mathbf{E} \cdot d\mathbf{l} = -\dfrac{\partial}{\partial t}\int_S \mathbf{B} \cdot d\mathbf{S}$	Faraday's law
$\nabla \times \mathbf{H} = \mathbf{J} + \dfrac{\partial \mathbf{D}}{\partial t}$	$\oint_L \mathbf{H} \cdot d\mathbf{l} = \int_S \left(\mathbf{J} + \dfrac{\partial \mathbf{D}}{\partial t}\right) \cdot d\mathbf{S}$	Ampère's circuit law

*This is also referred to as Gauss's law for magnetic fields.

---

[4] Maxwell's work can be found in his two-volume *Treatise on Electricity and Magnetism* (New York: Dover, 1954).

is associated with Maxwell's equations. Also the equation of continuity

$$\nabla \cdot \mathbf{J} = -\frac{\partial \rho_v}{\partial t} \tag{9.29}$$

is implicit in Maxwell's equations. The concepts of linearity, isotropy, and homogeneity of a material medium still apply for time-varying fields; in a linear, homogeneous, and isotropic medium characterized by σ, ε, and μ, the constitutive relations

$$\mathbf{D} = \varepsilon\mathbf{E} = \varepsilon_o\mathbf{E} + \mathbf{P} \tag{9.30a}$$

$$\mathbf{B} = \mu\mathbf{H} = \mu_o(\mathbf{H} + \mathbf{M}) \tag{9.30b}$$

$$\mathbf{J} = \sigma\mathbf{E} + \rho_v\mathbf{u} \tag{9.30c}$$

hold for time-varying fields. Consequently, the boundary conditions remain valid for time-varying fields, where $\mathbf{a}_n$ is the unit normal vector to the boundary.

$$E_{1t} - E_{2t} = 0 \quad \text{or} \quad (\mathbf{E}_1 - \mathbf{E}_2) \times \mathbf{a}_n = \mathbf{0} \tag{9.31a}$$

$$H_{1t} - H_{2t} = K \quad \text{or} \quad (\mathbf{H}_1 - \mathbf{H}_2) \times \mathbf{a}_n = \mathbf{K} \tag{9.31b}$$

$$D_{1n} - D_{2n} = \rho_s \quad \text{or} \quad (\mathbf{D}_1 - \mathbf{D}_2) \cdot \mathbf{a}_n = \rho_s \tag{9.31c}$$

$$B_{1n} - B_{2n} = 0 \quad \text{or} \quad (\mathbf{B}_2 - \mathbf{B}_1) \cdot \mathbf{a}_n = 0 \tag{9.31d}$$

However, for a perfect conductor ($\sigma \simeq \infty$) in a time-varying field,

$$\mathbf{E} = \mathbf{0}, \quad \mathbf{H} = \mathbf{0}, \quad \mathbf{J} = \mathbf{0} \tag{9.32}$$

and hence,

$$B_n = 0, \quad E_t = 0 \tag{9.33}$$

For a perfect dielectric ($\sigma \simeq 0$), eqs. (9.31) hold except that $\mathbf{K} = \mathbf{0}$. Though eqs. (9.28) to (9.33) are not Maxwell's equations, they are associated with them.

To complete this summary section, we present a structure linking the various potentials and vector fields of the electric and magnetic fields in Figure 9.11. This electromagnetic flow diagram helps with the visualization of the basic relationships between field quantities. It also shows that it is usually possible to find alternative formulations, for a given problem, in a relatively simple manner. It should be noted that in Figure 9.11(b) and (c), we introduce ρ^m as the free magnetic density (similar to ρ_v), which is, of course, zero, $\mathbf{A}_e$ as the electric vector potential (analogous to $\mathbf{A}$), and $\mathbf{J}_m$ as the magnetic current density (analogous to $\mathbf{J}$). Using terms from stress analysis, the principal relationships are typified as follows:

(a) compatibility equations

$$\nabla \cdot \mathbf{B} = \rho^m = 0 \tag{9.34}$$

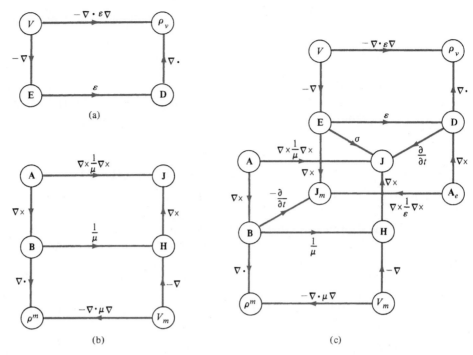

FIGURE 9.11 Electromagnetic flow diagrams showing the relationship between the potentials and vector fields: (**a**) electrostatic system, (**b**) magnetostatic system, (**c**) electromagnetic system. [Adapted with permission from the Publishing Department of the Institution of Electrical Engineers.]

and

$$\nabla \times \mathbf{E} = -\frac{\partial \mathbf{B}}{\partial t} = \mathbf{J}_m \tag{9.35}$$

(b) constitutive equations

$$\mathbf{B} = \mu \mathbf{H} \tag{9.36}$$

and

$$\mathbf{D} = \varepsilon \mathbf{E} \tag{9.37}$$

(c) equilibrium equations

$$\nabla \cdot \mathbf{D} = \rho_v \tag{9.38}$$

and

$$\nabla \times \mathbf{H} = \mathbf{J} + \frac{\partial \mathbf{D}}{\partial t} \tag{9.39}$$

†9.6 TIME-VARYING POTENTIALS

For static EM fields, we obtained the electric scalar potential as

$$V = \int_v \frac{\rho_v \, dv}{4\pi \varepsilon R} \tag{9.40}$$

and the magnetic vector potential as

$$\mathbf{A} = \int_v \frac{\mu \mathbf{J} \, dv}{4\pi R} \tag{9.41}$$

We would like to examine what happens to these potentials when the fields are time varying. Recall that $\mathbf{A}$ was defined from the fact that $\nabla \cdot \mathbf{B} = 0$, which still holds for time-varying fields. Hence the relation

$$\boxed{\mathbf{B} = \nabla \times \mathbf{A}} \tag{9.42}$$

holds for time-varying situations. Combining Faraday's law as expressed in eq. (9.8) with eq. (9.42) gives

$$\nabla \times \mathbf{E} = -\frac{\partial}{\partial t}(\nabla \times \mathbf{A}) \tag{9.43a}$$

or

$$\nabla \times \left(\mathbf{E} + \frac{\partial \mathbf{A}}{\partial t}\right) = \mathbf{0} \tag{9.43b}$$

Since the curl of the gradient of a scalar field is identically zero (see Practice Exercise 3.10), the solution to eq. (9.43b) is

$$\mathbf{E} + \frac{\partial \mathbf{A}}{\partial t} = -\nabla V \tag{9.44}$$

or

$$\boxed{\mathbf{E} = -\nabla V - \frac{\partial \mathbf{A}}{\partial t}} \tag{9.45}$$

From eqs. (9.42) and (9.45), we can determine the vector fields $\mathbf{B}$ and $\mathbf{E}$, provided the potentials $\mathbf{A}$ and V are known. However, we still need to find some expressions for $\mathbf{A}$ and V similar to those in eqs. (9.40) and (9.41) that are suitable for time-varying fields.

From Table 9.1 or eq. (9.38) we know that $\nabla \cdot \mathbf{D} = \rho_v$ is valid for time-varying conditions. By taking the divergence of eq. (9.45) and making use of eqs. (9.37) and (9.38), we obtain

$$\nabla \cdot \mathbf{E} = \frac{\rho_v}{\varepsilon} = -\nabla^2 V - \frac{\partial}{\partial t}(\nabla \cdot \mathbf{A})$$

or

$$\nabla^2 V + \frac{\partial}{\partial t}(\nabla \cdot \mathbf{A}) = -\frac{\rho_v}{\varepsilon} \tag{9.46}$$

Taking the curl of eq. (9.42) and incorporating eqs. (9.23) and (9.45) results in

$$\nabla \times \nabla \times \mathbf{A} = \mu \mathbf{J} + \varepsilon\mu \frac{\partial}{\partial t}\left(-\nabla V - \frac{\partial \mathbf{A}}{\partial t}\right)$$

$$= \mu \mathbf{J} - \mu\varepsilon \nabla\left(\frac{\partial V}{\partial t}\right) - \mu\varepsilon \frac{\partial^2 \mathbf{A}}{\partial t^2} \tag{9.47}$$

where $\mathbf{D} = \varepsilon\mathbf{E}$ and $\mathbf{B} = \mu\mathbf{H}$ have been assumed. By applying the vector identity

$$\nabla \times \nabla \times \mathbf{A} = \nabla(\nabla \cdot \mathbf{A}) - \nabla^2\mathbf{A} \tag{9.48}$$

to eq. (9.47),

$$\nabla^2\mathbf{A} - \nabla(\nabla \cdot \mathbf{A}) = -\mu \mathbf{J} + \mu\varepsilon \nabla\left(\frac{\partial V}{\partial t}\right) + \mu\varepsilon \frac{\partial^2 \mathbf{A}}{\partial t^2} \tag{9.49}$$

A vector field is uniquely defined when its curl and divergence are specified. The curl of **A** has been specified by eq. (9.42); for reasons that will be obvious shortly, we may choose the divergence of **A** as

$$\boxed{\nabla \cdot \mathbf{A} = -\mu\varepsilon \frac{\partial V}{\partial t}} \tag{9.50}$$

This choice relates **A** and *V*, and it is called the *Lorenz condition for potentials*.[5] We had this in mind when we chose $\nabla \cdot \mathbf{A} = 0$ for magnetostatic fields in eq. (7.59). By imposing the Lorenz condition of eq. (9.50), eqs. (9.46) and (9.49), respectively, become

$$\boxed{\nabla^2 V - \mu\varepsilon \frac{\partial^2 V}{\partial t^2} = -\frac{\rho_v}{\varepsilon}} \tag{9.51}$$

and

$$\boxed{\nabla^2\mathbf{A} - \mu\varepsilon \frac{\partial^2 \mathbf{A}}{\partial t^2} = -\mu \mathbf{J}} \tag{9.52}$$

which are *wave equations* to be discussed in the next chapter. The reason for choosing the Lorenz condition becomes obvious as we examine eqs. (9.51) and (9.52). The Lorenz condition uncouples eqs. (9.46) and (9.49) and also produces a symmetry between eqs. (9.51) and (9.52). It can be shown that the Lorenz condition can be obtained from the continuity equation; therefore, our choice of eq. (9.50) is not arbitrary. Notice that eqs. (6.4) and (7.60) are special static cases of eqs. (9.51) and (9.52), respectively. In other words, potentials *V* and **A** satisfy Poisson's equations for time-varying conditions. Just as

---

[5] Not to be confused with Hendrick A. Lorentz, Ludvig V. Lorenz (1829–1891) was a Danish mathematician and physicist.

eqs. (9.40) and (9.41) are the solutions, or the integral forms of eqs. (6.4) and (7.60), it can be shown that the solutions[6] to eqs. (9.51) and (9.52) are

$$V = \int_v \frac{[\rho_v]\,dv}{4\pi\varepsilon R} \tag{9.53}$$

and

$$\mathbf{A} = \int_v \frac{\mu[\mathbf{J}]\,dv}{4\pi R} \tag{9.54}$$

The term $[\rho_v]$ (or $[\mathbf{J}]$) means that the time t in $\rho_v(x, y, z, t)$ [or $\mathbf{J}(x, y, z, t)$] is replaced by the *retarded time* t' given by

$$t' = t - \frac{R}{u} \tag{9.55}$$

where $R = |\mathbf{r} - \mathbf{r}'|$ is the distance between the source point $\mathbf{r}'$ and the observation point $\mathbf{r}$ and

$$u = \frac{1}{\sqrt{\mu\varepsilon}} \tag{9.56}$$

is the velocity of wave propagation. In free space, $u = c \simeq 3 \times 10^8$ m/s is the speed of light in a vacuum. Potentials V and $\mathbf{A}$ in eqs. (9.53) and (9.54) are, respectively, called the *retarded electric scalar potential* and the *retarded magnetic vector potential*. Given ρ_v and $\mathbf{J}$, V and $\mathbf{A}$ can be determined by using eqs. (9.53) and (9.54); from V and $\mathbf{A}$, $\mathbf{E}$ and $\mathbf{B}$ can be determined by using eqs. (9.45) and (9.42), respectively.

9.7 TIME-HARMONIC FIELDS

So far, our time dependence of EM fields has been arbitrary. To be specific, we shall assume that the fields are *time harmonic*.

A **time-harmonic field** is one that varies periodically or sinusoidally with time.

Not only is sinusoidal analysis of practical value, but also it can be extended to most waveforms by Fourier analysis. Sinusoids are easily expressed in phasors, which are more convenient to work with. Before applying phasors to EM fields, it is worthwhile to have a brief review of the concept of phasor.

A *phasor* is a complex number that contains the amplitude and the phase of a sinusoidal oscillation. As a complex number, a phasor z can be represented as

$$z = x + jy = r\underline{/\phi} \tag{9.57}$$

[6] For example, see D. K. Cheng, *Fundamentals of Engineering Electromagnetics*. Reading, MA: Addison-Wesley, 1993, pp. 253–254.

or

$$z = r\, e^{j\phi} = r\,(\cos\phi + j\sin\phi) \qquad (9.58)$$

where $j = \sqrt{-1}$, x is the real part of z, y is the imaginary part of z, r is the magnitude of z, given by

$$r = |z| = \sqrt{x^2 + y^2} \qquad (9.59)$$

and ϕ is the phase of z, given by

$$\phi = \tan^{-1}\frac{y}{x} \qquad (9.60)$$

Here x, y, z, r, and ϕ should not be mistaken as the coordinate variables, although they look similar (different letters could have been used but it is hard to find better ones). The phasor z can be represented in *rectangular form* as $z = x + jy$ or in *polar form* as $z = r\,\underline{/\phi} = r\,e^{j\phi}$. The two forms of representing z are related in eqs. (9.57) to (9.60) and illustrated in Figure 9.12. Addition and subtraction of phasors are better performed in rectangular form; multiplication and division are better done in polar form.

Given complex numbers

$$z = x + jy = r\,\underline{/\phi}, \quad z_1 = x_1 + jy_1 = r_1\,\underline{/\phi_1}, \quad \text{and} \quad z_2 = x_2 + jy_2 = r_2\,\underline{/\phi_2}$$

the following basic properties should be noted.

addition:

$$z_1 + z_2 = (x_1 + x_2) + j(y_1 + y_2) \qquad (9.61a)$$

subtraction:

$$z_1 - z_2 = (x_1 - x_2) + j(y_1 - y_2) \qquad (9.61b)$$

multiplication:

$$z_1 z_2 = r_1 r_2\,\underline{/\phi_1 + \phi_2} \qquad (9.61c)$$

division:

$$\frac{z_1}{z_2} = \frac{r_1}{r_2}\,\underline{/\phi_1 - \phi_2} \qquad (9.61d)$$

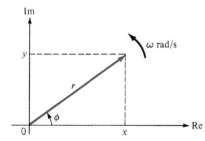

FIGURE 9.12 Representation of a phasor $z = x + jy = r\,\underline{/\phi}$.

Square root:

$$\sqrt{z} = \sqrt{r} \, \underline{/\phi/2} \qquad (9.61e)$$

Complex conjugate:

$$z^* = x - jy = r\underline{/-\phi} = re^{-j\phi} \qquad (9.61f)$$

Other properties of complex numbers can be found in Appendix A.2.

To introduce the time element, we let

$$\phi = \omega t + \theta \qquad (9.62)$$

where θ may be a function of time or space coordinates or a constant. The real (Re) and imaginary (Im) parts of

$$re^{j\phi} = re^{j(\omega t + \theta)} = re^{j\theta} \, e^{j\omega t} \qquad (9.63)$$

are respectively given by

$$\mathrm{Re}(re^{j\phi}) = r\cos(\omega t + \theta) \qquad (9.64a)$$

and

$$\mathrm{Im}(re^{j\phi}) = r\sin(\omega t + \theta) \qquad (9.64b)$$

Thus, a sinusoidal current $I(t) = I_o \cos(\omega t + \theta)$, for example, equals the real part of $I_o e^{j\theta} e^{j\omega t}$. The current $I'(t) = I_o \sin(\omega t + \theta)$, which is the imaginary part of $I_o e^{j\theta} e^{j\omega t}$, can be represented as the real part of $I_o e^{j\theta} e^{j\omega t} e^{-j90°}$ because $\sin \alpha = \cos(\alpha - 90°)$. However, in performing our mathematical operations, we must be consistent in our use of either the real part or the imaginary part of a quantity, but never both at the same time.

The complex term $I_o e^{j\theta}$, which results from dropping the time factor $e^{j\omega t}$ in $I(t)$, is called the *phasor* current, denoted by I_s; that is,

$$I_s = I_o e^{j\theta} = I_o \, \underline{/\theta} \qquad (9.65)$$

where the subscript s denotes the phasor form of $I(t)$. Thus $I(t) = I_o \cos(\omega t + \theta)$, the *instantaneous form*, can be expressed as

$$I(t) = \mathrm{Re}(I_s e^{j\omega t}) \qquad (9.66)$$

In general, a phasor is a complex quantity and could be a scalar or a vector. If a vector $\mathbf{A}(x, y, z, t)$ is a time-harmonic field, the *phasor form* of $\mathbf{A}$ is $\mathbf{A}_s(x, y, z)$; the two quantities are related as

$$\boxed{\mathbf{A}(x, y, z, t) = \mathrm{Re}[\mathbf{A}_s(x, y, z)e^{j\omega t}]} \qquad (9.67)$$

Note that the phasor is a function of position, not a function of time. For example, if $\mathbf{A} = A_o \cos(\omega t - \beta x)\,\mathbf{a}_y$, we can write $\mathbf{A}$ as

$$\mathbf{A} = \mathrm{Re}(A_o e^{-j\beta x}\mathbf{a}_y e^{j\omega t}) \qquad (9.68)$$

Comparing this with eq. (9.67) indicates that the phasor form of **A** is

$$\mathbf{A}_s = A_o e^{-j\beta x} \mathbf{a}_y \tag{9.69}$$

Notice from eq. (9.67) that

$$
\begin{aligned}
\frac{\partial \mathbf{A}}{\partial t} &= \frac{\partial}{\partial t} \operatorname{Re}(\mathbf{A}_s e^{j\omega t}) \\
&= \operatorname{Re}(j\omega \mathbf{A}_s e^{j\omega t})
\end{aligned} \tag{9.70}
$$

showing that taking the time derivative of the instantaneous quantity is equivalent to multiplying its phasor form by $j\omega$. That is,

$$\frac{\partial \mathbf{A}}{\partial t} \rightarrow j\omega \mathbf{A}_s \tag{9.71}$$

Similarly,

$$\int \mathbf{A} \, \partial t \rightarrow \frac{\mathbf{A}_s}{j\omega} \tag{9.72}$$

Note that the real part is chosen in eq. (9.67) as in circuit analysis; the imaginary part could equally have been chosen. Also notice the basic difference between the instantaneous form $\mathbf{A}(x, y, z, t)$ and its phasor form $\mathbf{A}_s(x, y, z)$: the former is time dependent and real, whereas the latter is time invariant and generally complex. It is easier to work with $\mathbf{A}_s$ and obtain **A** from $\mathbf{A}_s$ whenever necessary by using eq. (9.67).

We shall now apply the phasor concept to time-varying EM fields. The field quantities $\mathbf{E}(x, y, z, t)$, $\mathbf{D}(x, y, z, t)$, $\mathbf{H}(x, y, z, t)$, $\mathbf{B}(x, y, z, t)$, $\mathbf{J}(x, y, z, t)$, and $\rho_v(x, y, z, t)$ and their derivatives can be expressed in phasor form by using eqs. (9.67) and (9.71).

Let us see how we can write Maxwell's equations in phasor form. For example, consider

$$\nabla \times \mathbf{E}(x, y, z, t) = -\frac{\partial}{\partial t} \mathbf{B}(x, y, z, t) \tag{9.73}$$

We let

$$\mathbf{E}(x, y, z, t) = \operatorname{Re}\big(\mathbf{E}_s(x, y, z) e^{j\omega t}\big)$$

and

$$\mathbf{B}(x, y, z, t) = \operatorname{Re}\big(\mathbf{B}_s(x, y, z) e^{j\omega t}\big)$$

Substituting these in eq. (9.73) gives

$$\nabla \times \left\{ \operatorname{Re}[\mathbf{E}_s e^{j\omega t}] \right\} = -\frac{\partial}{\partial t} \left\{ \operatorname{Re}[\mathbf{B}_s e^{j\omega t}] \right\} \tag{9.74}$$

We consider the left-hand side of eq. (9.74). The curl operation operates only on (x, y, z),

$$\nabla \times \left\{ \operatorname{Re}[\mathbf{E}_s e^{j\omega t}] \right\} = \operatorname{Re}\left\{ [\nabla \times \mathbf{E}_s] e^{j\omega t} \right\} \tag{9.75}$$

We similarly consider the right-hand side of eq. (9.74), keeping in mind that $\mathbf{B}_s$ does not depend on time:

$$-\frac{\partial}{\partial t}\left\{\operatorname{Re}[\mathbf{B}_s e^{j\omega t}]\right\} = -\operatorname{Re}\left\{\mathbf{B}_s \frac{\partial}{\partial t} e^{j\omega t}\right\} = -\operatorname{Re}\{j\omega \mathbf{B}_s e^{j\omega t}\} \tag{9.76}$$

Comparing eqs. (9.75) and (9.76), we obtain

$$\nabla \times \mathbf{E}_s = -j\omega \mathbf{B}_s \tag{9.77}$$

which is the phasor form of eq. (9.73). Other Maxwell's equations can be treated in a similar manner, and we obtain Table 9.2. From Table 9.2, note that the time factor $e^{j\omega t}$ disappears because it is associated with every term and therefore factors out, resulting in time-independent equations. Herein lies the justification for using phasors: the time factor can be suppressed in our analysis of time-harmonic fields and inserted when necessary. Also note that in Table 9.2, the time factor $e^{j\omega t}$ has been assumed. It is equally possible to have assumed the time factor $e^{-j\omega t}$, in which case we would need to replace every j in Table 9.2 with $-j$.

TABLE 9.2 Time-Harmonic Maxwell's Equations Assuming Time Factor $e^{j\omega t}$

Point Form	Integral Form
$\nabla \cdot \mathbf{D}_s = \rho_{vs}$	$\oint \mathbf{D}_s \cdot d\mathbf{S} = \int \rho_{vs}\, dv$
$\nabla \cdot \mathbf{B}_s = 0$	$\oint \mathbf{B}_s \cdot d\mathbf{S} = 0$
$\nabla \times \mathbf{E}_s = -j\omega \mathbf{B}_s$	$\oint \mathbf{E}_s \cdot d\mathbf{l} = -j\omega \int \mathbf{B}_s \cdot d\mathbf{S}$
$\nabla \times \mathbf{H}_s = \mathbf{J}_s + j\omega \mathbf{D}_s$	$\oint \mathbf{H}_s \cdot d\mathbf{l} = \int (\mathbf{J}_s + j\omega \mathbf{D}_s) \cdot d\mathbf{S}$

EXAMPLE 9.5

Evaluate the complex numbers

(a) $z_1 = \dfrac{j(3 - j4)^*}{(-1 + j6)(2 + j)^2}$

(b) $z_2 = \left[\dfrac{1 + j}{4 - j8}\right]^{1/2}$

Solution:

(a) This can be solved in two ways: working with z in rectangular form or polar form.

Method 1 (working in rectangular form):

Let

$$z_1 = \frac{z_3 z_4}{z_5 z_6}$$

where

$$z_3 = j$$
$$z_4 = (3 - j4)^* = \text{the complex conjugate of } (3 - j4)$$
$$= 3 + j4$$

We note parenthetically that one can find the complex conjugate of a complex number simply by replacing every j with $-j$:

$$z_5 = -1 + j6$$

and

$$z_6 = (2 + j)^2 = 4 - 1 + j4 = 3 + j4$$

Hence,

$$z_3 z_4 = j(3 + j4) = -4 + j3$$
$$z_5 z_6 = (-1 + j6)(3 + j4) = -3 - j4 + j18 - 24$$
$$= -27 + j14$$

and

$$z_1 = \frac{-4 + j3}{-27 + j14}$$

Multiplying and dividing z_1 by $-27 - j14$ (rationalization), we have

$$z_1 = \frac{(-4 + j3)(-27 - j14)}{(-27 + j14)(-27 - j14)} = \frac{150 - j25}{27^2 + 14^2}$$
$$= 0.1622 - j0.027 = 0.1644 \underline{/-9.46°}$$

Method 2 (working in polar form):

$$z_3 = j = 1 \underline{/90°}$$
$$z_4 = (3 - j4)^* = (5 \underline{/-53.13})^* = 5 \underline{/53.13°}$$

$$z_5 = (-1 + j6) = \sqrt{37} \underline{/99.46°}$$
$$z_6 = (2 + j)^2 = (\sqrt{5} \underline{/26.56})^2 = 5 \underline{/53.13°}$$

Hence,

$$z_1 = \frac{(1 \underline{/90°})(5 \underline{/53.13°})}{(\sqrt{37} \underline{/99.46°})(5 \underline{/53.13°})}$$
$$= \frac{1}{\sqrt{37}} \underline{/90° - 99.46°} = 0.1644 \underline{/-9.46°}$$
$$= 0.1622 - j0.027$$

as obtained before.

(b) Let

$$z_2 = \left[\frac{z_7}{z_8}\right]^{1/2}$$

where

$$z_7 = 1 + j = \sqrt{2}\ \underline{/45°}$$

and

$$z_8 = 4 - j8 = 4\sqrt{5}\ \underline{/-63.4°}$$

Hence

$$\frac{z_7}{z_8} = \frac{\sqrt{2}\ \underline{/45°}}{4\sqrt{5}\ \underline{/-63.4°}} = \frac{\sqrt{2}}{4\sqrt{5}}\ \underline{/45° - -63.4°}$$

$$= 0.1581\ \underline{/108.4°}$$

and

$$z_2 = \sqrt{0.1581}\ \underline{/108.4°/2}$$

$$= 0.3976\ \underline{/54.2°}$$

PRACTICE EXERCISE 9.5

Evaluate these complex numbers:

(a) $j^3\left[\dfrac{1 + j}{2 - j}\right]^2$

(b) $6\ \underline{/30°} + j5 - 3 + e^{j45°}$

Answer: (a) $0.24 + j0.32$, (b) $2.03 + j8.707$.

EXAMPLE 9.6

Given that $\mathbf{A} = 10\cos(10^8 t - 10x + 60°)\,\mathbf{a}_z$ and $\mathbf{B}_s = (20/j)\,\mathbf{a}_x + 10\,e^{j2\pi x/3}\,\mathbf{a}_y$, express $\mathbf{A}$ in phasor form and $\mathbf{B}_s$ in instantaneous form.

Solution:

$$\mathbf{A} = \mathrm{Re}\left[10e^{j(\omega t - 10x + 60°)}\mathbf{a}_z\right]$$

where $\omega = 10^8$. Hence

$$\mathbf{A} = \mathrm{Re}\left[10e^{j(60° - 10x)}\,\mathbf{a}_z\,e^{j\omega t}\right] = \mathrm{Re}\left(\mathbf{A}_s e^{j\omega t}\right)$$

or

$$\mathbf{A}_s = 10\,e^{j(60° - 10x)}\mathbf{a}_z$$

If

$$\mathbf{B}_s = \frac{20}{j}\mathbf{a}_x + 10e^{j2\pi x/3}\mathbf{a}_y = -j20\mathbf{a}_x + 10e^{j2\pi x/3}\mathbf{a}_y$$

$$= 20e^{-j\pi/2}\mathbf{a}_x + 10e^{j2\pi x/3}\mathbf{a}_y$$

$$\mathbf{B} = \mathrm{Re}(\mathbf{B}_s e^{j\omega t})$$

$$= \mathrm{Re}\left[20e^{j(\omega t - \pi/2)}\mathbf{a}_x + 10e^{j(\omega t + 2\pi x/3)}\mathbf{a}_y\right]$$

$$= 20\cos(\omega t - \pi/2)\mathbf{a}_x + 10\cos\left(\omega t + \frac{2\pi x}{3}\right)\mathbf{a}_y$$

$$= 20\sin\omega t\,\mathbf{a}_x + 10\cos\left(\omega t + \frac{2\pi x}{3}\right)\mathbf{a}_y$$

PRACTICE EXERCISE 9.6

If $\mathbf{P} = 2\sin(10t + x - \pi/4)\mathbf{a}_y$ and $\mathbf{Q}_s = e^{jx}(\mathbf{a}_x - \mathbf{a}_z)\sin\pi y$, determine the phasor form of $\mathbf{P}$ and the instantaneous form of $\mathbf{Q}_s$.

Answer: $2e^{j(x-3\pi/4)}\mathbf{a}_y$, $\sin\pi y\cos(\omega t + x)(\mathbf{a}_x - \mathbf{a}_z)$.

EXAMPLE 9.7

The electric field and the magnetic field in free space are given by

$$\mathbf{E} = \frac{50}{\rho}\cos(10^6 t + \beta z)\mathbf{a}_\phi \text{ V/m}$$

$$\mathbf{H} = \frac{H_o}{\rho}\cos(10^6 t + \beta z)\mathbf{a}_\rho \text{ A/m}$$

Express these in phasor form and determine the constants H_o and β such that the fields satisfy Maxwell's equations.

Solution:

The instantaneous forms of $\mathbf{E}$ and $\mathbf{H}$ are written as

$$\mathbf{E} = \mathrm{Re}(\mathbf{E}_s e^{j\omega t}), \quad \mathbf{H} = \mathrm{Re}(\mathbf{H}_s e^{j\omega t}) \tag{9.7.1}$$

where $\omega = 10^6$ and phasors $\mathbf{E}_s$ and $\mathbf{H}_s$ are given by

$$\mathbf{E}_s = \frac{50}{\rho}e^{j\beta z}\mathbf{a}_\phi, \quad \mathbf{H}_s = \frac{H_o}{\rho}e^{j\beta z}\mathbf{a}_\rho \tag{9.7.2}$$

For free space, $\rho_v = 0$, $\sigma = 0$, $\varepsilon = \varepsilon_o$, and $\mu = \mu_o$, so Maxwell's equations become

$$\nabla \cdot \mathbf{D} = \varepsilon_o \nabla \cdot \mathbf{E} = 0 \quad \rightarrow \quad \nabla \cdot \mathbf{E}_s = 0 \tag{9.7.3}$$

$$\nabla \cdot \mathbf{B} = \mu_o \nabla \cdot \mathbf{H} = 0 \quad \rightarrow \quad \nabla \cdot \mathbf{H}_s = 0 \tag{9.7.4}$$

$$\nabla \times \mathbf{H} = \sigma \mathbf{E} + \varepsilon_o \frac{\partial \mathbf{E}}{\partial t} \quad \rightarrow \quad \nabla \times \mathbf{H}_s = j\omega\varepsilon_o \mathbf{E}_s \tag{9.7.5}$$

$$\nabla \times \mathbf{E} = -\mu_o \frac{\partial \mathbf{H}}{\partial t} \quad \rightarrow \quad \nabla \times \mathbf{E}_s = -j\omega\mu_o \mathbf{H}_s \tag{9.7.6}$$

Substituting eq (9.7.2) into eqs. (9.7.3) and (9.7.4), it is readily verified that two Maxwell's equations are satisfied; that is,

$$\nabla \cdot \mathbf{E}_s = \frac{1}{\rho} \frac{\partial}{\partial \phi} (E_{\phi s}) = 0$$

$$\nabla \cdot \mathbf{H}_s = \frac{1}{\rho} \frac{\partial}{\partial \rho} (\rho H_{\rho s}) = 0$$

Now

$$\nabla \times \mathbf{H}_s = \nabla \times \left(\frac{H_o}{\rho} e^{j\beta z} \mathbf{a}_\rho \right) = \frac{jH_o\beta}{\rho} e^{j\beta z} \mathbf{a}_\phi \tag{9.7.7}$$

Substituting eqs. (9.7.2) and (9.7.7) into eq. (9.7.5), we have

$$\frac{jH_o\beta}{\rho} e^{j\beta z} \mathbf{a}_\phi = j\omega\varepsilon_o \frac{50}{\rho} e^{j\beta z} \mathbf{a}_\phi$$

or

$$H_o\beta = 50\,\omega\varepsilon_o \tag{9.7.8}$$

Similarly, substituting eq. (9.7.2) into eq. (9.7.6) gives

$$-j\beta \frac{50}{\rho} e^{j\beta z} \mathbf{a}_\rho = -j\omega\mu_o \frac{H_o}{\rho} e^{j\beta z} \mathbf{a}_\rho$$

or

$$\frac{H_o}{\beta} = \frac{50}{\omega\mu_o} \tag{9.7.9}$$

Multiplying eq. (9.7.8) by eq. (9.7.9) yields

$$H_o^2 = (50)^2 \frac{\varepsilon_o}{\mu_o}$$

or

$$H_o = \pm 50 \sqrt{\varepsilon_o / \mu_o} = \pm \frac{50}{120\pi} = \pm 0.1326$$

Dividing eq. (9.7.8) by eq. (9.7.9), we get

$$\beta^2 = \omega^2 \mu_o \varepsilon_o$$

or

$$\beta = \pm \omega \sqrt{\mu_o \varepsilon_o} = \pm \frac{\omega}{c} = \pm \frac{10^6}{3 \times 10^8}$$
$$= \pm 3.33 \times 10^{-3}$$

In view of eq. (9.7.8), $H_o = 0.1326, \beta = 3.33 \times 10^{-3}$ or $H_o = -0.1326, \beta = -3.33 \times 10^{-3}$; only these will satisfy Maxwell's four equations.

PRACTICE EXERCISE 9.7

In air, $\mathbf{E} = \dfrac{\sin \theta}{r} \cos(6 \times 10^7 t - \beta r)\mathbf{a}_\phi$ V/m.

Find β and $\mathbf{H}$.

Answer: 0.2 rad/m, $-\dfrac{1}{12\pi r^2} \cos \theta \sin(6 \times 10^7 t - 0.2r)\mathbf{a}_r - \dfrac{1}{120\pi r} \sin \theta \times \cos(6 \times 10^7 t - 0.2r)\mathbf{a}_\theta$ /m.

EXAMPLE 9.8

In a medium characterized by $\sigma = 0, \mu = \mu_o, \varepsilon = 4\varepsilon_o$, and

$$\mathbf{E} = 20 \sin(10^8 t - \beta z)\mathbf{a}_y \text{ V/m}$$

calculate β and $\mathbf{H}$.

Solution:

This problem can be solved directly in time domain or by using phasors. As in Example 9.7, we find β and $\mathbf{H}$ by making $\mathbf{E}$ and $\mathbf{H}$ satisfy Maxwell's four equations.

Method 1 (time domain):

Let us solve this problem the harder way—in time domain. It is evident that Gauss's law for electric fields is satisfied; that is,

$$\nabla \cdot \mathbf{E} = \frac{\partial E_y}{\partial y} = 0$$

From Faraday's law,

$$\nabla \times \mathbf{E} = -\mu \frac{\partial \mathbf{H}}{\partial t} \quad \rightarrow \quad \mathbf{H} = -\frac{1}{\mu} \int (\nabla \times \mathbf{E}) \, dt$$

But

$$\nabla \times \mathbf{E} = \begin{vmatrix} \mathbf{a}_x & \mathbf{a}_y & \mathbf{a}_z \\ \dfrac{\partial}{\partial x} & \dfrac{\partial}{\partial y} & \dfrac{\partial}{\partial z} \\ 0 & E_y & 0 \end{vmatrix} = -\frac{\partial E_y}{\partial z} \mathbf{a}_x + \frac{\partial E_y}{\partial x} \mathbf{a}_z$$

$$= 20\beta \cos(10^8 t - \beta z) \, \mathbf{a}_x + \mathbf{0}$$

Hence,

$$\mathbf{H} = -\frac{20\beta}{\mu} \int \cos(10^8 t - \beta z) \, dt \, \mathbf{a}_x$$

$$= -\frac{20\beta}{\mu 10^8} \sin(10^8 t - \beta z) \, \mathbf{a}_x \tag{9.8.1}$$

It is readily verified that

$$\nabla \cdot \mathbf{H} = \frac{\partial H_x}{\partial x} = 0$$

showing that Gauss's law for magnetic fields is satisfied. Lastly, from Ampère's law

$$\nabla \times \mathbf{H} = \sigma \mathbf{E} + \varepsilon \frac{\partial \mathbf{E}}{\partial t} \quad \rightarrow \quad \mathbf{E} = \frac{1}{\varepsilon} \int (\nabla \times \mathbf{H}) \, dt \tag{9.8.2}$$

because $\sigma = 0$.

But

$$\nabla \times \mathbf{H} = \begin{vmatrix} \mathbf{a}_x & \mathbf{a}_y & \mathbf{a}_z \\ \dfrac{\partial}{\partial x} & \dfrac{\partial}{\partial y} & \dfrac{\partial}{\partial z} \\ H_x & 0 & 0 \end{vmatrix} = \frac{\partial H_x}{\partial z} \mathbf{a}_y - \frac{\partial H_x}{\partial y} \mathbf{a}_z$$

$$= \frac{20\beta^2}{\mu 10^8} \cos(10^8 t - \beta z) \, \mathbf{a}_y + \mathbf{0}$$

where **H** in eq. (9.8.1) has been substituted. Thus eq. (9.8.2) becomes

$$\mathbf{E} = \frac{20\beta^2}{\mu\varepsilon 10^8} \int \cos(10^8 t - \beta z)\, dt\, \mathbf{a}_y$$

$$= \frac{20\beta^2}{\mu\varepsilon 10^{16}} \sin(10^8 t - \beta z)\, \mathbf{a}_y$$

Comparing this with the given **E**, we have

$$\frac{20\beta^2}{\mu\varepsilon 10^{16}} = 20$$

or

$$\beta = \pm 10^8 \sqrt{\mu\varepsilon} = \pm 10^8 \sqrt{\mu_o \cdot 4\varepsilon_o} = \pm\frac{10^8(2)}{c} = \pm\frac{10^8(2)}{3 \times 10^8}$$

$$= \pm\frac{2}{3}$$

The β would be negative only in metamaterials, for an isotropic medium, $\beta = \frac{2}{3}$.

From eq. (9.8.1),

$$\mathbf{H} = +\frac{20\,(2/3)}{4\pi \cdot 10^{-7}(10^8)} \sin\left(10^8 t - \frac{2z}{3}\right) \mathbf{a}_x$$

or

$$\mathbf{H} = +\frac{1}{3\pi} \sin\left(10^8 t - \frac{2z}{3}\right)\mathbf{a}_x \text{ A/m}$$

Method 2 (using phasors):

$$\mathbf{E} = \text{Im}(E_s e^{j\omega t}) \quad \rightarrow \quad \mathbf{E}_s = 20 e^{-j\beta z}\, \mathbf{a}_y \tag{9.8.3}$$

where $\omega = 10^8$.

Again

$$\nabla \cdot \mathbf{E}_s = \frac{\partial E_{ys}}{\partial y} = 0$$

$$\nabla \times \mathbf{E}_s = -j\omega\mu\mathbf{H}_s \quad \rightarrow \quad \mathbf{H}_s = \frac{\nabla \times \mathbf{E}_s}{-j\omega\mu}$$

or

$$\mathbf{H}_s = \frac{1}{-j\omega\mu}\left[-\frac{\partial E_{ys}}{\partial z}\mathbf{a}_x\right] = -\frac{20\beta}{\omega\mu}e^{-j\beta z}\mathbf{a}_x \tag{9.8.4}$$

Notice that $\nabla \cdot \mathbf{H}_s = 0$ is satisfied.

$$\nabla \times \mathbf{H}_s = j\omega\varepsilon\mathbf{E}_s \quad \rightarrow \quad \mathbf{E}_s = \frac{\nabla \times \mathbf{H}_s}{j\omega\varepsilon} \tag{9.8.5}$$

Substituting $\mathbf{H}_s$ in eq. (9.8.4) into eq. (9.8.5) gives

$$\mathbf{E}_s = \frac{1}{j\omega\varepsilon}\frac{\partial H_{xs}}{\partial z}\mathbf{a}_y = \frac{20\beta^2 e^{-j\beta z}}{\omega^2\mu\varepsilon}\mathbf{a}_y$$

Comparing this with the given $\mathbf{E}_s$ in eq. (9.8.3), we have

$$20 = \frac{20\beta^2}{\omega^2\mu\varepsilon}$$

or

$$\beta = +\omega\sqrt{\mu\varepsilon} = +\frac{2}{3}$$

as obtained before. From eq. (9.8.4),

$$\mathbf{H}_s = +\frac{20(2/3)\,e^{-j\beta z}}{10^8(4\pi \times 10^{-7})}\mathbf{a}_x = +\frac{1}{3\pi}e^{-j\beta z}\mathbf{a}_x$$

$$\mathbf{H} = \text{Im}(\mathbf{H}_s e^{j\omega t})$$

$$= \pm\frac{1}{3\pi}\sin(10^8 t - \beta z)\mathbf{a}_x \text{ A/m}$$

as obtained before. It should be noticed that working with phasors is considerably simpler than working directly in time domain. Also, notice that we have used

$$\mathbf{A} = \text{Im}(\mathbf{A}_s e^{j\omega t})$$

because the given **E** is in sine form and not cosine. If we had used

$$\mathbf{A} = \text{Re}(\mathbf{A}_s e^{j\omega t})$$

sine would been expressed in terms of cosine, and eq. (9.8.3) would have been

$$E = 20\cos(10^8 t - \beta z - 90°)\mathbf{a}_y = \mathrm{Re}(E_s e^{j\omega t})$$

or

$$E_s = 20e^{-j\beta z - j90°}\mathbf{a}_y = -j20e^{-j\beta z}\mathbf{a}_y$$

and we follow the same procedure.

PRACTICE EXERCISE 9.8

A medium is characterized by $\sigma = 0$, $\mu = 2\mu_o$ and $\varepsilon = 5\varepsilon_o$. If $\mathbf{H} = 2\cos(\omega t - 3y)\mathbf{a}_z$ A/m, calculate ω and $\mathbf{E}$.

Answer: 2.846×10^8 rad/s, $-476.86\cos(2.846 \times 10^8 t - 3y)\mathbf{a}_x$ V/m.

---

†9.8 APPLICATION NOTE—MEMRISTOR

In 1971 Leon O. Chua of the University of California–Berkeley introduced the memristor (Figure 9.13) as one of the four basic circuit elements, coequal in importance with the other well-known circuit elements, namely, resistor (R), inductor (L), and capacitor (C). The new element had not been physically realized when Chua proposed it. However, he was the first to use this moniker. Not until 2008 was a physical approximation of such an element fabricated, as a TiO_2 nanodevice, by Stanley Williams's group at Hewlett-Packard (HP).

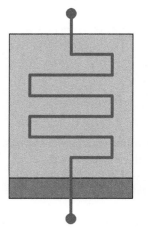

FIGURE 9.13 Schematic of a memristor.

Chua characterized the memristor in terms of the electric charge and the magnetic flux. He also linked this relationship with the quasi-static expansion of Maxwell's equations. A charge-controlled memristor can be defined as a two-terminal element satisfying the constitutive relation $\varphi = f_M(q)$, where φ and q are magnetic flux and electric charge, respectively, and $f_M(q_s)$ is a piecewise-differentiable function. Memristors have interesting circuit-theoretic properties different from those of the classical circuit elements R, L, and C. These properties, in turn, lead to remarkable applications not realizable with the earlier circuits.

A charge-controlled memristor behaves somewhat like a nonlinear resistor R_M satisfying a q-dependent Ohm's law, $v = R_M i$. The quantity R_M is *memresistance*, measured in webers per coulomb, and for all passive memristors $R_M = 0$. Since the voltage v is related to $\dfrac{d\varphi}{dt}$, we can express the memresistance as $R_M = \dfrac{df_M(q)}{dq}$. When current through a memristor is turned off at $t = t_0$, $\dfrac{dq}{dt} = 0$ implies $q = q(t_0)$. This allows us to view a memristor as a nonvolatile analog memory. In particular, it can be used as a nonvolatile binary memory, where two sufficiently different values of resistance are chosen to code binary states "\0" and "\1," respectively. The memristor reported by HP as well as many other nanodevices proposed recently can be scaled down to atomic dimensions. Thus the memristor offers immense potential for an ultra-low-power and ultradense nonvolatile memory technology that could replace flash memories and dynamic random-access memories (DRAMs).

The most important common property of a memristor is the pinched hysteresis loop; that is, the loci of $(v(t), i(t))$ due to any bipolar periodic current source $i(t)$ or periodic voltage source $v(t)$ must always be pinched at the origin in the sense that $(v(t), i(t)) = (0, 0)$ must always lie on the (v,i)-loci. The pinched hysteresis loop phenomenon of the memristor must hold for any bipolar periodic signal $v(t)$, or $i(t)$.

Although memristors have become popular only recently, they are known to abound in many other forms. For example, the electric arc, dating back to 1801, has been identified as a memristor. Also, a very interesting and scientifically significant example is the classic Hodgkin–Huxley axon circuit model of the squid giant axon. Chua showed that the Hodgkin–Huxley time-varying potassium conductance is in fact a first-order memristor, and the Hodgkin–Huxley time-varying sodium conductance is in fact a second-order memristor.

Besides serving as nonvolatile memories, locally passive memristors have been used for switching electromagnetic devices, for field-programmable logic arrays, for synaptic memories, and for learning. In addition, locally passive memristors have been found to exhibit many exotic dynamical phenomena, such as oscillations, chaos, Hamiltonian vortices, and autowaves.

†9.9 APPLICATION NOTE—OPTICAL NANOCIRCUITS

Circuit elements and electronic devices such as resistors, capacitors, inductors, switches, diodes, and transistors were developed at low frequencies; higher frequencies, even radio frequencies, were realized only later. With the development of metamaterials and nanotechnology, such elements have also been conceived at optical frequencies. Nader Engheta and his group at the University of Pennsylvania have recently proposed circuit elements at infrared and optical frequencies. The advantage of using lumped elements lies in their simplicity and modularity: when we want to use lumped elements by connecting them with one another, we need to know only their in-terminal behavior and the functionality of the overall circuit they comprise. Although it is difficult to pinpoint the frequency at which one might consider an element to be lumped, as long as it is smaller than the wavelength but larger than it is when the quantum effects begin to manifest, we can conveniently model an arbitrary particle as a lumped circuit element.

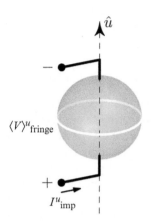

FIGURE 9.14 A nanoparticle used as a circuit element.

Optical lumped circuit elements with specific optical impedances have been realized by means of deep-subwavelength nanostructures. The use of gyroscopic nanospheres has permitted the identification of tunable circuit elements at infrared and optical frequencies. This new area is called *optical metatronics*—a portmanteau term derived from *metamaterials*, *optical*, and *electronics*. The three fields of electronics, photonics, and magnetics can be brought together seamlessly under one umbrella. In this paradigm, for information processing and data storage at the nanoscale, the optical electric displacement currents or the optical magnetic displacement currents play a more important role than the conventional drift of charged particles. In such optical circuitry, nanostructures with specific values of permittivity (or permeability) may act as lumped circuit elements (e.g., nanocapacitors, nanoinductors, nanoresistors).

By collecting properly arranging judiciously designed nanostructures, it is possible to achieve a new circuit platform in which optical signals can be tailored and manipulated, thus allowing optical information to be processed at the nanoscale. These nanostructures can be considered to be the modules and building blocks of metatronic circuits in which optical electrical fields and optical displacement currents are connected through the optical impedances of the nanoscale lumped elements. Such optical lumped circuit elements and metatronic circuitry afford the possibility of bringing many designs from RF electronics into the field of optics but with a much higher level of miniaturization and higher bandwidth. Just as electrons play the fundamental role in electronics, spins in spintronics, and photons in photonics, optical displacement current is of fundamental importance in the field of metatronics.

If the real part of the permittivity of the material forming a given nanoparticle is positive, its optical impedance is capacitive. If the imaginary part of the permittivity is nonzero, then the particle impedance arises from a lumped conductance in parallel with the lumped capacitance. Similarly, a plasmonic particle with negative real part for its permittivity may possess an optical impedance that resembles the impedance of an inductor at that frequency. These are fixed elements. But if we wish to make them variable, since it is not possible to mechanically change their physical size, we can consider gyrotropic materials impressed with a dc magnetic field and still realize tunable circuit elements. These nanocircuit elements play a vital role in metatronics for the design of various nanodevices. Actual fabrication of optical nanofilters, left-handed/right-handed nanotransmission lines, couplers, biosensors, information storage devices, and so on has become a reality, and more surprises are in the offing.

†9.10 APPLICATION NOTE—WIRELESS POWER TRANSFER AND QI STANDARD

Rapid growth in the area of high-speed wireless data transfer has resulted in the proliferation of cell-phones and various mobile devices that include even biomedical implants. In turn, rapid charging of batteries and remote powering of electric circuits have become a high priority and a pressing need. Especially the emergence of electric vehicles, aimed at reducing environmental pollution, became a greater impetus for more efficient ways of charging batteries. All along, recharging was done by connecting power cord battery, but to increase mobility and ease of handling, doing this task cordlessly or, if possible, dispensing of batteries altogether would be better. Toward this goal, operating mobile devices through wireless power transfer (WPT) became the preferred choice. The development in this area has been rather slows, although the idea dates as far back as a century to Tesla, who proposed that electric power can be transferred not only by means of radiation, but also by means of induction and resonant coupling. Induction machines, microwave heating, and similar power devices developed historically, are all based on WPT. Since the distance between the source and the receiver in these devices is usually small, the term *wireless* is not highlighted when we refer to them.

We might achieve WPT in three broad ways: (i) near-field resonant reactive coupling, (ii) far-field directive power beaming, and (iii) far-field nondirective power transfer. Near-field or non-radiative WPT is based on the near-field magnetic coupling of conductive loops and can be either short range or mid-range in its applications. Far-field or radiative WPT takes place from a transmitting antenna and propagates through a medium such as air over distances that are several wavelengths long to a receiver where power is used to energize the mobile device. This method of transferring power can be highly directive if the locations of the receiver are predetermined or nondirective otherwise. In the latter case, the efficiency of transmission is very low.

Wireless charging technology for portable electronic devices has escalated to the commercialization stage with the introduction of the Qi (pronounced "chee") Standard by the Wireless Power Consortium (WPC), now (at the time of this writing) growing with a membership of over 220 companies worldwide. The Qi system comprises a base station that appears as a charging pad, on which is placed a compatible device, which receives energy through resonant inductive coupling. The base station, connected to a power source, has planar coils that set up oscillating magnetic flux. Likewise, the mobile device has a receiver coil that harvests energy into a power receiver. Proper shielding of coils and selection of their parameters is done to ensure good inductive power transfer. To promote better coupling and higher power transfer, relative alignment of the device is made in a guided way by markings on the charging pad. Also, free positioning is allowed by the careful design of coil geometry or using a technique that employs multiple cooperative flux generators.

A typical WPT charging unit is depicted in Figure 9.15. It shows the base station with the charging pad on the top and a power transmitter section. It has a power conversion unit (PCU) and a communications and control unit (CCU). The transmitting coil of the PCU underneath the charging pad establishes the required oscillating magnetic flux. The Qi compatible mobile device is equipped with the power receiver section, which essentially has the power pickup unit (PPU) and a CCU similar to the one in the base station. A receiver coil above the charging pad collects the energy induced and conveys to the PPU, which then drives the load. The CCUs are designed to regulate the transferred power to the required level at the highest possible transfer efficiency. Although Figure 9.15 does not show, in practice, the base station has an array of transmitting coils to facilitate charging of numerous mobile devices. The system unit in the base station contains additional user interfaces. Between the receiver and the transmitter, communication is established with the aid of backscatter modulation.

In terms of the Qi standards, the low-power specification delivers up to 5 W, typically used to charge mobile devices, and the medium-power specification will deliver up to 120 W. Usually this is allocated for power displays and laptops. In 2015, WPC demonstrated a high-power specification that will deliver up to 1 kW, allowing the powering of kitchen utensils among other high-power utilities. The Qi logo is depicted in Figure 9.16. As the Qi standard gains popularity, it is forecast that Qi hotspots will begin to abound in all market places, coffee shops, airports, sports arenas, etc. The technology of WPT developed for electric vehicles and medical implants and other consumer power devices has begun to explode and readers are encouraged to consult additional references.

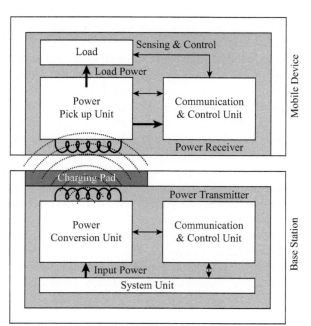

FIGURE 9.15 A typical wireless power transfer system for charging a mobile device.

FIGURE 9.16 Qi logo

MATLAB 9.1

```
% This script illustrates Matlab's complex arithmetic abilities
% and assists the user to solve Practice Exercise 9.5
%

clear

% Matlab recognizes the input of complex numbers using i or j
% for example z = 7 - 6*j sets variable z to the complex value of
% 7 plus sqrt(-1) times 6, thus it is interactive with respect to
% entering and displaying complex values
z = input('Enter the complex number z in the format a+j*b... \n > ');

disp(sprintf('The real part of z is %f', real(z)))
                                    % display the real part
disp(sprintf('The imaginary part of z is %f', imag(z)))
                                        % display the imag part
disp(sprintf('The magnitude of z is %f', abs(z)))
                                        % display the magnitude
disp(sprintf('The phase of z is %f degrees', angle(z)*180/pi))
                                    % display the phase (degrees)

% Matlab also recognizes complex  numbers in polar form
% the exponential function accepts imaginary arguments, however it
% interprets the value as being in radians, not degrees, so if
% degrees are desired a conversion must be made
disp('Enter the complex number z in the a*exp(j*b) where b is');
z = input(' in radians... \n >   ');

disp(sprintf('The real part of z is %f', real(z)))
disp(sprintf('The imaginary part of z is %f', imag(z)))
disp(sprintf('The magnitude of z is %f', abs(z)))
disp(sprintf('The phase of z is %f degrees', angle(z)*180/pi))

% part a
% complex numbers may be handled with the same math operators
% as real numbers in matlab....
z = j^3 * ((1+j)/(2-j))^2;

disp(sprintf('\nPart (a)\nz = %0.2f ', real(z)))
```

```
disp(sprintf(' + j%0.2f ', imag(z)))

% part b
% note the conversion from degrees to radians in the
% exponential
z = 6*exp(j*30*pi/180) + j*5 - 3 +exp(j*45*pi/180);

disp(sprintf('\nPart (b)\nz = %0.3f ', real(z)))
disp(sprintf(' + j%0.3f ', imag(z)))
```

SUMMARY

1. In this chapter, we have introduced two fundamental concepts: electromotive force (emf), based on Faraday's experiments, and displacement current, which resulted from Maxwell's hypothesis. These concepts call for modifications in Maxwell's curl equations obtained for static EM fields to accommodate the time dependence of the fields.

2. Faraday's law states that the induced emf is given by $(N = 1)$

$$V_{emf} = -\frac{\partial \Psi}{\partial t}$$

For transformer emf, $V_{emf} = -\displaystyle\int \frac{\partial \mathbf{B}}{\partial t} \cdot d\mathbf{S}$

and for motional emf, $V_{emf} = \displaystyle\int (\mathbf{u} \times \mathbf{B}) \cdot d\mathbf{l}$.

3. The displacement current

$$I_d = \int \mathbf{J}_d \cdot d\mathbf{S}$$

where $\mathbf{J}_d = \dfrac{\partial \mathbf{D}}{\partial t}$ (displacement current density) is a modification to Ampère's circuit law. This modification, attributed to Maxwell, predicted electromagnetic waves several years before the phenomenon was verified experimentally by Hertz.

4. In differential form, Maxwell's equations for dynamic fields are:

$$\nabla \cdot \mathbf{D} = \rho_v$$

$$\nabla \cdot \mathbf{B} = 0$$

$$\nabla \times \mathbf{E} = -\frac{\partial \mathbf{B}}{\partial t}$$

$$\nabla \times \mathbf{H} = \mathbf{J} + \frac{\partial \mathbf{D}}{\partial t}$$

Each differential equation has its integral counterpart (see Tables 9.1 and 9.2) that can be derived from the differential form by using Stokes's theorem or the divergence theorem. Any EM field must satisfy the four Maxwell's equations simultaneously.

5. Time-varying electric scalar potential $V(x, y, z, t)$ and magnetic vector potential $A(x, y, z, t)$ are shown to satisfy wave equations if Lorenz's condition is assumed.
6. Time-harmonic fields are those that vary sinusoidally with time. They are easily expressed in phasors, which are more convenient to work with. The cosine reference, can be used to show that the instantaneous vector quantity $A(x, y, z, t)$ is related to its phasor form $A_s(x, y, z)$ according to

$$A(x, y, z, t) = \text{Re}\left[A_s(x, y, z)\, e^{j\omega t}\right]$$

REVIEW QUESTIONS

9.1 The flux through each turn of a 100-turn coil is $(t^3 - 2t)$ mWb, where t is in seconds. The induced emf at $t = 2$ s is

(a) 1 V

(b) −1 V

(c) 4 mV

(d) 0.4 V

(e) −0.4 V

9.2 Assuming that each loop is stationary and the time-varying magnetic field B induces current I, which of the configurations in Figure 9.17 are incorrect?

9.3 Two conducting coils 1 and 2 (identical except that 2 is split) are placed in a uniform magnetic field that decreases at a constant rate as in Figure 9.18. If the plane of the coils is perpendicular to the field lines, which of the following statements is true?

(a) An emf is induced in both coils.

(b) An emf is induced in split coil 2.

(c) Equal Joule heating occurs in both coils.

(d) Joule heating does not occur in either coil.

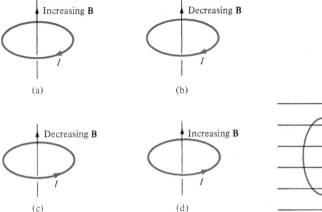

(a)   (b)

(c)   (d)

FIGURE 9.17 For Review Question 9.2.

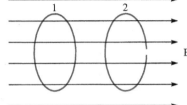

FIGURE 9.18 For Review Question 9.3.

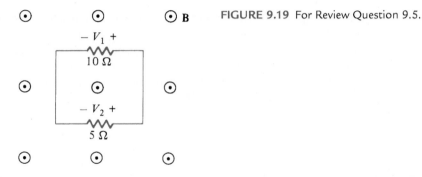

FIGURE 9.19 For Review Question 9.5.

9.4 A loop is rotating about the y-axis in a magnetic field $\mathbf{B} = B_o \sin \omega t \, \mathbf{a}_x$ Wb/m². The voltage induced in the loop is due to

(a) Motional emf

(b) Transformer emf

(c) A combination of motional and transformer emf

(d) None of the above

9.5 A rectangular loop is placed in the time-varying magnetic field $\mathbf{B} = 0.2 \cos 150\pi t \mathbf{a}_z$ Wb/m² as shown in Figure 9.19. V_1 is not equal to V_2.

(a) True   (b) False

9.6 The concept of displacement current was a major contribution attributed to

(a) Faraday

(b) Lenz

(c) Maxwell

(d) Lorenz

(e) Your professor

9.7 Identify which of the following expressions are not Maxwell's equations for time-varying fields:

(a) $\nabla \cdot \mathbf{J} + \dfrac{\partial \rho_v}{\partial t} = 0$

(b) $\nabla \cdot \mathbf{D} = \rho_v$

(c) $\nabla \cdot \mathbf{E} = -\dfrac{\partial \mathbf{B}}{\partial t}$

(d) $\oint_L \mathbf{H} \cdot d\mathbf{l} = \int_S \left(\sigma \mathbf{E} + \varepsilon \, \dfrac{\partial \mathbf{E}}{\partial t} \right) \cdot d\mathbf{S}$

(e) $\oint_S \mathbf{B} \cdot d\mathbf{S} = 0$

9.8 An EM field is said to be nonexistent or not Maxwellian if it fails to satisfy Maxwell's equations and the wave equations derived from them. Which of the following fields in free space are not Maxwellian?

(a) $\mathbf{H} = \cos x \cos 10^6 t \, \mathbf{a}_y$

(b) $\mathbf{E} = 100 \cos \omega t \, \mathbf{a}_x$

(c) $\mathbf{D} = e^{-10y} \sin(10^5 t - 10y) \, \mathbf{a}_z$

(d) $\mathbf{B} = 0.4 \sin 10^4 t \, \mathbf{a}_z$

(e) $\mathbf{H} = 10 \cos \left(10^5 t - \dfrac{z}{10} \right) \mathbf{a}_x$

(f) $\mathbf{E} = \dfrac{\sin \theta}{r} \cos (\omega t - r\omega \sqrt{\mu_o \varepsilon_o}) \, \mathbf{a}_\theta$

(g) $\mathbf{B} = (1 - \rho^2) \sin \omega t \, \mathbf{a}_z$

9.9 Which of the following statements is not true of a phasor?

(a) It may be a scalar or a vector.

(b) It is a time-dependent quantity.

(c) A phasor V_s may be represented as $V_o \, \underline{/\theta}$ or $V_o e^{j\omega}$ where $V_o = |V_s|$.

(d) It is a complex quantity.

9.10 If $\mathbf{E}_s = 10 \, e^{j4x} \, \mathbf{a}_y$, which of these is not a correct representation of E?

(a) $\mathrm{Re}(\mathbf{E}_s e^{j\omega t})$

(b) $\mathrm{Re}(\mathbf{E}_s e^{-j\omega t})$

(c) $\mathrm{Im}(\mathbf{E}_s e^{j\omega t})$

(d) $10 \cos(\omega t + j4x) \, \mathbf{a}_y$

(e) $10 \sin(\omega t + 4x) \, \mathbf{a}_y$

Answers: 9.1b, 9.2b, d, 9.3a, 9.4c, 9.5a, 9.6c, 9.7a,c, 9.8b, d, 9.9b, 9.10d.

PROBLEMS   **Sections 9.2 and 9.3—Faraday's Law and Electromotive Forces**

9.1 A conducting circular loop of radius 20 cm lies in the $z = 0$ plane in a magnetic field $\mathbf{B} = 10 \cos 377t \, \mathbf{a}_z$ mWb/m^2. Calculate the induced voltage in the loop.

9.2 The loop in Figure 9.20 exists in a magnetic field $\mathbf{B} = 4\cos(20t)\mathbf{a}_z$ Wb/m^2, where $\mathbf{a}_z$ is directed out of the page. If the area enclosed by the circuit is 2 cm^2, find the current $i(t)$.

9.3 A circuit conducting loop lies in the xy-plane as shown in Figure 9.21. The loop has a radius of 0.2 m and resistance $\mathbf{R} = 4 \, \Omega$. If $\mathbf{B} = 40 \sin 10^4 \, t\mathbf{a}_z$ mWb/m^2, find the currrent.

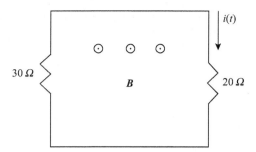

FIGURE 9.20 For Problem 9.2.

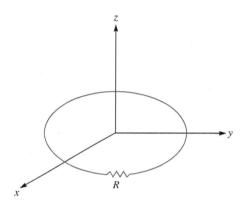

FIGURE 9.21 For Problem 9.3.

9.4 Two conducting bars slide over two stationary rails, as illustrated in Figure 9.22. If $\mathbf{B} = 0.2\mathbf{a}_z$ Wb/m², determine the induced emf in the loop thus formed.

9.5 A conductor located at $0 < y < 1.6$ m moves with velocity $2\mathbf{a}_x$ m/s in a magnetic field, $B = 10 \cos \beta y \mathbf{a}_z$ Wb/m² where β is a constant. Determine the induced voltage.

9.6 A square loop of side a recedes with a uniform velocity $u_o\mathbf{a}_y$ from an infinitely long filament carrying current I along $\mathbf{a}_z$ as shown in Figure 9.23. Assuming that $\rho = \rho_o$ at time $t = 0$, show that the emf induced in the loop at $t > 0$ is

$$V_{\text{emf}} = \frac{u_o a^2 \mu_o I}{2\pi\rho(\rho + a)}$$

9.7 A conducting rod moves with a constant velocity of 3 $\mathbf{a}_z$ m/s parallel to a long straight wire carrying a current of 15 A as in Figure 9.24. Calculate the emf induced in the rod and state which end is at the higher potential.

9.8 A conducting rod has one end grounded at the origin, while the other end is free to move in the $z = 0$ plane. The rod rotates at 30 rad/s in a static magnetic field $\mathbf{B} = 60\mathbf{a}_z$ mWb/m². If the rod is 8 cm long, find the voltage induced in the rod.

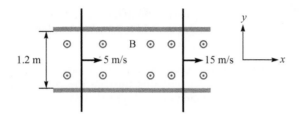

FIGURE 9.22 For Problem 9.4.

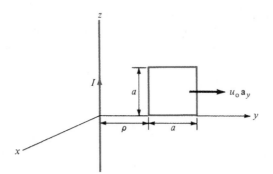

FIGURE 9.23 For Problem 9.6.

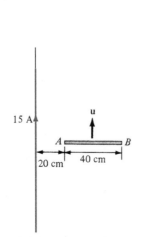

FIGURE 9.24 For Problem 9.7.

FIGURE 9.25 For Problem 9.10.

9.9 A rectangular coil has a cross-sectional area of 30 cm^2 and 50 turns. If the coil rotates at 60 rad/s in a magnetic field of 0.2 Wb/m^2 such that its axis of rotation is perpendicular to the direction of the field, determine the induced emf in the coil.

9.10 Determine the induced emf in the V-shaped loop of Figure 9.25. Take $\mathbf{B} = 0.6x\mathbf{a}_z$ Wb/m^2 and $\mathbf{u} = 5\mathbf{a}_x$ m/s. Assume that the sliding rod starts at the origin when $t = 0$.

9.11 A car travels at 120 km/hr. If the earth's magnetic field is 4.3×10^{-5} Wb/m^2, find the induced voltage in the car bumper of length 1.6 m. Assume that the angle between the earth's magnetic field and the normal to the car is 65°.

9.12 An airplane with a metallic wing of span 36 m flies at 410 m/s in a region where the vertical component of the earth's magnetic field is 0.4 μWb/m^2. Find the emf induced on the airplane wing.

9.13 As portrayed in Figure 9.26, a bar magnet is thrust toward the center of a coil of 10 turns and resistance 15 Ω. If the magnetic flux through the coil changes from 0.45 Wb to 0.64 Wb in 0.02 s, find the magnitude and direction (as viewed from the side near the magnet) of the induced current.

9.14 The cross section of a homopolar generator disk is shown in Figure 9.27. The disk has inner radius $\rho_1 = 2$ cm and outer radius $\rho_2 = 10$ cm and rotates in a uniform magnetic field 15 mWb/m^2 at a speed of 60 rad/s. Calculate the induced voltage.

Section 9.4—Displacement Current

9.15 A 50 V voltage generator at 20 MHz is connected to the plates of an air dielectric parallel-plate capacitor with a plate area of 2.8 cm^2 and a separation distance of 0.2 mm. Find the maximum value of displacement current density and displacement current.

9.16 A dielectric material with $\mu = \mu_o$, $\varepsilon = 9\varepsilon_o$, $\sigma = 4$ S/m is placed between the plates of a parallel-plate capacitor. Calculate the frequency at which the conduction and displacement currents are equal.

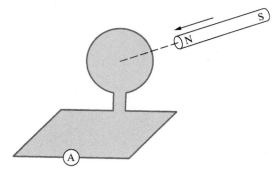

FIGURE 9.26 For Problem 9.13.

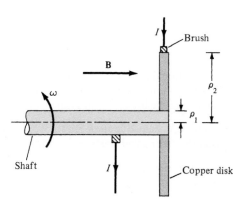

FIGURE 9.27 For Problem 9.14.

9.17 The ratio J/J_d (conduction current density to displacement current density) is very important at high frequencies. Calculate the ratio at 1 GHz for:

(a) distilled water $(\mu = \mu_o, \varepsilon = 81\varepsilon_o, \sigma = 2 \times 10^{-3}\,\text{S/m})$

(b) seawater $(\mu = \mu_o, \varepsilon = 81\varepsilon_o, \sigma = 25\,\text{S/m})$

(c) limestone $(\mu = \mu_o, \varepsilon = 5\varepsilon_o, \sigma = 2 \times 10^{-4}\,\text{S/m})$

9.18 In seawater $(\sigma = 4\,\text{S/m}, \varepsilon = 81\varepsilon_o, \mu = \mu_o)$, find the ratio of the conduction to the displacement currents at 10 MHz.

9.19 Assume that dry soil has $\sigma = 10^{-4}\,\text{S/m}$, $\varepsilon = 3\varepsilon_o$, and $\mu = \mu_o$. Determine the frequency at which the ratio of the magnitudes of the conduction current density and the displacement current density is unity.

9.20 In a dielectric $(\sigma = 10^{-4}\,\text{S/m}, \mu_r = 1, \varepsilon_r = 4.5)$, the conduction current density is given as $J_c = 0.4\cos(2\pi \times 10^8 t)\,\text{A/m}^2$. Determine the displacement current density.

9.21 In a source-free region, $\mathbf{H} = H_o\cos(wt - \beta z)\mathbf{a}_x$ A/m. Find the displacement current density.

9.22 An ac voltage source is connected across the plates of a parallel-plate capacitor so that $\mathbf{E} = 25\sin(10^3 t)\mathbf{a}_z$ V/m. Calculate the total current crossing a 2×5 m area placed perpendicular to the electric field. Assume that the capacitor is air filled.

Section 9.5—Maxwell's Equations

9.23 (a) Write Maxwell's equations for a linear, homogeneous medium in terms of $\mathbf{E}_s$ and $\mathbf{H}_s$, assuming only the time factor $e^{-j\omega t}$.

(b) In Cartesian coordinates, write the point form of Maxwell's equations in Table 9.2 as eight scalar equations.

9.24 Show that in a source-free region $(\mathbf{J} = 0, \rho_v = 0)$, Maxwell's equations can be reduced to two. Identify the two all-embracing equations.

9.25 Show that fields

$$\mathbf{E} = E_o\cos x \cos t\,\mathbf{a}_y \quad \text{and} \quad \mathbf{H} = \frac{E_o}{\mu_o}\sin x \sin t\,\mathbf{a}_z$$

do not satisfy all of Maxwell's equations.

9.26 Assuming a source-free region, derive the diffusion equation

$$\nabla^2\mathbf{E} = \mu\sigma\frac{\partial\mathbf{E}}{\partial t}$$

9.27 In a certain region,

$$\mathbf{J} = (2y\mathbf{a}_x + xz\mathbf{a}_y + z^3\mathbf{a}_z)\sin 10^4 t\,\text{A/m}$$

find ρ_v if $\rho_v(x, y, 0, t) = 0$.

9.28 In free space, the electric field is given by

$$\mathbf{E} = E_\mathrm{o} \cos z \cos t \mathbf{a}_z$$

Find the charge density ρ_v that will produce this field.

9.29 In free space,

$$\mathbf{H} = 10 \sin(10^8 t + \beta x)\mathbf{a}_y \text{ A/m}.$$

Find **E** and β.

9.30 In free space,

$$\mathbf{E} = \frac{50}{\rho} \cos(10^8 t - kz)\mathbf{a}_\rho \text{V/m}$$

Find k, $\mathbf{J}_d$, and **H**.

9.31 The electric field intensity of a spherical wave in free space is given by

$$\mathbf{E} = \frac{10}{r} \sin \theta \cos(\omega t - \beta r)\mathbf{a}_\theta \text{ V/m}$$

Find the corresponding magnetic field intensity **H**.

9.32 In a certain region for which $\sigma = 0$, $\mu = 2\mu_\mathrm{o}$, and $\varepsilon = 10\varepsilon_\mathrm{o}$

$$\mathbf{J} = 60 \sin(10^9 t - \beta z)\mathbf{a}_x \text{ mA/m}^2$$

(a) Find **D** and **H**.

(b) Determine β.

9.33 Use Maxwell's equations to derive the continuity equation.

9.34 In a source-free region, show that

$$\nabla^2 \mathbf{E} - \mu\sigma\frac{\partial \mathbf{E}}{\partial t} - \mu\varepsilon\frac{\partial^2 \mathbf{E}}{\partial t^2} = 0$$

9.35 Check whether the following fields are genuine EM fields (i.e., they satisfy Maxwell's equations). Assume that the fields exist in charge-free regions.

(a) $\mathbf{A} = 40 \sin(\omega t + 10x)\mathbf{a}_z$

(b) $\mathbf{B} = \dfrac{10}{\rho} \cos(\omega t - 2\rho)\mathbf{a}_\phi$

(c) $\mathbf{C} = \left(3\rho^2 \cot \phi \, \mathbf{a}_\rho + \dfrac{\cos \phi}{\rho} \mathbf{a}_\phi \right) \sin \omega t$

(d) $\mathbf{D} = \dfrac{1}{r} \sin \theta \sin(\omega t - 5r)\mathbf{a}_\theta$

9.36 Given the total electromagnetic energy

$$W = \frac{1}{2} \int (\mathbf{E} \cdot \mathbf{D} + \mathbf{H} \cdot \mathbf{B}) \, dv$$

show from Maxwell's equations that

$$\frac{\partial W}{\partial t} = - \oint_S (\mathbf{E} \times \mathbf{H}) \cdot d\mathbf{S} - \int_v \mathbf{E} \cdot \mathbf{J} \, dv$$

9.37 In air, $\mathbf{E} = \cos(12\pi x)\sin(10^{11}t - \alpha y)\mathbf{a}_z$ V/m. Find $\mathbf{H}$ and α.

9.38 An AM radio signal propagating in free space has

$$\mathbf{E} = \mathbf{E}_o \sin(1200\pi t - \beta z)\mathbf{a}_x$$

$$\mathbf{H} = \frac{E_o}{\eta} \sin(1200\pi t - \beta z)\mathbf{a}_y$$

Determine β and η.

9.39 An antenna radiates in free space and

$$\mathbf{H} = \frac{12 \sin \theta}{r} \cos(2\pi \times 10^8 t - \beta r)\mathbf{a}_\theta \text{ mA/m}$$

Find the corresponding $\mathbf{E}$ in terms of β.

Section 9.6—Time-Varying Potentials

9.40 In free space $(\rho_v = 0, \mathbf{J} = 0)$, show that

$$\mathbf{A} = \frac{\mu_0}{4\pi r} (\cos \theta \, \mathbf{a}_r - \sin \theta \, \mathbf{a}_\theta)e^{j\omega(t-r/c)}$$

satisfies the wave equation in eq. (9.52). Find the corresponding V. Take c as the speed of light in free space.

9.41 Retrieve Faraday's law in differential form from

$$\mathbf{E} = -\nabla V - \frac{\partial \mathbf{A}}{\partial t}$$

9.42 In free space, the retarded potentials are given by

$$V = x(z - ct)\text{V}, \qquad \mathbf{A} = x(z/c - t)\mathbf{a}_z \text{ Wb/m}$$

where $c = \dfrac{1}{\sqrt{\mu_o \varepsilon_o}}$

(a) Prove that $\nabla \cdot \mathbf{A} = \mu_o \varepsilon_o \dfrac{\partial V}{\partial t}$.

(b) Determine $\mathbf{E}$.

9.43 Let $\mathbf{A} = A_o \sin(\omega t - \beta z)\mathbf{a}_x$ Wb/m in free space. (a) Find V and $\mathbf{E}$. (b) Express β in terms of ω, ε_o, and μ_o.

Section 9.7—Time-Harmonic Fields

9.44 Evaluate the following complex numbers and express your answers in polar form:

(a) $(4 \underline{/30°} - 10 \underline{/50°})^{1/2}$

(b) $\dfrac{1 + j2}{6 + j8 - 7 \underline{/15°}}$

(c) $\dfrac{(3 + j4)^2}{12 - j7 + (-6 + j10)^*}$

(d) $\dfrac{(3.6 \underline{/-200°})^{1/2}}{(2.4 \underline{/45°})^2(-5 + j8)^*}$

9.45 Determine the phasor forms of the following instantaneous vector fields:

(a) $\mathbf{H} = -10\cos(10^6 t + \pi/3)\mathbf{a}_x$

(b) $\mathbf{E} = 4\cos(4y)\cos(10^4 t - 2x)\mathbf{a}_z$

(c) $\mathbf{D} = 5\sin(10^4 t + \pi/3)\mathbf{a}_x - 8\cos(10^4 t - \pi/4)\mathbf{a}_y$

9.46 Find the instantaneous form for each of the following phasors:

(a) $\mathbf{A}_s = j10\mathbf{a}_x + \dfrac{20}{j}\mathbf{a}_y$

(b) $\mathbf{B}_s = j4e^{-j2x}\mathbf{a}_x + 6e^{+j2x}\mathbf{a}_z$

(c) $\mathbf{C}_s = j2e^{-20z}e^{-j\pi/4}\mathbf{a}_z$

9.47 In a source-free vacuum region,

$$\mathbf{H} = \frac{1}{\rho}\cos(\omega t - 3z)\mathbf{a}_\phi \text{ A/m}$$

(a) Express $\mathbf{H}$ in phasor form.

(b) Find the associated $\mathbf{E}$ field.

(c) Determine ω.

9.48 In a certain homogeneous medium, $\varepsilon = 81\varepsilon_o$, and $\mu = \mu_o$,

$$\mathbf{E}_s = 10e^{j(\omega t + \beta z)}\mathbf{a}_y \text{ V/m}$$

$$\mathbf{H}_s = H_o e^{j(\omega t + \beta z)}\mathbf{a}_x \text{ A/m}$$

If $\omega = 2\pi \times 10^9$ rad/m, find β and H_o.

9.49 The magnetic phasor of a plane wave propagating in air is

$$\mathbf{H}_s(x) = 12e^{jax}\mathbf{a}_z$$

Determine α and $\mathbf{E}_s(x)$. Assume $\omega = 10^9$ rad/s.

9.50 Given that

$$\frac{d^2y}{dt^2} + 4\frac{dy}{dt} + y = 2\cos 3t$$

Solve for y by using phasors.

9.51 Show that in a linear homogeneous, isotropic source-free region, both $\mathbf{E}_s$ and $\mathbf{H}_s$ must satisfy the wave equation

$$\nabla^2\mathbf{A}_s + \gamma^2\mathbf{A}_s = 0$$

where $\gamma^2 = \omega^2\mu\varepsilon - j\omega\mu\sigma$ and $\mathbf{A}_s = \mathbf{E}_s$ or $\mathbf{H}_s$.

Hermann von Helmholtz (1821–1894), a German physicist, extended Joule's results to a general principle and derived the wave equation (to be discussed in this chapter).

Helmholtz was born in Potsdam, and his youth was marred by illness. He graduated from the Medical Institute in Berlin in 1843 and was assigned to a military regiment at Potsdam, but spent all his spare time doing research. In 1858 he became professor of anatomy and physiology at Bonn. In 1871 he became professor of physics at Berlin. Helmholtz made important contributions in all major fields of science, not only unifying the diverse fields of medicine, physiology, anatomy, and physics, but also relating this universal view to the fine arts. Helmholtz expressed the relationship between mechanics, heat, light, electricity, and magnetism by treating them all as manifestations of a single force. He sought to synthesize Maxwell's electromagnetic theory of light with the central force theorem.

Heinrich Rudolf Hertz (1857–1894), a German experimental physicist, demonstrated that electromagnetic waves obey the same fundamental laws that govern light. His work confirmed James Clerk Maxwell's celebrated theory and prediction that such waves existed.

Hertz was born into a prosperous family in Hamburg, Germany. He attended the University of Berlin and did his doctorate under Hermann von Helmholtz. He became a professor at Karlsruhe, where he began his quest for electromagnetic waves. Hertz successively generated and detected electromagnetic waves; he was first to show that light is electromagnetic energy. In 1887 Hertz noted for the first time the photoelectric effect of electrons in a molecular structure. Although Hertz died at the age of 37, his discovery of electromagnetic waves paved the way for the practical use of such waves in radio, television, and other communication systems. The unit of frequency, the hertz (Hz), bears his name.

CHAPTER 10

ELECTROMAGNETIC WAVE PROPAGATION

Young people tell what they are doing, old people what they have done, and fools what they wish to do.

—FRENCH PROVERB

10.1 INTRODUCTION

Our first application of Maxwell's equations will be in relation to electromagnetic wave propagation. The existence of EM waves, predicted by Maxwell's equations, was first investigated by Heinrich Hertz. After several calculations and experiments, Hertz succeeded in generating and detecting radio waves, which are sometimes called Hertzian waves in his honor.

In general, **waves** are means of transporting energy or information.

Typical examples of EM waves include radio waves, TV signals, radar beams, and light rays. All forms of EM energy share three fundamental characteristics: they all travel at high velocity; in traveling, they assume the properties of waves; and they radiate outward from a source, without benefit of any discernible physical vehicles. The problem of radiation will be addressed in Chapter 13.

In this chapter, our major goal is to solve Maxwell's equations and describe EM wave motion in the following media:

1. Free space $(\sigma = 0, \varepsilon = \varepsilon_o, \mu = \mu_o)$
2. Lossless dielectrics $(\sigma \simeq 0, \varepsilon = \varepsilon_r\varepsilon_o, \mu = \mu_r\mu_o, \text{ or } \sigma \ll \omega\varepsilon)$
3. Lossy dielectrics $(\sigma \neq 0, \varepsilon = \varepsilon_r\varepsilon_o, \mu = \mu_r\mu_o)$
4. Good conductors $(\sigma \simeq \infty, \varepsilon = \varepsilon_o, \mu = \mu_r\mu_o, \text{ or } \sigma \gg \omega\varepsilon)$

where ω is the angular frequency of the wave. Case 3, for lossy dielectrics, is the most general case and will be considered first. Once this general case has been solved, we simply derive the other cases (1, 2, and 4) from it as special cases by changing the values of σ, μ, and ε. However, before we consider wave motion in those different media, it is appropriate that we study the characteristics of waves in general. This is important for proper understanding of EM waves. The reader who is conversant with the concept of waves may skip Section 10.2. Power considerations, reflection, and transmission between two different media will be discussed later in the chapter.

†10.2 WAVES IN GENERAL

A clear understanding of EM wave propagation depends on a grasp of what waves are in general.

> A **wave** is a function of both space and time.

Wave motion occurs when a disturbance at point A, at time t_o, is related to what happens at point B, at time $t > t_o$. A wave equation, as exemplified by eqs. (9.51) and (9.52), is a partial differential equation of the second order. In one dimension, a scalar wave equation takes the form of

$$\frac{\partial^2 E}{\partial t^2} - u^2 \frac{\partial^2 E}{\partial z^2} = 0 \tag{10.1}$$

where u is the *wave velocity*. Equation (10.1) is a special case of eq. (9.51) in which the medium is source free $(\rho_v = 0, \mathbf{J} = 0)$. It can be solved by following a procedure similar to that in Example 6.5. Its solutions are of the form

$$E^+ = f(z - ut) \tag{10.2a}$$

$$E^- = g(z + ut) \tag{10.2b}$$

or

$$E = f(z - ut) + g(z + ut) \tag{10.2c}$$

where f and g denote any function of $z - ut$ and $z + ut$, respectively. Examples of such functions include $z \pm ut$, $\sin k(z \pm ut)$, $\cos k(z \pm ut)$, and $e^{jk(z \pm ut)}$, where k is a constant. It can easily be shown that these functions all satisfy eq. (10.1).

If we particularly assume harmonic (or sinusoidal) time dependence $e^{j\omega t}$, eq. (10.1) becomes

$$\frac{d^2 E_s}{dz^2} + \beta^2 E_s = 0 \tag{10.3}$$

where $\beta = \omega/u$ and E_s is the phasor form of E. The solution to eq. (10.3) is similar to Case 3 of Example 6.5 [see eq. (6.5.12)]. With the time factor inserted, the possible solutions to eq. (10.3) are

$$E^+ = A e^{j(\omega t - \beta z)} \tag{10.4a}$$

$$E^- = B e^{j(\omega t + \beta z)} \tag{10.4b}$$

where E^+ means positive z-travel and E^- means negative travel. Combining E^+ and E^- leads to

$$E = A e^{j(\omega t - \beta z)} + B e^{j(\omega t + \beta z)} \tag{10.4c}$$

where A and B are real constants.

For the moment, let us consider the solution in eq. (10.4a). Taking the imaginary part of this equation, we have

$$E = A \sin(\omega t - \beta z) \qquad (10.5)$$

This is a sine wave chosen for simplicity; a cosine wave would have resulted had we taken the real part of eq. (10.4a). Note the following characteristics of the wave in eq. (10.5):

1. It is time harmonic because we assumed time dependence of the form $e^{j\omega t}$ to arrive at eq. (10.5).
2. The *amplitude* of the wave A has the same units as E.
3. The *phase* (in radians) of the wave depends on time t and space variable z, it is the term $(\omega t - \beta z)$.
4. The *angular frequency* ω is given in radians per second; β, the *phase constant* or *wave number*, is given in radians per meter.

Because E varies with both time t and the space variable z, we may plot E as a function of t by keeping z constant and vice versa. The plots of $E(z, t = \text{constant})$ and $E(t, z = \text{constant})$ are shown in Figure 10.1(a) and (b), respectively. From Figure 10.1(a), we observe that the wave takes distance λ to repeat itself and hence λ is called the

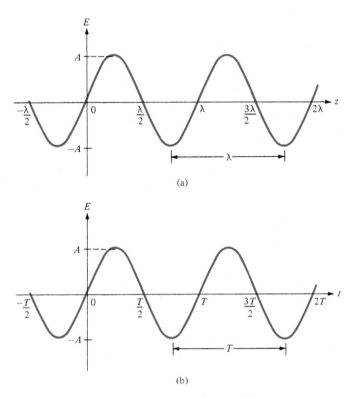

(a)

(b)

FIGURE 10.1 Plot of $E(z, t) = A \sin(\omega t - \beta z)$: (**a**) with constant t, (**b**) with constant z.

wavelength (in meters). From Figure 10.1(b), the wave takes time T to repeat itself; consequently T is known as the *period*, in seconds. Since it takes time T for the wave to travel distance λ at the speed u, we expect

$$\lambda = uT \tag{10.6a}$$

But $T = 1/f$, where f is the *frequency* (the number of cycles per second) of the wave in hertz (Hz). Hence,

$$\boxed{u = f\lambda} \tag{10.6b}$$

Because of this fixed relationship between wavelength and frequency, one can identify the position of a radio station within its band by either the frequency or the wavelength. Usually the frequency is preferred. Also, because

$$\omega = 2\pi f \tag{10.7a}$$

$$\beta = \frac{\omega}{u} \tag{10.7b}$$

and

$$T = \frac{1}{f} = \frac{2\pi}{\omega} \tag{10.7c}$$

we expect from eqs. (10.6) and (10.7) that

$$\boxed{\beta = \frac{2\pi}{\lambda} = \frac{\omega}{u}} \tag{10.8}$$

Equation (10.8) shows that for every wavelength of distance traveled, a wave undergoes a phase change of 2π radians.

We will now show that the wave represented by eq. (10.5) is traveling with a velocity u in the $+z$-direction. To do this, we consider a fixed point P on the wave. We sketch eq. (10.5) at times $t = 0$, $T/4$, and $T/2$ as in Figure 10.2. From the figure, it is evident that as the wave advances with time, point P moves along the $+z$-direction. Point P is a point of constant phase, therefore

$$\omega t - \beta z = \text{constant}$$

or

$$\frac{dz}{dt} = \frac{\omega}{\beta} = u \tag{10.9}$$

which is the same as eq. (10.7b). Equation (10.9) shows that the wave travels with velocity u in the $+z$-direction. Similarly, it can be shown that the wave $B \sin(\omega t + \beta z)$ in eq. (10.4b) is traveling with velocity u in the $-z$-direction.

In summary, we note the following:

1. A wave is a function of both time and space.
2. Though time $t = 0$ is arbitrarily selected as a reference for the wave, a wave is without beginning or end.

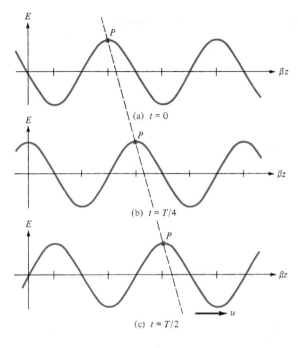

FIGURE 10.2 Plot of $E(z, t) = A \sin(\omega t - \beta z)$ at time (**a**) $t = 0$, (**b**) $t = T/4$, (**c**) $t = T/2$; P moves in the $+z$-direction with velocity u.

3. A negative sign in $(\omega t \pm \beta z)$ is associated with a wave propagating in the $+z$-direction (forward-traveling or positive-going wave), whereas a positive sign indicates that a wave is traveling in the $-z$-direction (backward-traveling or negative-going wave).

4. Since $\sin(-\psi) = -\sin \psi = \sin(\psi \pm \pi)$, whereas $\cos(-\psi) = \cos \psi$,

$$\sin(\psi \pm \pi/2) = \pm\cos \psi \tag{10.10a}$$

$$\sin(\psi \pm \pi) = -\sin \psi \tag{10.10b}$$

$$\cos(\psi \pm \pi/2) = \mp\sin \psi \tag{10.10c}$$

$$\cos(\psi \pm \pi) = -\cos \psi \tag{10.10d}$$

where $\psi = \omega t \pm \beta z$. One of the relations in eqs. (10.10) can be used to represent any time-harmonic wave in the form of sine or cosine.

5. E and H are called uniform waves if they lie in a plane and are constant over such planes.

A large number of frequencies visualized in numerical order constitute a *spectrum*. Table 10.1 shows the frequencies at which various types of energy in the EM spectrum occur. Frequencies usable for radio communication occur near the lower end of the EM spectrum. As frequency increases, the manifestation of EM energy becomes dangerous to human beings. Microwave ovens, for example, can pose a hazard if not properly shielded. The practical difficulties of using EM energy for communication purposes also increase as frequency increases, until finally it can no longer be used. As communication methods improve, the limit to usable frequency has been pushed higher. Today communication satellites use frequencies near 14 GHz. This is still far below light frequencies, but in the enclosed environment of fiber optics, light itself can be used for radio communication.

TABLE 10.1 Electromagnetic Spectrum

EM Phenomena	Examples of Uses	Approximate Frequency Range
Cosmic rays	Physics, astronomy	10^{14} GHz and above
Gamma rays	Cancer therapy	10^{10}–10^{13} GHz
X-rays	X-ray examination	10^{8}–10^{9} GHz
Ultraviolet radiation	Sterilization	10^{6}–10^{8} GHz
Visible light	Human vision	10^{5}–10^{6} GHz
Infrared radiation	Photography	10^{3}–10^{4} GHz
Microwave waves	Radar, microwave relays, satellite communication	3–300 GHz
Radio waves	UHF television	470–806 MHz
	VHF television, FM radio	54–216 MHz
	Short-wave radio	3–26 MHz
	AM radio	535–1605 kHz

EXAMPLE 10.1

An electric field in free space is given by

$$\mathbf{E} = 50 \cos(10^8 t + \beta x)\mathbf{a}_y \text{ V/m}$$

(a) Find the direction of wave propagation.

(b) Calculate β and the time it takes to travel a distance of $\lambda/2$.

(c) Sketch the wave at $t = 0$, $T/4$, and $T/2$.

Solution:

(a) From the positive sign in $(\omega t + \beta x)$, we infer that the wave is propagating along $-\mathbf{a}_x$. This will be confirmed in part (c) of this example.

(b) In free space, $u = c$:

$$\beta = \frac{\omega}{c} = \frac{10^8}{3 \times 10^8} = \frac{1}{3}$$

or

$$\beta = 0.3333 \text{ rad/m}$$

If T is the period of the wave, it takes T seconds to travel a distance λ at speed c. Hence to travel a distance of $\lambda/2$ will take

$$t_1 = \frac{T}{2} = \frac{1}{2}\frac{2\pi}{\omega} = \frac{\pi}{10^8} = 31.42 \text{ ns}$$

Alternatively, because the wave is traveling at the speed of light c,

$$\frac{\lambda}{2} = ct_1 \quad \text{or} \quad t_1 = \frac{\lambda}{2c}$$

But

$$\lambda = \frac{2\pi}{\beta} = 6\pi$$

Hence,

$$t_1 = \frac{6\pi}{2(3 \times 10^8)} = 31.42 \text{ ns}$$

as obtained before.

(c) At $t = 0, \quad E_y = 50 \cos \beta x$

At $t = T/4, E_y = 50 \cos\left(\omega \cdot \frac{2\pi}{4\omega} + \beta x \right) = 50 \cos(\beta x + \pi/2)$

$$= -50 \sin \beta x$$

At $t = T/2, E_y = 50 \cos\left(\omega \cdot \frac{2\pi}{2\omega} + \beta x \right) = 50 \cos(\beta x + \pi)$

$$= -50 \cos \beta x$$

E_y at $t = 0$, $T/4$, $T/2$ is plotted against x as shown in Figure 10.3. Notice that a point P (arbitrarily selected) on the wave moves along $-\mathbf{a}_x$ as t increases with time. This shows that the wave travels along $-\mathbf{a}_x$.

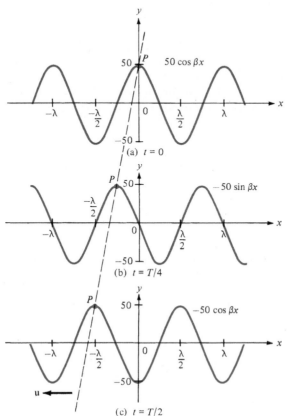

FIGURE 10.3 For Example 10.1; wave travels along $-\mathbf{a}_x$.

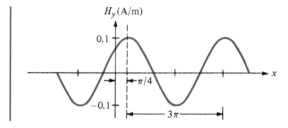

FIGURE 10.4 For Practice Exercise 10.1(c).

PRACTICE EXERCISE 10.1

In free space, $\mathbf{H} = 0.1 \cos(2 \times 10^8 t - kx)\mathbf{a}_y$ A/m.

(a) Calculate k, λ, and T.
(b) Calculate the time t_1 it takes the wave to travel $\lambda/8$.
(c) Sketch the wave at time t_1.

Answer: (a) 0.667 rad/m, 9.425 m, 31.42 ns, (b) 3.927 ns, (c) see Figure 10.4.

10.3 WAVE PROPAGATION IN LOSSY DIELECTRICS

As mentioned in Section 10.1, wave propagation in lossy dielectrics is a general case from which wave propagation in media of other types can be derived as special cases. Therefore, this section is foundational to the next three setions.

> A **lossy dielectric** is a medium in which an EM wave, as it propagates, loses power owing to imperfect dielectric.

In other words, a lossy dielectric is a partially conducting medium (imperfect dielectric or imperfect conductor) with $\sigma \neq 0$, as distinct from a lossless dielectric (perfect or good dielectric) in which $\sigma = 0$.

Consider a linear, isotropic, homogeneous, lossy dielectric medium that is charge free (macroscopic $\rho_v = 0$). Assuming and suppressing the time factor $e^{j\omega t}$, Maxwell's equations (see Table 9.2) become

$$\nabla \cdot \mathbf{E}_s = 0 \tag{10.11}$$

$$\nabla \cdot \mathbf{H}_s = 0 \tag{10.12}$$

$$\nabla \times \mathbf{E}_s = -j\omega\mu\mathbf{H}_s \tag{10.13}$$

$$\nabla \times \mathbf{H}_s = (\sigma + j\omega\varepsilon)\mathbf{E}_s \tag{10.14}$$

Taking the curl of both sides of eq. (10.13) gives

$$\nabla \times \nabla \times \mathbf{E}_s = -j\omega\mu(\nabla \times \mathbf{H}_s) \tag{10.15}$$

Applying the vector identity

$$\nabla \times (\nabla \times \mathbf{A}) = \nabla(\nabla \cdot \mathbf{A}) - \nabla^2\mathbf{A} \tag{10.16}$$

to the left-hand side of eq. (10.15) and invoking eqs. (10.11) and (10.14), we obtain

$$\nabla(\underset{0}{\underbrace{\nabla \cdot \mathbf{E}_s}}) - \nabla^2\mathbf{E}_s = -j\omega\mu(\sigma + j\omega\varepsilon)\mathbf{E}_s$$

or

$$\boxed{\nabla^2\mathbf{E}_s - \gamma^2\mathbf{E}_s = 0} \tag{10.17}$$

where

$$\gamma^2 = j\omega\mu(\sigma + j\omega\varepsilon) \tag{10.18}$$

and γ, in reciprocal meters, is called the *propagation constant* of the medium. By a similar procedure, it can be shown that for the **H** field,

$$\nabla^2\mathbf{H}_s - \gamma^2\mathbf{H}_s = 0 \tag{10.19}$$

Equations (10.17) and (10.19) are known as homogeneous vector *Helmholtz's equations* or simply vector *wave equations*. In Cartesian coordinates, eq. (10.17), for example, is equivalent to three scalar wave equations, one for each component of **E** along $\mathbf{a}_x$, $\mathbf{a}_y$, and $\mathbf{a}_z$.

Since γ in eqs. (10.17) to (10.19) is a complex quantity, we may let

$$\boxed{\gamma = \alpha + j\beta} \tag{10.20}$$

We obtain α and β from eqs. (10.18) and (10.20) by noting that

$$-\text{Re } \gamma^2 = \beta^2 - \alpha^2 = \omega^2\mu\varepsilon \tag{10.21}$$

and

$$|\gamma^2| = \beta^2 + \alpha^2 = \omega\mu\sqrt{\sigma^2 + \omega^2\varepsilon^2} \tag{10.22}$$

From eqs. (10.21) and (10.22), we obtain

$$\boxed{\alpha = \omega\sqrt{\frac{\mu\varepsilon}{2}\left[\sqrt{1 + \left[\frac{\sigma}{\omega\varepsilon}\right]^2} - 1\right]}} \tag{10.23}$$

$$\boxed{\beta = \omega\sqrt{\frac{\mu\varepsilon}{2}\left[\sqrt{1 + \left[\frac{\sigma}{\omega\varepsilon}\right]^2} + 1\right]}} \tag{10.24}$$

Without loss of generality, if we assume that a wave propagates along $+\mathbf{a}_z$ and that $\mathbf{E}_s$ has only an x-component, then

$$\mathbf{E}_s = E_{xs}(z)\mathbf{a}_x \qquad (10.25)$$

We then substitute into eq. (10.17), which yields

$$(\nabla^2 - \gamma^2)E_{xs}(z) = 0 \qquad (10.26)$$

Without loss of generality, if we assume that a wave propagates in an unbounded medium along $\mathbf{a}_z$ and that $\mathbf{E}$ has only an x-component that does not vary with x and y, then

$$\underbrace{\frac{\partial^2 E_{xs}(z)}{\partial x^2}}_{0} + \underbrace{\frac{\partial^2 E_{xs}(z)}{\partial y^2}}_{0} + \frac{\partial^2 E_{xs}(z)}{\partial z^2} - \gamma^2 E_{xs}(z) = 0$$

or

$$\left[\frac{d^2}{dz^2} - \gamma^2\right]E_{xs}(z) = 0 \qquad (10.27)$$

This is a scalar wave equation, a linear homogeneous differential equation, with solution (see eq. 6.5.13a in Case 3 of Example 6.5),

$$E_{xs}(z) = E_o e^{-\gamma z} + E_o' e^{\gamma z} \qquad (10.28)$$

where E_o and E_o' are constants. The fact that the field must be finite at infinity requires that $E_o' = 0$. Alternatively, because $e^{\gamma z}$ denotes a wave traveling along $-\mathbf{a}_z$, whereas we assume wave propagation along $\mathbf{a}_z$, $E_o' = 0$. Whichever way we look at it, $E_o' = 0$. Inserting the time factor $e^{j\omega t}$ into eq. (10.28) and using eq. (10.20), we obtain

$$\mathbf{E} + z, t) = \operatorname{Re}[E_{xs} + z)e^{j\omega t}\mathbf{a}_x] = \operatorname{Re} + E_o e^{-\alpha z}e^{j + \omega t - \beta z)}\mathbf{a}_x)$$

or

$$\boxed{\mathbf{E}(z, t) = E_o e^{-\alpha z}\cos(\omega t - \beta z)\mathbf{a}_x} \qquad (10.29)$$

A sketch of $|\mathbf{E}|$ at times $t = 0$ and $t = \Delta t$ is portrayed in Figure 10.5, where it is evident that $\mathbf{E}$ has only an x-component and it is traveling in the $+z$-direction. Having obtained $\mathbf{E}(z, t)$, we obtain $\mathbf{H}(z, t)$ either by taking similar steps to solve eq. (10.19) or by using eq. (10.29) in conjunction with Maxwell's equations, as we did in Example 9.8. We will eventually arrive at

$$\mathbf{H}(z, t) = \operatorname{Re}(H_o e^{-\alpha z}e^{j(\omega t - \beta z)}\mathbf{a}_y) \qquad (10.30)$$

where

$$H_o = \frac{E_o}{\eta} \qquad (10.31)$$

and η is a complex quantity known as the *intrinsic impedance*, in ohms, of the medium. It can be shown by following the steps taken in Example 9.8 that

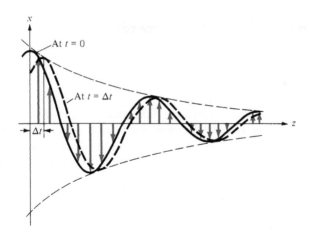

FIGURE 10.5 An *E*-field with an *x*-component traveling in the +*z*-direction at times $t = 0$ and $t = \Delta t$; arrows indicate instantaneous values of *E*.

$$\eta = \sqrt{\frac{j\omega\mu}{\sigma + j\omega\varepsilon}} = |\eta|\angle\theta_\eta = |\eta|e^{j\theta_\eta} \tag{10.32}$$

with

$$|\eta| = \frac{\sqrt{\mu/\varepsilon}}{\left[1 + \left(\dfrac{\sigma}{\omega\varepsilon}\right)^2\right]^{1/4}}, \qquad \tan 2\theta_\eta = \frac{\sigma}{\omega\varepsilon} \tag{10.33}$$

where $0 \le \theta_\eta \le 45°$. Substituting eqs. (10.31) and (10.32) into eq. (10.30) gives

$$\mathbf{H} = \mathrm{Re}\left[\frac{E_o}{|\eta|e^{j\theta_\eta}} e^{-\alpha z} e^{j(\omega t - \beta z)} \mathbf{a}_y\right]$$

or

$$\mathbf{H} = \frac{E_o}{|\eta|} e^{-\alpha z} \cos(\omega t - \beta z - \theta_\eta) \mathbf{a}_y \tag{10.34}$$

Notice from eqs. (10.29) and (10.34) that as the wave propagates along $\mathbf{a}_z$, it decreases or attenuates in amplitude by a factor $e^{-\alpha z}$, and hence α is known as the *attenuation constant*, or attenuation coefficient, of the medium. It is a measure of the spatial rate of decay of the wave in the medium, measured in nepers per meter (Np/m), and can be expressed in decibels per meter (dB/m). An attenuation of 1 neper denotes a reduction to e^{-1} of the original value, whereas an increase of 1 neper indicates an increase by a factor of *e*. Hence, for voltages

$$1 \text{ Np} = 20 \log_{10} e = 8.686 \text{ dB} \tag{10.35}$$

From eq. (10.23), we notice that if $\sigma = 0$, as is the case for a lossless medium and free space, $\alpha = 0$ and the wave is not attenuated as it propagates. The quantity β is a measure of the phase shift per unit length in radians per meter and is called the *phase constant* or *wave number.* In terms of β, the wave velocity u and wavelength λ are, respectively, given by [see eqs. (10.7b) and (10.8)]

$$u = \frac{\omega}{\beta}, \quad \lambda = \frac{2\pi}{\beta} \tag{10.36}$$

We also notice from eqs. (10.29) and (10.34) that **E** and **H** are out of phase by θ_η at any instant of time due to the complex intrinsic impedance of the medium. Thus at any time, **E** leads **H**(or **H** lags **E**) by θ_η. Finally, we notice that the ratio of the magnitude of the conduction current density $\mathbf{J}_c$ to that of the displacement current density $\mathbf{J}_d$ in a lossy medium is

$$\frac{|\mathbf{J}_{cs}|}{|\mathbf{J}_{ds}|} = \frac{|\sigma \mathbf{E}_s|}{|j\omega\varepsilon \mathbf{E}_s|} = \frac{\sigma}{\omega\varepsilon} = \tan\theta$$

or

$$\boxed{\tan\theta = \frac{\sigma}{\omega\varepsilon}} \tag{10.37}$$

where $\tan\theta$ is known as the *loss tangent* and θ is the *loss angle* of the medium as illustrated in Figure 10.6. Although a line of demarcation between good conductors and lossy dielectrics is not easy to make, $\tan\theta$ or θ may be used to determine how lossy a medium is. A medium is said to be a good (lossless or perfect) dielectric if $\tan\theta$ is very small ($\sigma \ll \omega\varepsilon$) or a good conductor if $\tan\theta$ is very large ($\sigma \gg \omega\varepsilon$). From the viewpoint of wave propagation, the characteristic behavior of a medium depends not only on its constitutive parameters σ, ε, and μ but also on the frequency of operation. A medium that is regarded as a good conductor at low frequencies may be a good dielectric at high frequencies. Note from eqs. (10.33) and (10.37) that

$$\theta = 2\theta_\eta \tag{10.38}$$

From eq. (10.14)

$$\nabla \times \mathbf{H}_s = (\sigma + j\omega\varepsilon)\mathbf{E}_s = j\omega\varepsilon\left[1 - \frac{j\sigma}{\omega\varepsilon}\right]\mathbf{E}_s \tag{10.39}$$

$$= j\omega\varepsilon_c\mathbf{E}_s$$

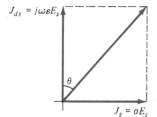

FIGURE 10.6 Loss angle of a lossy medium.

where

$$\varepsilon_c = \varepsilon\left[1 - j\frac{\sigma}{\omega\varepsilon}\right] = \varepsilon[1 - j\tan\theta]$$ (10.40a)

or

$$\varepsilon_c = \varepsilon' - j\varepsilon''$$ (10.40b)

with $\varepsilon' = \varepsilon$, $\varepsilon'' = \sigma/\omega$, $\varepsilon = \varepsilon_o\varepsilon_r$; ε_c is called the *complex permittivity* of the medium. We observe that the ratio of ε'' to ε' is the loss tangent of the medium; that is,

$$\tan\theta = \frac{\varepsilon''}{\varepsilon'} = \frac{\sigma}{\omega\varepsilon}$$ (10.41)

In subsequent sections, we will consider wave propagation in media of other types that may be regarded as special cases of what we have considered here. Thus we will simply deduce the governing formulas from those obtained for the general case treated in this section. The student is advised not just to memorize the formulas but to observe how they are easily obtained from the formulas for the general case.

EXAMPLE 10.2

A lossy dielectric has an intrinsic impedance of $200\ \underline{/30°}\ \Omega$ at a particular radian frequency ω. If, at that frequency, the plane wave propagating through the dielectric has the magnetic field component

$$\mathbf{H} = 10\ e^{-\alpha x}\cos\left(\omega t - \frac{1}{2}x\right)\mathbf{a}_y\ \text{A/m}$$

find $\mathbf{E}$ and α.

Solution:

The given wave travels along $\mathbf{a}_x$ so that $\mathbf{a}_k = \mathbf{a}_x$; $\mathbf{a}_H = \mathbf{a}_y$, so

$$-\mathbf{a}_E = \mathbf{a}_k \times \mathbf{a}_H = \mathbf{a}_x \times \mathbf{a}_y = \mathbf{a}_z$$

or

$$\mathbf{a}_E = -\mathbf{a}_z$$

Also $H_o = 10$, so

$$\frac{E_o}{H_o} = \eta = 200\ \underline{/30°} = 200\ e^{j\pi/6} \rightarrow E_o = 2000e^{j\pi/6}$$

Except for the amplitude and phase difference, $\mathbf{E}$ and $\mathbf{H}$ always have the same form. Hence

$$\mathbf{E} = \text{Re}\left(2000e^{j\pi/6}e^{-\gamma x}e^{j\omega t}\mathbf{a}_E\right)$$

or

$$\mathbf{E} = -2e^{-\alpha x} \cos\left(\omega t - \frac{x}{2} + \frac{\pi}{6}\right)\mathbf{a}_z \text{ kV/m}$$

Knowing that $\beta = 1/2$, we need to determine α. Since

$$\alpha = \omega\sqrt{\frac{\mu\varepsilon}{2}\left[\sqrt{1 + \left[\frac{\sigma}{\omega\varepsilon}\right]^2} - 1\right]}$$

and

$$\beta = \omega\sqrt{\frac{\mu\varepsilon}{2}\left[\sqrt{1 + \left[\frac{\sigma}{\omega\varepsilon}\right]^2} + 1\right]}$$

$$\frac{\alpha}{\beta} = \left[\frac{\sqrt{1 + \left[\frac{\sigma}{\omega\varepsilon}\right]^2} - 1}{\sqrt{1 + \left[\frac{\sigma}{\omega\varepsilon}\right]^2} + 1}\right]^{1/2}$$

But $\dfrac{\sigma}{\omega\varepsilon} = \tan 2\theta_\eta = \tan 60° = \sqrt{3}$. Hence,

$$\frac{\alpha}{\beta} = \left[\frac{2 - 1}{2 + 1}\right]^{1/2} = \frac{1}{\sqrt{3}}$$

or

$$\alpha = \frac{\beta}{\sqrt{3}} = \frac{1}{2\sqrt{3}} = 0.2887 \text{ Np/m}$$

PRACTICE EXERCISE 10.2

A plane wave propagating through a medium with $\varepsilon_r = 8$, $\mu_r = 2$ has $\mathbf{E} = 0.5$ $e^{-z/3} \sin(10^8 t - \beta z)\mathbf{a}_x$ V/m. Determine

(a) β (d) Wave velocity
(b) The loss tangent   (e) **H** field
(c) Intrinsic impedance

Answer: (a) 1.374 rad/m, (b) 0.5154, (c) 177.72 $\underline{/13.63°}$ Ω, (d) 7.278 × 10^7 m/s, (e) 2.817$e^{-z/3}$ sin($10^8 t - \beta z - 13.63°$)$\mathbf{a}_y$ mA/m.

10.4 PLANE WAVES IN LOSSLESS DIELECTRICS

In a lossless dielectric, $\sigma \ll \omega \varepsilon$. It is a special case of that in Section 10.3 except that

$$\boxed{\sigma \simeq 0, \quad \varepsilon = \varepsilon_o \varepsilon_r, \quad \mu = \mu_o \mu_r} \tag{10.42}$$

Substituting these into eqs. (10.23) and (10.24) gives

$$\alpha = 0, \quad \beta = \omega\sqrt{\mu\varepsilon} \tag{10.43a}$$

$$u = \frac{\omega}{\beta} = \frac{1}{\sqrt{\mu\varepsilon}}, \quad \lambda = \frac{2\pi}{\beta} \tag{10.43b}$$

Also

$$\eta = \sqrt{\frac{\mu}{\varepsilon}} \angle 0° \tag{10.44}$$

and thus **E** and **H** are in time phase with each other.

10.5 PLANE WAVES IN FREE SPACE

Plane waves in free space comprise a special case of what we considered in Section 10.3. In this case,

$$\boxed{\sigma = 0, \quad \varepsilon = \varepsilon_o, \quad \mu = \mu_o} \tag{10.45}$$

This may also be regarded as a special case of Section 10.4. Thus we simply replace ε by ε_o and μ by μ_o in eq. (10.43), or we substitute eq. (10.45) directly into eqs. (10.23) and (10.24). Either way, we obtain

$$\alpha = 0, \beta = \omega\sqrt{\mu_o\varepsilon_o} = \frac{\omega}{c} \tag{10.46a}$$

$$u = \frac{1}{\sqrt{\mu_o\varepsilon_o}} = c, \lambda = \frac{2\pi}{\beta} \tag{10.46b}$$

where $c \simeq 3 \times 10^8$ m/s, the speed of light in a vacuum. The fact that EM waves travel in free space at the speed of light is significant. It provides some evidence that light is the manifestation of an EM wave. In other words, light is characteristically electromagnetic.

By substituting the constitutive parameters in eq. (10.45) into eq. (10.33), $\theta_\eta = 0$ and $\eta = \eta_o$, where η_o is called the *intrinsic impedance of free space* and is given by

$$\eta_o = \sqrt{\frac{\mu_o}{\varepsilon_o}} = 120\pi \approx 377\ \Omega \tag{10.47}$$

$$\mathbf{E} = E_o \cos(\omega t - \beta z)\ \mathbf{a}_x \tag{10.48a}$$

then

$$\mathbf{H} = H_o \cos(\omega t - \beta z)\mathbf{a}_y = \frac{E_o}{\eta_o} \cos(\omega t - \beta z)\mathbf{a}_y \tag{10.48b}$$

The plots of **E** and **H** are shown in Figure 10.7(a). In general, if $\mathbf{a}_E$, $\mathbf{a}_H$, and $\mathbf{a}_k$ are unit vectors along the **E** field, the **H** field, and the direction of wave propagation; it can be shown that (see Problem 10.69).

$$\mathbf{a}_k \times \mathbf{a}_E = \mathbf{a}_H$$

or

$$\mathbf{a}_k \times \mathbf{a}_H = -\mathbf{a}_E$$

or

$$\boxed{\mathbf{a}_E \times \mathbf{a}_H = \mathbf{a}_k} \tag{10.49}$$

Both **E** and **H** fields (or EM waves) are everywhere normal to the direction of wave propagation, $\mathbf{a}_k$. That means that the fields lie in a plane that is transverse or orthogonal to the

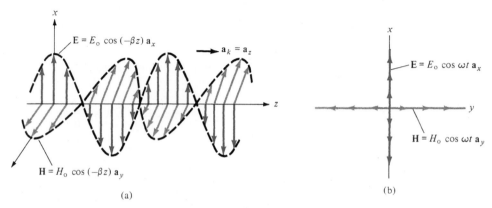

(a)   (b)

FIGURE 10.7 Plots of **E** and **H** (**a**) as functions of z at $t = 0$; and (**b**) at $z = 0$. The arrows indicate instantaneous values.

direction of wave propagation. They form an EM wave that has no electric or magnetic field components along the direction of propagation; such a wave is called a *transverse electromagnetic* (TEM) wave. A combination of **E** and **H** is called a *uniform plane wave* because **E** (or **H**) has the same magnitude throughout any transverse plane, defined by $z =$ constant. The direction in which the electric field points is the *polarization* of a TEM wave.[1] The wave in eq. (10.29), for example, is polarized in the x-direction. This should be observed in Figure 10.7(b), which illustrates uniform plane waves. A uniform plane wave cannot exist physically because it stretches to infinity and would represent an infinite energy. Such waves are characteristically simple and fundamentally important. They serve as approximations to practical waves such as those from a radio antenna at distances sufficiently far from radiating sources. Although our discussion after eq. (10.48) deals with free space, it also applies for any other isotropic medium.

10.6 PLANE WAVES IN GOOD CONDUCTORS

Plane waves in good conductors comprise another special case of that considered in Section 10.3. A perfect, or good conductor, is one in which $\sigma \gg \omega\varepsilon$, so that $\dfrac{\sigma}{\omega\varepsilon} \gg 1$; that is,

$$\sigma \simeq \infty, \quad \varepsilon = \varepsilon_o, \quad \mu = \mu_o\mu_r \tag{10.50}$$

Hence, eqs. (10.23) and (10.24) become

$$\alpha = \beta = \sqrt{\frac{\omega\mu\sigma}{2}} = \sqrt{\pi f\mu\sigma} \tag{10.51a}$$

$$u = \frac{\omega}{\beta} = \sqrt{\frac{2\omega}{\mu\sigma}}, \quad \lambda = \frac{2\pi}{\beta} \tag{10.51b}$$

Also, from eq. (10.32),

$$\eta = \sqrt{\frac{j\omega\mu}{\sigma}} = \sqrt{\frac{\omega\mu}{\sigma}}\,\angle 45° \tag{10.52}$$

and thus **E** leads **H** by 45°. If

$$\mathbf{E} = E_o e^{-\alpha z}\cos(\omega t - \beta z)\mathbf{a}_x \tag{10.53a}$$

then

$$\mathbf{H} = \frac{E_o}{\sqrt{\dfrac{\omega\mu}{\sigma}}} e^{-\alpha z}\cos(\omega t - \beta z - 45°)\mathbf{a}_y \tag{10.53b}$$

---

[1] Polarization will be covered in Section 10.7.

Therefore, as the **E** (or **H**) wave travels in a conducting medium, its amplitude is attenuated by the factor $e^{-\alpha z}$. The distance δ, shown in Figure 10.8, through which the wave amplitude decreases to a factor e^{-1} (about 37% of the original value) is called *skin depth* or *penetration depth* of the medium; that is,

$$E_o e^{-\alpha \delta} = E_o e^{-1}$$

or

$$\delta = \frac{1}{\alpha} \tag{10.54a}$$

The **skin depth** is a measure of the depth to which an **EM** wave can penetrate the medium.

Equation (10.54a) is generally valid for any material medium. For good conductors, eqs. (10.51a) and (10.54a) give

$$\boxed{\delta = \frac{1}{\sqrt{\pi f \mu \sigma}} = \frac{1}{\alpha}} \tag{10.54b}$$

The illustration in Figure 10.8 for a good conductor is exaggerated. However, for a partially conducting medium, the skin depth can be quite large. Note from eqs. (10.51a), (10.52), and (10.54b) that for a good conductor,

$$\eta = \frac{1}{\sigma \delta} \sqrt{2} \, e^{j\pi/4} = \frac{1+j}{\sigma \delta} \tag{10.55}$$

Noting that for good conductors we have $\alpha = \beta = \dfrac{1}{\delta}$, eq. (10.53a) can be written as

$$\mathbf{E} = E_o e^{-z/\delta} \cos\left(\omega t - \frac{z}{\delta}\right) \mathbf{a}_x$$

showing that δ measures the exponential damping of the wave as it travels through the conductor. The skin depth in copper at various frequencies is shown in Table 10.2. From Table 10.2, we notice that the skin depth decreases with increasing frequency. Thus, **E** and **H** can hardly propagate through good conductors.

The phenomenon whereby field intensity in a conductor rapidly decreases is known as the *skin effect*. It is a tendency of charges to migrate from the bulk of the conducting material to the surface, resulting in higher resistance. The fields and associated currents are confined to a very thin layer (the skin) of the conductor surface. For a wire of radius a, for example, it is a good approximation at high frequencies to assume that all of the current

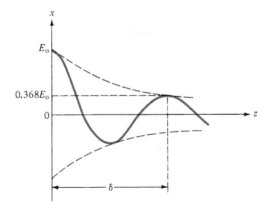

FIGURE 10.8 Illustration of skin depth.

flows in the circular ring of thickness δ as shown in Figure 10.9. The skin effect appears in different guises in such problems as attenuation in waveguides, effective or ac resistance of transmission lines, and electromagnetic shielding. It is used to advantage in many applications. For example, because the skin depth in silver is very small, the difference in performance between a pure silver component and a silver-plated brass component is negligible, so silver plating is often used to reduce the material cost of waveguide components. For the same reason, hollow tubular conductors are used instead of solid conductors in outdoor television antennas. Effective electromagnetic shielding of electrical devices can be provided by conductive enclosures a few skin depths in thickness.

The skin depth is useful in calculating the *ac resistance* due to skin effect. The resistance in eq. (5.16) is called the *dc resistance,* that is,

$$R_{dc} = \frac{\ell}{\sigma S}$$

We define the *surface or skin resistance* R_s (in Ω) as the real part of η for a good conductor. Thus from eq. (10.55)

$$R_s = \frac{1}{\sigma\delta} = \sqrt{\frac{\pi f \mu}{\sigma}} \tag{10.56}$$

This is the resistance of a unit width and unit length of the conductor. It is equivalent to the dc resistance for a unit length of the conductor having cross-sectional area $1 \times \delta$. Thus

TABLE 10.2 Skin Depth in Copper*

Frequency (Hz)	10	60	100	500	10^4	10^8	10^{10}
Skin depth (mm)	20.8	8.6	6.6	2.99	0.66	6.6×10^{-3}	6.6×10^{-4}

*For copper, $\sigma = 5.8 \times 10^7$ S/m, $\mu = \mu_o$, $\delta = 66.1/\sqrt{f}$ (in mm).

FIGURE 10.9 Skin depth at high frequencies, $\delta \ll a$.

for a given width w and length ℓ, the ac resistance is calculated by using the familiar dc resistance relation of eq. (5.16) and assuming a uniform current flow in the conductor of thickness δ; that is,

$$R_{ac} = \frac{\ell}{\sigma \delta w} = \frac{R_s \ell}{w} \tag{10.57}$$

where $S \simeq \delta w$. For a conductor wire of radius a (see Figure 10.9), $w = 2\pi a$, so

$$\frac{R_{ac}}{R_{dc}} = \frac{\dfrac{\ell}{\sigma 2\pi a \delta}}{\dfrac{\ell}{\sigma \pi a^2}} = \frac{a}{2\delta} = \frac{a}{2}\sqrt{\pi f \mu \sigma}$$

Since $\delta \ll a$ at high frequencies, this shows that R_{ac} is far greater than R_{dc}. In general, the ratio of the ac to the dc resistance starts at 1.0 for dc and very low frequencies and increases as the frequency increases. Also, although the bulk of the current is nonuniformly distributed over a thickness of 5δ of the conductor, the power loss is the same as though it were uniformly distributed over a thickness of δ and zero elsewhere. This is one more reason that δ is referred to as the skin depth. For easy reference, the formulas for propagation constants for different media are summarized in Table 10.3.

TABLE 10.3 Formulas, for α, β, η, v, and λ

	Lossy Medium	Lossless Medium	Free Space	Conductor
$\alpha =$	$\omega\left[\dfrac{\mu\varepsilon}{2}\left[\sqrt{1 + \left(\dfrac{\sigma}{\omega\varepsilon}\right)^2} - 1\right]\right]^{1/2}$	0	0	$\sqrt{\pi f \mu \sigma}$
$\beta =$	$\omega\left[\dfrac{\mu\varepsilon}{2}\left[\sqrt{1 + \left(\dfrac{\sigma}{\omega\varepsilon}\right)^2} + 1\right]\right]^{1/2}$	$\omega\sqrt{\mu\varepsilon}$	$\omega\sqrt{\mu_o\varepsilon_o}$	$\sqrt{\pi f \mu \sigma}$
$\eta =$	$\sqrt{\dfrac{j\omega\mu}{\sigma + j\omega\varepsilon}}$	$\sqrt{\dfrac{\mu}{\varepsilon}}$	$\sqrt{\dfrac{\mu_o}{\varepsilon_o}} \simeq 377$	$(1 + j)\dfrac{\alpha}{\sigma}$

$$v = \frac{\omega}{\beta}, \quad \lambda = \frac{2\pi}{\beta}$$

| EXAMPLE 10.3 | In a lossless dielectric for which $\eta = 60\pi$, $\mu_r = 1$, and $\mathbf{H} = -0.1 \cos(\omega t - z)\mathbf{a}_x + 0.5 \sin(\omega t - z)\mathbf{a}_y$ A/m, calculate ε_r, ω, and $\mathbf{E}$. |

Solution:

In this case, $\sigma = 0$, $\alpha = 0$, and $\beta = 1$, so

$$\eta = \sqrt{\mu/\varepsilon} = \sqrt{\frac{\mu_o}{\varepsilon_o}} \sqrt{\frac{\mu_r}{\varepsilon_r}} = \frac{120\pi}{\sqrt{\varepsilon_r}}$$

or

$$\sqrt{\varepsilon_r} = \frac{120\pi}{\eta} = \frac{120\pi}{60\pi} = 2 \quad \rightarrow \quad \varepsilon_r = 4$$

$$\beta = \omega\sqrt{\mu\varepsilon} = \omega\sqrt{\mu_o\varepsilon_o}\sqrt{\mu_r\varepsilon_r} = \frac{\omega}{c}\sqrt{4} = \frac{2\omega}{c}$$

or

$$\omega = \frac{\beta c}{2} = \frac{1\,(3 \times 10^8)}{2} = 1.5 \times 10^8 \text{ rad/s}$$

From the given $\mathbf{H}$ field, $\mathbf{E}$ can be calculated in two ways: by using the techniques (based on Maxwell's equations) developed in this chapter or directly, by using Maxwell's equations as in Chapter 9.

Method 1: To use the techniques developed in the present chapter, we let

$$\mathbf{H} = \mathbf{H}_1 + \mathbf{H}_2$$

where $\mathbf{H}_1 = -0.1 \cos(\omega t - z)\mathbf{a}_x$ and $\mathbf{H}_2 = 0.5 \sin(\omega t - z)\mathbf{a}_y$ and the corresponding electric field

$$\mathbf{E} = \mathbf{E}_1 + \mathbf{E}_2$$

where $\mathbf{E}_1 = E_{1o} \cos(\omega t - z)\mathbf{a}_{E_1}$ and $\mathbf{E}_2 = E_{2o} \sin(\omega t - z)\mathbf{a}_{E_2}$. Notice that although $\mathbf{H}$ has components along $\mathbf{a}_x$ and $\mathbf{a}_y$, it has no component along the direction of propagation; it is therefore a TEM wave.

For $\mathbf{E}_1$,

$$\mathbf{a}_{E_1} = -(\mathbf{a}_k \times \mathbf{a}_{H_1}) = -(\mathbf{a}_z \times -\mathbf{a}_x) = \mathbf{a}_y$$

$$E_{1o} = \eta H_{1o} = 60\pi\,(0.1) = 6\pi$$

Hence

$$\mathbf{E}_1 = 6\pi \cos(\omega t - z)\mathbf{a}_y$$

For $\mathbf{E}_2$,

$$\mathbf{a}_{E_2} = -(\mathbf{a}_k \times \mathbf{a}_{H_2}) = -(\mathbf{a}_z \times \mathbf{a}_y) = \mathbf{a}_x$$
$$E_{2o} = \eta H_{2o} = 60\pi(0.5) = 30\pi$$

Hence

$$\mathbf{E}_2 = 30\pi \sin(\omega t - z)\mathbf{a}_x$$

Adding $\mathbf{E}_1$ and $\mathbf{E}_2$ gives $\mathbf{E}$; that is,

$$\mathbf{E} = 94.25 \sin(1.5 \times 10^8 t - z)\mathbf{a}_x + 18.85 \cos(1.5 \times 10^8 t - z)\mathbf{a}_y \text{ V/m}$$

Method 2: We may apply Maxwell's equations directly

$$\nabla \times \mathbf{H} = \cancel{\sigma\mathbf{E}} + \varepsilon\frac{\partial \mathbf{E}}{\partial t} \quad \rightarrow \quad \mathbf{E} = \frac{1}{\varepsilon}\int \nabla \times \mathbf{H}\, dt$$

$$0$$

because $\sigma = 0$. But

$$\nabla \times \mathbf{H} = \begin{vmatrix} \mathbf{a}_x & \mathbf{a}_y & \mathbf{a}_z \\ \frac{\partial}{\partial x} & \frac{\partial}{\partial y} & \frac{\partial}{\partial z} \\ H_x(z) & H_y(z) & 0 \end{vmatrix} = -\frac{\partial H_y}{\partial z}\mathbf{a}_x + \frac{\partial H_x}{\partial z}\mathbf{a}_y$$

$$= H_{2o} \cos(\omega t - z)\mathbf{a}_x + H_{1o} \sin(\omega t - z)\mathbf{a}_y$$

where $H_{1o} = -0.1$ and $H_{2o} = 0.5$. Hence

$$\mathbf{E} = \frac{1}{\varepsilon}\int \nabla \times \mathbf{H}\, dt = \frac{H_{2o}}{\varepsilon\omega}\sin(\omega t - z)\mathbf{a}_x - \frac{H_{1o}}{\varepsilon\omega}\cos(\omega t - z)\mathbf{a}_y$$

$$= 94.25 \sin(\omega t - z)\mathbf{a}_x + 18.85 \cos(\omega t - z)\mathbf{a}_y \text{ V/m}$$

as expected.

PRACTICE EXERCISE 10.3

A plane wave in a nonmagnetic medium has $\mathbf{E} = 50 \sin(10^8 t + 2z)\mathbf{a}_y$ V/m. Find

(a) The direction of wave propagation
(b) λ, f, and ε_r
(c) $\mathbf{H}$

Answer: (a) in the $-z$-direction, (b) 3.142 m, 15.92 MHz, 36, (c) 0.7958 $\sin(10^8 t + 2z)\mathbf{a}_x$ A/m.

EXAMPLE 10.4

A uniform plane wave propagating in a medium has

$$\mathbf{E} = 2e^{-\alpha z} \sin(10^8 t - \beta z)\mathbf{a}_y \text{ V/m}$$

If the medium is characterized by $\varepsilon_r = 1$, $\mu_r = 20$, and $\sigma = 3$ S/m, find α, β, and $\mathbf{H}$.

Solution:

We need to determine the loss tangent to be able to tell whether the medium is a lossy dielectric or a good conductor.

$$\frac{\sigma}{\omega \varepsilon} = \frac{3}{10^8 \times 1 \times \dfrac{10^{-9}}{36\pi}} = 3393 \gg 1$$

showing that the medium may be regarded as a good conductor at the frequency of operation. Hence,

$$\alpha = \beta = \sqrt{\frac{\mu \omega \sigma}{2}} = \left[\frac{4\pi \times 10^{-7} \times 20(10^8)(3)}{2} \right]^{1/2}$$

$$= 61.4$$

$$\alpha = 61.4 \text{ Np/m}, \quad \beta = 61.4 \text{ rad/m}$$

Also

$$|\eta| = \sqrt{\frac{\mu \omega}{\sigma}} = \left[\frac{4\pi \times 10^{-7} \times 20(10^8)}{3} \right]^{1/2}$$

$$= \sqrt{\frac{800\pi}{3}}$$

$$\tan 2\theta_\eta = \frac{\sigma}{\omega \varepsilon} = 3393 \quad \rightarrow \quad \theta_\eta = 45° = \frac{\pi}{4}$$

Hence

$$\mathbf{H} = H_o e^{-\alpha z} \sin\left(\omega t - \beta z - \frac{\pi}{4} \right)\mathbf{a}_H$$

where

$$\mathbf{a}_H = \mathbf{a}_k \times \mathbf{a}_E = \mathbf{a}_z \times \mathbf{a}_y = -\mathbf{a}_x$$

and

$$H_o = \frac{E_o}{|\eta|} = 2\sqrt{\frac{3}{800\pi}} = 69.1 \times 10^{-3}$$

Thus

$$\mathbf{H} = -69.1 \, e^{-61.4z} \sin\left(10^8 t - 61.42z - \frac{\pi}{4} \right)\mathbf{a}_x \text{ mA/m}$$

PRACTICE EXERCISE 10.4

A plane wave traveling in the $+y$-direction in a lossy medium ($\varepsilon_r = 4$, $\mu_r = 1$, $\sigma = 10^{-2}$ S/m) has $\mathbf{E} = 30 \cos(10^9 \pi\, t + \pi/4)\mathbf{a}_z$ V/m at $y = 0$. Find

(a) $\mathbf{E}$ at $y = 1$ m, $t = 2$ ns
(b) The distance traveled by the wave to have a phase shift of $10°$
(c) The distance traveled by the wave to have its amplitude reduced by 40%
(d) $\mathbf{H}$ at $y = 2$ m, $t = 2$ ns

Answer: (a) $2.844\mathbf{a}_z$ V/m, (b) 8.349 mm, (c) 542 mm, (d) $-22.6\mathbf{a}_x$ mA/m.

EXAMPLE 10.5

A plane wave $\mathbf{E} = E_o \cos(\omega t - \beta z)\mathbf{a}_x$ is incident on a good conductor at $z \geq 0$. Find the current density in the conductor.

Solution:

Since the current density $\mathbf{J} = \sigma\mathbf{E}$, we expect $\mathbf{J}$ to satisfy the wave equation in eq. (10.17); that is, we expect to find

$$\nabla^2\mathbf{J}_s - \gamma^2\mathbf{J}_s = 0$$

Also the incident $\mathbf{E}$ has only an x-component and varies with z. Hence $\mathbf{J} = J_x(z, t)\mathbf{a}_x$ and

$$\frac{d^2}{dz^2}J_{sx} - \gamma^2 J_{sx} = 0$$

which is an ordinary differential equation with solution (see Case 2 of Example 6.5)

$$J_{sx} = Ae^{-\gamma z} + Be^{+\gamma z}$$

The constant B must be zero because J_{sx} is finite as $z \to \infty$. But in a good conductor, $\sigma \gg \omega\varepsilon$, so that $\alpha = \beta = 1/\delta$. Hence

$$\gamma = \alpha + j\beta = \alpha(1 + j) = \frac{(1 + j)}{\delta}$$

and

$$J_{sx} = Ae^{-z(1+j)/\delta}$$

or

$$J_{sx} = J_{sx}(0)\, e^{-z(1+j)/\delta}$$

where $J_{sx}(0)$ is the current density on the conductor surface.

EXAMPLE 10.6

For the copper coaxial cable of Figure 7.12, let $a = 2$ mm, $b = 6$ mm, and $t = 1$ mm. Calculate the resistance of a 2 m length of the cable at dc and at 100 MHz.

Solution:

Let

$$R = R_o + R_i$$

where R_o and R_i are the resistances of the inner and outer conductors.
At dc,

$$R_i = \frac{\ell}{\sigma S} = \frac{\ell}{\sigma \pi a^2} = \frac{2}{5.8 \times 10^7 \pi [2 \times 10^{-3}]^2} = 2.744 \text{ m}\Omega$$

$$R_o = \frac{\ell}{\sigma S} = \frac{\ell}{\sigma \pi [[b + t]^2 - b^2]} = \frac{\ell}{\sigma \pi [t^2 + 2bt]}$$

$$= \frac{2}{5.8 \times 10^7 \pi [1 + 12] \times 10^{-6}}$$

$$= 0.8429 \text{ m}\Omega$$

Hence $R_{dc} = 2.744 + 0.8429 = 3.587$ mΩ.
At $f = 100$ MHz,

$$R_i = \frac{R_s \ell}{w} = \frac{\ell}{\sigma \delta 2\pi a} = \frac{\ell}{2\pi a}\sqrt{\frac{\pi f \mu}{\sigma}}$$

$$= \frac{2}{2\pi \times 2 \times 10^{-3}}\sqrt{\frac{\pi \times 10^8 \times 4\pi \times 10^{-7}}{5.8 \times 10^7}}$$

$$= 0.41 \ \Omega$$

Since $\delta = 6.6 \ \mu$m $\ll t = 1$ mm, $w = 2\pi b$ for the outer conductor. Hence,

$$R_o = \frac{R_s \ell}{w} = \frac{\ell}{2\pi b}\sqrt{\frac{\pi f \mu}{\sigma}}$$

$$= \frac{2}{2\pi \times 6 \times 10^{-3}}\sqrt{\frac{\pi \times 10^8 \times 4\pi \times 10^{-7}}{5.8 \times 10^7}}$$

$$= 0.1384 \ \Omega$$

Hence,

$$R_{ac} = 0.41 + 0.1384 = 0.5484 \ \Omega$$

which is about 150 times greater than R_{dc}. Thus, for the same effective current i, the ohmic loss (i^2R) of the cable at 100 MHz is greater than the dc power loss by a factor of 150.

PRACTICE EXERCISE 10.6

For an aluminum wire having a diameter 2.6 mm, calculate the ratio of ac to dc resistance at

(a) 10 MHz
(b) 2 GHz

Answer: (a) 24.16, (b) 341.7.

10.7 WAVE POLARIZATION

It is a common practice to describe an EM wave by its polarization. Polarization is an important property of an EM wave, and the concept has been developed to describe the various types of electric field variation and orientation. The polarization of an EM wave depends on the transmitting antenna or source. It is determined by the direction of the electric field for fields having more than one component.

> **Polarization** may be regarded as the locus of the tip of the electric field (in a plane perpendicular to the direction of propagation) at a given point as a function of time.

There are three types of polarization: linear or plane, circular, and elliptical. That means that the tip of the electric field can describe a straight line, a circle, or an ellipse with time, as shown in Figure 10.10. Wave polarization is important for radio and TV broadcasting. Amplitude modulation (AM) radio broadcasting is with polarization vertical to the earth's surface, while frequency modulation (FM) broadcasting is generally circularly polarized.

A uniform plane wave is linearly polarized if it has only one component or when its transverse components are in phase. For a wave traveling in the +z-direction, we may have

$$E_x = E_{ox} \cos(\omega t - \beta z + \phi_x)$$
$$E_y = E_{oy} \cos(\omega t - \beta z + \phi_y)$$

(10.58)

where E_{ox} and E_{oy} are real. The composite wave

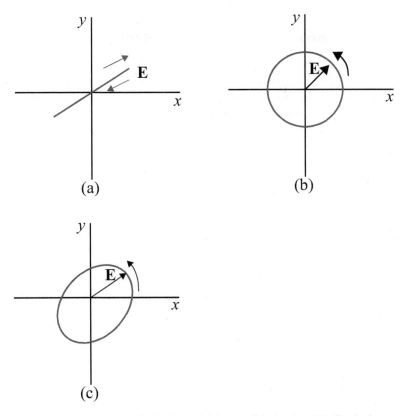

FIGURE 10.10 Wave polarizations: (**a**) linear, (**b**) circular, (**c**) elliptical.

$$\mathbf{E} = E_{ox}\cos(\omega t - \beta z + \phi_x)\mathbf{a}_x + E_{oy}\cos(\omega t - \beta z + \phi_y)\mathbf{a}_y \qquad (10.59)$$

is linearly polarized when the phase difference $\Delta\phi$ is

$$\Delta\phi = \phi_y - \phi_x = n\pi, \qquad n = 0, 1, 2, \ldots \qquad (10.60)$$

This allows the two components to maintain the same ratio at all times. If we observe the wave in the direction of propagation (z in this case), we will notice that the tip of the electric field follows a line—hence, the term *linear polarization*. Linearly polarized plane waves can be generated by simple antennas (such as dipole antennas) or lasers.

Circular polarization takes place when the x- and y-components are the same in magnitude ($E_{ox} = E_{oy} = E_o$) and the phase difference between them is an odd multiple of $\pi/2$; that is,

$$\Delta\phi = \phi_y - \phi_x = \pm(2n + 1)\pi/2, \qquad n = 0, 1, 2, \ldots \qquad (10.61)$$

For example, the x- and y-components may be of the form

$$E_x = E_o \cos(\omega t - \beta z) \tag{10.62}$$
$$E_y = E_o \cos(\omega t - \beta z + \pi/2)$$

The tip of the composite electric field as observed as a fixed point in the xy-plane moves along a circle as time progresses. Circularly polarized waves can be generated by a helically wound wire antenna or by two linear sources that are oriented perpendicular to each other and fed with currents that are out of phase by 90°. That the locus of total field traces a circle can be seen if we examine the components at a point, say $z = 0$,

$$E_x = E_o \cos(\omega t), \qquad E_y = E_o \cos(\omega t + \pi/2) = -E_o \sin(\omega t) \tag{10.63}$$
$$|E|^2 = E_x^2 + E_y^2 = E_o^2 \tag{10.64}$$

which is the equation of a circle.

Linear and circular polarizations are special cases of the more general case of the elliptical polarization. An elliptically polarized wave is one in which the tip of the field traces an elliptic locus in a fixed transverse plane as the field changes with time. Elliptical polarization is achieved when the x- and y-components are not equal in magnitude $E_{ox} \neq E_{oy}$ and the phase difference between them is an odd multiple of $\pi/2$; that is,

$$\Delta\phi = \phi_y - \phi_x = \pm (2n + 1)\pi/2, \qquad n = 0, 1, 2, \ldots \tag{10.65}$$

This allows the tip of the electric field to trace an ellipse in the xy-plane. To show that this is the case, consider eq. (10.58) when $z = 0$ and $\Delta\phi = \phi_y - \phi_x = \pi/2$,

$$E_x = E_{oz} \cos(\omega t) \longrightarrow \cos(\omega t) = \frac{E_x}{E_{ox}} \tag{10.66}$$

$$E_y = E_{oy} \cos(\omega t + \pi/2) = -E_o \sin(\omega t) \longrightarrow -\sin(\omega t) = \frac{E_y}{E_{oy}}$$

Squaring and adding these equations yields

$$\cos^2(\omega t) + \sin^2(\omega t) = 1 \longrightarrow \frac{E_x^2}{E_{ox}^2} + \frac{E_y^2}{E_{oy}^2} = 1 \tag{10.67}$$

which is the equation of an ellipse, as shown in Figure 10.10. Notice that if $E_{ox} = E_{oy}$, we have circular polarization. Thus, circular polarization is a special case of elliptical polarization. In fact, we can show that linear polarization is also a special case of elliptical polarization. Thus, the most general case is elliptical polarization.

EXAMPLE 10.7

Determine the polarization of a plane wave with:

(a) $\mathbf{E}(z, t) = 4e^{-0.25z} \cos(\omega t - 0.8z)\mathbf{a}_x + 3e^{-0.25z} \sin(\omega t - 0.8z)\mathbf{a}_y$ V/m

(b) $\mathbf{H}_s(z) = H_o e^{-j\beta z}\mathbf{a}_x - 2H_o e^{-j\beta z}\mathbf{a}_y$

Solution:

(a) From the given **E**,

$$E_x = 4e^{-0.25z} \cos(\omega t - 0.8z)$$

$$E_y = 3e^{-0.25z} \sin(\omega t - 0.8z)$$

In the $z = 0$ plane, we have

$$\frac{1}{4}E_x(0, t) = \cos(\omega t)$$

$$\frac{1}{3}E_x(0, t) = \sin(\omega t)$$

Squaring and adding gives

$$\frac{1}{16}E_x^2(0, t) + \frac{1}{9}E_y^2(0, t) = 1$$

which describes an ellipse. Hence, the wave is elliptically polarized.

(b) The two components of **H** are in phase; hence, the polarization is linear. For proper characterization, it is expedient to find the electric field component. This can be done in many ways. Using Maxwell's equation,

$$\nabla \times \mathbf{H}_s = j\omega\varepsilon\mathbf{E}_s \longrightarrow \mathbf{E}_s = \frac{1}{j\omega\varepsilon} \nabla \times \mathbf{H}_s$$

$$\nabla \times \mathbf{H}_s = \begin{vmatrix} \mathbf{a}_x & \mathbf{a}_y & \mathbf{a}_z \\ \dfrac{\partial}{\partial x} & \dfrac{\partial}{\partial y} & \dfrac{\partial}{\partial z} \\ H_o e^{-j\beta z} & -2H_o e^{-j\beta z} & 0 \end{vmatrix} = j2\beta H_o e^{-j\beta z}\mathbf{a}_x - j\beta H_o e^{-j\beta z}\mathbf{a}_y$$

Dividing both sides by $j\omega\varepsilon$ and setting $\eta = \beta/\omega\varepsilon$ yields

$$\mathbf{E}_s = 2\eta H_o e^{-j\beta z}\mathbf{a}_x - \eta H_o e^{-j\beta z}\mathbf{a}_y$$

In the time domain,

$$\mathbf{E}(z, t) = \text{Re}[\mathbf{E}_s e^{-j\omega t}] = 2\eta H_o\cos(\omega t - \beta z)\mathbf{a}_x - \eta H_o\cos(\omega t - \beta z)\mathbf{a}_y$$

If we set $z = 0$,

$$\mathbf{E}(0, t) = (2\eta H_o\mathbf{a}_x - \eta H_o\mathbf{a}_y) \cos(\omega t)$$

At $t = 0$, **E** has components $2\eta H_o$ in the x-direction and ηH_o in the y-direction. The ratio E_y/E_x remains the same as t changes. Hence, **E** is linearly polarized.

PRACTICE EXERCISE 10.7

Given that $\mathbf{E}_s = E_o(\mathbf{a}_y - j\mathbf{a}_y)e^{-j\beta z}$, determine the polarization.

Answer: Circular polarization.

10.8 POWER AND THE POYNTING VECTOR

As mentioned before, energy can be transported from one point (where a transmitter is located) to another point (with a receiver) by means of EM waves. The rate of such energy transportation can be obtained from Maxwell's equations:

$$\nabla \times \mathbf{E} = -\mu \frac{\partial \mathbf{H}}{\partial t} \qquad (10.68a)$$

$$\nabla \times \mathbf{H} = \sigma \mathbf{E} + \varepsilon \frac{\partial \mathbf{E}}{\partial t} \qquad (10.68b)$$

Dotting both sides of eq. (10.68b) with **E** gives

$$\mathbf{E} \cdot (\nabla \times \mathbf{H}) = \sigma E^2 + \mathbf{E} \cdot \varepsilon \frac{\partial \mathbf{E}}{\varepsilon t} \qquad (10.69)$$

But for any vector fields **A** and **B** (see Appendix A.10)

$$\nabla \cdot (\mathbf{A} \times \mathbf{B}) = \mathbf{B} \cdot (\nabla \times \mathbf{A}) - \mathbf{A} \cdot (\nabla \times \mathbf{B})$$

Applying this vector identity to eq. (10.69) (letting **A** = **H** and **B** = **E**) gives

$$\mathbf{H} \cdot (\nabla \times \mathbf{E}) + \nabla \cdot (\mathbf{H} \times \mathbf{E}) = \sigma E^2 + \mathbf{E} \cdot \varepsilon \frac{\partial \mathbf{E}}{\partial t} \qquad (10.70)$$

$$= \sigma E^2 + \frac{1}{2} \varepsilon \frac{\partial}{\partial t} E^2$$

Dotting both sides of eq. (10.68a) with **H**, we write

$$\mathbf{H} \cdot (\nabla \times \mathbf{E}) = \mathbf{H} \cdot \left(-\mu \frac{\partial \mathbf{H}}{\partial t} \right) = -\frac{\mu}{2} \frac{\partial}{\partial t} (\mathbf{H} \cdot \mathbf{H}) \qquad (10.71)$$

and thus eq. (10.70) becomes

$$-\frac{\mu}{2} \frac{\partial H^2}{\partial t} - \nabla \cdot (\mathbf{E} \times \mathbf{H}) = \sigma E^2 + \frac{1}{2} \varepsilon \frac{\partial E^2}{\partial t}$$

Rearranging terms and taking the volume integral of both sides,

$$\int_v \nabla \cdot (\mathbf{E} \times \mathbf{H}) dv = -\frac{\partial}{\partial t} \int_v \left[\frac{1}{2} \varepsilon E^2 + \frac{1}{2} \mu H^2 \right] dv - \int_v \sigma E^2 \, dv \qquad (10.72)$$

Applying the divergence theorem to the left-hand side gives

$$\oint_S (\mathbf{E} \times \mathbf{H}) \cdot d\mathbf{S} = -\frac{\partial}{\partial t} \int_v \left[\frac{1}{2} \varepsilon E^2 + \frac{1}{2} \mu H^2 \right] dv - \int_v \sigma E^2 dv \qquad (10.73)$$

$$\downarrow \qquad\qquad \downarrow \qquad\qquad \downarrow$$

$$\begin{matrix} \text{total power} \\ \text{leaving the volume} \end{matrix} = \begin{matrix} \text{rate of decrease in} \\ \text{energy stored in electric} \\ \text{and magnetic fields} \end{matrix} - \begin{matrix} \text{ohmic power} \\ \text{dissipated} \end{matrix} \qquad (10.74)$$

Equation (10.73) is referred to as *Poynting's theorem*.[2] The various terms in the equation are identified using energy-conservation arguments for EM fields. The first term on the right-hand side of eq. (10.73) is interpreted as the rate of decrease in energy stored in the electric and magnetic fields. The second term is the power dissipated because the medium is conducting ($\sigma \neq 0$). The quantity $\mathbf{E} \times \mathbf{H}$ on the left-hand side of eq. (10.73) is known as the *Poynting vector* $\mathscr{P}$, measured in watts per square meter (W/m²); that is,

$$\boxed{\mathscr{P} = \mathbf{E} \times \mathbf{H}} \tag{10.75}$$

It represents the instantaneous power density vector associated with the EM field at a given point. The integration of the Poynting vector over any closed surface gives the net power flowing out of that surface.

> **Poynting's theorem** states that the net power flowing out of a given volume v is equal to the time rate of decrease in the energy stored within v minus the ohmic losses.

The theorem is illustrated in Figure 10.11.

It should be noted that $\mathscr{P}$ is normal to both $\mathbf{E}$ and $\mathbf{H}$ and is therefore along the direction of wave propagation $\mathbf{a}_k$ for uniform plane waves. Thus

$$\mathbf{a}_k = \mathbf{a}_E \times \mathbf{a}_H \tag{10.49}$$

The fact that $\mathscr{P}$ points along $\mathbf{a}_k$ causes $\mathscr{P}$ to be regarded as a "pointing" vector.

Again, if we assume that

$$\mathbf{E}(z, t) = E_o e^{-\alpha z} \cos(\omega t - \beta z)\mathbf{a}_x$$

then

$$\mathbf{H}(z, t) = \frac{E_o}{|\eta|} e^{-\alpha z} \cos(\omega t - \beta z - \theta_\eta)\mathbf{a}_y$$

and

$$\mathscr{P}(z, t) = \frac{E_o^2}{|\eta|} e^{-2\alpha z} \cos(\omega t - \beta z) \cos(\omega t - \beta z - \theta_\eta)\mathbf{a}_z$$

$$= \frac{E_o^2}{2|\eta|} e^{-2\alpha z} \left[\cos \theta_\eta + \cos(2\omega t - 2\beta z - \theta_\eta)\right]\mathbf{a}_z \tag{10.76}$$

---

[2] After J. H. Poynting, "On the transfer of energy in the electromagnetic field," *Philosophical Transactions*, vol. 174, 1883, p. 343.

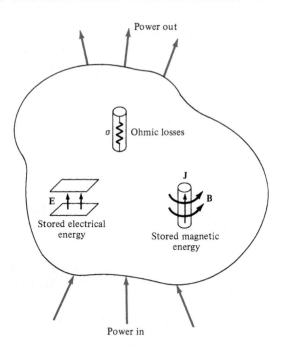

FIGURE 10.11 Illustration of power balance for EM fields.

since $\cos A \cos B = \frac{1}{2}[\cos(A - B) + \cos(A + B)]$. To determine the time-average Poynting vector $\mathcal{P}_{ave}(z)$ (in W/m²), which is of more practical value than the instantaneous Poynting vector $\mathcal{P}(z, t)$, we integrate eq. (10.76) over the period $T = 2\pi/\omega$; that is,

$$\mathcal{P}_{ave}(z) = \frac{1}{T} \int_0^T \mathcal{P}(z, t) \, dt \tag{10.77}$$

It can be shown (see Problem 10.42) that this is equivalent to

$$\boxed{\mathcal{P}_{ave}(z) = \frac{1}{2} \text{Re}(\mathbf{E}_s \times \mathbf{H}_s^*)} \tag{10.78}$$

By substituting eq. (10.76) into eq. (10.77), we obtain

$$\boxed{\mathcal{P}_{ave}(z) = \frac{E_o^2}{2|\eta|} e^{-2\alpha z} \cos \theta_\eta \, \mathbf{a}_z} \tag{10.79}$$

The total time-average power crossing a given surface S is given by

$$\boxed{P_{ave} = \int_S \mathcal{P}_{ave} \cdot d\mathbf{S}} \tag{10.80}$$

We should note the difference between $\mathscr{P}$, $\mathscr{P}_{ave}$, and P_{ave}: whereas $\mathscr{P}(x, y, z, t)$ is the Poynting vector in watts per square meter and is time varying, $\mathscr{P}_{ave}(x, y, z)$, also in watts per square meter, is the time average of the Poynting vector $\mathscr{P}$; it is a vector but is time invariant. Finally, P_{ave} is a total time-average power through a surface in watts; it is a scalar.

EXAMPLE 10.8

In a nonmagnetic medium

$$\mathbf{E} = 4 \sin(2\pi \times 10^7 t - 0.8x)\mathbf{a}_z \text{ V/m}$$

Find

(a) ε_r, η

(b) The time-average power carried by the wave

(c) The total power crossing 100 cm^2 of plane $2x + y = 5$

Solution:

(a) Since $\alpha = 0$ and $\beta \neq \omega/c$, the medium is not free space but a lossless medium:

$$\beta = 0.8, \quad \omega = 2\pi \times 10^7, \quad \mu = \mu_o \text{ (nonmagnetic)}, \quad \varepsilon = \varepsilon_o \varepsilon_r$$

Hence

$$\beta = \omega\sqrt{\mu\varepsilon} = \omega\sqrt{\mu_o\varepsilon_o\varepsilon_r} = \frac{\omega}{c}\sqrt{\varepsilon_r}$$

or

$$\sqrt{\varepsilon_r} = \frac{\beta c}{\omega} = \frac{0.8(3 \times 10^8)}{2\pi \times 10^7} = \frac{12}{\pi}$$

$$\varepsilon_r = 14.59$$

$$\eta = \sqrt{\frac{\mu}{\varepsilon}} = \sqrt{\frac{\mu_o}{\varepsilon_o\varepsilon_r}} = \frac{120\pi}{\sqrt{\varepsilon_r}} = 120\pi \cdot \frac{\pi}{12} = 10\pi^2$$

$$= 98.7 \ \Omega$$

(b) $\mathscr{P} = \mathbf{E} \times \mathbf{H} = \dfrac{E_o^2}{\eta}\sin^2(\omega t - \beta x)\mathbf{a}_x$

$$\mathscr{P}_{ave} = \frac{1}{T}\int_0^T \mathscr{P} \, dt = \frac{E_o^2}{2\eta}\mathbf{a}_x = \frac{16}{2 \times 10\pi^2}\mathbf{a}_x$$

$$= 81\mathbf{a}_x \text{ mW/m}^2$$

(c) On plane $2x + y = 5$ (see Example 3.5 or 8.5),

$$\mathbf{a}_n = \frac{2\mathbf{a}_x + \mathbf{a}_y}{\sqrt{5}}$$

Hence the total power is

$$P_{ave} = \int \mathscr{P}_{ave} \cdot d\mathbf{S} = \mathscr{P}_{ave} \cdot S\,\mathbf{a}_n$$

$$= (81 \times 10^{-3}\mathbf{a}_x) \cdot (100 \times 10^{-4}) \left[\frac{2\mathbf{a}_x + \mathbf{a}_y}{\sqrt{5}}\right]$$

$$= \frac{162 \times 10^{-5}}{\sqrt{5}} = 724.5 \ \mu\text{W}$$

PRACTICE EXERCISE 10.8

In free space, $\mathbf{H} = 0.2 \cos(\omega t - \beta x)\mathbf{a}_z$ A/m. Find the total power passing through:

(a) A square plate of side 10 cm on plane $x + y = 1$
(b) A circular disk of radius 5 cm on plane $x = 1$.

Answer: (a) 53.31 mW, (b) 59.22 mW.

10.9 REFLECTION OF A PLANE WAVE AT NORMAL INCIDENCE

So far, we have considered uniform plane waves traveling in unbounded, homogeneous, isotropic media. When a plane wave from one medium meets a different medium, it is partly reflected and partly transmitted. The proportion of the incident wave that is reflected or transmitted depends on the constitutive parameters $(\varepsilon, \mu, \sigma)$ of the two media involved. Here we will assume that the incident plane wave is normal to the boundary between the media; oblique incidence of plane waves will be covered in the next section after we have presented the simpler case of normal incidence.

Suppose that a plane wave propagating along the $+z$-direction is incident normally on the boundary $z = 0$ between medium $1(z < 0)$ characterized by $\sigma_1, \varepsilon_1, \mu_1$ and medium $2(z > 0)$ characterized by $\sigma_2, \varepsilon_2, \mu_2$, as shown in Figure 10.12. In Figure 10.12, subscripts i, r, and t denote incident, reflected, and transmitted waves, respectively. The incident, reflected, and transmitted waves shown in Figure 10.12 are obtained as follows.

Incident Wave

$(\mathbf{E}_i, \mathbf{H}_i)$ is traveling along $+\mathbf{a}_z$ in medium 1. If we suppress the time factor $e^{j\omega t}$ and assume that

$$\mathbf{E}_{is}(z) = E_{io}e^{-\gamma_1 z}\mathbf{a}_x \tag{10.81}$$

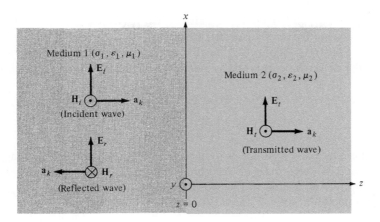

FIGURE 10.12 A plane wave incident normally on an interface between two different media.

then

$$\mathbf{H}_{is}(z) = H_{io}e^{-\gamma_1 z}\mathbf{a}_y = \frac{E_{io}}{\eta_1}e^{-\gamma_1 z}\mathbf{a}_y \qquad (10.82)$$

Reflected Wave

$(\mathbf{E}_r, \mathbf{H}_r)$ is traveling along $-\mathbf{a}_z$ in medium 1. If

$$\mathbf{E}_{rs}(z) = E_{ro}e^{\gamma_1 z}\mathbf{a}_x \qquad (10.83)$$

then

$$\mathbf{H}_{rs}(z) = H_{ro}\, e^{\gamma_1 z}(-\mathbf{a}_y) = -\frac{E_{ro}}{\eta_1}e^{\gamma_1 z}\mathbf{a}_y \qquad (10.84)$$

where $\mathbf{E}_{rs}$ has been assumed to be along $\mathbf{a}_x$. To satisfy the necessary boundary conditions at the interface, we will consistently assume that for normal incidence $\mathbf{E}_i$, $\mathbf{E}_r$, and $\mathbf{E}_t$ have the same polarization.

Transmitted Wave

$(\mathbf{E}_t, \mathbf{H}_t)$ is traveling along $+\mathbf{a}_z$ in medium 2. If

$$\mathbf{E}_{ts}(z) = E_{to}\, e^{-\gamma_2 z}\mathbf{a}_x \qquad (10.85)$$

then

$$\mathbf{H}_{ts}(z) = H_{to}\, e^{-\gamma_2 z}\, \mathbf{a}_y = \frac{E_{to}}{\eta_2}e^{-\gamma_2 z}\mathbf{a}_y \qquad (10.86)$$

In eqs. (10.81) to (10.86), E_{io}, E_{ro}, and E_{to} are, respectively, the magnitudes of the incident, reflected, and transmitted electric fields at $z = 0$.

Notice from Figure 10.12 that the total field in medium 1 comprises both the incident and the reflected fields, whereas medium 2 has only the transmitted field; that is,

$$\mathbf{E}_1 = \mathbf{E}_i + \mathbf{E}_r, \quad \mathbf{H}_1 = \mathbf{H}_i + \mathbf{H}_r$$

$$\mathbf{E}_2 = \mathbf{E}_t, \quad \mathbf{H}_2 = \mathbf{H}_t$$

At the interface $z = 0$, the boundary conditions require that the tangential components of **E** and **H** fields must be continuous. Since the waves are transverse, **E** and **H** fields are entirely tangential to the interface. Hence at $z = 0$, $\mathbf{E}_{1\text{tan}} = \mathbf{E}_{2\text{tan}}$ and $\mathbf{H}_{1\text{tan}} = \mathbf{H}_{2\text{tan}}$ imply that

$$\mathbf{E}_i(0) + \mathbf{E}_r(0) = \mathbf{E}_t(0) \quad \rightarrow \quad E_{io} + E_{ro} = E_{to} \tag{10.87}$$

$$\mathbf{H}_i(0) + \mathbf{H}_r(0) = \mathbf{H}_t(0) \quad \rightarrow \quad \frac{1}{\eta_1}(E_{io} - E_{ro}) = \frac{E_{to}}{\eta_2} \tag{10.88}$$

From eqs. (10.87) and (10.88), we obtain

$$E_{ro} = \frac{\eta_2 - \eta_1}{\eta_2 + \eta_1} E_{io} \tag{10.89}$$

and

$$E_{to} = \frac{2\eta_2}{\eta_2 + \eta_1} E_{io} \tag{10.90}$$

We now define the *reflection coefficient* Γ and the *transmission coefficient* τ from eqs. (10.89) and (10.90) as

$$\boxed{\Gamma = \frac{E_{ro}}{E_{io}} = \frac{\eta_2 - \eta_1}{\eta_2 + \eta_1}} \tag{10.91a}$$

or

$$E_{ro} = \Gamma E_{io} \tag{10.91b}$$

and

$$\boxed{\tau = \frac{E_{to}}{E_{io}} = \frac{2\eta_2}{\eta_2 + \eta_1}} \tag{10.92a}$$

or

$$E_{to} = \tau E_{io} \tag{10.92b}$$

Note that

1. $1 + \Gamma = \tau$
2. Both Γ and τ are dimensionless and may be complex.
3. $0 \le |\Gamma| \le 1$ (10.93)

The case just considered is the general case. Let us now consider the following special case: medium 1 is a perfect dielectric (lossless, $\sigma_1 = 0$) and medium 2 is a perfect conductor ($\sigma_2 \simeq \infty$). For this case, $\eta_2 = 0$; hence, $\Gamma = -1$, and $\tau = 0$, showing that the wave is totally reflected. This should be expected because fields in a perfect conductor must vanish, so there can be no transmitted wave ($\mathbf{E}_2 = 0$). The totally reflected wave combines with the incident wave to form a *standing wave*. A standing wave "stands" and does not travel; it consists of two traveling waves ($\mathbf{E}_i$ and $\mathbf{E}_r$) of equal amplitudes but in opposite directions. Combining eqs. (10.81) and (10.83) gives the standing wave in medium 1 as

$$\mathbf{E}_{1s} = \mathbf{E}_{is} + \mathbf{E}_{rs} = \left(E_{io}e^{-\gamma_1 z} + E_{ro}e^{\gamma_1 z}\right)\mathbf{a}_x \qquad (10.94)$$

But

$$\Gamma = \frac{E_{ro}}{E_{io}} = -1, \sigma_1 = 0, \alpha_1 = 0, \gamma_1 = j\beta_1$$

Hence,

$$\mathbf{E}_{1s} = -E_{io}\left(e^{j\beta_1 z} - e^{-j\beta_1 z}\right)\mathbf{a}_x$$

or

$$\mathbf{E}_{1s} = -2jE_{io}\sin\beta_1 z\,\mathbf{a}_x \qquad (10.95)$$

Thus

$$\mathbf{E}_1 = \mathrm{Re}\left(\mathbf{E}_{1s}e^{j\omega t}\right)$$

or

$$\boxed{\mathbf{E}_1 = 2E_{io}\sin\beta_1 z\,\sin\omega t\,\mathbf{a}_x} \qquad (10.96)$$

By taking similar steps, it can be shown that the magnetic field component of the wave is

$$\boxed{\mathbf{H}_1 = \frac{2E_{io}}{\eta_1}\cos\beta_1 z\,\cos\omega t\,\mathbf{a}_y} \qquad (10.97)$$

A sketch of the standing wave in eq. (10.96) is presented in Figure 10.13 for $t = 0, T/8, T/4,$ $3T/8, T/2,$ and so on, where $T = 2\pi/\omega$. From Figure 10.13, we notice that the wave does not travel but oscillates.

When media 1 and 2 are both lossless, we have another special case: $\sigma_1 = 0 = \sigma_2$. In this case, η_1 and η_2 are real and so are Γ and τ. Let us consider two more cases:

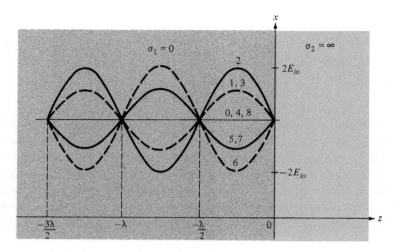

FIGURE 10.13 Standing waves $E = 2E_{io} \sin \beta_1 z \sin \omega t \, \mathbf{a}_x$. The curves 0, 1, 2, 3, 4, ..., are, respectively, at times $t = 0, T/8, T/4, 3T/8, T/2, ...$; $\lambda = 2\pi/\beta_1$.

CASE 1.

If $\eta_2 > \eta_1$, $\Gamma > 0$. Again there is a standing wave in medium 1, but there is also a transmitted wave in medium 2. However, the incident and reflected waves have amplitudes that are not equal in magnitude. It can be shown that a relative maximum of $|\mathbf{E}_1|$ occurs at

$$-\beta_1 z_{\max} = n\pi$$

or

$$z_{\max} = -\frac{n\pi}{\beta_1} = -\frac{n\lambda_1}{2}, \quad n = 0, 1, 2, \ldots \tag{10.98}$$

and the minimum values of $|\mathbf{E}_1|$ occur at

$$-\beta_1 z_{\min} = (2n + 1)\frac{\pi}{2}$$

or

$$z_{\min} = -\frac{(2n + 1)\pi}{2\beta_1} = -\frac{(2n + 1)}{4}\lambda_1, \quad n = 0, 1, 2, \ldots \tag{10.99}$$

CASE 2.

If $\eta_2 < \eta_1$, $\Gamma < 0$. For this case, the locations of $|\mathbf{E}_1|$ maximum are given by eq. (10.99), whereas those of $|\mathbf{E}_1|$ minimum are given by eq. (10.98). All these are illustrated in Figure 10.14. Note that

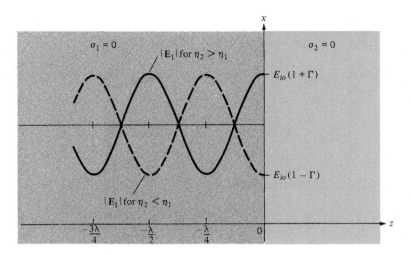

FIGURE 10.14 Standing waves due to reflection at an interface between two lossless media; $\lambda = 2\pi/\beta_1$.

1. $|\mathbf{H}_1|$ minimum occurs whenever there is $|\mathbf{E}_1|$ maximum, and vice versa.
2. The transmitted wave (not shown in Figure 10.14) in medium 2 is a purely traveling wave, and consequently there are no maxima or minima in this region.

The ratio of $|\mathbf{E}_1|_{max}$ to $|\mathbf{E}_1|_{min}$ (or $|\mathbf{H}_1|_{max}$ to $|\mathbf{H}_1|_{min}$) is called the *standing wave ratio s*; that is,

$$s = \frac{|\mathbf{E}_1|_{max}}{|\mathbf{E}_1|_{min}} = \frac{|\mathbf{H}_1|_{max}}{|\mathbf{H}_1|_{min}} = \frac{1 + |\Gamma|}{1 - |\Gamma|} \qquad (10.100)$$

or

$$|\Gamma| = \frac{s - 1}{s + 1} \qquad (10.101)$$

Since $|\Gamma| \le 1$, it follows that $1 \le s \le \infty$. The standing wave ratio is dimensionless, and it is customarily expressed in decibels (dB) as

$$s\, dB = 20 \log_{10} s \qquad (10.102)$$

EXAMPLE 10.9

In free space $(z \le 0)$, a plane wave with

$$\mathbf{H}_i = 10 \cos(10^8 t - \beta z)\mathbf{a}_x\ mA/m$$

is incident normally on a lossless medium $(\varepsilon = 2\varepsilon_o, \mu = 8\mu_o)$ in region $z \ge 0$. Determine the reflected wave $\mathbf{H}_r$, $\mathbf{E}_r$ and the transmitted wave $\mathbf{H}_t$, $\mathbf{E}_t$.

Solution:

This problem can be solved in two different ways.

Method 1: Consider the problem as illustrated in Figure 10.15. For free space,

$$\beta_1 = \frac{\omega}{c} = \frac{10^8}{3 \times 10^8} = \frac{1}{3}$$

$$\eta_1 = \eta_o = 120\pi$$

For the lossless dielectric medium,

$$\beta_2 = \omega\sqrt{\mu\varepsilon} = \omega\sqrt{\mu_o\varepsilon_o}\sqrt{\mu_r\varepsilon_r} = \frac{\omega}{c} \cdot (4) = 4\beta_1 = \frac{4}{3}$$

$$\eta_2 = \sqrt{\frac{\mu}{\varepsilon}} = \sqrt{\frac{\mu_o}{\varepsilon_o}}\sqrt{\frac{\mu_r}{\varepsilon_r}} = 2\,\eta_o$$

Given that $\mathbf{H}_i = 10\cos(10^8 t - \beta_1 z)\mathbf{a}_x$ mA/m, we expect that

$$\mathbf{E}_i = E_{io}\cos(10^8 t - \beta_1 z)\mathbf{a}_{E_i}$$

where

$$\mathbf{a}_{E_i} = \mathbf{a}_{H_i} \times \mathbf{a}_{k_i} = \mathbf{a}_x \times \mathbf{a}_z = -\mathbf{a}_y$$

and

$$E_{io} = \eta_1 H_{io} = 10\,\eta_o$$

Hence,

$$\mathbf{E}_i = -10\eta_o\cos(10^8 t - \beta_1 z)\mathbf{a}_y \text{ mV/m}$$

FIGURE 10.15 For Example 10.9.

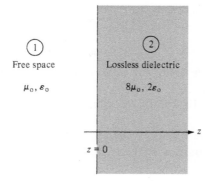

Now

$$\frac{E_{ro}}{E_{io}} = \Gamma = \frac{\eta_2 - \eta_1}{\eta_2 + \eta_1} = \frac{2\eta_o - \eta_o}{2\eta_o + \eta_o} = \frac{1}{3}$$

$$E_{ro} = \frac{1}{3} E_{io}$$

Thus

$$\mathbf{E}_r = -\frac{10}{3} \eta_o \cos\left(10^8 t + \frac{1}{3} z\right) \mathbf{a}_y \, \text{mV/m}$$

from which we easily obtain $\mathbf{H}_r$ as

$$\mathbf{H}_r = -\frac{10}{3} \cos\left(10^8 t + \frac{1}{3} z\right) \mathbf{a}_x \, \text{mA/m}$$

Similarly,

$$\frac{E_{to}}{E_{io}} = \tau = 1 + \Gamma = \frac{4}{3} \quad \text{or} \quad E_{to} = \frac{4}{3} E_{io}$$

Thus

$$\mathbf{E}_t = E_{to} \cos(10^8 t - \beta_2 z) \mathbf{a}_{E_t}$$

where $\mathbf{a}_{E_t} = \mathbf{a}_{E_i} = -\mathbf{a}_y$. Hence,

$$\mathbf{E}_t = -\frac{40}{3} \eta_o \cos\left(10^8 t - \frac{4}{3} z\right) \mathbf{a}_y \, \text{mV/m}$$

from which we obtain

$$\mathbf{H}_t = \frac{20}{3} \cos\left(10^8 t - \frac{4}{3} z\right) \mathbf{a}_x \, \text{mA/m}$$

Method 2: Alternatively, we can obtain $\mathbf{H}_r$ and $\mathbf{H}_t$ directly from $\mathbf{H}_i$ by using

$$\frac{H_{ro}}{H_{io}} = -\Gamma \quad \text{and} \quad \frac{H_{to}}{H_{io}} = \tau \frac{\eta_1}{\eta_2}$$

Thus

$$H_{ro} = -\frac{1}{3} H_{io} = -\frac{10}{3}$$

$$H_{to} = \frac{4}{3} \frac{\eta_o}{2\eta_o} \cdot H_{io} = \frac{2}{3} H_{io} = \frac{20}{3}$$

and

$$\mathbf{H}_r = -\frac{10}{3}\cos(10^8 t + \beta_1 z)\mathbf{a}_x \text{ mA/m}$$

$$\mathbf{H}_t = \frac{20}{3}\cos(10^8 t - \beta_2 z)\mathbf{a}_x \text{ mA/m}$$

as obtained by Method 1.

Notice that the boundary conditions at $z = 0$, namely,

$$\mathbf{E}_i(0) + \mathbf{E}_r(0) = \mathbf{E}_t(0) = -\frac{40}{3}\eta_0 \cos(10^8 t)\mathbf{a}_y$$

and

$$\mathbf{H}_i(0) + \mathbf{H}_r(0) = \mathbf{H}_t(0) = \frac{20}{3}\cos(10^8 t)\mathbf{a}_x$$

are satisfied. The boundary conditions can always be used to cross-check $\mathbf{E}$ and $\mathbf{H}$.

PRACTICE EXERCISE 10.9

A 5 GHz uniform plane wave $\mathbf{E}_{is} = 10\, e^{-j\beta z}\, \mathbf{a}_x$ V/m in free space is incident normally on a large, plane, lossless dielectric slab $(z > 0)$ having $\varepsilon = 4\varepsilon_0$, $\mu = \mu_0$. Find the reflected wave $\mathbf{E}_{rs}$ and the transmitted wave $\mathbf{E}_{ts}$.

Answer: $-3.333 \exp(j\beta_1 z)\mathbf{a}_x$ V/m, $\quad 6.667 \exp(-j\beta_2 z)\mathbf{a}_x$ V/m, $\quad$ where $\quad \beta_2 = 2\beta_1 = 200\pi/3$.

EXAMPLE 10.10

Given a uniform plane wave in air as

$$\mathbf{E}_i = 40\cos(\omega t - \beta z)\mathbf{a}_x + 30\sin(\omega t - \beta z)\mathbf{a}_y \text{ V/m}$$

(a) Find $\mathbf{H}_i$.
(b) If the wave encounters a perfectly conducting plate normal to the z-axis at $z = 0$, find the reflected wave $\mathbf{E}_r$ and $\mathbf{H}_r$.
(c) What are the total $\mathbf{E}$ and $\mathbf{H}$ fields for $z \leq 0$?
(d) Calculate the time-average Poynting vectors for $z \leq 0$ and $z \geq 0$.

Solution:

(a) This is similar to the problem in Example 10.3. We may treat the wave as consisting of two waves $\mathbf{E}_{i1}$ and $\mathbf{E}_{i2}$, where

$$\mathbf{E}_{i1} = 40\cos(\omega t - \beta z)\mathbf{a}_x, \quad \mathbf{E}_{i2} = 30\sin(\omega t - \beta z)\mathbf{a}_y$$

At atmospheric pressure, air has $\varepsilon_r = 1.0006 \simeq 1$. Thus air may be regarded as free space. Let $\mathbf{H}_i = \mathbf{H}_{i1} + \mathbf{H}_{i2}$.

$$\mathbf{H}_{i1} = H_{i1o} \cos(\omega t - \beta z)\mathbf{a}_{H_1}$$

where

$$H_{i1o} = \frac{E_{i1o}}{\eta_o} = \frac{40}{120\pi} = \frac{1}{3\pi}$$

$$\mathbf{a}_{H_1} = \mathbf{a}_k \times \mathbf{a}_E = \mathbf{a}_z \times \mathbf{a}_x = \mathbf{a}_y$$

Hence

$$\mathbf{H}_{i1} = \frac{1}{3\pi} \cos(\omega t - \beta z)\mathbf{a}_y$$

Similarly,

$$\mathbf{H}_{i2} = H_{i2o} \sin(\omega t - \beta z)\mathbf{a}_{H_2}$$

where

$$H_{i2o} = \frac{E_{i2o}}{\eta_o} = \frac{30}{120\pi} = \frac{1}{4\pi}$$

$$\mathbf{a}_{H_2} = \mathbf{a}_k \times \mathbf{a}_E = \mathbf{a}_z \times \mathbf{a}_y = -\mathbf{a}_x$$

Hence

$$\mathbf{H}_{i2} = -\frac{1}{4\pi} \sin(\omega t - \beta z)\mathbf{a}_x$$

and

$$\mathbf{H}_i = \mathbf{H}_{i1} + \mathbf{H}_{i2}$$
$$= -\frac{1}{4\pi} \sin(\omega t - \beta z)\mathbf{a}_x + \frac{1}{3\pi} \cos(\omega t - \beta z)\mathbf{a}_y \text{ mA/m}$$

This problem can also be solved using Method 2 of Example 10.3.

(b) Since medium 2 is perfectly conducting,

$$\frac{\sigma_2}{\omega \varepsilon_2} \gg 1 \;\; \rightarrow \;\; \eta_2 \ll \eta_1$$

that is,

$$\Gamma \simeq -1, \;\; \tau = 0$$

showing that the incident $\mathbf{E}$ and $\mathbf{H}$ fields are totally reflected:

$$E_{ro} = \Gamma E_{io} = -E_{io}$$

Hence,

$$\mathbf{E}_r = -40 \cos(\omega t + \beta z)\mathbf{a}_x - 30 \sin(\omega t + \beta z)\mathbf{a}_y \text{ V/m}$$

We can find $\mathbf{H}_r$ from $\mathbf{E}_r$ just as we did in part (a) of this example or by using Method 2 of Example 10.9, starting with $\mathbf{H}_i$. Whichever approach is taken, we obtain

$$\mathbf{H}_r = \frac{1}{3\pi}\cos(\omega t + \beta z)\mathbf{a}_y - \frac{1}{4\pi}\sin(\omega t + \beta z)\mathbf{a}_x \text{ A/m}$$

(c)   The total fields in air

$$\mathbf{E}_1 = \mathbf{E}_i + \mathbf{E}_r \quad \text{and} \quad \mathbf{H}_1 = \mathbf{H}_i + \mathbf{H}_r$$

can be shown to be standing waves. The total fields in the conductor are

$$\mathbf{E}_2 = \mathbf{E}_t = 0, \quad \mathbf{H}_2 = \mathbf{H}_t = 0$$

(d)   For $z \leq 0$,

$$\mathcal{P}_{1\text{ave}} = \frac{|\mathbf{E}_{1s}|^2}{2\eta_1}\mathbf{a}_k = \frac{1}{2\eta_0}\left[E_{io}^2\mathbf{a}_z - E_{ro}^2\mathbf{a}_z\right]$$

$$= \frac{1}{240\pi}\left[(40^2 + 30^2)\mathbf{a}_z - (40^2 + 30^2)\mathbf{a}_z\right]$$
$$= 0$$

For $z \geq 0$,

$$\mathcal{P}_{2\text{ave}} = \frac{|\mathbf{E}_{2S}|^2}{2\eta_2}\mathbf{a}_k = \frac{E_{to}^2}{2\eta_2}\mathbf{a}_z = 0$$

PRACTICE EXERCISE 10.10

The plane wave $\mathbf{E} = 50 \sin(\omega t - 5x)\mathbf{a}_y$ V/m in a lossless medium ($\mu = 4\mu_0$, $\varepsilon = \varepsilon_0$) encounters a lossy medium ($\mu = \mu_0$, $\varepsilon = 4\varepsilon_0$, $\sigma = 0.1$ S/m) normal to the x-axis at $x = 0$. Find

(a) Γ, τ, and s
(b) $\mathbf{E}_r$ and $\mathbf{H}_r$
(c) $\mathbf{E}_t$ and $\mathbf{H}_t$
(d) The time-average Poynting vectors in both regions

Answer: (a) $0.8186 \underline{/171.1°}$, $0.2295 \underline{/33.56°}$, 10.025, (b) $40.93 \sin(\omega t + 5x + 171.9°)\mathbf{a}_y$ V/m, $-54.3 \sin(\omega t + 5x + 171.9°)\mathbf{a}_z$ mA/m,

(c) $11.47\, e^{-6.021x}\sin(\omega t - 7.826x + 33.56°)\mathbf{a}_y$ V/m,
$120.2\, e^{-6.021x} \sin(\omega t - 7.826x - 4.01°)\mathbf{a}_z$ mA/m,

(d) $0.5469\, \mathbf{a}_x$ W/m², $0.5469 \exp(-12.04x)\mathbf{a}_x$ W/m².

†10.10 REFLECTION OF A PLANE WAVE AT OBLIQUE INCIDENCE

We now consider a more general situation than that in Section 10.9. To simplify the analysis, we will assume that we are dealing with lossless media. (We may extend our analysis to that of lossy media by merely replacing ε by ε_c.) It can be shown (see Problems 10.69 and 10.72) that a uniform plane wave takes the general form of

$$
\begin{aligned}
\mathbf{E}(\mathbf{r}, t) &= \mathbf{E}_o \cos(\mathbf{k} \cdot \mathbf{r} - \omega t) \\
&= \mathrm{Re}\left[\mathbf{E}_o e^{j(\mathbf{k} \cdot \mathbf{r} - \omega t)}\right]
\end{aligned}
\tag{10.103}
$$

where $\mathbf{r} = x\mathbf{a}_x + y\mathbf{a}_y + z\mathbf{a}_z$ is the radius or position vector and $\mathbf{k} = k_x\mathbf{a}_x + k_y\mathbf{a}_y + k_z\mathbf{a}_z$ is the *wave number vector* or the *propagation vector*; $\mathbf{k}$ is always in the direction of wave propagation. The magnitude of $\mathbf{k}$ is related to ω according to the dispersion relation:[3]

$$
k^2 = k_x^2 + k_y^2 + k_z^2 = \omega^2 \mu \varepsilon
\tag{10.104}
$$

Thus, for lossless media, k is essentially the same as β in the preceding sections. With the general form of $\mathbf{E}$ as in eq. (10.103), Maxwell's equations for a source-free region reduce to

$$
\mathbf{k} \times \mathbf{E} = \omega \mu \mathbf{H}
\tag{10.105a}
$$

$$
\mathbf{k} \times \mathbf{H} = -\omega \varepsilon \mathbf{E}
\tag{10.105b}
$$

$$
\mathbf{k} \cdot \mathbf{H} = 0
\tag{10.105c}
$$

$$
\mathbf{k} \cdot \mathbf{E} = 0
\tag{10.105d}
$$

showing that (i) $\mathbf{E}$, $\mathbf{H}$, and $\mathbf{k}$ are mutually orthogonal, and (ii) $\mathbf{E}$ and $\mathbf{H}$ lie on the plane

$$
\mathbf{k} \cdot \mathbf{r} = k_x x + k_y y + k_z z = \text{constant}
$$

From eq. (10.105a), the $\mathbf{H}$ field corresponding to the $\mathbf{E}$ field in eq. (10.103) is

$$
\mathbf{H} = \frac{1}{\omega\mu}\mathbf{k} \times \mathbf{E} = \frac{\mathbf{a}_k \times \mathbf{E}}{\eta}
\tag{10.106}
$$

Having expressed $\mathbf{E}$ and $\mathbf{H}$ in the general form, we can now consider the oblique incidence of a uniform plane wave at a plane boundary as illustrated in Figure 10.16(a). The plane defined by the propagation vector $\mathbf{k}$ and a unit normal vector $\mathbf{a}_n$ to the boundary is called the *plane of incidence*. The angle θ_i between $\mathbf{k}$ and $\mathbf{a}_n$ is the *angle of incidence*.

---

[3] The phenomenon of signal distortion due to a dependence of the phase velocity on frequency is known as dispersion.

Again, both the incident and the reflected waves are in medium 1, while the transmitted (or refracted) wave is in medium 2. Let

$$\mathbf{E}_i = \mathbf{E}_{io} \cos(k_{ix}x + k_{iy}y + k_{iz}z - \omega_i t) \tag{10.107a}$$

$$\mathbf{E}_r = \mathbf{E}_{ro} \cos(k_{rx}x + k_{ry}y + k_{rz}z - \omega_r t) \tag{10.107b}$$

$$\mathbf{E}_t = \mathbf{E}_{to} \cos(k_{tx}x + k_{ty}y + k_{tz}z - \omega_t t) \tag{10.107c}$$

where k_i, k_r, and k_t with their normal and tangential components are shown in Figure 10.16(b). Since the tangential component of $\mathbf{E}$ must be continuous across the boundary $z = 0$,

$$\mathbf{E}_i(z = 0) + \mathbf{E}_r(z = 0) = \mathbf{E}_t(z = 0) \tag{10.108}$$

This boundary condition can be satisfied by the waves in eq. (10.107) for all x and y only if

1. $\omega_i = \omega_r = \omega_t = \omega$
2. $k_{ix} = k_{rx} = k_{tx} = k_x$
3. $k_{iy} = k_{ry} = k_{ty} = k_y$

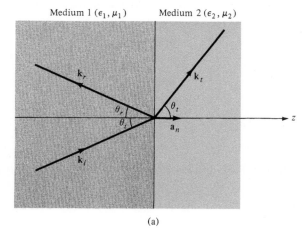

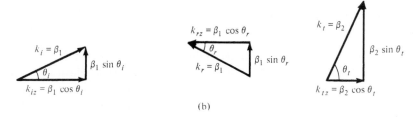

FIGURE 10.16 Oblique incidence of a plane wave: (**a**) illustration of θ_i, θ_r, and θ_t; (**b**) illustration of the normal and tangential components of **k**.

Condition 1 implies that the frequency is unchanged. Conditions 2 and 3 require that the tangential components of the propagation vectors be continuous (called the *phase-matching conditions*). This means that the propagation vectors $\mathbf{k}_i$, $\mathbf{k}_t$, and $\mathbf{k}_r$ must all lie in the plane of incidence. Thus, by conditions 2 and 3,

$$k_i \sin \theta_i = k_r \sin \theta_r \tag{10.109}$$

$$k_i \sin \theta_i = k_t \sin \theta_t \tag{10.110}$$

where θ_r is *the angle of reflection* and θ_t is the *angle of transmission*. But for lossless media,

$$k_i = k_r = \beta_1 = \omega \sqrt{\mu_1 \varepsilon_1} \tag{10.111a}$$

$$k_t = \beta_2 = \omega \sqrt{\mu_2 \varepsilon_2} \tag{10.111b}$$

From eqs. (10.109) and (10.111a), it is clear that

$$\boxed{\theta_r = \theta_i} \tag{10.112}$$

so that the angle of reflection θ_r equals the angle of incidence θ_i, as in optics. Also from eqs. (10.110) and (10.111),

$$\frac{\sin \theta_t}{\sin \theta_i} = \frac{k_i}{k_t} = \frac{u_2}{u_1} = \sqrt{\frac{\mu_1 \varepsilon_1}{\mu_2 \varepsilon_2}} \tag{10.113}$$

where $u = \omega/k$ is the phase velocity. Equation (10.113) is the well-known *Snell's law*, which can be written as

$$\boxed{n_1 \sin \theta_i = n_2 \sin \theta_t} \tag{10.114}$$

where $n_1 = c\sqrt{\mu_1 \varepsilon_1} = c/u_1$ and $n_2 = c\sqrt{\mu_2 \varepsilon_2} = c/u_2$ are the *refractive indices* of the media.

Based on these general preliminaries on oblique incidence, we will now consider two special cases: one with the **E** field perpendicular to the plane of incidence and the other with the **E** field parallel to it. Any other polarization may be considered as a linear combination of these two cases.

A. Parallel Polarization

Figure 10.17, where the **E** field lies in the *xz*-plane, the plane of incidence, illustrates the case of parallel polarization. In medium 1, we have both incident and reflected fields given by

$$\mathbf{E}_{is} = E_{io}(\cos \theta_i \, \mathbf{a}_x - \sin \theta_i \, \mathbf{a}_z) \, e^{-j\beta_1(x \sin \theta_i + z \cos \theta_i)} \tag{10.115a}$$

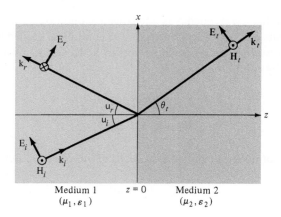

FIGURE 10.17 Oblique incidence with **E** parallel to the plane of incidence.

Medium 1
(μ_1, ε_1)

$z = 0$

Medium 2
(μ_2, ε_2)

$$\mathbf{H}_{is} = \frac{E_{io}}{\eta_1} e^{-j\beta_1(x \sin \theta_i + z \cos \theta_i)} \mathbf{a}_y \tag{10.115b}$$

$$\mathbf{E}_{rs} = E_{ro}(\cos \theta_r \, \mathbf{a}_x + \sin \theta_r \, \mathbf{a}_z) \, e^{-j\beta_1(x \sin \theta_r - z \cos \theta_r)} \tag{10.116a}$$

$$\mathbf{H}_{rs} = -\frac{E_{ro}}{\eta_1} e^{-j\beta_1(x \sin \theta_r - z \cos \theta_r)} \mathbf{a}_y \tag{10.116b}$$

where $\beta_1 = \omega\sqrt{\mu_1\varepsilon_1}$. Notice carefully how we arrive at each field component. The trick in deriving the components is to first get the propagation vector **k** as shown in Figure 10.16(b) for incident, reflected, and transmitted waves. Once **k** is known, we define $\mathbf{E}_s$ such that $\nabla \cdot \mathbf{E}_s = 0$ or $\mathbf{k} \cdot \mathbf{E}_s = 0$ and then $\mathbf{H}_s$ is obtained from $\mathbf{H}_s = \frac{\mathbf{k}}{\omega\mu} \times \mathbf{E}_s = \mathbf{a}_k \times \frac{\mathbf{E}}{\eta}$.

The transmitted fields exist in medium 2 and are given by

$$\mathbf{E}_{ts} = E_{to}(\cos \theta_t \, \mathbf{a}_x - \sin \theta_t \, \mathbf{a}_z) \, e^{-j\beta_2(x \sin \theta_t + z \cos \theta_t)} \tag{10.117a}$$

$$\mathbf{H}_{ts} = \frac{E_{to}}{\eta_2} e^{-j\beta_2(x \sin \theta_t + z \cos \theta_t)} \mathbf{a}_y \tag{10.117b}$$

where $\beta_2 = \omega\sqrt{\mu_2\varepsilon_2}$. Should our assumption about the relative directions in eqs. (10.115) to (10.117) be wrong, the final result will show us this by means of its sign.

Requiring that $\theta_r = \theta_i$ and that the tangential components of **E** and **H** be continuous at the boundary $z = 0$, we obtain

$$(E_{io} + E_{ro}) \cos \theta_i = E_{to} \cos \theta_t \tag{10.118a}$$

$$\frac{1}{\eta_1} (E_{io} - E_{ro}) = \frac{1}{\eta_2} E_{to} \tag{10.118b}$$

Expressing E_{ro} and E_{to} in terms of E_{io}, we obtain

$$\boxed{\Gamma_{\|} = \frac{E_{ro}}{E_{io}} = \frac{\eta_2 \cos \theta_t - \eta_1 \cos \theta_i}{\eta_2 \cos \theta_t + \eta_1 \cos \theta_i}} \qquad (10.119a)$$

or

$$E_{ro} = \Gamma_{\|} E_{io} \qquad (10.119b)$$

and

$$\boxed{\tau_{\|} = \frac{E_{to}}{E_{io}} = \frac{2\eta_2 \cos \theta_i}{\eta_2 \cos \theta_t + \eta_1 \cos \theta_i}} \qquad (10.120a)$$

or

$$E_{to} = \tau_{\|} E_{io} \qquad (10.120b)$$

Equations (10.119) and (10.120) are called *Fresnel's equations*. $\Gamma_{\|}$ and $\tau_{\|}$ are known as Fresnel coefficients. Note that the equations reduce to eqs. (10.91) and (10.92) when $\theta_i = \theta_t = 0$ as expected. Since θ_i and θ_t are related according to Snell's law of eq. (10.113), eqs. (10.119) and (10.120) can be written in terms of θ_i by substituting

$$\cos \theta_t = \sqrt{1 - \sin^2 \theta_t} = \sqrt{1 - (u_2/u_1)^2 \sin^2 \theta_i} \qquad (10.121)$$

From eqs. (10.119) and (10.120), it is easily shown that

$$1 + \Gamma_{\|} = \tau_{\|} \left(\frac{\cos \theta_t}{\cos \theta_i} \right) \qquad (10.122)$$

From eq. (10.119a), it is evident that it is possible that $\Gamma_{\|} = 0$ because the numerator is the difference of two terms. Under this condition, there is no reflection ($E_{ro} = 0$), and the incident angle at which this takes place is called the *Brewster angle* $\theta_{B_{\|}}$. The Brewster angle is also known as the *polarizing angle* because an arbitrarily polarized incident wave will be reflected with only the component of **E** perpendicular to the plane of incidence. The Brewster effect is utilized in a laser tube where quartz windows are set at the Brewster angle to control polarization of emitted light. The Brewster angle is obtained by setting $\theta_i = \theta_{B_{\|}}$ when $\Gamma_{\|} = 0$ in eqs. (10.119); that is,

$$\eta_2 \cos \theta_t = \eta_1 \cos \theta_{B_{\|}}$$

or

$$\eta_2^2 (1 - \sin^2 \theta_t) = \eta_1^2 (1 - \sin^2 \theta_{B_{\|}})$$

Introducing eq. (10.113) or (10.114) gives

$$\boxed{\sin^2 \theta_{B_{\|}} = \frac{1 - \mu_2 \varepsilon_1 / \mu_1 \varepsilon_2}{1 - (\varepsilon_1 / \varepsilon_2)^2}} \qquad (10.123)$$

It is of practical value to consider the case when the dielectric media are not only lossless but nonmagnetic as well—that is, $\mu_1 = \mu_2 = \mu_0$. For this situation, eq. (10.123) becomes

$$\sin^2 \theta_{B_\parallel} = \frac{1}{1 + \varepsilon_1/\varepsilon_2} \quad \rightarrow \quad \sin \theta_{B_\parallel} = \sqrt{\frac{\varepsilon_2}{\varepsilon_1 + \varepsilon_2}}$$

or

$$\tan \theta_{B_\parallel} = \sqrt{\frac{\varepsilon_2}{\varepsilon_1}} = \frac{n_2}{n_1} \tag{10.124}$$

showing that there is a Brewster angle for any combination of ε_1 and ε_2.

B. Perpendicular Polarization

When the **E** field is perpendicular to the plane of incidence (the xz-plane) as shown in Figure 10.18, we have perpendicular polarization. This may also be viewed as the case in which the **H** field is parallel to the plane of incidence. The incident and reflected fields in medium 1 are given by

$$\mathbf{E}_{is} = E_{io} e^{-j\beta_1(x \sin \theta_i + z \cos \theta_i)} \mathbf{a}_y \tag{10.125a}$$

$$\mathbf{H}_{is} = \frac{E_{io}}{\eta_1} \left(-\cos \theta_i \, \mathbf{a}_x + \sin \theta_i \, \mathbf{a}_z \right) e^{-j\beta_1(x \sin \theta_i + z \cos \theta_i)} \tag{10.125b}$$

$$\mathbf{E}_{rs} = E_{ro} e^{-j\beta_1(x \sin \theta_r - z \cos \theta_r)} \mathbf{a}_y \tag{10.126a}$$

$$\mathbf{H}_{rs} = \frac{E_{ro}}{\eta_1} \left(\cos \theta_r \, \mathbf{a}_x + \sin \theta_r \, \mathbf{a}_z \right) e^{-j\beta_1(x \sin \theta_r - z \cos \theta_r)} \tag{10.126b}$$

while the transmitted fields in medium 2 are given by

$$\mathbf{E}_{ts} = E_{to} e^{-j\beta_2(x \sin \theta_t + z \cos \theta_t)} \mathbf{a}_y \tag{10.127a}$$

$$\mathbf{H}_{ts} = \frac{E_{to}}{\eta_2} \left(-\cos \theta_t \, \mathbf{a}_x + \sin \theta_t \, \mathbf{a}_z \right) e^{-j\beta_2(x \sin \theta_t + z \cos \theta_t)} \tag{10.127b}$$

Notice that in defining the field components in eqs. (10.125) to (10.127), Maxwell's equations (10.105) are always satisfied. Again, requiring that the tangential components of **E** and **H** be continuous at $z = 0$ and setting θ_r equal to θ_i, we get

$$E_{io} + E_{ro} = E_{to} \tag{10.128a}$$

$$\frac{1}{\eta_1} \left(E_{io} - E_{ro} \right) \cos \theta_i = \frac{1}{\eta_2} E_{to} \cos \theta_t \tag{10.128b}$$

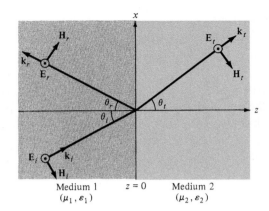

Medium 1 $z = 0$ Medium 2
(μ_1, ε_1) (μ_2, ε_2)

Expressing E_{ro} and E_{to} in terms of E_{io} leads to

$$\Gamma_\perp = \frac{E_{ro}}{E_{io}} = \frac{\eta_2 \cos \theta_i - \eta_1 \cos \theta_t}{\eta_2 \cos \theta_i + \eta_1 \cos \theta_t} \tag{10.129a}$$

or

$$E_{ro} = \Gamma_\perp E_{io} \tag{10.129b}$$

and

$$\tau_\perp = \frac{E_{to}}{E_{io}} = \frac{2\eta_2 \cos \theta_i}{\eta_2 \cos \theta_i + \eta_1 \cos \theta_t} \tag{10.130a}$$

or

$$E_{to} = \tau_\perp E_{io} \tag{10.130b}$$

which are the *Fresnel's equations* for perpendicular polarization $\Gamma_\parallel$ and $\tau_\parallel$ are known as Fresnel coefficients. From eqs. (10.129) and (10.130), it is easy to show that

$$1 + \Gamma_\perp = \tau_\perp \tag{10.131}$$

which is similar to eq. (10.93) for normal incidence. Also, when $\theta_i = \theta_t = 0$, eqs. (10.129) and (10.130) become eqs. (10.91) and (10.92), as they should.

For no reflection, $\Gamma_\perp = 0$ (or $E_r = 0$). This is the same as the case of total transmission ($\tau_\perp = 1$). By replacing θ_i with the corresponding Brewster angle $\theta_{B_\perp}$, we obtain

$$\eta_2 \cos \theta_{B_\perp} = \eta_1 \cos \theta_t$$

or

$$\eta_2^2 (1 - \sin^2 \theta_{B_\perp}) = \eta_1^2 (1 - \sin^2 \theta_t)$$

Incorporating eq. (10.114) yields

$$\sin^2 \theta_{B_\perp} = \frac{1 - \mu_1 \varepsilon_2 / \mu_2 \varepsilon_1}{1 - (\mu_1 / \mu_2)^2} \tag{10.132}$$

Note that for nonmagnetic media ($\mu_1 = \mu_2 = \mu_o$), $\sin^2 \theta_{B_\perp} \rightarrow \infty$ in eq. (10.132), so $\theta_{B_\perp}$ does not exist because the sine of an angle is never greater than unity. Also if $\mu_1 \neq \mu_2$ and $\varepsilon_1 = \varepsilon_2$, eq. (10.132) reduces to

$$\sin \theta_{B_\perp} = \sqrt{\frac{\mu_2}{\mu_1 + \mu_2}}$$

or

$$\tan \theta_{B_\perp} = \sqrt{\frac{\mu_2}{\mu_1}} \tag{10.133}$$

Although this situation is theoretically possible, it rarely occurs in practice.

EXAMPLE 10.11

An EM wave travels in free space with the electric field component

$$\mathbf{E}_s = 100 \, e^{j(0.866y + 0.5z)} \mathbf{a}_x \text{ V/m}$$

Determine

(a) ω and λ
(b) The magnetic field component
(c) The time average power in the wave

Solution:

(a) Comparing the given **E** with

$$\mathbf{E}_s = \mathbf{E}_o \, e^{j\mathbf{k} \cdot \mathbf{r}} = E_o e^{j(k_x x + k_y y + k_z z)} \mathbf{a}_x$$

it is clear that

$$k_x = 0, \quad k_y = 0.866, \quad k_z = 0.5$$

Thus

$$k = \sqrt{k_x^2 + k_y^2 + k_z^2} = \sqrt{(0.866)^2 + (0.5)^2} = 1$$

But in free space,

$$k = \beta = \omega \sqrt{\mu_o \varepsilon_o} = \frac{\omega}{c} = \frac{2\pi}{\lambda}$$

Hence,

$$\omega = kc = 3 \times 10^8 \text{ rad/s}$$

$$\lambda = \frac{2\pi}{k} = 2\pi = 6.283 \text{ m}$$

(b) From eq. (10.106), the corresponding magnetic field is given by

$$\mathbf{H}_s = \frac{1}{\mu\omega}\mathbf{k} \times \mathbf{E}_s$$

$$= \frac{(0.866\mathbf{a}_y + 0.5\mathbf{a}_z)}{4\pi \times 10^{-7} \times 3 \times 10^8} \times 100\, \mathbf{a}_x e^{jk \cdot r}$$

or

$$\mathbf{H}_s = (132.63\mathbf{a}_y - 229.7\mathbf{a}_z)\, e^{j(0.866y + 0.5z)}\, \text{mA/m}$$

(c) The time-average power is

$$\mathcal{P}_{ave} = \frac{1}{2}\text{Re}(\mathbf{E}_s \times \mathbf{H}_s^*) = \frac{E_o^2}{2\eta}\mathbf{a}_k$$

$$= \frac{(100)^2}{2(120\pi)}(0.866\mathbf{a}_y + 0.5\mathbf{a}_z)$$

$$= 11.49\mathbf{a}_y + 6.631\mathbf{a}_z \text{ W/m}^2$$

PRACTICE EXERCISE 10.11

Rework Example 10.11 if

$$\mathbf{E} = (10\mathbf{a}_y + 5\mathbf{a}_z)\cos(\omega t + 2y - 4z) \text{ V/m}$$

in free space.

Answer: (a) 1.342×10^9 rad/s, 1.405 m, (b) $-29.66\cos(1.342 \times 10^9 t + 2y - 4z)\mathbf{a}_x$ mA/m, (c) $-0.07415\mathbf{a}_y + 0.489\mathbf{a}_z$ W/m².

EXAMPLE 10.12

A uniform plane wave in air with

$$\mathbf{E} = 8\cos(\omega t - 4x - 3z)\mathbf{a}_y \text{ V/m}$$

is incident on a dielectric slab ($z \geq 0$) with $\mu_r = 1.0$, $\varepsilon_r = 2.5$, $\sigma = 0$. Find

(a) The polarization of the wave

(b) The angle of incidence

(c) The reflected **E** field

(d) The transmitted **H** field

Solution:

(a) From the incident **E** field, it is evident that the propagation vector is

$$\mathbf{k}_i = 4\mathbf{a}_x + 3\mathbf{a}_z \rightarrow k_i = 5 = \omega\sqrt{\mu_o\varepsilon_o} = \frac{\omega}{c}$$

Hence,

$$\omega = 5c = 15 \times 10^8 \text{ rad/s}$$

A unit vector normal to the interface $(z = 0)$ is $\mathbf{a}_z$. The plane containing $\mathbf{k}$ and $\mathbf{a}_z$ is $y = $ constant, which is the xz-plane, the plane of incidence. Since $\mathbf{E}_i$ is normal to this plane, we have perpendicular polarization (similar to Figure 10.18).

(b) The propagation vectors are illustrated in Figure 10.19 where it is clear that

$$\tan \theta_i = \frac{k_{ix}}{k_{iz}} = \frac{4}{3} \rightarrow \theta_i = 53.13°$$

Alternatively, without Figure 10.19, we can obtain θ_i from the fact that θ_i is the angle between $\mathbf{k}$ and $\mathbf{a}_n$; that is,

$$\cos \theta_i = \mathbf{a}_k \cdot \mathbf{a}_n = \left(\frac{4\mathbf{a}_x + 3\mathbf{a}_z}{5}\right) \cdot \mathbf{a}_z = \frac{3}{5}$$

or

$$\theta_i = 53.13°$$

(c) An easy way to find $\mathbf{E}_r$ is to use eq. (10.126a) because we have noticed that this problem is similar to that considered in Section 10.10B. Suppose we are not aware of this. Let

$$\mathbf{E}_r = E_{ro} \cos(\omega t - \mathbf{k}_r \cdot \mathbf{r})\mathbf{a}_y$$

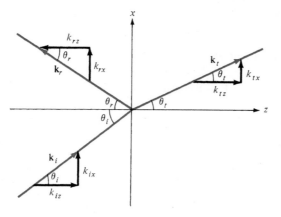

FIGURE 10.19 Propagation vectors of Example 10.12.

which is similar to the form of the given $\mathbf{E}_i$. The unit vector $\mathbf{a}_y$ is chosen in view of the fact that the tangential component of $\mathbf{E}$ must be continuous at the interface. From Figure 10.19,

$$\mathbf{k}_r = k_{rx}\,\mathbf{a}_x - k_{rz}\,\mathbf{a}_z$$

where

$$k_{rx} = k_r \sin\theta_r, \quad k_{rz} = k_r \cos\theta_r$$

But $\theta_r = \theta_i$ and $k_r = k_i = 5$ because both k_r and k_i are in the same medium. Hence,

$$\mathbf{k}_r = 4\mathbf{a}_x - 3\mathbf{a}_z$$

To find E_{ro}, we need θ_t. From Snell's law

$$\sin\theta_t = \frac{n_1}{n_2}\sin\theta_i = \frac{c\sqrt{\mu_1\varepsilon_1}}{c\sqrt{\mu_2\varepsilon_2}}\sin\theta_i$$

$$= \frac{\sin 53.13°}{\sqrt{2.5}}$$

or

$$\theta_t = 30.39°$$

$$\Gamma_\perp = \frac{E_{ro}}{E_{io}}$$

$$= \frac{\eta_2 \cos\theta_i - \eta_1 \cos\theta_t}{\eta_2 \cos\theta_i + \eta_1 \cos\theta_t}$$

where $\eta_1 = \eta_o = 377\ \Omega$, $\eta_2 = \sqrt{\dfrac{\mu_o\mu_{r_2}}{\varepsilon_o\varepsilon_{r_2}}} = \dfrac{377}{\sqrt{2.5}} = 238.4\ \Omega$

$$\Gamma_\perp = \frac{238.4 \cos 53.13° - 377 \cos 30.39°}{238.4 \cos 53.13° + 377 \cos 30.39°} = -0.389$$

Hence,

$$E_{ro} = \Gamma_\perp E_{io} = -0.389\,(8) = -3.112$$

and

$$\mathbf{E}_r = -3.112 \cos(15 \times 10^8 t - 4x + 3z)\mathbf{a}_y \text{ V/m}$$

(d) Similarly, let the transmitted electric field be

$$\mathbf{E}_t = E_{to}\cos(\omega t - \mathbf{k}_t \cdot \mathbf{r})\mathbf{a}_y$$

where

$$k_t = \beta_2 = \omega\sqrt{\mu_2\varepsilon_2} = \frac{\omega}{c}\sqrt{\mu_{r_2}\varepsilon_{r_2}}$$

$$= \frac{15\times10^8}{3\times10^8}\sqrt{1\times2.5} = 7.906$$

From Figure 10.19,

$$k_{tx} = k_t\sin\theta_t = 4$$

$$k_{tz} = k_t\cos\theta_t = 6.819$$

or

$$\mathbf{k}_t = 4\mathbf{a}_x + 6.819\,\mathbf{a}_z$$

Notice that $k_{ix} = k_{rx} = k_{tx}$ as expected.

$$\tau_\perp = \frac{E_{to}}{E_{io}} = \frac{2\,\eta_2\cos\theta_i}{\eta_2\cos\theta_i + \eta_1\cos\theta_t}$$

$$= \frac{2\times238.4\cos53.13°}{238.4\cos53.13° + 377\cos30.39°}$$

$$= 0.611$$

The same result could be obtained from the relation $\tau_\perp = 1 + \Gamma_\perp$. Hence,

$$E_{to} = \tau_\perp E_{io} = 0.611\times8 = 4.888$$

$$\mathbf{E}_t = 4.888\cos(15\times10^8t - 4x - 6.819z)\mathbf{a}_y \text{ V/m}$$

From $\mathbf{E}_t$, $\mathbf{H}_t$ is easily obtained as

$$\mathbf{H}_i = \frac{1}{\mu_2\omega}\mathbf{k}_t\times\mathbf{E}_t = \frac{\mathbf{a}_{k_t}\times\mathbf{E}_t}{\eta_2}$$

$$= \frac{4\mathbf{a}_x + 6.819\,\mathbf{a}_z}{7.906\,(238.4)}\times4.888\,\mathbf{a}_y\cos(\omega t - \mathbf{k}\cdot\mathbf{r})$$

$$\mathbf{H}_t = (-17.69\mathbf{a}_x + 10.37\mathbf{a}_z)\cos(15\times10^8t - 4x - 6.819z) \text{ mA/m}$$

PRACTICE EXERCISE 10.12

If the plane wave of Practice Exercise 10.11 is incident on a dielectric medium having $\sigma = 0$, $\varepsilon = 4\varepsilon_o$, $\mu = \mu_o$ and occupying $z \geq 0$, calculate

(a) The angles of incidence, reflection, and transmission
(b) The reflection and transmission coefficients

(c) The total **E** field in free space
(d) The total **E** field in the dielectric
(e) The Brewster angle

Answer: (a) 26.56°, 26.56°, 12.92°, (b) -0.295, 0.647, (c) $(10\mathbf{a}_y + 5\mathbf{a}_z) \times$
$\cos(\omega t + 2y - 4z) + (-2.946\mathbf{a}_y + 1.473\mathbf{a}_z)\cos(\omega t + 2y + 4z)$ V/m,
(d) $(7.055\mathbf{a}_y + 1.618\mathbf{a}_z)\cos(\omega t + 2y - 8.718z)$ V/m, (e) 63.43°.

†10.11 APPLICATION NOTE—MICROWAVES

At the moment, there are three means for carrying thousands of channels over long distances: (a) microwave links, (b) coaxial cables, and (c) fiber optic, a technology to be covered later in Section 12.9.

> **Microwaves** are EM waves whose frequencies range from approximately 300 MHz to 1000 GHz.

For comparison, the signal from an AM radio station is about 1 MHz, while that from an FM station is about 100 MHz. The higher-frequency edge of microwaves borders on the optical spectrum. This accounts for why microwaves behave more like rays of light than ordinary radio waves. You may be familiar with microwave appliances such as the microwave oven, which operates at 2.4 GHz, the satellite television receiver, which operates at about 4 GHz, and the police radar, which works at about 22 GHz.

Features that make microwaves attractive for communications include wide available bandwidths (capacities to carry information) and directive properties of short wavelengths. Since the amount of information that can be transmitted is limited by the available bandwidth, the microwave spectrum provides more communication channels than the radio and TV bands. With the ever-increasing demand for channel allocation, microwave communications has become more common.

A microwave system[4] normally consists of a transmitter (including a microwave oscillator, waveguides, and a transmitting antenna) and a receiver subsystem (including a receiving antenna, transmission line or waveguide, microwave amplifiers, and a receiver). A microwave network is usually an interconnection of various microwave components and devices. There are several microwave components and variations of these components. Common microwave components include the following:

- Coaxial cables, which are transmission lines for interconnecting microwave components
- Resonators, which are usually cavities in which EM waves are stored

---

[4] For a comprehensive treatment of microwaves, see D. M. Pozar, *Microwave Engineering,* 3rd ed. Hoboken, NJ: John Wiley & Sons, 2004.

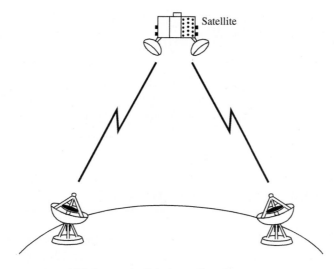

(a) Point-to-point link via satellite microwave

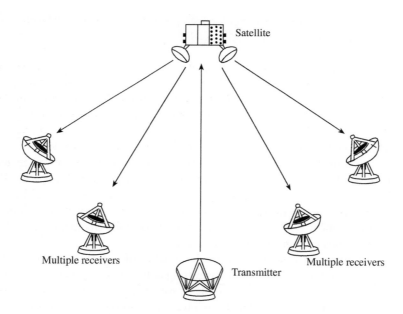

(b) Broadcast link via satellite microwave

FIGURE 10.20 Satellite communications configurations. (From W. Stallings, *Data and Computer Communications*, 5th ed. Upper Saddle River, NJ: Prentice Hall, 1997, p. 90.)

- Waveguide sections, which may be straight, curved, or twisted
- Antennas, which transmit or receive EM waves efficiently
- Terminators, which are designed to absorb the input power and therefore act as one-port networks
- Attenuators, which are designed to absorb some of the EM power passing through the device, thereby decreasing the power level of the microwave signal
- Directional couplers, which consist of two waveguides and a mechanism for coupling signals between them
- Isolators, which allow energy flow in only one direction
- Circulators, which are designed to establish various entry/exit points where power can be either fed or extracted
- Filters, which suppress unwanted signals and/or separate signals of different frequencies.

The use of microwaves has greatly expanded. Examples include telecommunications, radio astronomy, land surveying, radar, meteorology, UHF television, terrestrial microwave links, solid-state devices, heating devices, medical therapeutic and diagnostic equipment, and identification systems. We will consider only three of these.

1. **Telecommunications** (the transmission of analog or digital information from one point to another): This is the largest application of microwave frequencies. Microwaves propagate along a straight line like light rays and are not bent by the ionosphere as are signals of lower frequency. This makes communication satellites possible. In essence, a communication satellite is a microwave relay station that is used to link two or more ground-based transmitters and receivers. The satellite receives signals at one frequency, repeats or amplifies them, and transmits at another frequency. Two common modes of operation for satellite communication are portrayed in Figure 10.20. The satellite provides a point-to-point link in Figure 10.20(a), while it is being used to provide multiple links between one ground-based transmitter and several ground-based receivers in Figure 10.20(b).

2. **Radar systems:** Radar systems provided the major incentive for the development of microwave technology because they give better resolution for radar instruments at higher frequencies. Only the microwave region of the spectrum could provide the required resolution with antennas of reasonable size. The ability to focus a radiated wave sharply is what makes microwaves so useful in radar applications. Radar is used to detect aircraft, guide supersonic missiles, observe and track weather patterns, and control flight traffic at airports. It is also used in burglar alarms, garage-door openers, and police speed detectors.

3. **Heating:** Microwave energy is more easily directed, controlled, and concentrated than low-frequency EM waves. Also, various atomic and molecular resonances occur at microwave frequencies, creating diverse application areas in basic science, remote sensing, and heating methods. The heating properties of microwave power are useful in a wide variety of commercial and industrial applications. The microwave oven, shown in Figure 10.21, is a typical example. When the magnetron oscillates, microwave energy is extracted from the resonant cavities. The reflections from the stationary walls and the motion of the stirring fan cause the microwave energy to be well distributed. Thus the microwave enables the

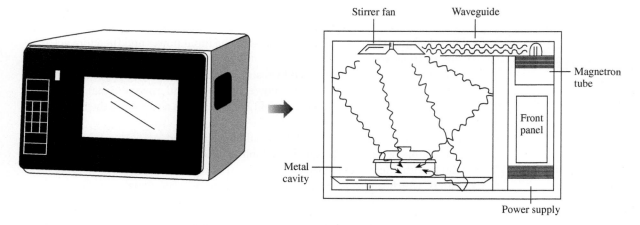

FIGURE 10.21 Microwave oven. (From N. Schlager, ed., *How Products Are Made.* Detroit: Gale Research, 1994, p. 289.)

cooking process to be fast and even. Microwave heating properties also are used in physical diathermy and in drying potato chips, paper, cloth, and so on.

A microwave circuit consists of microwave components such as sources, transmission lines, waveguides, attenuators, resonators, circulators, and filters. One way of analyzing, for example, a circuit, is to relate the input and output variables of each component. Several sets of parameters can be used for relating input and output variables; but at high frequencies, such as microwave frequencies, where voltage and current are not well defined, S-parameters are often used to analyze microwave circuits. The *scattering* or S-*parameters* are defined in terms of wave variables, which are more easily measured at microwave frequencies than voltage and current.

Consider the two-port network shown in Figure 10.22. The traveling waves are related to the scattering parameters according to

$$b_1 = S_{11}a_1 + S_{12}a_2$$
$$b_2 = S_{21}a_1 + S_{22}a_2 \qquad (10.134)$$

or in matrix form

$$\begin{bmatrix} b_1 \\ b_2 \end{bmatrix} = \begin{bmatrix} S_{11} & S_{12} \\ S_{21} & S_{22} \end{bmatrix} \begin{bmatrix} a_1 \\ a_2 \end{bmatrix} \qquad (10.135)$$

where a_1 and a_2 represent the incident waves at ports 1 and 2, respectively, while b_1 and b_2 represent the reflected waves, as shown in Figure 10.22. For the S matrix, the off-diagonal terms represent voltage wave transmission coefficients, while the diagonal terms represent reflection coefficients. If the network is *reciprocal*, it will have the same transmission characteristics in either direction; that is,

$$S_{12} = S_{21} \qquad (10.136)$$

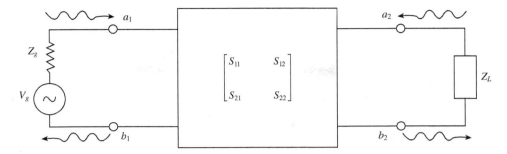

FIGURE 10.22 A two-port network.

If the network is *symmetrical,* then

$$S_{11} = S_{22} \qquad (10.137)$$

For a matched two-port network, the reflection coefficients are zero and

$$S_{11} = S_{22} = 0 \qquad (10.138)$$

The input reflection coefficient can be expressed in terms of the S-parameters and the load Z_L as

$$\Gamma_i = \frac{b_1}{a_1} = S_{11} + \frac{S_{12}S_{21}\Gamma_L}{1 - S_{22}\Gamma_L} \qquad (10.139)$$

where

$$\Gamma_L = \frac{Z_L - Z_o}{Z_L + Z_o} \qquad (10.140)$$

Similarly, the output reflection coefficient (with $V_g = 0$) can be expressed in terms of the generator impedance Z_g and the S-parameters as

$$\Gamma_o = \frac{b_2}{a_2}\bigg|_{V_g=0} = S_{22} + \frac{S_{12}S_{21}\Gamma_g}{1 - S_{11}\Gamma_g} \qquad (10.141)$$

where

$$\Gamma_g = \frac{Z_g - Z_o}{Z_g + Z_o} \qquad (10.142)$$

EXAMPLE 10.13   The following S-parameters are obtained for a microwave transistor operating at 2.5 GHz: $S_{11} = 0.85\underline{/-30°}$, $S_{12} = 0.07\underline{/56°}$, $S_{21} = 1.68\underline{/120°}$, $S_{22} = 0.85\underline{/-40}$. Determine the input reflection coefficient when $Z_L = Z_o = 75\ \Omega$.

Solution:

From eq. (10.140),

$$\Gamma_L = \frac{Z_L - Z_o}{Z_L + Z_o} = 0$$

Hence, using eq. (10.139) leads to

$$\Gamma_i = S_{11} = 0.85\underline{/-30°}$$

PRACTICE EXERCISE 10.13

For an hybrid coupler, the voltage standing wave ratios (VSWRs) for the input and output ports are, respectively, given as

$$s_i = \frac{1 + |S_{11}|}{1 - |S_{11}|}$$

$$s_o = \frac{1 + |S_{22}|}{1 - |S_{22}|}$$

Calculate s_i and s_o for the following scattering matrix:

$$S = \begin{bmatrix} 0.4 & j0.6 \\ j0.6 & 0.2 \end{bmatrix}$$

Answer: 2.333, 1.5.

10.12 APPLICATION NOTE—60 GHz TECHNOLOGY

The next-generation wireless technology—namely, 60 GHz—can provide wireless connectivity for short distances between electronic devices at speeds in the multi-gigabyte-per-second range, as typically shown in Figure 10.23. At such a high frequency, the wavelength is nearly 5 mm. Although millimeter wave (mmWave) technology has been known for a long time, it was used initially only for military applications. With the strides in process technologies and low-cost integration solutions, academia, industry, and standardization bodies also turned to mmWave technology. Broadly speaking, mmWave refers to the electromagnetic spectrum that spans 30 to 300 GHz, corresponding to wavelengths from 10 mm down to 1 mm. At these smaller wavelengths, the data rates are expected to be 40 to 100 times faster than current wireless technologies for local-area networks. Also, IBM's high-speed signal processing and coding techniques,

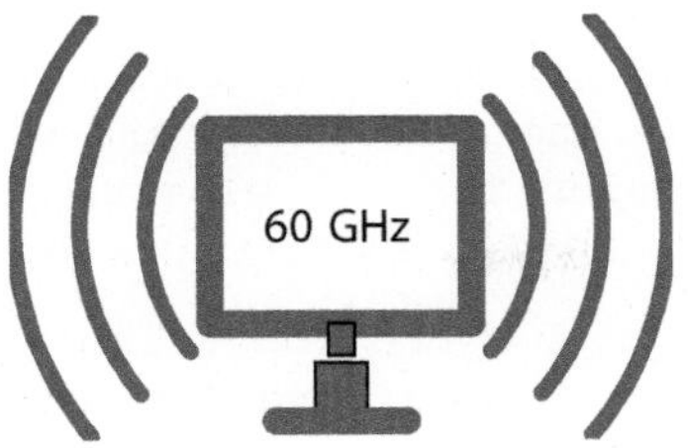

FIGURE 10.23 Predicting the use of 60 GHz Wi-Fi.

such as error and packet recovery, enable efficient and robust digital data transmission for small additional cost and low power consumption. Some of the factors that make this technology attractive are the following.

- There is a large bandwidth (up to 7 GHz) available worldwide.
- The 60 GHz technology is less restricted in terms of power limits.
- At 60 GHz, the path loss is higher, but higher transmitting power overcomes this, especially when the operation is restricted to indoor environments. The effective interference levels for 60 GHz are less severe than those systems located in the congested 2–2.5 and 5–5.8 GHz bands. In addition, higher frequency reuse can also be achieved in indoor environments, allowing a very high throughput.
- The compact size of the 60 GHz radio band permits the use of multiple antenna arrays, which can be conveniently integrated into consumer electronic products.
- Operators at these bands are exempt from license fees.
- Narrow beamwidth is possible.
- For example, this technology can be applied to an in-flight entertainment distribution system without causing interference with flight controls or navigation equipment.
- Oxygen absorption does not pose a problem when a 60 GHz system is used between satellites.

While the advantages of 60 GHz technology are very attractive, there are also some challenges to be met. At the data rates and range of communication offered by 60 GHz technology, ensuring a reliable communication link with sufficient power margin is not a trivial task. Delay spread of the channel under study, another limiting factor for high-speed transmissions, necessitates sophisticated coding techniques. Large delay spread values can easily increase the complexity of the system beyond the practical limit for channel equalization. The technology permits instant wireless downloading of multimedia content. Transmitting a large amount of data across remotely controlled miniature robots or vehicles without cable connections would be possible. Rescuing people in case of emergency or accidents becomes less problematic. Applications of mmWave technology to automobiles are also attractive. Overall, this technology could significantly affect the way computers and electronic devices communicate with each other.

Intense efforts are under way to expedite the commercialization of this fascinating technology. For example, industrial alliances and regulatory bodies are working to draft standards for mmWave.

MATLAB 10.1

```
% This script assists with the solution and graphing of Example 10.1
% We use symbolic variables in the creation of the waveform equation
% that describes the expression for the electric field

clear
syms E omega Beta t x          % symbolic variables for fields,
                               % time, and frequency

% Enter the frequency (in rad/s)
w = input('Enter the angular frequency\n >  ');

% The expression for the y-component of the electric field
E = 50*cos(w*t+Beta*x);

% part (b)
% solve for Beta
B = w/3e8;          % B is the numeric variable for Beta,
                    % with value as calculated here
E = subs(E,Beta,B); % substitute the value B in for
                    % variable Beta

% Generate numerical sequence
xfinal=ceil(6*2*pi/B);  % we will compute spatial values
                        % out to 3 wavelengths
dx=xfinal/1000;         % the discrete distance
space=0:dx:xfinal; % create a vector of 1000 discrete space segments
unityvec=ones(1,length(space));   % create a vector of 1s that is
        % the same length as the spatial vector discrete space
        segments

% Plot
figure
f = w/(2*pi);      % determine the frequency

for time=0:1/(20*f):1/f,  % time loop - each interation of this loop
                          % will plot the e-field waveform for
                          % a different increment of time we will loop
                          % through exactly one wavelength
    En = subs(E,{x,t},{space,unityvec*time});  % substitute
                 % the time and space vector into the variable
                 % to get a vector of the field as a function of
    plot(space, En)
    axis([0 6*2*pi/B min(En)-10 max(En)+10])
                         % add buffer space of 5 units to graph
    xlabel('x-axis (m)')
    ylabel('y-axis (m)')
    str=strcat('time = ', num2str(time), ' (s)');
            % concatenate string "time = " to the actual time
    text(1.5, max(En)+5, str)  % put annotation on figure
                         % to show time
    pause(0.5)     % pause for half a second then re-draw
    hold off
end
```

SUMMARY

1. The wave equation is of the form

$$\frac{\partial^2 \Phi}{\partial t^2} - u^2 \frac{\partial^2 \Phi}{\partial z^2} = 0$$

with the solution

$$\Phi = A \sin(\omega t - \beta z)$$

where u = wave velocity, A = wave amplitude, ω = angular frequency $(=2\pi f)$, and β = phase constant. Also, $\beta = \omega/u = 2\pi/\lambda$ or $u = f\lambda = \lambda/T$, where λ = wavelength and T = period.

2. In a lossy, charge-free medium, the wave equation based on Maxwell's equations is of the form

$$\nabla^2 \mathbf{A}_s - \gamma^2 \mathbf{A}_s = 0$$

where $\mathbf{A}_s$ is either $\mathbf{E}_s$ or $\mathbf{H}_s$ and $\gamma = \alpha + j\beta$ is the propagation constant. If we assume $\mathbf{E}_s = E_{xs}(z)\,\mathbf{a}_x$, we obtain EM waves of the form

$$\mathbf{E}(z, t) = E_o e^{-\alpha z} \cos(\omega t - \beta z)\mathbf{a}_x$$

$$\mathbf{H}(z, t) = H_o e^{-\alpha z} \cos(\omega t - \beta z - \theta_\eta)\mathbf{a}_y$$

where α = attenuation constant, β = phase constant, $\eta = |\eta|\underline{/\theta_\eta}$ = intrinsic impedance of the medium, also called wave impedance. The reciprocal of α is the skin depth $(\delta = 1/\alpha)$. The relations between β, ω, and λ as stated in item 1 remain valid for EM waves.

3. Wave propagation in media of other types can be derived from that for lossy media as special cases. For free space, set $\sigma = 0$, $\varepsilon = \varepsilon_o$, $\mu = \mu_o$; for lossless dielectric media, set $\sigma = 0$, $\varepsilon = \varepsilon_o \varepsilon_r$, and $\mu = \mu_o \mu_r$; and for good conductors, set $\sigma \simeq \infty$, $\varepsilon = \varepsilon_o$, $\mu = \mu_o$, or $\sigma/\omega\varepsilon \to \infty$.

4. A medium is classified as a lossy dielectric, a lossless dielectric, or a good conductor depending on its loss tangent, given by

$$\tan \theta = \frac{|\mathbf{J}_s|}{|\mathbf{J}_{d_s}|} = \frac{\sigma}{\omega\varepsilon} = \frac{\varepsilon''}{\varepsilon'}$$

where $\varepsilon_c = \varepsilon' - j\varepsilon''$ is the complex permittivity of the medium. For lossless dielectrics $\tan \theta \ll 1$, for good conductors $\tan \theta \gg 1$, and for lossy dielectrics $\tan \theta$ is of the order of unity.

5. In a good conductor, the fields tend to concentrate within the initial distance δ from the conductor surface. This phenomenon is called the skin effect. For a conductor of width w and length ℓ, the effective or ac resistance is

$$R_{ac} = \frac{\ell}{\sigma w \delta}$$

where δ is the skin depth.

6. The Poynting vector $\mathscr{P}$ is the power-flow vector whose direction is the same as the direction of wave propagation; its magnitude is the same as the amount of power flowing through a unit area normal to the propagation direction.

$$\mathscr{P} = \mathbf{E} \times \mathbf{H}, \quad \mathscr{P}_{ave} = 1/2\,\text{Re}(\mathbf{E}_s \times \mathbf{H}_s^*)$$

7. If a plane wave is incident normally from medium 1 to medium 2, the reflection coefficient Γ and transmission coefficient τ are given by

$$\Gamma = \frac{E_{ro}}{E_{io}} = \frac{\eta_2 - \eta_1}{\eta_2 + \eta_1}, \quad \tau = \frac{E_{to}}{E_{io}} = 1 + \Gamma$$

The standing wave ratio, s, is defined as

$$s = \frac{1 + |\Gamma|}{1 - |\Gamma|}$$

8. For oblique incidence from lossless medium 1 to lossless medium 2, we have the Fresnel coefficients as

$$\Gamma_{\|} = \frac{\eta_2 \cos\theta_t - \eta_1 \cos\theta_i}{\eta_2 \cos\theta_t + \eta_1 \cos\theta_i}, \quad \tau_{\|} = \frac{2\eta_2 \cos\theta_i}{\eta_2 \cos\theta_t + \eta_1 \cos\theta_i}$$

for parallel polarization and

$$\Gamma_{\perp} = \frac{\eta_2 \cos\theta_i - \eta_1 \cos\theta_t}{\eta_2 \cos\theta_i + \eta_1 \cos\theta_t} \quad \tau_{\perp} = \frac{2\eta_2 \cos\theta_i}{\eta_2 \cos\theta_i + \eta_1 \cos\theta_t}$$

for perpendicular polarization. As in optics,

$$\theta_r = \theta_i$$

$$\frac{\sin\theta_t}{\sin\theta_i} = \frac{\beta_1}{\beta_2} = \sqrt{\frac{\mu_1 \varepsilon_1}{\mu_2 \varepsilon_2}}$$

Total transmission or no reflection $(\Gamma = 0)$ occurs when the angle of incidence θ_i is equal to the Brewster angle.

9. Microwaves are EM waves of very short wavelengths. They propagate along a straight line like light rays and can therefore be focused easily in one direction by antennas. They are used in radar, guidance, navigation, and heating.

REVIEW QUESTIONS

10.1 Which of these is not a correct form of the wave $E_x = \cos(\omega t - \beta z)$?

(a) $\cos(\beta z - \omega t)$

(b) $\sin(\beta z - \omega t - \pi/2)$

(c) $\cos\left(\dfrac{2\pi t}{T} - \dfrac{2\pi z}{\lambda}\right)$

(d) $\text{Re}(e^{j(\omega t - \beta z)})$

(e) $\cos \beta(z - ut)$

10.2 Which of these functions does not satisfy the wave equation?

(a) $50e^{j\omega(t - 3z)}$
(b) $\sin \omega(10z + 5t)$
(c) $(x + 2t)^2$
(d) $\cos^2(y + 5t)$
(e) $\sin x \cos t$
(f) $\cos(5y + 2x)$

10.3 Which of the following statements is not true of waves in general?

(a) The phenomenon may be a function of time only.
(b) The phenomenon may be sinusoidal or cosinusoidal.
(c) The phenomenon must be a function of time and space.
(d) For practical reasons, it must be finite in extent.

10.4 The electric field component of a wave in free space is given by
$\mathbf{E} = 10 \cos(10^7 t + kz)\mathbf{a}_y$ V/m. It can be inferred that

(a) The wave propagates along $\mathbf{a}_y$.
(b) The wavelength $\lambda = 188.5$ m.
(c) The wave amplitude is 10 V/m.
(d) The wave number $k = 0.33$ rad/m.
(e) The wave attenuates as it travels.

10.5 Given that $\mathbf{H} = 0.5 \, e^{-0.1x} \sin(10^6 t - 2x)\mathbf{a}_z$ A/m, which of these statements are incorrect?

(a) $\alpha = 0.1$ Np/m
(b) $\beta = -2$ rad/m
(c) $\omega = 10^6$ rad/s
(d) The wave travels along $\mathbf{a}_x$.
(e) The wave is polarized in the z-direction.
(f) The period of the wave is 1 μs.

10.6 What is the major factor for determining whether a medium is free space, a lossless dielectric, a lossy dielectric, or a good conductor?

(a) Attenuation constant
(b) Constitutive parameters $(\sigma, \varepsilon, \mu)$
(c) Loss tangent
(d) Reflection coefficient

10.7 In a certain medium, $\mathbf{E} = 10 \cos(10^8 t - 3y)\mathbf{a}_x$ V/m. What type of medium is it?

(a) Free space
(b) Lossy dielectric
(c) Lossless dielectric
(d) Perfect conductor

10.8 Electromagnetic waves travel faster in conductors than in dielectrics.

(a) True
(b) False

10.9 In a good conductor, **E** and **H** are in time phase.

(a) True
(b) False

10.10 The Poynting vector physically denotes the power density leaving or entering a given volume in a time-varying field.

(a) True
(b) False

Answers: 10.1b, 10.2f, 10.3a, 10.4b,c 10.5b,e,f, 10.6c, 10.7c, 10.8b, 10.9b, 10.10a.

PROBLEMS

Section 10.2—Waves in General

10.1 An EM wave propagating in a certain medium is described by

$$\mathbf{E} = 25 \sin(2\pi \times 10^6 t - 6x)\mathbf{a}_z \text{ V/m}$$

(a) Determine the direction of wave propagation.
(b) Compute the period T, the wavelength λ, and the velocity u.
(c) Sketch the wave at $t = 0$, $T/8$, $T/4$, $T/2$.

10.2 Calculate the wavelength for plane waves in vacuum at the following frequencies:

(a) 60 Hz (power line)
(b) 2 MHz (AM radio)
(c) 120 MHz (FM radio)
(d) 2.4 GHz (microwave oven)

10.3 An EM wave in free space is described by

$$H = 0.4 \cos(10^8 t + \beta y) \text{ A/M}$$

Determine (a) the angular frequency ω, (b) the wave number β, (c) the wavelength λ, (d) the direction of wave propagation, (e) the value of $H(2, 3, 4, 10 \text{ ns})$.

10.4 (a) Show that $E(x, t) = \cos(x + \omega t) + \cos(x - \omega t)$ satisfies the scalar wave equation.
(b) Determine the velocity of wave propagation.

Section 10.3—Wave Propagation in Lossy Dielectrics

10.5 (a) Derive eqs. (10.23) and (10.24) from eqs. (10.18) and (10.20).
(b) Using eq. (10.29) in conjunction with Maxwell's equations, show that

$$\eta = \frac{j\omega\mu}{\gamma}$$

(c) From part (b), derive eqs. (10.32) and (10.33).

10.6   Show that the phase constant in eq. (10.24) can be approximated as

$$\beta = \omega\sqrt{\mu\varepsilon}\left(1 + \frac{\sigma}{8\omega^2\varepsilon^2}\right)$$

for $\dfrac{\sigma}{\omega\varepsilon} = 1$.

10.7   At 50 MHz, a lossy dielectric material is characterized by $\varepsilon = 3.6\varepsilon_0$, $\mu = 2.1\mu_0$, and $\sigma = 0.08$ S/m. If $\mathbf{E}_s = 6e^{-\gamma x}\, \mathbf{a}_z$ V/m, compute (a) γ, (b) λ, (c) u, (d) η, (e) $\mathbf{H}_s$.

10.8   Determine the loss tangent for each of the following nonmagnetic media at 12 MHz.

(a) wet earth $(\varepsilon = 10\varepsilon_0, \sigma = 10^{-2}$ S/m$)$
(b) dry earth $(\varepsilon = 4\varepsilon_0, \sigma = 10^{-4}$ S/m$)$
(c) seawater $(\varepsilon = 81\varepsilon_0, \sigma = 4$ S/m$)$

10.9   Alumina is a ceramic material used in making printed circuit boards. At 15 GHz, $\varepsilon = 9.6\varepsilon_0$, $\mu = \mu_0$, $\tan\theta = 3 \times 10^{-4}$. Calculate (a) the penetration depth, (b) the total attenuation over a thickness of 5 mm.

10.10   The properties of a medium are $\sigma = \varepsilon\omega$, $\varepsilon = 4\varepsilon_0$, $\mu = \mu_0$. If the wavelength in free space is 12 cm, find α, β, and u in the medium.

10.11   At $f = 100$ MHz, show that silver $(\sigma = 6.1 \times 10^7$ S/m, $\mu_r = 1, \varepsilon_r = 1)$ is a good conductor, while rubber $(\sigma = 10^{-15}$ S/m, $\mu_r = 1, \varepsilon_r = 3.1)$ is a good insulator.

10.12   Seawater plays a vital role in the study of submarine communications. Assuming that for seawater, $\sigma = 4$ S/m, $\varepsilon_r = 80$, $\mu_r = 1$, and $f = 100$ kHz, calculate (a) the phase velocity, (b) the wavelength, (c) the skin depth, (d) the intrinsic impedance.

10.13   In a certain medium with $\mu = \mu_0$, $\varepsilon = 4\varepsilon_0$,

$$\mathbf{H} = 12e^{-0.1y}\sin(\pi \times 10^8 t - \beta y)\mathbf{a}_x \text{ A/m}$$

Find (a) the wave period T, (b) the wavelength λ, (c) the electric field $\mathbf{E}$, (d) the phase difference between $\mathbf{E}$ and $\mathbf{H}$.

10.14   In a nonmagnetic medium,

$$\mathbf{H} = 50e^{-100x}\cos(2\pi \times 10^9 t - 200x)\mathbf{a}_y \text{ mA/m}$$

Find $\mathbf{E}$.

10.15   A certain medium has $\sigma = 1$ S/m, $\varepsilon = 4\varepsilon_0$, and $\mu = 9\,\mu_0$ at a frequency of 1 GHz. Determine the (a) attenuation constant, (b) phase constant, (c) intrinsic impedance, and (d) wave velocity.

Sections 10.4 and 10.5—Waves in Lossless Dielectrics and Free Space

10.16 The electric field of a TV broadcast signal progagating in air is given by

$$E (z,t) = 0.2 \cos(\omega t - 6.5z)a_x \text{ V/m}$$

(a) Determine the wave frequency ω and the wavelength λ.
(b) Sketch E_x as a function of t at $z = 0$ and $z = \lambda/2$.
(c) Find the corresponding $H(z, t)$.

10.17 A uniform plane wave has a wavelength of 6.4 cm in free space and 2.8 cm in a dielectric $(\sigma = 0, \varepsilon = \varepsilon_o\varepsilon_r, \mu = \mu_o)$. Find ε_r.

10.18 The magnetic field component of an EM wave propagating through a nonmagnetic medium $(\mu = \mu_o)$ is

$$H = 25 \sin(2 \times 10^8 t + 6x)a_y \text{ mA/m}$$

Determine:

(a) The direction of wave propagation
(b) The permittivity of the medium
(c) The electric field intensity

10.19 A manufacturer produces a ferrite material with $\mu = 750\mu_o$, $\varepsilon = 5\varepsilon_o$, and $\sigma = 10^{-6}$ S/m at 10 MHz.

(a) Would you classify the material as lossless, lossy, or conducting?
(b) Calculate β and λ.
(c) Determine the phase difference between two points separated by 2 m.
(d) Find the intrinsic impedance.

10.20 The electric field intensity of a uniform plane wave in air is given by

$$E = 50 \sin(10^8\pi t - \beta x)a_z \text{ mV/m}$$

(a) Calculate β.
(b) Determine the location(s) where E vanishes at $t = 50$ ns.
(c) Find H.

10.21 For a uniform plane wave at 4 GHz, the intrinsic impedance and phase velocity of an unknown material are measured as 105 Ω and 7.6×10^7 m/s, respectively. Find ε_r and μ_r of the material.

10.22 In a lossless medium $(\varepsilon_r = 4.5, \mu_r = 1)$, a uniform plane wave

$$E = 8 \cos(\omega t - \beta z)a_x - 6 \sin(\omega t - \beta z)a_y \text{ V/m}$$

propagates at 40 MHz. (a) Find H. (b) Determine β, λ, η, and u.

10.23 A uniform plane wave in a lossy nonmagnetic medium has

$$E_s = (5a_x + 12a_y)e^{-\gamma z}, \quad \gamma = (0.2 + j3.4) \text{ /m}$$

(a) Compute the magnitude of the wave at $z = 4$ m, $t = T/8$.
(b) Find the loss in decibels suffered by the wave in the interval $0 < z < 3$ m.
(c) Calculate the intrinsic impedance.

Section 10.6—Plane Waves in Good Conductors

10.24 The magnet field intensity of a uniform plane wave in a good conductor ($\varepsilon = \varepsilon_o, \mu = \mu_o$) is

$$\mathbf{H} = 20e^{-12z}\cos(2\pi \times 10^6 t + 12z)\mathbf{a}_y \text{ mA/m}$$

Find the conductivity and the corresponding **E** field.

10.25 Which of the following media may be treated as conducting at 8 MHz?

    (a) Wet marshy soil ($\varepsilon = 15\varepsilon_o, \mu = \mu_o, \sigma = 10^{-2}$ S/m)

    (b) Intrinsic germanium ($\varepsilon = 16\varepsilon_o, \mu = \mu_o, \sigma = 0.025$ S/m)

    (c) Seawater ($\varepsilon = 81\varepsilon_o, \mu = \mu_o, \sigma = 25$ S/m)

10.26 A uniform plane wave impinges normally on a conducting medium. If the frequency is 100 MHz and the skin depth is 0.02 mm, determine the velocity of the wave in the conducting medium.

10.27 (a) Determine the dc resistance of a round copper wire ($\sigma = 5.8 \times 10^7$ S/m, $\mu_r = 1, \varepsilon_r = 1$) of radius 1.2 mm and length 600 m.

    (b) Find the ac resistance at 100 MHz.

    (c) Calculate the approximate frequency at which dc and ac resistances are equal.

10.28 A 10 GHz wave passes through a medium made of copper ($\varepsilon = \varepsilon_o, \mu = \mu_o, \sigma = 5.8 \times 10^7$ S/m); find: (a) attenuation constant, (b) the skin depth, (c) the intrinsic impedance.

10.29 For silver, $\sigma = 6.1 \times 10^7$ S/m, $\mu_r = 1, \varepsilon_r = 1$, determine the frequency at which the penetration depth is 2 mm.

10.30 Compute the penetration depth of copper at the power frequency of 60 Hz.

10.31 By measurements conducted at 12 MHz on a certain material, it is found that the intrinsic impedance is $24.6\angle 45°$ Ω with $\mu = \mu$. Find α, β, λ, and u.

10.32 Fat tissue at 2.42 GHz has the following properties:

$$\sigma = 0.12 \text{ S/m}, \varepsilon = 5.5\,\varepsilon_o, \text{ and } \mu = \mu_o. \text{ Find the penetration depth.}$$

10.33 Brass waveguides are often silver plated to reduce losses. If the thickness of silver ($\mu = \mu_o, \varepsilon = \varepsilon_o, \sigma = 6.1 \times 10^7$ S/m) must be 5δ, find the minimum thickness required for a waveguide operating at 12 GHz.

10.34 How deep does a radar wave at 2 GHz travel in seawater before its amplitude is reduced to 10^{-5} of its amplitude just below the surface? Assume that $\mu = \mu_o, \varepsilon = 24\varepsilon_o, \sigma = 4$ S/m.

Section 10.7—Wave Polarization

10.35 The electric field intensity of a uniform plane wave in a medium ($\sigma = 0, \mu = \mu_o, \varepsilon = \varepsilon_o\varepsilon_r$) is

$$\mathbf{E} = 12\sin(2\pi \times 10^7 t - 3y)\mathbf{a}_z \text{ V/m}$$

    (a) Determine the polarization of the wave.

    (b) Find the frequency.

    (c) Calculate ε_r.

    (d) Obtain the magnetic field intensity **H**.

10.36 Let $\mathbf{E} = 2 \sin(\omega t - \beta x)\mathbf{a}_y - 5 \sin(\omega t - \beta x)\mathbf{a}_z$ V/m. What is the wave polarization?

10.37 Determine the wave polarization of each of the following waves:

(a) $E_o \cos(\omega t + \beta y)\mathbf{a}_x + E_o \sin(\omega t + \beta y)\mathbf{a}_z$ V/m

(b) $E_o \cos(\omega t - \beta y)\mathbf{a}_x - 3E_o \sin(\omega t + \beta y)\mathbf{a}_z$ V/m

10.38 Determine the polarization of the following waves:

(a) $\mathbf{E}_s = 40e^{j10z}\mathbf{a}_x + 60e^{j10z}\mathbf{a}_y$ V/m

(b) $\mathbf{E}_s = 12e^{j\pi/3}e^{-j10x}\mathbf{a}_y + 5e^{-j\pi/3}e^{-j10x}\mathbf{a}_z$ V/m

10.39 The electric field intensity of a uniform plane wave in free space is given by

$$\mathbf{E} = 40 \cos(\omega t - \beta z)\mathbf{a}_x + 60 \sin(\omega t - \beta z)\mathbf{a}_y \text{ V/m}$$

(a) What is the wave polarization?

(b) Determine the magnetic field intensity.

10.40 Show that a linearly polarized plane wave of the form $\mathbf{E}_s = E_o e^{-j\beta z}\mathbf{a}_x$ can be expressed as the sum of two circularly polarized waves.

10.41 Suppose $\mathbf{E}(y,t) = E_{o1} \cos(\omega t - \beta y)\mathbf{a}_x + E_{o2} \cos(\omega t - \beta y + \phi)\mathbf{a}_z$ V/m. Determine the polarization when (a) $\phi = 0$, (b) $\phi = \pi/2$, (c) $\phi = \pi$.

Section 10.8—Power and the Poynting Vector

10.42 Show that eqs. (10.77) and (10.78) are equivalent.

10.43 The electric field intensity in a dielectric medium ($\mu = \mu_o$, $\varepsilon = \varepsilon_o\varepsilon_r$) is given by

$$\mathbf{E} = 150 \cos(10^9 t + 8x)\mathbf{a}_z \text{ V/m}$$

Calculate

(a) The dielectric constant ε_r

(b) The intrinsic impedance

(c) The velocity of propagation

(d) The magnetic field intensity

(e) The Poynting vector $\mathscr{P}$

10.44 In free space, $\mathbf{E} = 40 \cos(\omega t - 10z)\mathbf{a}_y$ V/m. Find the total average power passing through a circular disk of radius 1.5 m in the $z = 0$ plane.

10.45 The electric field due a short dipole antenna located in free space is

$$\mathbf{E}_s = \frac{10}{r} \sin \theta e^{-j3r}\mathbf{a}_\theta \text{ V/m}$$

Find (a) $\mathbf{H}_s$, (b) the average power crossing the surface $r = 2$, $0 < \theta < \pi/6$, $0 < \phi < \pi$.

10.46 The electric field component of a uniform plane wave traveling in seawater $(\sigma = 4 \text{ S/m}, \varepsilon = 81 \, \varepsilon_o, \mu = \mu_o)$ is

$$\mathbf{E} = 8e^{-0.1z} \cos(\omega t - 0.3z)\mathbf{a}_x \text{ V/m}$$

(a) Determine the average power density. (b) Find the depth at which the power density is reduced by 20 dB.

10.47 In a coaxial transmission line filled with a lossless dielectric $(\varepsilon = 4.5\varepsilon_o, \mu = \mu_o)$,

$$\mathbf{E} = \frac{40}{\rho} \sin(\omega t - 2z)\mathbf{a}_\rho \text{ V/m}$$

Find (a) ω and $\mathbf{H}$, (b) the Poynting vector, (c) the total time-average power crossing the surface $z = 1$ m, 2 mm $< \rho <$ 3 mm, $0 < \phi < 2\pi$.

10.48 An antenna is located at the origin of a spherical coordinate system. The fields produced by the antenna in free space are

$$\mathbf{E} = \frac{E_o}{r} \sin \theta \sin \omega(t - r/c)\mathbf{a}_\theta$$

$$\mathbf{H} = \frac{E_o}{\eta r} \sin \theta \sin \omega(t - r/c)\mathbf{a}_\phi$$

where $c = \dfrac{1}{\sqrt{\mu_o \varepsilon_o}}$ and $\eta = \sqrt{\dfrac{\mu_o}{\mu_o}}$. Determine the time-average power radiated by the antenna.

10.49 A plane wave in free space has

$$\mathbf{H}(x, t) = (10\mathbf{a}_y - 20\mathbf{a}_z) \sin(\omega t - 40x) \text{ A/m}$$

Find ω, $\mathbf{E}$, and $\mathscr{P}_{ave}$.

10.50 Human exposure to the electromagnetic radiation in air is regarded as safe if the power density is less than 10 mW/m². What is the corresponding electric field intensity?

10.51 Given that $\mathbf{E} = \cos(\omega t - \beta z)\mathbf{a}_x + \sin(\omega t - \beta z)\mathbf{a}_y$ V/m, show that the Poynting vector is constant everywhere.

10.52 At the bottom of a microwave oven, $E = 2.4$ kV/m. If this value is found uniformly over the entire area of the oven, which is 450 cm², determine the power delivered by the oven. Assume $\mu = \mu_o$, $\varepsilon = \varepsilon_o$.

10.53 A coaxial cable consists of two conducting cylinders of radii a and b. The electric and magnetic fields in the cable are

$$\mathbf{E} = \frac{V_o}{\rho \ln(b/a)} \sin(\omega t - \beta z)\mathbf{a}_\rho, \quad a < \rho < b$$

$$\mathbf{H} = \frac{I_o}{2\pi\rho}\sin(\omega t - \beta z)\mathbf{a}_\phi, \ a < \rho < b$$

where V_o and I_o are constants. (a) Determine the time-average Poynting vector. (b) Find the time-average power flowing through the cable.

Section 10.9—Reflection at Normal Incidence

10.54 (a) For a normal incidence upon the dielectric–dielectric interface for which $\mu_1 = \mu_2 = \mu_o$, we define R and T as the reflection and transmission coefficients for average powers; that is, $P_{r,\text{ave}} = RP_{i,\text{ave}}$ and $P_{t,\text{ave}} = TP_{i,\text{ave}}$. Prove that

$$R = \left(\frac{n_1 - n_2}{n_1 + n_2}\right)^2 \quad \text{and} \quad T = \frac{4n_1 n_2}{(n_1 + n_2)^2}$$

where n_1 and n_2 are the refractive indices of the media.

(b) Determine the ratio n_1/n_2 so that the reflected and the transmitted waves have the same average power.

10.55 A plane wave in a lossless medium $(\sigma = 0, \mu = 2\mu_o, \varepsilon = 8\varepsilon_o)$ is given as

$$\mathbf{E} = 60\sin(\omega t - 10z)\mathbf{a}_x + 30\sin(\omega t - 10z + \pi/6)\mathbf{a}_y \text{ V/m}.$$

This wave is incident on a lossless medium $(\sigma = 0, \mu = \mu_o, \varepsilon = 16\varepsilon_o)$. Determine the reflected and transmitted electric fields.

10.56 A uniform plane wave in free space impinges perpendicularly on a lossless nonmagnetic material with $\varepsilon = 9\varepsilon_o$. Calculate the fraction of the incident average power that is transmitted.

10.57 A uniform plane wave in air with

$$\mathbf{H} = 4\sin(\omega t - 5x)\mathbf{a}_y \text{ A/m}$$

is normally incident on a plastic region $(x \geq 0)$ with the parameters $\mu = \mu_o, \varepsilon = 4\varepsilon_o$, and $\sigma = 0$. (a) Obtain the total electric field in air. (b) Calculate the time-average power density in the plastic region. (c) Find the standing wave ratio.

10.58 Region 1 is a lossless medium for which $y \geq 0, \mu = \mu_o, \varepsilon = 4\varepsilon_o$, whereas region 2 is free space, $y \leq 0$. If a plane wave $\mathbf{E}_i = 5\cos(10^8 t + \beta y)\mathbf{a}_z$ V/m exists in region 1, find (a) the total electric field component of the wave in region 1, (b) the time-average Poynting vector in region 1, (c) the time-average Poynting vector in region 2.

10.59 A uniform plane wave propagates in a medium for which $\sigma = 0, \varepsilon = 16\varepsilon_o, \mu = \mu_o$. The electric field in the medium is

$$\mathbf{E} = 60\cos(\omega t - \beta z)\mathbf{a}_x \text{ V/m}$$

where $\omega = 90$ Mrad/s. If the wave hits free space normally at $z = 0$, determine the power densities in both media.

10.60 A uniform plane wave in air is normally incident on an infinite lossless dielectric material occupying $z > 0$ and having $\varepsilon = 3\varepsilon_o$ and $\mu = \mu_o$. If the incident wave is $\mathbf{E}_i = 10 \cos(\omega t - z)\mathbf{a}_y$ V/m, find

(a) λ and ω of the wave in air and the transmitted wave in the dielectric medium

(b) The incident $\mathbf{H}_i$ field

(c) Γ and τ

(d) The total electric field and the time-average power in both regions

10.61 A 100 MHz plane wave is normally incident from air to the sea surface, which may be assumed to be calm and smooth. If $\sigma = 4$ S/m, $\mu_r = 1$, and $\varepsilon_r = 81$ for seawater, calculate the fractions of the incident power that are transmitted and reflected.

10.62 A uniform plane wave in a certain medium ($\mu = \mu_o$, $\varepsilon = 4\varepsilon_o$) is given by

$$\mathbf{E} = 12 \cos(\omega t - 40\pi x)\mathbf{a}_z \text{ V/m}$$

(a) Find ω.

(b) If the wave is normally incident on a dielectric ($\mu = \mu_o$, $\varepsilon = 3.2\varepsilon_o$), determine $\mathbf{E}_r$ and $\mathbf{E}_t$.

*10.63 A signal in air ($z \geq 0$) with the electric field component

$$\mathbf{E} = 10 \sin(\omega t + 3z)\mathbf{a}_x \text{ V/m}$$

hits normally the ocean surface at $z = 0$ as in Figure 10.24. Assuming that the ocean surface is smooth and that $\varepsilon = 80\varepsilon_o$, $\mu = \mu_o$, $\sigma = 4$ S/m in ocean, determine

(a) ω

(b) The wavelength of the signal in air

(c) The loss tangent and intrinsic impedance of the ocean

(d) The reflected and transmitted $\mathbf{E}$ field

10.64 Sketch the standing wave in eq. (10.97) at $t = 0$, $T/8$, $T/4$, $3T/8$, $T/2$, and so on, where $T = 2\pi/\omega$.

10.65 A uniform plane wave is incident at an angle $\theta_i = 45°$ on a pair of dielectric slabs joined together as shown in Figure 10.25. Determine the angles of transmission θ_{t1} and θ_{t2} in the slabs.

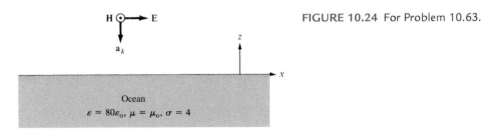

FIGURE 10.24 For Problem 10.63.

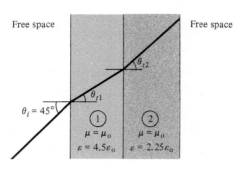

FIGURE 10.25 For Problem 10.65.

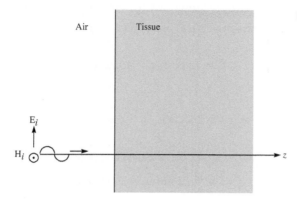

FIGURE 10.26 For Problem 10.67.

10.66 Show that the field

$$E_s = 20 \sin(k_x x) \cos(k_y y) a_z$$

where $k_x^2 + k_y^2 = \omega^2 \mu_o \varepsilon_o$, can be represented as the superposition of four propagating plane waves. Find the corresponding H_s.

10.67 Electromagnetic radiation can be used to heat cancerous tumors. If a plane wave is normally incident on the tissue surface at 1.2 GHz as shown in Figure. 10.26, determine the reflection coefficient. At 1.2 GHz, the electrical properties of the tissue are $\varepsilon_r = 50$, $\mu_r = 1$, $\sigma = 4$ S/m.

10.68 An EM plane wave in a lossless medium impinges normally on a lossy medium.

(a) Determine the ratio of transmitted to incident power in terms of the standing wave ratio s.

(b) Express the ratio of reflected to incident power in terms of s.

Section 10.10—Reflection at Oblique Incidence

*10.69 By assuming the time-dependent fields $\mathbf{E} = \mathbf{E}_o e^{j(\mathbf{k} \cdot \mathbf{r} - \omega t)}$ and $\mathbf{H} = \mathbf{H}_o e^{j(\mathbf{k} \cdot \mathbf{r} - \omega t)}$ where $\mathbf{k} = k_x \mathbf{a}_x + k_y \mathbf{a}_y + k_z \mathbf{a}_z$ is the wave number vector and $\mathbf{r} = x \mathbf{a}_x + y \mathbf{a}_y + z \mathbf{a}_z$ is the radius vector, show that $\nabla \times \mathbf{E} = -\partial \mathbf{B}/\partial t$ can be expressed as $\mathbf{k} \times \mathbf{E} = \mu \omega \mathbf{H}$ and deduce $\mathbf{a}_k \times \mathbf{a}_E = \mathbf{a}_H$.

10.70 A plane wave in free space has a propagation vector

$$\mathbf{k} = 124\mathbf{a}_x + 124\mathbf{a}_y + 263\mathbf{a}_z$$

Find the wavelength, frequency, and angles $\mathbf{k}$ makes with the x-, y-, and z-axes.

10.71 In free space,

$$\mathbf{E}_s = [E_o\mathbf{a}_x + \mathbf{a}_y + (3 + j4)\mathbf{a}_z]e^{-j(3.4x - 4.2y)} \text{ V/m}$$

Determine E_o, $\mathbf{H}_s$, and frequency.

10.72 Assume the same fields as in Problem 10.69 and show that Maxwell's equations in a source-free region can be written as

$$\mathbf{k} \cdot \mathbf{E} = 0$$
$$\mathbf{k} \cdot \mathbf{H} = 0$$
$$\mathbf{k} \times \mathbf{E} = \omega\mu\mathbf{H}$$
$$\mathbf{k} \times \mathbf{H} = -\omega\varepsilon\mathbf{E}$$

From these equations deduce

$$\mathbf{a}_k \times \mathbf{a}_E = \mathbf{a}_H \quad \text{and} \quad \mathbf{a}_k \times \mathbf{a}_H = -\mathbf{a}_E$$

10.73 Show that for nonmagnetic dielectric media, the reflection and transmission coefficients for oblique incidence become

$$\Gamma_\parallel = \frac{\tan(\theta_t - \theta_i)}{\tan(\theta_t + \theta_i)}, \quad \tau_\parallel = \frac{2\cos\theta_i \sin\theta_t}{\sin(\theta_t + \theta_i)\cos(\theta_t - \theta_i)}$$

$$\Gamma_\perp = \frac{\sin(\theta_t - \theta_i)}{\sin(\theta_t + \theta_i)}, \quad \tau_\perp = \frac{2\cos\theta_i \sin\theta_t}{\sin(\theta_t + \theta_i)}$$

10.74 If region 1 is in free space, while region 2 is a nonmagnetic dielectric medium ($\sigma_2 = 0$, $\varepsilon_{r2} = 6.4$), compute E_{ro}/E_{io} and E_{to}/E_{io} for oblique incidence at $\theta_i = 12°$. Assume parallel polarization.

10.75 A parallel-polarized wave in air with

$$\mathbf{E} = (8\mathbf{a}_y - 6\mathbf{a}_z)\sin(\omega t - 4y - 3z) \text{ V/m}$$

impinges a dielectric half-space as shown in Figure 10.27. Find (a) the incidence angle θ_i, (b) the time-average power in air ($\mu = \mu_o$, $\varepsilon = \varepsilon_o$), (c) the reflected and transmitted $\mathbf{E}$ fields.

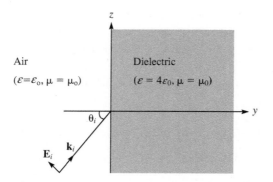

FIGURE 10.27 For Problem 10.75.

10.76 In a dielectric medium $(\varepsilon = 9\varepsilon_0, \mu = \mu_0)$, a plane wave with

$$\mathbf{H} = 0.2\cos(10^9 t - kx - k\sqrt{8}z)\mathbf{a}_y \text{ A/m}$$

is incident on an air boundary at $z = 0$. Find

(a) θ_r and θ_t

(b) k

(c) The wavelength in the dielectric and in air

(d) The incident $\mathbf{E}$

(e) The transmitted and reflected $\mathbf{E}$

(f) The Brewster angle

10.77 A plane wave in free space has

$$\mathbf{E}_i = (4\mathbf{a}_x + 5\mathbf{a}_y - 3\mathbf{a}_z)\cos(\omega_t - 0.5\pi x - 0.866\pi z)$$

Determine: (a) the perpendicular- and parallel-polarized components of the wave, (b) the angle of incidence. Assume that the xy-plane is the boundary between the two media.

10.78 A parallel-polarized wave in free space impinges on a dielectric medium $(\sigma = 0, \varepsilon = \varepsilon_0\varepsilon_r, \mu = \mu_0)$. If the Brewster angle is 68°, find ε_r.

10.79 If u is the phase velocity of an EM wave in a given medium, the index of refraction of the medium is $n = c/u$, where c is the speed of light in vacuum.

(a) Paraffin has $\mu_r = 1$, $\varepsilon_r = 2.1$. Determine n for unbounded medium of paraffin.

(b) Distilled water has $\mu_r = 1$, $\varepsilon_r = 81$. Find n.

(c) Polystyrene has $\mu_r = 1$, $\varepsilon_r = 2.7$. Calculate n.

Section 10.11—Application Note—Microwaves

10.80 Discuss briefly some applications of microwaves other than those discussed in the text.

10.81 A useful set of parameters, known as the *scattering transfer parameters,* is related to the incident and reflected waves as

$$\begin{bmatrix} a_1 \\ b_1 \end{bmatrix} = \begin{bmatrix} T_{11} & T_{12} \\ T_{21} & T_{22} \end{bmatrix} \begin{bmatrix} b_2 \\ a_2 \end{bmatrix}$$

(a) Express the T-parameters in terms of the S-parameters.

(b) Find T when

$$S = \begin{bmatrix} 0.2 & 0.4 \\ 0.4 & 0.2 \end{bmatrix}$$

10.82 The S-parameters of a two-port network are:

$$S_{11} = 0.33 - j0.16, S_{12} = S_{21} = 0.56, S_{22} = 0.44 - j0.62$$

Find the input and output reflection coefficients when $Z_L = Z_o = 50 \ \Omega$ and $Z_g = 2Z_o$.

10.83 Why can't regular lumped circuit components such as resistors, inductors, and capacitors be used at microwave frequencies?

10.84 In free space, a microwave signal has a frequency of 8.4 GHz. Calculate the wavelength of the signal.

BIOELECTROMAGNETICS

Is it safe to live close to power transmission lines? Are cellular phones safe? Can video display terminals cause problems for pregnant women? These and other questions are addressed by bioelectromagnetics (BEM), which is the branch of electromagnetics that deals with the biological effects of man-made EM fields with respect to humans and the environment. BEM may also be regarded as the emerging science that studies how living organisms interact with EM fields. It combines the investigative efforts of scientists from various disciplines.

A major challenge the industry faces as a result of the immense spread of wireless technology is the growth of health concerns on the part of the public and health agencies alike. The recent focus has been on the design criteria for transmitters operating in the closest proximity of the human body. When a person is exposed to an EM field, energy incident on the person may be scattered, reflected, transmitted, or absorbed into the body depending on the field strength, the frequency, the dimensions of the body, and the electrical properties of the tissue. The heat produced by radiation may affect live tissue. If the body cannot dissipate this heat energy as fast as it is produced, the internal temperature of the body will rise. This may result in damage to tissues and organs and in death if the rise is sufficiently high.

An EM field is classified as ionizing if its energy is high enough to dislodge electrons from an atom or molecule. High-energy forms of EM radiation, such as gamma rays and X-rays, are strongly ionizing in biological matter. For this reason, prolonged exposure to such rays is harmful. Radiation in the middle portion of the frequency and energy spectrum—such as visible, especially ultraviolet, light—is weakly ionizing. Although it has long been known that exposure to strongly ionizing EM radiation can cause extreme damage to biological tissues, only recently has evidence implicated long-term exposure to nonionizing EM fields, such as those emitted by power lines, in increased health hazards.

Researchers have reported that to prevent deep-tissue burning, individuals wearing or carrying metal objects such as hairpins, metal implants, buckles, coins, or metal-framed eyeglasses, any of which may concentrate the EM field and cause burning, should not be exposed to radio-frequency (RF) radiation. Researchers have also pointed out that radiation can be absorbed deeply and is actually greater in tissue such as muscle or the brain than in regions of poorer absorption in the bone and fatty layers near the body surface. At this time, there are fewer scientists who will say that there is positively no possibility of nonthermal hazards of low-level EM fields. Ongoing research is aimed at determining whether there is a hazard and if so, at what levels.

CHAPTER 11

TRANSMISSION LINES

Kind hearts are the garden. Kind thoughts are the roots. Kind words are the flowers. Kind deeds are the fruits. Take care of your garden, And keep out the weeds. Fill it up with sunshine, Kind words and kind deeds.

—LONGFELLOW

11.1 INTRODUCTION

Our discussion in Chapter 10 was essentially on wave propagation in unbounded media, media of infinite extent. Such wave propagation is said to be unguided in that the uniform plane wave exists throughout all space, and EM energy associated with the wave spreads over a wide area. Wave propagation in unbounded media is used in radio or TV broadcasting, where the information being transmitted is meant for everyone who may be interested.

Another means of transmitting power or information is by guided structures. Guided structures serve to guide (or direct) the propagation of energy from the source to the load. Typical examples of such structures are transmission lines and waveguides. Waveguides are discussed in the next chapter; transmission lines are considered in this chapter.

Transmission lines are commonly used in power distribution (at low frequencies) and in communications (at high frequencies). Transmission lines such as twisted-pair and coaxial cables (thinnet and thicknet) are used in computer networks such as the Ethernet and the Internet.

A transmission line basically consists of two or more parallel conductors used to connect a source to a load. The source may be a hydroelectric generator, a transmitter, or an oscillator; the load may be a factory, an antenna, or an oscilloscope. Typical transmission lines include coaxial cable, a two-wire line, a parallel-plate or planar line, a wire above the conducting plane, and a microstrip line. These lines are portrayed in Figure 11.1. Notice that each of these lines consists of two conductors in parallel. Coaxial cables are routinely used in electrical laboratories and in connecting TV sets to antennas. Microstrip lines (similar to that in Figure 11.1e) are particularly important in integrated circuits, where metallic strips connecting electronic elements are deposited on dielectric substrates.

Transmission line problems are usually solved by means of EM field theory and electric circuit theory, the two major theories on which electrical engineering is based. In this chapter, we use circuit theory because it is easier to deal with mathematically. The basic concepts of wave propagation (such as propagation constant, reflection coefficient, and standing wave ratio) covered in the preceding chapter apply here.

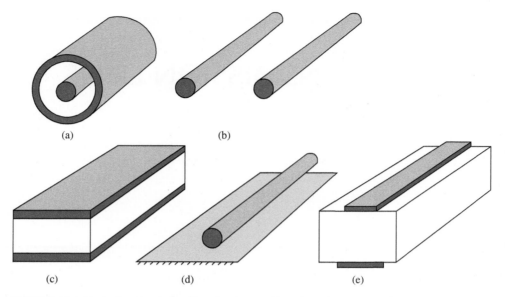

FIGURE 11.1 Typical transmission lines in cross-sectional view: **(a)** coaxial line, **(b)** two-wire line, **(c)** planar line, **(d)** wire above conducting plane, **(e)** microstrip line.

Our analysis of transmission lines will include the derivation of the transmission line equations and characteristic quantities, the use of the Smith chart, various practical applications of transmission lines, and transients on transmission lines.

11.2 TRANSMISSION LINE PARAMETERS

It is customary and convenient to describe a transmission line in terms of its line parameters, which are its resistance per unit length R, inductance per unit length L, conductance per unit length G, and capacitance per unit length C. Each of the lines shown in Figure 11.1 has specific formulas for finding R, L, G, and C. For coaxial, two-wire, and planar lines, the formulas for calculating the values of R, L, G, and C are provided in Table 11.1. The dimensions of the lines are as shown in Figure 11.2. Some of the formulas[1] in Table 11.1 were derived in Chapters 6 and 8. It should be noted that

1. The line parameters R, L, G, and C are not discrete or lumped. Rather, they are distributed as shown in Figure 11.3. By this we mean that the parameters are uniformly distributed along the entire length of the line.

---

[1]Similar formulas for other transmission lines can be obtained from engineering handbooks or data books—for example, M. A. R. Gunston, *Microwave Transmission-Line Impedance Data*. London: Van Nostrand Reinhold, 1972.

TABLE 11.1 Distributed Line Parameters at High Frequencies*

Parameters	Coaxial Line	Two-Wire Line	Planar Line
$R\ (\Omega/\text{m})$	$\dfrac{1}{2\pi\delta\sigma_c}\left[\dfrac{1}{a}+\dfrac{1}{b}\right]$	$\dfrac{1}{\pi a\delta\sigma_c}$	$\dfrac{2}{w\delta\sigma_c}$
	$(\delta \ll a, c - b)$	$(\delta \ll a)$	$(\delta \ll t)$
$L\ (\text{H/m})$	$\dfrac{\mu}{2\pi}\ln\dfrac{b}{a}$	$\dfrac{\mu}{\pi}\cosh^{-1}\dfrac{d}{2a}$	$\dfrac{\mu d}{w}$
$G\ (\text{S/m})$	$\dfrac{2\pi\sigma}{\ln\dfrac{b}{a}}$	$\dfrac{\pi\sigma}{\cosh^{-1}\dfrac{d}{2a}}$	$\dfrac{\sigma w}{d}$
$C\ (\text{F/m})$	$\dfrac{2\pi\varepsilon}{\ln\dfrac{b}{a}}$	$\dfrac{\pi\varepsilon}{\cosh^{-1}\dfrac{d}{2a}}$	$\dfrac{\varepsilon w}{d}$ $(w \gg d)$

$^{*}\ \delta = \dfrac{1}{\sqrt{\pi f \mu_c \sigma_c}}$ = skin depth of the conductor; $\cosh^{-1}\dfrac{d}{2a} \simeq \ln\dfrac{d}{a}$ if $\left[\dfrac{d}{2a}\right]^2 \gg 1$.

2. For each line, the conductors are characterized by σ_c, μ_c, $\varepsilon_c = \varepsilon_o$, and the homogeneous dielectric separating the conductors is characterized by σ, μ, ε.

3. $G \neq 1/R$; R is the ac resistance per unit length of the conductors comprising the line, and G is the conductance per unit length due to the dielectric medium separating the conductors.

4. The value of L shown in Table 11.1 is the external inductance per unit length, that is, $L = L_{\text{ext}}$. The effects of internal inductance $L_{\text{in}}\ (= R/\omega)$ are negligible at the high frequencies at which most communication systems operate.

5. For each line,

$$LC = \mu\varepsilon \quad \text{and} \quad \frac{G}{C} = \frac{\sigma}{\varepsilon} \tag{11.1}$$

As a way of preparing for the next section, let us consider how an EM wave propagates through a two-conductor transmission line. For example, consider the coaxial line connecting the generator or source to the load as in Figure 11.4(a). When switch S is closed, the inner conductor is made positive with respect to the outer one so that

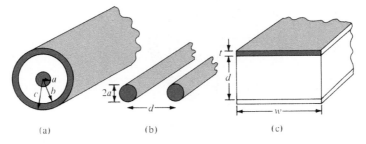

(a) (b) (c)

FIGURE 11.2 Common transmission lines: **(a)** coaxial line, **(b)** two-wire line, **(c)** planar line.

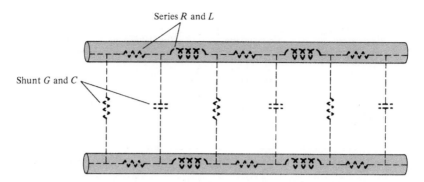

FIGURE 11.3 Distributed parameters of a two-conductor transmission line.

the **E** field is radially outward as in Figure 11.4(b). According to Ampère's law, the **H** field encircles the current-carrying conductor as in Figure 11.4(b). The Poynting vector $(\mathbf{E} \times \mathbf{H})$ points along the transmission line. Thus, closing the switch simply establishes a disturbance, which appears as a transverse electromagnetic (TEM) wave propagating along the line. This wave is a nonuniform plane wave, and by means of it, power is transmitted through the line.

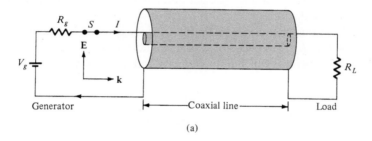

(a)

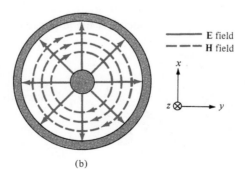

(b)

FIGURE 11.4 **(a)** Coaxial line connecting the generator to the load; **(b) E** and **H** fields on the coaxial line.

11.3 TRANSMISSION LINE EQUATIONS

As mentioned in Section 11.2, a two-conductor transmission line supports a TEM wave; that is, the electric and magnetic fields on the line are perpendicular to each other and transverse to the direction of wave propagation. An important property of TEM waves is that the fields **E** and **H** are uniquely related to voltage V and current I, respectively:

$$V = -\int_L \mathbf{E} \cdot d\mathbf{l}, \quad I = \oint_L \mathbf{H} \cdot d\mathbf{l} \qquad (11.2)$$

In view of this, we will use circuit quantities V and I in solving the transmission line problem instead of solving field quantities **E** and **H** (i.e., solving Maxwell's equations and boundary conditions). The circuit model is simpler and more convenient.

Let us examine an incremental portion of length Δz of a two-conductor transmission line. We intend to find an equivalent circuit for this line and derive the line equations. From Figure 11.3, we expect the equivalent circuit of a portion of the line to be as in Figure 11.5. The model in Figure 11.5 is in terms of the line parameters R, L, G, and C, and may represent any of the two-conductor lines of Figure 11.2. The model is called the L-type equivalent circuit; there are other possible types. In the model of Figure 11.5, we assume that the wave propagates along the $+z$-direction, from the generator to the load.

By applying Kirchhoff's voltage law to the outer loop of the circuit in Figure 11.5, we obtain

$$V(z, t) = R \, \Delta z \, I(z, t) + L \, \Delta z \frac{\partial I(z, t)}{\partial t} + V(z + \Delta z, t)$$

or

$$-\frac{V(z + \Delta z, t) - V(z, t)}{\Delta z} = RI(z, t) + L \frac{\partial I(z, t)}{\partial t} \qquad (11.3)$$

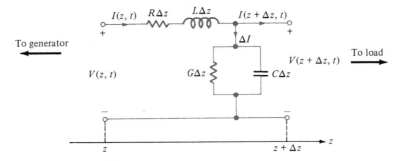

FIGURE 11.5 An L-type equivalent circuit model of a two-conductor transmission line of differential length Δz.

Taking the limit of eq. (11.3) as $\Delta z \to 0$ leads to

$$-\frac{\partial V(z, t)}{\partial z} = RI(z, t) + L\frac{\partial I(z, t)}{\partial t}$$
(11.4)

Similarly, applying Kirchhoff's current law to the main node of the circuit in Figure 11.5 gives

$$I(z, t) = I(z + \Delta z, t) + \Delta I$$

$$= I(z + \Delta z, t) + G\Delta z\, V(z + \Delta z, t) + C\Delta z\frac{\partial V(z + \Delta z, t)}{\partial t}$$

or

$$-\frac{I(z + \Delta z, t) - I(z, t)}{\Delta z} = GV(z + \Delta z, t) + C\frac{\partial V(z + \Delta z, t)}{\partial t}$$
(11.5)

As $\Delta z \to 0$, eq. (11.5) becomes

$$-\frac{\partial I(z, t)}{\partial z} = GV(z, t) + C\frac{\partial V(z, t)}{\partial t}$$
(11.6)

If we assume harmonic time dependence so that

$$V(z, t) = \text{Re}[V_s(z)\, e^{j\omega t}]$$
(11.7a)

$$I(z, t) = \text{Re}[I_s(z)\, e^{j\omega t}]$$
(11.7b)

where $V_s(z)$ and $I_s(z)$ are the phasor forms of $V(z, t)$ and $I(z, t)$, respectively, eqs. (11.4) and (11.6) become

$$-\frac{dV_s}{dz} = (R + j\omega L)I_s$$
(11.8)

$$-\frac{dI_s}{dz} = (G + j\omega C)V_s$$
(11.9)

The differential eqs. (11.8) and (11.9) are coupled. To separate them, we take the second derivative of V_s in eq. (11.8) and employ eq. (11.9) so that we obtain

$$\frac{d^2V_s}{dz^2} = (R + j\omega L)(G + j\omega C)V_s$$

or

$$\frac{d^2V_s}{dz^2} - \gamma^2 V_s = 0$$
(11.10)

where

$$\boxed{\gamma = \alpha + j\beta = \sqrt{(R + j\omega L)(G + j\omega C)}}$$

(11.11)

By taking the second derivative of I_s in eq. (11.9) and employing eq. (11.8), we get

$$\frac{d^2 I_s}{dz^2} - \gamma^2 I_s = 0$$

(11.12)

We notice that eqs. (11.10) and (11.12) are, respectively, the wave equations for voltage and current similar in form to the wave equations obtained for plane waves in eqs. (10.17) and (10.19). Thus, in our usual notations, γ in eq. (11.11) is the propagation constant (in per meter), α is the attenuation constant (in nepers per meter or decibels per meter),[2] and β is the phase constant (in radians per meter). The wavelength λ and wave velocity u are, respectively, given by

$$\lambda = \frac{2\pi}{\beta}$$

(11.13)

$$\boxed{u = \frac{\omega}{\beta} = f\lambda}$$

(11.14)

The solutions of the linear homogeneous differential equations (11.10) and (11.12) are similar to Case 2 of Example 6.5, namely,

$$V_s(z) = \underset{\longrightarrow \ +z}{V_o^+ e^{-\gamma z}} + \underset{-z \ \longleftarrow}{V_o^- e^{\gamma z}}$$

(11.15)

and

$$I_s(z) = \underset{\longrightarrow \ +z}{I_o^+ e^{-\gamma z}} + \underset{-z \ \longleftarrow}{I_o^- e^{\gamma z}}$$

(11.16)

where V_o^+, V_o^-, I_o^+, and I_o^- are wave amplitudes; the $+$ and $-$ signs, respectively, denote waves traveling along $+z$- and $-z$-directions, as is also indicated by the arrows. We obtain the instantaneous expression for voltage as

$$V(z, t) = \text{Re}[V_s(z) e^{j\omega t}]$$
$$= V_o^+ e^{-\alpha z} \cos(\omega t - \beta z) + V_o^- e^{\alpha z} \cos(\omega t + \beta z)$$

(11.17)

> The **characteristic impedance** Z_o of the line is the ratio of the positively traveling voltage wave to the current wave at any point on the line.

---

[2]Recall from eq. (10.35) that 1 Np $= 8.686$ dB.

The characteristic impedance Z_o is analogous to η, the intrinsic impedance of the medium of wave propagation. By substituting eqs. (11.15) and (11.16) into eqs. (11.8) and (11.9) and equating coefficients of terms $e^{\gamma z}$ and $e^{-\gamma z}$, we obtain

$$Z_o = \frac{V_o^+}{I_o^+} = -\frac{V_o^-}{I_o^-} = \frac{R + j\omega L}{\gamma} = \frac{\gamma}{G + j\omega C} \tag{11.18}$$

or

$$\boxed{Z_o = \sqrt{\frac{R + j\omega L}{G + j\omega C}} = R_o + jX_o} \tag{11.19}$$

where R_o and X_o are the real and imaginary parts of Z_o. Do not mistake R_o for R—while R is in ohms per meter, R_o is in ohms. The propagation constant γ and the characteristic impedance Z_o are important properties of the line because both depend on the line parameters R, L, G, and C and the frequency of operation. The reciprocal of Z_o is the characteristic admittance Y_o, that is, $Y_o = 1/Z_o$.

The transmission line considered thus far in this section is the *lossy* type in that the conductors comprising the line are imperfect ($\sigma_c \neq \infty$) and the dielectric in which the conductors are embedded is lossy ($\sigma \neq 0$). Having considered this general case, we may now consider two special cases: the lossless transmission line and the distortionless line.

A. Lossless Line ($R = 0 = G$)

A transmission line is said to be **lossless** if the conductors of the line are perfect ($\sigma_c \approx \infty$) and the dielectric medium separating them is lossless ($\sigma \approx 0$).

For such a line, it is evident from Table 11.1 that when $\sigma_c \simeq \infty$ and $\sigma \simeq 0$,

$$\boxed{R = 0 = G} \tag{11.20}$$

This is a necessary condition for a line to be lossless. Thus for such a line, eq. (11.20) forces eqs. (11.11), (11.14), and (11.19) to become

$$\alpha = 0, \quad \gamma = j\beta = j\omega\sqrt{LC} \tag{11.21a}$$

$$u = \frac{\omega}{\beta} = \frac{1}{\sqrt{LC}} = f\lambda \tag{11.21b}$$

$$X_o = 0, \quad Z_o = R_o = \sqrt{\frac{L}{C}} \tag{11.21c}$$

B. Distortionless Line ($R/L = G/C$)

A signal normally consists of a band of frequencies; wave amplitudes of different frequency components will be attenuated differently in a lossy line because α is frequency dependent. Since, in general, the phase velocity of each frequency component is also frequency dependent, this will result in distortion.

> A **distortionless line** is one in which the attenuation constant α is frequency independent while the phase constant β is linearly dependent on frequency.

From the general expression for α and β [in eq. (11.11)], a distortionless line results if the line parameters are such that

$$\boxed{\frac{R}{L} = \frac{G}{C}} \tag{11.22}$$

Thus, for a distortionless line,

$$\gamma = \sqrt{RG\left(1 + \frac{j\omega L}{R}\right)\left(1 + \frac{j\omega C}{G}\right)}$$

$$= \sqrt{RG}\left(1 + \frac{j\omega C}{G}\right) = \alpha + j\beta$$

or

$$\alpha = \sqrt{RG}, \quad \beta = \omega\sqrt{LC} \tag{11.23a}$$

showing that α does not depend on frequency, whereas β is a linear function of frequency. Also

$$Z_0 = \sqrt{\frac{R(1 + j\omega L/R)}{G(1 + j\omega C/G)}} = \sqrt{\frac{R}{G}} = \sqrt{\frac{L}{C}} = R_0 + jX_0$$

or

$$R_0 = \sqrt{\frac{R}{G}} = \sqrt{\frac{L}{C}}, \quad X_0 = 0 \tag{11.23b}$$

and

$$u = \frac{\omega}{\beta} = \frac{1}{\sqrt{LC}} = f\lambda \tag{11.23c}$$

Note the following important properties.

1. The phase velocity is independent of frequency because the phase constant β linearly depends on frequency. We have shape distortion of signals unless α and u are independent of frequency.

TABLE 11.2 Transmission Line Characteristics

Case	Propagation Constant $\gamma = \alpha + j\beta$	Characteristic Impedance $Z_o = R_o + jX_o$
General	$\sqrt{(R + j\omega L)(G + j\omega C)}$	$\sqrt{\dfrac{R + j\omega L}{G + j\omega C}}$
Lossless	$0 + j\omega\sqrt{LC}$	$\sqrt{\dfrac{L}{C}} + j0$
Distortionless	$\sqrt{RG} + j\omega\sqrt{LC}$	$\sqrt{\dfrac{L}{C}} + j0$

2. Both u and Z_o remain the same as for lossless lines.
3. A lossless line is also a distortionless line, but a distortionless line is not necessarily lossless. Although lossless lines are desirable in power transmission, telephone lines are required to be distortionless.

A summary of our discussion in this section is in Table 11.2. For the greater part of our analysis, we shall restrict our discussion to lossless transmission lines.

EXAMPLE 11.1

An air line has a characteristic impedance of 70 Ω and a phase constant of 3 rad/m at 100 MHz. Calculate the inductance per meter and the capacitance per meter of the line.

Solution:

An air line can be regarded as a lossless line because $\sigma \simeq 0$ and $\sigma_c \to \infty$. Hence

$$R = 0 = G \quad \text{and} \quad \alpha = 0$$

$$Z_o = R_o = \sqrt{\frac{L}{C}} \tag{11.1.1}$$

$$\beta = \omega\sqrt{LC} \tag{11.1.2}$$

Dividing eq. (11.1.1) by eq. (11.1.2) yields

$$\frac{R_o}{\beta} = \frac{1}{\omega C}$$

or

$$C = \frac{\beta}{\omega R_o} = \frac{3}{2\pi \times 100 \times 10^6 (70)} = 68.2 \text{ pF/m}$$

From eq. (11.1.1),

$$L = R_o^2 C = (70)^2 (68.2 \times 10^{-12}) = 334.2 \text{ nH/m}$$

EXAMPLE 11.2

A distortionless line has $Z_o = 60\ \Omega$, $\alpha = 20$ mNp/m, $u = 0.6c$, where c is the speed of light in a vacuum. Find R, L, G, C, and λ at 100 MHz.

Solution:

For a distortionless line,

$$RC = GL \quad \text{or} \quad G = \frac{RC}{L}$$

and hence

$$Z_o = \sqrt{\frac{L}{C}} \tag{11.2.1}$$

$$\alpha = \sqrt{RG} = R\sqrt{\frac{C}{L}} = \frac{R}{Z_o} \tag{11.2.2a}$$

or

$$R = \alpha Z_o \tag{11.2.2b}$$

But

$$u = \frac{\omega}{\beta} = \frac{1}{\sqrt{LC}} \tag{11.2.3}$$

From eq. (11.2.2b),

$$R = \alpha Z_o = (20 \times 10^{-3})(60) = 1.2\ \Omega/\text{m}$$

Dividing eq. (11.2.1) by eq. (11.2.3) results in

$$L = \frac{Z_o}{u} = \frac{60}{0.6\,(3 \times 10^8)} = 333\ \text{nH/m}$$

From eq. (11.2.2a),

$$G = \frac{\alpha^2}{R} = \frac{400 \times 10^{-6}}{1.2} = 333\ \mu\text{S/m}$$

Multiplying eqs. (11.2.1) and (11.2.3) together gives

$$uZ_o = \frac{1}{C}$$

or

$$C = \frac{1}{uZ_o} = \frac{1}{0.6\,(3 \times 10^8)\,60} = 92.59 \text{ pF/m}$$

$$\lambda = \frac{u}{f} = \frac{0.6\,(3 \times 10^8)}{10^8} = 1.8 \text{ m}$$

PRACTICE EXERCISE 11.2

A telephone line has $R = 30\ \Omega$/km, $L = 100$ mH/km, $G = 0$, and $C = 20\ \mu$F/km. At $f = 1$ kHz, obtain:

(a) The characteristic impedance of the line
(b) The propagation constant
(c) The phase velocity

Answer: (a) $70.75\underline{/-1.367°}\ \Omega$, (b) $2.121 \times 10^{-4} + j8.888 \times 10^{-3}$/m,
(c) 7.069×10^5 m/s.

11.4 INPUT IMPEDANCE, STANDING WAVE RATIO, AND POWER

Consider a transmission line of length ℓ, characterized by γ and Z_o, connected to a load Z_L as shown in Figure 11.6(a). Looking into the line, the generator sees the line with the load as an input impedance Z_{in}. It is our intention in this section to determine the input impedance, the standing wave ratio (SWR), and the power flow on the line.

Let the transmission line extend from $z = 0$ at the generator to $z = \ell$ at the load. First of all, we need the voltage and current waves in eqs. (11.15) and (11.16), that is,

$$V_s(z) = V_o^+ e^{-\gamma z} + V_o^- e^{\gamma z} \tag{11.24}$$

$$I_s(z) = \frac{V_o^+}{Z_o} e^{-\gamma z} - \frac{V_o^-}{Z_o} e^{\gamma z} \tag{11.25}$$

where eq. (11.18) has been incorporated. To find V_o^+ and V_o^-, the terminal conditions must be given. For example, if we are given the conditions at the input, say

$$V_o = V(z = 0), \quad I_o = I(z = 0) \tag{11.26}$$

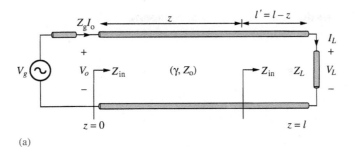

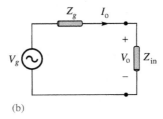

(b)

FIGURE 11.6 (a) Input impedance due to a line terminated by a load. (b) Equivalent circuit for finding V_o and I_o in terms of Z_{in} at the input.

substituting these into eqs. (11.24) and (11.25) results in

$$V_o^+ = \frac{1}{2}(V_o + Z_o I_o)$$ (11.27a)

$$V_o^- = \frac{1}{2}(V_o - Z_o I_o)$$ (11.27b)

If the input impedance at the input terminals is Z_{in}, the input voltage V_o and the input current I_o are easily obtained from Figure 11.6(b) as

$$V_o = \frac{Z_{in}}{Z_{in} + Z_g} V_g, \quad I_o = \frac{V_g}{Z_{in} + Z_g}$$ (11.28)

On the other hand, if we are given the conditions at the load, say

$$V_L = V(z = \ell), \quad I_L = I(z = \ell)$$ (11.29)

Substituting these into eqs. (11.24) and (11.25) gives

$$V_o^+ = \frac{1}{2}(V_L + Z_o I_L)e^{\gamma \ell}$$ (11.30a)

$$V_o^- = \frac{1}{2}(V_L - Z_o I_L)e^{-\gamma \ell}$$ (11.30b)

Next, we determine the input impedance $Z_{in} = V_s(z)/I_s(z)$ at any point on the line. At the generator, for example, eqs. (11.24) and (11.25) yield

$$Z_{in} = \frac{V_s(z)}{I_s(z)} = \frac{Z_o(V_o^+ + V_o^-)}{V_o^+ - V_o^-} \tag{11.31}$$

Substituting eq. (11.30) into (11.31) and utilizing the fact that

$$\frac{e^{\gamma\ell} + e^{-\gamma\ell}}{2} = \cosh\gamma\ell, \quad \frac{e^{\gamma\ell} - e^{-\gamma\ell}}{2} = \sinh\gamma\ell \tag{11.32a}$$

or

$$\tanh\gamma\ell = \frac{\sinh\gamma\ell}{\cosh\gamma\ell} = \frac{e^{\gamma\ell} - e^{-\gamma\ell}}{e^{\gamma\ell} + e^{-\gamma\ell}} \tag{11.32b}$$

we get

$$\boxed{Z_{in} = Z_o\left[\frac{Z_L + Z_o\tanh\gamma\ell}{Z_o + Z_L\tanh\gamma\ell}\right]} \quad \text{(lossy)} \tag{11.33}$$

Although eq. (11.33) has been derived for the input impedance Z_{in} at the generation end, it is a general expression for finding Z_{in} at any point on the line. To find Z_{in} at a distance ℓ' from the load as in Figure 11.6(a), we replace ℓ by ℓ'. A formula for calculating the hyperbolic tangent of a complex number, required in eq. (11.33), is found in Appendix A.3.

For a lossless line, $\gamma = j\beta$, $\tanh j\beta\ell = j\tan\beta\ell$, and $Z_o = R_o$, so eq. (11.33) becomes

$$\boxed{Z_{in} = Z_o\left[\frac{Z_L + jZ_o\tan\beta\ell}{Z_o + jZ_L\tan\beta\ell}\right]} \quad \text{(lossless)} \tag{11.34}$$

showing that the input impedance varies periodically with distance ℓ from the load. The quantity $\beta\ell$ in eq. (11.34) is usually referred to as the *electrical length* of the line and can be expressed in degrees or radians.

We now define Γ_L as the *voltage reflection coefficient* (at the load). The reflection coefficient Γ_L is the ratio of the voltage reflection wave to the incident wave at the load; that is,

$$\Gamma_L = \frac{V_o^- e^{\gamma\ell}}{V_o^+ e^{-\gamma\ell}} \tag{11.35}$$

Substituting V_o^- and V_o^+ in eq. (11.30) into eq. (11.35) and incorporating $V_L = Z_L I_L$ gives

$$\boxed{\Gamma_L = \frac{Z_L - Z_o}{Z_L + Z_o}} \tag{11.36}$$

The **voltage reflection coefficient** at any point on the line is the ratio of the reflected voltage wave to that of the incident wave.

That is,

$$\Gamma(z) = \frac{V_o^- e^{\gamma z}}{V_o^+ e^{-\gamma z}} = \frac{V_o^-}{V_o^+} e^{2\gamma z}$$

But $z = \ell - \ell'$. Substituting and combining with eq. (11.35), we get

$$\Gamma(z) = \frac{V_o^-}{V_o^+} e^{2\gamma \ell} e^{-2\gamma \ell'} = \Gamma_L e^{-2\gamma \ell'} \qquad (11.37)$$

The **current reflection coefficient** at any point on the line is the negative of the voltage reflection coefficient at that point.

Thus, the current reflection coefficient at the load is $I_o^- e^{\gamma \ell}/I_o^+ e^{-\gamma \ell} = -\Gamma_L$.

Just as we did for plane waves, we define the *standing wave ratio s* (otherwise denoted by SWR) as

$$s = \frac{V_{max}}{V_{min}} = \frac{I_{max}}{I_{min}} = \frac{1 + |\Gamma_L|}{1 - |\Gamma_L|} \qquad (11.38a)$$

$$|\Gamma_L| = \frac{S - 1}{S + 1} \qquad (11.38b)$$

It is easy to show that $I_{max} = V_{max}/Z_o$ and $I_{min} = V_{min}/Z_o$. The input impedance Z_{in} in eq. (11.34) has maxima and minima that occur, respectively, at the maxima and minima of the voltage standing wave. It can also be shown that

$$|Z_{in}|_{max} = \frac{V_{max}}{I_{min}} = sZ_o \qquad (11.39a)$$

and

$$|Z_{in}|_{min} = \frac{V_{min}}{I_{max}} = \frac{Z_o}{s} \qquad (11.39b)$$

As a way of demonstrating these concepts, consider a lossless line with characteristic impedance of $Z_o = 50\ \Omega$. For the sake of simplicity, we assume that the line is terminated in a pure resistive load $Z_L = 100\ \Omega$ and the voltage at the load is 100 V (rms). The conditions on the line are displayed in Figure 11.7. Note from Figure 11.7 that conditions on the line repeat themselves every half-wavelength.

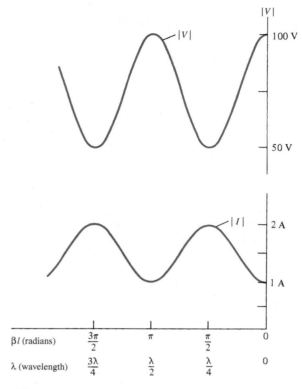

FIGURE 11.7 Voltage and current standing wave patterns on a lossless line terminated by a resistive load.

As mentioned at the beginning of this chapter, a transmission line is used in transferring power from the source to the load. The average input power at a distance ℓ from the load is given by an equation similar to eq. (10.78); that is,

$$P_{\text{ave}} = \frac{1}{2} \text{Re}\left[V_s(\ell)I_s^*(\ell)\right]$$

where the factor $\frac{1}{2}$ is needed because we are dealing with the peak values instead of the rms values. Assuming a lossless line, we substitute eqs. (11.24) and (11.25) to obtain

$$P_{\text{ave}} = \frac{1}{2} \text{Re}\left[V_o^+(e^{j\beta\ell} + \Gamma e^{-j\beta\ell}) \frac{V^{+*}}{Z_o}(e^{-j\beta\ell} - \Gamma^* e^{j\beta\ell})\right]$$

$$= \frac{1}{2} \text{Re}\left[\frac{|V_o^+|^2}{Z_o}(1 - |\Gamma|^2 + \Gamma e^{-2j\beta\ell} - \Gamma^* e^{2j\beta\ell})\right]$$

Since the last two terms together become purely imaginary, we have

$$\boxed{P_{\text{ave}} = \frac{|V_o^+|^2}{2Z_o}(1 - |\Gamma|^2)} \tag{11.40}$$

The first term is the incident power P_i, while the second term is the reflected power P_r. Thus eq. (11.40) may be written as

$$P_t = P_i - P_r$$

where P_t is the input or transmitted power and the negative sign is due to the negative-going wave (since we take the reference direction as that of the voltage/current traveling toward the right). We should notice from eq. (11.40) that the power is constant and does not depend on ℓ, since it is a lossless line. Also, we should notice that maximum power is delivered to the load when $\Gamma = 0$, as expected.

We now consider special cases when the line is connected to load $Z_L = 0$, $Z_L = \infty$, and $Z_L = Z_o$. These special cases can easily be derived from the general case.

A. Shorted Line (Z_L = 0)

For this case, eq. (11.34) becomes

$$Z_{sc} = Z_{in}\Big|_{Z_L=0} = jZ_o \tan \beta\ell \qquad (11.41a)$$

Also, from eqs. (11.36) and (11.38)

$$\Gamma_L = -1, \quad s = \infty \qquad (11.41b)$$

We notice from eq. (11.41a) that Z_{in} is a pure reactance, which could be capacitive or inductive depending on the value of ℓ. The variation of Z_{in} with ℓ is shown in Figure 11.8(a).

B. Open-Circuited Line (Z_L = ∞)

In this case, eq. (11.34) becomes

$$Z_{oc} = \lim_{Z_L \to \infty} Z_{in} = \frac{Z_o}{j \tan \beta\ell} = -jZ_o \cot \beta\ell \qquad (11.42a)$$

and from eqs. (11.36) and (11.38),

$$\Gamma_L = 1, \quad s = \infty \qquad (11.42b)$$

The variation of Z_{in} with ℓ is shown in Figure 11.8(b). Notice from eqs. (11.41a) and (11.42a) that

$$Z_{sc}Z_{oc} = Z_o^2 \qquad (11.43)$$

C. Matched Line (Z_L = Z_o)

The most desired case from the practical point of view is the matched line i.e., $Z_L = Z_o$. For this case, eq. (11.34) reduces to

$$Z_{in} = Z_o \qquad (11.44a)$$

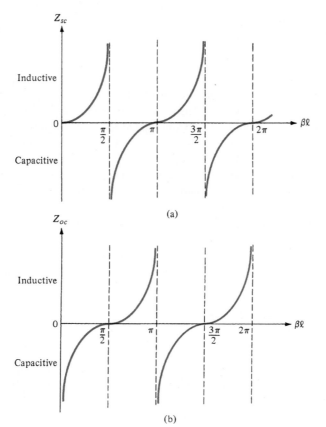

and from eqs. (11.36) and (11.38),

$$\Gamma_L = 0, \quad s = 1 \tag{11.44b}$$

that is, $V_o^- = 0$; the whole wave is transmitted, and there is no reflection. The incident power is fully absorbed by the load. Thus maximum power transfer is possible when a transmission line is matched to the load.

EXAMPLE 11.3

A certain transmission line 2 m long operating at $\omega = 10^6$ rad/s has $\alpha = 8$ dB/m, $\beta = 1$ rad/m, and $Z_o = 60 + j40 \ \Omega$. If the line is connected to a source of $10\underline{/0°}$ V, $Z_g = 40 \ \Omega$ and terminated by a load of $20 + j50 \ \Omega$, determine

(a) The input impedance
(b) The sending-end current
(c) The current at the middle of the line

Solution:

(a) Since 1 Np = 8.686 dB,

$$\alpha = \frac{8}{8.686} = 0.921 \text{ Np/m}$$

$$\gamma = \alpha + j\beta = 0.921 + j1 \text{ /m}$$

$$\gamma\ell = 2(0.921 + j1) = 1.84 + j2$$

Using the formula for $\tanh(x + jy)$ in Appendix A.3, we obtain

$$\tanh \gamma\ell = 1.033 - j0.03929$$

$$Z_{in} = Z_o\left(\frac{Z_L + Z_o \tanh \gamma\ell}{Z_o + Z_L \tanh \gamma\ell}\right)$$

$$= (60 + j40)\left[\frac{20 + j50 + (60 + j40)(1.033 - j0.03929)}{60 + j40 + (20 + j50)(1.033 - j0.03929)}\right]$$

$$Z_{in} = 60.25 + j38.79 \text{ } \Omega$$

(b) The sending-end current is $I(z = 0) = I_o$. From eq. (11.28),

$$I(z = 0) = \frac{V_g}{Z_{in} + Z_g} = \frac{10}{60.25 + j38.79 + 40}$$

$$= 93.03\underline{/-21.15°} \text{ mA}$$

(c) To find the current at any point, we need V_o^+ and V_o^-. But

$$I_o = I(z = 0) = 93.03\underline{/-21.15°} \text{ mA}$$

$$V_o = Z_{in}I_o = (71.66\underline{/32.77°})(0.09303\underline{/-21.15°}) = 6.667\underline{/11.62°} \text{ V}$$

From eq. (11.27),

$$V_o^+ = \frac{1}{2}(V_o + Z_oI_o)$$

$$= \frac{1}{2}[6.667\underline{/11.62°} + (60 + j40)(0.09303\underline{/-21.15°})] = 6.687\underline{/12.08°}$$

$$V_o^- = \frac{1}{2}(V_o - Z_oI_o) = 0.0518\underline{/260°}$$

At the middle of the line, $z = \ell/2$, $\gamma z = 0.921 + j1$. Hence, the current at this point is

$$I_s(z = \ell/2) = \frac{V_o^+}{Z_o}e^{-\gamma z} - \frac{V_o^-}{Z_o}e^{\gamma z}$$

$$= \frac{(6.687e^{j12.08°})e^{-0.921-j1}}{60 + j40} - \frac{(0.0518e^{j260°})e^{0.921+j1}}{60 + j40}$$

Note that $j1$ is in radians and is equivalent to $j57.3°$. Thus,

$$I_s(z = \ell/2) = \frac{6.687e^{j12.08°}e^{-0.921}e^{-j57.3°}}{72.1e^{j33.69°}} - \frac{0.0518e^{j260°}e^{0.921}e^{j57.3°}}{72.1e^{33.69°}}$$

$$= 0.0369e^{-j78.91°} - 0.001805e^{j283.61°}$$

$$= 6.673 - j34.456 \text{ mA}$$

$$= 35.10\underline{/281°} \text{ mA}$$

PRACTICE EXERCISE 11.3

The transmission line shown in Figure 11.9 is 40 m long and has $V_g = 15\underline{/0°} \text{ V}_{rms}$, $Z_o = 30 + j60 \text{ }\Omega$, and $V_L = 5\underline{/-48°} \text{ V}_{rms}$. If the line is matched to the load and $Z_g = 0$, calculate:

(a) The input impedance Z_{in}
(b) The sending-end current I_{in} and voltage V_{in}
(c) The propagation constant γ

Answer: (a) $30 + j60 \text{ }\Omega$, (b) $0.2236\underline{/-63.43°} \text{ A}$, $7.5\underline{/0°} \text{ V}_{rms}$, (c) $0.0101 + j0.02094 \text{ /m}$.

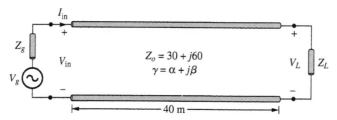

FIGURE 11.9 For Practice Exercise 11.3.

11.5 THE SMITH CHART

Prior to the advent of digital computers and calculators, engineers developed all sorts of aids (slide rules, tables, charts, graphs, etc.) to facilitate their calculations for design and analysis. To reduce the tedious manipulations involved in calculating the characteristics of transmission lines, graphical means were then developed. The Smith chart[3] is the most commonly

---

[3]Devised by Phillip H. Smith in 1939. See P. H. Smith, "Transmission line calculator," *Electronics*, vol. 12, pp. 29–31, 1939, and "An improved transmission line calculator," *Electronics*, vol. 17, pp. 130–133, 318–325, 1944.

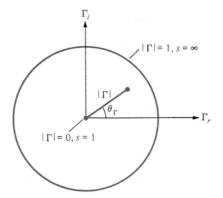

FIGURE 11.10 Unit circle on which the Smith chart is constructed.

used of the graphical techniques. It is basically a graphical indication of the impedance of a transmission line and of the corresponding reflection coefficient as one moves along the line. It becomes easy to use after a small amount of experience. We will first examine how the Smith chart is constructed and later employ it in our calculations of transmission line characteristics such as Γ_L, s, and Z_{in}. We will assume that the transmission line to which the Smith chart will be applied is lossless ($Z_o = R_o$), although this is not fundamentally required.

The Smith chart is constructed within a circle of unit radius ($|\Gamma| \leq 1$) as shown in Figure 11.10. The construction of the chart is based on the relation in eq. (11.36)[4]; that is,

$$\Gamma = \frac{Z_L - Z_o}{Z_L + Z_o} \tag{11.45}$$

or

$$\Gamma = |\Gamma| \underline{/\theta_\Gamma} = \Gamma_r + j\Gamma_i \tag{11.46}$$

where Γ_r and Γ_i are the real and imaginary parts of the reflection coefficient Γ.

Instead of having separate Smith charts for transmission lines with different characteristic impedances (e.g., $Z_o = 60, 100, 120 \ \Omega$), we prefer to have just one that can be used for any line. We achieve this by using a normalized chart in which all impedances are normalized with respect to the characteristic impedance Z_o of the particular line under consideration. For the load impedance Z_L, for example, the *normalized impedance* z_L is given by

$$z_L = \frac{Z_L}{Z_o} = r + jx \tag{11.47}$$

Substituting eq. (11.47) into eqs. (11.45) and (11.46) gives

$$\Gamma = \Gamma_r + j\Gamma_i = \frac{z_L - 1}{z_L + 1} \tag{11.48a}$$

---

[4] Whenever a subscript is not attached to Γ, we simply mean voltage reflection coefficient at the load ($\Gamma_L = \Gamma$).

or

$$z_L = r + jx = \frac{(1 + \Gamma_r) + j\Gamma_i}{(1 - \Gamma_r) - j\Gamma_i} \qquad (11.48b)$$

Normalizing and equating real and imaginary components, we obtain

$$r = \frac{1 - \Gamma_r^2 - \Gamma_i^2}{(1 - \Gamma_r)^2 + \Gamma_i^2} \qquad (11.49a)$$

$$x = \frac{2\,\Gamma_i}{(1 - \Gamma_r)^2 + \Gamma_i^2} \qquad (11.49b)$$

Rearranging terms in eqs. (11.49) leads to

$$\left[\Gamma_r - \frac{r}{1 + r}\right]^2 + \Gamma_i^2 = \left[\frac{1}{1 + r}\right]^2 \qquad (11.50)$$

and

$$[\Gamma_r - 1]^2 + \left[\Gamma_i - \frac{1}{x}\right]^2 = \left[\frac{1}{x}\right]^2 \qquad (11.51)$$

Each of eqs. (11.50) and (11.51) is similar to

$$(x - h)^2 + (y - k)^2 = a^2 \qquad (11.52)$$

which is the general equation of a circle of radius a, centered at (h, k). Thus eq. (11.50) is an *r-circle (resistance circle)* with

$$\text{center at } (\Gamma_r, \Gamma_i) = \left(\frac{r}{1 + r}, 0\right) \qquad (11.53a)$$

$$\text{radius} = \frac{1}{1 + r} \qquad (11.53b)$$

TABLE 11.3 Radii and Centers of *r*-Circles for Typical Values of *r*

Normalized Resistance (r)	Radius $\left(\dfrac{1}{1 + r}\right)$	Center $\left(\dfrac{r}{1 + r}, 0\right)$
0	1	(0, 0)
1/2	2/3	(1/3, 0)
1	1/2	(1/2, 0)
2	1/3	(2/3, 0)
5	1/6	(5/6, 0)
∞	0	(1, 0)

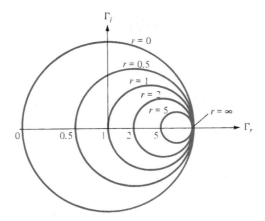

FIGURE 11.11 Typical r-circles for
$r = 0, 0.5, 1, 2, 5, \infty$.

For typical values of the normalized resistance r, the corresponding centers and radii of the r-circles are presented in Table 11.3. Typical examples of the r-circles based on the data in Table 11.3 are shown in Figure 11.11. Similarly, eq. (11.51) is an *x-circle (reactance circle)* with

$$\text{center at } (\Gamma_r, \Gamma_i) = \left(1, \frac{1}{x} \right) \tag{11.54a}$$

$$\text{radius} = \frac{1}{x} \tag{11.54b}$$

Table 11.4 presents centers and radii of the x-circles for typical values of x, and Figure 11.12 shows the corresponding plots. Notice that while r is always positive, x can be positive (for inductive impedance) or negative (for capacitive impedance).

If we superpose the r-circles and x-circles, what we have is the Smith chart shown in Figure 11.13. On the chart, we locate a normalized impedance $z = 2 + j$, for example, as the point of intersection of the $r = 2$ circle and the $x = 1$ circle. This is point P_1 in Figure 11.13. Similarly, $z = 1 - j0.5$ is located at P_2, where the $r = 1$ circle and the $x = -0.5$ circle intersect.

Apart from the r- and x-circles (shown on the Smith chart), we can draw the *s-circles* or *constant standing wave ratio circles* (always not shown on the Smith chart), which are

TABLE 11.4 Radii and Centers of x-Circles
for Typical Values of x

Normalized Reactance (x)	Radius $\left(\dfrac{1}{x} \right)$	Center $\left(1, \dfrac{1}{x} \right)$
0	∞	$(1, \infty)$
$\pm 1/2$	2	$(1, \pm 2)$
± 1	1	$(1, \pm 1)$
± 2	1/2	$(1, \pm 1/2)$
± 5	1/5	$(1, \pm 1/5)$
$\pm \infty$	0	$(1, 0)$

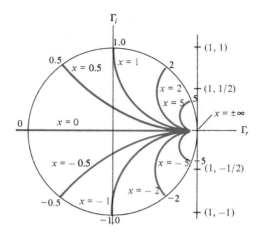

FIGURE 11.12 Typical x-circles for $x = 0, \pm 0.5, \pm 1, \pm 2, \pm 5, \pm\infty$.

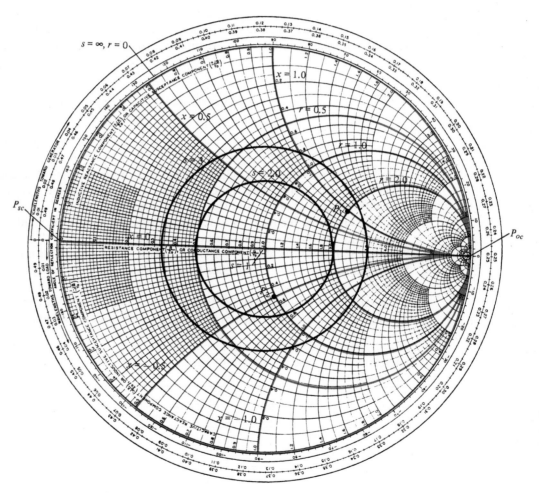

FIGURE 11.13 Illustration of the r-, x-, and s-circles on the Smith chart.

centered at the origin with s varying from 1 to ∞. The value of the standing wave ratio s is determined by locating where an s-circle crosses the Γ_r axis. Typical examples of s-circles for $s = 1, 2, 3,$ and ∞ are shown in Figure 11.13. Since $|\Gamma|$ and s are related according to eq. (11.38), the s-circles are sometimes referred to as $|\Gamma|$-circles, with $|\Gamma|$ varying linearly from 0 to 1 as we move away from the center O toward the periphery of the chart, while s varies nonlinearly from 1 to ∞.

The following points should be noted about the Smith chart.

1. At point P_{sc} on the chart $r = 0, x = 0$; that is, $Z_L = 0 + j0$, showing that P_{sc} represents a short circuit on the transmission line. At point $P_{oc}, r = \infty$ and $x = \infty$, or $Z_L = \infty + j\infty$, which implies that P_{oc} corresponds to an open circuit on the line. Also at $P_{oc}, r = 0$ and $x = 0$, showing that P_{oc} is another location of a short circuit on the line.

2. A complete revolution (360°) around the Smith chart represents a distance of $\lambda/2$ on the line. Clockwise movement on the chart is regarded as moving toward the generator (or away from the load) as shown by the arrow G in Figure 11.14(a) and (b). Similarly, counterclockwise movement on the chart corresponds to moving toward the load (or away from the generator) as indicated by the arrow L in Figure 11.14. Notice from Figure 11.14(b) that at the load, moving toward the load does not make sense (because we are already at the load). The same can be said of the case when we are at the generator end.

3. There are three scales around the periphery of the Smith chart as illustrated in Figure 11.14(a). The three scales are included for the sake of convenience but they are actually meant to serve the same purpose; one scale should be sufficient. The scales are used in determining the distance from the load or generator in degrees or wavelengths. The outermost scale is used to determine the distance on the line from the generator end in terms of wavelengths, and the next scale determines the distance from the load end in terms of wavelengths. The innermost scale is a protractor (in degrees) and is primarily used in determining θ_Γ; it can also be used to determine the distance from the load or generator. Since a $\lambda/2$ distance on the line corresponds to a movement of 360° on the chart, λ *distance on the line corresponds to a 720° movement on the chart.*

$$\boxed{\lambda \rightarrow 720°}$$ (11.55)

Thus we may ignore the other outer scales and use the protractor (the innermost scale) for all our θ_Γ and distance calculations.

4. The voltage V_{max} occurs where $Z_{in, max}$ is located on the chart [see eq. (11.39a)], and that is on the positive Γ_r-axis or on OP_{oc} in Figure 11.14(a). The voltage V_{min} is located at the same point where we have $Z_{in, min}$ on the chart, that is, on the negative Γ_r-axis or on OP_{sc} in Figure 11.14(a). Notice that V_{max} and V_{min} (or $Z_{in, max}$ and $Z_{in, min}$) are $\lambda/4$ (or 180°) apart.

5. The Smith chart is used both as impedance chart and admittance chart ($Y = 1/Z$). As admittance chart (normalized admittance $y = Y/Y_o = g + jb$), the g- and b-circles correspond to r- and x-circles, respectively.

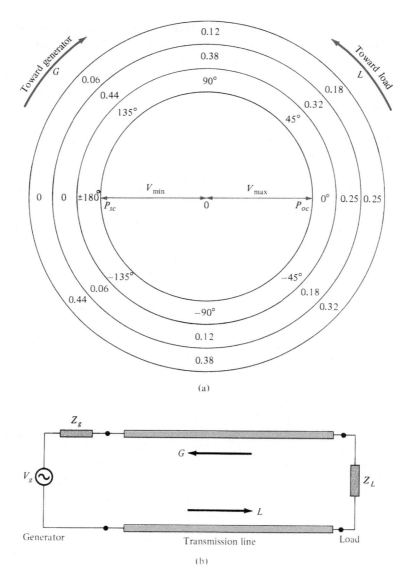

FIGURE 11.14 **(a)** Smith chart illustrating scales around the periphery and movements around the chart. **(b)** Corresponding movements along the transmission line.

Based on these important properties, the Smith chart may be used to determine, among other things, (a) $\Gamma = |\Gamma|\underline{/\theta_\Gamma}$ and s; (b) Z_{in} or Y_{in}; and (c) the locations of V_{max} and V_{min} provided that we are given Z_o, Z_L, λ, and the length of the line. Some examples will clearly show how we can find all these and much more with the aid of the Smith chart, a compass, and a plain straightedge. A complete Smith chart is available in Appendix D. You may copy this.

EXAMPLE 11.4

A lossless transmission line with $Z_o = 50\ \Omega$ is 30 m long and operates at 2 MHz. The line is terminated with a load $Z_L = 60 + j40\ \Omega$. If $u = 0.6c$ on the line, find

(a) The reflection coefficient Γ

(b) The standing wave ratio s

(c) The input impedance

Solution:

This problem will be solved with and without using the Smith chart.

Method 1 (without the Smith chart):

(a) $\Gamma = \dfrac{Z_L - Z_o}{Z_L + Z_o} = \dfrac{60 + j40 - 50}{50 + j40 + 50} = \dfrac{10 + j40}{110 + j40}$

$= 0.3523\underline{/56°}$

(b) $s = \dfrac{1 + |\Gamma|}{1 - |\Gamma|} = \dfrac{1 + 0.3523}{1 - 0.3523} = 2.088$

(c) Since $u = \omega/\beta$, or $\beta = \omega/u$,

$$\beta\ell = \frac{\omega\ell}{u} = \frac{2\pi(2 \times 10^6)(30)}{0.6(3 \times 10^8)} = \frac{2\pi}{3} = 120°$$

Note that $\beta\ell$ is the electrical length of the line.

$$Z_{in} = Z_o\left[\frac{Z_L + jZ_o \tan \beta\ell}{Z_o + jZ_L \tan \beta\ell}\right]$$

$$= \frac{50(60 + j40 + j50 \tan 120°)}{[50 + j(60 + j40) \tan 120°]}$$

$$= \frac{50(6 + j4 - j5\sqrt{3})}{(5 + 4\sqrt{3} - j6\sqrt{3})} = 24.01\underline{/3.22°}$$

$$= 23.97 + j1.35\ \Omega$$

Method 2 (using the Smith chart):

(a) Calculate the normalized load impedance

$$z_L = \frac{Z_L}{Z_o} = \frac{60 + j40}{50}$$

$$= 1.2 + j0.8$$

Locate z_L on the Smith chart of Figure 11.15 at point P, where the $r = 1.2$ circle and the $x = 0.8$ circle meet. To get Γ at z_L, extend OP to meet the $r = 0$ circle at Q and measure OP and OQ. Since OQ corresponds to $|\Gamma| = 1$, then at P,

$$|\Gamma| = \frac{OP}{OQ} = \frac{3.2\ \text{cm}}{9.1\ \text{cm}} = 0.3516$$

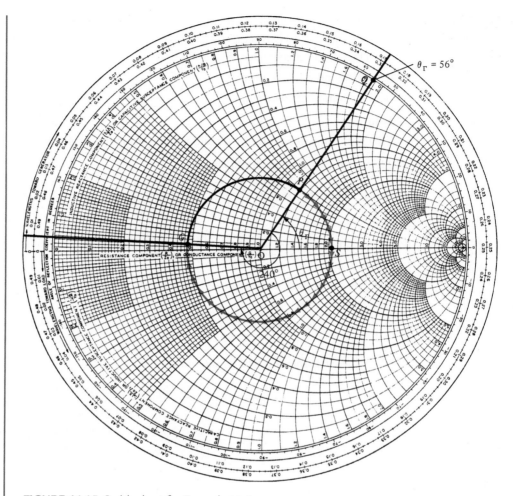

FIGURE 11.15 Smith chart for Example 11.4.

Note that $OP = 3.2$ cm and $OQ = 9.1$ cm were taken from the Smith chart used by the author; the Smith chart in Figure 11.15 is reduced, but the ratio of OP/OQ remains the same.

Angle θ_Γ is read directly on the chart as the angle between OS and OP; that is,

$$\theta_\Gamma = \text{angle } POS = 56°$$

Thus

$$\Gamma = 0.3516 \underline{/56°}$$

(b) To obtain the standing wave ratio s, draw a circle with radius OP and center at O. This is the constant s or $|\Gamma|$ circle. Locate point S where the s-circle meets the Γ_r-axis. [This is easily shown by setting $\Gamma_i = 0$ in eq. (11.49a).] The value of r at this point is s; that is,

$$s = r(\text{for } r \geq 1)$$
$$= 2.1$$

(c) To obtain Z_{in}, first express ℓ in terms of λ or in degrees:

$$\lambda = \frac{u}{f} = \frac{0.6(3 \times 10^8)}{2 \times 10^6} = 90 \text{ m}$$

$$\ell = 30 \text{ m} = \frac{30}{90}\lambda = \frac{\lambda}{3} \rightarrow \frac{720°}{3} = 240°$$

Since λ corresponds to an angular movement of $720°$ on the chart, the length of the line corresponds to an angular movement of $240°$. That means we move toward the generator (or away from the load, in the clockwise direction) $240°$ on the s-circle from point P to point G. At G, we obtain

$$z_{in} = 0.47 + j0.03$$

Hence

$$Z_{in} = Z_o z_{in} = 50(0.47 + j0.03) = 23.5 + j1.5 \ \Omega$$

Although the results obtained using the Smith chart are only approximate, for engineering purposes they are close enough to the exact ones obtained by Method 1. However, an inexpensive modern calculator can handle the complex algebra in less time and with much less effort than are needed to use the Smith chart. The value of the Smith chart is that it allows us to observe the variation of Z_{in} with ℓ.

PRACTICE EXERCISE 11.4

A 70 Ω lossless line has $s = 1.6$ and $\theta_\Gamma = 300°$. If the line is 0.6λ long, obtain

(a) Γ, Z_L, Z_{in}
(b) The distance of the first minimum voltage from the load

Answer: (a) $0.228 \ \underline{/300°}$, $80.5 - j33.6 \ \Omega$, $47.6 - j17.5 \ \Omega$, (b) $\lambda/6$.

EXAMPLE 11.5

A load of $100 + j150 \ \Omega$ is connected to a 75 Ω lossless line. Find:

(a) Γ
(b) s
(c) The load admittance Y_L
(d) Z_{in} at 0.4λ from the load
(e) The locations of V_{max} and V_{min} with respect to the load if the line is 0.6λ long
(f) Z_{in} at the generator.

Solution:

(a) We can use the Smith chart to solve this problem. The normalized load impedance is

$$z_L = \frac{Z_L}{Z_o} = \frac{100 + j150}{75} = 1.33 + j2$$

We locate this at point P on the Smith chart of Figure 11.16. At P, we obtain

$$|\Gamma| = \frac{OP}{OQ} = \frac{6 \text{ cm}}{9.1 \text{ cm}} = 0.659$$

$$\theta_\Gamma = \text{angle } POS = 40°$$

Hence,

$$\Gamma = 0.659 \ \underline{/40°}$$

Check:

$$\Gamma = \frac{Z_L - Z_o}{Z_L + Z_o} = \frac{100 + j150 - 75}{100 + j150 + 75}$$

$$= 0.6598 \ \underline{/39.94°}$$

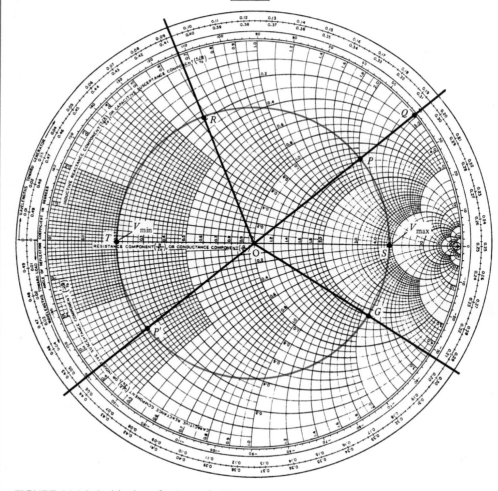

FIGURE 11.16 Smith chart for Example 11.5.

(b) Draw the constant s-circle passing through P and obtain

$$s = 4.82$$

Check:

$$s = \frac{1 + |\Gamma|}{1 - |\Gamma|} = \frac{1 + 0.659}{1 - 0.659} = 4.865$$

(c) To obtain Y_L, extend PO to POP' and note point P' where the constant s-circle meets POP'. At P', obtain

$$y_L = 0.228 - j0.35$$

The load admittance is

$$Y_L = Y_o y_L = \frac{1}{75}(0.228 - j0.35) = 3.04 - j4.67 \text{ mS}$$

Check:

$$Y_L = \frac{1}{Z_L} = \frac{1}{100 + j150} = 3.07 - j4.62 \text{ mS}$$

(d) The 0.4λ corresponds to an angular movement of $0.4 \times 720° = 288°$ on the constant s-circle. From P, we move $288°$ toward the generator (clockwise) on the s-circle to reach point R. At R,

$$z_{\text{in}} = 0.3 + j0.63$$

Hence

$$Z_{\text{in}} = Z_o z_{\text{in}} = 75(0.3 + j0.63)$$
$$= 22.5 + j47.25 \ \Omega$$

Check:

$$\beta\ell = \frac{2\pi}{\lambda}(0.4\lambda) = 360°\,(0.4) = 144°$$

$$Z_{\text{in}} = Z_o \left[\frac{Z_L + jZ_o \tan \beta\ell}{Z_o + jZ_L \tan \beta\ell}\right]$$

$$= \frac{75(100 + j150 + j75 \tan 144°)}{[75 + j(100 + j150) \tan 144°]}$$

$$= 54.41\underline{/65.25°}$$

or

$$Z_{in} = 21.9 + j47.6 \ \Omega$$

(e) The 0.6λ corresponds to an angular movement of

$$0.6 \times 720° = 432° = 1 \text{ revolution} + 72°$$

Thus, we start from P (load end), move clockwise along the s-circle 432°, or one revolution plus 72°, and reach the generator at point G. Note that to reach G from P, we have passed through point T (location of V_{min}) once and point S (location of V_{max}) twice. Thus, from the load,

$$\text{1st } V_{max} \text{ is located at } \frac{40°}{720°}\lambda = 0.055\lambda$$

$$\text{2nd } V_{max} \text{ is located at } 0.0555\lambda + \frac{\lambda}{2} = 0.555\lambda$$

and the only V_{min} is located at $0.055\lambda + \lambda/4 = 0.3055\lambda$
(f) At G (generator end),

$$z_{in} = 1.8 - j2.2$$

$$Z_{in} = 75(1.8 - j2.2) = 135 - j165 \ \Omega$$

This can be checked by using eq. (11.34), where $\beta\ell = \dfrac{2\pi}{\lambda}(0.6\lambda) = 216°$.

We can see how much time and effort are saved by using the Smith chart.

PRACTICE EXERCISE 11.5

A lossless 60 Ω line is terminated by a load of $60 + j60 \ \Omega$.

(a) Find Γ and s. If $Z_{in} = 120 - j60 \ \Omega$, how far (in terms of wavelengths) is the load from the generator? Solve this without using the Smith chart.

(b) Use the Smith chart to solve the problem in part (a). Calculate Z_{max} and $Z_{in, min}$. How far (in terms of λ) is the first maximum voltage from the load?

Answer: (a) 0.4472 $\underline{/63.43°}$, 2.618, $\dfrac{\lambda}{8}(1 + 4n)$, $n = 0, 1, 2, \ldots$, (b) 0.4457 $\underline{/62°}$, 2.612, $\dfrac{\lambda}{8}(1 + 4n)$, 157.1 Ω, 22.92 Ω, 0.0861 λ.

11.6 SOME APPLICATIONS OF TRANSMISSION LINES

Transmission lines are used to serve different purposes. Here we consider how transmission lines are used for load matching and impedance measurements.

A. Quarter-Wave Transformer (Matching)

When $Z_o \neq Z_L$, we say that the load is *mismatched* and a reflected wave exists on the line. However, for maximum power transfer, it is desired that the load be matched to the transmission line $(Z_o = Z_L)$ so that there is no reflection $(|\Gamma| = 0$ or $s = 1)$. The matching is achieved by using shorted sections of transmission lines.

We recall from eq. (11.34) that when $\ell = \lambda/4$ or $\beta\ell = (2\pi/\lambda)(\lambda/4) = \pi/2$,

$$Z_{in} = Z_o \left[\frac{Z_L + jZ_o \tan \pi/2}{Z_o + jZ_L \tan \pi/2} \right] = \frac{Z_o^2}{Z_L} \tag{11.56}$$

that is,

$$\frac{Z_{in}}{Z_o} = \frac{Z_o}{Z_L}$$

or

$$z_{in} = \frac{1}{z_L} \rightarrow y_{in} = z_L \tag{11.57}$$

Thus by adding a $\lambda/4$ line on the Smith chart, we obtain the input admittance corresponding to a given load impedance.

Also, a mismatched load Z_L can be properly matched to a line (with characteristic impedance Z_o) by inserting prior to the load a transmission line $\lambda/4$ long (with characteristic impedance Z_o') as shown in Figure 11.17. The $\lambda/4$ section of the transmission line is called a *quarter-wave transformer* because it is used for impedance matching like an ordinary transformer. From eq. (11.56), Z_o' is selected such that $(Z_{in} = Z_o)$

$$Z_o' = \sqrt{Z_o Z_L} \tag{11.58}$$

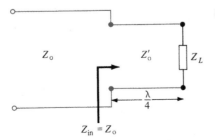

FIGURE 11.17 Load matching using a $\lambda/4$ transformer.

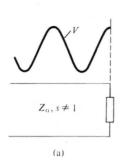
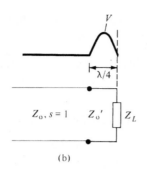

FIGURE 11.18 Voltage standing wave pattern of mismatched load: **(a)** without a λ/4 transformer, **(b)** with a λ/4 transformer.

where Z_o', Z_o and Z_L are all real. If, for example, a 120 Ω load is to be matched to a 75 Ω line, the quarter-wave transformer must have a characteristic impedance of $\sqrt{(75)(120)} \simeq$ 95 Ω. This 95 Ω quarter-wave transformer will also match a 75 Ω load to a 120 Ω line. The voltage standing wave patterns without and with the λ/4 transformer are shown in Figure 11.18(a) and (b), respectively. From Figure 11.18, we observe that although a standing wave still exists between the transformer and the load, there is no standing wave to the left of the transformer due to the matching. However, the reflected wave (or standing wave) is eliminated only at the desired wavelength (or frequency f); there will be reflection at a slightly different wavelength. Thus, the main disadvantage of the quarter-wave transformer is that it is a narrow-band or frequency-sensitive device.

B. Single-Stub Tuner (Matching)

The major drawback of using a quarter-wave transformer as a line-matching device is eliminated by using a *single-stub* tuner. The tuner consists of an open or shorted section of transmission line of length d connected in parallel with the main line at some distance ℓ from the load, as in Figure 11.19. Notice that the stub has the same characteristic impedance as the main line, although stubs may be designed with different values of Z_o. It is more difficult to use a series stub although it is theoretically feasible. An open-circuited stub radiates some energy at high frequencies. Consequently, shunt short-circuited parallel stubs are preferred.

Since we intend that $Z_{in} = Z_o$, that is, $z_{in} = 1$ or $y_{in} = 1$ at point A on the line, we first draw the locus $y = 1 + jb(r = 1$ circle$)$ on the Smith chart as shown in Figure 11.20. If a shunt stub of admittance $y_s = -jb$ is introduced at A, then

$$y_{in} = 1 + jb + y_s = 1 + jb - jb = 1 + j0 \tag{11.59}$$

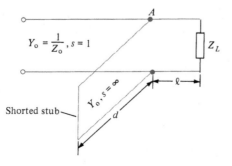

FIGURE 11.19 Matching with a single-stub tuner.

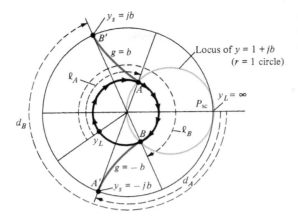

FIGURE 11.20 Using the Smith chart to determine ℓ and d of a shunt-shorted single-stub tuner.

as desired. Since b could be positive or negative, two possible values of $\ell(<\lambda/2)$ can be found on the line. At A, $y_s = -jb$, $\ell = \ell_A$ and at B, $y_s = jb$, $\ell = \ell_B$ as in Figure 11.20. Because the stub is shorted $(y'_L = \infty)$, we determine the length d of the stub by finding the distance from P_{sc} (at which $z'_L = 0 + j0$) to the required stub admittance y_s. For the stub at A, we obtain $d = d_A$ as the distance from P_{sc} to A', where A' corresponds to $y_s = -jb$ located on the periphery of the chart as in Figure 11.20. Similarly, we obtain $d = d_B$ as the distance from P_{sc} to $B'(y_s = jb)$.

Thus we obtain $d = d_A$ and $d = d_B$, corresponding to A and B, respectively, as shown in Figure 11.20. Note that $d_A + d_B = \lambda/2$ always. Since we have two possible shunted stubs, we normally choose to match the shorter stub or one at a position closer to the load. Instead of having a single stub shunted across the line, we may have two stubs. This arrangement, which is called *double-stub matching*, allows for the adjustment of the load impedance.

C. Slotted Line (Impedance Measurement)

At high frequencies, it is very difficult to measure current and voltage because measuring devices become significant in size and every circuit becomes a transmission line. The slotted line is a simple device used in determining the impedance of an unknown load at high frequencies up into the region of gigahertz. It consists of a section of an air (lossless) line with a slot in the outer conductor as shown in Figure 11.21. The line has a probe, along the E field (see Figure 11.4), which samples the E field and consequently measures the potential difference between the probe and its outer shield.

The slotted line is primarily used in conjunction with the Smith chart to determine the standing wave ratio s (the ratio of maximum voltage to the minimum voltage) and the load impedance Z_L. The value of s is read directly on the detection meter when the load is connected. To determine Z_L, we first replace the load by a short circuit and note the locations of voltage minima (which are more accurately determined than the maxima because of the sharpness of the turning point) on the scale. Since impedances repeat every half-wavelength, any of the minima may be selected as the load reference point. We now determine the distance from the selected reference point to the load by replacing the short circuit by the load and noting the locations of voltage minima. The distance ℓ (distance of

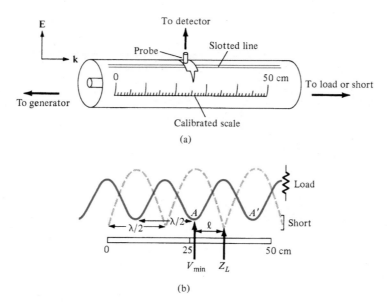

(a)

Load

Short

(b)

FIGURE 11.21 (a) Typical slotted line; **(b)** Determining the location of the load Z_L and V_{min} on the line.

V_{min} toward the load) expressed in terms of λ is used to locate the position of the load of an s-circle on the chart as shown in Figure 11.22.

The procedure for using the slotted line can be summarized as follows.

1. With the load connected, read s on the detection meter. With the value of s, draw the s-circle on the Smith chart.
2. With the load replaced by a short circuit, locate a reference position for Z_L at a voltage minimum point.
3. With the load on the line, note the position of V_{min} and determine ℓ.
4. On the Smith chart, move toward the load a distance ℓ from the location of V_{min}. Find Z_L at that point.

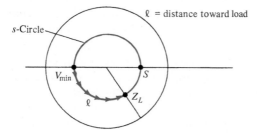

FIGURE 11.22 Determining the load impedance from the Smith chart by using the data obtained from the slotted line.

EXAMPLE 11.6

With an unknown load connected to a slotted air line, $s = 2$ is recorded by a standing wave indicator, and minima are found at 11 cm, 19 cm, . . . , on the scale. When the load is replaced by a short circuit, the minima are at 16 cm, 24 cm, ... If $Z_o = 50\ \Omega$, calculate $\lambda, f,$ and Z_L.

Solution:

Consider the standing wave patterns as in Figure 11.23(a). From this, we observe that

$$\frac{\lambda}{2} = 19 - 11 = 8\ \text{cm} \quad \text{or} \quad \lambda = 16\ \text{cm}$$

$$f = \frac{u}{\lambda} = \frac{3 \times 10^8}{16 \times 10^{-2}} = 1.875\ \text{GHz}$$

Electrically speaking, the load can be located at 16 cm or 24 cm. If we assume that the load is at 24 cm, the load is at a distance ℓ from $V_{\min}$, where

$$\ell = 24 - 19 = 5\ \text{cm} = \frac{5}{16}\lambda = 0.3125\ \lambda$$

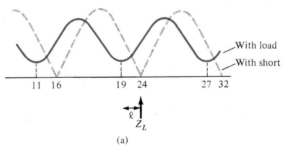

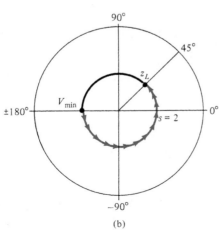

FIGURE 11.23 Determining Z_L by using the slotted line: **(a)** wave pattern, **(b)** Smith chart for Example 11.6.

This corresponds to an angular movement of $0.3125 \times 720° = 225°$ on the $s = 2$ circle. By starting at the location of $V_{\min}$ and moving 225° toward the load (counterclockwise), we reach the location of z_L as illustrated in Figure 11.23(b). Thus

$$z_L = 1.4 + j0.75$$

and

$$Z_L = Z_o z_L = 50(1.4 + j0.75) = 70 + j37.5 \ \Omega$$

PRACTICE EXERCISE 11.6

The following measurements were taken by means of the slotted line technique: with load, $s = 1.8$, $V_{\max}$ occurred at 23 cm, 33.5 cm, . . . ; with short, $s = \infty$, $V_{\max}$ occurred at 25 cm, 37.5 cm, ... If $Z_0 = 50 \ \Omega$, determine Z_L.

Answer: $32.5 - j17.5 \ \Omega$.

EXAMPLE 11.7

An antenna with an impedance of $40 + j30 \ \Omega$ is to be matched to a 100 Ω lossless line with a shorted stub. Determine

(a) The required stub admittance

(b) The distance between the stub and the antenna

(c) The stub length

(d) The standing wave ratio on each segment of the system

Solution:

(a) $z_L = \dfrac{Z_L}{Z_o} = \dfrac{40 + j30}{100} = 0.4 + j0.3$

Locate z_L on the Smith chart as in Figure 11.24 and from this draw the s-circle so that y_L can be located diametrically opposite z_L. Thus $y_L = 1.6 - j1.2$. Alternatively, we may find y_L by using

$$y_L = \frac{Z_o}{Z_L} = \frac{100}{40 + j30} = 1.6 - j1.2$$

Locate points A and B where the s-circle intersects the $g = 1$ circle. At A, $y_s = -j1.04$ and at B, $y_s = +j1.04$. Thus the required stub admittance is

$$Y_s = Y_o y_s = \pm j1.04 \frac{1}{100} = \pm j10.4 \text{ mS}$$

Both $j10.4$ mS and $-j10.4$ mS are possible values.

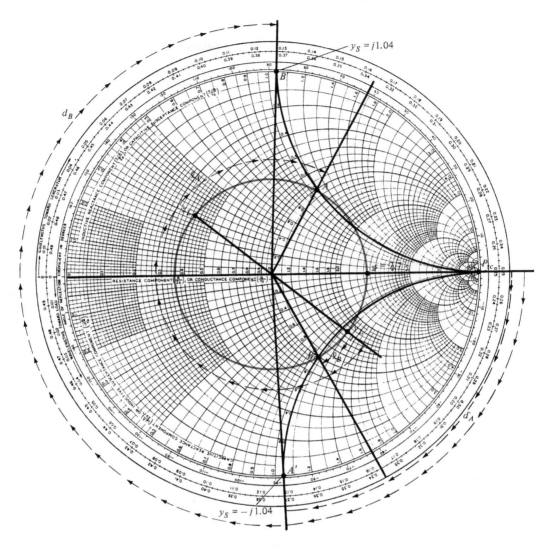

FIGURE 11.24 Smith chart for Example 11.7.

(b) From Figure 11.24, we determine the distance between the load (antenna in this case) y_L and the stub. At A,

$$\ell_A = \frac{\lambda}{2} - \frac{(62° - {-39°})\lambda}{720°} = 0.36\lambda$$

and at B:

$$\ell_B = \frac{(62° - 39°)\lambda}{720°} = 0.032\lambda$$

(c) Locate points A' and B' corresponding to stub admittance $-j1.04$ and $j1.04$, respectively. Determine the stub length (distance from P_{sc} to A' and B'):

$$d_A = \frac{88°}{720°}\lambda = 0.1222\lambda$$

$$d_B = \frac{272°}{720°}\lambda = 0.3778\lambda$$

Notice that $d_A + d_B = 0.5\lambda$ as expected.

(d) From Figure 11.24, $s = 2.7$. This is the standing wave ratio on the line segment between the stub and the load (see Figure 11.18); $s = 1$ to the left of the stub because the line is matched, and $s = \infty$ along the stub because the stub is shorted.

PRACTICE EXERCISE 11.7

A 75 Ω lossless line is to be matched to a load of $100 - j80$ Ω with a shorted stub. Calculate the stub length, its distance from the load, and the necessary stub admittance.

Answer: $\ell_A = 0.093\lambda$, $\ell_B = 0.272\lambda$, $d_A = 0.126\lambda$, $d_B = 0.374\lambda$, $\pm j12.67$ mS.

---

†11.7 TRANSIENTS ON TRANSMISSION LINES

In our discussion so far, we have assumed that a transmission line operates at a single frequency. In computer networks and in certain other practical applications, pulsed signals may be sent through the line. From Fourier analysis, a pulse can be regarded as a superposition of waves of many frequencies. Thus, sending a pulsed signal on the line may be regarded as the same as simultaneously sending waves of different frequencies.

As in circuit analysis, when a pulse generator or battery connected to a transmission line is switched on, it takes some time for the current and voltage on the line to reach steady values. This transitional period is called the *transient*. The transient behavior just after closing the switch (or due to lightning strokes) is usually analyzed in the frequency domain by using Laplace transformation. For the sake of convenience, we treat the problem in the time domain.

Consider a lossless line of length ℓ and characteristic impedance Z_o as shown in Figure 11.25(a). Suppose that the line is driven by a pulse generator of voltage V_g with internal impedance Z_g at $z = 0$ and terminated with a purely resistive load Z_L. At the instant $t = 0$ that the switch is closed, the starting current "sees" only Z_g and Z_o, so the initial situation can be described by the equivalent circuit of Figure 11.25(b). From the figure, the starting current at $z = 0$, $t = 0^+$ is given by

$$I(0, 0^+) = I_o = \frac{V_g}{Z_g + Z_o} \qquad (11.60)$$

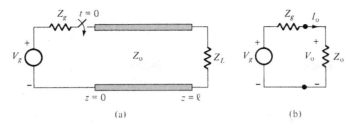

FIGURE 11.25 Transients on a transmission line: **(a)** a line driven by a pulse generator, **(b)** the equivalent circuit at $z = 0, t = 0^+$.

and the initial voltage is

$$V(0, 0^+) = V_0 = I_0 Z_0 = \frac{Z_0}{Z_g + Z_0} V_g \tag{11.61}$$

After the switch is closed, waves $I^+ = I_0$ and $V^+ = V_0$ propagate toward the load at the speed

$$u = \frac{1}{\sqrt{LC}} \tag{11.62}$$

Since this speed is finite, it takes some time for the waves traveling in the positive direction to reach the load and interact with it. The presence of the load has no effect on the waves before the transit time given by

$$t_1 = \frac{\ell}{u} \tag{11.63}$$

After t_1 seconds, the waves reach the load. The voltage (or current) at the load is the sum of the incident and reflected voltages (or currents). Thus

$$V(\ell, t_1) = V^+ + V^- = V_0 + \Gamma_L V_0 = (1 + \Gamma_L) V_0 \tag{11.64}$$

and

$$I(\ell, t_1) = I^+ + I^- = I_0 - \Gamma_L I_0 = (1 - \Gamma_L) I_0 \tag{11.65}$$

where Γ_L is the load reflection coefficient given in eq. (11.36); that is,

$$\Gamma_L = \frac{Z_L - Z_0}{Z_L + Z_0} \tag{11.66}$$

The reflected waves $V^- = \Gamma_L V_0$ and $I^- = -\Gamma_L I_0$ travel back toward the generator in addition to the waves V_0 and I_0 already on the line. At time $t = 2t_1$, the reflected waves have reached the generator, so

$$V(0, 2t_1) = V^+ + V^- = \Gamma_G \Gamma_L V_0 + (1 + \Gamma_L) V_0$$

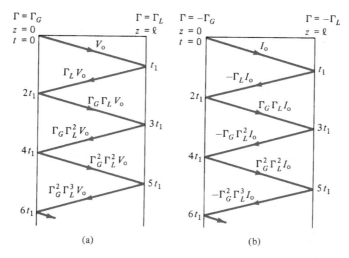

FIGURE 11.26 Bounce diagram for **(a)** a voltage wave and **(b)** a current wave.

or

$$V(0, 2t_1) = (1 + \Gamma_L + \Gamma_G\Gamma_L)V_o \qquad (11.67)$$

and

$$I(0, 2t_1) = I^+ + I^- = -\Gamma_G(-\Gamma_L I_o) + (1 - \Gamma_L)I_o$$

or

$$I(0, 2t_1) = (1 - \Gamma_L + \Gamma_L\Gamma_G)I_o \qquad (11.68)$$

where Γ_G is the generator reflection coefficient given by

$$\Gamma_G = \frac{Z_g - Z_o}{Z_g + Z_o} \qquad (11.69)$$

Again the reflected waves (from the generator end) $V^+ = \Gamma_G\Gamma_L V_o$ and $I^+ = \Gamma_G\Gamma_L I_o$ propagate toward the load, and the process continues until the energy of the pulse is actually absorbed by the resistors Z_g and Z_L.

Instead of tracing the voltage and current waves back and forth, it is easier to keep track of the reflections using a *bounce diagram*, otherwise known as a *lattice diagram*. The bounce diagram consists of a zigzag line indicating the position of the voltage (or current) wave with respect to the generator end, as shown in Figure 11.26. On the bounce diagram, the voltage (or current) at any time may be determined by adding those values that appear on the diagram above that time.

EXAMPLE 11.8

For the transmission line of Figure 11.27, calculate and sketch

(a) The voltage at the load and generator ends for $0 < t < 6\ \mu s$

(b) The current at the load and generator ends for $0 < t < 6\ \mu s$

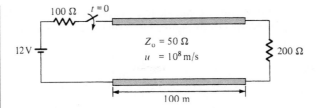

FIGURE 11.27 For Example 11.8.

Solution:

(a) We first calculate the voltage reflection coefficients at the generator and load ends:

$$\Gamma_G = \frac{Z_g - Z_o}{Z_g + Z_o} = \frac{100 - 50}{100 + 50} = \frac{1}{3}$$

$$\Gamma_L = \frac{Z_L - Z_o}{Z_L + Z_o} = \frac{200 - 50}{200 + 50} = \frac{3}{5}$$

The transit time $t_1 = \dfrac{\ell}{u} = \dfrac{100}{10^8} = 1 \ \mu s$.

The initial voltage at the generator end is

$$V_o = \frac{Z_o}{Z_o + Z_g} V_g = \frac{50}{150}(12) = 4 \ V$$

The 4 V is sent out to the load. The leading edge of the pulse arrives at the load at $t = t_1 = 1 \ \mu s$. A portion of it, $4(3/5) = 2.4 \ V$, is reflected back and reaches the generator at $t = 2t_1 = 2 \ \mu s$. At the generator, $2.4(1/3) = 0.8$ is reflected and the process continues. The whole process is best illustrated in the voltage bounce diagram of Figure 11.28.

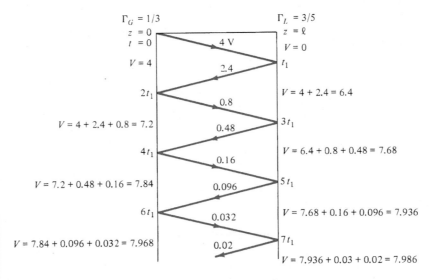

FIGURE 11.28 Voltage bounce diagram for Example 11.8.

From the bounce diagram, we can sketch $V(0, t)$ and $V(\ell, t)$ as functions of time as shown in Figure 11.29. Notice from Figure 11.29 that as $t \to \infty$, the voltages approach an asymptotic value of

$$V_\infty = \frac{Z_L}{Z_L + Z_g} V_g = \frac{200}{300}(12) = 8 \text{ V}$$

This should be expected because the equivalent circuits at $t = 0$ and $t = \infty$ are as shown in Figure 11.30.

(b) The current reflection coefficients at the generator and load ends are $-\Gamma_G = -1/3$ and $-\Gamma_L = -3/5$, respectively. The initial current is

$$I_o = \frac{V_o}{Z_o} = \frac{4}{50} = 80 \text{ mA}$$

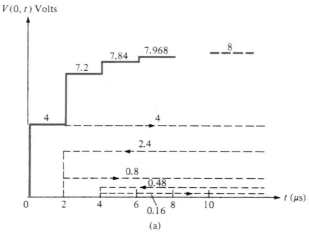

$V(0, t)$ Volts

(a)

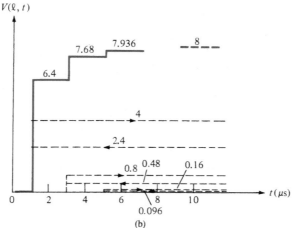

$V(\ell, t)$

(b)

FIGURE 11.29 Voltage (not to scale) for Example 11.8: **(a)** at the generator end, **(b)** at the load end.

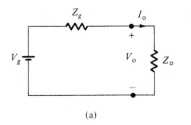 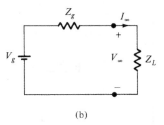

(a)   (b)

FIGURE 11.30 Equivalent circuits for the line in Figure 11.27 for **(a)** $t = 0$ and **(b)** $t = \infty$.

Again, $I(0, t)$ and $I(\ell, t)$ are easily obtained from the current bounce diagram shown in Figure 11.31. These currents are sketched in Figure 11.32. Note that $I(\ell, t) = V(\ell, t)/Z_L$. Hence, Figure 11.32(b) can be obtained either from the current bounce diagram of Figure 11.31 or by scaling Figure 11.29(b) by a factor of $1/Z_L = 1/200$. Notice from Figures 11.30(b) and 11.32 that the currents approach an asymptotic value of

$$I_\infty = \frac{V_g}{Z_g + Z_L} = \frac{12}{300} = 40 \text{ mA}$$

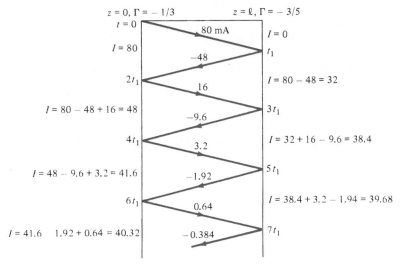

FIGURE 11.31 Current bounce diagram for Example 11.8.

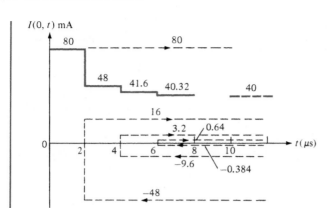

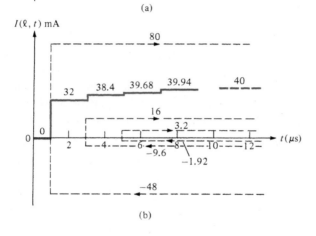

(a)

(b)

FIGURE 11.32 Current (not to scale) for Example 11.8: **(a)** at the generator end, **(b)** at the load end.

PRACTICE EXERCISE 11.8

Repeat Example 11.8 if the transmission line is

(a) Short-circuited
(b) Open-circuited

Answer: (a) See Figure 11.33, (b) See Figure 11.34.

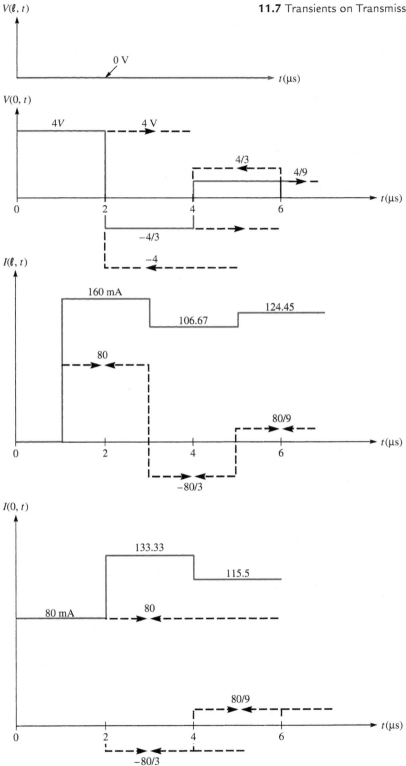

FIGURE 11.33 For Practice Exercise 11.8(a).

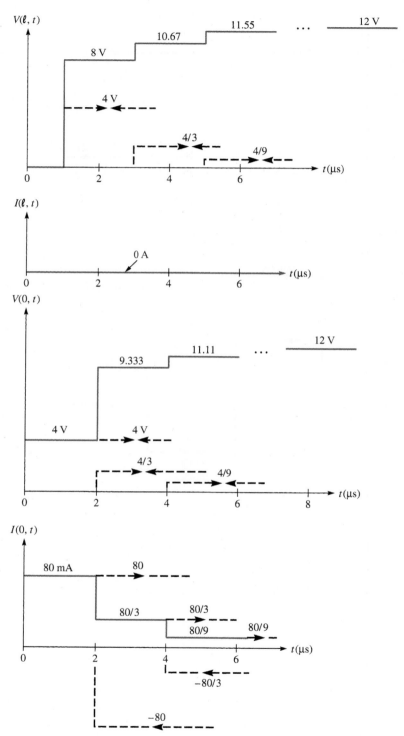

FIGURE 11.34 For Practice Exercise 11.8(b).

EXAMPLE 11.9

A 75 Ω transmission line of length 60 m is terminated by a 100 Ω load. If a rectangular pulse of width 5 μs and magnitude 4 V is sent out by the generator connected to the line, sketch $I(0, t)$ and $I(\ell, t)$ for $0 < t < 15$ μs. Take $Z_g = 25$ Ω and $u = 0.1c$.

Solution:

In the previous example, the switching on of a battery created a step function, a pulse of infinite width. In this example, the pulse is of finite width of 5 μs. We first calculate the voltage reflection coefficients:

$$\Gamma_G = \frac{Z_g - Z_o}{Z_g + Z_o} = \frac{25 - 75}{25 + 75} = \frac{-1}{2}$$

$$\Gamma_L = \frac{Z_L - Z_o}{Z_L + Z_o} = \frac{100 - 75}{100 + 75} = \frac{1}{7}$$

The initial voltage and transit time are given by

$$V_o = \frac{Z_o}{Z_o + Z_g} V_g = \frac{75}{100}(4) = 3 \text{ V}$$

$$t_1 = \frac{\ell}{u} = \frac{60}{0.1 (3 \times 10^8)} = 2 \text{ } \mu\text{s}$$

The time taken by V_o to go forth and back is $2t_1 = 4$ μs, which is less than the pulse duration of 5 μs. Hence, there will be overlapping.

The current reflection coefficients are

$$-\Gamma_L = -\frac{1}{7} \quad \text{and} \quad -\Gamma_G = \frac{1}{2}$$

The initial current $I_o = \dfrac{V_g}{Z_g + Z_o} = \dfrac{4}{100} = 40 \text{ mA}$.

Let i and r denote incident and reflected pulses, respectively. At the generator end:

$$0 < t < 5 \text{ } \mu\text{s}, \qquad I_r = I_o = 40 \text{ mA}$$

$$4 < t < 9, \qquad I_i = -\frac{1}{7}(40) = -5.714$$

$$I_r = \frac{1}{2}(-5.714) = -2.857$$

$$8 < t < 13, \qquad I_i = -\frac{1}{7}(-2.857) = 0.4082$$

$$I_r = \frac{1}{2}(0.4082) = 0.2041$$

$$12 < t < 17, \qquad I_i = -\frac{1}{7}(0.2041) = -0.0292$$

$$I_r = \frac{1}{2}(-0.0292) = -0.0146$$

and so on. Hence, the plot of $I(0, t)$ versus t is as shown in Figure 11.35(a).

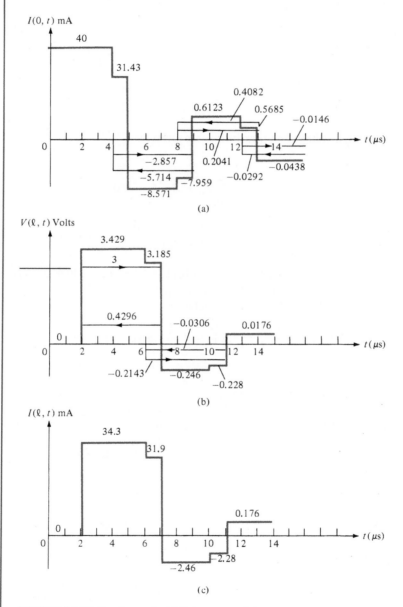

FIGURE 11.35 For Example 11.9 (not to scale).

At the load end:

$$0 < t < 2\ \mu s, \qquad\qquad V = 0$$

$$2 < t < 7, \qquad\qquad V_i = 3$$

$$V_r = \frac{1}{7}(3) = 0.4296$$

$$6 < t < 11, \qquad\qquad V_i = -\frac{1}{2}(0.4296) = -0.2143$$

$$V_r = \frac{1}{7}(-0.2143) = -0.0306$$

$$10 < t < 14, \qquad\qquad V_i = -\frac{1}{2}(-0.0306) = 0.0154$$

$$V_r = \frac{1}{7}(0.0154) = 0.0022$$

and so on. From $V(\ell, t)$, we can obtain $I(\ell, t)$ as

$$I(\ell, t) = \frac{V(\ell, t)}{Z_L} = \frac{V(\ell, t)}{100}$$

The plots of $V(\ell, t)$ and $I(\ell, t)$ are shown in Figure 11.35(b) and (c).

PRACTICE EXERCISE 11.9

Repeat Example 11.9, replacing the rectangular pulse by the triangular pulse of Figure 11.36.

Answer: $(I_o)_{max} = 100$ mA. See Figure 11.37 for the current waveforms.

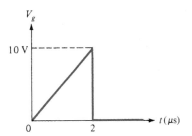

FIGURE 11.36 Triangular pulse for Practice Exercise 11.9.

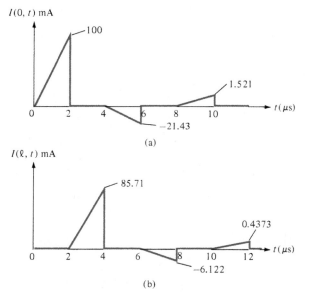

FIGURE 11.37 Current waves for Practice Exercise 11.9.

†11.8 APPLICATION NOTE—MICROSTRIP LINES AND CHARACTERIZATION OF DATA CABLES

†A. Microstrip Transmission Lines

Microstrip lines belong to a group of lines known as parallel-plate transmission lines. They are widely used in present-day electronics. Apart from being the most commonly used form of transmission lines for microwave integrated circuits, microstrips are used for circuit components such as filters, couplers, resonators, and antennas. In comparison with coaxial lines, microstrip lines allow for greater flexibility and compactness of design.

A microstrip line consists of a single ground plane and an open strip conductor separated by dielectric substrate as shown in Figure 11.38. It is constructed by the

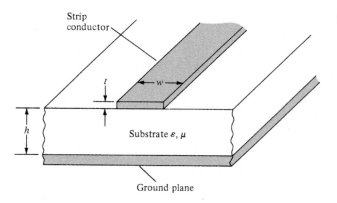

FIGURE 11.38 Microstrip line.

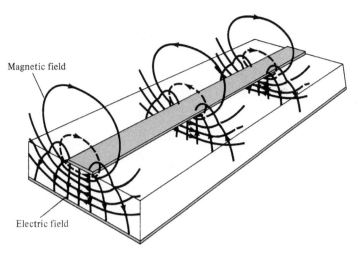

Magnetic field

Electric field

FIGURE 11.39 Pattern of the EM field of a microstrip line. (From D. Roddy, Microwave Technology, 1986, by permission of Prentice-Hall.)

photographic processes used for integrated circuits. Analytical derivation of the characteristic properties of the line is cumbersome. We will consider only some basic, valid empirical formulas necessary for calculating the phase velocity, impedance, and losses of the line.

Owing to the open structure of the microstrip line, the EM field is not confined to the dielectric, but is partly in the surrounding air as in Figure 11.39. Provided the frequency is not too high, the microstrip line will propagate a wave that, for all practical purposes, is a TEM wave. Because of the fringing, the *effective relative permittivity* ε_{eff} is less than the relative permittivity ε_r of the substrate. If w is the line width and h is the substrate thickness, an approximate value of ε_{eff} is given by

$$\varepsilon_{\text{eff}} = \frac{(\varepsilon_r + 1)}{2} + \frac{(\varepsilon_r - 1)}{2\sqrt{1 + 12h/w}} \tag{11.70}$$

The characteristic impedance is given by the following approximate formulas:

$$Z_0 = \begin{cases} \dfrac{60}{\sqrt{\varepsilon_{\text{eff}}}} \ln\!\left(\dfrac{8h}{w} + \dfrac{w}{4h}\right) & w/h \leq 1 \\[4mm] \dfrac{1}{\sqrt{\varepsilon_{\text{eff}}}} \dfrac{120\pi}{[w/h + 1.393 + 0.667\ln(w/h + 1.444)]} & w/h \geq 1 \end{cases} \tag{11.71}$$

The characteristic impedance of a wide strip is often low, while that of a narrow strip is high.

For design purposes, if ε_r and Z_o are known, the ratio w/h necessary to achieve Z_o is given by

$$\frac{w}{h} = \begin{cases} \dfrac{8e^A}{e^{2A} - 2}, & w/h \leq 2 \\[2ex] \dfrac{2}{\pi} \left\{ B - 1 - \ln(2B - 1) \right. \\[2ex] \left. + \dfrac{\varepsilon_r - 1}{2\varepsilon_r} \left[\ln(B - 1) + 0.39 - \dfrac{0.61}{\varepsilon_r} \right] \right\}, & w/h > 2 \end{cases} \tag{11.72}$$

where

$$A = \frac{Z_o}{60} \sqrt{\frac{\varepsilon_r + 1}{2}} + \frac{\varepsilon_r - 1}{\varepsilon_r + 1} \left(0.23 + \frac{0.11}{\varepsilon_r} \right) \tag{11.73a}$$

$$B = \frac{60\pi^2}{Z_o \sqrt{\varepsilon_r}} \tag{11.73b}$$

From the knowledge of ε_{eff} and Z_o, the phase constant and the phase velocity of a wave propagating on the microstrip are given, respectively, by

$$\beta = \frac{\omega \sqrt{\varepsilon_{\text{eff}}}}{c} \tag{11.74a}$$

$$u = \frac{c}{\sqrt{\varepsilon_{\text{eff}}}} \tag{11.74b}$$

where c is the speed of light in a vacuum. The attenuation due to conduction (or ohmic) loss is (in dB/m)

$$\alpha_c \simeq 8.686 \frac{R_s}{wZ_o} \tag{11.75}$$

where $R_s = \dfrac{1}{\sigma_c \delta}$ is the skin resistance of the conductor. The attenuation due to dielectric loss is (in dB/m)

$$\alpha_d \simeq 27.3 \frac{(\varepsilon_{\text{eff}} - 1)\, \varepsilon_r}{(\varepsilon_r - 1)\, \sqrt{\varepsilon_{\text{eff}}}} \frac{\tan \theta}{\lambda} \tag{11.76}$$

where $\lambda = u/f$ is the line wavelength and $\tan \theta = \sigma/\omega\varepsilon$ is the loss tangent of the substrate. The total attenuation constant is the sum of the ohmic attenuation constant α_c and the dielectric attenuation constant α_d; that is,

$$\alpha = \alpha_c + \alpha_d \tag{11.77}$$

Sometimes α_d is negligible in comparison with α_c. Although they offer an advantage of flexibility and compactness, microstrip lines are not useful for long transmission because attenuation is excessive.

EXAMPLE 11.10

A certain microstrip line has fused quartz ($\varepsilon_r = 3.8$) as a substrate. If the ratio of line width to substrate thickness is $w/h = 4.5$, determine

(a) The effective relative permittivity of the substrate
(b) The characteristic impedance of the line
(c) The wavelength of the line at 10 GHz

Solution:

(a) For $w/h = 4.5$, we have a wide strip. From eq. (11.70),

$$\varepsilon_{\text{eff}} = \frac{4.8}{2} + \frac{2.8}{2}\left[1 + \frac{12}{4.5}\right]^{-1/2} = 3.131$$

(b) From eq. (11.71),

$$Z_o = \frac{120\pi}{\sqrt{3.131}\left[4.5 + 1.393 + 0.667 \ln(4.5 + 1.444)\right]}$$

$$= 30.08 \ \Omega$$

(c) $\lambda = \dfrac{u}{f} = \dfrac{c}{f\sqrt{\varepsilon_{\text{eff}}}} = \dfrac{3 \times 10^8}{10^{10}\sqrt{3.131}}$

$\qquad = 1.69 \times 10^{-2} \text{ m} = 16.9 \text{ mm}$

PRACTICE EXERCISE 11.10

Repeat Example 11.10 for $w/h = 0.8$.

Answer: (a) 2.75, b) 84.03 Ω, (c) 18.09 mm.

EXAMPLE 11.11

At 10 GHz, a microstrip line has the following parameters:

$$h = 1 \text{ mm}$$

$$w = 0.8 \text{ mm}$$

$$\varepsilon_r = 6.6$$

$$\tan \theta = 10^{-4}$$

$$\sigma_c = 5.8 \times 10^7 \text{ S/m}$$

Calculate the attenuation due to conduction loss and dielectric loss.

Solution:

The ratio $w/h = 0.8$. Hence from eqs. (11.70) and (11.71)

$$\varepsilon_{eff} = \frac{7.6}{2} + \frac{5.6}{2}\left(1 + \frac{12}{0.8}\right)^{-1/2} = 4.5$$

$$Z_o = \frac{60}{\sqrt{4.5}}\ln\left(\frac{8}{0.8} + \frac{0.8}{4}\right)$$

$$= 65.69\ \Omega$$

The skin resistance of the conductor is

$$R_s = \frac{1}{\sigma_c\delta} = \sqrt{\frac{\pi f \mu_o}{\sigma_c}} = \sqrt{\frac{\pi \times 10 \times 10^9 \times 4\pi \times 10^{-7}}{5.8 \times 10^7}}$$

$$= 2.609 \times 10^{-2}\ \Omega/m^2$$

Using eq. (11.75), we obtain the conduction attenuation constant as

$$\alpha_c = 8.686 \times \frac{2.609 \times 10^{-2}}{0.8 \times 10^{-3} \times 65.69}$$

$$= 4.31\ dB/m$$

To find the dielectric attenuation constant, we need λ:

$$\lambda = \frac{u}{f} = \frac{c}{f\sqrt{\varepsilon_{eff}}} = \frac{3 \times 10^8}{10 \times 10^9\sqrt{4.5}}$$

$$= 1.414 \times 10^{-2}\ m$$

Applying eq. (11.76), we have

$$\alpha_d = 27.3 \times \frac{3.5 \times 6.6 \times 10^{-4}}{5.6 \times \sqrt{4.5} \times 1.414 \times 10^{-2}}$$

$$= 0.3754\ dB/m$$

PRACTICE EXERCISE 11.11

Calculate the attenuation due to ohmic losses at 20 GHz for a microstrip line constructed of copper conductor having a width of 2.5 mm on an alumina substrate. Take the characteristic impedance of the line as 50 Ω.

Answer: 2.564 dB/m.

B. Characterization of Data Cables

Data communication has become a vital part of our daily life, the educational system, and business enterprises. Cables (copper or optical fiber) play an important role in data communication because they constitute the vehicle that transmits electrical signals from one point to another. Before such cables are installed, they must meet certain requirements specified in terms of parameters including insertion loss (or attenuation), return loss (RL), near-end crosstalk (NEXT), far-end crosstalk (FEXT), attenuation-to-crosstalk ratio (ACR), power sum NEXT (PSNEXT), propagation delay, propagation delay skew, equal level far-end crosstalk (ELFEXT), and power sum ELFEXT (PSELFEXT). In this section, we focus on the most popular measures: attenuation, RL, NEXT, and ELFEXT.

Attenuation

Attenuation (also known as insertion loss) is one of the greatest concerns of any cabling infrastructure. It is the reduction of signal strength during transmission. It is the opposite of amplification. Although it is normal to expect attenuation, a signal that attenuates too much becomes unintelligible, which is why most networks require repeaters at regular intervals.

The factors that contribute to a cable's attenuation include conductor size, material, insulation, frequency (bandwidth), speed, and distance.

Attenuation describes the phenomenon of reduction of power intensity according to the law

$$\frac{dP}{dz} = -2\alpha P \rightarrow P = P_o e^{-2\alpha z} \tag{11.78}$$

where it is assumed that signal propagates along z, α is the attenuation coefficient, and P_o is the power at $z = 0$. Thus, attenuation describes how energy is lost or dissipated. Energy loss occurs as a transformation from one type of energy to another. Attenuation increases with both frequency and length. Attenuation is usually expressed in decibels. For a cable of length L, attenuation (or loss) through the cable is

$$A = 8.686 L \alpha \text{ dB} \tag{11.79}$$

Since it is a loss, it is usually expressed as a negative value. Thus, -12 dB is a weaker signal than -10 dB.

Return Loss

Return loss (RL) is a measure of the reflected energy caused by impedance mismatches in a cabling system. It is a measure of the dissimilarity between impedances in metallic transmission lines and loads. It may also be regarded as the ratio, at the junction of a transmission line and a terminating impedance or other discontinuity, of the amplitude of the reflected wave to the amplitude of the incident wave. Return loss is important in applications that use simultaneous bidirectional transmission. Possible causes of excessive return loss include fluctuation in characteristic impedance, cable kinks, excessive bends, and cable jacket.

Return loss is defined as the ratio of the incident power to the reflected power:

$$RL = 10 \log_{10} \frac{P^+}{P^-} = -20 \log_{10} |\Gamma| \text{ dB} \tag{11.80}$$

since $P^- = |\Gamma|^2 P^+$ and the reflection coefficient is given by

$$\Gamma = \frac{Z_L - Z_\text{o}}{Z_L + Z_\text{o}} \tag{11.81}$$

where Z_o is the characteristic impedance of the cable and Z_L is the load impedance. Thus, return loss is a number that indicates the amount of signal that is reflected back into the cable from the terminating equipment. It is generally specified in decibels, and larger values are better because they indicate less reflection. Ideally, there would be no reflection, and return loss would have a value of infinity. Generally, values of 35 to 40 dB or higher are considered acceptable. A value of 40 dB indicates that only 1% of the signal is reflected.

NEXT

Crosstalk is a major impairment in any two-wire transmission system. Within a cable, there are usually several active pairs. Because these pairs are in close physical proximity over long distances, coupling takes place, and the pairs "crosstalk" in each other. Thus, the idea of crosstalk refers to interference that enters a communication channel through some coupling path. There are two types of crosstalk in multipair cables: near-end crosstalk (NEXT) and far-end crosstalk (FEXT).

When current flows in a wire, an electromagnetic field is created which can interfere with signals on adjacent wires. As frequency increases, this effect becomes stronger. Each pair is twisted because this allows opposing fields in the wire pair to cancel each other. The tighter the twist, the more effective the cancellation and the higher the data rate supported by the cable. If wires are not tightly twisted, the result is near-end crosstalk (NEXT). If you have ever been talking on the telephone and could hear another conversation faintly in the background, you have experienced crosstalk. In fact, the term derives from telephony applications where 'talk' came 'across.' In local-area networks, NEXT occurs when a strong signal on one pair of wires is picked up by an adjacent pair of wires. NEXT is the portion of the transmitted signal that is electromagnetically coupled back into the received signal, as illustrated in Figure 11.40. In many cases, excessive crosstalk is due to poorly twisted terminations at connection points.

Since NEXT is a measure of difference in signal strength between a disturbing pair and a disturbed pair, a larger number (less crosstalk) is more desirable than a smaller number (more crosstalk). Because NEXT varies significantly with frequency, it is important to measure it across a range of frequencies, typically 1–250 MHz. Twisted-pair coupling becomes less effective for higher frequencies.

ELFEXT

Far-end crosstalk (FEXT) is similar to NEXT, except that the signal is sent from the local end as shown in Figure 11.41, and crosstalk is measured at the far end. As a result of attenuation, signals that induce FEXT can be much weaker, especially for longer cable lengths.

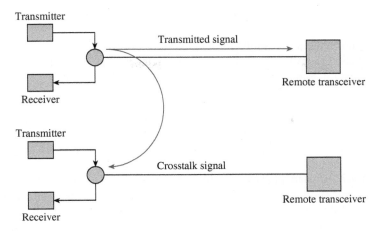

FIGURE 11.40 Near-end crosstalk (NEXT) in a paired cable.

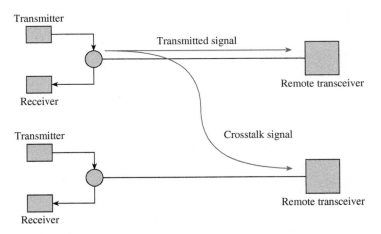

FIGURE 11.41 Far-end crosstalk (FEXT) in a paired cable.

For that reason, FEXT results are not meaningful without an indication of the corresponding attenuation on the link. Thus, FEXT is measured but rarely reported. FEXT results are used to derive equal-level far-end crosstalk (ELFEXT).

Noise occurring at the far end can be difficult to measure. It is common in practice to eliminate the attenuation effects and look at the pure noise taking place. ELFEXT is used when one is looking at the noise minus the effects of attenuation.

ELFEXT is a measure of the unwanted signal coupling from a transmitter at the near end into a neighboring pair measured at the far end relative to the received signal level measured on the same pair. Unlike attenuation, return loss, NEXT, and FEXT, ELFEXT is a calculated rather than a measured quantity. It is derived by subtracting the attenuation of the disturbing pair from the FEXT this pair induces in an adjacent pair. That is, if the disturbing pair is i and the disturbed pair is j,

$$ELFEXT_{ij} = FEXT_{ij} - attenuation_j \qquad (11.82)$$

This normalizes the results for length. Since both FEXT and attenuation are measured in decibels, ELFEXT is also measured in decibels. High ELFEXT is indicative of excessive attenuation, higher than expected FEXT, or both.

11.9 APPLICATION NOTE—METAMATERIALS

In 1967, Russian physicist Viktor Veselago studied theoretically the problem of time-harmonic monochromatic plane wave propagation in a material whose permittivity and permeability he assumed to be simultaneously negative at the frequency of interest. He showed that such a material could possess interesting electromagnetic features such as anomalous refraction. At the turn of the century, American physicists Richard Shelby, David Smith, Sheldon Schultz, and their group constructed such a composite medium by embedding split rings in a host medium. They experimentally showed in the microwave region the presence of anomalous refraction in the medium. These complex materials, called metamaterials, have gained considerable attention.

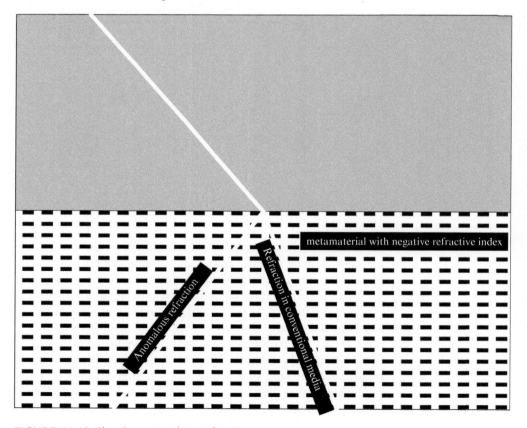

FIGURE 11.42 Showing anomalous refraction.

Metamaterials are synonymously called left-handed materials, negative refractive index materials, and double-negative materials. A conventional material inherits its electromagnetic properties from the constituent atoms. But metamaterials derive their properties from the constituent units embedded in a host medium. They can be engineered as we desire.

The phenomena of radiation, scattering, guidance in metamaterials, and complex structures made by pairing them with layers of conventional materials are being actively investigated. Researchers envision exciting possibilities for the development of improved computer chips, DVDs with vastly increased memory, cheaper and better-performing wireless communication systems, and improved medical imaging equipment. Several important ideas and developments that should lead to fabrication of components such as phase conjugators, unconventional guided-wave structures, compact thin cavities, high-impedance surfaces, and perfect lens are expected to emerge from these studies. The concepts underlying metamaterials also led to the development of artificial left-handed transmission lines.

Metamaterials are artificial materials engineered to have specific properties that are not found in nature. Unlike naturally occurring materials, the behavior of a metamaterial is determined not only by the properties of its components, but also by the size, geometry, orientation, and arrangement of those components. Metamaterials have gained attention owing to their unique ability to bend waves rather than reflect them, rendering an object surrounded by a metamaterial *invisible*. Section 11.10 describes some applications of these unusual materials to microwave imaging.

11.10 APPLICATION NOTE—MICROWAVE IMAGING

Section 10.2 introduced mmWave technology, which takes advantage of the properties of electromagnetic waves in the 30–300 GHz frequency band, or mm-waves, which correspond to wavelengths in the range of 10–1 mm. Many optically opaque objects are rendered transparent when they are imaged with signals only a few millimeters in wavelength. This makes mm-wave imaging attractive for a wide variety of commercial, defense, and scientific applications. Examples include nondestructive testing (NDT) for structural integrity, material characterization, security scanning, and medical screening. Microwaves and millimeter waves have been used extensively to image dielectric bodies. They can penetrate into many optically opaque media such as living tissues, wood, ceramics, plastics, clothing, concrete, soil, fog, and foliage. In the past, many people believed that microwaves could be used just for target detection and tracking. But they have also been used for decades in remote sensing, that is, for the imaging of weather patterns or the surfaces of remote planets, for underground surveillance, and so on. More recently, microwave and millimeter-wave systems have been deployed for a variety of short-range applications, most notably in concealed-weapon detection (Figure 11.43) and through-the-wall imaging. Additional information about cloaking applications for metamaterials is presented in Section 12.10.

FIGURE 11.43 A concealed weapon revealed.

The main measures of performance of an imaging system are considered to be the spatial resolution in lateral and range directions and the image dynamic range offered. With the availability of more channels, combined with the powerful digital signal processing (DSP) capabilities of modern computers, the performance of mm-wave imaging systems is advancing rapidly.

The most commonly known imaging systems are based on X-ray technology. Their applications are in, for example, computed tomography (CT) for medical diagnostics, NDT applications, and luggage inspection at security checkpoints. These systems work in a transmission setup. Furthermore, backscatter X-ray systems, which work in a reflection setup, have been investigated over the past few years, especially for the screening of airline passengers for concealed objects. On the one hand, X-ray images have an inherent high lateral resolution owing to the extremely short wavelength (0.01–10 nm). But on the other hand, the energy of the photons is high enough to ionize organic and inorganic matter. Therefore, health aspects are critical with respect to the imaging of humans, especially people who must be imaged frequently, such as airport personnel.

In contrast, electromagnetic mm-waves offer a contactless inspection of materials with nonionizing radiation and a high spatial resolution. Since spatial resolution and penetration depth are conflicting parameters regarding the wavelength, the E-band (60–90 GHz with wavelengths of 5–3.3 mm) is a good compromise for NDT applications to detect flaws, material inhomogeneities, and inclusions in dielectrics. A lateral resolution of about 2 mm is sufficient for many applications (e.g., personnel screening at airport security checkpoints).

The application of microwaves in biomedical imaging and diagnostics, however, remains a field with many uncharted territories. So far, the contributions of microwave technology to medical imaging have been limited except for radio-frequency component design for magnetic resonance imaging (MRI) systems. From the imaging of isolated organs to the imaging of the brain, the bones, or the breast, the microwave community is striving to make an entrance in the highly competitive world of medical imaging.

MATLAB 11.1

```
% This script computes the voltage and current waveforms as functions
% of length along
% a transmission line terminated with a complex load

clear
syms w Z0 ZL ZG VG gamma z Zin % symbolic variables for frequency,
                               % characteristic impedance,
% load impedance, source impedance, source voltage, propagation
% constant, distance

% Enter the frequency (in rad/s)
wn = input('Enter the angular frequency\n >  ');
% Enter the propagation constant gamma (in a+j*b format)
gamman = input('Enter the propagation constant\n >  ');
% Enter the length (m)
L = input('Enter the length\n >  ');
% Enter the characteristic impedance
Z0n = input('Enter the characteristic impedance\n >  ');
% Enter the load impedance (in a+j*b format)
disp('Enter the complex load impedance ');
ZLn = input('(in a+j*b format)\n >  ');
% Enter the source impedance (in a+j*b format)
disp('Enter the complex source impedance ');
ZGn = input('(in a+j*b format)\n >  ');
% Enter the source voltage (in a*exp(j*b) format)
disp('Enter the source voltage ');
VGn = input(' (in a*exp(j*b) format)\n >  ');

% The expression for the input impedance as a function of
% length along the line
% This expression at this point is purely symbolic and
% contains no numerical data, it will be used in line 37
Zin = Z0*(ZL - Z0 * tanh(gamma *z)) / (Z0 - ...
                ZL * tanh(gamma * z));

% Output the key parameters
% Reflection coefficient at load
GammaL = (ZLn - Z0n) / (ZLn + Z0n);
disp('\nThe reflection coefficient at the load is');
disp(sprintf(' %0.2f+j%0.2f\n',real(GammaL), imag(GammaL)))
```

```
% SWR
SWR=(1+abs(GammaL))/(1-abs(GammaL));
disp(sprintf('The SWR at the load is %0.2f\n', SWR))
% Input impedance seen at source
ZinG=subs(Zin,{Z0,ZL,gamma,z},{Z0n,ZLn, gamman,L});
disp('The input impedance seen by the generator ')
disp(sprintf(' is %0.2f+j%0.2f\n', real(ZinG), imag(ZinG)))

% now determine the forward-traveling voltage coefficient
% V0plus at the load (z = 0) from the equation
% V(z) = (V0+) e^(-j B z) (1 + GammaL exp (2 j B z))
% by setting z = -L
% First Determine the voltage at the source
V0G = VGn * ZinG / (ZGn + ZinG);
V0plus = V0G * exp(-gamman*L)/(1 + GammaL *exp(-2*gamman*L));

% Generate vector of voltage over length
z=-L:L/1000:0;   % vector of length
Vz = V0plus*exp(gamman*z).*(1 + GammaL*exp(2*gamman*z));

% voltage vector
Iz = V0plus/Z0n*exp(gamman*z).*(1 - GammaL*exp(2*gamman*z));

% voltage vector
%(notice the .* for multiplying two vectors
% together element-by-element)

% Plot
figure
subplot(2,1,1)  % generate a subplot within one figure window
% the two plots will be one on top of the other
% (2,`,1) means there are two rows, one column, and this plot
% goes into the top subplot
plot(z, abs(Vz))  % plot only voltage magnitude
axis([-L 0 0 abs(2*V0plus)])
title('Voltage on transmission line as a function of length')
xlabel('distance from load (m)')
ylabel('magnitude of voltage (V)')
subplot(2,1,2)
plot(z, abs(Iz))  % plot only current magnitude
axis([-L 0 0 abs(2*V0plus/Z0n)])
title('Current on transmission line as a function of length')
xlabel('distance from load (m)')
ylabel('magnitude of current (A)')
```

SUMMARY

1. A transmission line is commonly described by its distributed parameters R (in Ω/m), L (in H/m), G (in S/m), and C (in F/m). Formulas for calculating R, L, G, and C are provided in Table 11.1 for coaxial, two-wire, and planar lines.

2. The distributed parameters are used in an equivalent circuit model to represent a differential length of the line. The transmission line equations are obtained by applying Kirchhoff's laws and allowing the length of the line to approach zero. The voltage and current waves on the line are

$$V(z, t) = V_o^+ e^{-\alpha z} \cos(\omega t - \beta z) + V_o^- e^{\alpha z} \cos(\omega t + \beta z)$$

$$I(z, t) = \frac{V_o^+}{Z_o} e^{-\alpha z} \cos(\omega t - \beta z) - \frac{V_o^-}{Z_o} e^{\alpha z} \cos(\omega t + \beta z)$$

showing that there are two waves traveling in opposite directions on the line.

3. The characteristic impedance Z_o (analogous to the intrinsic impedance η of plane waves in a medium) of a line is given by

$$Z_o = \sqrt{\frac{R + j\omega L}{G + j\omega C}}$$

and the propagation constant γ (per meter) is given by

$$\gamma = \alpha + j\beta = \sqrt{(R + j\omega L)(G + j\omega C)}$$

The wavelength and wave velocity are

$$\lambda = \frac{2\pi}{\beta}, \quad u = \frac{\omega}{\beta} = f\lambda$$

4. The general case is that of the lossy transmission line ($G \neq 0 \neq R$) considered earlier. For a lossless line, $R = 0 = G$; for a distortionless line, $R/L = G/C$. It is desirable that power lines be lossless and telephone lines be distortionless.

5. The voltage reflection coefficient at the load end is defined as

$$\Gamma_L = \frac{V_o^-}{V_o^+} = \frac{Z_L - Z_o}{Z_L + Z_o}$$

and the standing wave ratio is

$$s = \frac{1 + |\Gamma_L|}{1 - |\Gamma_L|}$$

where Z_L is the load impedance.

6. At any point on the line, the ratio of the phasor voltage to phasor current is the impedance at that point looking toward the load and would be the input impedance to the line if the line were that long. For a lossy line,

$$Z(z) = \frac{V_s(z)}{I_s(z)} = Z_{in} = Z_o \left[\frac{Z_L + Z_o \tanh \gamma\ell}{Z_o + Z_L \tanh \gamma\ell} \right]$$

where ℓ is the distance from load to the point. For a lossless line ($\alpha = 0$), tanh $\gamma\ell = j \tan \beta\ell$; for a shorted line, $Z_L = 0$; for an open-circuited line, $Z_L = \infty$; and for a matched line, $Z_L = Z_o$.

7. The Smith chart is a graphical means of obtaining line characteristics such as Γ, s, and Z_{in}. It is constructed within a circle of unit radius and based on the formula for Γ_L given in eq. (11.36). For each r and x, there are two explicit circles (the resistance and reactance circles) and one implicit circle (the constant s-circle). The Smith chart is conveniently used in determining the location of a stub tuner and its length. It is also used with the slotted line to determine the value of the unknown load impedance.

8. When a dc voltage is suddenly applied at the sending end of a line, a pulse moves forth and back on the line. The transient behavior is conveniently analyzed by using bounce diagrams.

9. Microstrip transmission lines are useful in microwave integrated circuits. Useful formulas for constructing microstrip lines and determining losses on the lines have been presented.

10. Some parameters that are commonly used in characterizing data communication cables are presented. These parameters include attenuation, return loss, NEXT, and ELFEXT.

REVIEW QUESTIONS

11.1 Which of the following statements are not true of the line parameters R, L, G, and C?

(a) R and L are series elements.

(b) G and C are shunt elements.

(c) $G = \dfrac{1}{R}$.

(d) $LC = \mu\varepsilon$ and $RG = \sigma\varepsilon$.

(e) Both R and G depend on the conductivity of the conductors forming the line.

(f) Only R depends explicitly on frequency.

(g) The parameters are not lumped but distributed.

11.2 For a lossy transmission line, the characteristic impedance does not depend on

(a) The operating frequency of the line

(b) The length of the line

(c) The load terminating the line

(d) The conductivity of the conductors

(e) The conductivity of the dielectric separating the conductors

11.3 Which of the following conditions will not guarantee a distortionless transmission line?

(a) $R = 0 = G$

(b) $RC = GL$

(c) Very low frequency range $(R \gg \omega L, G \gg \omega C)$

(d) Very high frequency range $(R \ll \omega L, G \ll \omega C)$

11.4 Which of these is not true of a lossless line?

(a) $Z_{\text{in}} = -jZ_{\text{o}}$ for a shorted line with $\ell = \lambda/8$.

(b) $Z_{\text{in}} = j\infty$ for a shorted line with $\ell = \lambda/4$.

(c) $Z_{\text{in}} = jZ_{\text{o}}$ for an open line with $\ell = \lambda/2$.

(d) $Z_{\text{in}} = Z_{\text{o}}$ for a matched line.

(e) At a half-wavelength from a load, $Z_{\text{in}} = Z_L$ and repeats for every half-wavelength thereafter.

11.5 A lossless transmission line of length 50 cm with $L = 10 \, \mu\text{H/m}, C = 40 \, \text{pF/m}$ is operated at 30 MHz. Its electrical length is

(a) 20λ

(b) 0.2λ

(c) $108°$

(d) 40π

(e) None of the above

11.6 Match the following normalized impedances with points A, B, C, D, and E on the Smith chart of Figure 11.44.

(i) $0 + j0$ (ii) $1 + j0$

(iii) $0 - j1$ (iv) $0 + j1$

(v) $\infty + j\infty$ (vi) $\left[\dfrac{Z_{\text{in}}}{Z_{\text{o}}}\right]_{\text{min}}$

(vii) $\left[\dfrac{Z_{\text{in}}}{Z_{\text{o}}}\right]_{\text{max}}$ (viii) Matched load $(\Gamma = 0)$

11.7 A 500 m lossless transmission line is terminated by a load that is located at P on the Smith chart of Figure 11.45. If $\lambda = 150$ m, how many voltage maxima exist on the line?

(a) 7   (b) 6

(c) 5   (d) 3

(e) None

11.8 Write true (T) or false (F) for each of the following statements.

(a) All r- and x-circles pass through point $(\Gamma_r, \Gamma_i) = (1, 0)$.

(b) Any impedance repeats itself every $\lambda/4$ on the Smith chart.

(c) An $s = 2$ circle is the same as $|\Gamma| = 0.5$ circle on the Smith chart.

(d) The basic principle of any matching scheme is to eliminate the reflected wave between the source and the matching device.

(e) The slotted line is used to determine Z_L only.

(f) At any point on a transmission line, the current reflection coefficient is the reciprocal of the voltage reflection coefficient at that point.

11.9 In an air line, adjacent maxima are found at 12.5 cm and 37.5 cm. The operating frequency is

(a) 1.5 GHz   (b) 600 MHz

(c) 300 MHz   (d) 1.2 GHz

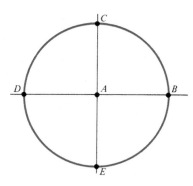

FIGURE 11.44 Smith chart for Review Question 11.6.

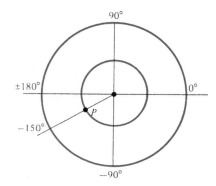

FIGURE 11.45 Smith chart for Review Question 11.7.

11.10 Two identical pulses each of magnitude 12 V and width 2 μs are incident at $t = 0$ on a lossless transmission line of length 400 m terminated with a load. If the two pulses are separated 3 μs and $u = 2 \times 10^8$ m/s, when does the contribution to $V_L(\ell, t)$ by the second pulse start overlapping that of the first?

(a) $t = 0.5 \, \mu$s

(b) $t = 2 \, \mu$s

(c) $t = 5 \, \mu$s

(d) $t = 5.5 \, \mu$s

(e) $t = 6 \, \mu$s

Answers: 11.1c,d,e, 11.2b,c, 11.3c, 11.4a,c, 11.5c, 11.6 (i) D,e (ii) A, (iii) E, (iv) C, (v) B, (vi) D, (vii) B, (viii) A, 11.7a, 11.8 (a) T, (b) F, (c) F, (d) T, (e) F, (f) F, 11.9b, 11.10e.

PROBLEMS

Section 11.2—Transmission Line Parameters

11.1 An air-filled planar line with $w = 30$ cm, $d = 1.2$ cm, $t = 3$ mm has conducting plates with $\sigma_c = 7 \times 10^7$ S/m. Calculate R, L, C, and G at 500 MHz.

11.2 A coaxial cable has an inner conductor of radius $a = 0.8$ mm and an outer conductor of radius $b = 2.6$ mm. The conductors have $\sigma_c = 5.28 \times 10^7$ S/m, $\mu_c = \mu_o$, and $\epsilon_c = \epsilon_o$; they are separated by a dielectric material having $\sigma = 10^{-5}$ S/m, $\mu = \mu_o$, $\epsilon = 3.5\,\epsilon_o$. At 80 MHz, calculate the line parameters L, C, G, and R.

11.3 A coaxial cable has inner radius a and outer radius b. If the inner and outer conductors are separated by a material with conductivity σ, show that the conductance per unit length is

$$G = \frac{2\pi\sigma}{\ln\dfrac{b}{a}}$$

11.4 A planar transmission line is made of copper strips of width 30 mm and are separated by a dielectric of thickness 2 mm, $\sigma = 10^{-3}$ S/m, $\varepsilon = 4\varepsilon_o$, $\mu = \mu_o$. The conductivity of copper is 5.8×10^7 S/m. Assuming that the line operates at 200 MHz, (a) find R, L, G, and C, (b) determine γ and Z_o.

11.5 The copper leads of a diode are 16 mm in length and have a radius of 0.3 mm. They are separated by a distance of 2 mm as shown in Figure 11.46. Find the capacitance between the leads and the ac resistance at 10 MHz.

Section 11.3—Transmission Line Equations

*11.6 A TV twin-lead is made of two parallel copper wires with $a = 1.2$ mm. The wires are separated by 1.5 cm of a dielectric material with $\varepsilon_r = 4$. Calculate L, C, and Z_o.

11.7 A small section Δz of a transmission line may be represented by the equivalent circuit in Figure 11.47. Determine the voltage–current relationship for the section.

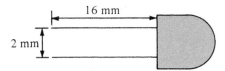

FIGURE 11.46 The diode of Problem 11.5.

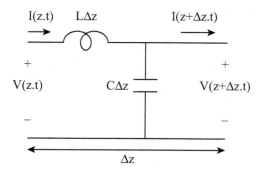

FIGURE 11.47 For Problem 11.7.

11.8 (a) Show that at high frequencies ($R \ll \omega L, G \ll \omega L$),

$$\gamma \simeq \left(\frac{R}{2}\sqrt{\frac{C}{L}} + \frac{G}{2}\sqrt{\frac{L}{C}} \right) + j\omega\sqrt{LC}$$

(b) Obtain a similar formula for Z_0.

11.9 A transmission line operates at 12 MHz and has the following parameters:

$$R = 0.2 \ \Omega/\text{m}, G = 4 \ \text{mS/m}, L = 40 \ \mu\text{H/m}, C = 25 \ \mu\text{H/m}$$

(a) Compute γ using eq. (11.11).

(b) Compute γ using the result of Problem 11.8(a).

11.10 The current along a transmission line is given by

$$I(z,t) = I_1(\omega t - \beta z) + I_2(\omega t + \beta z)$$

(a) Determine which of the two components represents a wave traveling from source to load.

(b) Find the corresponding voltage $V(z, t)$.

11.11 At 60 MHz, the following characteristics of a lossy line are measured:

$$Z_0 = 50 \ \Omega, \quad \alpha = 0.04 \ \text{dB/m}, \quad \beta = 2.5 \ \text{rad/m}$$

Calculate R, L, C, and G of the line.

11.12 A 78 Ω lossless planar line was designed but did not meet a requirement. What fraction of the widths of the strip should be added or removed to get the characteristic impedance of 75 Ω?

11.13 A telephone line operating at 1 kHz has $R = 6.8 \ \Omega/\text{mi}, L = 3.4 \ \text{mH/mi}, C = 8.4 \ \text{nF/mi}$, and $G = 0.42 \ \mu\text{S/mi}$. Find (a) Z and γ, (b) phase velocity, (c) wavelength.

11.14 A TV antenna lead-in wire 10 cm long has a characteristic impedance of 250 Ω and is open-circuited at its end. If the line operates at 400 MHz, determine its input impedance.

11.15 A coaxial cable has its conductors made of copper ($\sigma_c = 5.8 \times 10^7$ S/m) and its dielectric made of polyethylene ($\varepsilon_r = 2.25, \mu_r = 1$). If the radius of the outer conductor is 3 mm, determine the radius of the inner conductor so that $Z_o = 75\ \Omega$.

11.16 For a lossless two-wire transmission line, show that

(a) The phase velocity $u = c = \dfrac{1}{\sqrt{LC}}$

(b) The characteristic impedance $Z_o = \dfrac{120}{\sqrt{\varepsilon_r}} \cosh^{-1} \dfrac{d}{2a}$

Is part (a) true of other lossless lines?

11.17 A twisted line, which may be approximated by a two-wire line, is very useful in the telephone industry. Consider a line comprising two copper wires of diameter 0.12 cm that have a 0.32 cm center-to-center spacing. If the wires are separated by a dielectric material with $\varepsilon = 3.5\varepsilon_o$, find L, C, and Z_o.

11.18 A distortionless cable is 4 m long and has a characteristic impedance of 60 Ω. An attenuation of 0.24 dB is observed at the receiving end. Also, a signal applied to the cable is delayed by 80 μs before it is measured at the receiving end. Find R, G, L, and C for the cable.

11.19 A distortionless line operating at 120 MHz has $R = 20\ \Omega/$m, $L = 0.3\ \mu$H/m, and $C = 63$ pF/m. (a) Determine γ, u, and Z_o. (b) How far will a voltage wave travel before it is reduced to 20% of its initial magnitude? (c) How far will it travel to suffer a 45° phase shift?

11.20 On a distortionless line, the voltage wave is given by

$$V(\ell') = 120e^{0.0025\ell'}\cos(10^8 t + 2\ell') + 60e^{-0.0025\ell'}\cos(10^8 t - 2\ell')$$

where ℓ' is the distance from the load. If $Z_L = 300\ \Omega$, find (a) α, β, and u, (b) Z_o and $I(\ell')$.

11.21 The voltage on a line is given by

$$V(\ell) = 80e^{10^{-3\ell}}\cos(2\pi \times 10^4 t + 0.01\ell) + 60e^{-10^{-3\ell}}\cos(2\pi \times 10^4 t + 0.01\ell)\text{V}$$

where ℓ is the distance from the load. Calculate γ and u.

11.22 A distortionless transmission line satisfies $RC = LG$. If the line has $R = 10$ m$\Omega/$m, $C = 82$ pF/m, and $L = 0.6\ \mu$H/m, calculate its characteristic impedance and propagation constant. Assume that the line operates at 80 MHz.

11.23 A coaxial line 5.6 m long has distributed parameters $R = 6.5\ \Omega/$m, $L = 3.4\ \mu$H/m, $G = 8.4$ mS/m, and $C = 21.5$ pF/m. If the line operates at 2 MHz, calculate the characteristic impedance and the end-to-end propagation time delay.

11.24 A lossy transmission line of length 2.1 m has characteristic impedance of $80 + j60\ \Omega$. When the line is short-circuited, the input impedance is $30 - j12\ \Omega$. (a) Determine α and β. (b) Find the input impedance when the short circuit is replaced by $Z_L = 40 + j30\ \Omega$.

11.25 A lossy transmission line with characteristic impedance of $75 + j60\ \Omega$ is connected to a 200 Ω load. If attenuation is 1.4 Np/m and phase constant is 2.6 rad/m, find the input impedance for $\ell = 0.5$ m.

Section 11.4—Input Impedance, Standing Wave Ratio, and Power

11.26 (a) Show that a transmission coefficient may be defined as

$$\tau_L = \frac{V_L}{V_o^+} = 1 + \Gamma_L = \frac{2Z_L}{Z_L + Z_o}$$

(b) Find τ_L when the line is terminated by (i) a load whose value is nZ_o, (ii) an open circuit, (iii) a short circuit, (iv) $Z_L = Z_o$ (matched line).

11.27 (a) Show that

$$Z_L = \frac{1 + \Gamma}{1 - \Gamma} Z_o$$

(b) A 50 Ω line has a reflection coefficient of 0.6∠45°. Find the load impedance.

11.28 A 120 Ω lossless line is terminated at a load impedance $200 - j240$ Ω. Find Γ_L and s.

11.29 A lossy transmission line has $R = 3.5$ Ω/m, $L = 2$ μH/m, $C = 120$ pF/m, and $G \approx 0$. At 400 MHz, determine α, β, Z_o, and u.

11.30 Find the input impedance of a short-circuited coaxial transmission line of Figure 11.48 if $Z_o = 65 + j38$ Ω, $\gamma = 0.7 + j2.5$/m, $\ell = 0.8$ m.

11.31 Calculate the reflection coefficient due to $Z_L = (1 + 2j)Z_o$.

11.32 Refer to the lossless transmission line shown in Figure 11.49. (a) Find Γ and s. (b) Determine Z_{in} at the generator.

11.33 A 60 Ω lossless line is connected to a source with $V_g = 10\angle0°$ V_{rms} and $Z_g = 50 - j40$ Ω and terminated with a load of $j40$ Ω. If the line is 100 m long and $\beta = 0.25$ rad/m, calculate Z_{in} and V at

(a) The sending end   (c) 4 m from the load

(b) The receiving end   (d) 3 m from the source

FIGURE 11.48 For Problem 11.30.

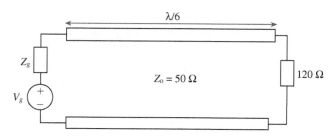

FIGURE 11.49 For Problem 11.32.

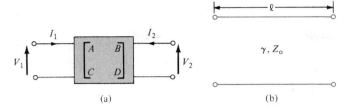

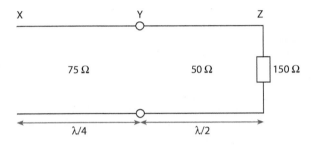

FIGURE 11.50 For Problem 11.34: **(a)** network, **(b)** lossy line.

FIGURE 11.51 For Problem 11.38.

*11.34 Consider the two-port network shown in Figure 11.50(a). The relation between the input and output variables can be written in matrix form as

$$\begin{bmatrix} V_1 \\ I_1 \end{bmatrix} = \begin{bmatrix} A & B \\ C & D \end{bmatrix} \begin{bmatrix} V_2 \\ -I_2 \end{bmatrix}$$

For the lossy line in Figure 11.50(b), show that the *ABCD* matrix is

$$\begin{bmatrix} \cosh \gamma \ell & Z_o \sinh \gamma \ell \\ \dfrac{1}{Z_o} \sinh \gamma \ell & \cosh \gamma \ell \end{bmatrix}$$

Section 11.5—The Smith Chart

11.35 Normalize the following impedances with respect to 50 Ω and locate them on the Smith chart: (a) $Z_a = 80\ \Omega$, (b) $Z_b = 60 + j40\ \Omega$, (c) $Z_c = 30 - j120\ \Omega$.

11.36 A quarter-wave lossless 100 Ω line is terminated by a load $Z_L = 210\ \Omega$. If the voltage at the receiving end is 80 V, what is the voltage at the sending end?

11.37 Determine the impedance at a point $\lambda/4$ distant from a load of impedance $(1 + j2)Z_o$.

11.38 Two lines are cascaded as shown in Figure 11.51. Determine:
(a) The input impedance
(b) The standing wave ratio for sections *XY* and *YZ*
(c) The reflection coefficient at *Z*

11.39 A 50 Ω lossless line operates at 600 MHz and is terminated by a load of Z_L. If the line is 0.1 m long and $Z_{in} = 100 - j120$, find Z_L and s. Assume $u = 0.8\ c$.

11.40 A lossless transmission line, with characteristic impedance of 50 Ω and electrical length of $\ell = 0.27\lambda$, is terminated by a load impedance $40 - j25\ \Omega$. Determine Γ_L, s, and Z_{in}.

11.41 A lossless 100 Ω transmission line is terminated in an unknown impedance Z_L. The standing wave ratio is 2.4, and the nearest voltage minimum is 0.2 λ from the load. Find Z_L and Γ.

11.42 The distance from the load to the first minimum voltage in a 50 Ω line is 0.12λ, and the standing wave ratio $s = 4$.

(a) Find the load impedance Z_L.

(b) Is the load inductive or capacitive?

(c) How far from the load is the first maximum voltage?

11.43 A lossless 50 Ω line is terminated by a load $Z_L = 75 + j60\ \Omega$. Using a Smith chart, determine (a) the reflection coefficient Γ, (b) the standing wave ratio s, (c) the input impedance at 0.2λ from the load, (d) the location of the first minimum voltage from the load, (e) the shortest distance from the load at which the input impedance is purely resistive.

11.44 A transmission line is terminated by a load with admittance $Y_L = (0.6 + j0.8)/Z_0$. Find the normalized input impedance at $\lambda/6$ from the load.

11.45 Using the Smith chart, determine the admittance of $Z = 100 + j60\ \Omega$ with respect to $Z_0 = 50\ \Omega$.

11.46 A 50 Ω transmission line operates at 160 MHz and is terminated by a load of $50 + j30$ Ω. If its wave speed is $c/2$ and the input impedance is to be made real, calculate the minimum possible length of the line and the corresponding input impedance.

11.47 A 50 Ω transmission line, $\lambda/4$ in length, is connected to a $\lambda/2$ section of 100 Ω line terminated by a 60 Ω resistor. Calculate the input impedance to the 50 Ω line.

11.48 (a) Calculate the reflection coefficient corresponding to $Z_L = (0.5 - j)Z_0$.

(b) Determine the load impedance corresponding to the reflection coefficient 0.4 $\underline{/25°}$.

11.49 An 80 Ω transmission line operating at 12 MHz is terminated by a load Z_L. At 22 m from the load, the input impedance is $100 - j120\ \Omega$. If $u = 0.8c$,

(a) Calculate Γ_L, $Z_{in,\ max}$, and $Z_{in,\ min}$.

(b) Find Z_L, s, and the input impedance at 28 m from the load.

(c) How many $Z_{in,\ max}$ and $Z_{in,\ min}$ are there between the load and the $100 - j120\ \Omega$ input impedance?

11.50 An antenna, connected to a 150 Ω lossless line, produces a standing wave ratio of 2.6. If measurements indicate that voltage maxima are 120 cm apart and that the last maximum is 40 cm from the antenna, calculate

(a) The operating frequency

(b) The antenna impedance

(c) The reflection coefficient (assume that $u = c$).

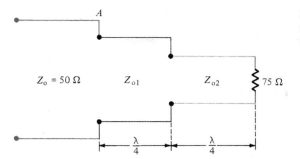

FIGURE 11.52 Double section transformer of Problem 11.53.

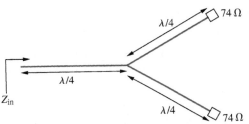

FIGURE 11.53 For Problems 11.54 and 11.55.

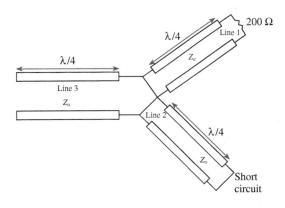

FIGURE 11.54 For Problem 11.56.

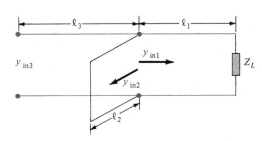

FIGURE 11.55 For Problem 11.57.

11.51 An 80 Ω lossless line has $Z_L = j60\ \Omega$ and $Z_{\text{in}} = j40\ \Omega$. (a) Determine the shortest length of the line. (b) Calculate s and Γ_L.

11.52 A 50 Ω air-filled line is terminated in a mismatched load of $40 + j25\ \Omega$. Find the shortest distance from the load at which the voltage has the smallest magnitude.

11.53 Two $\lambda/4$ transformers in tandem are to connect a 50 Ω line to a 75 Ω load as in Figure 11.52.
   (a) Determine the characteristic impedance Z_{o1} if $Z_{o2} = 30\ \Omega$ and there is no reflected wave to the left of A.
   (b) If the best results are obtained when

$$\left[\frac{Z_o}{Z_{o1}}\right]^2 = \frac{Z_{o1}}{Z_{o2}} = \left[\frac{Z_{o2}}{Z_L}\right]^2$$

determine Z_{o1} and Z_{o2} for this case.

11.54 Two identical antennas, each with input impedance 74 Ω, are fed with three identical 50 Ω quarter-wave lossless transmission lines as shown in Figure 11.53. Calculate the input impedance at the source end.

11.55 If the lines in Figure 11.53 are connected to a voltage source of 120 V with an internal impedance of 80 Ω, calculate the average power delivered to either antenna.

11.56 Consider the three lossless lines in Figure 11.54. If $Z_o = 50\ \Omega$, calculate:
   (a) Z_{in} looking into line 1
   (b) Z_{in} looking into line 2
   (c) Z_{in} looking into line 3

11.57 A section of lossless transmission line is shunted across the main line as in Figure 11.55. If $\ell_1 = \lambda/4$, $\ell_2 = \lambda/8$, and $\ell_3 = 7\lambda/8$, find y_{in_1}, y_{in_2}, and y_{in_3} given that $Z_o = 100\ \Omega$, $Z_L = 200 + j150\ \Omega$. Repeat the calculations as if the shorted section were open.

Section 11.6—Some Applications of Transmission Lines

11.58 A load $Z_L = 75 + j100\ \Omega$ is to be matched to a 50 Ω line. A shorted shunt-stub tuner is preferred. Find the length of the stub in terms of λ.

11.59 A stub of length 0.12λ is used to match a 60 Ω lossless line to a load. If the stub is located at 0.3λ from the load, calculate
   (a) The load impedance Z_L
   (b) The length of an alternative stub and its location with respect to the load
   (c) The standing wave ratio between the stub and the load

11.60 A 50 Ω lossless transmission line that is 20 m long is terminated into a $120 + j220\ \Omega$ load. To perfectly match, what should be the length and location of a short-circuited stub line? Assume an operating frequency of 10 MHz.

11.61 On a lossless line, measurements indicate $s = 4.2$ with the first maximum voltage at $\lambda/4$ from the load. Determine how far from the load a short-circuited stub should be located and calculate its length.

11.62 A 60 Ω lossless line terminated by load Z_L has a voltage wave as shown in Figure 11.56. Find s, Γ, and Z_L.

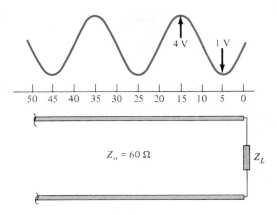

FIGURE 11.56 For Problem 11.62.

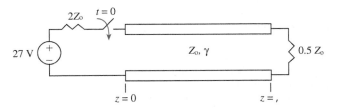

FIGURE 11.57 For Problem 11.65.

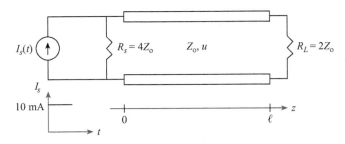

FIGURE 11.58 For Problem 11.66.

11.63 A 50 Ω air-filled slotted line is applied in measuring a load impedance. Adjacent minima are found at 14 cm and 22.5 cm from the load when the unknown load is connected, and $V_{max} = 0.95$ V and $V_{min} = 0.45$ V. When the load is replaced by a short circuit, the minima are 3.2 cm to the load. Determine s, f, Γ, and Z_L.

Section 11.7—Transients on Transmission Lines

11.64 A 50 Ω coaxial cable is connected to an 80 Ω resistive load and a dc source with zero internal resistance. Calculate the voltage reflection coefficients at the source and at the load.

11.65 The switch in Figure 11.57 is closed at $t = 0$. Sketch the voltage and current at the right side of the switch for $0 < t < 6\ell/u$. Take $Z_o = 50 \ \Omega$ and $\ell/u = 2 \ \mu s$. Assume a lossless transmission line.

11.66 A step current of 10 mA is turned on at $t = 0$, at $z = 0$ of a transmission line as shown in Figure 11.58. Determine the load voltage and current as functions of time. Let $Z_o = 50 \ \Omega$, $\mu = 2 \times 10^8$ m/s. You may Thévenin equivalent at $z = 0$. Carry out your analysis for $0 < t < 6\ell/u$.

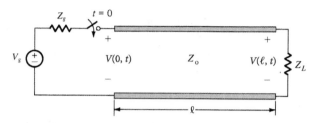

FIGURE 11.59 For Problem 11.68.

11.67 A 50-cm-long cable, of characteristic impedance 75 Ω and wave velocity 2×10^8 m/s, is used to connect a source of internal impedance 32 Ω to an amplifier with an input impedance of 2 M Ω. If the source voltage changes from 0 to 8 V at $t = 0$, obtain the voltage seen by the amplifier at $t = 20$ ns.

*11.68 Refer to Figure 11.59, where $Z_o = 50 \ \Omega$, $Z_g = 40 \ \Omega$, $V_g = 12V$, $Z_t = 100\Omega$, $l = 40$ cm Assuming wave speed $u = 2.5 \times 10^8$ m/s, find $V(0, t)$ and $V(l, t)$.

11.69 Using a time-domain reflectometer, the voltage waveform shown in Figure 11.60 was observed at the sending end of a lossless transmission line in response to a step voltage. If $Z_g = 60 \ \Omega$ and $Z_o = 50 \ \Omega$, calculate the generator voltage and the length of the line. Assume $u = c$.

11.70 A 12 V battery with an internal resistance of 10 Ω is connected to a 20 m length of 50 Ω coaxial cable with phase velocity of 2×10^8 m/s. If the receiving end is short-circuited, sketch the sending voltage $V(0, t)$ and the receiving-end voltage $V(\ell, t)$.

11.71 Suppose $Z_L = Z_G$ and a dc voltage is turned on at $t = 0$ (i.e., a unit step function) of amplitude V_o. The voltage is launched on a lossless line with characteristic impedance Z_o. Find the voltage across the line after a very long time.

Section 11.8—Application Note: Microstrip and Characterization of Data Cables

11.72 A microstrip line is 1 cm thick and 1.5 cm wide. The conducting strip is made of brass $(\sigma_c = 1.1 \times 10^7 \text{S/m})$, while the substrate is a dielectric material with $\varepsilon_r = 2.2$ and $\tan \theta = 0.02$. If the line operates at 2.5 GHz, find (a) Z_o and ε_{eff}, (b) α_c and α_d, (c) the distance down the line before the wave drops by 20 dB.

11.73 A 50 Ω microstrip line has a phase shift of 45° at 8 GHz. If the substrate thickness is $h = 8$ mm with $\varepsilon_r = 4.6$, find (a) the width of the conducting strip, (b) the length of the microstrip line.

11.74 An alumina substrate ($\varepsilon = 9.6\varepsilon_o$) of thickness 2 mm is used for the construction of a microstrip circuit. If the circuit designer can choose a line width between 0.4 mm and 8.0 mm, what is the range of characteristic impedance of the line?

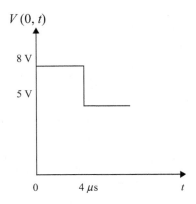

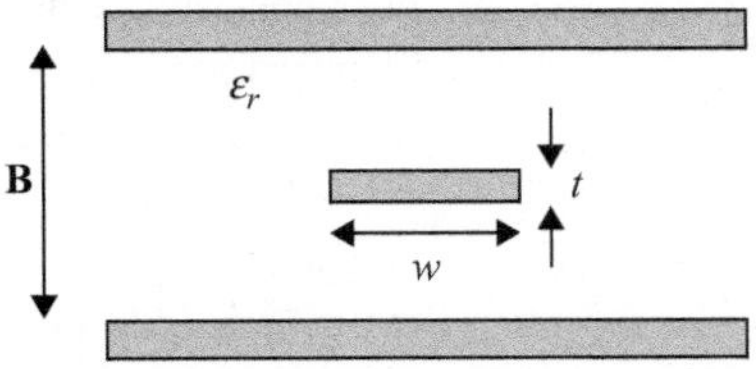

FIGURE 11.61 For Problem 11.75.

FIGURE 11.60 For Problem 11.69.

11.75 A strip transmission line is shown in Figure 11.61. An approximate expression for the characteristic impedance is

$$Z_o = \frac{377}{2\pi\sqrt{\varepsilon_r}} \ln\left\{1 + \frac{4b}{\pi w'}\left[\frac{8b}{\pi w'} + \sqrt{6.27 + \left(\frac{8b}{\pi w'}\right)^2}\right]\right\}$$

where $w' = w + \dfrac{t}{3.2} \ln\left(\dfrac{5b}{t}\right)$. Determine Z_o for $w = 0.5$ cm, $t = 0.1$ cm, $b = 1.2$ cm, $\varepsilon_r = 2$.

11.76 Design a 75 Ω microstrip line on a 1.2 mm thick duroid ($\varepsilon_r = 2.3$) substrate. Find the width of the conducting strip and the phase velocity.

11.77 Find the return loss due to a 150 Ω cable terminated by a 100 Ω load.

11.78 The effective relative permittivity ε_e is given by eq. (11.70). Use MATLAB to plot ε_e for $0.1 < w/h < 10$. Assume $\varepsilon_r = 2.2$ (Teflon).

RF MEMS

MEMS stands for **m**icro **e**lectro **m**echanical **s**ystems. These devices and systems range in size from a few micrometers to a few millimeters. The field of MEMS (or nanotechnology) encompasses all aspects of science and technology and is involved with things on a smaller scale. MEMS technology gives scientists and engineers the tools to build things that have been impossible or prohibitively expensive with other technologies.

MEMS technology has emerged with a major application area in telecommunications, particularly in optical switching and wireless communication. Wireless communication is expanding at an incredible pace for applications ranging from mobile phones to satellite communications. Radiofrequency MEMS technologies are helping fuel this expansion. The integration of MEMS into traditional RF circuits has resulted in systems with superior performance levels and lower manufacturing costs. RF MEMS are providing critical reductions in power consumption and signal loss to extend battery life and reduce weight. RF MEMS devices have a broad range of potential applications in military and commercial wireless communication and in navigation and sensor systems. Though RF MEMS devices are small, they can be very complex, commonly encompassing multiple interdependent engineering disciplines.

RF MEMS application areas are in phased arrays and reconfigurable apertures for defense and communication systems and in switching networks for satellite communications. As a breakthrough technology, allowing unparalleled synergy between apparently unrelated fields of endeavor such as biology and microelectronics, many new MEMS applications will emerge, expanding beyond that which is currently identified or known.

Traditionally, the training of MEMS engineers and scientists has entailed a graduate education at one of a few research universities, with the student working under the direction of an experienced faculty member to design, fabricate, and test a MEMS device. A graduate education in MEMS technology is very costly and comparatively time-consuming. Consequently, the current output from our universities of technical persons trained in MEMS technology does not meet the requirement for personnel to support the projected growth of MEMS industry. If your university offers classes in MEMS, take as many as possible. Better still, if your university has a MEMS laboratory, consider doing your senior design or your graduate thesis in that area. That should prepare you well enough for the job market. (*Note*: For an additional discussion of RF MEMS, see Section 6.8.)

CHAPTER 12

WAVEGUIDES

Reading makes a full man; conference makes a ready man; and writing makes an accurate man.

—ANONYMOUS

12.1 INTRODUCTION

As mentioned in the preceding chapter, a transmission line can be used to guide EM energy from one point (generator) to another (load). A waveguide is another means of achieving the same goal. However, a waveguide differs from a transmission line in some respects, although we may regard the latter as a special case of the former. In the first place, a transmission line can support only a transverse electromagnetic (TEM) wave, whereas a waveguide can support many possible field configurations. Second, at microwave frequencies (roughly 3–300 GHz), transmission lines become inefficient as a result of skin effect and dielectric losses; waveguides are used at that range of frequencies to obtain larger bandwidth and lower signal attenuation. Moreover, a transmission line may operate from dc ($f = 0$ Hz) to a very high frequency; a waveguide can operate only above a certain frequency called the *cutoff frequency* and therefore acts as a high-pass filter. Thus, waveguides cannot transmit dc, and they become excessively large at frequencies below microwave frequencies.

Although a waveguide may assume any arbitrary but uniform cross section, common waveguides are either rectangular or circular. Typical waveguides[1] are shown in Figure 12.1. Analysis of circular waveguides is involved and requires familiarity with Bessel functions, which are beyond our scope.[2] We will consider only hollow rectangular waveguides. By assuming lossless waveguides ($\sigma_c \simeq \infty, \sigma \approx 0$), we shall apply Maxwell's equations with the appropriate boundary conditions to obtain different modes of wave propagation and the corresponding **E** and **H** fields. When we close both ends of a waveguide, a cavity is formed. We will also consider optical fiber guide, which is basic to optical communications.

---

[1] For other types of waveguides, see J. A. Seeger, *Microwave Theory, Components and Devices.* Englewood Cliffs, NJ: Prentice-Hall, 1986, pp. 128–133.

[2] Analysis of circular waveguides can be found in advanced EM or EM-related texts (e.g., S. Y. Liao, *Microwave Devices and Circuits,* 3rd ed. Englewood Cliffs, NJ: Prentice-Hall, 1990, pp. 119–141).

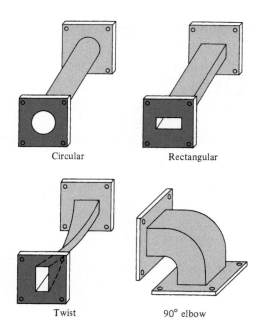

FIGURE 12.1 Typical waveguides.

Circular

Rectangular

Twist

90° elbow

12.2 RECTANGULAR WAVEGUIDES

Consider the rectangular waveguide shown in Figure 12.2, where a and b are the inner dimensions of the waveguide. We shall assume that the waveguide is filled with a source-free $(\rho_v = 0, \mathbf{J} = \mathbf{0})$ lossless dielectric material $(\sigma \simeq 0)$ and that its walls are perfectly conducting $(\sigma_c \simeq \infty)$. From eqs. (10.17) and (10.19), we recall that for a lossless medium, Maxwell's equations in phasor form become

$$\nabla^2 \mathbf{E}_s + k^2 \mathbf{E}_s = 0 \tag{12.1}$$

$$\nabla^2 \mathbf{H}_s + k^2 \mathbf{H}_s = 0 \tag{12.2}$$

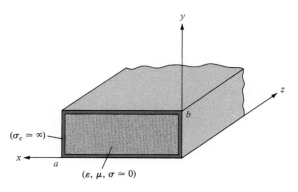

FIGURE 12.2 A rectangular waveguide with perfectly conducting walls, filled with a lossless material.

where

$$k = \omega \sqrt{\mu \varepsilon} \tag{12.3}$$

and the time factor $e^{j\omega t}$ is assumed. If we let

$$\mathbf{E}_s = (E_{xs}, E_{ys}, E_{zs}) \quad \text{and} \quad \mathbf{H}_s = (H_{xs}, H_{ys}, H_{zs})$$

each of eqs. (12.1) and (12.2) comprises three scalar Helmholtz equations. In other words, to obtain the **E** and **H** fields, we have to solve six scalar equations. For the z-component, for example, eq. (12.1) becomes

$$\frac{\partial^2 E_{zs}}{\partial x^2} + \frac{\partial^2 E_{zs}}{\partial y^2} + \frac{\partial^2 E_{zs}}{\partial z^2} + k^2 E_{zs} = 0 \tag{12.4}$$

which is a partial differential equation. From Example 6.5, we know that eq. (12.4) can be solved by separation of variables (product solution). So we let

$$E_{zs}(x, y, z) = X(x)\, Y(y)\, Z(z) \tag{12.5}$$

where $X(x)$, $Y(y)$, and $Z(z)$ are functions of x, y, and z, respectively. Substituting eq. (12.5) into eq. (12.4) and dividing by XYZ gives

$$\frac{X''}{X} + \frac{Y''}{Y} + \frac{Z''}{Z} = -k^2 \tag{12.6}$$

Since the variables are independent, each term in eq. (12.6) must be constant, so the equation can be written as

$$-k_x^2 - k_y^2 + \gamma^2 = -k^2 \tag{12.7}$$

where $-k_x^2$, $-k_y^2$, and γ^2 are separation constants. Thus, eq. (12.6) is separated as

$$X'' + k_x^2 X = 0 \tag{12.8a}$$

$$Y'' + k_y^2 Y = 0 \tag{12.8b}$$

$$Z'' - \gamma^2 Z = 0 \tag{12.8c}$$

The choice of γ^2 is due to the realization that the guided waves propagate along the guide axis z in the positive or negative direction, and the propagation may result in E_{zs} and H_{zs} that approach zero as $z \rightarrow \pm\infty$.

By following the same argument as in Example 6.5, we obtain the solution to eq. (12.8) as

$$X(x) = c_1 \cos k_x x + c_2 \sin k_x x \tag{12.9a}$$

$$Y(y) = c_3 \cos k_y y + c_4 \sin k_y y \tag{12.9b}$$

$$Z(z) = c_5 e^{\gamma z} + c_6 e^{-\gamma z} \tag{12.9c}$$

Substituting eq. (12.9) into eq. (12.5) gives

$$E_{zs}(x, y, z) = (c_1 \cos k_x x + c_2 \sin k_x x)(c_3 \cos k_y y + c_4 \sin k_y y) (c_5 e^{\gamma z} + c_6 e^{-\gamma z}) \tag{12.10}$$

As usual, if we assume that the wave propagates along the waveguide in the $+z$-direction, the multiplicative constant $c_5 = 0$ because the wave has to be finite at infinity [i.e., $E_{zs}(x, y, z = \infty) = 0$]. Hence eq. (12.10) is reduced to

$$E_{zs}(x, y, z) = (A_1 \cos k_x x + A_2 \sin k_x x)(A_3 \cos k_y y + A_4 \sin k_y y)e^{-\gamma z} \tag{12.11}$$

where $A_1 = c_1 c_6$, $A_2 = c_2 c_6$, $A_3 = c_3 c_6$, and $A_4 = c_4 c_6$. By taking similar steps, we get the solution of the z-component of eq. (12.2) as

$$H_{zs}(x, y, z) = (B_1 \cos k_x x + B_2 \sin k_x x)(B_3 \cos k_y y + B_4 \sin k_y y)e^{-\gamma z} \tag{12.12}$$

Instead of solving for other field components E_{xs}, E_{ys}, H_{xs}, and H_{ys} in eqs. (12.1) and (12.2) in the same manner, it is more convenient to use Maxwell's equations to determine them from E_{zs} and H_{zs}. From

$$\nabla \times \mathbf{E}_s = -j\omega\mu\mathbf{H}_s$$

and

$$\nabla \times \mathbf{H}_s = j\omega\varepsilon\mathbf{E}_s$$

we obtain

$$\frac{\partial E_{zs}}{\partial y} - \frac{\partial E_{ys}}{\partial z} = -j\omega\mu H_{xs} \tag{12.13a}$$

$$\frac{\partial H_{zs}}{\partial y} - \frac{\partial H_{ys}}{\partial z} = j\omega\varepsilon E_{xs} \tag{12.13b}$$

$$\frac{\partial E_{xs}}{\partial z} - \frac{\partial E_{zs}}{\partial x} = -j\omega\mu H_{ys} \tag{12.13c}$$

$$\frac{\partial H_{xs}}{\partial z} - \frac{\partial H_{zs}}{\partial x} = j\omega\varepsilon E_{ys} \tag{12.13d}$$

$$\frac{\partial E_{ys}}{\partial x} - \frac{\partial E_{xs}}{\partial y} = -j\omega\mu H_{zs} \tag{12.13e}$$

$$\frac{\partial H_{ys}}{\partial x} - \frac{\partial H_{xs}}{\partial y} = j\omega\varepsilon E_{zs} \tag{12.13f}$$

We will now express E_{xs}, E_{ys}, H_{xs}, and H_{ys} in terms of E_{zs} and H_{zs}. For E_{xs}, for example, we combine eqs. (12.13b) and (12.13c) and obtain

$$jωεE_{xs} = \frac{∂H_{zs}}{∂y} + \frac{1}{jωμ}\left(\frac{∂^2E_{xs}}{∂z^2} - \frac{∂^2E_{zs}}{∂x∂z}\right) \qquad (12.14)$$

From eqs. (12.11) and (12.12), it is clear that all field components vary with z according to $e^{-γz}$, that is,

$$E_{zs} \sim e^{-γz}, \quad E_{xs} \sim e^{-γz}$$

Hence

$$\frac{∂E_{zs}}{∂z} = -γE_{zs}, \quad \frac{∂^2E_{xs}}{∂z^2} = γ^2E_{xs}$$

and eq. (12.14) becomes

$$jωεE_{xs} = \frac{∂H_{zs}}{∂y} + \frac{1}{jωμ}\left(γ^2E_{xs} + γ\frac{∂E_{zs}}{∂x}\right)$$

or

$$-\frac{1}{jωμ}(γ^2 + ω^2με)\,E_{xs} = \frac{γ}{jωμ}\frac{∂E_{zs}}{∂x} + \frac{∂H_{zs}}{∂y}$$

Thus, if we let $h^2 = γ^2 + ω^2με = γ^2 + k^2$,

$$E_{xs} = -\frac{γ}{h^2}\frac{∂E_{zs}}{∂x} - \frac{jωμ}{h^2}\frac{∂H_{zs}}{∂y}$$

Similar manipulations of eqs. (12.13) yield expressions for E_{ys}, H_{xs}, and H_{ys} in terms of E_{zs} and H_{zs}. Thus,

$$E_{xs} = -\frac{γ}{h^2}\frac{∂E_{zs}}{∂x} - \frac{jωμ}{h^2}\frac{∂H_{zs}}{∂y} \qquad (12.15a)$$

$$E_{ys} = -\frac{γ}{h^2}\frac{∂E_{zs}}{∂y} + \frac{jωμ}{h^2}\frac{∂H_{zs}}{∂x} \qquad (12.15b)$$

$$H_{xs} = \frac{jωε}{h^2}\frac{∂E_{zs}}{∂y} - \frac{γ}{h^2}\frac{∂H_{zs}}{∂x} \qquad (12.15c)$$

$$H_{ys} = -\frac{jωε}{h^2}\frac{∂E_{zs}}{∂x} - \frac{γ}{h^2}\frac{∂H_{zs}}{∂y} \qquad (12.15d)$$

where

$$h^2 = γ^2 + k^2 = k_x^2 + k_y^2 \qquad (12.16)$$

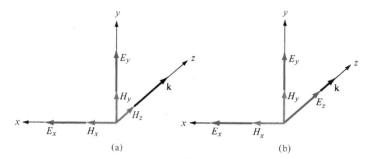

FIGURE 12.3 Components of EM fields in a rectangular waveguide: (**a**) TE mode $E_z = 0$, (**b**) TM mode, $H_z = 0$.

Thus we can use eqs. (12.15) in conjunction with eqs. (12.11) and (12.12) to obtain E_{xs}, E_{ys}, H_{xs}, and H_{ys}.

From eqs. (12.11), (12.12), and (12.15), we notice that the field patterns or configurations come in different types. Each of these distinct field patterns is called a *mode*. Four different mode categories can exist, namely:

1. $E_{zs} = 0 = H_{zs}$ (TEM mode): In the *transverse electromagnetic* mode, both the **E** and **H** fields are transverse to the direction of wave propagation. From eq. (12.15), all field components vanish for $E_{zs} = 0 = H_{zs}$. Consequently, we conclude that a hollow rectangular waveguide cannot support TEM mode.
2. $E_{zs} = 0$, $H_{zs} \neq 0$ (TE modes): For this case, the remaining components (E_{xs} and E_{ys}) of the electric field are transverse to the direction of propagation $\mathbf{a}_z$. Under this condition, fields are said to be in *transverse electric* (TE) modes. See Figure 12.3(a).
3. $E_{zs} \neq 0$, $H_{zs} = 0$ (TM modes): In this case, the **H** field is transverse to the direction of wave propagation. Thus we have *transverse magnetic* (TM) modes. See Figure 12.3(b).
4. $E_{zs} \neq 0$, $H_{zs} \neq 0$ (HE modes): In this case neither the **E** nor the **H** field is transverse to the direction of wave propagation. Sometimes these modes are referred to as *hybrid* modes.

We should note the relationship between k in eq. (12.3) and β of eq. (10.43a). The phase constant β in eq. (10.43a) was derived for TEM mode. For the TEM mode, $h = 0$, so from eq. (12.16), $\gamma^2 = -k^2 \rightarrow \gamma = \alpha + j\beta = jk$; that is, $\beta = k$. For other modes, $\beta \neq k$. In the subsequent sections, we shall examine the TM and TE modes of propagation separately.

12.3 TRANSVERSE MAGNETIC MODES

For the TM case, the magnetic field has its components transverse (or normal) to the direction of wave propagation. This implies that we set $H_z = 0$ and determine E_x, E_y, E_z, H_x, and H_y by using eqs. (12.11) and (12.15) and the boundary conditions. We shall solve for E_z and later determine other field components from E_z. At the walls (perfect conductors) of the waveguide in Figure 12.2, the tangential components of the **E** field must be continuous; that is,

$$E_{zs} = 0 \quad \text{at} \quad y = 0 \qquad \text{(bottom wall)} \qquad \text{(12.17a)}$$

$$E_{zs} = 0 \quad \text{at} \quad y = b \qquad \text{(top wall)} \qquad \text{(12.17b)}$$

$$E_{zs} = 0 \quad \text{at} \quad x = 0 \qquad \text{(right wall)} \qquad \text{(12.17c)}$$

$$E_{zs} = 0 \quad \text{at} \quad x = a \qquad \text{(left wall)} \qquad \text{(12.17d)}$$

Equations (12.17a) and (12.17c) require that $A_1 = 0 = A_3$ in eq. (12.11), so eq. (12.11) becomes

$$E_{zs} = E_o \sin k_x x \sin k_y y \, e^{-\gamma z} \qquad \text{(12.18)}$$

where $E_o = A_2 A_4$. Also eqs. (12.17d) and (12.17b) when applied to eq. (12.18) require, respectively, that

$$\sin k_x a = 0, \quad \sin k_y b = 0 \qquad \text{(12.19)}$$

This implies that

$$k_x a = m\pi, \quad m = 1, 2, 3, \ldots \qquad \text{(12.20a)}$$

$$k_y b = n\pi, \quad n = 1, 2, 3, \ldots \qquad \text{(12.20b)}$$

or

$$\boxed{k_x = \frac{m\pi}{a}, \quad k_y = \frac{n\pi}{b}} \qquad \text{(12.21)}$$

The negative integers are not chosen for m and n in eq. (12.20a) for the reason given in Example 6.5. Substituting eq. (12.21) into eq. (12.18) gives

$$\boxed{E_{zs} = E_o \sin\left(\frac{m\pi x}{a}\right) \sin\left(\frac{n\pi y}{b}\right) e^{-\gamma z}} \qquad \text{(12.22)}$$

We obtain other field components from eqs. (12.22) and (12.15), bearing in mind that $H_{zs} = 0$. Thus

$$E_{xs} = -\frac{\gamma}{h^2}\left(\frac{m\pi}{a}\right) E_o \cos\left(\frac{m\pi x}{a}\right) \sin\left(\frac{n\pi y}{b}\right) e^{-\gamma z} \qquad \text{(12.23a)}$$

$$E_{ys} = -\frac{\gamma}{h^2}\left(\frac{n\pi}{b}\right) E_o \sin\left(\frac{m\pi x}{a}\right) \cos\left(\frac{n\pi y}{b}\right) e^{-\gamma z} \qquad \text{(12.23b)}$$

$$H_{xs} = \frac{j\omega\varepsilon}{h^2}\left(\frac{n\pi}{b}\right) E_o \sin\left(\frac{m\pi x}{a}\right) \cos\left(\frac{n\pi y}{b}\right) e^{-\gamma z} \qquad \text{(12.23c)}$$

$$H_{ys} = -\frac{j\omega\varepsilon}{h^2}\left(\frac{m\pi}{a}\right) E_o \cos\left(\frac{m\pi x}{a}\right) \sin\left(\frac{n\pi y}{b}\right) e^{-\gamma z} \qquad \text{(12.23d)}$$

where

$$h^2 = k_x^2 + k_y^2 = \left[\frac{m\pi}{a}\right]^2 + \left[\frac{n\pi}{b}\right]^2 \qquad (12.24)$$

which is obtained from eqs. (12.16) and (12.21). Notice from eqs. (12.22) and (12.23) that each set of integers m and n gives a different field pattern or mode, referred to as TM_{mn} mode, in the waveguide. Integer m equals the number of half-cycle variations in the x-direction, and integer n is the number of half-cycle variations in the y-direction. We also notice from eqs. (12.22) and (12.23) that if (m, n) is $(0, 0)$, $(0, n)$, or $(m, 0)$, all field components vanish. Thus neither m nor n can be zero. Consequently, TM_{11} is the lowest-order mode of all the TM_{mn} modes.

By substituting eq. (12.21) into eq. (12.16), we obtain the propagation constant

$$\gamma = \sqrt{\left[\frac{m\pi}{a}\right]^2 + \left[\frac{n\pi}{b}\right]^2 - k^2} \qquad (12.25)$$

where $k = \omega\sqrt{\mu\varepsilon}$ as in eq. (12.3). We recall that, in general, $\gamma = \alpha + j\beta$. In the case of eq. (12.25), we have three possibilities depending on k (or ω), m, and n:

CASE 1 (cutoff)

If

$$k^2 = \omega^2\mu\varepsilon = \left[\frac{m\pi}{a}\right]^2 + \left[\frac{n\pi}{b}\right]^2$$

$$\gamma = 0 \qquad \text{or} \qquad \alpha = 0 = \beta$$

The value of ω that causes this is called the *cutoff angular frequency* ω_c; that is,

$$\omega_c = \frac{1}{\sqrt{\mu\varepsilon}}\sqrt{\left[\frac{m\pi}{a}\right]^2 + \left[\frac{n\pi}{b}\right]^2} \qquad (12.26)$$

No propagation takes place at this frequency.

CASE 2 (evanescent)

If

$$k^2 = \omega^2\mu\varepsilon < \left[\frac{m\pi}{a}\right]^2 + \left[\frac{n\pi}{b}\right]^2$$

$$\gamma = \alpha, \qquad \beta = 0$$

In this case, we have no wave propagation at all. These nonpropagating modes are said to be *evanescent*.

CASE 3 (propagation)

If

$$k^2 = \omega^2 \mu \varepsilon > \left[\frac{m\pi}{a}\right]^2 + \left[\frac{n\pi}{b}\right]^2$$

$$\gamma = j\beta, \qquad \alpha = 0$$

that is, from eq. (12.25) the phase constant β becomes

$$\beta = \sqrt{k^2 - \left[\frac{m\pi}{a}\right]^2 - \left[\frac{n\pi}{b}\right]^2} \qquad (12.27)$$

This is the only case in which propagation takes place, because all field components will have the factor $e^{-\gamma z} = e^{-j\beta z}$.

Thus for each mode, characterized by a set of integers m and n, there is a corresponding *cutoff frequency* f_c.

> The **cutoff frequency** is the operating frequency below which attenuation occurs and above which propagation takes place.

The waveguide therefore operates as a high-pass filter. The cutoff frequency is obtained from eq. (12.26) as

$$f_c = \frac{\omega_c}{2\pi} = \frac{1}{2\pi\sqrt{\mu\varepsilon}} \sqrt{\left[\frac{m\pi}{a}\right]^2 + \left[\frac{n\pi}{b}\right]^2}$$

or

$$\boxed{f_c = \frac{u'}{2} \sqrt{\left(\frac{m}{a}\right)^2 + \left(\frac{n}{b}\right)^2}} \qquad (12.28)$$

where $u' = \dfrac{1}{\sqrt{\mu\varepsilon}}$ = phase velocity of uniform plane wave in the lossless dielectric medium $(\sigma = 0, \mu, \varepsilon)$ filling the waveguide. The *cutoff wavelength* λ_c is given by

$$\lambda_c = \frac{u'}{f_c}$$

or

$$\boxed{\lambda_c = \frac{2}{\sqrt{\left(\frac{m}{a}\right)^2 + \left(\frac{n}{b}\right)^2}}} \qquad (12.29)$$

Note from eqs. (12.28) and (12.29) that TM_{11} has the lowest cutoff frequency (or the longest cutoff wavelength) of all the TM modes. The phase constant β in eq. (12.27) can be written in terms of f_c as

$$\beta = \omega \sqrt{\mu\varepsilon} \sqrt{1 - \left[\frac{f_c}{f}\right]^2}$$

or

$$\beta = \beta' \sqrt{1 - \left[\frac{f_c}{f}\right]^2} \tag{12.30}$$

where $\beta' = \omega/u' = \omega\sqrt{\mu\varepsilon}$ = phase constant of uniform plane wave in the dielectric medium. It should be noted that γ for evanescent mode can be expressed in terms of f_c, namely,

$$\gamma = \alpha = \beta' \sqrt{\left(\frac{f_c}{f}\right)^2 - 1} \tag{12.30'}$$

The phase velocity u_p and the wavelength in the guide are, respectively, given by

$$u_p = \frac{\omega}{\beta'}, \lambda = \frac{2\pi}{\beta} = \frac{u_p}{f} \tag{12.31}$$

The intrinsic wave impedance of the mode is obtained from eq. (12.23) as $(\gamma = j\beta)$

$$\eta_{TM} = \frac{E_x}{H_y} = -\frac{E_y}{H_x}$$

$$= \frac{\beta}{\omega\varepsilon} = \sqrt{\frac{\mu}{\varepsilon}} \sqrt{1 - \left[\frac{f_c}{f}\right]^2}$$

or

$$\eta_{TM} = \eta' \sqrt{1 - \left[\frac{f_c}{f}\right]^2} \tag{12.32}$$

where $\eta' = \sqrt{\mu/\varepsilon}$ is the intrinsic impedance of a uniform plane wave in the medium. Note the difference between u', β', and η', and u, β, and η. The primed quantities are wave characteristics of the dielectric medium unbounded by the waveguide, as discussed in Chapter 10 (i.e., for TEM mode). For example, u' would be the velocity of the wave if the waveguide were removed and the entire space were filled with the dielectric. The unprimed quantities are the wave characteristics of the medium bounded by the waveguide.

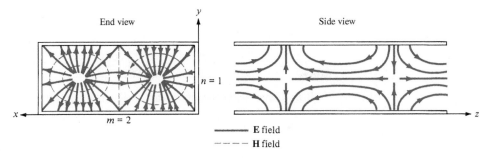

FIGURE 12.4 Field configuration for TM_{21} mode.

As mentioned before, the integers m and n indicate the number of half-cycle variations in the x–y cross section of the guide. Thus for a fixed time, the field configuration of Figure 12.4 results for TM_{21} mode, for example.

12.4 TRANSVERSE ELECTRIC MODES

In the TE modes, the electric field is transverse (or normal) to the direction of wave propagation. We set $E_z = 0$ and determine other field components E_x, E_y, H_x, H_y, and H_z from eqs. (12.12) and (12.15) and the boundary conditions just as we did for the TM modes. The boundary conditions are obtained from the requirement that the tangential components of the electric field be continuous at the walls (perfect conductors) of the waveguide; that is,

$$E_{xs} = 0 \quad \text{at} \quad y = 0 \tag{12.33a}$$

$$E_{xs} = 0 \quad \text{at} \quad y = b \tag{12.33b}$$

$$E_{ys} = 0 \quad \text{at} \quad x = 0 \tag{12.33c}$$

$$E_{ys} = 0 \quad \text{at} \quad x = a \tag{12.33d}$$

From eqs. (12.15) and (12.33), the boundary conditions can be written as

$$\frac{\partial H_{zs}}{\partial y} = 0 \quad \text{at} \quad y = 0 \tag{12.34a}$$

$$\frac{\partial H_{zs}}{\partial y} = 0 \quad \text{at} \quad y = b \tag{12.34b}$$

$$\frac{\partial H_{zs}}{\partial x} = 0 \quad \text{at} \quad x = 0 \tag{12.34c}$$

$$\frac{\partial H_{zs}}{\partial x} = 0 \quad \text{at} \quad x = a \tag{12.34d}$$

Imposing these boundary conditions on eq. (12.12) yields

$$H_{zs} = H_o \cos\left(\frac{m\pi x}{a}\right) \cos\left(\frac{n\pi y}{b}\right) e^{-\gamma z} \qquad (12.35)$$

where $H_o = B_1 B_3$. Other field components are easily obtained from eqs. (12.35) and (12.15) as

$$E_{xs} = \frac{j\omega\mu}{h^2}\left(\frac{n\pi}{b}\right) H_o \cos\left(\frac{m\pi x}{a}\right) \sin\left(\frac{n\pi y}{b}\right) e^{-\gamma z} \qquad (12.36a)$$

$$E_{ys} = -\frac{j\omega\mu}{h^2}\left(\frac{m\pi}{a}\right) H_o \sin\left(\frac{m\pi x}{a}\right) \cos\left(\frac{n\pi y}{b}\right) e^{-\gamma z} \qquad (12.36b)$$

$$H_{xs} = \frac{\gamma}{h^2}\left(\frac{m\pi}{a}\right) H_o \sin\left(\frac{m\pi x}{a}\right) \cos\left(\frac{n\pi y}{b}\right) e^{-\gamma z} \qquad (12.36c)$$

$$H_{ys} = \frac{\gamma}{h^2}\left(\frac{n\pi}{b}\right) H_o \cos\left(\frac{m\pi x}{a}\right) \sin\left(\frac{n\pi y}{b}\right) e^{-\gamma z} \qquad (12.36d)$$

where $m = 0, 1, 2, 3, \ldots$; and $n = 0, 1, 2, 3, \ldots$; h and γ remain as defined for the TM modes. Again, m and n denote the number of half-cycle variations in the x–y cross section of the guide. For TE_{32} mode, for example, the field configuration is in Figure 12.5. The cutoff frequency f_c, the cutoff wavelength λ_c, the phase constant β, the phase velocity u_p, and the wavelength λ for TE modes are the same as for TM modes [see eqs. (12.28) to (12.31)].

For TE modes, (m, n) may be $(0, 1)$ or $(1, 0)$ but not $(0, 0)$. Both m and n cannot be zero at the same time because this will force the field components in eq. (12.36) to vanish. This implies that the lowest mode can be TE_{10} or TE_{01} depending on the values of a and b, the dimensions of the guide. It is standard practice to have $a > b$ so that $1/a^2 < 1/b^2$ in eq. (12.28). Thus TE_{10} is the lowest mode because $f_{c_{TE_{10}}} = \frac{u'}{2a} < f_{c_{TE_{01}}} = \frac{u'}{2b}$. This mode is

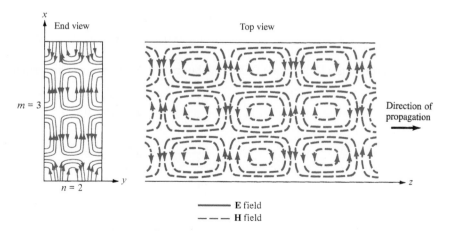

FIGURE 12.5 Field configuration for TE_{32} mode.

called the *dominant mode* of the waveguide and is of practical importance. The cutoff frequency for the TE_{10} mode is obtained from eq. (12.28) as $(m = 1, n = 0)$

$$f_{c_{10}} = \frac{u'}{2a} \tag{12.37}$$

and the cutoff wavelength for TE_{10} mode is obtained from eq. (12.29) as

$$\lambda_{c_{10}} = 2a \tag{12.38}$$

Note that from eq. (12.28) the cutoff frequency for TM_{11} is

$$\frac{u'[a^2 + b^2]^{1/2}}{2ab}$$

which is greater than the cutoff frequency for TE_{10}. Hence, TM_{11} cannot be regarded as the dominant mode.

> The **dominant mode** is the mode with the lowest cutoff frequency (or longest cutoff wavelength).

Also note that any EM wave with frequency $f < f_{c_{10}}$ (or $\lambda > \lambda_{c_{10}}$) will not be propagated in the guide.

The intrinsic impedance for the TE mode is not the same as for TM modes. From eq. (12.36), it is evident that $(\gamma = j\beta)$

$$\eta_{TE} = \frac{E_x}{H_y} = -\frac{E_y}{H_x} = \frac{\omega\mu}{\beta}$$

$$= \sqrt{\frac{\mu}{\varepsilon}} \frac{1}{\sqrt{1 - \left[\frac{f_c}{f}\right]^2}}$$

or

$$\boxed{\eta_{TE} = \frac{\eta'}{\sqrt{1 - \left[\frac{f_c}{f}\right]^2}}} \tag{12.39}$$

Note from eqs. (12.32) and (12.39) that η_{TE} and η_{TM} are purely resistive and vary with frequency, as shown in Figure 12.6. Also note that

$$\eta_{TE}\, \eta_{TM} = \eta'^2 \tag{12.40}$$

Important equations for TM and TE modes are listed in Table 12.1 for convenience and quick reference.

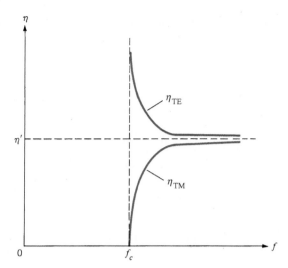

FIGURE 12.6 Variation of wave impedance with frequency for TE and TM modes.

TABLE 12.1 Important Equations for TM and TE Modes

TM Modes	TE Modes
$E_{xs} = -\dfrac{j\beta}{h^2}\left(\dfrac{m\pi}{a}\right)E_o \cos\left(\dfrac{m\pi x}{a}\right)\sin\left(\dfrac{n\pi y}{b}\right)e^{-\gamma z}$	$E_{xs} = \dfrac{j\omega\mu}{h^2}\left(\dfrac{n\pi}{b}\right)H_o \cos\left(\dfrac{m\pi x}{a}\right)\sin\left(\dfrac{n\pi y}{b}\right)e^{-\gamma z}$
$E_{ys} = -\dfrac{j\beta}{h^2}\left(\dfrac{n\pi}{b}\right)E_o \sin\left(\dfrac{m\pi x}{a}\right)\cos\left(\dfrac{n\pi y}{b}\right)e^{-\gamma z}$	$E_{ys} = -\dfrac{j\omega\mu}{h^2}\left(\dfrac{m\pi}{a}\right)H_o \sin\left(\dfrac{m\pi x}{a}\right)\cos\left(\dfrac{n\pi y}{b}\right)e^{-\gamma z}$
$E_{zs} = E_o \sin\left(\dfrac{m\pi x}{a}\right)\sin\left(\dfrac{n\pi y}{b}\right)e^{-\gamma z}$	$E_{zs} = 0$
$H_{xs} = \dfrac{j\omega\varepsilon}{h^2}\left(\dfrac{n\pi}{b}\right)E_o \sin\left(\dfrac{m\pi x}{a}\right)\cos\left(\dfrac{n\pi y}{b}\right)e^{-\gamma z}$	$H_{xs} = \dfrac{j\beta}{h^2}\left(\dfrac{m\pi}{a}\right)H_o \sin\left(\dfrac{m\pi x}{a}\right)\cos\left(\dfrac{n\pi y}{b}\right)e^{-\gamma z}$
$H_{ys} = -\dfrac{j\omega\varepsilon}{h^2}\left(\dfrac{m\pi}{a}\right)E_o \cos\left(\dfrac{m\pi x}{a}\right)\sin\left(\dfrac{n\pi y}{b}\right)e^{-\gamma z}$	$H_{ys} = \dfrac{j\beta}{h^2}\left(\dfrac{n\pi}{b}\right)H_o \cos\left(\dfrac{m\pi x}{a}\right)\sin\left(\dfrac{n\pi y}{b}\right)e^{-\gamma z}$
$H_{zs} = 0$	$H_{zs} = H_o \cos\left(\dfrac{m\pi x}{a}\right)\cos\left(\dfrac{n\pi y}{b}\right)e^{-\gamma z}$
$\eta = \eta'\sqrt{1 - \left(\dfrac{f_c}{f}\right)^2}$	$\eta = \dfrac{\eta'}{\sqrt{1 - \left(\dfrac{f_c}{f}\right)^2}}$

$$f_c = \frac{u'}{2}\sqrt{\left(\frac{m}{a}\right)^2 + \left(\frac{n}{b}\right)^2}$$

$$\lambda_c = \frac{u'}{f_c}$$

$$\beta = \beta'\sqrt{1 - \left(\frac{f_c}{f}\right)^2}$$

$$u_p = \frac{\omega}{\beta} = f\lambda$$

where $h^2 = \left(\dfrac{m\pi}{a}\right)^2 + \left(\dfrac{n\pi}{b}\right)^2$, $u' = \dfrac{1}{\sqrt{\mu\varepsilon}}$, $\beta' = \dfrac{\omega}{u'}$, $\eta' = \sqrt{\dfrac{\mu}{\varepsilon}}$

From eqs. (12.22), (12.23), (12.35), and (12.36), we obtain the field patterns for the TM and TE modes. For the dominant TE_{10} mode, $m = 1$ and $n = 0$, so eq. (12.35) becomes

$$H_{zs} = H_o \cos\left(\frac{\pi x}{a}\right) e^{-j\beta z} \tag{12.41}$$

In the time domain,

$$H_z = \text{Re}(H_{zs} e^{j\omega t})$$

or

$$H_z = H_o \cos\left(\frac{\pi x}{a}\right) \cos(\omega t - \beta z) \tag{12.42}$$

Similarly, from eq. (12.36),

$$E_y = \frac{\omega \mu a}{\pi} H_o \sin\left(\frac{\pi x}{a}\right) \sin(\omega t - \beta z) \tag{12.43a}$$

$$H_x = -\frac{\beta a}{\pi} H_o \sin\left(\frac{\pi x}{a}\right) \sin(\omega t - \beta z) \tag{12.43b}$$

$$E_z = E_x = H_y = 0 \tag{12.43c}$$

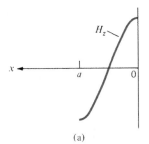

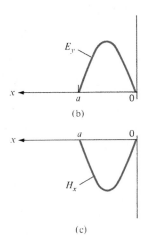

FIGURE 12.7 Variation of the field components with x for TE_{10} mode.

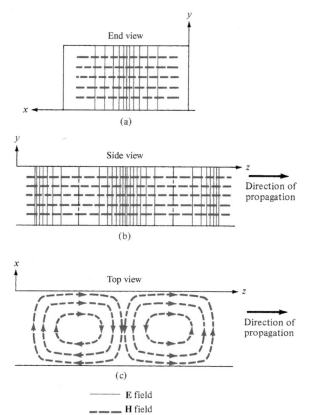

FIGURE 12.8 Field lines for TE_{10} mode, corresponding to components (**a**), (**b**), and (**c**) in Figure 12.7.

The variation of the **E** and **H** fields with x in an xy-plane, say plane $\cos(\omega t - \beta z) = 1$ for H_z, and plane $\sin(\omega t - \beta z) = 1$ for E_y and H_x, is shown in Figure 12.7 for the TE_{10} mode. The corresponding field lines are shown in Figure 12.8.

EXAMPLE 12.1

A rectangular waveguide with dimensions $a = 2.5$ cm, $b = 1$ cm is to operate below 15.1 GHz. How many TE and TM modes can the waveguide transmit if the guide is filled with a medium characterized by $\sigma = 0$, $\varepsilon = 4\,\varepsilon_o$, $\mu_r = 1$? Calculate the cutoff frequencies of the modes.

Solution:

The cutoff frequency is given by

$$f_{c_{mn}} = \frac{u'}{2}\sqrt{\frac{m^2}{a^2} + \frac{n^2}{b^2}}$$

where $a = 2.5b$ or $a/b = 2.5$, and

$$u' = \frac{1}{\sqrt{\mu\varepsilon}} = \frac{c}{\sqrt{\mu_r\varepsilon_r}} = \frac{c}{2}$$

Hence,

$$f_{c_{mn}} = \frac{c}{4a}\sqrt{m^2 + \frac{a^2}{b^2}n^2}$$

$$= \frac{3 \times 10^8}{4(2.5 \times 10^{-2})}\sqrt{m^2 + 6.25n^2}$$

or

$$f_{c_{mn}} = 3\sqrt{m^2 + 6.25n^2} \text{ GHz} \tag{12.1.1}$$

We are looking for $f_{c_{mn}} < 15.1$ GHz. A systematic way of doing this is to fix m or n and increase the other until $f_{c_{mn}}$ is greater than 15.1 GHz. From eq. (12.1.1), it is evident that fixing m and increasing n will quickly give us an $f_{c_{mn}}$ that is greater than 15.1 GHz.

For TE$_{01}$ mode $(m = 0, n = 1)$, $f_{c_{01}} = 3(2.5) = 7.5$ GHz

$$\text{TE}_{02} \text{ mode } (m = 0, n = 2), f_{c_{02}} = 3(5) = 15 \text{ GHz}$$

$$\text{TE}_{03} \text{ mode}, f_{c_{03}} = 3(7.5) = 22.5 \text{ GHz}$$

Thus for $f_{c_{mn}} < 15.1$ GHz, the maximum $n = 2$. We now fix n and increase m until $f_{c_{mn}}$ is greater than 15.1 GHz.

For TE$_{10}$ mode $(m = 1, n = 0)$, $f_{c_{10}} = 3$ GHz

$$\text{TE}_{20} \text{ mode}, f_{c_{20}} = 6 \text{ GHz}$$

$$\text{TE}_{30} \text{ mode}, f_{c_{30}} = 9 \text{ GHz}$$

$$\text{TE}_{40} \text{ mode}, f_{c_{40}} = 12 \text{ GHz}$$

$$\text{TE}_{50} \text{ mode}, f_{c_{50}} = 15 \text{ GHz (the same as for TE}_{02})$$

$$\text{TE}_{60} \text{ mode}, f_{c_{60}} = 18 \text{ GHz}$$

That is, for $f_{c_{mn}} < 15.1$ GHz, the maximum $m = 5$. Now that we know the maximum m and n, we try other possible combinations in between these maximum values.

For TE$_{11}$, TM$_{11}$ (degenerate modes), $f_{c_{11}} = 3\sqrt{7.25} = 8.078$ GHz

$$\text{TE}_{21}, \text{TM}_{21}, f_{c_{21}} = 3\sqrt{10.25} = 9.6 \text{ GHz}$$

$$\text{TE}_{31}, \text{TM}_{31}, f_{c_{31}} = 3\sqrt{15.25} = 11.72 \text{ GHz}$$

$$\text{TE}_{41}, \text{TM}_{41}, f_{c_{41}} = 3\sqrt{22.25} = 14.14 \text{ GHz}$$

$$\text{TE}_{12}, \text{TM}_{12}, f_{c_{12}} = 3\sqrt{26} = 15.3 \text{ GHz}$$

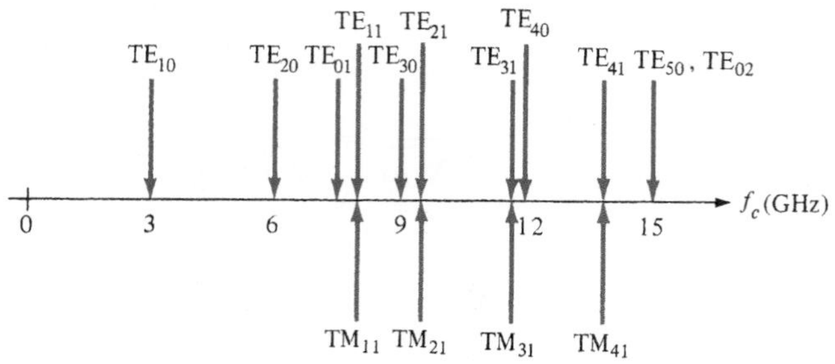

FIGURE 12.9 Cutoff frequencies of rectangular waveguide with $a = 2.5b$; for Example 12.1.

Those modes, whose cutoff frequencies are less than or equal to 15.1 GHz, will be transmitted—that is, 11 TE modes and 4 TM modes (all the foregoing modes except TE_{12}, TM_{12}, TE_{60}, and TE_{03}). The cutoff frequencies for the 15 modes are illustrated in Figure 12.9.

PRACTICE EXERCISE 12.1

Consider the waveguide of Example 12.1. Calculate the phase constant, phase velocity, and wave impedance for TE_{10} and TM_{11} modes at the operating frequency of 15 GHz.

Answer: For TE_{10}, $\beta = 615.6$ rad/m, $u = 1.531 \times 10^8$ m/s, $\eta_{TE} = 192.4\ \Omega$; for TM_{11}, $\beta = 529.4$ rad/m, $u = 1.78 \times 10^8$ m/s, $\eta_{TM} = 158.8\ \Omega$.

EXAMPLE 12.2

Write the general instantaneous field expressions for the TM and TE modes. Deduce those for TE_{01} and TM_{12} modes.

Solution:

The instantaneous field expressions are obtained from the phasor forms by using

$$\mathbf{E} = \text{Re}\left(\mathbf{E}_s e^{j\omega t}\right) \quad \text{and} \quad \mathbf{H} = \text{Re}\left(\mathbf{H}_s e^{j\omega t}\right)$$

Applying these to eqs. (12.22) and (12.23) while replacing γ with $j\beta$ gives the following field components for the TM modes:

$$E_x = \frac{\beta}{h^2}\left[\frac{m\pi}{a}\right] E_o \cos\left(\frac{m\pi x}{a}\right) \sin\left(\frac{n\pi y}{b}\right) \sin\left(\omega t - \beta z\right)$$

$$E_y = \frac{\beta}{h^2}\left[\frac{n\pi}{b}\right] E_o \sin\left(\frac{m\pi x}{a}\right) \cos\left(\frac{n\pi y}{b}\right) \sin\left(\omega t - \beta z\right)$$

$$E_z = E_o \sin\left(\frac{m\pi x}{a}\right) \sin\left(\frac{n\pi y}{b}\right) \cos\left(\omega t - \beta z\right)$$

$$H_x = -\frac{\omega\varepsilon}{h^2}\left[\frac{n\pi}{b}\right]E_o \sin\left(\frac{m\pi x}{a}\right)\cos\left(\frac{n\pi y}{b}\right)\sin(\omega t - \beta z)$$

$$H_y = \frac{\omega\varepsilon}{h^2}\left[\frac{m\pi}{a}\right]E_o \cos\left(\frac{m\pi x}{a}\right)\sin\left(\frac{n\pi y}{b}\right)\sin(\omega t - \beta z)$$

$$H_z = 0$$

Similarly, for the TE modes, eqs. (12.35) and (12.36) become

$$E_x = -\frac{\omega\mu}{h^2}\left[\frac{n\pi}{b}\right]H_o \cos\left(\frac{m\pi x}{a}\right)\sin\left(\frac{n\pi y}{b}\right)\sin(\omega t - \beta z)$$

$$E_y = \frac{\omega\mu}{h^2}\left[\frac{m\pi}{a}\right]H_o \sin\left(\frac{m\pi x}{a}\right)\cos\left(\frac{n\pi y}{b}\right)\sin(\omega t - \beta z)$$

$$E_z = 0$$

$$H_x = -\frac{\beta}{h^2}\left[\frac{m\pi}{a}\right]H_o \sin\left(\frac{m\pi x}{a}\right)\cos\left(\frac{n\pi y}{b}\right)\sin(\omega t - \beta z)$$

$$H_y = -\frac{\beta}{h^2}\left[\frac{n\pi}{b}\right]H_o \cos\left(\frac{m\pi x}{a}\right)\sin\left(\frac{n\pi y}{b}\right)\sin(\omega t - \beta z)$$

$$H_z = H_o \cos\left(\frac{m\pi x}{a}\right)\cos\left(\frac{n\pi y}{b}\right)\cos(\omega t - \beta z)$$

For the TE_{01} mode, we set $m = 0$, $n = 1$ to obtain

$$h^2 = \left[\frac{\pi}{b}\right]^2$$

$$E_x = -\frac{\omega\mu b}{\pi}H_o \sin\left(\frac{\pi y}{b}\right)\sin(\omega t - \beta z)$$

$$E_y = 0 = E_z = H_x$$

$$H_y = -\frac{\beta b}{\pi}H_o \sin\left(\frac{\pi y}{b}\right)\sin(\omega t - \beta z)$$

$$H_z = H_o \cos\left(\frac{\pi y}{b}\right)\cos(\omega t - \beta z)$$

For the TM_{12} mode, we set $m = 1$, $n = 2$ to obtain

$$E_x = \frac{\beta}{h^2}\left(\frac{\pi}{a}\right)E_o \cos\left(\frac{\pi x}{a}\right)\sin\left(\frac{2\pi y}{b}\right)\sin(\omega t - \beta z)$$

$$E_y = \frac{\beta}{h^2}\left(\frac{2\pi}{b}\right)E_o \sin\left(\frac{\pi x}{a}\right)\cos\left(\frac{2\pi y}{b}\right)\sin(\omega t - \beta z)$$

$$E_z = E_o \sin\left(\frac{\pi x}{a}\right)\sin\left(\frac{2\pi y}{b}\right)\cos(\omega t - \beta z)$$

$$H_x = -\frac{\omega\varepsilon}{h^2}\left(\frac{2\pi}{b}\right) E_o \sin\left(\frac{\pi x}{a}\right) \cos\left(\frac{2\pi y}{b}\right) \sin(\omega t - \beta z)$$

$$H_y = \frac{\omega\varepsilon}{h^2}\left(\frac{\pi}{a}\right) E_o \cos\left(\frac{\pi x}{a}\right) \sin\left(\frac{2\pi y}{b}\right) \sin(\omega t - \beta z)$$

$$H_z = 0$$

where

$$h^2 = \left[\frac{\pi}{a}\right]^2 + \left[\frac{2\pi}{b}\right]^2$$

PRACTICE EXERCISE 12.2

At 15 GHz, an air-filled 5 cm × 2 cm waveguide has

$$E_{zs} = 20 \sin 40\pi x \sin 50\pi y \, e^{-j\beta z} \, \text{V/m}$$

(a) What mode is being propagated?
(b) Find β.
(c) Determine E_y/E_x.

Answer: (a) TM_{21}, (b) 241.3 rad/m, (c) 1.25 tan 40πx cot 50πy.

EXAMPLE 12.3

In a rectangular waveguide for which $a = 1.5$ cm, $b = 0.8$ cm, $\sigma = 0$, $\mu = \mu_o$, and $\varepsilon = 4\varepsilon_o$,

$$H_x = 2 \sin\left(\frac{\pi x}{a}\right) \cos\left(\frac{3\pi y}{b}\right) \sin(\pi \times 10^{11}t - \beta z) \, \text{A/m}$$

Determine

(a) The mode of operation
(b) The cutoff frequency
(c) The phase constant β
(d) The propagation constant γ
(e) The intrinsic wave impedance η

Solution:

(a) It is evident from the given expression for H_x and the field expressions in Example 12.2 that $m = 1$, $n = 3$; that is, the guide is operating at TM_{13} or TE_{13}. Suppose we choose TM_{13} mode (the possibility of having TE_{13} mode is left as an exercise in Practice Exercise 12.3).

(b)
$$f_{c_{mn}} = \frac{u'}{2} \sqrt{\frac{m^2}{a^2} + \frac{n^2}{b^2}}$$

$$u' = \frac{1}{\sqrt{\mu\varepsilon}} = \frac{c}{\sqrt{\mu_r \varepsilon_r}} = \frac{c}{2}$$

Hence

$$f_{c_{13}} = \frac{c}{4} \sqrt{\frac{1}{[1.5 \times 10^{-2}]^2} + \frac{9}{[0.8 \times 10^{-2}]^2}}$$

$$= \frac{3 \times 10^8}{4} (\sqrt{0.444 + 14.06}) \times 10^2 = 28.57 \text{ GHz}$$

(c)
$$\beta = \omega\sqrt{\mu\varepsilon} \sqrt{1 - \left[\frac{f_c}{f}\right]^2} = \frac{\omega\sqrt{\varepsilon_r}}{c} \sqrt{1 - \left[\frac{f_c}{f}\right]^2}$$

$$\omega = 2\pi f = \pi \times 10^{11} \quad \text{or} \quad f = \frac{10^{11}}{2} = 50 \text{ GHz}$$

$$\beta = \frac{\pi \times 10^{11}(2)}{3 \times 10^8} \sqrt{1 - \left[\frac{28.57}{50}\right]^2} = 1718.81 \text{ rad/m}$$

(d)
$$\gamma = j\beta = j1718.81 \text{ /m}$$

(e)
$$\eta_{TM_{13}} = \eta' \sqrt{1 - \left[\frac{f_c}{f}\right]^2} = \frac{377}{\sqrt{\varepsilon_r}} \sqrt{1 - \left[\frac{28.57}{50}\right]^2}$$

$$= 154.7 \ \Omega$$

PRACTICE EXERCISE 12.3

Repeat Example 12.3 if TE_{13} mode is assumed. Determine other field components for this mode.

Answer: $f_c = 28.57 \text{ GHz}$, $\beta = 1718.81 \text{ rad/m}$, $\gamma = j\beta$, $\eta_{TE_{13}} = 229.69 \ \Omega$

$$E_x = 2584.1 \cos\left(\frac{\pi x}{a}\right) \sin\left(\frac{3\pi y}{b}\right) \sin(\omega t - \beta z) \text{ V/m}$$

$$E_y = -459.4 \sin\left(\frac{\pi x}{a}\right) \cos\left(\frac{3\pi y}{b}\right) \sin(\omega t - \beta z) \text{ V/m}, \quad E_z = 0$$

$$H_y = 11.25 \cos\left(\frac{\pi x}{a}\right) \sin\left(\frac{3\pi y}{b}\right) \sin(\omega t - \beta z) \text{ A/m}$$

$$H_z = -7.96 \cos\left(\frac{\pi x}{a}\right) \cos\left(\frac{3\pi y}{b}\right) \cos(\omega t - \beta z) \text{ A/m}$$

12.5 WAVE PROPAGATION IN THE GUIDE

Examination of eq. (12.23) or (12.36) shows that the field components all involve the terms sine or cosine of $(m\pi/a)x$ or $(n\pi/b)y$ times $e^{-\gamma z}$. Since

$$\sin \theta = \frac{1}{2j}\left(e^{j\theta} - e^{-j\theta}\right) \tag{12.44a}$$

$$\cos \theta = \frac{1}{2}\left(e^{j\theta} + e^{-j\theta}\right) \tag{12.44b}$$

a wave within the waveguide can be resolved into a combination of plane waves reflected from the waveguide walls. For the TE_{10} mode, for example,

$$
\begin{aligned}
E_{ys} &= -\frac{j\omega\mu a}{\pi} \sin\left(\frac{\pi x}{a}\right) e^{-j\beta z} \\
&= -\frac{\omega\mu a}{2\pi}\left(e^{j\pi x/a} - e^{-j\pi x/a}\right) e^{-j\beta z} \\
&= \frac{\omega\mu a}{2\pi}\left[e^{-j\beta(z + \pi x/\beta a)} - e^{-j\beta(z - \pi x/\beta a)}\right]
\end{aligned}
\tag{12.45}
$$

where $H_o = 1$. The first term of eq. (12.45) represents a wave traveling in the positive z-direction at an angle

$$\theta = \tan^{-1}\left(\frac{\pi}{\beta a}\right) \tag{12.46}$$

with the z-axis. The second term of eq. (12.45) represents a wave traveling in the positive z-direction at an angle $-\theta$. The field may be depicted as a sum of two plane TEM waves propagating along zigzag paths between the guide walls at $x = 0$ and $x = a$ as illustrated in Figure 12.10(a). The decomposition of the TE_{10} mode into two plane waves can be extended to any TE and TM mode. When n and m are both different from zero, four plane waves result from the decomposition.

The wave component in the z-direction has a different wavelength from that of the plane waves. This wavelength along the axis of the guide is called the *waveguide wavelength* and is given by

$$\lambda = \frac{\lambda'}{\sqrt{1 - \left[\dfrac{f_c}{f}\right]^2}} \tag{12.47}$$

where $\lambda' = u'/f$.

As a consequence of the zigzag paths, we have three types of velocity: the *medium velocity* u', the *phase velocity* u_p, and the *group velocity* u_g. Figure 12.10(b) illustrates the relationship between the three different velocities. The medium velocity $u' = 1/\sqrt{\mu\varepsilon}$ is as

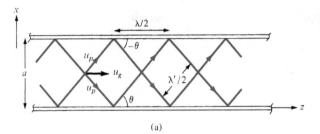

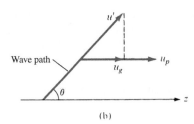

FIGURE 12.10 (**a**) Decomposition of the TE_{10} mode into two plane waves. (**b**) Relationship between u', u_p, and u_g.

explained in the preceding sections. The phase velocity u_p is the velocity at which loci of constant phase are propagated down the guide and is given by eq. (12.31); that is,

$$u_p = \frac{\omega}{\beta} \tag{12.48a}$$

or

$$u_p = \frac{u'}{\cos\theta} = \frac{u'}{\sqrt{1 - \left[\dfrac{f_c}{f}\right]^2}} \tag{12.48b}$$

This shows that $u_p \geq u'$, since $\cos\theta \leq 1$. If $u' = c$, then u_p is greater than the speed of light in vacuum. Does this violate Einstein's relativity theory that messages cannot travel faster than the speed of light? Not really, because information (or energy) in a waveguide generally does not travel at the phase velocity. Information travels at the group velocity, which must be less than the speed of light. The group velocity u_g is the velocity with which the resultant repeated reflected waves are traveling down the guide and is given by

$$\boxed{u_g = \frac{1}{\partial\beta/\partial\omega}} \tag{12.49a}$$

or

$$u_g = u' \cos\theta = u' \sqrt{1 - \left[\frac{f_c}{f}\right]^2} \tag{12.49b}$$

Although the concept of group velocity is fairly complex and is beyond the scope of this chapter, a group velocity is essentially the velocity of propagation of the wave-packet envelope of a group of frequencies. It is the energy propagation velocity in the guide and is always less than or equal to u'. From eqs. (12.48) and (12.49), it is evident that

$$u_p u_g = u'^2 \tag{12.50}$$

This relation is similar to eq. (12.40). Hence the variation of u_p and u_g with frequency is similar to that in Figure 12.6 for η_{TE} and η_{TM}.

EXAMPLE 12.4

A standard air-filled rectangular waveguide with dimensions $a = 8.636$ cm, $b = 4.318$ cm is fed by a 4 GHz carrier from a coaxial cable. Determine whether a TE_{10} mode will be propagated. If so, calculate the phase velocity and the group velocity.

Solution:

For the TE_{10} mode, $f_c = u'/2a$. Since the waveguide is air filled, $u' = c = 3 \times 10^8$. Hence,

$$f_c = \frac{3 \times 10^8}{2 \times 8.636 \times 10^{-2}} = 1.737 \text{ GHz}$$

As $f = 4$ GHz $> f_c$, the TE_{10} mode will propagate.

$$u_p = \frac{u'}{\sqrt{1 - (f_c/f)^2}} = \frac{3 \times 10^8}{\sqrt{1 - (1.737/4)^2}}$$
$$= 3.33 \times 10^8 \text{ m/s}$$
$$u_g = \frac{u'^2}{u_p} = \frac{9 \times 10^{16}}{3.33 \times 10^8} = 2.702 \times 10^8 \text{ m/s}$$

PRACTICE EXERCISE 12.4

Repeat Example 12.4 for the TM_{11} mode.

Answer: 12.5×10^8 m/s, 7.2×10^7 m/s.

12.6 POWER TRANSMISSION AND ATTENUATION

To determine power flow in the waveguide, we first find the average Poynting vector [given earlier as eq. (10.78)],

$$\mathcal{P}_{ave} = \frac{1}{2} \text{Re}(\mathbf{E}_s \times \mathbf{H}_s^*) \tag{12.51}$$

In this case, the Poynting vector is along the z-direction so that

$$\mathcal{P}_{\text{ave}} = \frac{1}{2}\,\text{Re}\!\left(E_{xs}H_{ys}^* - E_{ys}H_{xs}^*\right)\mathbf{a}_z$$

$$= \frac{|E_{xs}|^2 + |E_{ys}|^2}{2\eta}\,\mathbf{a}_z \tag{12.52}$$

where $\eta = \eta_{\text{TE}}$ for TE modes or $\eta = \eta_{\text{TM}}$ for TM modes. The total average power transmitted across the cross section of the waveguide is

$$P_{\text{ave}} = \int \mathcal{P}_{\text{ave}} \cdot d\mathbf{S}$$

$$= \int_{x=0}^{a} \int_{y=0}^{b} \frac{|E_{xs}|^2 + |E_{ys}|^2}{2\eta}\, dy\, dx \tag{12.53}$$

Of practical importance is the attenuation in a lossy waveguide. In our analysis thus far, we have assumed lossless waveguides ($\sigma = 0, \sigma_c \simeq \infty$) for which $\alpha = 0$, $\gamma = j\beta$. When the dielectric medium is lossy ($\sigma \neq 0$) and the guide walls are not perfectly conducting ($\sigma_c \neq \infty$), there is a continuous loss of power as a wave propagates along the guide. According to eq. (10.79), the power flow in the guide is of the form

$$P_{\text{ave}} = P_o e^{-2\alpha z} \tag{12.54}$$

In general,

$$\alpha = \alpha_c + \alpha_d \tag{12.55}$$

where α_c and α_d are attenuation constants due to ohmic or conduction losses ($\sigma_c \neq \infty$) and dielectric losses ($\sigma \neq 0$), respectively.

To determine α_d, recall that we started with eq. (12.1) assuming a lossless dielectric medium ($\sigma = 0$). For a lossy dielectric, we need to incorporate the fact that $\sigma \neq 0$. All our equations still hold except that $\gamma = j\beta$ needs to be modified. This is achieved by replacing ε in eq. (12.25) by the complex permittivity of eq. (10.40). Thus, we obtain

$$\gamma = \alpha_d + j\beta_d = \sqrt{\left(\frac{m\pi}{a}\right)^2 + \left(\frac{n\pi}{b}\right)^2 - \omega^2\mu\varepsilon_c} \tag{12.56}$$

where

$$\varepsilon_c = \varepsilon' - j\varepsilon'' = \varepsilon - j\frac{\sigma}{\omega} \tag{12.57}$$

Substituting eq. (12.57) into eq. (12.56) and squaring both sides of the equation, we obtain

$$\gamma^2 = \alpha_d^2 - \beta_d^2 + 2j\alpha_d\beta_d = \left(\frac{m\pi}{a}\right)^2 + \left(\frac{n\pi}{b}\right)^2 - \omega^2\mu\varepsilon + j\omega\mu\sigma$$

Equating real and imaginary parts, we have

$$\alpha_d^2 - \beta_d^2 = \left(\frac{m\pi}{a}\right)^2 + \left(\frac{n\pi}{b}\right)^2 - \omega^2\mu\varepsilon \tag{12.58a}$$

$$2\alpha_d\beta_d = \omega\mu\sigma \quad \text{or} \quad \alpha_d = \frac{\omega\mu\sigma}{2\beta_d} \tag{12.58b}$$

Assuming that $\alpha_d^2 \ll \beta_d^2$, $\alpha_d^2 - \beta_d^2 \simeq -\beta_d^2$, so eq. (12.58a) gives

$$\beta_d = \sqrt{\omega^2\mu\varepsilon - \left(\frac{m\pi}{a}\right)^2 - \left(\frac{n\pi}{b}\right)^2}$$

$$= \omega\sqrt{\mu\varepsilon}\sqrt{1 - \left(\frac{f_c}{f}\right)^2} \tag{12.59}$$

which is the same as β in eq. (12.30). Substituting eq. (12.59) into eq. (12.58b) gives

$$\boxed{\alpha_d = \frac{\sigma\eta'}{2\sqrt{1 - \left(\frac{f_c}{f}\right)^2}}} \tag{12.60}$$

where $\eta' = \sqrt{\mu/\varepsilon}$.

The determination of α_c for TM_{mn} and TE_{mn} modes is time-consuming and tedious. We shall illustrate the procedure by finding α_c for the TE_{10} mode. For this mode, only E_y, H_x, and H_z exist. Substituting eq. (12.43a) into eq. (12.53) yields

$$P_{\text{ave}} = \int_{x=0}^{a}\int_{y=0}^{b}\frac{|E_{ys}|^2}{2\eta}\,dx\,dy = \frac{\omega^2\mu^2a^2H_o^2}{2\pi^2\eta}\int_0^b dy\int_0^a \sin^2\frac{\pi x}{a}\,dx$$

$$P_{\text{ave}} = \frac{\omega^2\mu^2a^3H_o^2b}{4\pi^2\eta} \tag{12.61}$$

The total power loss per unit length in the walls is

$$P_L = P_L\big|_{y=0} + P_L\big|_{y=b} + P_L\big|_{x=0} + P_L\big|_{x=a}$$

$$= 2(P_L\big|_{y=0} + P_L\big|_{x=0}) \tag{12.62}$$

since the same amount is dissipated in the walls $y = 0$ and $y = b$ or $x = 0$ and $x = a$. For the wall $y = 0$,

$$P_L\big|_{y=0} = \frac{1}{2}\text{Re}\left[\eta_c\int\left(|H_{xs}|^2 + |H_{zs}|^2\right)dx\right]\Bigg|_{y=0}$$

$$= \frac{1}{2}R_s\left[\int_0^a\frac{\beta^2a^2}{\pi^2}H_o^2\sin^2\frac{\pi x}{a}\,dx + \int_0^a H_o^2\cos^2\frac{\pi x}{a}\,dx\right] \tag{12.63}$$

$$= \frac{R_saH_o^2}{4}\left(1 + \frac{\beta^2a^2}{\pi^2}\right)$$

where R_s is the real part of the intrinsic impedance η_c of the conducting wall. From eq. (10.56), we write

$$R_s = \frac{1}{\sigma_c \delta} = \sqrt{\frac{\pi f \mu}{\sigma_c}} \tag{12.64}$$

where δ is the skin depth. The skin resistance of the wall R_s may be regarded as the resistance of 1 m by δ by 1 m of the conducting material. For the wall $x = 0$,

$$P_L \, |_{x=0} = \frac{1}{2} \text{Re} \left[\eta_c \int (|H_{zs}|^2) \, dy \right] \, |_{x=0} = \frac{1}{2} R_s \int_0^b H_o^2 \, dy$$
$$= \frac{R_s b H_o^2}{2} \tag{12.65}$$

Substituting eqs. (12.63) and (12.65) into eq. (12.62) gives

$$P_L = R_s H_o^2 \left[b + \frac{a}{2} \left(1 + \frac{\beta^2 a^2}{\pi^2} \right) \right] \tag{12.66}$$

For energy to be conserved, the rate of decrease in P_{ave} must equal the time-average power loss P_L per unit length; that is,

$$P_L = -\frac{dP_{\text{ave}}}{dz} = 2\alpha P_{\text{ave}}$$

or

$$\alpha = \frac{P_L}{2P_{\text{ave}}} \tag{12.67}$$

Finally, substituting eqs. (12.61) and (12.66) into eq. (12.67), we have

$$\alpha_c = \frac{R_s H_o^2 \left[b + \frac{a}{2} \left(1 + \frac{\beta^2 a^2}{\pi^2} \right) \right] 2\pi^2 \eta}{\omega^2 \mu^2 a^3 H_o^2 b} \tag{12.68a}$$

It is convenient to express α_c in terms of f and f_c. After some manipulations, we obtain for the TE_{10} mode

$$\boxed{\alpha_c = \frac{2R_s}{b\eta' \sqrt{1 - \left[\frac{f_c}{f} \right]^2}} \left(0.5 + \frac{b}{a} \left[\frac{f_c}{f} \right]^2 \right)} \tag{12.68b}$$

By following the same procedure, the attenuation constant for the TE_{mn} modes $(n \neq 0)$ can be obtained as

$$\boxed{\alpha_c \, |_{\text{TE}} = \frac{2R_s}{b\eta' \sqrt{1 - \left[\frac{f_c}{f} \right]^2}} \left[\left(1 + \frac{b}{a} \right) \left[\frac{f_c}{f} \right]^2 + \frac{\frac{b}{a} \left(\frac{b}{a} m^2 + n^2 \right)}{\frac{b^2}{a^2} m^2 + n^2} \left(1 - \left[\frac{f_c}{f} \right]^2 \right) \right]} \tag{12.69}$$

and for the TM_{mn} modes as

$$\alpha_c \big|_{TM} = \frac{2R_s}{b\eta' \sqrt{1 - \left[\frac{f_c}{f}\right]^2}} \left[\frac{(b/a)^3 m^2 + n^2}{(b/a)^2 m^2 + n^2}\right] \tag{12.70}$$

The total attenuation constant α is obtained by substituting eqs. (12.60) and (12.69) or (12.70) into eq. (12.55).

†12.7 WAVEGUIDE CURRENT AND MODE EXCITATION

For either TM or TE modes, the surface current density **K** on the walls of the waveguide may be found by using

$$\mathbf{K} = \mathbf{a}_n \times \mathbf{H} \tag{12.71}$$

where $\mathbf{a}_n$ is the unit outward normal to the wall and **H** is the field intensity evaluated on the wall. The current flow on the guide walls for TE_{10} mode propagation can be found by using eq. (12.71) with eqs. (12.42) and (12.43). The result is sketched in Figure 12.11.

The surface charge density ρ_S on the walls is given by

$$\rho_S = \mathbf{a}_n \cdot \mathbf{D} = \mathbf{a}_n \cdot \varepsilon\mathbf{E} \tag{12.72}$$

where **E** is the electric field intensity evaluated on the guide wall.

A waveguide is usually fed or excited by a coaxial line or another waveguide. Most often, a probe (central conductor of a coaxial line) is used to establish the field intensities of the desired mode and achieve a maximum power transfer. The probe is located so as to produce **E** and **H** fields that are roughly parallel to the lines of **E** and **H** fields of the desired mode. To excite the TE_{10} mode, for example, we know from eq. (12.43a) that E_y has maximum value at $x = a/2$. Hence, the probe is located at $x = a/2$ to excite the TE_{10} mode as

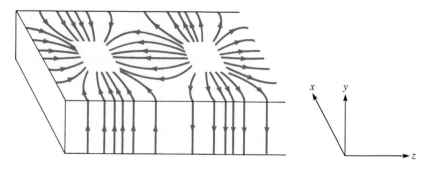

FIGURE 12.11 Surface current on guide walls for TE_{10} mode.

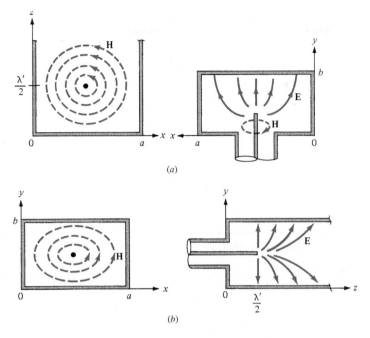

FIGURE 12.12 Excitation of modes in a rectangular waveguide:
(**a**) TE$_{10}$ mode, (**b**) TM$_{11}$ mode.

shown in Figure 12.12(a), where the field lines are similar to those of Figure 12.8. Similarly, the TM$_{11}$ mode is launched by placing the probe along the z-direction as in Figure 12.12(b).

EXAMPLE 12.5

An air-filled rectangular waveguide of dimensions $a = 4$ cm, $b = 2$ cm transports energy in the dominant mode at a rate of 2 mW. If the frequency of operation is 10 GHz, determine the peak value of the electric field in the waveguide.

Solution:

The dominant mode for $a > b$ is TE$_{10}$ mode. The field expressions corresponding to this mode ($m = 1, n = 0$) are in eq. (12.36) or (12.43), namely,

$$E_{xs} = 0, \quad E_{ys} = -jE_o \sin\left(\frac{\pi x}{a}\right) e^{-j\beta z}, \quad \text{where } E_o = \frac{\omega\mu a}{\pi} H_o$$

$$f_c = \frac{u'}{2a} = \frac{3 \times 10^8}{2(4 \times 10^{-2})} = 3.75 \text{ GHz}$$

$$\eta = \eta_{\text{TE}} = \frac{\eta'}{\sqrt{1 - \left[\frac{f_c}{f}\right]^2}} = \frac{377}{\sqrt{1 - \left[\frac{3.75}{10}\right]^2}} = 406.7 \ \Omega$$

From eq. (12.53), the average power transmitted is

$$P_{ave} = \int_{y=0}^{b} \int_{x=0}^{a} \frac{|E_{ys}|^2}{2\eta} \, dx \, dy = \frac{E_o^2}{2\eta} \int_0^b dy \int_0^a \sin^2\left(\frac{\pi x}{a}\right) dz$$

$$= \frac{E_o^2 \, ab}{4\eta}$$

Hence,

$$E_o^2 = \frac{4\eta P_{ave}}{ab} = \frac{4(406.7) \times 2 \times 10^{-3}}{8 \times 10^{-4}} = 4067$$

$$E_o = 63.77 \text{ V/m}$$

PRACTICE EXERCISE 12.5

In Example 12.5, calculate the peak value H_o of the magnetic field in the guide if $a = 2$ cm, $b = 4$ cm, while other things remain the same.

Answer: 63.34 mA/m.

EXAMPLE 12.6

A copper-plated waveguide ($\sigma_c = 5.8 \times 10^7$ S/m) operating at 4.8 GHz is supposed to deliver a minimum power of 1.2 kW to an antenna. If the guide is filled with polystyrene ($\sigma = 10^{-17}$ S/m, $\varepsilon = 2.55\varepsilon_o$) and its dimensions are $a = 4.2$ cm, $b = 2.6$ cm, calculate the power dissipated in a length 60 cm of the guide in the TE_{10} mode.

Solution:

Let

P_d = power loss or dissipated
P_a = power delivered to the antenna
P_o = input power to the guide

so that $P_o = P_d + P_a$
From eq. (12.54),

$$P_a = P_o e^{-2\alpha z}$$

Hence,

$$P_a = (P_d + P_a)e^{-2\alpha z}$$

or

$$P_d = P_a(e^{2\alpha z} - 1)$$

Now we need to determine α from

$$\alpha = \alpha_d + \alpha_c$$

From eq. (12.60),

$$\alpha_d = \frac{\sigma \eta'}{2\sqrt{1 - \left[\dfrac{f_c}{f}\right]^2}}$$

Since the loss tangent

$$\frac{\sigma}{\omega \varepsilon} = \frac{10^{-17}}{2\pi \times 4.8 \times 10^9 \times \dfrac{10^{-9}}{36\pi} \times 2.55}$$

$$= 1.47 \times 10^{-17} \ll 1 \quad (\text{lossless dielectric medium})$$

then

$$\eta' \simeq \sqrt{\frac{\mu}{\varepsilon}} = \frac{377}{\sqrt{\varepsilon_r}} = 236.1$$

$$u' = \frac{1}{\sqrt{\mu\varepsilon}} = \frac{c}{\sqrt{\varepsilon_r}} = 1.879 \times 10^8 \text{ m/s}$$

$$f_c = \frac{u'}{2a} = \frac{1.879 \times 10^8}{2 \times 4.2 \times 10^{-2}} = 2.234 \text{ GHz}$$

$$\alpha_d = \frac{10^{-17} \times 236.1}{2\sqrt{1 - \left[\dfrac{2.234}{4.8}\right]^2}}$$

$$\alpha_d = 1.334 \times 10^{-15} \text{ Np/m}$$

For the TE$_{10}$ mode, eq. (12.68b) gives

$$\alpha_c = \frac{2R_s}{b\eta'\sqrt{1 - \left[\dfrac{f_c}{f}\right]^2}}\left(0.5 + \frac{b}{a}\left[\frac{f_c}{f}\right]^2\right)$$

where

$$R_s = \frac{1}{\sigma_c \delta} = \sqrt{\frac{\pi f \mu}{\sigma_c}} = \sqrt{\frac{\pi \times 4.8 \times 10^9 \times 4\pi \times 10^{-7}}{5.8 \times 10^7}}$$

$$= 1.808 \times 10^{-2} \, \Omega$$

Hence

$$\alpha_c = \frac{2 \times 1.808 \times 10^{-2}\left(0.5 + \dfrac{2.6}{4.2}\left[\dfrac{2.234}{4.8}\right]^2\right)}{2.6 \times 10^{-2} \times 236.1 \sqrt{1 - \left[\dfrac{2.234}{4.8}\right]^2}}$$

$$= 4.218 \times 10^{-3}\,\text{Np/m}$$

Note that $\alpha_d \ll \alpha_c$, showing that the loss due to the finite conductivity of the guide walls is more important than the loss due to the dielectric medium. Thus

$$\alpha = \alpha_d + \alpha_c \simeq \alpha_c = 4.218 \times 10^{-3}\,\text{Np/m}$$

and the power dissipated is

$$P_d = P_a\left(e^{2\alpha z} - 1\right) = 1.2 \times 10^3\left(e^{2 \times 4.218 \times 10^{-3} \times 0.6} - 1\right)$$

$$= 6.089\,\text{W}$$

PRACTICE EXERCISE 12.6

A brass waveguide $(\sigma_c = 1.1 \times 10^7\,\text{S/m})$ of dimensions $a = 4.2$ cm, $b = 1.5$ cm is filled with Teflon $(\varepsilon_r = 2.6, \sigma = 10^{-15}\,\text{S/m})$. The operating frequency is 9 GHz. For the TE_{10} mode:

(a) Calculate α_d and α_c.
(b) Find the loss in decibels in the guide if it is 40 cm long.

Answer: (a) $1.205 \times 10^{-13}\,\text{Np/m}$, $2 \times 10^{-2}\,\text{Np/m}$, (b) 0.06945 dB.

EXAMPLE 12.7

Sketch the field lines for the TM_{11} mode. Derive the instantaneous expressions for the surface current density of this mode.

Solution:

From Table 12.1, we obtain the fields for TM_{11} mode $(m = 1, n = 1)$ as

$$E_x = \frac{\beta}{h^2}\left(\frac{\pi}{a}\right)E_o \cos\left(\frac{\pi x}{a}\right)\sin\left(\frac{\pi y}{b}\right)\sin(\omega t - \beta z)$$

$$E_y = \frac{\beta}{h^2}\left(\frac{\pi}{b}\right)E_o \sin\left(\frac{\pi x}{a}\right)\cos\left(\frac{\pi y}{b}\right)\sin(\omega t - \beta z)$$

$$E_z = E_o \sin\left(\frac{\pi x}{a}\right)\sin\left(\frac{\pi y}{b}\right)\cos(\omega t - \beta z)$$

$$H_x = -\frac{\omega\varepsilon}{h^2}\left(\frac{\pi}{b}\right)E_o \sin\left(\frac{\pi x}{a}\right)\cos\left(\frac{\pi y}{b}\right)\sin(\omega t - \beta z)$$

$$H_y = \frac{\omega\varepsilon}{h^2}\left(\frac{\pi}{a}\right) E_{\mathrm{o}} \cos\left(\frac{\pi x}{a}\right) \sin\left(\frac{\pi y}{b}\right) \sin(\omega t - \beta z)$$

$$H_z = 0$$

For the electric field lines,

$$\frac{dy}{dx} = \frac{E_y}{E_x} = \frac{a}{b}\tan\left(\frac{\pi x}{a}\right)\cot\left(\frac{\pi y}{b}\right)$$

For the magnetic field lines,

$$\frac{dy}{dx} = \frac{H_y}{H_x} = -\frac{b}{a}\cot\left(\frac{\pi x}{a}\right)\tan\left(\frac{\pi y}{b}\right)$$

Notice that $(E_y/E_x)(H_y/H_x) = -1$, showing that electric and magnetic field lines are mutually orthogonal. This should also be observed in Figure 12.13, where the field lines are sketched.

The surface current density on the walls of the waveguide is given by

$$\mathbf{K} = \mathbf{a}_n \times \mathbf{H} = \mathbf{a}_n \times (H_x, H_y, 0)$$

At $x = 0$, $\mathbf{a}_n = \mathbf{a}_x$, $\mathbf{K} = H_y(0, y, z, t)\,\mathbf{a}_z$; that is,

$$\mathbf{K} = \frac{\omega\varepsilon}{h^2}\left(\frac{\pi}{a}\right) E_{\mathrm{o}} \sin\left(\frac{\pi y}{b}\right) \sin(\omega t - \beta z)\,\mathbf{a}_z$$

At $x = a$, $\mathbf{a}_n = -\mathbf{a}_x$, $\mathbf{K} = -H_y(a, y, z, t)\,\mathbf{a}_z$, or

$$\mathbf{K} = \frac{\omega\varepsilon}{h^2}\left(\frac{\pi}{a}\right) E_{\mathrm{o}} \sin\left(\frac{\pi y}{b}\right) \sin(\omega t - \beta z)\,\mathbf{a}_z$$

At $y = 0$, $\mathbf{a}_n = \mathbf{a}_y$, $\mathbf{K} = -H_x(x, 0, z, t)\,\mathbf{a}_z$, or

$$\mathbf{K} = \frac{\omega\varepsilon}{h^2}\left(\frac{\pi}{b}\right) E_{\mathrm{o}} \sin\left(\frac{\pi x}{a}\right) \sin(\omega t - \beta z)\,\mathbf{a}_z$$

At $y = b$, $\mathbf{a}_n = -\mathbf{a}_y$, $\mathbf{K} = H_x(x, b, z, t)\,\mathbf{a}_z$, or

$$\mathbf{K} = \frac{\omega\varepsilon}{h^2}\left(\frac{\pi}{b}\right) E_{\mathrm{o}} \sin\left(\frac{\pi x}{a}\right) \sin(\omega t - \beta z)\,\mathbf{a}_z$$

FIGURE 12.13 Field lines for TM$_{11}$ mode; for Example 12.7.

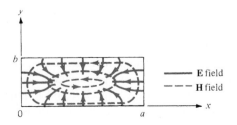

— E field

--- H field

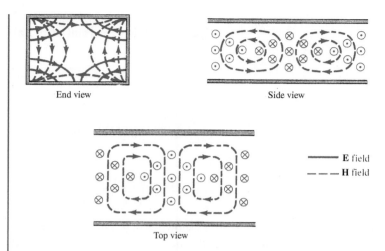

End view   Side view

——— **E** field
— — — **H** field

Top view

FIGURE 12.14 Field lines for TE$_{11}$ mode; for Practice Exercise 12.7.

PRACTICE EXERCISE 12.7

Sketch the field lines for the TE$_{11}$ mode.

Answer: See Figure 12.14. The strength of the field at any point is indicated by the density of the lines; the field is strongest (or weakest) where the lines are closest together (or farthest apart).

12.8 WAVEGUIDE RESONATORS

Resonators are primarily used for energy storage. At high frequencies (≥ 100 MHz) the *RLC* circuit elements are inefficient when used as resonators because the dimensions of the circuits are comparable to the operating wavelength, and consequently, there is unwanted radiation. Therefore, at high frequencies the *RLC* resonant circuits are replaced by electromagnetic cavity resonators. Such resonator cavities are used in klystron tubes, bandpass filters, and wave meters. The microwave oven essentially consists of a power supply, a waveguide feed, and an oven cavity.

Consider the rectangular cavity (or closed conducting box) shown in Figure 12.15. We notice that the cavity is simply a rectangular waveguide shorted at both ends. We therefore expect to have standing wave and also TM and TE modes of wave propagation. Depending on how the cavity is excited, the wave can propagate in the x-, y-, or z-direction. We will choose the $+z$-direction as the "direction of wave propagation." In fact, there is no wave propagation. Rather, there are standing waves. We recall from Section 10.9 that a standing wave is a combination of two waves traveling in opposite directions.

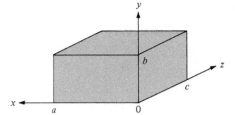

FIGURE 12.15 Rectangular cavity.

A. TM Mode to z

For propagation to z in TM mode, $H_z = 0$ and we let

$$E_{zs}(x, y, z) = X(x)\, Y(y)\, Z(z) \tag{12.73}$$

be the product solution of eq. (12.1). We follow the same procedure taken in Section 12.2 and obtain

$$X(x) = c_1 \cos k_x x + c_2 \sin k_x x \tag{12.74a}$$

$$Y(y) = c_3 \cos k_y y + c_4 \sin k_y y \tag{12.74b}$$

$$Z(z) = c_5 \cos k_z z + c_6 \sin k_z z \tag{12.74c}$$

where

$$k^2 = k_x^2 + k_y^2 + k_z^2 = \omega^2 \mu \varepsilon \tag{12.75}$$

The boundary conditions are:

$$E_z = 0 \qquad \text{at} \quad x = 0, a \tag{12.76a}$$

$$E_z = 0 \qquad \text{at} \quad y = 0, b \tag{12.76b}$$

$$E_y = 0, E_x = 0 \quad \text{at} \quad z = 0, c \tag{12.76c}$$

As shown in Section 12.3, the conditions in eqs. (12.76a,b) are satisfied when $c_1 = 0 = c_3$ and

$$k_x = \frac{m\pi}{a}, \quad k_y = \frac{n\pi}{b} \tag{12.77}$$

where $m = 1, 2, 3, \ldots$, $n = 1, 2, 3, \ldots$. To invoke the conditions in eq. (12.76c), we notice that eq. (12.14) (with $H_{zs} = 0$) yields

$$j\omega \varepsilon E_{xs} = \frac{1}{j\omega \mu} \left(\frac{\partial^2 E_{xs}}{\partial z^2} - \frac{\partial^2 E_{zs}}{\partial z\, \partial x} \right) \tag{12.78}$$

Similarly, combining eqs. (12.13a) and (12.13d) $\left(\text{with } H_{zs} = 0\right)$ results in

$$j\omega\varepsilon E_{ys} = \frac{1}{-j\omega\mu}\left(\frac{\partial^2 E_{zs}}{\partial y\, \partial z} - \frac{\partial^2 E_{ys}}{\partial z^2}\right) \tag{12.79}$$

From eqs. (12.78) and (12.79), it is evident that eq. (12.76c) is satisfied if

$$\frac{\partial E_{zs}}{\partial z} = 0 \quad \text{at} \quad z = 0, c \tag{12.80}$$

This implies that $c_6 = 0$ and $\sin k_z c = 0 = \sin p\pi$. Hence,

$$k_z = \frac{p\pi}{c} \tag{12.81}$$

where $p = 0, 1, 2, 3, \ldots$. Substituting eqs. (12.77) and (12.81) into eq. (12.74) yields

$$\boxed{E_{zs} = E_o \sin\left(\frac{m\pi x}{a}\right) \sin\left(\frac{n\pi y}{b}\right) \cos\left(\frac{p\pi z}{c}\right)} \tag{12.82}$$

where $E_o = c_2 c_4 c_5$. Other field components are obtained from eqs. (12.82) and (12.13). The phase constant β is obtained from eqs. (12.75), (12.77), and (12.81) as

$$\beta^2 = k^2 = \left[\frac{m\pi}{a}\right]^2 + \left[\frac{n\pi}{b}\right]^2 + \left[\frac{p\pi}{c}\right]^2 \tag{12.83}$$

Since $\beta^2 = \omega^2 \mu\varepsilon$, from eq. (12.83), we obtain the *resonant frequency* f_r

$$2\pi f_r = \omega_r = \frac{\beta}{\sqrt{\mu\varepsilon}} = \beta u'$$

or

$$\boxed{f_r = \frac{u'}{2}\sqrt{\left[\frac{m}{a}\right]^2 + \left[\frac{n}{b}\right]^2 + \left[\frac{p}{c}\right]^2}} \tag{12.84}$$

The corresponding resonant wavelength is

$$\boxed{\lambda_r = \frac{u'}{f_r} = \frac{2}{\sqrt{\left[\dfrac{m}{a}\right]^2 + \left[\dfrac{n}{b}\right]^2 + \left[\dfrac{p}{c}\right]^2}}} \tag{12.85}$$

From eq. (12.84), we notice that the lowest-order TM mode is TM_{110}.

B. TE Mode to z

For propagation to z in TE mode, $E_z = 0$ and

$$H_{zs} = (b_1 \cos k_x x + b_2 \sin k_x x)(b_3 \cos k_y y + b_4 \sin k_y y)$$

$$(b_5 \cos k_z z + b_6 \sin k_z z) \tag{12.86}$$

The boundary conditions in eq. (12.76c) combined with eq. (12.13) yields

$$H_{zs} = 0 \quad \text{at} \quad z = 0, c \tag{12.87a}$$

$$\frac{\partial H_{zs}}{\partial x} = 0 \quad \text{at} \quad x = 0, a \tag{12.87b}$$

$$\frac{\partial H_{zs}}{\partial y} = 0 \quad \text{at} \quad y = 0, b \tag{12.87c}$$

Imposing the conditions in eq. (12.87) on eq. (12.86) in the same manner as for TM mode to z leads to

$$H_{zs} = H_o \cos\left(\frac{m\pi x}{a}\right) \cos\left(\frac{n\pi y}{b}\right) \sin\left(\frac{p\pi z}{c}\right) \tag{12.88}$$

where $m = 0, 1, 2, 3, \ldots$, $n = 0, 1, 2, 3, \ldots$, and $p = 1, 2, 3, \ldots$. Other field components can be obtained from eqs. (12.13) and (12.88). The resonant frequency is the same as that of eq. (12.84) except that m or n (but not both at the same time) can be zero for TE modes. It is impossible for m and n to be zero at the same time because the field components will be zero if m and n are zero. The mode that has the lowest resonant frequency for a given cavity size (a, b, c) is the *dominant mode*. If $a > b < c$, it implies that $1/a < 1/b > 1/c$; hence, the dominant mode is TE_{101}. Note that for $a > b < c$, the resonant frequency of TM_{110} mode is higher than that for TE_{101} mode; hence, TE_{101} is dominant. When different modes have the same resonant frequency, we say that the modes are *degenerate;* one mode will dominate others depending on how the cavity is excited.

A practical resonant cavity has walls with finite conductivity σ_c and is, therefore, capable of losing stored energy. The *quality factor Q* is a means of determining the loss.

The **quality factor** is also a measure of the bandwidth of the cavity resonator.

It may be defined as

$$Q = 2\pi \cdot \frac{\text{time average energy stored}}{\text{energy loss per cycle of oscillation}}$$

$$= 2\pi \cdot \frac{W}{P_L T} = \omega \frac{W}{P_L} \tag{12.89}$$

where $T = 1/f =$ the period of oscillation, P_L is the time-average power loss in the cavity, and W is the total time-average energy stored in electric and magnetic fields in the cavity. Usually Q is very high for a cavity resonator compared with Q for an RLC resonant circuit. By following a procedure similar to that used in deriving α_c in Section 12.6, it can be shown that the quality factor for the dominant TE_{101} is given by[3]

$$Q_{TE_{101}} = \frac{(a^2 + c^2)abc}{\delta[2b(a^3 + c^3) + ac(a^2 + c^2)]} \tag{12.90}$$

where $\delta = \dfrac{1}{\sqrt{\pi f_{101}\mu_o\sigma_c}}$ is the skin depth of the cavity walls.

EXAMPLE 12.8

An air-filled resonant cavity with dimensions $a = 5$ cm, $b = 4$ cm, and $c = 10$ cm is made of copper ($\sigma_c = 5.8 \times 10^7$ S/m). Find

(a) The five lowest-order modes

(b) The quality factor for TE_{101} mode

Solution:

(a) The resonant frequency is given by

$$f_r = \frac{u'}{2}\sqrt{\left[\frac{m}{a}\right]^2 + \left[\frac{n}{b}\right]^2 + \left[\frac{p}{c}\right]^2}$$

where

$$u' = \frac{1}{\sqrt{\mu\varepsilon}} = c$$

Hence

$$f_r = \frac{3 \times 10^8}{2}\sqrt{\left[\frac{m}{5 \times 10^{-2}}\right]^2 + \left[\frac{n}{4 \times 10^{-2}}\right]^2 + \left[\frac{p}{10 \times 10^{-2}}\right]^2}$$

$$= 15\sqrt{0.04m^2 + 0.0625n^2 + 0.01p^2} \text{ GHz}$$

Since $c > a > b$ or $1/c < 1/a < 1/b$, the lowest-order mode is TE_{101}. Notice that TM_{101} and TE_{100} do not exist because $m = 1, 2, 3, \ldots, n = 1, 2, 3, \ldots,$ and $p = 0, 1, 2, 3, \ldots$ for the TM modes, and $m = 0, 1, 2, \ldots, n = 0, 1, 2, \ldots,$ and $p = 1, 2, 3, \ldots$ for the TE modes. The resonant frequency for the TE_{101} mode is

$$f_{r_{101}} = 15\sqrt{0.04 + 0 + 0.01} = 3.354 \text{ GHz}$$

---

[3] For the proof, see S. V. Marshall and G. G. Skitek, *Electromagnetic Concepts and Applications*, 3rd ed. Englewood Cliffs, NJ: Prentice-Hall, 1990, pp. 440–442.

The next higher mode is TE_{011} (TM_{011} does not exist), with

$$f_{r_{011}} = 15\sqrt{0 + 0.0625 + 0.01} = 4.04 \text{ GHz}$$

The next mode is TE_{102} (TM_{102} does not exist), with

$$f_{r_{102}} = 15\sqrt{0.04 + 0 + 0.04} = 4.243 \text{ GHz}$$

The next mode is TM_{110} (TE_{110} does not exist), with

$$f_{r_{110}} = 15\sqrt{0.04 + 0.0625 + 0} = 4.8 \text{ GHz}$$

The next two modes are TE_{111} and TM_{111} (degenerate modes), with

$$f_{r_{111}} = 15\sqrt{0.04 + 0.0625 + 0.01} = 5.031 \text{ GHz}$$

The next mode is TM_{103} with

$$f_{r_{103}} = 15\sqrt{0.04 + 0 + 0.09} = 5.408 \text{ GHz}$$

Thus the five lowest order modes in ascending order are

TE_{101} (3.35 GHz)
TE_{011} (4.04 GHz)
TE_{102} (4.243 GHz)
TM_{110} (4.8 GHz)
TE_{111} or TM_{111} (5.031 GHz)

(b) The quality factor for TE_{101} is given by

$$\begin{aligned}
Q_{TE_{101}} &= \frac{(a^2 + c^2)\, abc}{\delta[2b(a^3 + c^3) + ac(a^2 + c^2)]} \\[2mm]
&= \frac{(25 + 100)\, 200 \times 10^{-2}}{\delta[8(125 + 1000) + 50(25 + 100)]} \\[2mm]
&= \frac{1}{61\delta} = \frac{\sqrt{\pi f_{101}\, \mu_o \sigma_c}}{61} \\[2mm]
&= \frac{\sqrt{\pi(3.35 \times 10^9)\, 4\pi \times 10^{-7}\,(5.8 \times 10^7)}}{61} \\[2mm]
&= 14{,}358
\end{aligned}$$

PRACTICE EXERCISE 12.8

If the resonant cavity of Example 12.8 is filled with a lossless material $(\mu_r = 1, \varepsilon_r = 3)$, find the resonant frequency f_r and the quality factor for TE_{101} mode.

Answer: 1.936 GHz, 1.093×10^4.

†12.9 APPLICATION NOTE—OPTICAL FIBER

In the mid-1970s, it was recognized that the existing copper technology would be unsuitable for future communication networks. In view of this, the telecommunication industry invested heavily in research into optical fibers. Optical fiber provides an attractive alternative to wire transmission lines such as twisted pair and coaxial cable (or coax). Optical fiber[4] has the following advantages over copper:

- *Bandwidth:* It provides a very high capacity for carrying information. It has sufficient bandwidth that bit-serial transmission can be used, thereby considerably reducing the size, cost, and complexity of the hardware.
- *Attenuation:* It provides low attenuation and is therefore capable of transmitting over a long distance without the need of repeaters.
- *Noise susceptibility:* It neither radiates nor is affected by electromagnetic interference. The immunity from EMI is due to the absence of metal parts, which means that there can be no conduction currents.
- *Security:* It is more secure from malicious interception because it is not easy to tap a fiber-optic cable without interrupting communication.
- *Cost:* The cost of optical fibers has fallen considerably since the turn of the century and will continue to fall. The cost of related components such as optical transmitters and receivers also is falling.

These impressive advantages over electrical media have made fiber optics a popular transmission medium in recent times. Although optical fiber is more expensive and is used mainly for point-to-point links, there has been a rapid changeover from coax and twisted pair to optical fibers for telecommunication systems, instrumentation, cable TV networks, industrial automation, and data transmission systems.

> An **optical fiber** is a dielectric waveguide operating at optical frequency.

Optical frequencies are on the order of 100 THz. As shown in Figure 12.16, an optical fiber consists of three concentric cylindrical sections: the core, the cladding, and the jacket. The core consists of one or more thin strands made of glass or plastic. The cladding is the glass or plastic coating surrounding the core, which may be step index or graded index. In the step-index core, the refractive index is uniform but undergoes an abrupt change at the core–cladding interface, while the graded-index core has a refractive index that varies with the radial distance from the center of the fiber. The jacket surrounds one or a bundle of cladded fibers. The jacket is made of plastic or other materials to protect against moisture, crushing, and other forms of damage.

A ray of light entering the core will be internally reflected when incident in the denser medium and the angle of incidence is greater than a critical value. Thus a light ray is

---

[4] There are several excellent books that can provide further exposition on optical fiber. See, for example, S. L. W. Meardon, *The Elements of Fiber Optics*, Englewood Cliffs, NJ: Regents/Prentice Hall, 1993.

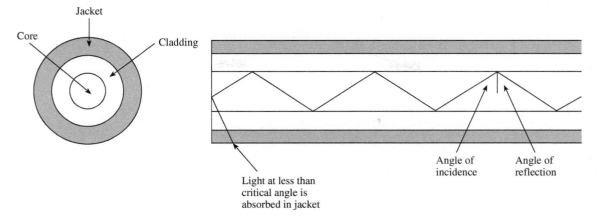

FIGURE 12.16 Optical fiber.

reflected back into the original medium and the process is repeated as light passes down the core. This form of propagation is multimode, referring to the variety of angles that will reflect, as shown in Figure 12.17. It causes the signal to spread out in time and limits the rate at which data can be accurately received. When the radius of the core is reduced a single-mode propagation occurs. This eliminates distortion.

A fiber-optic system is similar to a conventional transmission system. As shown in Figure 12.18, a fiber-optic system consists of a transmitter, a transmission medium, and a receiver. The transmitter accepts and converts to optical signals electrical signals input in analog or digital form. The transmitter sends the optical signal by modulating the output of a light source (usually an LED or a laser) by varying its intensity. The optical signal is transmitted over the optical fiber to a receiver. At the receiver, the optical signal is converted back into an electrical signal by a photodiode.

The performance of a fiber-optic link depends on the numerical aperture (NA), attenuation, and dispersion characteristics of the fiber. As signals propagate through the fiber, they become distorted owing to attenuation and dispersion.

Numerical Aperture

The most important parameter of an optical fiber is its numerical aperture (NA). The value of NA is dictated by the refractive indices of the core and cladding. By definition, the refractive index n of a medium is defined as

$$n = \frac{\text{speed of light in a vacuum}}{\text{speed of light in the medium}}$$

$$= \frac{c}{u_m} = \frac{\dfrac{1}{\sqrt{\mu_o \varepsilon_o}}}{\dfrac{1}{\sqrt{\mu_m \varepsilon_m}}} \tag{12.91}$$

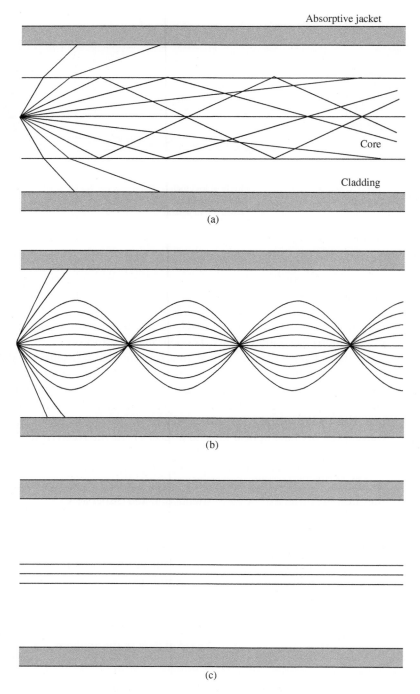

FIGURE 12.17. Optical fiber transmission modes: (**a**) multimode, (**b**) multimode graded index, (**c**) single mode. (From W. Stallings, *Local and Metropolitan Area Networks*, 4th ed. New York: Macmillan, 1993, p. 85.)

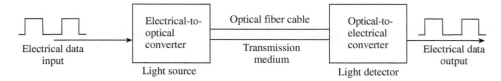

FIGURE 12.18 A typical fiber-optic system.

Since $\mu_m = \mu_o$ in most practical cases,

$$n = \sqrt{\frac{\varepsilon_m}{\varepsilon_o}} = \sqrt{\varepsilon_r} \tag{12.92}$$

indicating that the refractive index is essentially the square root of the dielectric constant. Keep in mind that ε_r can be complex, as discussed in Chapter 10. For common materials, $n = 1$ for air, $n = 1.33$ for water, and $n = 1.5$ for glass.

As a light ray propagates from medium 1 to medium 2, Snell's law must be satisfied.

$$n_1 \sin \theta_1 = n_2 \sin \theta_2 \tag{12.93}$$

where θ_1 is the incident angle in medium 1 and θ_2 is the transmission angle in medium 2. The total reflection occurs when $\theta_2 = 90°$, resulting in

$$\theta_1 = \theta_c = \sin^{-1} \frac{n_2}{n_1} \tag{12.94}$$

where θ_c is the *critical angle* for total internal reflection. Note that eq. (12.94) is valid only if $n_1 > n_2$, since the value of $\sin \theta_c$ must be less than or equal to 1.

Another way of looking at the light-guiding capability of a fiber is to measure the *acceptance angle* θ_a, which is the maximum angle over which light rays entering the fiber will be trapped in its core. We know that the maximum angle occurs when θ_c is the critical angle, thereby satisfying the condition for total internal reflection. Thus, for a step-index fiber,

$$\boxed{NA = \sin \theta_a = n_1 \sin \theta_c = \sqrt{n_1^2 - n_2^2}} \tag{12.95}$$

where n_1 is the refractive index of the core and n_2 is the refractive index of the cladding, as shown in Figure 12.19. Since most fiber cores are made of silica, $n_1 = 1.48$. Typical values of NA range between 0.19 and 0.25. The larger the value of NA, the more optical power the fiber can capture from a source.

Because such optical fibers may support the numerous modes, they are called a *multimode step-index* fibers. The mode volume V is given by

$$V = \frac{\pi d}{\lambda} \sqrt{n_1^2 - n_2^2} \tag{12.96}$$

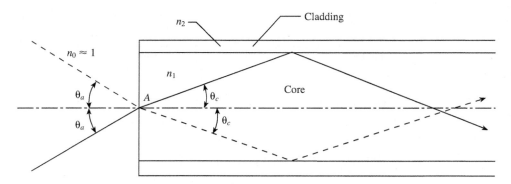

FIGURE 12.19 Numerical aperture and acceptance angle.

where d is the fiber core diameter and λ is the wavelength of the optical source. From eq. (12.96), the number N of modes propagating in a step-index fiber can be estimated as

$$N = \frac{V^2}{2} \tag{12.97}$$

Attenuation

As discussed in Chapter 10, attentuation is the reduction in the power of the optical signal. Power attenuation (or fiber loss) in an optical fiber is governed by

$$\frac{dP}{dz} = -\alpha P \tag{12.98}$$

where α is the attenuation and P is the optical power. In eq. (12.98), it is assumed that a wave propagates along z. By solving eq. (12.98), the power $P(0)$ at the input of the fiber and the power $P(\ell)$ of the light after ℓ are related as

$$P(\ell) = P(0)e^{-\alpha\ell} \tag{12.99}$$

It is customary to express attenuation α in decibels per kilometer and length ℓ of the fiber in kilometers. In this case, eq. (12.99) becomes

$$\boxed{\alpha\ell = 10 \log_{10} \frac{P(0)}{P(\ell)}} \tag{12.100}$$

Thus, the power of the light reduces by α decibels per kilometer as it propagates through the fiber. Equation (12.100) may be written as

$$P(\ell) = P(0) \cdot 10^{-\alpha\ell/10} \tag{12.101}$$

For $\ell = 100$ km,

$$\frac{P(\ell)}{P(0)} \sim \begin{cases} 10^{-100} & \text{for coaxial cable} \\ 10^{-2} & \text{for optical fiber} \end{cases} \qquad (12.102)$$

indicating that much more power is lost in the coaxial cable than in optical fiber.

Dispersion

The spreading of pulses of light as they propagate down a fiber is called dispersion. As the pulses representing 0s spread, they overlap epochs that represent 1s. If dispersion is beyond a certain limit, it may confuse the receiver. The dispersive effects in single-mode fibers are much smaller than in multimode fibers.

EXAMPLE 12.9

A step-index fiber has a core diameter of 80 μm, a core refractive index of 1.62, and a numerical aperture of 0.21. Calculate (a) the acceptance angle, (b) the refractive index that the fiber can propagate at a wavelength of 0.8 μm, (c) the number of modes that the fiber can propagate at a wavelength of 0.8 μm.

Solution:

(a) Since $\sin \theta_a = \text{NA} = 0.21$, then

$$\theta_a = \sin^{-1} 0.21 = 12.12°$$

(b) From $\text{NA} = \sqrt{n_1^2 - n_2^2}$, we obtain

$$n_2 = \sqrt{n_1^2 - \text{NA}^2} = \sqrt{1.62^2 - 0.21^2} = 1.606$$

(c)

$$V = \frac{\pi d}{\lambda}\sqrt{n_1^2 - n_2^2} = \frac{\pi d\,\text{NA}}{\lambda}$$

$$= \frac{\pi(80 \times 10^{-6}) \times 0.21}{0.8 \times 10^{-6}} = 65.973$$

Hence

$$N = \frac{V^2}{2} = 2176 \text{ modes}$$

PRACTICE EXERCISE 12.9

A silica fiber has a refractive index of 1.48. It is surrounded by a cladding material with a refractive index of 1.465. Find (a) the critical angle above which total internal reflection occurs, (b) the numerical aperture of the fiber.

Answer: (a) 81.83°, (b) 0.21.

EXAMPLE 12.10

Light pulses propagate through a fiber cable with an attenuation of 0.25 dB/km. Determine the distance through which the power of pulses is reduced by 40%.

Solution:

If the power is reduced by 40%, it means that

$$\frac{P(\ell)}{P(0)} = 1 - 0.4 = 0.6$$

Hence

$$\ell = \frac{10}{\alpha} \log_{10} \frac{P(0)}{P(\ell)}$$
$$= \frac{10}{0.25} \log_{10} \frac{1}{0.6}$$
$$= 8.874 \text{ km}$$

PRACTICE EXERCISE 12.10

A 10 km fiber with an attenuation of 0.2 dB/km serves as an optical link between two cities. How much of input power is received?

Answer: 63.1%.

12.10 APPLICATION NOTE—CLOAKING AND INVISIBILITY

The practice of using metamaterials to *hide* an object is called *metamaterial cloaking*. Metamaterials are ideal for cloaking because they are designed to have a negative refractive index. All materials have an index of refraction, a number that describes that amount of light, or electromagnetic wave, that is reflected as the wave passes through the material. All materials that are found in nature have a positive refraction index, allowing the reflected light to hit an observer's eye, making the object visible. However, the negative refraction index of metamaterials can bend the wave around an object instead of reflecting the light, thus making the object invisible.

Many attempts at cloaking an object have been made and have been successful to some degree, leaving only small reflections of the cloaked object. Recently, however, researchers at Duke University discovered a method of cloaking an object completely, making it perfectly invisible. The research at Duke began in 2006, but the cloaking models suffered from the common problem of reflected light. In 2011, David Smith and graduate student Nathan Landy modified the models by altering the arrangement of the metamaterial to a diamond-like configuration and shifting the metamaterial so that the reflections were canceled by its mirror image at each intersection. With this adjustment, illustrated schematically in Figure 12.20, perfect invisibility was achieved.

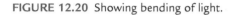

FIGURE 12.20 Showing bending of light.

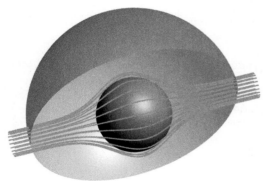

This perfect invisibility, however, comes at the price of a few caveats. An invisibility cloak has been created only on the centimeter scale. Also, the object surrounded by this metamaterial cloak is invisible only to microwaves. In other words, the researchers have been able to bend microwaves but have not yet achieved the bending of light waves, which are more difficult to bend because they have a higher frequency. Finally, the invisibility is unidirectional: that is, the object cloaked is invisible from only one specific angle. Nonetheless, this research at Duke University marks a breakthrough in metamaterial cloaking. These researchers, who were the first to bend waves without any reflection, now plan to further develop the cloak to make it omnidirectional, hiding the object from every angle.

While using metamaterials to render objects invisible to the human eye may be decades away, invisibility to microwaves has many practical applications in telecommunications and defense. Potential applications include radar and sensor detection, battlefield communication, and infrastructure monitoring.

MATLAB 12.1

```
% This script computes the cutoff frequencies of the first
% 10 waveguide modes, allowing the user to enter the
% dimensions (assuming a > b) and relative material properties.
% The script first finds the lowest 100 modes by cutoff frequency
% for both TE and TM, creating a list of 200 total modes, from
% which the lowest 10 of all (TE and TM) are found

clear

% Enter the frequency (in rad/s)
a = input('Enter the waveguide width\n >  ');
% Enter the propagation constant gamma (in a+j*b format)
b = input('Enter the waveguide height\n >  ');
% Enter the relative permittivity
er = input('Enter the relative permittivity \n >  ');
% Enter the propagation constant gamma (in a+j*b format)
ur = input('Enter the relative permeability\n >  ');

% Determine the first 100 TM modes
```

```
index=1; % start a count
for m=1:10,
    for n=1:10,
        modes(index,1)=1; % store a 1 in row <index>, and
                          % column 1 for TM modes
        modes(index,2)=m; % store m in row <index>, and column 2
        modes(index,3)=n; % store n in row <index>, and column 3
        modes(index,4)=3e8/sqrt(er*ur)*sqrt((m*pi/a)^2+(n*pi/b)^2);
                  % store cutoff in row <index>, and column 4
        index=index+1; % increment counter
    end
end

% Determine the first 100 TE modes
for m=0:9,
    for n=0:9,
        if m | n  % check if either m or n is nonzero and
                          % compute mode
            modes(index,1)=2;
            modes(index,2)=m;
            modes(index,3)=n;
            modes(index,4)=3e8/sqrt(er*ur)*sqrt((m*pi/
            a)^2+(n*pi/b)^2);
            index=index+1;
        else
            % do nothing, because m = n = 0
        end
    end
end

% Sort these 100 modes by lowest cutoff
% this command sorts the matrix by grouping the fourth
% column (the frequencies) in ascending order)
modes=sortrows(modes,4);

% Print out the lowest 10 modes of the lowest 100 modes
mode_string='ME'; % 'M' is the first character, 'E' is the second
disp(sprintf('\n'));  % format extra line
for k = 1:10
    disp(sprintf('Mode: T%c%d%d, ',...
        mode_string(modes(k,1)),modes(k,2), modes(k,3)))
    disp(sprintf('Cutoff frequency = %0.3f GHz\n',...
                modes(k,4)/(2*pi*1e9)))
end
```

SUMMARY

1. Waveguides are structures used in guiding EM waves at high frequencies. Assuming a lossless rectangular waveguide ($\sigma_c \simeq \infty, \sigma \simeq 0$), we apply Maxwell's equations in analyzing EM wave propagation through the guide. The resulting partial differential

equation is solved by using the method of separation of variables. On applying the boundary conditions on the walls of the guide, the basic formulas for the guide are obtained for different modes of operation.

2. Two modes of propagation (or field patterns) are the TM_{mn} and TE_{mn}, where m and n are positive integers. For TM modes, $m = 1, 2, 3, \ldots$, and $n = 1, 2, 3, \ldots$ and for TE modes, $m = 0, 1, 2, \ldots$, and $n = 0, 1, 2, \ldots, n = m \neq 0$.

3. Each mode of propagation has an associated propagation constant and cutoff frequency. The propagation constant $\gamma = \alpha + j\beta$ does not depend only on the constitutive parameters $(\varepsilon, \mu, \sigma)$ of the medium as in the case of plane waves in an unbounded space; it depends on the cross-sectional dimensions (a, b) of the guide. The cutoff frequency is the frequency at which γ changes from being purely real (attenuation) to purely imaginary (propagation). The dominant mode of operation is the lowest mode possible. It is the mode with the lowest cutoff frequency. If $a > b$, the dominant mode is TE_{10}.

4. The basic equations for calculating the cutoff frequency f_c, the phase constant β, and the phase velocity u_p are summarized in Table 12.1. Formulas for calculating the attenuation constants due to lossy dielectric medium and imperfectly conducting walls are also provided.

5. The group velocity (or velocity of energy flow) u_g is related to the phase velocity u_p of the wave propagation by

$$u_p u_g = u'^2$$

where $u' = 1/\sqrt{\mu\varepsilon}$ is the medium velocity (i.e., the velocity of the wave in the dielectric medium unbounded by the guide). Although u_p is greater than u', u_g does not exceed u'.

6. The mode of operation for a given waveguide is dictated by the method of excitation.

7. A waveguide resonant cavity is used for energy storage at high frequencies. It is nothing but a waveguide shorted at both ends. Hence its analysis is similar to that of a waveguide. The resonant frequency for both the TE and the TM modes to z is given by

$$f_r = \frac{u'}{2} \sqrt{\left[\frac{m}{a}\right]^2 + \left[\frac{n}{b}\right]^2 + \left[\frac{p}{c}\right]}$$

For TM modes, $m = 1, 2, 3, \ldots, n = 1, 2, 3, \ldots$, and $p = 0, 1, 2, 3, \ldots$, and for TE modes, $m = 0, 1, 2, 3, \ldots, n = 0, 1, 2, 3, \ldots$, and $p = 1, 2, 3, \ldots, m = n \neq 0$. If $a > b < c$, the dominant mode (one with the lowest resonant frequency) is TE_{101}.

8. The quality factor, a measure of the energy loss in the cavity, is given by

$$Q = \omega \frac{W}{P_L}$$

9. An optical fiber is a dielectric waveguiding structure operating at optical frequencies; it consists of a core region and a cladding region.

10. Advantages of optical fiber over copper wire include large bandwidth, low attenuation, immunity to electromagnetic intererence, and low cost.

12.1 At microwave frequencies, we prefer waveguides to transmission lines for transporting EM energy because of all the following *except* that

(a) losses in transmission lines are prohibitively large.

(b) waveguides have larger bandwidths and lower signal attenuation.

(c) transmission lines are larger than waveguides.

(d) transmission lines support only TEM mode.

12.2 An evanescent mode occurs when

(a) a wave is attenuated rather than propagated.

(b) the propagation constant is purely imaginary.

(c) $m = 0 = n$ so that all field components vanish.

(d) the wave frequency is the same as the cutoff frequency.

12.3 The dominant mode for rectangular waveguides is

(a) TE_{11} (c) TE_{101}

(b) TM_{11} (d) TE_{10}

12.4 The TM_{10} mode can exist in a rectangular waveguide.

(a) True   (b) False

12.5 For TE_{30} mode, which of the following field components exist?

(a) E_x (d) H_x

(b) E_y (e) H_y

(c) E_z

12.6 If in a rectangular waveguide for which $a = 2b$, the cutoff frequency for TE_{02} mode is 12 GHz, the cutoff frequency for TM_{11} mode is

(a) 3 GHz   (d) $6\sqrt{5}$ GHz

(b) $3\sqrt{5}$ GHz   (e) None of the above

(c) 12 GHz

12.7 If a tunnel is 4 m by 7 m in cross section, a car in the tunnel will not receive an AM radio signal (e.g., $f = 10$ MHz).

(a) True   (b) False

12.8 When the electric field is at its maximum value, the magnetic energy of a cavity is

(a) at its maximum value

(b) at $\sqrt{2}$ of its maximum value

(c) at $\dfrac{1}{\sqrt{2}}$ of its maximum value

(d) at 1/2 of its maximum value

(e) zero

12.9 Which of these modes does not exist in a rectangular resonant cavity?

(a) TE_{110} (c) TM_{110}

(b) TE_{011} (d) TM_{111}

12.10 How many degenerate dominant modes exist in a rectangular resonant cavity for which $a = b = c$?

(a) 0   (d) 5

(b) 2   (e) ∞

(c) 3

Answers: 12.1c, 12.2a,d, 12.3d, 12.4b, 12.5b,d, 12.6b, 12.7a, 12.8e, 12.9a, 12.10c.

PROBLEMS

Sections 12.3 and 12.4—TM and TE Modes

12.1 An air-filled rectangular waveguide has a cross section of 6 cm × 4 cm.

(a) Calculate the cutoff frequency of the dominant mode.

(b) Determine how many modes are passed at three times cutoff frequency of dominant mode.

12.2 A square waveguide (a by a) can propagate only TE_{10} and not TE_{11} or higher modes. In order to achieve this, what must be the size a?

12.3 A rectangular waveguide (2.28 cm × 1.01 cm) is filled with polyethylene ($\varepsilon_r = 2.25$). Calculate the cutoff frequencies for the following modes:

TE_{01}, TE_{10}, TE_{11}, TE_{02}, TE_{22}, TM_{11}, TM_{12}, TM_{21}. Assume that polyethylene is lossless.

12.4 An air-filled waveguide has a cross section of 2.4 cm × 1.2 cm. A microwave signal of 12 GHz propagates down the guide. (a) Calculate the cutoff frequencies of TE_{10}, TE_{01}, TE_{20}, and TE_{02} modes. (b) Which modes will propagate?

12.5 Design a rectangular waveguide with an aspect ratio of 3 to 1 for use in the K-band (18–26.5 GHz). Assume that the guide is air filled.

12.6 A tunnel is modeled as an air-filled metallic rectangular waveguide with dimensions $a = 8$ m and $b = 16$ m. Determine whether the tunnel will pass (a) a 1.5 MHz AM broadcast signal, (b) a 120 MHz FM broadcast signal.

12.7 In an air-filled rectangular waveguide, the cutoff frequency of a TE_{10} mode is 5 GHz, whereas that of a TE_{01} mode is 12 GHz. Calculate

(a) The dimensions of the guide

(b) The cutoff frequencies of the next three higher TE modes

(c) The cutoff frequency for TE_{11} mode if the guide is filled with a lossless material having $\varepsilon_r = 2.25$ and $\mu_r = 1$

12.8 An air-filled rectangular waveguide operates at 40 GHz. If the cutoff frequency of the TE_{12} mode is 25 GHz, calculate the wavelength, phase constant, phase velocity, and intrinsic impedance of this mode.

12.9 An air-filled rectangular waveguide of dimension 5 cm $\times$ 3 cm operates on the TE_{10} mode at a frequency of 12.5 GHz. Find the phase constant, phase velocity, and the wave impedance.

12.10 An air-filled waveguide has $a = 2b = 4$ cm and operates at the TE_{10} mode. Determine f_c, β, and λ at 24 MHz.

12.11 An air-filled hollow rectangular waveguide is 150 m long and is capped at the end with a metal plate. If a short pulse of frequency 7.2 GHz is introduced into the input end of the guide, how long will it take the pulse to return to the input end? Assume that the cutoff frequency of the guide is 6.5 GHz.

12.12 A section of an air-filled rectangular waveguide ($a = 2.4$ cm, $b = 1.2$ cm) operates in the TE_{10} mode. The operating frequency is 25% higher than the cutoff frequency. Determine f_c, f, and η.

12.13 A K-band waveguide (1.067 cm $\times$ 0.533 cm) is filled by a dielectric material with $\varepsilon_r = 6.8$. If it operates in the dominant TE_{10} mode at 6 GHz, determine the following:

(a) The cutoff frequency

(b) The phase velocity

(c) The waveguide wavelength

12.14 Show that the attenuation due to a waveguide operating below cutoff is

$$\alpha = 2\pi \sqrt{\mu\varepsilon}\, f_c \sqrt{1 - \left(\frac{f}{f_c}\right)^2}$$

12.15 An air-filled rectangular waveguide has cross-sectional dimensions $a = 6$ cm and $b = 3$ cm. Given that

$$E_z = 5 \sin\left(\frac{2\pi x}{a}\right) \sin\left(\frac{3\pi y}{b}\right) \cos(10^{12}t - \beta z) \text{ V/m}$$

calculate the intrinsic impedance of this mode and the average power flow in the guide.

12.16 In an air-filled rectangular waveguide, a TE mode operating at 6 GHz has

$$E_y = 5 \sin\left(\frac{2\pi x}{a}\right) \cos\left(\frac{\pi y}{b}\right) \sin(\omega t - 12z) \text{ V/m}$$

Determine (a) the mode of operation, (b) the cutoff frequency, (c) the intrinsic impedance, (d) H_x.

12.17 In an air-filled rectangular waveguide with $a = 2.286$ cm and $b = 1.016$ cm, the y-component of the TE mode is given by

$$E_y = \sin\left(\frac{2\pi x}{a}\right) \cos\left(\frac{3\pi y}{b}\right) \sin(10\pi \times 10^{10}t - \beta z) \text{ V/m}$$

Find (a) the operating mode, (b) the propagation constant γ, (c) the intrinsic impedance η.

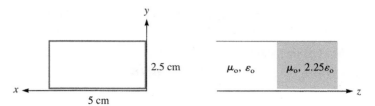

FIGURE 12.21 For Problem 12.18.

12.18 A rectangular waveguide with cross sections shown in Figure 12.21 has dielectric discontinuity. Calculate the standing wave ratio if the guide operates at 8 GHz in the dominant mode.

12.19 Analysis of a circular waveguide requires solution of the scalar Helmholtz equation in cylindrical coordinates, namely,

$$\nabla^2 E_{zs} + k^2 E_{zs} = 0$$

or

$$\frac{1}{\rho}\frac{\partial}{\partial \rho}\left(\rho \frac{\partial E_{zs}}{\partial \rho}\right) + \frac{1}{\rho^2}\frac{\partial^2 E_{zs}}{\partial \phi^2} + \frac{\partial^2 E_{zs}}{\partial z^2} + k^2 E_{zs} = 0$$

By assuming the product solution

$$E_{zs}(\rho, \phi, z) = R(\rho)\,\Phi(\phi)\,Z(z)$$

show that the separated equations are

$$Z'' - k_z^2\, Z = 0$$
$$\Phi'' + k_\phi^2\, \Phi = 0$$
$$\rho^2 R'' + \rho R' + (k_\rho^2\, \rho^2 - k_\phi^2)\, R = 0$$

where

$$k_\rho^2 = k^2 + k_z^2$$

12.20 For an air-filled waveguide, use MATLAB to plot u_p and u_g for 10 GHz $< f <$ 100 GHz. Assume that $f_c = 8$ GHz.

Section 12.5—Wave Propagation in the Guide

12.21 Determine the values of β, μ_p, μ_g, and η_{TE10} for a 7.2 cm $\times$ 3.4 cm rectangular waveguide operating at 6.2 GHz (a) if the waveguide is air filled, (b) if the waveguide is filled with a material having $\varepsilon = 2.25\,\varepsilon_o$, $\mu = \mu_o$, $\sigma = 0$.

12.22 Consider a WR284 waveguide ($a = 7.214$ cm, $b = 3.404$ cm). If it is filled with polyethylene ($\varepsilon_r = 2.5$) and operates at 4 GHz, determine u_p and u_g.

12.23 In a certain medium, the phase velocity is

$$u_p = c\left[\frac{\lambda_o}{\lambda}\right]^2$$

where $c = 3 \times 10^8$ m/s. Obtain the expression for the group velocity.

12.24 The group velocity of a dielectric-filled rectangular waveguide operating at 12 GHz is $c/4$. When the frequency becomes 15 GHz, the group velocity is $c/3$ for the same mode. Determine f_c and ε_r.

12.25 A square waveguide operates at 4.5 GHz in the dominant mode. If the group velocity is determined to be 1.8×10^8 m/s, calculate the largest dimension of the waveguide. Assume that the waveguide is filled with oil ($\varepsilon = 2.2\varepsilon_o$).

12.26 A rectangular waveguide is filled with polyethylene ($\varepsilon = 2.25\varepsilon_o$) and operates at 24 GHz. If the cutoff frequency of a certain TE mode is 16 GHz, find the group velocity and intrinsic impedance of the mode.

Section 12.6—Power Transmission and Attenuation

12.27 The average power density is given by

$$\mathcal{P}_{ave} = \frac{1}{2}\operatorname{Re}[\mathbf{E}_s \times \mathbf{H}_s^*]$$

Show that for a rectangular waveguide operating in the TE_{10} mode,

$$\mathcal{P}_{ave} = \frac{\omega\mu\beta a^2}{2\pi^2}H_o^2\sin^2\left(\frac{\pi x}{a}\right)\mathbf{a}_z$$

12.28 For TE_{01} mode,

$$E_{xs} = \frac{j\omega\mu\pi}{bh^2}H_o\sin(\pi y/b)e^{-\gamma z}, \quad E_{ys} = 0$$

Find $\mathcal{P}_{ave}$ and P_{ave}.

12.29 A 1 cm $\times$ 2 cm waveguide is made of copper ($\sigma_c = 5.8 \times 10^7$ S/m) and filled with a dielectric material for which $\varepsilon = 2.6\varepsilon_o$, $\mu = \mu_o$, $\sigma_d = 10^{-4}$ S/m. If the guide operates at 12 GHz, evaluate α_c and α_d for (a) TE_{10} and (b) TM_{11}.

12.30 A 4 cm square waveguide is filled with a dielectric with complex permittivity $\varepsilon_c = 16\varepsilon_o(1 - j10^{-4})$ and is excited with the TM_{21} mode. If the waveguide operates at 10% above the cutoff frequency, calculate attenuation α_d. How far can the wave travel down the guide before its magnitude is reduced by 20%?

12.31 If the walls of the square waveguide in Problem 12.30 are made of brass ($\sigma_c = 1.5 \times 10^7$ S/m), find α_c and the distance over which the wave is attenuated by 30%.

12.32 An air-filled waveguide with dimensions $a = 6$ cm and $b = 3$ cm is excited at the level of $|\mathbf{E}| = 2.2$ kV/m. If the dominant mode propagates at 4 GHz, determine the power transmitted.

12.33 A rectangular waveguide with $a = 2b = 4.8$ cm is filled with Teflon ($\varepsilon_r = 2.11$, loss tangent of 3×10^{-4}). Assume that the walls of the waveguide are coated with gold ($\sigma_c = 4.1 \times 10^7$ S/m) and that a TE_{10} wave at 4 GHz propagates down the waveguide. Find (a) α_d, (b) α_c.

12.34 Use MATLAB to plot the attenuation for the TE_{10} mode a of waveguide with copper walls as a function of frequency. Do this for frequencies above cutoff. Keep in mind that R_s varies with frequency. Take $a = 2b = 1$ cm, $f_c = 10$ GHz, and assume that the waveguide is filled with a dielectric having $\varepsilon_r = 2.25$.

12.35 An air-filled X-band rectangular waveguide has dimensions $a = 2.286$ cm and $b = 1.016$ cm. If the waveguide has copper walls ($\varepsilon = \varepsilon_o, \mu = \mu_o, \sigma = 5.8 \times 10^7$ S/m), find the attenuation in dB/m due to the wall loss when the dominant mode is propagating at 8.4 GHz.

12.36 A rectangular, air-filled waveguide has dimensions $a = 3.8$ cm and $b = 1.6$ cm, and walls are made of copper. For the dominant mode at $f = 10$ GHz, calculate

(a) the group velocity

(b) the attenuation dB/m

12.37 A rectangular waveguide has transverse dimensions $a = 2.5$ cm and $b = 1.5$ cm and operates at 7.5 GHz in the dominant mode. If the waveguide is filled with a lossy dielectric material with $\varepsilon_r = 2.26, \mu_r = 1, \sigma = 10^{-4}$ S/m and the walls are made of brass ($\sigma_o = 1.1 \times 10^7$ S/m), calculate $\beta, \alpha_d, \alpha_c, u_p, u_g,$ and λ_c.

12.38 A rectangular brass ($\sigma_c = 1.37 \times 10^7$ S/m) waveguide with dimensions $a = 2.25$ cm and $b = 1.5$ cm operates in the dominant mode at frequency 5 GHz. If the waveguide is filled with Teflon ($\mu_r = 1, \varepsilon_r = 2.11, \sigma \simeq 0$), determine (a) the cutoff frequency for the dominant mode, (b) the attenuation constant due to the loss in the guide walls.

12.39 For a square waveguide, show that attenuation α_c is minimum for the TE_{10} mode when $f = 2.962 f_c$.

Section 12.8—Waveguide Resonators

12.40 Show that for propagation from the TE mode to z in a rectangular cavity,

$$E_{ys} = -\frac{j\omega\mu}{h^2}\left(\frac{m\pi}{a}\right)H_o \sin\left(\frac{m\pi x}{a}\right)\cos\left(\frac{n\pi y}{b}\right)\sin\left(\frac{p\pi z}{c}\right)$$

Find H_{xs}.

12.41 For a rectangular cavity, show that

$$H_{xs} = \frac{j\omega\varepsilon}{h^2}\left(\frac{n\pi}{b}\right)E_o \sin\left(\frac{m\pi x}{a}\right)\cos\left(\frac{n\pi y}{b}\right)\cos\left(\frac{p\pi z}{c}\right)$$

for propagation from the TM mode to z. Determine E_{ys}.

12.42 In a rectangular resonant cavity, identify the dominant made when

(a) $a < b < c$

(b) $a > b > c$

(c) $a = c > b$

12.43 A rectangular cavity has dimensions $a = 3$ cm, $b = 4$ cm, and $c = 6$ cm. The cavity is filled with a lossless dielectric with $\varepsilon_r = 4.6$. Calculate: (a) the resonant frequency of the dominant mode, (b) the quality factor.

12.44 An air-filled waveguide has dimension 3 cm $\times$ 2.5 cm. The guide is 4 cm long. It is shorted at each end, forming a cavity. Determine the lowest three resonance frequency.

12.45 A rectangular cavity has dimension $a = 1$ cm, $b = 2$ cm, $c = 3$ cm. If it is filled with polyethylene ($\varepsilon = 2.5\varepsilon_o$), find the first five resonant frequencies.

12.46 For a cubical cavity ($a = b = c$) in the TE_{101} mode, show that

$$Q = \frac{a}{3\delta}$$

where δ is the skin depth.

12.47 An air-filled cavity has dimensions 20 mm $\times$ 8 mm $\times$ 10 mm. If the walls are silver-plated, find (a) dominant resonant frequency, (b) Q for the TE_{101} mode.

12.48 Design an air-filled cubical cavity to have its dominant resonant frequency at 3 GHz.

12.49 Design a cubical resonant cavity with a dominant frequency of 5.6 GHz. Assume that the cavity is filled with (a) air, (b) Teflon having $\varepsilon_r = 2.05$.

12.50 An air-filled cubical cavity of size 10 cm has

$$\mathbf{E} = 200 \sin(30\pi x) \sin(30\pi y) \cos(6 \times 10^9 t)\mathbf{a}_z \text{ V/m}$$

(a) Find $\mathbf{H}$. (b) Show that $\mathbf{E} \cdot \mathbf{H} = 0$.

12.51 (a) Determine the size of an air-filled cubical cavity made of copper that it will give a dominant resonant frequency of 12 GHz.

(b) Calculate the quality factor Q at that frequency.

12.52 Shielded rooms act as resonant cavities. We must avoid operating equipment in any such room at a resonant frequency of the cavity. If an air-filled shielded room has the dimensions 10.2 m by 8.7 m by 3.6 m, find all resonant frequencies below 50 MHz.

Section 12.9—Application Note—Optical Fiber

12.53 The speed of light in a given medium is measured as 2.1×10^8 m/s. Find its refractive index.

12.54 Determine the numerical aperture of an optical fiber which has $n_1 = 1.51$ and $n_2 = 1.45$.

12.55 A glass fiber has a core diameter of 50 μm, a core refractive index of 1.62, and a cladding with a refractive index of 1.604. If light having a wavelength of 1300 nm is used, find

(a) The numerical aperture

(b) The acceptance angle

(c) The number of transmission modes

12.56 A silicon fiber has a core index of 1.48 and a cladding index of 1.46. If the core radius is 5 μm, find the number of propagating modes for the source wavelength of 1300 nm.

12.57 An optical fiber with an attenuation of 0.4 dB/km is 5 km long. The fiber has $n_1 = 1.53$, $n_2 = 1.45$, and a diameter of 50 μm. Find:

(a) The maximum angle at which rays will enter the fiber and be trapped

(b) The percentage of input power received

12.58 A laser diode is capable of coupling 10 mW into a fiber with attenuation of 0.5 dB/km. If the fiber is 850 m long, calculate the power received at the end of the fiber.

12.59 Attenuation α_{10} in Chapter 10 is in nepers per meter (Np/m), whereas attenuation α_{12} in this chapter is in decibels per kilometer (dB/km). What is the relationship between the two?

12.60 A power of 1.25 mW is launched into an optical fiber that has a 0.4 dB/km attenuation. Determine the fiber length such that a power of 1 μW is received at the other end of the fiber.

12.61 A lightwave system uses a 30 km fiber link with a loss of 0.4 dB/km. If the system requires at least 0.2 mW at the receiver, calculate the minimum power that must be launched into the fiber.

12.62 (a) Discuss the advantages derived from using a fiber-optic cable.

(b) What is pulse dispersion?

SMART ANTENNAS

Just as we hear better with two ears than with one, a communications system with two or more antennas can outperform a system with a single antenna. Smart antennas (also known as adaptive antennas) basically consist of an antenna array combined with signal processing in both time and space. They are different from common antennas in that they have adaptive (nonfixed) lobe patterns. They exploit the fact that interferers and users rarely have the same location.

There are basically two types of smart antenna: switched beam (a finite number of fixed, predefined patterns) and adaptive array (an infinite number of patterns that are adjusted in real time). The switched-beam type is the simplest technique. It is simply a controlled RF switch connected to many fixed antennas. It employs a grid of beams and usually chooses the beam that gives the best signal-to-noise ratio. It is easily deployed but has low gain between beams.

The wireless personal communications market, especially the cellular telephone segment, has been growing exponentially for years and will continue to grow. But there are some challenges along the way. These include quality of service, traffic capacity, and cost of service. The smart antenna technology is a promising approach to these problems, offering increased capacity, extended range, better link quality, and longer battery life in mobile units. Smart antenna systems enable operators of PCs, cellular phones, wireless local-area networks, and wireless local-loop networks to realize significant increases in channel capacity, signal quality, spectrum efficiency, and coverage.

Although smart antennas make wireless systems more complex, they provide real improvements in areas that are critical for making wireless service more universal and reliable. It has been rightly argued that the performance requirements of future wireless systems cannot be met without the use of smart antennas. To know about smart antennas, one should take a class on antennas to acquire general background information. Unfortunately, most electrical engineering departments do not offer courses on antennas at the undergraduate level.

Source: Adapted with permission from M. Chryssomallis, "Smart antennas," *IEEE Antennas and Propagation Magazine,* vol. 42, no. 3, June 2000, pp. 129–136.

CHAPTER **13**

ANTENNAS

A committee is a group of the unwilling, chosen from the unfit, to do the unnecessary.
—ANONYMOUS

13.1 INTRODUCTION

Up until now, we have not asked ourselves how EM waves are produced. Recall that electric charges are the sources of EM fields. If the sources are time varying, EM waves propagate away from the sources and radiation is said to have taken place. Radiation may be thought of as the process of transmitting electric energy. The radiation or launching of the waves into space is efficiently accomplished with the aid of conducting or dielectric structures called *antennas*. Theoretically, any structure can radiate EM waves, but not all structures can serve as efficient radiation mechanisms.

An antenna may also be viewed as a transducer used in matching the transmission line or as a waveguide (used in guiding the wave to be launched) to the surrounding medium, or vice versa. Figure 13.1 shows how an antenna is used to accomplish a match between the line or guide and the medium. The antenna is needed for two main reasons: for efficient radiation and for matching wave impedances to minimize reflection. The antenna uses voltage and current from the transmission line (or the EM fields from the waveguide) to launch an EM wave into the medium. An antenna may be used for either transmitting or receiving EM energy.

Typical antennas are illustrated in Figure 13.2. The dipole antenna in Figure 13.2(a) consists of two straight wires lying along the same axis. The loop antenna, exemplified in Figure 13.2(b), consists of one or more turns of wire. The helical antenna in Figure 13.2(c) consists of a wire in the form of a helix backed by a ground plane. Antennas in Figure 13.2(a–c) are called *wire antennas;* they are used in automobiles, buildings, aircraft, ships, and so on. The horn antenna in Figure 13.2(d), an example of an *aperture antenna,* is a tapered section of waveguide providing a transition between a waveguide and the surroundings. Since it is conveniently flush mounted, it is useful in various applications such as aircraft communications. The parabolic dish reflector in Figure 13.2(e) utilizes the fact that EM waves are reflected by a conducting sheet. When used as a transmitting antenna, a feed antenna such as a dipole or horn is placed at the focal point. The radiation from the source is reflected by the dish (acting like a mirror), and a parallel beam results. Parabolic dish antennas are used in communications, radar, and astronomy.

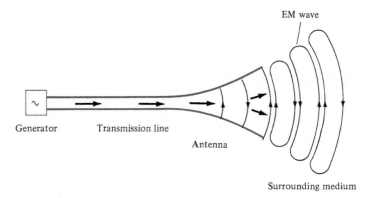

FIGURE 13.1 An antenna as a matching device between the guiding structure and the surrounding medium.

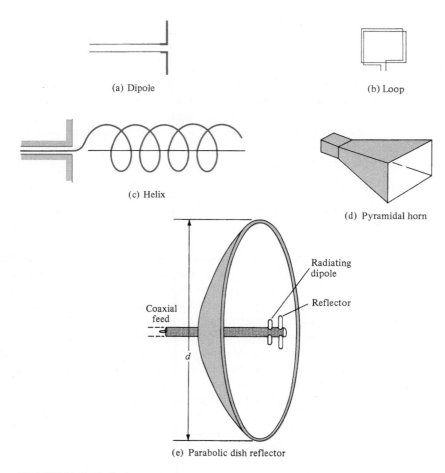

FIGURE 13.2 Typical antennas.

The phenomenon of radiation is rather complicated, so we have intentionally delayed its discussion until this chapter. We will not attempt a broad coverage of antenna theory; our discussion will be limited to the basic types of antennas such as the Hertzian dipole, the half-wave dipole, the quarter-wave monopole, and the small loop. For each of these types, we will determine the radiation fields by taking the following steps:

1. Select an appropriate coordinate system and determine the magnetic vector potential $\mathbf{A}$.
2. Find $\mathbf{H}$ from $\mathbf{B} = \mu\mathbf{H} = \nabla \times \mathbf{A}$.
3. Determine $\mathbf{E}$ from $\nabla \times \mathbf{H} = \varepsilon \dfrac{\partial \mathbf{E}}{\partial t}$ or $\mathbf{E} = \eta\mathbf{H} \times \mathbf{a}_k$ assuming a lossless medium $(\sigma = 0)$.
4. Find the far field and determine the time-average power radiated by using

$$P_{\text{rad}} = \int \mathscr{P}_{\text{ave}} \cdot d\mathbf{S}$$

where

$$\mathscr{P}_{\text{ave}} = \frac{1}{2}\,\mathrm{Re}\,(\mathbf{E}_s \times \mathbf{H}_s^*)$$

Note that P_{rad} throughout this chapter is the same as P_{ave} in eq. (10.80).

We will consider antenna arrays which produce particular directional properties of the radiated field. We will derive the Friis transmission equation for coupling between two antennas. Finally, we will consider the problem of electromagnetic interference (EMI).

13.2 HERTZIAN DIPOLE

By "Hertzian dipole" we mean an infinitesimal current element $I\,dl$, where $dl \le \lambda/10$. Although such a current element does not exist in real life, it serves as a building block from which the field of a practical antenna can be calculated by integration.

Consider the Hertzian dipole shown in Figure 13.3. We assume that it is located at the origin of a coordinate system and that it carries a uniform current (constant throughout the

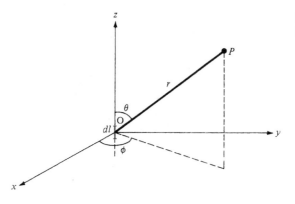

FIGURE 13.3 A Hertzian dipole carrying current $I = I_o \cos \omega t$.

dipole), $I = I_o \cos \omega t$. From eq. (9.54), the retarded magnetic vector potential at the field point P, due to the dipole, is given by

$$\mathbf{A} = \frac{\mu[I]\,dl}{4\pi\,r}\,\mathbf{a}_z \tag{13.1}$$

where $[I]$ is the retarded current given by

$$[I] = I_o \cos \omega \left(t - \frac{r}{u} \right) = I_o \cos (\omega t - \beta r)$$
$$= \text{Re}\left[I_o e^{j(\omega t - \beta r)} \right] \tag{13.2}$$

where $\beta = \omega/u = 2\pi/\lambda$, and $u = 1/\sqrt{\mu\varepsilon}$. The current is said to be *retarded* at point P because there is a propagation time delay r/u or phase delay βr from O to P. By substituting eq. (13.2) into eq. (13.1), we may write $\mathbf{A}$ in phasor form as

$$A_{zs} = \frac{\mu I_o dl}{4\pi r} e^{-j\beta r} \tag{13.3}$$

Transforming this vector from Cartesian to spherical coordinates yields

$$\mathbf{A}_s = (A_{rs}, A_{\theta s}, A_{\phi s})$$

where

$$A_{rs} = A_{zs} \cos \theta, \quad A_{\theta s} = -A_{zs} \sin \theta, \quad A_{\phi s} = 0 \tag{13.4}$$

Since $\mathbf{B}_s = \mu \mathbf{H}_s = \nabla \times \mathbf{A}_s$, we obtain the $\mathbf{H}$ field as

$$H_{\phi s} = \frac{I_o dl}{4\pi} \sin \theta \left[\frac{j\beta}{r} + \frac{1}{r^2} \right] e^{-j\beta r} \tag{13.5a}$$

$$H_{rs} = 0 = H_{\theta s} \tag{13.5b}$$

We find the $\mathbf{E}$ field by using $\nabla \times \mathbf{H} = \varepsilon\, \partial\mathbf{E}/\partial t$ or $\nabla \times \mathbf{H}_s = j\omega\varepsilon\mathbf{E}_s$,

$$E_{rs} = \frac{\eta I_o dl}{2\pi} \cos \theta \left[\frac{1}{r^2} - \frac{j}{\beta r^3} \right] e^{-j\beta r} \tag{13.6a}$$

$$E_{\theta s} = \frac{\eta I_o dl}{4\pi} \sin \theta \left[\frac{j\beta}{r} + \frac{1}{r^2} - \frac{j}{\beta r^3} \right] e^{-j\beta r} \tag{13.6b}$$

$$E_{\phi s} = 0 \tag{13.6c}$$

where

$$\eta = \frac{\beta}{\omega\varepsilon} = \sqrt{\frac{\mu}{\varepsilon}}$$

A close observation of the field equations in eqs. (13.5) and (13.6) reveals that we have terms varying as $1/r^3$, $1/r^2$, and $1/r$. The $1/r^3$ term is called the *electrostatic field*, since it corresponds to the field of an electric dipole [see eq. (4.82)]. This term dominates other terms in a region very close to the Hertzian dipole. The $1/r^2$ term is called the *inductive field*, and it is predictable from the Biot–Savart law [see eq. (7.3)]. The term is important only at near field, that is, at distances close to the current element. The $1/r$ term is called the *far field or radiation field* because it is the only term that remains at the far zone, that is, at a point very far from the current element. Here, we are mainly concerned with the far field or radiation zone ($\beta r \gg 1$ or $2\pi r \gg \lambda$), where the terms in $1/r^3$ and $1/r^2$ can be neglected in favor of the $1/r$ term. Also note that near-zone and far-zone fields are determined, respectively, to be the inequalities $\beta r \ll 1$ and $\beta r \gg 1$. More specifically, we define the boundary between the near and the far zones by the value of r given by

$$r = \frac{2d^2}{\lambda} \tag{13.7}$$

where d is the largest dimension of the antenna. Thus at far field,

$$\boxed{H_{\phi s} = \frac{jI_o\beta dl}{4\pi r}\sin\theta\, e^{-j\beta r}, \quad E_{\theta s} = \eta\, H_{\phi s}} \tag{13.8a}$$

$$H_{rs} = H_{\theta s} = E_{rs} = E_{\phi s} = 0 \tag{13.8b}$$

Note from eq. (13.8a) that the radiation terms of $H_{\phi s}$ and $E_{\theta s}$ are in time phase and orthogonal just as the fields of a uniform plane wave.

The time-average power density is obtained as

$$\mathscr{P}_{\text{ave}} = \frac{1}{2}\operatorname{Re}(\mathbf{E}_s \times \mathbf{H}_s^*) = \frac{1}{2}\operatorname{Re}(E_{\theta s}\, H_{\phi s}^* \mathbf{a}_r)$$

$$= \frac{1}{2}\eta\, |H_{\phi s}|^2\, \mathbf{a}_r \tag{13.9}$$

Substituting eq. (13.8a) into eq. (13.9) yields the time-average radiated power as

$$P_{\text{rad}} = \int \mathscr{P}_{\text{ave}} \cdot d\mathbf{S}$$

$$= \int_{\phi=0}^{2\pi}\left[\int_{\theta=0}^{\pi} \frac{I_o^2\eta\beta^2\, dl^2}{32\pi^2 r^2}\sin^2\theta\, r^2 \sin\theta\, d\theta\right] d\phi \tag{13.10}$$

$$= \frac{I_o^2\eta\beta^2\, dl^2}{32\pi^2}\, 2\pi \int_0^{\pi} \sin^3\theta\, d\theta$$

But

$$\int_0^{\pi} \sin^3\theta\, d\theta = \int_0^{\pi} (1 - \cos^2\theta)\, d(-\cos\theta)$$

$$= \left.\frac{\cos^3\theta}{3} - \cos\theta\right|_0^{\pi} = \frac{4}{3}$$

and $\beta^2 = 4\pi^2/\lambda^2$. Hence eq. (13.10) becomes

$$P_{rad} = \frac{I_o^2 \pi \eta}{3} \left[\frac{dl}{\lambda} \right]^2 \tag{13.11a}$$

If free space is the medium of propagation, $\eta = 120\pi$ and

$$P_{rad} = 40\pi^2 \left[\frac{dl}{\lambda} \right]^2 I_o^2 \tag{13.11b}$$

This power is equivalent to the power dissipated in a fictitious resistance R_{rad} by current $I = I_o \cos \omega t$; that is,

$$P_{rad} = I_{rms}^2 R_{rad}$$

or

$$\boxed{P_{rad} = \frac{1}{2} I_o^2 R_{rad}} \tag{13.12}$$

where I_{rms} is the root-mean-square value of I. From eqs. (13.11) and (13.12), we obtain

$$R_{rad} = \frac{2P_{rad}}{I_o^2} \tag{13.13a}$$

or

$$\boxed{R_{rad} = 80\pi^2 \left[\frac{dl}{\lambda} \right]^2} \tag{13.13b}$$

The resistance R_{rad} is a characteristic property of the Hertzian dipole antenna and is called its *radiation resistance*. From eqs. (13.12) and (13.13), we observe antennas with large radiation resistances are required to deliver large amounts of power to space. For example, if $dl = \lambda/20$, $R_{rad} \simeq 2\ \Omega$, which is small in that it can deliver relatively small amounts of power. It should be noted that R_{rad} in eq. (13.13b) is for a Hertzian dipole in free space. If the dipole is in a different, lossless medium, $\eta = \sqrt{\mu/\varepsilon}$ is substituted in eq. (13.11a) and R_{rad} is determined by using eq. (13.13a).

Note that the Hertzian dipole is assumed to be infinitesimally small ($\beta\ dl \ll 1$ or $dl \leq \lambda/10$). Consequently, its radiation resistance is very small, and it is in practice difficult to match it with a real transmission line. We have also assumed that the dipole has a uniform current; this requires that the current be nonzero at the end points of the dipole. This is practically impossible because the surrounding medium is not conducting. However, our analysis will serve as a useful, valid approximation for an antenna with $dl \leq \lambda/10$. A more practical (and perhaps the most important) antenna is the half-wave dipole, considered in the next section.

13.3 HALF-WAVE DIPOLE ANTENNA

The half-wave dipole derives its name from the fact that its length is half a wavelength ($\ell = \lambda/2$). As shown in Figure 13.4(a), it consists of a thin wire fed or excited at the midpoint by a voltage source connected to the antenna via a transmission line (e.g., a two-wire line). The field due to the dipole can be easily obtained if we consider it as consisting of a chain of Hertzian dipoles. The magnetic vector potential at P due to a differential length $dl\,(= dz)$ of the dipole [see Figure. 13.4(b)] carrying a phasor current $I_s = I_o \cos \beta z$ is

$$dA_{zs} = \frac{\mu I_o \cos \beta z \, dz}{4\pi r'} e^{-j\beta r'} \tag{13.14}$$

Notice that to obtain eq. (13.14), we have assumed a sinusoidal current distribution for two reasons. First, the sinusoidal current assumption is based on the transmission line model of the dipole. Second, the current must vanish at the ends of the dipole. A triangular current distribution is also possible (see Problem 13.5) but would give less accurate results. The actual current distribution on the antenna is not precisely known. It is determined by solving Maxwell's equations subject to the boundary conditions on the antenna, but

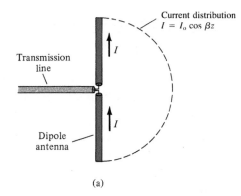

(a)

FIGURE 13.4 (a) A half-wave dipole.
(b) Geometry for calculating the fields.

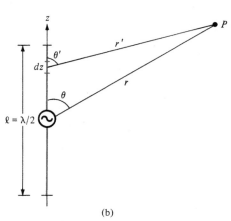

(b)

the procedure is mathematically complex. However, the sinusoidal current assumption approximates the distribution obtained by solving the boundary-value problem and is commonly used in antenna theory.

If $r \gg \ell$, as explained in Section 4.9 on electric dipoles (see Figure 4.20), then

$$r - r' = z \cos \theta \quad \text{or} \quad r' = r - z \cos \theta$$

Thus we may substitute $r' \simeq r$ in the denominator of eq. (13.14), where the magnitude of the distance is needed. For the phase term in the numerator of eq. (13.14), the difference between βr and $\beta r'$ is significant, so we replace r' by $r - z \cos \theta$ and not r. In other words, we maintain the cosine term in the exponent while neglecting it in the denominator because the exponent involves the phase constant while the denominator does not. Thus,

$$
\begin{aligned}
A_{zs} &= \frac{\mu I_o}{4\pi r} \int_{-\lambda/4}^{\lambda/4} e^{-j\beta(r - z \cos \theta)} \cos \beta z \, dz \\
&= \frac{\mu I_o}{4\pi r} e^{-j\beta r} \int_{-\lambda/4}^{\lambda/4} e^{j\beta z \cos \theta} \cos \beta z \, dz
\end{aligned}
\tag{13.15}
$$

From the integral tables of Appendix A.8,

$$\int e^{az} \cos bz \, dz = \frac{e^{az} (a \cos bz + b \sin bz)}{a^2 + b^2} + c$$

Applying this to eq. (13.15) gives

$$A_{zs} = \frac{\mu I_o e^{-j\beta r} e^{j\beta z \cos \theta} (j\beta \cos \theta \cos \beta z + \beta \sin \beta z)}{4\pi r} \frac{}{-\beta^2 \cos^2 \theta + \beta^2} \Bigg|_{-\lambda/4}^{\lambda/4} \tag{13.16}$$

Since $\beta = 2\pi/\lambda$ or $\beta \lambda/4 = \pi/2$ and $-\cos^2 \theta + 1 = \sin^2 \theta$, eq. (13.16) becomes

$$A_{zs} = \frac{\mu I_o e^{-j\beta r}}{4\pi r \beta^2 \sin^2 \theta} \left[e^{j(\pi/2) \cos \theta}(0 + \beta) - e^{-j(\pi/2) \cos \theta}(0 - \beta) \right] \tag{13.17}$$

Using the identity $e^{jx} + e^{-jx} = 2 \cos x$, we obtain

$$A_{zs} = \frac{\mu I_o e^{-j\beta r} \cos \left(\dfrac{\pi}{2} \cos \theta \right)}{2\pi r \beta \sin^2 \theta} \tag{13.18}$$

We use eq. (13.4) in conjunction with the fact that $\mathbf{B}_s = \mu \mathbf{H}_s = \nabla \times \mathbf{A}_s$ and $\nabla \times \mathbf{H}_s = j\omega\varepsilon\mathbf{E}_s$ to obtain the magnetic and electric fields at far zone (discarding the $1/r^3$ and $1/r^2$ terms) as

$$\boxed{ H_{\phi s} = \frac{j I_o e^{-j\beta r} \cos \left(\dfrac{\pi}{2} \cos \theta \right)}{2\pi r \sin \theta}, \quad E_{\theta s} = \eta H_{\phi s} } \tag{13.19}$$

Notice again that the radiation terms $H_{\phi s}$ and $E_{\theta s}$ are in time phase and orthogonal.

By using eqs. (13.9) and (13.19), we obtain the time-average power density as

$$
\begin{aligned}
\mathcal{P}_{ave} &= \frac{1}{2}\eta \, |H_{\phi s}|^2 \, \mathbf{a}_r \\[2mm]
&= \frac{\eta I_o^2 \cos^2\left(\dfrac{\pi}{2}\cos\theta\right)}{8\pi^2 r^2 \sin^2\theta} \, \mathbf{a}_r
\end{aligned}
\tag{13.20}
$$

The time-average radiated power can be determined as

$$
\begin{aligned}
P_{rad} &= \int_S \mathcal{P}_{ave} \cdot d\mathbf{S} \\[2mm]
&= \int_{\phi=0}^{2\pi}\left[\int_{\theta=0}^{\pi}\frac{\eta I_o^2 \cos^2\left(\dfrac{\pi}{2}\cos\theta\right)}{8\pi^2 r^2 \sin^2\theta}\, r^2 \sin\theta\, d\theta\right] d\phi \\[2mm]
&= \frac{\eta I_o^2}{8\pi^2}\, 2\pi \int_0^{\pi}\frac{\cos^2\left(\dfrac{\pi}{2}\cos\theta\right)}{\sin\theta}\, d\theta \\[2mm]
&= 30\, I_o^2 \int_0^{\pi}\frac{\cos^2\left(\dfrac{\pi}{2}\cos\theta\right)}{\sin\theta}\, d\theta
\end{aligned}
\tag{13.21}
$$

where $\eta = 120\pi$ has been substituted assuming free space as the medium of propagation. Due to the nature of the integrand in eq. (13.21),

$$
\int_0^{\pi/2}\frac{\cos^2\left(\dfrac{\pi}{2}\cos\theta\right)}{\sin\theta}\, d\theta = \int_{\pi/2}^{\pi}\frac{\cos^2\left(\dfrac{\pi}{2}\cos\theta\right)}{\sin\theta}\, d\theta
$$

This is easily illustrated by a rough sketch of the variation of the integrand with θ. Hence

$$
P_{rad} = 60 I_o^2 \int_0^{\pi/2}\frac{\cos^2\left(\dfrac{\pi}{2}\cos\theta\right)d\theta}{\sin\theta}
\tag{13.22}
$$

Changing variables, $u = \cos\theta$, and using partial fraction reduces eq. (13.22) to

$$
\begin{aligned}
P_{rad} &= 60 I_o^2 \int_0^{1}\frac{\cos^2\dfrac{1}{2}\pi u}{1 - u^2}\, du \\[2mm]
&= 30 I_o^2 \left[\int_0^{1}\frac{\cos^2\dfrac{1}{2}\pi u}{1 + u}\, du + \int_0^{1}\frac{\cos^2\dfrac{1}{2}\pi u}{1 - u}\, du\right]
\end{aligned}
\tag{13.23}
$$

Replacing $1 + u$ with v in the first integrand and $1 - u$ with v in the second results in

$$P_{rad} = 30I_o^2 \left[\int_0^1 \frac{\sin^2 \frac{1}{2}\pi v}{v} \, dv + \int_1^2 \frac{\sin^2 \frac{1}{2}\pi v}{v} \, dv \right]$$

$$= 30I_o^2 \int_0^2 \frac{\sin^2 \frac{1}{2}\pi v}{v} \, dv$$

(13.24)

Changing variables, $w = \pi v$, yields

$$P_{rad} = 30I_o^2 \int_0^{2\pi} \frac{\sin^2 \frac{1}{2} w}{w} \, dw$$

$$= 15I_o^2 \int_0^{2\pi} \frac{(1 - \cos w)}{w} \, dw$$

(13.25)

$$= 15I_o^2 \int_0^{2\pi} \left[\frac{w}{2!} - \frac{w^3}{4!} + \frac{w^5}{6!} - \frac{w^7}{8!} + \cdots \right] dw$$

since $\cos w = 1 - \dfrac{w^2}{2!} + \dfrac{w^4}{4!} - \dfrac{w^6}{6!} + \dfrac{w^8}{8!} - \cdots$. Integrating eq. (13.25) term by term and evaluating at the limit leads to

$$P_{rad} = 15I_o^2 \left[\frac{(2\pi)^2}{2(2!)} - \frac{(2\pi)^4}{4(4!)} + \frac{(2\pi)^6}{6(6!)} - \frac{(2\pi)^8}{8(8!)} + \cdots \right]$$

$$\approx 36.56 \, I_o^2$$

(13.26)

The radiation resistance R_{rad} for the half-wave dipole antenna is readily obtained from eqs. (13.12) and (13.26) as

$$\boxed{R_{rad} = \frac{2P_{rad}}{I_o^2} \approx 73 \, \Omega}$$

(13.27)

Note the significant increase in the radiation resistance of the half-wave dipole over that of the Hertzian dipole. Thus the half-wave dipole is capable of delivering greater amounts of power to space than the Hertzian dipole.

The total input impedance Z_{in} of the antenna is the impedance seen at the terminals of the antenna and is given by

$$Z_{in} = R_{in} + jX_{in}$$

(13.28)

where $R_{in} = R_{rad}$ for a lossless antenna. Deriving the value of the reactance X_{in} involves a complicated procedure beyond the scope of this text. It is found that $X_{in} = 42.5 \, \Omega$, so $Z_{in} = 73 + j42.5 \, \Omega$ for a dipole length $\ell = \lambda/2$. The inductive reactance drops rapidly to zero as the length of the dipole is slightly reduced. For $\ell = 0.485 \, \lambda$, the dipole is resonant, with $X_{in} = 0$. Thus in practice, a $\lambda/2$ dipole is designed such that X_{in} approaches zero and $Z_{in} \simeq 73 \, \Omega$. This value of the radiation resistance of the $\lambda/2$ dipole antenna is the reason for the standard 75 Ω coaxial cable. Also, the value is easy to match to transmission lines. These factors in addition to the resonance property are the reasons for the dipole antenna's popularity and its extensive use.

13.4 QUARTER-WAVE MONOPOLE ANTENNA

Basically, the quarter-wave monopole antenna consists of half of a half-wave dipole antenna located on a conducting ground plane, as in Figure 13.5. The monopole antenna is perpendicular to the plane, which is usually assumed to be infinite and perfectly conducting. It is fed by a coaxial cable connected to its base.

Using image theory of Section 6.6, we replace the infinite, perfectly conducting ground plane with the image of the monopole. The field produced in the region above the ground plane due to the $\lambda/4$ monopole with its image is the same as the field due to a $\lambda/2$ wave dipole. Thus eq. (13.19) holds for the $\lambda/4$ monopole. However, the integration in eq. (13.21) is only over the hemispherical surface above the ground plane (i.e., $0 \leq \theta \leq \pi/2$) because the monopole radiates only through that surface. Hence, the monopole radiates only half as much power as the dipole with the same current. Thus for a $\lambda/4$ monopole,

$$P_{rad} \simeq 18.28 \, I_o^2 \qquad (13.29)$$

and

$$R_{rad} = \frac{2P_{rad}}{I_o^2}$$

or

$$\boxed{R_{rad} \simeq 36.5 \, \Omega} \qquad (13.30)$$

By the same token, the total input impedance for a $\lambda/4$ monopole is $Z_{in} = 36.5 + j21.25 \, \Omega$.

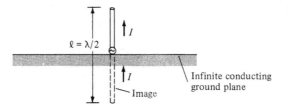

FIGURE 13.5 The monopole antenna.

13.5 SMALL-LOOP ANTENNA

The loop antenna is of practical importance. It is used as a directional finder (or search loop) in radiation detection and as a TV antenna for ultrahigh frequencies. The term "small" implies that the dimensions (such as ρ_o) of the loop are much smaller than λ.

Consider a small filamentary circular loop of radius ρ_o carrying a uniform current, $I_o \cos \omega t$, as in Figure 13.6. The loop may be regarded as an elemental magnetic dipole. The magnetic vector potential at the field point P due to the loop is

$$\mathbf{A} = \oint_L \frac{\mu[I]\,d\mathbf{l}}{4\pi r'} \tag{13.31}$$

where $[I] = I_o \cos(\omega t - \beta r') = \operatorname{Re}\left[I_o e^{j(\omega t - \beta r')}\right]$. Substituting $[I]$ into eq. (13.31), we obtain $\mathbf{A}$ in phasor form as

$$\mathbf{A}_s = \frac{\mu I_o}{4\pi} \oint_L \frac{e^{-j\beta r'}}{r'}\,d\mathbf{l} \tag{13.32}$$

Evaluating this integral requires a lengthy procedure. It can be shown that for a small loop ($\rho_o \ll \lambda$), r' can be replaced by r in the denominator of eq. (13.32) and $\mathbf{A}_s$ has only a ϕ-component given by

$$A_{\phi s} = \frac{\mu I_o S}{4\pi r^2}(1 + j\beta r)e^{-j\beta r}\sin\theta \tag{13.33}$$

where $S = \pi\rho_o^2 =$ loop area. For a loop with N turns, $S = N\pi\rho_o^2$. Using the fact that $\mathbf{B}_s = \mu\mathbf{H}_s = \nabla \times \mathbf{A}_s$ and $\nabla \times \mathbf{H}_s = j\omega\varepsilon\mathbf{E}_s$, we obtain the electric and magnetic fields from eq. (13.33) as

$$E_{\phi s} = \frac{-j\omega\mu I_o S}{4\pi}\sin\theta\left[\frac{j\beta}{r} + \frac{1}{r^2}\right]e^{-j\beta r} \tag{13.34a}$$

$$H_{rs} = \frac{j\omega\mu I_o S}{2\pi\eta}\cos\theta\left[\frac{1}{r^2} - \frac{j}{\beta r^3}\right]e^{-j\beta r} \tag{13.34b}$$

$$H_{\theta s} = \frac{j\omega\mu I_o S}{4\pi\eta}\sin\theta\left[\frac{j\beta}{r} + \frac{1}{r^2} - \frac{j}{\beta r^3}\right]e^{-j\beta r} \tag{13.34c}$$

$$E_{rs} = E_{\theta s} = H_{\phi s} = 0 \tag{13.34d}$$

Comparing eqs. (13.5) and (13.6) with eq. (13.34), we observe the dual nature of the field due to an electric dipole of Figure 13.3 and the elemental magnetic dipole of Figure 13.6 (see Table 8.2 also). At far field, only the $1/r$ term (the radiation term) in eq. (13.34) remains. Thus at far field,

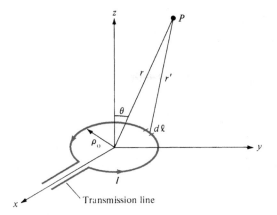

FIGURE 13.6 The small-loop antenna.

$$E_{\phi s} = \frac{\omega \mu I_o S}{4 \pi r} \beta \sin \theta \, e^{-j\beta r}$$

$$= \frac{\eta \pi I_o S}{r \lambda^2} \sin \theta \, e^{-j\beta r}$$

or

$$E_{\phi s} = \frac{120 \pi^2 I_o}{r} \frac{S}{\lambda^2} \sin \theta \, e^{-j\beta r}, \quad H_{\theta s} = -\frac{E_{\phi s}}{\eta} \qquad (13.35a)$$

$$E_{rs} = E_{\theta s} = H_{rs} = H_{\phi s} = 0 \qquad (13.35b)$$

where $\eta = 120\pi$ for free space has been assumed. Though the far-field expressions in eqs. (13.35) are obtained for a small circular loop, they can be used for a small square loop with one turn $(S = a^2)$ or with N turns $(S = Na^2)$, or for any small loop, provided the loop dimensions are small $(d \leq \lambda/10$, where d is the largest dimension of the loop). It is left as an exercise to show that using eqs. (13.13a) and (13.35) gives the radiation resistance of a small loop antenna as

$$R_{\text{rad}} = \frac{320 \, \pi^4 S^2}{\lambda^4} \qquad (13.36)$$

EXAMPLE 13.1

A magnetic field strength of 5 μA/m is required at a point on $\theta = \pi/2$, which is 2 km from an antenna in air. Neglecting ohmic loss, how much power must the antenna transmit if it is

(a) A Hertzian dipole of length $\lambda/25$?

(b) A half-wave dipole?

(c) A quarter-wave monopole?

(d) A 10-turn loop antenna of radius $\rho_o = \lambda/20$?

Solution:

(a) For a Hertzian dipole,

$$|H_{\phi s}| = \frac{I_o \beta \, dl \sin \theta}{4\pi r}$$

where $dl = \lambda/25$ or $\beta \, dl = \dfrac{2\pi}{\lambda} \cdot \dfrac{\lambda}{25} = \dfrac{2\pi}{25}$. Hence,

$$5 \times 10^{-6} = \frac{I_o \cdot \dfrac{2\pi}{25}(1)}{4\pi \, (2 \times 10^3)} = \frac{I_o}{10^5}$$

or

$$I_o = 0.5 \text{ A}$$

$$P_{rad} = 40\pi^2 \left[\frac{dl}{\lambda}\right]^2 I_o^2 = \frac{40\pi^2(0.5)^2}{(25)^2}$$

$$= 158 \text{ mW}$$

(b) For a $\lambda/2$ dipole,

$$|H_{\phi s}| = \frac{I_o \cos\left(\dfrac{\pi}{2} \cos \theta\right)}{2\pi r \sin \theta}$$

$$5 \times 10^{-6} = \frac{I_o \cdot 1}{2\pi \, (2 \times 10^3) \cdot (1)}$$

or

$$I_o = 20\pi \text{ mA}$$

$$P_{rad} = \frac{1}{2} I_o^2 R_{rad} = \frac{1}{2} (20\pi)^2 \times 10^{-6}(73)$$

$$= 144 \text{ mW}$$

(c) For a $\lambda/4$ monopole,

$$I_o = 20\pi \text{ mA}$$

as in part (b).

$$P_{rad} = \frac{1}{2} I_o^2 R_{rad} = \frac{1}{2} (20\pi)^2 \times 10^{-6}(36.56)$$

$$= 72 \text{ mW}$$

(d) For a loop antenna,

$$|H_{\theta s}| = \frac{\pi I_o}{r} \frac{S}{\lambda^2} \sin \theta$$

For a single turn, $S = \pi \rho_o^2$. For N-turn, $S = N\pi \rho_o^2$. Hence,

$$5 \times 10^{-6} = \frac{\pi I_o 10\pi}{2 \times 10^3} \left[\frac{\rho_o}{\lambda} \right]^2$$

or

$$I_o = \frac{10}{10\pi^2} \left[\frac{\lambda}{\rho_o} \right]^2 \times 10^{-3} = \frac{20^2}{\pi^2} \times 10^{-3}$$

$$= 40.53 \text{ mA}$$

$$R_{\text{rad}} = \frac{320 \pi^4 S^2}{\lambda^4} = 320 \pi^6 N^2 \left[\frac{\rho_o}{\lambda} \right]^4$$

$$= 320 \pi^6 \times 100 \left[\frac{1}{20} \right]^4 = 192.3 \, \Omega$$

$$P_{\text{rad}} = \frac{1}{2} I_o^2 R_{\text{rad}} = \frac{1}{2} (40.53)^2 \times 10^{-6} (192.3)$$

$$= 158 \text{ mW}$$

PRACTICE EXERCISE 13.1

A Hertzian dipole of length $\lambda/100$ is located at the origin in free space and fed with a current of $0.25 \sin 10^8 t$ A. Determine the magnetic field at

(a) $r = \lambda/5, \theta = 30°$
(b) $r = 200\lambda, \theta = 60°$

Answer: (a) $0.2119 \sin (10^8 t - 20.5°) \mathbf{a}_\phi$ mA/m, (b) $0.2871 \sin (10^8 t + 90°) \mathbf{a}_\phi$ μA/m.

EXAMPLE 13.2

An electric field strength of 10 μV/m is to be measured at an observation point $\theta = \pi/2$, 500 km from a half-wave (resonant) dipole antenna operating in air at 50 MHz.

(a) What is the length of the dipole?
(b) Calculate the current that must be fed to the antenna.
(c) Find the average power radiated by the antenna.
(d) If a transmission line with $Z_o = 75 \, \Omega$ is connected to the antenna, determine the standing wave ratio.

Solution:

(a) The wavelength $\lambda = \dfrac{c}{f} = \dfrac{3 \times 10^8}{50 \times 10^6} = 6$ m.

Hence, the length of the half-dipole is $\ell = \dfrac{\lambda}{2} = 3$ m.

(b) From eq. (13.19),

$$|E_{\theta s}| = \frac{\eta_o I_o \cos\left(\dfrac{\pi}{2}\cos\theta\right)}{2\pi r \sin\theta}$$

or

$$I_o = \frac{|E_{\theta s}|\, 2\pi r \sin\theta}{\eta_o \cos\left(\dfrac{\pi}{2}\cos\theta\right)}$$

$$= \frac{10 \times 10^{-6}\, 2\pi\, (500 \times 10^3) \cdot (1)}{120\pi\,(1)}$$

$$= 83.33 \text{ mA}$$

(c) $\qquad R_{\text{rad}} \simeq 73\ \Omega$

$$P_{\text{rad}} = \frac{1}{2}I_o^2 R_{\text{rad}} = \frac{1}{2}(83.33)^2 \times 10^{-6} \times 73$$

$$= 253.5 \text{ mW}$$

(d) $\qquad \Gamma = \dfrac{Z_L - Z_o}{Z_L + Z_o}\quad (Z_L = Z_{\text{in}} \text{ in this case})$

$$= \frac{73 + j42.5 - 75}{73 + j42.5 + 75} = \frac{-2 + j42.5}{148 + j42.5}$$

$$= \frac{42.55\underline{/92.69^\circ}}{153.98\underline{/16.02^\circ}} = 0.2763\underline{/76.67^\circ}$$

$$s = \frac{1 + |\Gamma|}{1 - |\Gamma|} = \frac{1 + 0.2763}{1 - 0.2763} = 1.763$$

PRACTICE EXERCISE 13.2

Repeat Example 13.2 with the dipole antenna replaced by a $\lambda/4$ monopole.

Answer: (a) 1.5 m, (b) 83.33 mA, (c) 126.8 mW, (d) 2.265.

13.6 ANTENNA CHARACTERISTICS

Having considered the basic elementary antenna types, we now discuss some important characteristics of antennas as radiators of electromagnetic energy. These characteristics include (a) antenna patterns, (b) radiation intensity, (c) directive gain, (d) power gain.

A. Antenna Patterns

When the amplitude of a specified component of the **E** field is plotted, it is called the *field pattern* or *voltage pattern*. When the square of the amplitude of **E** is plotted, it is called the *power pattern*. A three-dimensional plot of an antenna pattern is avoided by plotting separately the normalized $|E_s|$ versus θ for a constant ϕ (this is called an *E-plane pattern* or *vertical pattern*) and the normalized $|E_s|$ versus ϕ for $\theta = \pi/2$ (called the *H-plane pattern* or *horizontal pattern*). The normalization of $|E_s|$ is with respect to the maximum value of the $|E_s|$ so that the maximum value of the normalized $|E_s|$ is unity.

For the Hertzian dipole, for example, the normalized $|E_s|$ is obtained from eq. (13.8a) as

$$f(\theta) = |\sin \theta| \tag{13.37}$$

which is independent of ϕ. From eq. (13.37), we obtain the *E*-plane pattern as the polar plot of $f(\theta)$ with θ varying from 0° to 180°. The result is shown in Figure 13.7(a). Note that the plot is symmetric about the *z*-axis ($\theta = 0$). For the *H*-plane pattern, we set $\theta = \pi/2$ so that $f(\theta) = 1$, which is circle of radius 1 as shown in Figure 13.7(b). When the two

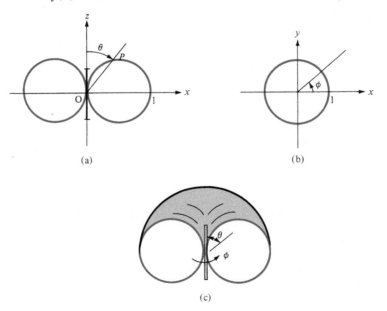

FIGURE 13.7 Field patterns of the Hertzian dipole: (**a**) normalized *E*-plane or vertical pattern ($\phi = $ constant $= 0$), (**b**) normalized *H*-plane or horizontal pattern ($\theta = \pi/2$), (**c**) three-dimensional pattern.

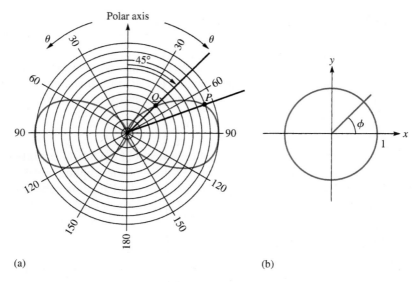

FIGURE 13.8 Power patterns of the Hertzian dipole: (**a**) (ϕ = constant = 0), (**b**) θ = constant = $\pi/2$.

plots of Figure 13.7(a) and (b) are combined, we have the three-dimensional field pattern of Figure 13.7(c), which has the shape of a doughnut.

A plot of the time-average power, $|\mathcal{P}_{ave}| = \mathcal{P}_{ave}$, for a fixed distance r is the power pattern of the antenna. It is obtained by plotting separately $\mathcal{P}_{ave}$ versus θ for constant ϕ and $\mathcal{P}_{ave}$ versus ϕ for constant θ.

For the Hertzian dipole, the normalized power pattern is easily obtained from eqs. (13.37) or (13.9) as

$$f^2(\theta) = \sin^2 \theta \tag{13.38}$$

which is sketched in Figure 13.8. Notice that Figures 13.7(b) and 13.8(b) show circles because $f(\theta)$ is independent of ϕ and that the value of OP in Figure 13.8(a) is the relative average power for that particular θ. Thus, at point Q ($\theta = 45°$), the average power is half the maximum average power (the maximum average power is at $\theta = \pi/2$).

An **antenna pattern** (or **radiation pattern**) is a three-dimensional plot of its radiation at far field.

B. Radiation Intensity

The radiation intensity of an antenna is defined as

$$\boxed{U(\theta, \phi) = r^2 \mathcal{P}_{ave}} \tag{13.39}$$

From eq. (13.39), the total average power radiated can be expressed as

$$P_{rad} = \oint_S \mathcal{P}_{ave} \, dS = \oint_S \mathcal{P}_{ave} \, r^2 \sin\theta \, d\theta \, d\phi$$

$$= \int_S U(\theta, \phi) \sin\theta \, d\theta \, d\phi \qquad (13.40)$$

$$= \int_{\phi=0}^{2\pi} \int_{\theta=0}^{\pi} U(\theta, \phi) \, d\Omega$$

where $d\Omega = \sin\theta \, d\theta \, d\phi$ is the *differential solid angle* in steradian (sr). Hence the radiation intensity $U(\theta, \phi)$ is measured in watts per steradian (W/sr). The average value of $U(\theta, \phi)$ is the total radiated power divided by 4π sr; that is,

$$U_{ave} = \frac{P_{rad}}{4\pi} \qquad (13.41)$$

C. Directive Gain

Besides the antenna patterns just described, we are often interested in measurable quantities such as gain and directivity to determine the radiation characteristics of an antenna.

> **The directive gain** $G_d(\theta, \phi)$ of an antenna is a measure of the concentration of the radiated power in a particular direction (θ, ϕ).

It may be regarded as the ability of the antenna to direct radiated power in a given direction. It is usually obtained as the ratio of radiation intensity in a given direction (θ, ϕ) to the average radiation intensity, that is,

$$G_d(\theta, \phi) = \frac{U(\theta, \phi)}{U_{ave}} = \frac{4\pi \, U(\theta, \phi)}{P_{rad}} \qquad (13.42)$$

By substituting eq. (13.39) into eq. (13.42), $\mathcal{P}_{ave}$ may be expressed in terms of directive gain as

$$\mathcal{P}_{ave} = \frac{G_d}{4\pi r^2} P_{rad} \qquad (13.43)$$

The directive gain $G_d(\theta, \phi)$ depends on antenna pattern. For the Hertzian dipole (as well as for $\lambda/2$ dipole and $\lambda/4$ monopole), we notice from Figure 13.8 that $\mathcal{P}_{ave}$ is maximum at $\theta = \pi/2$ and minimum (zero) at $\theta = 0$ or π. Thus the Hertzian dipole radiates power in a direction broadside to its length. For an *isotropic* antenna (one that radiates equally in all directions), $G_d = 1$. However, such an antenna is not a physicality but an ideality.

> The **directivity** D of an antenna is the ratio of the maximum radiation intensity to the average radiation intensity.

Obviously, D is the maximum directive gain $G_{d\,max}$. Thus

$$\boxed{D = \frac{U_{max}}{U_{ave}} = G_{d\,max}}$$ (13.44a)

or, from eq. (13.41),

$$D = \frac{4\pi\,U_{max}}{P_{rad}}$$ (13.44b)

For an isotropic antenna, $D = 1$; this is the smallest value D can have. For the Hertzian dipole,

$$G_d(\theta, \phi) = 1.5 \sin^2 \theta, \quad D = 1.5$$ (13.45)

For the $\lambda/2$ dipole,

$$G_d(\theta, \phi) = \frac{\eta}{\pi R_{rad}} f^2(\theta), \quad D = 1.64$$ (13.46)

where $\eta = 120\pi$, $R_{rad} \simeq 73\ \Omega$, and

$$f(\theta) = \frac{\cos\left(\dfrac{\pi}{2}\cos\theta\right)}{\sin\theta}$$ (13.47)

D. Power Gain

Our definition of the directive gain in eq. (13.42) does not account for the ohmic power loss P_ℓ of the antenna. This power loss P_ℓ occurs because the antenna is made of a conductor with finite conductivity. As illustrated in Figure 13.9, if P_{in} is the total input power to the antenna,

$$P_{in} = P_\ell + P_{rad}$$

$$= \frac{1}{2}|I_{in}|^2 (R_\ell + R_{rad})$$ (13.48)

where I_{in} is the current at the input terminals and R_ℓ is the *loss* or *ohmic resistance* of the antenna. In other words, P_{in} is the power accepted by the antenna at its terminals during the radiation process, and P_{rad} is the power radiated by the antenna; the difference between the two powers is P_ℓ, the power dissipated within the antenna.

FIGURE 13.9 Relating P_{in}, P_ℓ, and P_{rad}.

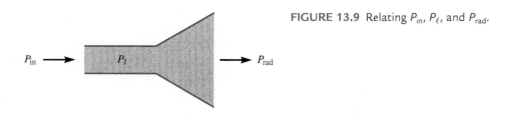

We define the *power gain* $G_p(\theta, \phi)$ of the antenna as

$$G_p(\theta, \phi) = \frac{4\pi \, U(\theta, \phi)}{P_{in}}$$

(13.49)

The ratio of the power gain in any specified direction (θ, ϕ) to the directive gain in that direction is referred to as the *radiation efficiency* η_r of the antenna; that is,

$$\eta_r = \frac{G_P}{G_d} = \frac{P_{rad}}{P_{in}}$$

Introducing eq. (13.48) leads to

$$\eta_r = \frac{P_{rad}}{P_{in}} = \frac{R_{rad}}{R_{rad} + R_\ell}$$

(13.50)

For many antennas, η_r is close to 100% so that $G_P \simeq G_d$. It is customary to express directivity and gain in decibels. Thus

$$D\,(\text{dB}) = 10 \log_{10} D$$

(13.51a)

$$G\,(\text{dB}) = 10 \log_{10} G$$

(13.51b)

It should be mentioned at this point that the radiation patterns of an antenna are measured in the far-field region. The far-field region of an antenna is commonly taken to exist at a distance $r \geq r_{min}$, where

$$r_{min} = \frac{2d^2}{\lambda}$$

(13.52)

and d is the largest dimension of the antenna. For example, $d = \ell$ for the electric dipole antenna and $d = 2\rho_o$ for the small-loop antenna.

EXAMPLE 13.3

Show that the directive gain of the Hertzian dipole is

$$G_d(\theta, \phi) = 1.5 \sin^2 \theta$$

and that of the half-wave dipole is

$$G_d(\theta, \phi) = 1.64 \, \frac{\cos^2\left(\dfrac{\pi}{2} \cos \theta\right)}{\sin^2 \theta}$$

Solution:

Starting from eq. (13.42) and introducing the expressions for $U(\theta, \phi)$ and P_{rad}, we obtain

$$G_d(\theta, \phi) = \frac{4\pi f^2(\theta)}{\int f^2(\theta)\, d\Omega}$$

(a) For the Hertzian dipole,

$$G_d(\theta, \phi) = \frac{4\pi \sin^2 \theta}{\int_{\phi=0}^{2\pi} \int_{\theta=0}^{\pi} \sin^3 \theta\, d\theta\, d\phi} = \frac{4\pi \sin^2 \theta}{2\pi (4/3)}$$

$$= 1.5 \sin^2 \theta$$

as required.

(b) For the half-wave dipole,

$$G_d(\theta, \phi) = \frac{\dfrac{4\pi \cos^2\left(\dfrac{\pi}{2} \cos \theta\right)}{\sin^2 \theta}}{\displaystyle\int_{\phi=0}^{2\pi} \int_{\theta=0}^{\pi} \frac{\cos^2\left(\dfrac{\pi}{2} \cos \theta\right)}{\sin \theta}\, d\theta\, d\phi}$$

From eq. (13.26), the integral in the denominator gives $2\pi(1.2188)$. Hence,

$$G_d(\theta, \phi) = \frac{4\pi \cos^2\left(\dfrac{\pi}{2} \cos \theta\right)}{\sin^2 \theta} \cdot \frac{1}{2\pi (1.2188)}$$

$$= 1.64 \frac{\cos^2\left(\dfrac{\pi}{2} \cos \theta\right)}{\sin^2 \theta}$$

as required.

PRACTICE EXERCISE 13.3

Calculate the directivity of

(a) The Hertzian monopole

(b) The quarter-wave monopole

Answer: (a) 3, (b) 3.28.

EXAMPLE 13.4

Determine the electric field intensity at a distance of 10 km from an antenna having a directive gain of 5 dB and radiating a total power of 20 kW.

Solution:

$$5 = G_d \, (\text{dB}) = 10 \log_{10} G_d$$

or

$$0.5 = \log_{10} G_d \rightarrow G_d = 10^{0.5} = 3.162$$

From eq. (13.43),

$$\mathcal{P}_{\text{ave}} = \frac{G_d P_{\text{rad}}}{4\pi r^2}$$

But

$$\mathcal{P}_{\text{ave}} = \frac{|E_s|^2}{2\eta}$$

Hence,

$$|E_s|^2 = \frac{\eta G_d P_{\text{rad}}}{2\pi r^2} = \frac{120\pi (3.162)(20 \times 10^3)}{2\pi \left[10 \times 10^3 \right]^2}$$

$$|E_s| = 0.1948 \text{ V/m}$$

PRACTICE EXERCISE 13.4

A certain antenna with an efficiency of 95% has maximum radiation intensity of 0.5 W/sr. Calculate its directivity when

(a) The input power is 0.4 W
(b) The radiated power is 0.3 W

Answer: (a) 16.53, (b) 20.94.

EXAMPLE 13.5

The radiation intensity of a certain antenna is

$$U(\theta, \phi) = \begin{cases} 2 \sin \theta \sin^3 \phi, & 0 \leq \theta \leq \pi, 0 \leq \phi \leq \pi \\ 0, & \text{elsewhere} \end{cases}$$

Determine the directivity of the antenna.

Solution:

The directivity is defined as

$$D = \frac{U_{\text{max}}}{U_{\text{ave}}}$$

From the given U,

$$U_{\text{max}} = 2$$

From eqs. (13.40) and (13.41), we get the expression for the average radiated intensity.

$$U_{\text{ave}} = \frac{1}{4\pi} \int U(\theta, \phi) \, d\Omega$$

$$= \frac{1}{4\pi} \int_{\phi=0}^{\pi} \int_{\theta=0}^{\pi} 2 \sin \theta \sin^3 \phi \sin \theta \, d\theta \, d\phi$$

$$= \frac{1}{2\pi} \int_0^\pi \sin^2 \theta \, d\theta \int_0^\pi \sin^3 \phi \, d\phi$$

$$= \frac{1}{2\pi} \int_0^\pi \frac{1}{2} (1 - \cos 2\theta) \, d\theta \int_0^\pi (1 - \cos^2 \phi) \, d(-\cos \phi)$$

$$= \frac{1}{2\pi} \frac{1}{2} \left(\theta - \frac{\sin 2\theta}{2} \right) \Big|_0^\pi \left(\frac{\cos^3 \phi}{3} - \cos \phi \right) \Big|_0^\pi$$

$$= \frac{1}{2\pi} \left(\frac{\pi}{2} \right) \left(\frac{4}{3} \right) = \frac{1}{3}$$

Hence

$$D = \frac{2}{(1/3)} = 6$$

PRACTICE EXERCISE 13.5

Evaluate the directivity of an antenna with normalized radiation intensity

$$U(\theta, \phi) = \begin{cases} \sin \theta, & 0 \le \theta \le \pi/2, 0 \le \phi \le 2\pi \\ 0, & \text{otherwise} \end{cases}$$

Answer: 2.546.

13.7 ANTENNA ARRAYS

In many practical applications (e.g., in an AM broadcast station), it is necessary to design antennas with more energy radiated in some particular directions and less in other directions. This is tantamount to requiring that the radiation pattern be concentrated in the direction of interest. This is hardly achievable with a single antenna element. An antenna array is used to obtain greater directivity than can be obtained with a single antenna element.

> An **antenna array** is a group of radiating elements arranged to produce particular radiation characteristics.

It is practical and convenient that the array consists of identical elements, but this is not fundamentally required. We shall consider the simplest case of a two-element array and extend our results to the more complicated, general case of an N-element array.

Consider an antenna consisting of two Hertzian dipoles placed in free space along the z-axis but oriented parallel to the x-axis as depicted in Figure 13.10. We assume that the dipole at $(0, 0, d/2)$ carries current $I_{1s} = I_o\underline{/\alpha}$ and the one at $(0, 0, -d/2)$ carries current $I_{2s} = I_o\underline{/0}$, where α is the phase difference between the two currents. By varying the spacing d and phase difference α, the fields from the array can be made to interfere constructively (add) in certain directions of interest and interfere destructively (cancel) in other directions. The total electric field at point P is the vector sum of the fields due to the individual elements. If P is in the far-field zone, we obtain the total electric field at P from eq. (13.8a) as

$$\mathbf{E}_s = \mathbf{E}_{1s} + \mathbf{E}_{2s}$$

$$= \frac{j\eta\beta I_o dl}{4\pi}\left[\cos\theta_1\frac{e^{-j\beta r_1}}{r_1}e^{j\alpha}\,\mathbf{a}_{\theta_1} + \cos\theta_2\frac{e^{-j\beta r_2}}{r_2}\,\mathbf{a}_{\theta_2}\right] \qquad (13.53)$$

Note that $\sin\theta$ in eq. (13.8a) has been replaced by $\cos\theta$ because the element of Figure 13.3 is z-directed, whereas those in Figure 13.10 are x-directed. Since P is far from the array, $\theta_1 \simeq \theta \simeq \theta_2$ and $\mathbf{a}_{\theta_1} \simeq \mathbf{a}_\theta \simeq \mathbf{a}_{\theta_2}$. In the amplitude, we can set $r_1 \simeq r \simeq r_2$ but in the phase, we use

$$r_1 \simeq r - \frac{d}{2}\cos\theta \qquad (13.54a)$$

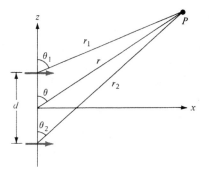

FIGURE 13.10 A two-element array.

$$r_2 \simeq r + \frac{d}{2} \cos \theta \qquad\qquad (13.54\text{b})$$

Thus eq. (13.53) becomes

$$\mathbf{E}_s = \frac{j\eta\beta I_o \, dl}{4\pi \, r} \cos \theta \; e^{-j\beta r} e^{j\alpha/2} \big[e^{j(\beta d \cos \theta)/2} e^{j\alpha/2} + e^{-j(\beta d \cos \theta)/2} e^{-j\alpha/2} \big] \, \mathbf{a}_\theta$$

$$= \frac{j\eta\beta I_o \, dl}{4\pi \, r} \cos \theta \; e^{-j\beta r} e^{j\alpha/2} 2 \cos \left[\frac{1}{2} (\beta d \cos \theta + \alpha) \right] \mathbf{a}_\theta \qquad (13.55)$$

Comparing this with eq. (13.8a) shows that the total field of an array is equal to the field of single element located at the origin multiplied by an *array factor* given by

$$\boxed{AF = 2 \cos \left[\frac{1}{2} (\beta d \cos \theta + \alpha) \right] e^{j\alpha/2}} \qquad\qquad (13.56)$$

Thus, in general, the far field due to a two-element array is given by

$$\mathbf{E} \, (\text{total}) = (\mathbf{E} \text{ due to single element at origin}) \times (\text{array factor}) \qquad (13.57)$$

Also, from eq. (13.55), note that $|\cos \theta|$ is the radiation pattern due to a single element, whereas the normalized array factor, $|\cos [1/2(\beta d \cos \theta + \alpha)]|$, is the radiation pattern the array would have if the elements were isotropic. These may be regarded as "unit pattern" and "group pattern," respectively. Thus the "resultant pattern" is the product of the unit pattern and the group pattern; that is,

$$\boxed{\text{resultant pattern} = \text{unit pattern} \times \text{group pattern}} \qquad\qquad (13.58)$$

This is known as *pattern multiplication,* and it can be used to sketch, almost by inspection, the pattern of an array. Therefore, pattern multiplication is a useful tool in the design of an array. We should note that while the unit pattern depends on the type of elements comprising the array, the group pattern is independent of the element type as long as the spacing d, the phase difference α, and the orientation of the elements remain the same.

Let us now extend the results on the two-element array to the general case of an N-element array shown in Figure 13.11. We assume that the array is *linear* in that the elements are spaced equally along a straight line and lie along the z-axis. Also, we assume that the array is *uniform* so that each element is fed with current of the same magnitude but of progressive phase shift α; that is, $I_{1s} = I_o\underline{/0}$, $I_{2s} = I_o\underline{/\alpha}$, $I_{3s} = I_o\underline{/2\alpha}$, and so on. We are mainly interested in finding the array factor; the far field can easily be found from eq. (13.57) once the array factor is known. For the uniform linear array, the array factor is the sum of the contributions by all the elements. Thus,

$$AF = 1 + e^{j\psi} + e^{j2\psi} + e^{j3\psi} + \cdots + e^{j(N-1)\psi} \qquad\qquad (13.59)$$

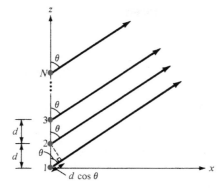

FIGURE 13.11 An *N*-element uniform linear array.

where

$$\psi = \beta d \cos \theta + \alpha \tag{13.60}$$

In eq. (13.60), $\beta = 2\pi/\lambda$, and d and α are, respectively, the spacing, and the interelement phase shift. Notice that the right-hand side of eq. (13.59) is a geometric series of the form

$$1 + x + x^2 + x^3 + \cdots + x^{N-1} = \frac{1 - x^N}{1 - x} \tag{13.61}$$

Hence eq. (13.59) becomes

$$AF = \frac{1 - e^{jN\psi}}{1 - e^{j\psi}} \tag{13.62}$$

which can be written as

$$AF = \frac{e^{jN\psi} - 1}{e^{j\psi} - 1} = \frac{e^{jN\psi/2}}{e^{j\psi/2}} \frac{e^{jN\psi/2} - e^{-jN\psi/2}}{e^{j\psi/2} - e^{-j\psi/2}}$$

$$= e^{j(N-1)\psi/2} \frac{\sin (N\psi/2)}{\sin (\psi/2)} \tag{13.63}$$

The phase factor $e^{j(N-1)\psi/2}$ would not be present if the array were centered about the origin. Neglecting this unimportant term, we have

$$\boxed{|AF| = \left| \frac{\sin \dfrac{N\psi}{2}}{\sin \dfrac{\psi}{2}} \right|, \quad \psi = \beta d \cos \theta + \alpha} \tag{13.64}$$

Note that this equation reduces to eq. (13.56) when $|AF|$ is considered and $N = 2$ as expected. Also, note the following:

1. Since $|AF|$ has the maximum value of N, the normalized $|AF|$ is obtained by dividing $|AF|$ by N. The principal maximum occurs when $\psi = 0$; that is,

$$0 = \beta d \cos\theta + \alpha \quad \text{or} \quad \cos\theta = -\frac{\alpha}{\beta d} \tag{13.65}$$

2. When $|AF| = 0$, $|AF|$ has *nulls* (or *zeros*); that is,

$$\frac{N\psi}{2} = \pm k\pi, \quad k = 1, 2, 3, \ldots \tag{13.66}$$

   where k is not a multiple of N.

3. A *broadside* array has its maximum radiation directed normal to the axis of the array; that is, $\psi = 0$, $\theta = 90°$ so that $\alpha = 0$.

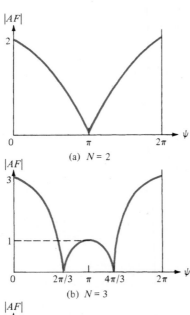

(a) $N = 2$

(b) $N = 3$

(c) $N = 4$

FIGURE 13.12 Array factors for uniform linear arrays.

4. An *end-fire* array has its maximum radiation directed along the axis of the array,

$$\text{that is, } \psi = 0, \theta = \begin{bmatrix} 0 \\ \pi \end{bmatrix} \text{ so that } \alpha = \begin{bmatrix} -\beta d \\ \beta d \end{bmatrix}.$$

These points are helpful in plotting $|AF|$. For $N = 2, 3$, and 4, the plots of $|AF|$ are sketched in Figure 13.12.

EXAMPLE 13.6

For the two-element antenna array of Figure 13.10, sketch the normalized field pattern when the currents are:

(a) Fed in phase ($\alpha = 0$), $d = \lambda/2$
(b) Fed 90° out of phase ($\alpha = \pi/2$), $d = \lambda/4$

Solution:

The normalized field of the array is obtained from eqs. (13.55) to (13.57) as

$$f(\theta) = \left| \cos \theta \cos \left[\frac{1}{2} (\beta d \cos \theta + \alpha) \right] \right|$$

(a) If $\alpha = 0, d = \lambda/2, \beta d = \dfrac{2\pi}{\lambda} \dfrac{\lambda}{2} = \pi$. Hence,

$$f(\theta) \qquad = \quad |\cos \theta| \qquad \left| \cos \frac{\pi}{2} (\cos \theta) \right|$$

$$\downarrow \qquad\qquad \downarrow \qquad\qquad\qquad \downarrow$$

$$\text{resultant} \; = \quad \text{unit} \quad \times \quad \text{group}$$
$$\text{pattern} \qquad \text{pattern} \qquad\quad \text{pattern}$$

The sketch of the unit pattern is straightforward. It is merely a rotated version of that in Figure 13.7(a) for the Hertzian dipole and is shown in Figure 13.13(a). To sketch a group pattern, we must first determine nulls and maxima. For the nulls (or zeros),

$$\cos \left(\frac{\pi}{2} \cos \theta \right) = 0 \rightarrow \frac{\pi}{2} \cos \theta = \pm \frac{\pi}{2}$$

or

$$\theta = 0°, 180°$$

For the maxima,

$$\cos \left(\frac{\pi}{2} \cos \theta \right) = 1 \rightarrow \cos \theta = 0$$

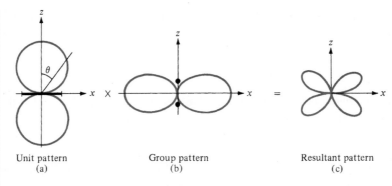

Unit pattern   Group pattern   Resultant pattern
(a)   (b)   (c)

FIGURE 13.13 For part (**a**) of Example 13.6: field patterns in the plane containing the axes of the elements.

or

$$\theta = 90°$$

The group pattern, shown in Figure 13.12(b), is the polar plot obtained by sketching $\left|\cos\left(\dfrac{\pi}{2}\cos\theta\right)\right|$ for $\theta = 0, 5°, 10°, 15°, \ldots, 180°$ and incorporating the nulls and maxima at $\theta = 0°, 180°$ and $\theta = 90°$, respectively. Multiplying Figure 13.13(a) with Figure 13.13 (b) gives the resultant pattern in Figure 13.13(c). MATLAB can easily be used to do this. It should be observed that the field patterns in Figure 13.13 are in the plane containing the axes of the elements. Note the following: (1) In the yz-plane, which is normal to the axes of the elements, the unit pattern ($= 1$) is a circle [see Figure 13.7(b)] while the group pattern remains as in Figure 13.13(b); therefore, the resultant pattern is the same as the group pattern in this case. (2) In the xy-plane, $\theta = \pi/2$, so the unit pattern vanishes while the group pattern ($= 1$) is a circle.

(b) If $\alpha = \pi/2$, $d = \lambda/4$, and $\beta d = \dfrac{2\pi}{\lambda}\dfrac{\lambda}{4} = \dfrac{\pi}{2}$

$$f(\theta) \quad = \quad |\cos\theta| \quad \left|\cos\dfrac{\pi}{4}(\cos\theta + 1)\right|$$

$$\downarrow \qquad\qquad \downarrow \qquad\qquad\qquad \downarrow$$

resultant   =   unit   ×   group
pattern   pattern   pattern

The unit pattern remains as in Figure 13.13(a). For the group pattern, the null occurs when

$$\cos\dfrac{\pi}{4}(1 + \cos\theta) = 0 \rightarrow \dfrac{\pi}{4}(1 + \cos\theta) = \dfrac{\pi}{2}$$

or

$$\cos\theta = 1 \rightarrow \theta = 0$$

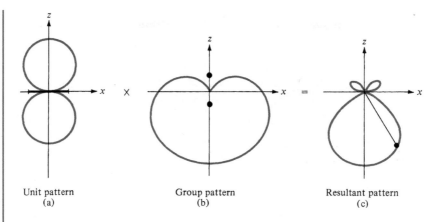

Unit pattern   Group pattern   Resultant pattern
   (a)   (b)   (c)

FIGURE 13.14 For part (**b**) of Example 13.6; field patterns in the plane containing the axes of the elements.

The maxima and minima occur when

$$\frac{d}{d\theta}\left[\cos\frac{\pi}{4}(1 + \cos\theta)\right] = 0 \rightarrow \sin\theta\sin\frac{\pi}{4}(1 + \cos\theta) = 0$$

$$\sin\theta = 0 \rightarrow \theta = 0°, 180°$$

and

$$\sin\frac{\pi}{4}(1 + \cos\theta) = 0 \rightarrow \cos\theta = -1 \text{ or } \theta = 180°$$

Each field pattern is obtained by varying $\theta = 0°, 5°, 10°, 15°, \ldots, 180°$. Note that $\theta = 180°$ corresponds to the maximum value of $|AF|$, whereas $\theta = 0°$ corresponds to the null. Thus the unit, group, and resultant patterns in the plane containing the axes of the elements are shown in Figure 13.14. Observe from the group patterns that the broadside array ($\alpha = 0$) in Figure 13.13 is bidirectional, while the end-fire array ($\alpha = \beta d$) in Figure 13.14 is unidirectional.

PRACTICE EXERCISE 13.6

Repeat Example 13.6 for the following cases: (a) $\alpha = \pi$, $d = \lambda/2$, (b) $\alpha = -\pi/2$, $d = \lambda/4$.

Answer: See Figure 13.15.

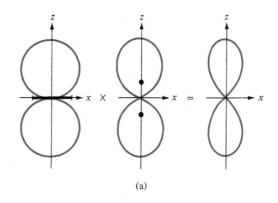

(a)

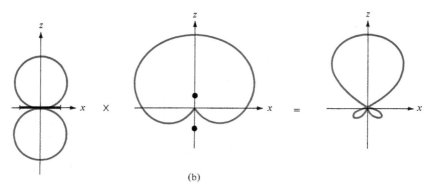

(b)

FIGURE 13.15 For Practice Exercise 13.6.

EXAMPLE 13.7

Consider a three-element array that has current ratios $1:2:1$ as in Figure 13.16(a). Sketch the group pattern in the plane containing the axes of the elements.

Solution:

For the purpose of analysis, we split the middle element in Figure 13.16(a) carrying current $2I\underline{/0°}$ into two elements each carrying current $I\underline{/0°}$. This results in four elements instead of three, as shown in Figure 13.16(b). If we consider elements 1 and 2 as a group and elements 3 and 4 as another group, we have a two-element array of Figure 13.16(c). Each group is a two-element array with $d = \lambda/2$, $\alpha = 0$, so that the group pattern of the two-element array (or the unit pattern for the three-element array) is as shown in Figure 13.13(b). The two groups form a two-element array similar to Example 13.6(a) with $d = \lambda/2$, $\alpha = 0$, so that the group pattern is the same as that in Figure 13.13(b). Thus, in this case, both the unit and group patterns are the same pattern in Figure 13.13(b). The resultant group pattern is obtained in Figure 13.17(c). We should note that the pattern in Figure 13.17(c) is not the resultant pattern but the group pattern of the three-element array. The resultant group pattern of the array is Figure 13.17(c) multiplied by the field pattern of the element type.

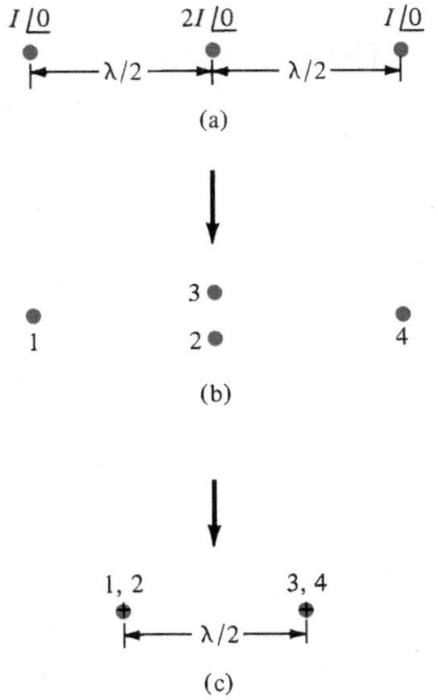

FIGURE 13.16 For Example 13.7: (**a**) a three-element array with current ratios 1:2:1; (**b**) and (**c**) equivalent two-element arrays.

An alternative method of obtaining the resultant group pattern of the three-element array of Figure 13.16 is by following steps similar to those taken to obtain eq. (13.59). We obtain the normalized array factor (or the group pattern) as

$$(AF)_n = \frac{1}{4}|1 + 2e^{j\psi} + e^{j2\psi}|$$

$$= \frac{1}{4}|e^{j\psi}||2 + e^{-j\psi} + e^{j\psi}|$$

$$= \frac{1}{2}|1 + \cos\psi| = \left|\cos\frac{\psi}{2}\right|^2$$

where $\psi = \beta d \cos\theta + \alpha$ if the elements are placed along the z-axis but oriented parallel to the x-axis. Since $\alpha = 0$, $d = \lambda/2$, $\beta d = \frac{2\pi}{\lambda} \cdot \frac{\lambda}{2} = \pi$, and

$$(AF)_n = \left|\cos\left(\frac{\pi}{2}\cos\theta\right)\right|^2$$

$$(AF)_n = \left|\cos\left(\frac{\pi}{2}\cos\theta\right)\right| \qquad \left|\cos\left(\frac{\pi}{2}\cos\theta\right)\right|$$

$$\downarrow \qquad\qquad\qquad \downarrow \qquad\qquad\qquad \downarrow$$

resultant   unit $\times$ group
group pattern   pattern   pattern

The sketch of these patterns is exactly what is in Figure 13.17.

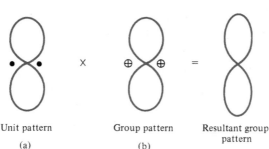

FIGURE 13.17 For Example 13.7; obtaining the resultant group pattern of the three-element array of Figure 13.16(a).

Unit pattern
(a)

Group pattern
(b)

Resultant group pattern
(c)

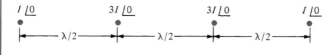

$I \underline{/0}$ $3I \underline{/0}$ $3I \underline{/0}$ $I \underline{/0}$

$\lambda/2$ $\lambda/2$ $\lambda/2$

FIGURE 13.18 For Example 13.7 and Practice Exercise 13.7: four-element array with current ratios 1:3:3:1.

If two three-element arrays in Figure 13.16(a) are displaced by $\lambda/2$, we obtain a four-element array with current ratios 1:3:3:1 as in Figure 13.18. Two of such four-element arrays, displaced by $\lambda/2$, give a five-element array with current ratios 1:4:6:4:1. Continuing this process results in an N-element array, spaced $\lambda/2$ and $(N-1)\lambda/2$ long, whose current ratios are the binomial coefficients. Such an array is called a linear *binomial array*.

PRACTICE EXERCISE 13.7

(a) Sketch the resultant group pattern for the four-element array with current ratios 1:3:3:1 shown in Figure 13.18.

(b) Derive an expression for the group pattern of a linear binomial array of N elements. Assume that the elements are placed along the z-axis, oriented parallel to the x-axis with spacing d and interelement phase shift α.

Answer: (a) See Figure 13.19, (b) $\left| \cos \dfrac{\psi}{2} \right|^{N-1}$, where $\psi = \beta d \cos \theta + \alpha$.

FIGURE 13.19 For part (**a**) of Practice Exercise 13.7.

†13.8 EFFECTIVE AREA AND THE FRIIS EQUATION

When the incoming EM wave is normal to the entire surface of a receiving antenna, the power received is

$$P_r = \int_S \mathscr{P}_{\text{ave}} \cdot d\mathbf{S} = \mathscr{P}_{\text{ave}} S \qquad (13.67)$$

But in most cases, the incoming EM wave is not normal to the entire surface of the antenna. This necessitates the idea of the effective area of a receiving antenna.

The concept of effective area or effective aperture (receiving cross section of an antenna) is usually employed in the anaysis of receiving antennas.

> The **effective area** A_e of a receiving antenna is the ratio of the time-average power received P_r (or delivered to the load, to be strict) to the time-average power density $\mathscr{P}_{\text{ave}}$ of the incident wave at the antenna.

That is,

$$A_e = \frac{P_r}{\mathscr{P}_{\text{ave}}} \qquad (13.68)$$

From eq. (13.68), we notice that the effective area is a measure of the ability of the antenna to extract energy from a passing EM wave.

Let us derive the formula for calculating the effective area of the Hertzian dipole acting as a receiving antenna. The Thévenin equivalent circuit for the receiving antenna is shown in Figure 13.20, where V_{oc} is the open-circuit voltage induced on the antenna terminals by a remote transmitter, $Z_{\text{in}} = R_{\text{rad}} + jX_{\text{in}}$ is the antenna impedance, and $Z_L = R_L + jX_L$ is the external load impedance, which might be the input impedance to the transmission line feeding the antenna. For maximum power transfer, $Z_L = Z_{\text{in}}^*$ and $X_L = -X_{\text{in}}$. The time-average power delivered to the matched load is therefore

$$\begin{aligned} P_r &= \frac{1}{2} \left[\frac{|V_{\text{oc}}|}{2R_{\text{rad}}} \right]^2 R_{\text{rad}} \\ &= \frac{|V_{\text{oc}}|^2}{8\,R_{\text{rad}}} \end{aligned} \qquad (13.69)$$

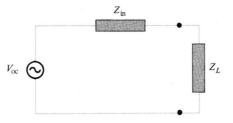

FIGURE 13.20 Thévenin equivalent of a receiving antenna.

For the Hertzian dipole, eq. (13.13b) gives $R_{rad} = 80\pi^2(dl/\lambda)^2$ and $V_{oc} = E\,dl$, where E is the effective field strength parallel to the dipole axis. Hence, eq. (13.69) becomes

$$P_r = \frac{E^2\lambda^2}{640\pi^2} \tag{13.70}$$

The time-average power at the antenna is

$$\mathcal{P}_{ave} = \frac{E^2}{2\eta} = \frac{E^2}{240\pi} \tag{13.71}$$

Inserting eqs. (13.70) and (13.71) in eq. (13.68) gives

$$A_e = \frac{3\lambda^2}{8\pi} = 1.5\frac{\lambda^2}{4\pi}$$

or

$$A_e = \frac{\lambda^2}{4\pi}D \tag{13.72}$$

where $D = 1.5$ is the directivity of the Hertzian dipole. Although eq. (13.72) was derived for the Hertzian dipole, it holds for any antenna if D is replaced by $G_d(\theta, \phi)$. Thus, in general

$$\boxed{A_e = \frac{\lambda^2}{4\pi}G_d(\theta, \phi)} \tag{13.73}$$

Now suppose we have two antennas separated by distance r in free space as shown in Figure 13.21. The transmitting antenna has effective area A_{et} and directive gain G_{dt} and transmits a total power $P_t \,(= P_{rad})$. The receiving antenna has effective area of A_{er} and directive gain G_{dr} and receives a total power of P_r. At the transmitter,

$$G_{dt} = \frac{4\pi U}{P_t} = \frac{4\pi r^2\mathcal{P}_{ave}}{P_t}$$

or

$$\mathcal{P}_{ave} = \frac{P_t}{4\pi r^2}G_{dt} \tag{13.74}$$

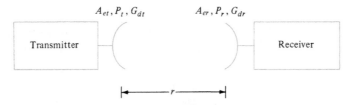

FIGURE 13.21 Transmitting and receiving antennas in free space.

By applying eqs. (13.68) and (13.73), we obtain the time-average power received as

$$P_r = \mathcal{P}_{ave} A_{er} = \frac{\lambda^2}{4\pi} G_{dr} \mathcal{P}_{ave} \tag{13.75}$$

Substituting eq. (13.74) into eq. (13.75) results in

$$\boxed{P_r = G_{dr} G_{dt} \left[\frac{\lambda}{4\pi r} \right]^2 P_t} \tag{13.76}$$

This is referred to as the *Friis transmission formula*. It relates the power received by one antenna to the power transmitted by the other, provided the two antennas are separated by $r \geq 2d^2/\lambda$, where d is the largest dimension of either antenna [see eq. (13.52)]. Therefore, to apply the Friis equation, we must make sure that each antenna is in the far field of the other.

EXAMPLE 13.8

Find the maximum effective area of a $\lambda/2$ wire dipole operating at 30 MHz. How much power is received with an incident plane wave of strength 2 mV/m?

Solution:

$$A_e = \frac{\lambda^2}{4\pi} G_d(\theta, \phi)$$

$$\lambda = \frac{c}{f} = \frac{3 \times 10^8}{30 \times 10^6} = 10 \text{ m}$$

$$G_d(\theta, \phi) = \frac{\eta}{\pi R_{rad}} f^2(\theta) = \frac{120\pi}{73\pi} f^2(\theta) = 1.64 f^2(\theta)$$

$$G_{d\,max} = 1.64$$

$$A_{e\,max} = \frac{10^2}{4\pi} (1.64) = 13.05 \text{ m}^2$$

$$P_r = \mathcal{P}_{ave} A_e = \frac{E_0^2}{2\eta} A_e$$

$$= \frac{(2 \times 10^{-3})^2}{240\pi} 13.05 = 69.23 \text{ nW}$$

PRACTICE EXERCISE 13.8

Determine the maximum effective area of a Hertzian dipole of length 10 cm operating at 100 MHz. If the antenna receives 3 μW of power, what is the power density of the incident wave?

Answer: 1.074 m², 2.793 μW/m².

EXAMPLE 13.9

The transmitting and receiving antennas are separated by a distance of 200 λ and have directive gains of 25 dB and 18 dB, respectively. If 5 mW of power is to be received, calculate the minimum transmitted power.

Solution:

Given that $G_{dt} \, (\text{dB}) = 25 \, \text{dB} = 10 \log_{10} G_{dt}$,

$$G_{dt} = 10^{2.5} = 316.23$$

Similarly,

$$G_{dr} \, (\text{dB}) = 18 \, \text{dB} \quad \text{or} \quad G_{dr} = 10^{1.8} = 63.1$$

Using the Friis equation, we have

$$P_r = G_{dr} G_{dt} \left[\frac{\lambda}{4\pi r} \right]^2 P_t$$

or

$$P_t = P_r \left[\frac{4\pi r}{\lambda} \right]^2 \frac{1}{G_{dr} G_{dt}}$$

$$= 5 \times 10^{-3} \left[\frac{4\pi \times 200 \, \lambda}{\lambda} \right]^2 \frac{1}{(63.1)(316.23)}$$

$$= 1.583 \, \text{W}$$

PRACTICE EXERCISE 13.9

An antenna in air radiates a total power of 100 kW so that a maximum radiated electric field strength of 12 mV/m is measured 20 km from the antenna. Find (a) its directivity in decibels, (b) its maximum power gain if $\eta_r = 98\%$.

Answer: (a) -20.18 dB, (b) 9.408×10^{-3}.

---

†13.9 THE RADAR EQUATION

Radars are electromagnetic devices used for detection and location of objects. The term *radar* is derived from the phrase **ra**dio **d**etection **a**nd **r**anging. In a typical radar system, as shown in Figure 13.22(a), pulses of EM energy are transmitted to a distant object. The same antenna is used for transmitting and receiving, so the time interval between the transmitted and reflected pulses is used to determine the distance of the target. If r is the distance between the radar and target and c is the speed of light, the elapsed time between the transmitted and received pulse is $2r/c$. By measuring the elapsed time, we determine r.

The ability of a target to scatter (or reflect) energy is characterized by its *scattering cross section* σ (also called the *radar cross section*). The scattering cross section has the units of area and can be measured experimentally.

> The **scattering cross section** is the equivalent area intercepting the amount of power that, when scattering isotropically, produces at the radar a power density that is equal to that scattered (or reflected) by the actual target.

That is,

$$\mathcal{P}_s = \lim_{r \to \infty} \left[\frac{\sigma \mathcal{P}_i}{4\pi r^2} \right]$$

or

$$\sigma = \lim_{r \to \infty} 4\pi r^2 \frac{\mathcal{P}_s}{\mathcal{P}_i} \tag{13.77}$$

where $\mathcal{P}_i$ is the incident power density at the target T while $\mathcal{P}_s$ is the scattered power density at the transceiver O as in Figure 13.22(b).

From eq. (13.43), the incident power density $\mathcal{P}_i$ at the target T is

$$\mathcal{P}_i = \mathcal{P}_{\text{ave}} = \frac{G_d}{4\pi r^2} P_{\text{rad}} \tag{13.78}$$

The power received at transreceiver O is

$$P_r = A_{er} \mathcal{P}_s$$

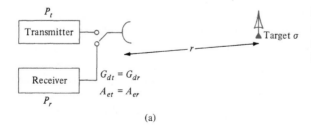

(a)

(b)

FIGURE 13.22 (a) Typical radar system. (b) Simplification of the system in (a) for calculating the target cross section σ.

or

$$\mathcal{P}_s = \frac{P_r}{A_{er}} \tag{13.79}$$

Note that $\mathcal{P}_i$ and $\mathcal{P}_s$ are the time-average power densities in watts per square meter, and P_{rad} and P_r are the total time-average powers in watts. Since $G_{dr} = G_{dt} = G_d$ and $A_{er} = A_{et} = A_e$, substituting eqs. (13.78) and (13.79) into eq. (13.77) gives

$$\sigma = (4\pi r^2)^2 \frac{P_r}{P_{rad}} \frac{1}{A_e \, G_d} \tag{13.80a}$$

or

$$P_r = \frac{A_e \sigma G_d P_{rad}}{(4\pi r^2)^2} \tag{13.80b}$$

From eq. (13.73), $A_e = \lambda^2 G_d/4\pi$. Hence,

$$\boxed{P_r = \frac{(\lambda G_d)^2 \sigma P_{rad}}{(4\pi)^3 r^4}} \tag{13.81}$$

This is the *radar transmission equation* for free space. It is the basis for measurement of scattering cross section of a target. Solving for r in eq. (13.81) results in

$$\boxed{r = \left[\frac{\lambda^2 \, G_d^2 \sigma}{(4\pi)^3} \cdot \frac{P_{rad}}{P_r} \right]^{1/4}} \tag{13.82}$$

Equation (13.82) is called the *radar range equation*. Given the minimum detectable power of the receiver, the equation determines the maximum range for a radar. It is also useful for obtaining engineering information concerning the effects of the various parameters on the performance of a radar system.

The radar considered so far is the *monostatic* type because of the predominance of this type of radar in practical applications. A *bistatic radar* is one in which the transmitter and receiver are separated. If the transmitting and receiving antennas are at distances r_1 and r_2 from the target and $G_{dr} \neq G_{dt}$, eq. (13.81) for bistatic radar becomes

$$P_r = \frac{G_{dt} G_{dr}}{4\pi} \left[\frac{\lambda}{4\pi r_1 r_2} \right]^2 \sigma P_{rad} \tag{13.83}$$

Radar transmission frequencies range from 25 to 70,000 MHz. Table 13.1 shows radar frequencies and their designations as commonly used by radar engineers.

TABLE 13.1 Designations of Radar Frequencies

Designation	Frequency
UHF	300–1000 MHz
L	1000–2000 MHz
S	2000–4000 MHz
C	4000–8000 MHz
X	8000–12,500 MHz
Ku	12.5–18 GHz
K	18–26.5 GHz
Millimeter	>35 GHz

EXAMPLE 13.10

An S-band radar transmitting at 3 GHz radiates 200 kW. Determine the signal power density at ranges 100 and 400 nautical miles if the effective area of the radar antenna is 9 m². With a 20 m² target at 300 nautical miles, calculate the power of the reflected signal at the radar.

Solution:

The nautical mile is a common unit in radar communications.

$$1 \text{ nautical mile (nm)} = 1852 \text{ m}$$

$$\lambda = \frac{c}{f} = \frac{3 \times 10^8}{3 \times 10^9} = 0.1 \text{ m}$$

$$G_{dt} = \frac{4\pi}{\lambda^2} A_{et} = \frac{4\pi}{(0.1)^2} 9 = 3600\pi$$

For $r = 100$ nm $= 1.852 \times 10^5$ m

$$\mathcal{P} = \frac{G_{dt}P_{\text{rad}}}{4\pi r^2} = \frac{3600\pi \times 200 \times 10^3}{4\pi (1.852)^2 \times 10^{10}}$$

$$= 5.248 \text{ mW/m}^2$$

For $r = 400$ nm $= 4 (1.852 \times 10^5)$ m

$$\mathcal{P} = \frac{5.248}{(4)^2} = 0.328 \text{ mW/m}^2$$

Using eq. (13.80b)

$$P_r = \frac{A_e \sigma \, G_d \, P_{\text{rad}}}{[4\pi r^2]^2}$$

where $r = 300$ nm $= 5.556 \times 10^5$ m

$$P_r = \frac{9 \times 20 \times 3600\pi \times 200 \times 10^3}{[4\pi \times 5.556^2]^2 \times 10^{20}} = 2.706 \times 10^{-14}\,\text{W}$$

The same result can be obtained by using eq. (13.81).

PRACTICE EXERCISE 13.10

A C-band radar with an antenna 1.8 m in radius transmits 60 kW at a frequency of 6000 MHz. If the minimum detectable power is 0.26 mW, for a target cross section of 5 m^2, calculate the maximum range in nautical miles and the signal power density at half this range. Assume unity efficiency and that the effective area of the antenna is 70% of the actual area.

Answer: 0.031 nm, 501 W/m^2.

---

†13.10 APPLICATION NOTE—ELECTROMAGNETIC INTERFERENCE AND COMPATIBILITY

Every electronic device is a source of radiated electromagnetic fields called *radiated emissions*. These are often an accidental by-product of the design.

Electromagnetic interference (EMI) is the degradation in the performance of a device due to the fields making up the electromagnetic environment.

The electromagnetic environment consists of various apparatuses such as radio and TV broadcast stations, radar, and navigational aids that radiate EM energy as they operate. Every electronic device is susceptible to EMI. Its influence can be seen all around us. The results include "ghosts" in TV picture reception, taxicab radio interference with police radio systems, power line transient interference with personal computers, and self-oscillation of a radio receiver or transmitter circuit.

Electromagnetic compatibility (EMC) is achieved when a device functions satisfactorily without introducing intolerable disturbances to the electromagnetic environment or to other devices in its neighborhood.

EMC[1] is achieved when electronic devices coexist in harmony, such that each device functions according to its intended purpose in the presence of, and in spite of, the others. EMI

---

[1] For an in-depth treatment of EMC, see C. R. Paul, *Introduction to Electromagnetic Compatibility*, 2nd ed. Hoboken, NJ: John Wiley & Sons, 2006.

is the problem that occurs when unwanted voltages or currents are present to influence the performance of a device, while EMC is the solution to the problem. The goal of EMC is system or subsystem compatibility, and this is achieved by applying proven design techniques, the use of which ensures a system relatively free of EMI problems.

EMC is a growing field because of the ever-increasing density of electronic circuits in modern systems for computation, communication, control, and so on. It is a concern not only to electrical and computer engineers, but also to automotive engineers. The increasing application of automotive electronic systems to improve fuel economy, reduce exhaust emissions, ensure vehicle safety, and provide assistance to the driver has resulted in a growing need to ensure compatibility during normal operation. We will consider the sources and characteristics of EMI. Later, we will examine EMI control techniques.

A. Source and Characteristics of EMI

First, let us classify EMI in terms of its causes and sources. The classification will facilitate recognition of sources and assist in determining means of control. As mentioned earlier, any electronic device may be the source of EMI, although this is not the intention of the designer. The cause of the EMI problem may be either within the system, in which case it is termed an *intrasystem problem,* or from the outside, in which case it is called an *intersystem problem.* Figure 13.23 shows intersystem EMI problems. The term "emitter" is commonly used to denote the source of EMI, while the term "susceptor" is used to designate a victim device. Tables 13.2 and 13.3 present typical causes of intrasystem and intersystem problems. Both intrasystem and intersystem EMI generally can be controlled by the system design engineer by following some design guidelines and techniques. For intrasystem EMI problems, for example, the design engineer may apply proper grounding and wiring arrangements, shielding of circuits and devices, and filtering.

The sources of EMI can be classified as natural or artificial (manmade). The origins of EMI are basically undesired conducted emissions (voltages and/or currents) or radiated emissions (electric and/or magnetic fields). Conducted emissions are currents that are carried by metallic paths (the unit's power cord) and placed on the common power network, where they may cause interference with other devices that are connected to the network. Radiated emissions concern the electric fields radiated by the device that may be received by other electronic devices causing interference in those devices. Figure 13.24 illustrates the conceptual difference between conducted and radiated paths.

No single operating agency has jurisdiction over all systems to dictate actions necessary to achieve EMC. Thus, EMC is usually achieved by industrial association, voluntary regulation, government-enforced regulation, and negotiated agreements between the affected parties. Frequency plays a significant role in EMC. Frequency allocations and assignments are made according to the constraints established by international treaties. The Radio Regulations resulting from such international treaties are published by the International Telecommunication Union (ITU). The Federal Communications Commission (FCC) has the authority over radio and wire communications in the United States. The FCC has set limits on the radiated and conducted emissions of electronic devices including calculators, televisions, printers, modems, and personal computers. It is illegal to market an electronic device in the United States unless its radiated and conducted emissions have been measured and do not exceed the limits of FCC regulations. Therefore, any electronic device designed today that is designed without incorporating EMC design principles will probably fail to comply with the FCC limits.

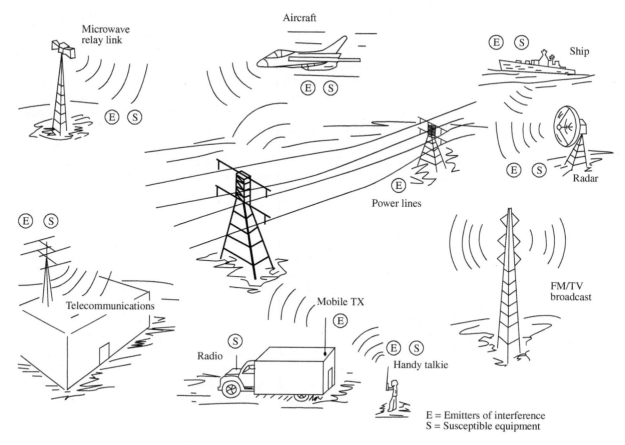

FIGURE 13.23 Typical examples of intersystem EMI problems. *Source:* J. I. N. Violette et al., *Electromagnetic Compatibility Handbook.* New York: Van Nostrand Reinhold, 1987, p. 4.

TABLE 13.2 Intrasystem EMI Causes

Emitters	Susceptors
Power supplies	Relays
Radar transmitters	Radar receivers
Mobile radio transmitters	Mobile radio receivers
Fluorescent lights	Ordnance
Car ignition systems	Car radio receivers

TABLE 13.3 Intersystem EMI Causes

Emitters	Susceptors
Lightning strokes	Radio receivers
Computers	TV sets
Power lines	Heart pacers
Radar transmitters	Aircraft navigation systems
Police radio transmitters	Taxicab radio receivers
Fluorescent lights	Industrial controls
Aircraft transmitters	Ship receivers

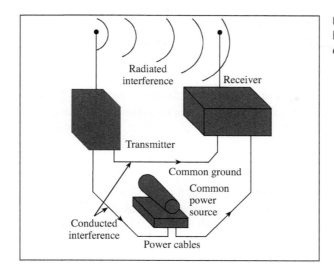

FIGURE 13.24 Differences between conducted and radiated emissions.

B. EMI Control Techniques

The three common design approaches used to control or suppress EMI are grounding, shielding, and filtering. Although each technique has a distinct role in system design, proper grounding may sometimes minimize the need for shielding and filtering; also proper shielding may minimize the need for filtering. Therefore, we discuss the three techniques, grounding, shielding, and filtering, in that order.

Grounding

Grounding is the establishment of an electrically conductive path between two points to connect electrical and electronic elements of a system to one another or to some reference point, which may be designated the *ground*. An ideal ground plane is a zero-potential, zero-impedance body that can be used as a reference for all signals in associated circuitry and to which any undesired current can be transferred for the elimination of its effects.

The purpose of the floating ground is to isolate circuits or equipment electrically from a common ground plane. This type of grounding technique may cause a hazard. Single-point grounding is used to minimize the effects of facility ground currents. Multiple-point grounding minimizes ground lead lengths. The ground plane might be a ground wire that is carried throughout the system or a large conductive body.

Bonding is the establishment of a low-impedance path between two metal surfaces. Grounding is a circuit concept, while bonding denotes the physical implementation of that concept. The purpose of a bond is to make a structure homogeneous with respect to the flow of electrical currents, thus avoiding the development of potentials between the metallic parts, since such potentials may result in EMI. Bonds provide protection from electrical shock, power circuit current return paths, and antenna ground plane connections and also minimize the potential difference between the devices. They have the ability to carry large fault current.

There are two types of bonds: direct and indirect. The direct bond is a metal-to-metal contact between the elements connected, while the indirect bond is a contact through the use of conductive jumpers.

The dc resistance R_{dc} of a bond is often used as an indication of bond quality. It is given by

$$R_{dc} = \frac{\ell}{\sigma S} \tag{13.84}$$

where ℓ is the length of the bond, σ is its conductivity, and S is its cross-sectional area. As frequency increases, the bond resistance increases due to skin effect. Thus the ac resistance R_{ac} is given as

$$R_{ac} = \frac{\ell}{\sigma \delta w} \tag{13.85}$$

where w is the width of the bond and δ is the skin depth.

Bonding effectiveness can be expressed as the difference (in dB) between the induced voltages on an equipment case with and without the bond trap.

Shielding

The purpose of shielding is to confine radiated energy to a specific region or to prevent radiated energy from entering a specific region. Shields may be in the form of partitions and boxes as well as in the form of cable and connector shields.

Shield types include solid, nonsolid (e.g., screen), and braid, as is used on cables. In all cases, a shield can be characterized by its *shielding effectiveness*. The shielding effectiveness (*SE*) is defined as

$$SE = 10 \log_{10} \frac{\text{incident power density}}{\text{transmitted power density}} \tag{13.86}$$

where the incident power density is the power density at a measuring point before a shield is installed and the transmitted power is the power density at the same point after the shield is in place. In terms of the field strengths, the shielding effectiveness may also be defined as the ratio of the field E_t transmitted through to the inside to the incident field E_i. Thus, *SE* is given by

$$SE = 20 \log_{10} \frac{E_i}{E_t} \tag{13.87}$$

For magnetic fields,

$$SE = 20 \log_{10} \frac{H_i}{H_t} \tag{13.88}$$

For example, aluminum has $\sigma = 3.5 \times 10^7$ S/m, $\varepsilon = \varepsilon_o$, $\mu = \mu_o$. An aluminum sheet at 100 MHz has an *SE* of 100 dB at a thickness of 0.01 mm. Since an aluminum sheet for a computer cabinet is much thicker than this, an aluminum case is considered a highly effective shield. A cabinet that effectively shields the circuits inside from external fields is also highly effective in preventing radiation from those circuits to the external world. Because of the effective shield, radiated emission from the computer system is caused by openings in the cabinet such as cracks or holes from disk drives and from wires that penetrate the cabinet such as power cords and cables to external devices.

Filtering

An electrical filter is a network of lumped or distributed constant resistors, inductors, and capacitors that offers comparatively little opposition to certain frequencies, while blocking the passage of other frequencies. A filter provides the means whereby levels of conducted interference are substantially reduced.

The most significant characteristic of a filter is the *insertion loss* it provides as a function of frequency. Insertion loss (*IL*) is defined as

$$IL = 20 \log_{10} \frac{V_1}{V_2} \tag{13.89}$$

where V_1 is the output voltage of a signal source with the filter in the circuit, and V_2 is the output voltage of the signal source without the use of the filter. Low-pass filters are commonly used in EMC work. The insertion loss for the low-pass filters is given by

$$IL = 10 \log_{10}\left(1 + F^2\right) \text{ dB} \tag{13.90}$$

where

$$F = \begin{cases} \pi f R C, & \text{for a capacitive filter} \\ \pi f L / R, & \text{for an inductive filter} \end{cases} \tag{13.91}$$

and *f* is the frequency.

13.11 APPLICATION NOTE—TEXTILE ANTENNAS AND SENSORS

Antennas for body-centric communication (Figure 13.25) have been introduced in recent years. Antennas in this new class can be sewn directly onto clothes. Weaving antennas and other electronic sensors into textiles heralds a new era for the apparel industry. The garments of the future will not only cover the human body and protect against the extremes of nature, but also collect and transmit crucial information about the wearer's vital signs and current environment. These capabilities will be achieved by seamlessly tailoring biomedical and environmental monitoring systems into fabric.

Researchers at the Ohio State University created a prototype using plastic film and metallic thread. Some of the novel body-worn antennas and medical sensors they have developed are based on embroidered conductive polymer fibers called e-fibers on textiles. The flexible conductors are constructed from silver-coated *p*-phenylene-2,6-benzobisoxazole (PBO) fibers. The e-fibers are composed of high-strength and flexible polymer cores that incorporate conductive metallic coatings. They are readily embroidered onto regular textiles and can also be laminated onto polymer dielectric substrates.

The e-fiber textiles exhibit an insertion loss of only 0.07 dB/cm at 1 GHz and 0.15 dB/cm at 2 GHz. They provide inherent mechanical strength that is due to their polymer core, together with high electrical conductivity resulting from the silver coating. These e-fibers are twisted together to improve their conductivity. For instance, the 332-strand e-fibers have a low resistivity of only 0.8 Ω/m. More importantly, e-fibers are suitable for automatic

FIGURE 13.25 Textile antenna.

embroidery onto textiles to realize various antenna and circuit designs. The embroidered e-fiber textile electronics exhibit both mechanical and electrical advantages. Also, because of their high conductivity, e-fibers provide much better antenna performance than other textile antennas utilizing less conductive materials or embedded metal wires.

The e-fibers are sewn onto textiles via computerized embroidery processes to form antennas or RF circuits. Because the fibers are so thin, they can be bundled to form much thicker threads (664 strands per thread) for improved conductivity. During embroidery, an "assistant" yarn is used to couch the e-fibers onto one side of the textile's surface. This procedure avoids abrasion damage to the silver coatings on the e-fiber's polymer core. The antenna and sensor designs are translated into embroidery software, followed by digitizing stitches of the assistant yarn. As the sewing machine carries out each stitch, the e-fibers are firmly and precisely placed onto the textile. To improve surface conductivity by minimizing physical discontinuities and thread gaps, a second layer can be embroidered right on top of the first (double embroidery). It is recommended that the resultant e-fiber surface discontinuities be kept to less than $\lambda/20$, where λ is the free-space wavelength of the operational frequency. This precaution is critical to realizing high-performance antennas and circuits.

Although primarily designed for military use, the e-fiber technology could potentially be applied to the manufacture of gear for police officers, firefighters, and astronauts—anybody who needs to keep the hands free for important work. The European Integrated Project *Proetex* aims at developing such wearable textile systems chiefly for professional firefighters and other first responders. A variety of sensors are being sewn inside and outside the firefighter's outfit, and the signals from them are processed in a wearable electronic unit and transmitted to a base station. Suitable antennas that combine flexibility with robustness and reliability are needed for this purpose. In other applications substrates at least 2 mm thick are used to print the antennas. But since the clothes are usually thinner, a flexible protective foam available in a variety of thicknesses and easily layered with garments such as firefighter suits is employed in the design. Proetex also contemplates designing wearable textile systems for civilian victims of natural and other disasters.

Textile antennas find applications not only for continuous monitoring, but also for therapeutic regimes. For instance, they can be made to produce hyperthermia for the treatment of tumors and to monitor various physiological parameters. In addition to medical applications, textile antennas serve as part of a biotelemetry system to establish wireless communication links between implantable devices and exterior instruments.

13.12 APPLICATION NOTE—FRACTAL ANTENNAS

Historically, antennas were called aerials; but they were so named because of their resemblance to insect antennae. Standard antennas come in various shapes, almost all of which have mathematical description. Many shapes occurring in nature cannot be described in terms of euclidean geometry. For example, a fern leaf, a thin snowflake, the shoreline, and a class of crustaceans possess self-similarity, a property that the whole is deterministically or statistically similar to a part thereof. These are characterized by fractal geometry. A fractal is an iteratively generated geometry that has fractal dimensions. The underlying notions were conceived by Benoit B. Mandelbrot. He investigated the relationship between these iterated function systems and the nature around us using previous contributions of Gaston Julia, Pierre Fatou, and Felix Hausdorff. He depicted many fractals existing in nature and was able to accurately model certain phenomena. Also, he introduced new fractal models for more complex structures, including trees, clouds, and mountains, that possess an inherent self-similarity and self-affnity by way of geometrical continuation in terms of non-euclidean elements.

Antennas whose shape is inherited from fractals are called fractal antennas. The widespread use of wireless communication systems posed the necessity for the design of wideband, or multiband, low-profile, small antennas. Their role became important in personal communication systems, satellite communication terminals, RFID, unmanned aerial vehicles, and so on. The central idea in their design involves optimal appropriation of the physical space either in planar or in 3-dimension. This accomplishes greater bandwidth of operation from a low quality factor. Fractal concepts have been applied to many branches of science and engineering, including fractal electrodynamics for radiation, propagation, and scattering. They have been extended to antenna theory and design resulting in the implementations of different fractal antenna elements and arrays.

There are broadly two categories of fractals: deterministic and random. Deterministic fractals are generated from several scaled-down and rotated copies of themselves. Examples are the von Koch snowflake and the Sierpinski gaskets. Recursive algorithms are used to generate such fractals. Random fractals evince some degree of randomness such as is found in natural phenomena. Fractal geometries can best be characterized and generated using an iterative process suitable for self-similar and self-affinity structures. Figure 13.26 illustrates this iterative process in generating few iterations of a Koch loop and a Koch loop antenna at a chosen iteration. In a similar fashion, other shapes can be generated. Figure 13.27 shows the stages of fractal tree dipole antenna. Figures 13.28 and 13.29 depict a fern leaf and the Sierpinski triangle, which can also be rendered as antennas.

A useful feature of fractal antennas arises from their space-filling property that helps in miniaturization while increasing the length. This permits low values of Q factors and higher bandwidths. At resonance, the impedance is higher compared to that of traditional antennas. A small circular loop of quarter wavelength perimeter has low radiation resistance, but a Koch loop occupying similar space has 35 times higher resistance. Higher iterative geometries cause longer electrical lengths and exhibit lower resonant frequencies. The meanderings of fractal contour impart distributive loading. For example, in the Koch fractal loop inductance adds in a distributive way. The increase in inductance allows the monopole to be smaller than the corresponding linear monopole and still be resonant. Most of the miniaturization benefits of the fractal dipoles occur within the first five iterations, with meager marginal changes in the characteristics at greater complexities. Similar to several fractal antennas, Sierpinski gaskets possess desirable radiation

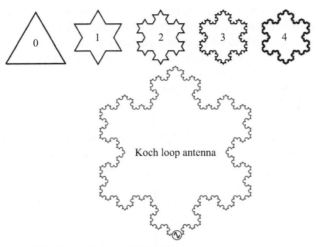

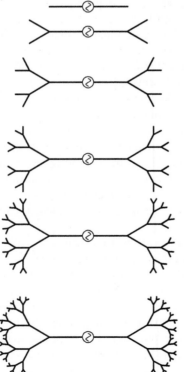

FIGURE 13.26 Generating Koch loop antenna.

FIGURE 13.27 Fractal tree dipole antenna iteration after iteration.

and impedance characteristics and can be used as monopoles and dipoles with similar cross-sectional contours. Although fractal antennas have several advantages, they do not necessarily outperform other classical categories.

FIGURE 13.28 Fern leaf.

FIGURE 13.29 Sierpinski triangle.

13.13 APPLICATION NOTE—RFID

Radio-frequency identification (RFID) is a relatively new technology for tracking and access applications that have been under development since the 1980s. For example, with RFID, one can use electromagnetic signals to read data from electronic labels without actual contact. The label to be read does not have be located in a line-of-sight region. This technology is very attractive for tracking and tagging purposes (Figure 13.30). For a long time, the universal product code (UPC), also known as the bar code, has been the foremost means of identifying products. Designed to provide an open standard for product labeling, bar codes not only reduced costs but also increased efficiency and favored innovations that benefited consumers, manufacturers, wholesalers, and retailers alike. However, bar codes suffer drawbacks such as the need for line-of-sight from the scanner to the bar code. They are also limited in their capacity for storing data and thus are employed only to track product categories.

The Electronic Product Code (EPC) is one of the industrial standards for global RFID usage, and a core element of the EPCglobal network, an architecture of open standards developed by the GS1 EPCglobal community. Most currently deployed EPC RFID tags comply with ISO/IEC 18000-6C for the RFID air interface standard. In ambient conditions that do not permit bar code labels as effective identifiers, RFID tags could serve the purpose. Thanks to the technology's ability to track moving objects, use of RFID has become an established practice in a wide range of markets.

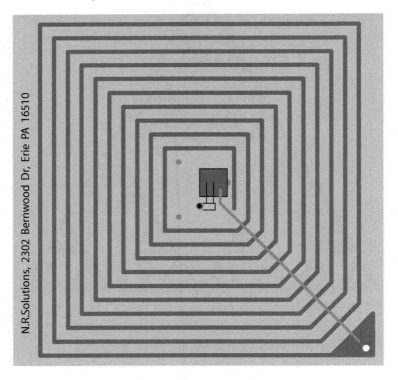

N.R.Solutions, 2302 Bernwood Dr, Erie PA 16510

FIGURE 13.30 Schematic view of an RFID tag in use.

RFID also greatly facilitates identification of and automated data collection from livestock, moving trucks, or other nonstationary objects. Pharmaceuticals can be tracked through warehouses. Livestock and pets may have tags injected, allowing positive identification of the animals.

Essentially, an RFID system works as follows. Two-way radio transmitter–receivers called interrogators or readers send a signal to the tag and read its response. The readers generally transmit their observations to a computer system running RFID software or RFID middleware. RFID systems typically come in three configurations. One is a passive reader/active tag (PRAT) system in which a passive reader does no more than receive radio signals from active tags that are battery operated and do no more than transmit. The reception range of a PRAT system reader can be adjusted from 1 to 2000 feet, thereby allowing for great flexibility in applications such as asset protection and supervision. Another configuration is an active reader/passive tag (ARPT) system, in which an active reader transmits interrogator signals and also receives authentication replies from passive tags. Finally, there is the active reader/active tag (ARAT) system in which active tags are awakened with an interrogator signal from the active reader. A variation of this system could also use a battery-assisted passive (BAP) tag, which acts like a passive tag but has a small battery to power the tag's return reporting signal. RFID tags can be passive, active, or battery-assisted passive. An active tag has an onboard battery and periodically transmits its ID signal. A BAP tag has a small battery onboard and is activated in the presence of an RFID reader. A passive tag is cheaper and smaller because it has no battery. Instead, the tag uses the radio energy transmitted by the reader as its energy source. The interrogator must be close for the RF field to be strong enough to transfer sufficient power to the tag. Since tags have individual serial numbers, the RFID system design can discriminate several tags that might be within the range of the RFID reader and read them simultaneously.

The tag's information is stored electronically in a nonvolatile memory. The RFID tag includes a small RF transmitter and receiver. An RFID reader transmits an encoded radio signal to interrogate the tag. The tag receives the message and responds with its identification information. This may be only a unique tag serial number or perhaps such product-related information as a stock number, lot or batch number, production date, or other specific information. RFID tags contain at least two parts: an integrated circuit (for storing and processing information, modulating and demodulating an RF signal, collecting dc power from the incident reader signal, and other specialized functions) and an antenna (for receiving and transmitting the signal). Fixed readers are set up to create a specific interrogation zone, which can be tightly controlled. This allows a highly defined reading area to accommodate movement of the tags in and out of the interrogation zone. Mobile readers may be handheld or mounted on carts or vehicles.

Printed antenna technology, electronic circuitry, the middleware for data collection and filtering, and other areas of innovation are constantly evolving. Some of the complexities are not understood by the general public. Since RFID tags can be attached to clothing, possessions, or even implanted within people, the possibility of reading personally linked information without consent has raised privacy concerns.

```
% This script allows the user to calculate and plot the far-field
% radiation pattern for an array of isotropic point sources,
% placed anywhere around the "origin".
% The main assumption is that the array elements are clustered
% together relative to the far-field, so that the origin is roughly
% in the center of the cluster. The user can still enter [x y]
% coordinates and magnitude/phase for each point source. Also the
% calculation assumes free space.

% Prompt user for number of antennas and frequency
n=input('Enter number of point sources \n> ');
f=input('Enter frequency \n> ');

% Begin loop prompting the user to enter each coefficient and each
% coordinate
for i=1:n
    disp('Enter source coefficient for point ')
    disp(sprintf(' %d source coefficient (a*exp(j*b) format)',i))
    A(i)=input('> ');  % store complex coefficients in A
    disp('Enter the (x,y) coordinate for point ')
    disp(sprintf(' %d source coefficient (in format [x y])',i))
    P(i,:)=input('> '); % store coordinates in P, where i is
                                   % the ith element (row)
    % and column 1 contains the x values, column 2 the y values
end

Beta = 2*pi*f/3e8;  % solve for free space

phi = 0:2*pi/1000:2*pi;   % create a vector of 1001 points around
                                   % a circle
Etotal=zeros(1,length(phi));   % the total electric field is also
                             % a vector of 1001 points

% loop through the antennas
for i=1:n
    % first determine how much the point source is offset from the
    % phi = 0
    % axis (x-axis)
    phi_offset(i) = atan2(P(i,2), P(i,1));
    % now find the distance from the ith point source to the origin
    rho(i) = sqrt(P(i,1)^2+P(i,2)^2);
end

% loop through the phi array and calculate the far field as a
% function of phi
for phi_index = 1:length(phi)
    for i=1:n  % need to add
        % The apparent angle between a point in the far field is the
        % difference between phi and the offset of the antenna
          from the
        % x-axis
```

```
            phi_apparent = phi(phi_index) - phi_offset(i);
            Etotal(phi_index)=Etotal(phi_index) + A(i)*...
            exp(j*Beta*rho(i)*cos(phi_apparent));
        end
end

% Polar plot of far-field field pattern
polar(phi, abs(Etotal))
```

SUMMARY

1. We have discussed the fundamental ideas and definitions in antenna theory. The basic types of antenna considered were the Hertzian (or differential length) dipole, the half-wave dipole, the quarter-wave monopole, and the small loop.

2. Theoretically, if we know the current distribution on an antenna, we can find the retarded magnetic vector potential **A**, and from it we can find the retarded electromagnetic fields **H** and **E** by using

$$\mathbf{H} = \nabla \times \frac{\mathbf{A}}{\mu}, \quad \mathbf{E} = \eta\, \mathbf{H} \times \mathbf{a}_k$$

The far-zone fields are obtained by retaining only $1/r$ terms.

3. The analysis of the Hertzian dipole serves as a stepping-stone for other antennas. The radiation resistance of the dipole is very small. This limits the practical usefulness of the Hertzian dipole.

4. The half-wave dipole has a length equal to $\lambda/2$. It is more popular and of more practical use than the Hertzian dipole. Its input impedance is $73 + j42.5\ \Omega$.

5. The quarter-wave monopole is essentially half a half-wave dipole placed on a conducting plane.

6. The radiation patterns commonly used are the field intensity, power intensity, and radiation intensity patterns. The field pattern is usually a plot of $|E_s|$ or its normalized form $f(\theta)$. The power pattern is the plot of $\mathscr{P}_{\text{ave}}$ or its normalized form $f^2(\theta)$.

7. The directive gain is the ratio of $U(\theta, \phi)$ to its average value. The directivity is the maximum value of the directive gain.

8. An antenna array is a group of radiating elements arranged to produce particular radiation characteristics. Its radiation pattern is obtained by multiplying the unit pattern (due to a single element in the group) with the group pattern, which is the plot of the normalized array factor. For an N-element linear uniform array,

$$|AF| = \left| \frac{\sin\left(\dfrac{N\psi}{2}\right)}{\sin\left(\dfrac{\psi}{2}\right)} \right|$$

where $\psi = \beta d \cos\theta + \alpha$, $\beta = 2\pi/\lambda$, $d =$ spacing between the elements, and $\alpha =$ interelement phase shift.

9. The Friis transmission formula characterizes the coupling between two antennas in terms of their directive gains, separation distance, and frequency of operation.

10. For a bistatic radar (one in which the transmitting and receiving antennas are separated), the power received is given by

$$P_r = \frac{G_{dt}G_{dr}}{4\pi} \left[\frac{\lambda}{4\pi r_1 r_2} \right]^2 \sigma\, P_{rad}$$

For a monostatic radar, $r_1 = r_2 = r$ and $G_{dt} = G_{dr}$.

11. Electromagnetic compatibility (EMC) is the capability of electrical and electronic devices to operate in their intended electromagnetic environment without suffering or causing unacceptable degradation as a result of EMI.

12. Electromagnetic interference (EMI) is the disturbance generated by one electronic device that affects another device. It can be suppressed by grounding, shielding, and filtering.

REVIEW QUESTIONS

13.1 An antenna located in a city is a source of radio waves. How much time does it take the waves to reach a town 12,000 km away?

(a) 36 s   (d) 40 ms

(b) 20 μs   (e) None of the above

(c) 20 ms

13.2 In eq. (13.34a–c), which term is the radiation term?

(a) $1/r$ term   (c) $1/r^3$ term

(b) $1/r^2$ term   (d) All of the above

13.3 A very small, thin wire of length $\lambda/100$ has a radiation resistance of

(a) $\approx 0\ \Omega$ (c) $7.9\ \Omega$

(b) $0.08\ \Omega$ (d) $790\ \Omega$

13.4 A quarter-wave monopole antenna operating in air at frequency 1 MHz must have an overall length of

(a) $\ell \gg \lambda$ (c) 150 m   (e) $\ell \ll \lambda$

(b) 300 m   (d) 75 m

13.5 If a small single-turn loop antenna has a radiation resistance of 0.04 Ω, how many turns are needed to produce a radiation resistance of 1 Ω?

(a) 150   (c) 50   (e) 5

(b) 125   (d) 25

13.6 At a distance of 8 km from a differential antenna, the field strength is 12 μV/m. The field strength at a location 20 km from the antenna is

(a) 75 μV/m   (c) 4.8 μV/m

(b) 30 μV/m   (d) 1.92 μV/m

13.7 An antenna has $U_{max} = 10$ W/sr, $U_{ave} = 4.5$ W/sr, and $\eta_r = 95\%$. The input power to the antenna is

(a) 2.222 W   (c) 55.55 W

(b) 12.11 W   (d) 59.52 W

13.8 A receiving antenna in an airport has a maximum dimension of 3 m and operates at 100 MHz. An aircraft approaching the airport is 0.5 km from the antenna. The aircraft is in the far-field region of the antenna.

(a) True   (b) False

13.9 A receiving antenna is located 100 m away from the transmitting antenna. If the effective area of the receiving antenna is 500 cm^2 and the power density at the receiving location is 2 mW/m^2, the total power received is:

(a) 10 nW   (c) 1 μW   (e) 100 μW

(b) 100 nW   (d) 10 μW

13.10 Let R be the maximum range of a monostatic radar. If a target with radar cross section of 5 m^2 exists at $R/2$, what should be the target cross section at $3R/2$ to result in an equal signal strength at the radar?

(a) 0.0617 m^2 (c) 15 m^2 (e) 405 m^2

(b) 0.555 m^2 (d) 45 m^2

Answers: 13.1d, 13.2a, 13.3b, 13.4d, 13.5e, 13.6c, 13.7d, 13.8a, 13.9e, 13.10e.

PROBLEMS   **Section 13.2—Hertzian Dipole**

13.1 The magnetic vector potential at point $P(r, \theta, \phi)$ due to a small antenna located at the origin is given by

$$\mathbf{A}_s = \frac{50\, e^{-j\beta r}}{r}\, \mathbf{a}_x$$

where $r^2 = x^2 + y^2 + z^2$. Find $\mathbf{E}(r, \theta, \phi, t)$ and $\mathbf{H}(r, \theta, \phi, t)$ at the far field.

13.2 A dipole antenna has the following parameters :

Antenna length $\ell = 0.02\lambda$

Current magnitude $I_o = 3$ A

Operating frequency $f = 400$ MHz

Radiation range $r = 60$ m

Determine the following:

(a) The magnitude of the electric field intensity at $\theta = 90°$

(b) The magnitude of the magnetic field intensity at $\theta = 90°$

(c) The radiation resistance

(d) The radiated power

13.3 A Hertzian antenna in free space is 10 cm long. It is fed by a current of 20 A at frequency of 50 MHz. Find the electric and magnetic fields at far zone.

13.4 Determine the current necessary for a 2 cm Hertzian dipole to radiate 12 W at 140 MHz.

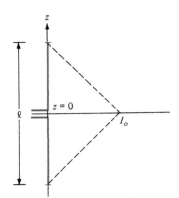

FIGURE 13.31 Short dipole antenna with triangular current distribution; for Problem 13.5.

13.5 (a) Instead of a constant current distribution assumed for the short dipole of Section 13.2, assume a triangular current distribution $I_s = I_o \left(1 - \dfrac{2|z|}{\ell} \right)$ shown in Figure 13.31.

Show that

$$R_{\text{rad}} = 20\,\pi^2 \left[\frac{\ell}{\lambda} \right]^2$$

which is one-fourth of that in eq. (13.13). Thus R_{rad} depends on the current distribution.

(b) Calculate the length of the dipole that will result in a radiation resistance of 0.5 Ω.

13.6 A short dipole antenna operates at the AM broadcast band at 1.2 MHz. To achieve a radiation resistance of 0.5 Ω, how long must the antenna be?

Section 13.3—Half-Wave Dipole Antenna

13.7 A half-wave dipole antenna is driven by a 24 V, 200 MHz source having an internal impedance of 40 Ω. Find the average power radiated by the antenna, given that $Z_{\text{in}} = 73 + j42\,\Omega$.

13.8 A car radio antenna 1 m long operates in the AM frequency band of 1.5 MHz. How much current is required to transmit 4 W of power?

*13.9 (a) Show that the generated far-field expressions for a thin dipole of length ℓ carrying sinusoidal current $I_o \cos \beta z$ are

$$H_{\phi s} = \frac{jI_o e^{-\beta r}}{2\pi r} \frac{\cos \left(\dfrac{\beta \ell}{2} \cos \theta \right) - \cos \dfrac{\beta \ell}{2}}{\sin \theta}, \quad E_{\theta s} = \eta H_{\phi s}$$

[*Hint:* Use Figure 13.4 and start with eq. (13.14).]

(b) On a polar coordinate sheet, plot $f(\theta)$ in part (a) for $\ell = \lambda$, $3\lambda/2$, and 2λ.

13.10 An antenna engineer is asked to design a $\lambda/2$ dipole antenna to operate at 450 MHz.

(a) Calculate the length of the antenna if it is located in free space.

(b) Determine the length of the antenna if it is located under seawater ($\varepsilon = 81\varepsilon_0$, $\mu = \mu_0$, $\sigma = 4$ S/m).

Section 13.4—Quarter-Wave Monopole Antenna

13.11 Quarter-wavelength antennas are used to transmit:
  (a) AM signals at 1150 kHz
  (b) FM signals at 90 MHz
  (c) VHF-TV signals at 80 MHz
  (d) UHF-TV signals at 600 MHz

  Calculate the respective antenna lengths.

13.12 A dipole antenna ($\ell = \lambda/8$) operating at 400 MHz is used to send a message to a satellite in space. Find the radiation resistance of the antenna.

Section 13.5—Small-Loop Antenna

13.13 A 100-turn loop antenna of radius 20 cm operating at 10 MHz in air is to give a 50 mV/m field strength at a distance 3 m from the loop. Determine
  (a) The current that must be fed to the antenna
  (b) The average power radiated by the antenna

13.14 A circular loop antenna has a mean radius of 1.2 cm and N turns. If it operates at 80 MHz, find N that will produce a radiation resistance of 8 Ω.

13.15 A loop antenna with loop radius of 0.4 m is made of copper wire of radius 4 mm. If the loop radiates at 6 MHz and carries a current of 50 A, find
  (a) The radiation resistance of the loop
  (b) The power radiated
  (c) The radiation efficiency

Section 13.6—Antenna Characteristics

13.16 Sketch the normalized E-field and H-field patterns for
  (a) A half-wave dipole
  (b) A quarter-wave monopole

13.17 The radiation efficiency is given by eq. (13.50) as

$$\eta_r = \frac{P_{rad}}{P_{rad} + P_{ohm}}$$

where P_{rad} is the radiation power and P_{ohm} is the power loss due to ohmic resistance of the antenna. For a cylindrical conductor of length Δz that carries current I,

$$P_{ohm} = \frac{1}{2}I^2 R_s \Delta z$$

where $R_s = \dfrac{1}{\sigma\delta 2\pi a}$ and δ is the skin depth. A short antenna of length Δz has an efficiency of 20%. Assume that the efficiency is improved by increasing the length to $2\Delta z$, while keeping the same current. What is the new efficiency?

13.18   Determine the radiation efficiency of a half-wave dipole operating at 6 MHz. The wire is made of copper $(\sigma = 58$ MS/m) and is 1.2 mm in radius.

13.19   An antenna located at the origin has a far-zone electric field as

$$\mathbf{E}_s = \frac{\cos 2\theta}{r} e^{-j\beta r} \mathbf{a}_\theta \text{ V/m}$$

(a) Obtain the corresponding $\mathbf{H}_s$ field.
(b) Determine the power radiated.
(c) What fraction of the total power is radiated in the belt $60° < \theta < 120°$?

13.20   In the far field of a particular antenna located at the origin, the magnetic field intensity is

$$\mathbf{H}_s = \frac{j\beta I_o}{4\pi r} (\sin \phi \, \mathbf{a}_\theta + \cos \theta \cos \phi \, \mathbf{a}_\phi)$$

where I_o is the peak value of the input current. Show that the radiation resistance is given by $R_{\text{rad}} = 20 \, \beta^2$.

***13.21**   An antenna located on the surface of a flat earth transmits an average power of 200 kW. Assuming that all the power is radiated uniformly over the surface of a hemisphere with the antenna at the center, calculate

(a) The time-average Poynting vector at 50 km
(b) The maximum electric field at that location.

13.22   The radiated power density of an antenna at far-zone is given by

$$\mathscr{P}_{ave} = \frac{\alpha \sin^2 \theta \sin^3 \phi}{r^2} \mathbf{a}_\phi, \quad 0 < \theta < \pi, 0 < \phi < \pi$$

where α is a constant. Determine the directive gain and directivity.

13.23   For a one-and-a-half-wave antenna, show that the normalized field pattern is

$$f(\theta) = \frac{\cos(1.5 \, \pi \cos \theta)}{\sin \theta}$$

Use MATLAB to plot $f(\theta)$.

13.24   Divide the interval $0 < \theta < 2\pi$ into 20 equal parts and use MATLAB to show that

$$\int_0^{2\pi} \frac{(1 - \cos \theta)}{\theta} \, d\theta = 2.438$$

13.25 For a thin dipole $\lambda/16$ long, find

(a) The directive gain   (b) The directivity

(c) The effective area   (d) The radiation resistance

13.26 Find the directive gain and directivity of the small loop antenna.

13.27 A quarter-wavelength monopole antenna is used at 1.2 MHz for AM transmission. The antenna is vertically placed above a conducting surface. Determine

(a) The length of the antenna

(b) The radiation resistance

(c) The directivity of the antenna

13.28 Find U_{ave}, U_{max}, and D if:

(a) $U(\theta, \phi) = \sin^2 2\theta$, $0 < \theta < \pi, 0 < \phi < 2\pi$

(b) $U(\theta, \phi) = 4 \csc^2 \theta$, $\pi/3 < \theta < \pi/2, 0 < \phi < \pi$

(c) $U(\theta, \phi) = 2 \sin^2 \theta \sin^2 \phi$, $0 < \theta < \pi, 0 < \phi < \pi$

13.29 For each of the following radiation intensities, calculate the directive gain and directivity.

(a) $U(\theta, \phi) = 10 \sin \theta \sin^2 \phi$, $0 < \theta < \pi, 0 < \phi < 2\pi$

(b) $U(\theta, \phi) = 2 \sin^2 \theta \sin^3 \phi$, $0 < \theta < \pi, 0 < \phi < \pi$

(c) $U(\theta, \phi) = 5 (1 + \sin^2 \theta \sin^2 \phi)$, $0 < \theta < \pi, 0 < \phi < 2\pi$

13.30 The radiation intensity of an antenna is given by

$$U(\theta, \phi) = \begin{cases} 4 \sin^2\theta \sin \phi/2, & 0 < \phi < \pi, 0 < \theta < \pi \\ 0, & \text{otherwise} \end{cases}$$

Calculate the directivity of the antenna.

13.31 An antenna has a far-field electric field given by

$$\mathbf{E}_s = \frac{I_o}{r} e^{-j\beta r} \sin \theta \, \mathbf{a}_\theta$$

where I_o is the maximum input current. Determine the value of I_o to radiate a power of 50 mW.

13.32 Find the radiation intensity of a small loop antenna.

13.33 The field due to an isotropic (or omnidirectional) antenna is given by

$$E = \frac{\alpha l}{r}$$

where α is a constant. Determine the radiation resistance of the antenna.

13.34 Determine the fraction of the total power radiated by the elemental (Hertzian) antenna over $0 < \theta < 60°$.

Section 13.7—Antenna Array

13.35 Derive E_s at far field due to the two-element array shown in Figure 13.32. Assume that the Hertzian dipole elements are fed in phase with uniform current $I_o \cos \omega t$.

13.36 An array comprises two dipoles that are separated by one wavelength. If the dipoles are fed by currents of the same magnitude and phase,

(a) Find the array factor.

(b) Calculate the angles where the nulls of the pattern occur.

(c) Determine the angles where the maxima of the pattern occur.

(d) Sketch the group pattern in the plane containing the elements.

13.37 Sketch the group pattern in the xz-plane of the two-element array of Figure 13.10 with

(a) $d = \lambda, \alpha = \pi/2$

(b) $d = \lambda/4, \alpha = 3\pi/4$

(c) $d = 3\lambda/4, \alpha = 0$

13.38 An antenna array consists of N identical Hertzian dipoles uniformly located along the z-axis and polarized in the z-direction. If the spacing between the dipoles is $\lambda/4$, sketch the group pattern when (a) $N = 2$, (b) $N = 4$.

13.39 An array of isotropic elements has the group pattern

$$F(\psi) = \left| \frac{\sin(2\pi \cos \psi)}{\sin\left(\dfrac{\pi}{2} \cos \psi\right)} \cos^2\left(\frac{\pi}{2} \cos \psi\right) \right|$$

Use MATLAB to plot $F(\psi)$ for $0° < \psi < 180°$.

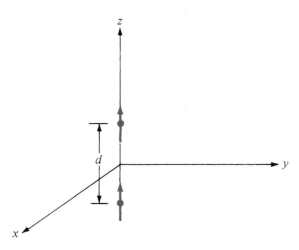

FIGURE 13.32 Two-element array of Problem 13.35.

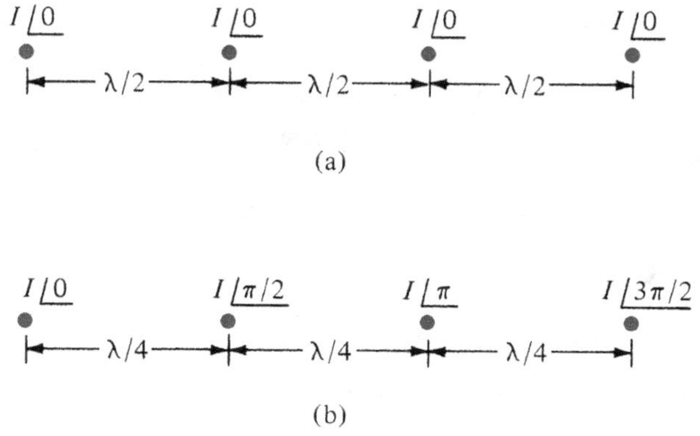

(a)

(b)

FIGURE 13.33 For Problem 13.40.

13.40 Sketch the resultant group patterns for the four-element arrays shown in Figure 13.33.

Section 13.8—Effective Area and the Friis Equation

13.41 A microwave antenna has a power gain of 20 dB at 10 GHz. Determine its effective area.

13.42 An antenna receives a power of 2 μW from a radio station. Calculate its effective area if the antenna is located in the far zone of the station where $E = 50$ mV/m.

13.43 A telemetry transmitter situated on the moon transmits 120 mW at 200 MHz. The gain of the transmitting antenna is 15 dB. Calculate the gain (in dB) of the receiving antenna (situated on earth) in order to receive 4 nW. Assume that the moon is 238,857 miles away from the earth and that 1 mile = 1.609 km.

13.44 In a communication system, suppose the transmitting and receiving antennas have gains 25 dB and 20 dB, respectively, and are 42 km apart. Find the minimum power that must be transmitted in order to deliver a minimum of 3 μW. The channel frequency is 3 GHz.

13.45 The power transmitted by a synchronous orbit satellite antenna is 320 W. If the antenna has a gain of 40 dB at 15 GHz, calculate the power received by another antenna with a gain of 32 dB at the range of 24,567 km.

13.46 A communication link uses a half-wave dipole antenna for transmission and another half-wave dipole antenna for reception. The link operates at 20 MHz and the two antennas are separated by a distance of 80 km. If the receiving antenna must receive an average power of 0.5 μW, determine the minimum required current of the transmitting antenna.

13.47 Two identical antennas in an anechoic chamber are separated by 12 m and are oriented for maximum directive gain. At a frequency of 5 GHz, the power received by one is 30 dB down from that transmitted by the other. Calculate the gain of the antennas in dB.

Section 13.9—The Radar Equation

13.48 An L-band pulse radar with a common transmitting and receiving antenna having a directive gain of 3500 operates at 1500 MHz and transmits 200 kW. If an object is 120 km from the radar and its scattering cross section is 8 m^2, find

(a) The magnitude of the incident electric field intensity of the object

(b) The magnitude of the scattered electric field intensity at the radar

(c) The amount of power captured by the object

(d) The power absorbed by the antenna from the scattered wave

13.49 A monostatic radar operates at 4 GHz and has a directive gain of 30 dB. The radar is used to track a target 10 km away, and the radar cross section of the target is 12 m^2. If the antenna of the radar transmits 80 kW, calculate the power intercepted by the target.

13.50 A 4 GHz radar antenna with effective area of 2 m^2 transmits 60 kW. A target with cross section of 5 m^2 is located 160 km away. Calculate: (a) the round trip travel time, (b) the power received, (c) the maximum detectable range, assuming that the minimum detectable power is 8 pW.

13.51 It is required to double the range capacity of a radar. What percentage increase in transmitter power is necessary to achieve this?

13.52 In the bistatic radar system of Figure 13.34, the ground-based antennas are separated by 4 km and the 2.4 m^2 target is at a height of 3 km. The system operates at 5 GHz. For G_{dt} of 36 dB and G_{dr} of 20 dB, determine the minimum necessary radiated power to obtain a return power of 8×10^{-12} W.

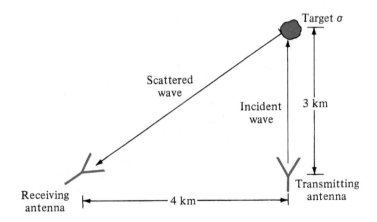

FIGURE 13.34 For Problem 13.52.

Section 13.10—Application Note—Electromagnetic Interference and Compatibility

13.53 (a) For an RL filter with $L = 50$ nH and $R = 20$ Ω, find the insertion loss in dB at 300 MHz.

(b) Repeat part (a) if an RC filter with $C = 60$ pF and $R = 10$ kΩ is used instead.

*13.54 The insertion loss of a filter circuit can be calculated in terms of its A, B, C, and D parameters when terminated by Z_g and Z_L as shown in Figure 13.35. Show that

$$IL = 20 \log_{10} \left| \frac{AZ_L + B + CZ_gZ_L + DZ_g}{Z_g + Z_L} \right|$$

13.55 A silver rod has rectangular cross section with height 0.8 cm and width 1.2 cm. Find

(a) The dc resistance per kilometer of the conductor

(b) The ac resistance per kilometer of the conductor at 6 MHz

13.56 Within a shielded enclosure, the electric field is 6 V/m. It is required that the electric field outside the shield be no more than 20 μV/m. Find the shielding effectiveness in dB.

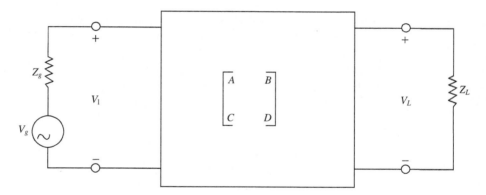

FIGURE 13.35 For Problem 13.54.

COMPUTATIONAL ELECTROMAGNETICS

Until the 1940s, most electromagnetic problems were solved by using the classical methods of separation of variables and integral equations. Application of those methods, however, required a high degree of ingenuity, experience, and effort, and only a narrow range of practical problems could be investigated, owing to the complex geometries defining the problems. While theory and experiment remain the two conventional pillars of science and engineering, modeling and simulation represent the third pillar that complements them.

Computational electromagnetics (CEM) is the theory and practice of solving EM field problems on digital computers. It offers the key to comprehensive solutions of Maxwell's equations. In this chapter, the common methods used in computational electromagnetics will be presented. CEM techniques can be used to model electromagnetic interaction phenomena in circuits, devices, and systems.

Numerical modeling and simulation have revolutionized all aspects of engineering design to the extent that several software packages have been developed to aid designing and modeling. Widely used software packages for CEM include the *Numerical Electromagnetics Code* (NEC) based on the method of moments and developed at Lawrence Livermore National Laboratory, the *High Frequency Structure Simulator* (HFSS) based on the finite element method and developed by Ansoft, *Microwave Office* based on the method of moments and developed by Applied Wave Research, *Sonnet* developed by Sonnet, COMSOL based on the finite element method by Sonnet, COMSOL based on the finite element method, and QuickField, which is another finite element modeling software. These software packages put powerful tools and techniques, previously available only to full-time theorists, into the hands of engineers not formally trained in CEM. The best method or software package to use depends on the particular problem you are trying to solve.

CHAPTER 14

NUMERICAL METHODS

Young men think old men are fools, but old men know young men are fools.

—GEORGE CHAPMAN

14.1 INTRODUCTION

In the preceding chapters we considered various analytic techniques for solving EM problems and obtaining solutions in closed form. A *closed-form solution* is one in the form of an explicit, algebraic equation in which values of the problem parameters can be substituted. Some of these analytic solutions were obtained assuming certain situations, thereby making the solutions applicable to those idealized situations. For example, in deriving the formula for calculating the capacitance of a parallel-plate capacitor, we assumed that the fringing effect was negligible and that the separation distance was very small compared with the width and length of the plates. Also, our application of Laplace's equation in Chapter 6 was restricted to problems with boundaries coinciding with coordinate surfaces. Analytic solutions have an inherent advantage of being exact. They also make it easy to observe the behavior of the solution when there is variation in the problem parameters. However, analytic solutions are available only for problems with simple configurations.

When the complexities of theoretical formulas make analytic solution intractable, we resort to nonanalytic methods, which include (1) graphical methods, (2) experimental methods, (3) analog methods, and (4) numerical methods. Graphical, experimental, and analog methods are applicable to solving relatively few problems. Numerical methods have come into prominence and have become more attractive with the advent of fast digital computers. The three most commonly used simple numerical techniques in EM are the moment method, the finite difference method, and the finite element method. Most EM problems involve either partial differential equations or integral equations. Partial differential equations are usually solved by using the finite difference method or the finite element method; integral equations are solved conveniently by using the moment method. Although numerical methods give approximate solutions, the solutions are sufficiently accurate for engineering purposes. We should not get the impression that analytic techniques are outdated because of numerical methods; rather, they are complementary. As will be observed later, every numerical method involves analytic simplification until the method can be easily applied.

The MATLAB codes developed for computer implementation of the concepts developed in this chapter are simplified and self-explanatory for instructional purposes. (Appendix C provides a short tutorial on MATLAB.) The notations used in the programs are as close as possible to those used in the main text; some are defined wherever necessary. These programs are by no means unique; there are several ways of writing a computer program. Therefore, users may decide to modify the programs to suit their objectives.

†14.2 FIELD PLOTTING

In Section 4.9, we used field lines and equipotential surfaces for visualizing an electrostatic field. However, the graphical representations in Figure 4.21 for electrostatic fields and in Figures 7.8(b) and 7.16 for magnetostatic fields are very simple, trivial, and qualitative. Accurate pictures of more complicated charge distributions would be more helpful. This section presents a numerical technique that may be developed into an interactive computer program. It generates data points for electric field lines and equipotential lines for arbitrary configuration of point sources.

Electric field lines and equipotential lines can be plotted for coplanar point sources with simple programs. Suppose we have N point charges located at position vectors $\mathbf{r}_1, \mathbf{r}_2, \ldots, \mathbf{r}_N$, the electric field intensity $\mathbf{E}$ and potential V at position vector $\mathbf{r}$ are given, respectively, by

$$\mathbf{E} = \sum_{k=1}^{N} \frac{Q_k (\mathbf{r} - \mathbf{r}_k)}{4\pi\varepsilon \, |\mathbf{r} - \mathbf{r}_k|^3} \tag{14.1}$$

and

$$V = \sum_{k=1}^{N} \frac{Q_k}{4\pi\varepsilon \, |\mathbf{r} - \mathbf{r}_k|} \tag{14.2}$$

If the charges are on the same plane ($z = $ constant), eqs. (14.1) and (14.2) become

$$\mathbf{E} = \sum_{k=1}^{N} \frac{Q_k[(x - x_k)\mathbf{a}_x + (y - y_k)\mathbf{a}_y]}{4\pi\varepsilon[(x - x_k)^2 + (y - y_k)^2]^{3/2}} \tag{14.3}$$

$$V = \sum_{k=1}^{N} \frac{Q_k}{4\pi\varepsilon[(x - x_k)^2 + (y - y_k)^2]^{1/2}} \tag{14.4}$$

To plot the electric field lines, follow these steps:

1. Choose a starting point on the field line.
2. Calculate E_x and E_y at that point using eq. (14.3).
3. Take a small step along the field line to a new point in the plane. As shown in Figure 14.1, a movement $\Delta\ell$ along the field line corresponds to movements Δx and Δy in the x- and y-directions, respectively. From Figure 14.1, it is evident that

$$\frac{\Delta x}{\Delta\ell} = \frac{E_x}{E} = \frac{E_x}{[E_x^2 + E_y^2]^{1/2}}$$

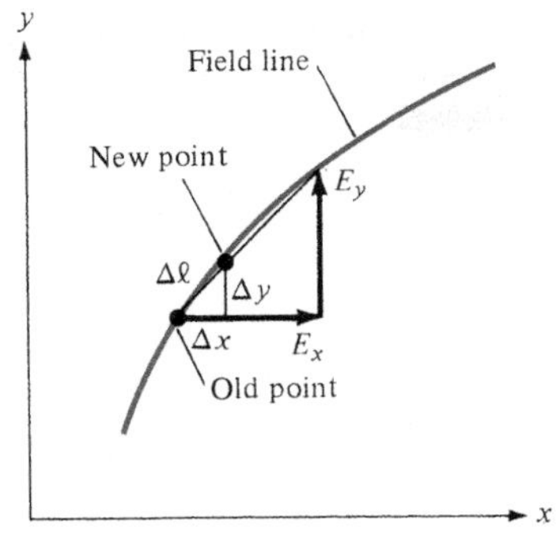

FIGURE 14.1 A small displacement on a field line.

or

$$\Delta x = \frac{\Delta\ell \cdot E_x}{[E_x^2 + E_y^2]^{1/2}} \tag{14.5}$$

Similarly,

$$\Delta y = \frac{\Delta\ell \cdot E_y}{[E_x^2 + E_y^2]^{1/2}} \tag{14.6}$$

Move along the field line from the old point (x, y) to a new point $x' = x + \Delta x$, $y' = y + \Delta y$.

4. Go back to steps 2 and 3 and repeat the calculations. Continue to generate new points until a line is completed within a given range of coordinates. On completing the line, go back to step 1 and choose another starting point. Note that since there are an infinite number of field lines, any starting point is likely to be on a field line. The points generated can be plotted by a plotter as illustrated in Figure 14.2.

To plot the equipotential lines, follow these steps:

1. Choose a starting point.
2. Calculate the electric field (E_x, E_y) at that point by using eq. (14.3).
3. Move a small step along the line perpendicular to the E-field line at that point. Utilize the fact that if a line has slope m, a perpendicular line must have slope $-1/m$. Since an E-field line and an equipotential line meeting at a given point are mutually orthogonal there,

$$\Delta x = \frac{-\Delta\ell \cdot E_y}{[E_x^2 + E_y^2]^{1/2}} \tag{14.7}$$

$$\Delta y = \frac{\Delta\ell \cdot E_x}{[E_x^2 + E_y^2]^{1/2}} \tag{14.8}$$

move along the equipotential line from the old point (x, y) to a new point $(x + \Delta x, y + \Delta y)$. As a way of checking the new point, calculate the potential at

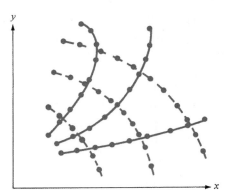

FIGURE 14.2 Generated points on *E*-field lines (shown thick) and equipotential lines (dotted).

the new and old points using eq. (14.4); the potentials must be equal because the points are on the same equipotential line.

4. Go back to steps 2 and 3 and repeat the calculations. Continue to generate new points until a line is completed within the given range of *x* and *y*. After completing the line, go back to step 1 and choose another starting point. Join the points generated by a plotter as illustrated in Figure 14.2.

By following the same reasoning, we can use the Biot–Savart law to plot the magnetic field line due to various current distributions. Programs for determining the magnetic field line due to line current, a current loop, a Helmholtz pair, and a solenoid can be developed. Programs for drawing the electric and magnetic field lines inside a rectangular waveguide or the power radiation pattern produced by a linear array of vertical half-wave electric dipole antennas can also be written.

EXAMPLE 14.1

Write a program to plot the electric field and equipotential lines due to:

(a) Two point charges Q and $-4Q$, located at $(x, y) = (-1, 0)$ and $(1, 0)$, respectively.

(b) Four point charges Q, $-Q$, Q, and $-Q$, located at $(x, y) = (-1, -1)$, $(1, -1)$, $(1, 1)$, and $(-1, 1)$, respectively. Take $Q/4\pi\varepsilon = 1$ and $\Delta\ell = 0.1$. Consider the range $-5 < x < 5$, $-5 < y < 5$.

Solution:

Based on the steps given in Section 14.2, the program in Figure 14.3 was developed. Enough comments are inserted to make the program as self-explanatory as possible. For example, to use the program to generate the plot in Figure 14.4(a), load program **plotit** in your MATLAB directory. At the command prompt in MATLAB, type

$$\textbf{plotit} \left(\begin{bmatrix} 1 & -4 \end{bmatrix}, \begin{bmatrix} -1 & 0; 1 & 0 \end{bmatrix}, 1, 1, 0.1, 0.01, 8, 2, 5\right)$$

where the numbers have meanings provided in the program. Further explanation of the program is provided in the following paragraphs.

Since the E-field lines emanate from positive charges and terminate on negative charges, it seems reasonable to generate starting points (x_s, y_s) for the E-field lines on small circles centered at charge locations (x_Q, y_Q); that is,

$$x_s = x_Q + r\cos\theta \tag{14.1.1a}$$

$$y_s = y_Q + r\sin\theta \tag{14.1.1b}$$

```
function plotit(charges,location,ckEField,ckEq,DLE,DLV,NLE,NLV,PTS)
figure;
hold on;
% Program for plotting the electric field lines
% and equipotential lines due to coplanar point charges
% the plot is to be within the range -5<x,y<5
%
% This is the correct usage:
% function plotit(charges,location,ckEField,ckEq,DLE,DLV,NLE,NLV,PTS)
%
% where,
%    charges = a vector containing the charges
%   location = a matrix where each row is a charge location
%   ckEField = Flag set to 1 plots the Efield lines
%       ckEq = Flag set to 1 plots the Equipotential lines
% DLE or DLV = the increment along E & V lines
%        NLE = No. of E-Field lines per charge
%        NLV = No. of Equipotential lines per charge
%        PTS => Plots every PTS point (i.e. if PTS = 5 then plot
every 5th point)
% note that constant Q/4*Pie*ErR is set equal to 1.0

% Determine the E-Field LInes
% For convenience, the starting points (XS,YS) are radially
distributed about charge locations
Q=charges;
XQ = location(:,1);
YQ = location(:,2);
JJ=1;
NQ = length(charges);
if (ckEField)
for K=1:NQ
    for I =1:NLE
        THETA = 2*pi*(I-1)/(NLE);
        XS=XQ(K) + 0.1*cos(THETA);
        YS=YQ(K) + 0.1*sin(THETA);
        XE=XS;
        YE=YS;
        JJ=JJ+1;
        if (~mod(JJ,PTS))
```

FIGURE 14.3 Computer program for Example 14.1.

```
                plot (XE,YE);
        end
        while(1)
        % FIND INCREMENT AND NEW POINT (X,Y)
            EX=0;
            EY=0;
            for J=1:NQ
                R =sqrt((XE-XQ(J))^2 + (YE - YQ(J))^2);
                EX = EX +Q(J)*(XE-XQ(J))/(R^3);
                EY = EY +Q(J)*(YE-YQ(J))/(R^3);
            end
            E = sqrt(EX^2 + EY^2);

            % CHECK FOR A SINGULAR POINT
            if (E <=.00005)
                break;
            end
            DX = DLE*EX/E;
            DY = DLE*EY/E;
            % FOR NEGATIVE CHARGE, NEGATE DX & DY SO THAT INCREMENT
            % IS AWAY FROM THE CHARGE
            if (Q(K) < 0)
                DX = -DX;
                DY = -DY;
            end
            XE = XE + DX;
            YE = YE + DY;
            % CHECK WHETHER NEW POINT IS WITHIN THE GIVEN RANGE OR
            TOO
            % CLOSE TO ANY OF THE POINT CHARGES - TO AVOID SINGULAR
            POINT
            if ((abs(XE) >= 5) | (abs(YE) >= 5))
                break;
            end

            if (sum(abs(XE-XQ) < .05 & abs(YE-YQ) < .05) >0)
                break;
            end
            JJ=JJ+1;
            if (~mod(JJ,PTS))
                plot (XE,YE);
            end
        end % while loop
    end % I =1:NLE
end   % K = 1:NQ
end % if
% NEXT, DETERMINE THE EQUIPOTENTIAL LINES
% FOR CONVENIENCE, THE STARTING POINTS (XS,YS) ARE
% CHOSEN LIKE THOSE FOR THE E-FIELD LINES
if (ckEq)
```

FIGURE 14.3 (Continued)

```
JJ=1;
DELTA = .2;
ANGLE = 45*pi/180;
for K =1:NQ
    FACTOR = .5;
    for KK = 1:NLV
        XS = XQ(K) + FACTOR*cos(ANGLE);
        YS = YQ(K) + FACTOR*sin(ANGLE);
        if ( abs(XS) >= 5 | abs(YS) >=5)
            break;
        end
        DIR = 1;
        XV = XS;
        YV = YS;
        JJ=JJ+1;
        if (~mod(JJ,PTS))
            plot(XV,YV);
        end
% FIND INCREMENT AND NEW POINT (XV,YV)
        N=1;
        while (1)
            EX = 0;
            EY = 0;
            for J = 1:NQ
                R = sqrt((XV-XQ(J))^2 + (YV-YQ(J))^2);
                EX = EX + Q(J)*(XV-XQ(J))/(R^3);
                EY = EY + Q(J)*(YV-YQ(J))/(R^3);
            end
            E=sqrt(EX^2 + EY^2);
            if (E <= .00005)
                FACTOR = 2*FACTOR;
                break;
            end
            DX = -DLV*EY/E;
            DY = DLV*EX/E;
            XV = XV + DIR*DX;
            YV = YV + DIR*DY;
            % CHECK IF THE EQUIPOTENTIAL LINE LOOPS BACK TO (X,YS)
            R0 = sqrt((XV - XS)^2 + (YV - YS)^2);
            if (R0 < DELTA & N < 50)
                FACTOR = 2*FACTOR;
                break;
            end
            % CHECK WHETHER NEW POINT IS WITHIN THE GIVEN RANGE
            % IF FOUND OUT OF RANGE, GO BACK TO THE STARTING POINT
            % (XS,YS)BUT INCREMENT IN THE OPPOSITE DIRECTION
            if (abs(XV) > 5 | abs(YV) > 5)
                DIR = DIR -2;
                XV = XS;
```

FIGURE 14.3 (Continued)

```
                    YV = YS;
                    if (abs(DIR) > 1)
                        FACTOR = 2*FACTOR;
                        break;
                    end
              else
                    if (sum(abs(XV-XQ) < .005 & abs(YV-YQ) < .005) >0)
                            break;
                    end
              end
              JJ=JJ+1;
              if (~mod(JJ,PTS))
                  N=N+1;
                  plot(XV,YV);
              end
        end % WHILE loop
    end  % KK
end    % K

end % if
```

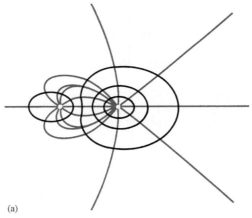

(a)

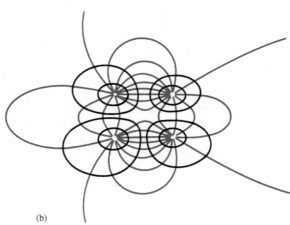

(b)

FIGURE 14.4 For Example 14.1; plots of *E*-field lines and equipotential lines due to (**a**) two point charges and (**b**) four point charges (a two-dimensional quadrupole).

where r is the radius of the small circle (e.g., $r = 0.1$ or 0.05), and θ is a prescribed angle chosen for each E-field line. The starting points for the equipotential lines can be generated in different ways: along the x- and y-axes, along line $y = x$, and so on. However, to make the program as general as possible, the starting points should depend on the charge locations like those for the E-field lines. They could be chosen by using eq. (14.1.1) but with fixed θ (e.g., $45°$) and variable r (e.g., 0.5, 1.0, 2.0, . . .).

The value of incremental length $\Delta\ell$ is crucial for accurate plots. Although the smaller the value of $\Delta\ell$, the more accurate the plots, we must keep in mind that the smaller the value of $\Delta\ell$, the more points we generate, and memory storage may be a problem. For example, a line may consist of more than 1000 generated points. In view of the large number of points to be plotted, the points are usually stored in a data file and a graphics routine is used to plot the data.

For both the E-field and equipotential lines, different checks are inserted in the program in Figure 14.3:

(a) Check for singular point ($\mathbf{E} = \mathbf{0}$?).
(b) Check whether the point generated is too close to a charge location.
(c) Check whether the point is within the given range of $-5 < x < 5$, $-5 < y < 5$,
(d) Check whether the (equipotential) line loops back to the starting point.

The plot of the points generated for the cases of two point charges and four point charges are shown in Figure 14.4(a) and (b), respectively.

PRACTICE EXERCISE 14.1

Write a complete program for plotting the electric field lines and equipotential lines due to coplanar point charges. Run the program for $N = 3$; that is, there are three point charges $-Q$, $+Q$, and $-Q$, located at $(x, y) = (-1, 0)$, $(0, 1)$, and $(1, 0)$, respectively. Take $Q/4\pi\varepsilon = 1$, $\Delta\ell = 0.1$ or 0.01 for greater accuracy and limit your plot to $-5 < x < 5$, $-5 < y < 5$.

Answer: See Figure 14.5.

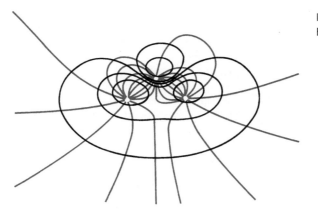

FIGURE 14.5 For Practice Exercise 14.1.

14.3 THE FINITE DIFFERENCE METHOD

The finite difference method[1] (FDM) is a simple numerical technique used in solving problems like those solved analytically in Chapter 6. A problem is uniquely defined by three things:

1. A partial differential equation such as Laplace's or Poisson's equation
2. A solution region
3. Boundary and/or initial conditions

A finite difference solution to Poisson's or Laplace's equation, for example, proceeds in three steps: (1) dividing the solution region into a grid of nodes, (2) approximating the differential equation and boundary conditions by a set of linear algebraic equations (called *difference equations*) on grid points within the solution region, and (3) solving this set of algebraic equations.

Step 1: Suppose we intend to apply the finite difference method to determine the electric potential in a region shown in Figure 14.6(a). The solution region is divided into rectangular meshes with *grid points* or *nodes* as in Figure 14.6(a). A node on the boundary of the region where the potential is specified is called a *fixed node* (fixed by the problem), and interior points in the region are called *free points* (free in that the potential is unknown).

Step 2: Our objective is to obtain the finite difference approximation to Poisson's equation and use this to determine the potentials at all the free points. We recall that Poisson's equation is given by

$$\nabla^2 V = -\frac{\rho_v}{\varepsilon} \tag{14.9a}$$

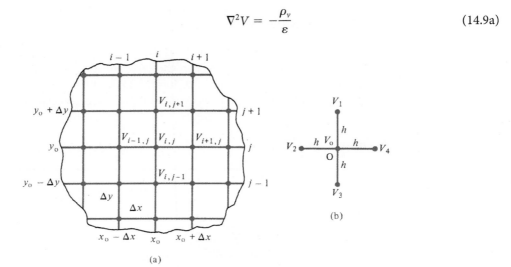

FIGURE 14.6 Finite difference solution pattern: (**a**) division of the solution into grid points, (**b**) finite difference five-node molecule.

---

[1] For an extensive treatment of the finite difference method, see G. D. Smith, *Numerical Solution of Partial Differential Equations: Finite Difference Methods*, 3rd ed. Oxford: Oxford Univ. Press, 1985.

For two-dimensional solution region such as in Figure 14.6(a), $\dfrac{\partial^2 V}{\partial z^2} = 0$, so

$$\frac{\partial^2 V}{\partial x^2} + \frac{\partial^2 V}{\partial y^2} = -\frac{\rho_v}{\varepsilon} \tag{14.9b}$$

From the definition of the derivative of $V(x, y)$ at point (x_o, y_o),

$$V' = \frac{\partial V}{\partial x}\bigg|_{x=x_o} \simeq \frac{V(x_o + \Delta x, y_o) - V(x_o - \Delta x, y_o)}{2\Delta x}$$

$$= \frac{V_{i+1,j} - V_{i-1,j}}{2\,\Delta x} \tag{14.10}$$

where Δx is a sufficiently small increment along x. For the second derivative, which is the derivative of the first derivative V',

$$V'' = \frac{\partial^2 V}{\partial x^2}\bigg|_{x=x_o} = \frac{\partial V'}{\partial x} \simeq \frac{V'(x_o + \Delta x/2, y_o) - V'(x_o - \Delta x/2, y_o)}{\Delta x}$$

$$= \frac{V(x_o + \Delta x, y_o) - 2V(x_o, y_o) + V(x_o - \Delta x, y_o)}{(\Delta x)^2}$$

$$= \frac{V_{i+1,j} - 2V_{i,j} + V_{i-1,j}}{(\Delta x)^2} \tag{14.11}$$

Equations (14.10) and (14.11) are the finite difference approximations for the first and second partial derivatives of V with respect to x, evaluated at $x = x_o$. The approximation in eq. (14.10) is associated with an error of the order of the Δx while that of eq. (14.11) has an associated error on the order of $(\Delta x)^2$. Similarly,

$$\frac{\partial^2 V}{\partial y^2}\bigg|_{y=y_o} \simeq \frac{V(x_o, y_o + \Delta y) - 2V(x_o, y_o) + V(x_o, y_o - \Delta y)}{(\Delta y)^2}$$

$$= \frac{V_{i,j+1} - 2V_{i,j} + V_{i,j-1}}{(\Delta y)^2} \tag{14.12}$$

Substituting eqs. (14.11) and (14.12) into eq. (14.9b) and letting $\Delta x = \Delta y = h$ gives

$$V_{i+1,j} + V_{i-1,j} + V_{i,j+1} + V_{i,j-1} - 4V_{i,j} = -\frac{h^2 \rho_v}{\varepsilon}$$

or

$$\boxed{V_{i,j} = \frac{1}{4}\left(V_{i+1,j} + V_{i-1,j} + V_{i,j+1} + V_{i,j-1} + \frac{h^2 \rho_v}{\varepsilon}\right)} \tag{14.13}$$

where h is called the *mesh size*. Equation (14.13) is the finite difference approximation to Poisson's equation. If the solution region is charge free ($\rho_v = 0$), eq. (14.9) becomes Laplace's equation:

$$\nabla^2 V = \frac{\partial^2 V}{\partial x^2} + \frac{\partial^2 V}{\partial y^2} = 0 \tag{14.14}$$

The finite difference approximation to this equation is obtained from eq. (14.13) by setting $\rho_v = 0$; that is,

$$V_{i,j} = \frac{1}{4}\left(V_{i+1,j} + V_{i-1,j} + V_{i,j+1} + V_{i,j-1}\right) \tag{14.15}$$

This equation is essentially a five-node finite difference approximation for the potential at the central point of a square mesh. Figure 14.6(b) illustrates what is called the finite difference *five-node molecule*. The molecule in Figure 14.6(b) is taken out of Figure 14.6(a). Thus eq. (14.15) applied to the molecule becomes

$$V_o = \frac{1}{4}\left(V_1 + V_2 + V_3 + V_4\right) \tag{14.16}$$

This equation clearly shows the average-value property of Laplace's equation. In other words, Laplace's equation can be interpreted as a differential means of stating the fact that the potential at a specific point is the average of the potentials at the neighboring points.
Step 3: To apply eq. (14.16) [or eq. (14.13)] to a given problem, one of the following two methods is commonly used.

A. Iteration Method

We start by setting initial values of the potentials at the free nodes equal to zero or to any reasonable guessed value. Keeping the potentials at the fixed nodes unchanged at all times, we apply eq. (14.16) to every free node in turn until the potentials at all free nodes have been calculated. The potentials obtained at the end of this first iteration are just approximate. To increase the accuracy of the potentials, we repeat the calculation at every free node, using old values to determine new ones. The iterative or repeated modification of the potential at each free node is continued until a prescribed degree of accuracy is achieved or until the old and the new values at each node are satisfactorily close.

B. Band Matrix Method

Equation (14.16) applied to all free nodes results in a set of simultaneous equations of the form

$$[A][V] = [B] \tag{14.17}$$

where $[A]$ is a *sparse* matrix (i.e., one having many zero terms), $[V]$ consists of the unknown potentials at the free nodes, and $[B]$ is another column matrix formed by the known potentials at the fixed nodes. Matrix $[A]$ is also *banded* in that its nonzero terms appear clustered near the main diagonal because only nearest neighboring nodes affect the potential at each node. The sparse, band matrix is easily inverted to determine $[V]$. Thus we obtain the potentials at the free nodes from matrix $[V]$ as

$$[V] = [A]^{-1}[B] \tag{14.18}$$

The finite difference method can be applied to solve time-varying problems. For example, consider the one-dimensional wave equation of eq. (10.1), then

$$u^2 \frac{\partial^2 \Phi}{\partial x^2} = \frac{\partial^2 \Phi}{\partial t^2} \tag{14.19}$$

where u is the wave velocity and Φ is the E- or H-field component of the EM wave. The difference approximations of the derivatives at the (x_o, t_o) or (i, j)th node shown in Figure 14.7 are

$$\frac{\partial^2 \Phi}{\partial x^2}\bigg|_{x=x_o} \simeq \frac{\Phi_{i-1,j} - 2\Phi_{i,j} + \Phi_{i+1,j}}{(\Delta x)^2} \tag{14.20}$$

$$\frac{\partial^2 \Phi}{\partial t^2}\bigg|_{t=t_o} \simeq \frac{\Phi_{i,j-1} - 2\Phi_{i,j} + \Phi_{i,j+1}}{(\Delta t)^2} \tag{14.21}$$

Where Δx and Δt are increments along x and t. Inserting eqs. (14.20) and (14.21) in eq. (14.19) and solving for $\Phi_{i,j+1}$ gives

$$\Phi_{i,j+1} \simeq \alpha(\Phi_{i-1,j} + \Phi_{i+1,j}) + 2(1 - \alpha)\Phi_{i,j} - \Phi_{i,j-1} \tag{14.22}$$

where

$$\alpha = \left[\frac{u\,\Delta t}{\Delta x}\right]^2 \tag{14.23}$$

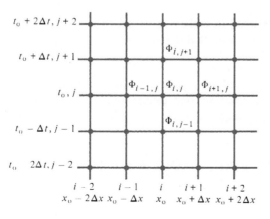

FIGURE 14.7 Finite difference solution pattern for the wave equation: eq. (14.19).

It can be shown that for the solution in eq. (14.22) to be stable, $\alpha \leq 1$. To start the finite difference algorithm in eq. (14.22), we use the initial conditions. We assume that at $t = 0$, $\partial \Phi_{i,0} / \partial t = 0$ and use (central) difference approximation (see Review Question 14.2) to get

$$\frac{\partial \Phi_{i,0}}{\partial t} \simeq \frac{\Phi_{i,1} - \Phi_{i,-1}}{2\Delta t} = 0$$

or

$$\Phi_{i,1} = \Phi_{i,-1} \tag{14.24}$$

Substituting eq. (14.24) into eq. (14.22) and taking $j = 0$ ($t = 0$), we obtain

$$\Phi_{i,1} \simeq \alpha(\Phi_{i-1,0} + \Phi_{i+1,0}) + 2(1 - \alpha)\Phi_{i,0} - \Phi_{i,1}$$

or

$$\Phi_{i,1} \simeq \frac{1}{2}\left[\alpha(\Phi_{i-1,0} + \Phi_{i+1,0}) + 2(1 - \alpha)\Phi_{i,0}\right] \tag{14.25}$$

With eq. (14.25) as the "starting" formula, the value of Φ at any point on the grid can be obtained directly from eq. (14.22). Note that the three methods discussed for solving eq. (14.16) do not apply to eq. (14.22) because eq. (14.22) can be used directly with eq. (14.25) as the starting formula. In other words, we do not have a set of simultaneous equations; eq. (14.22) is an explicit formula.

The FDM concept can be extended to Poisson's, Laplace's, or wave equations in other coordinate systems. The accuracy of the method depends on the fineness of the grid and the amount of time spent in refining the potentials. We can reduce computer time and increase the accuracy and convergence rate by the method of successive overrelaxation, by making reasonable guesses at initial values, by taking advantage of symmetry if possible, by making the mesh size as small as possible, and by using more complex finite difference molecules. One limitation of the finite difference method is that interpolation of some kind must be used to determine solutions at points not on the grid. One obvious way to overcome this is to use a finer grid, but this requires a greater number of computations and a larger amount of computer storage.

EXAMPLE 14.2

Solve the one-dimensional boundary-value problem $-\Phi'' = x^2$, $0 \leq x \leq 1$, subject to $\Phi(0) = 0 = \Phi(1)$. Use the finite difference method.

Solution:

First, we obtain the finite difference approximation to the differential equation $\Phi'' = -x^2$, which is Poisson's equation in one dimension. Next, we divide the entire domain $0 \leq x \leq 1$ into N equal segments each of length h ($= 1/N$) as in Figure 14.8(a) so that there are $(N + 1)$ nodes.

$$-x_o^2 = \left.\frac{d^2\Phi}{dx^2}\right|_{x=x_o} \simeq \frac{\Phi(x_o + h) - 2\Phi(x_o) + \Phi(x_o - h)}{h^2}$$

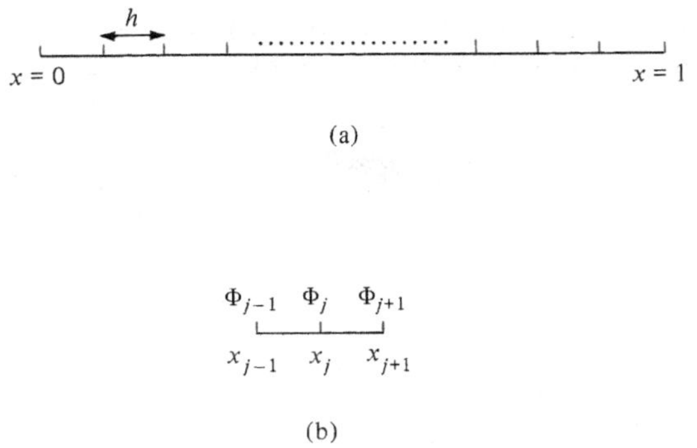

FIGURE 14.8 For Example 14.2.

(a)

(b)

or

$$-x_j^2 = \frac{\Phi_{j+1} - 2\Phi_j + \Phi_{\xi-1}}{h^2}$$

Thus

$$-2\Phi_j = -x_j^2 h^2 - \Phi_{j+1} - \Phi_{j-1}$$

or

$$\Phi_j = \frac{1}{2}(\Phi_{j+1} + \Phi_{j-1} + x_j^2 h^2)$$

Using this finite difference scheme, we obtain an approximate solution for various values of N. The MATLAB code is shown in Figure 14.9. The number of iterations ni depends on the degree of accuracy desired. For a one-dimensional problem such as this, $ni = 50$ may suffice. For two- or three-dimensional problems, larger values of ni would be required (see later: Table 14.1). It should be noted that the values of Φ at end points (fixed nodes) are held fixed. The solutions for $N = 4$ and 10 are shown in Figure 14.10.

We may compare this with the exact solution obtained as follows. Given that $d^2\Phi/dx^2 = -x^2$, integrating twice gives

$$\Phi = -\frac{x^4}{12} + Ax + B$$

where A and B are integration constants. From the boundary conditions,

$$\Phi(0) = 0 \rightarrow B = 0$$
$$\Phi(1) = 0 \rightarrow 0 = -\frac{1}{12} + A \quad \text{or} \quad A = \frac{1}{12}$$

Hence, the exact solution is $\Phi = x(1 - x^3)/12$, which is calculated in Figure 14.9 and found to be very close to case $N = 10$.

```
% ONE-DIMENSIONAL PROBLEM OF EXAMPLE 14.2
% SOLVED USING FINITE DIFFERENCE METHOD
%
% h = MESH SIZE
% ni = NO. OF ITERATIONS DESIRED

      P = [ ];
      n=20;
      ni=500;
      l=1.0;
      h = 1/n;
      phi=zeros(n+1,1);
      x=h*[0:n]';
      x1=x(2:n);
      for k=1:ni
          phi([2:n])=[phi(3:n+1)+phi(1:n-1)+x1.^2*h^2]/2;
      end
      %   CALCULATE THE EXACT VALUE ALSO
      phiex=x.*(1.0-x.^3)/12.0;
      diary a:test.out
      [[1:n+1]' phi phiex]
      diary off
```

FIGURE 14.9 Computer program for Example 14.2.

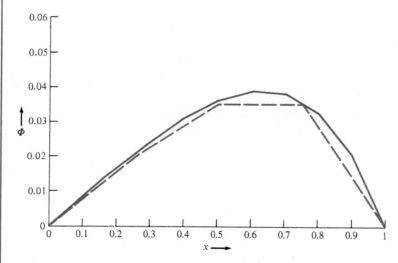

FIGURE 14.10 For Example 14.2: plot of $\Phi(x)$. Continuous curve is for $N = 10$; dashed curve is for $N = 4$.

PRACTICE EXERCISE 14.2

Solve the differential equation $d^2y/dx^2 + y = 0$ with the boundary conditions $y(0) = 0$, $y(1) = 1$ by using the finite difference method. Take $\Delta x = 1/4$.

Answer: Compare your result with the exact solution $y(x) = \dfrac{\sin(x)}{\sin(1)}$.

EXAMPLE 14.3

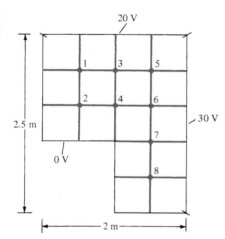

FIGURE 14.11 For Example 14.3.

Determine the potential at the free nodes in the potential system of Figure 14.11 using the finite difference method.

Solution:

This problem will be solved by using the iteration method first, and then the band matrix method.

Method 1 (Iteration Method): We first set the initial values of the potential at the free nodes equal to zero. We apply eq. (14.16) to each free node, using the newest surrounding potentials each time the potential at that node is calculated. For the first iteration:

$$V_1 = 1/4(0 + 20 + 0 + 0) = 5$$
$$V_2 = 1/4(5 + 0 + 0 + 0) = 1.25$$
$$V_3 = 1/4(5 + 20 + 0 + 0) = 6.25$$
$$V_4 = 1/4(1.25 + 6.25 + 0 + 0) = 1.875$$

and so on. To avoid confusion, each time a new value at a free node is calculated, we cross out the old value as shown in Figure 14.12. After V_8 is calculated, we start the second iteration at node 1:

$$V_1 = 1/4(0 + 20 + 1.25 + 6.25) = 6.875$$

$$V_2 = 1/4(6.875 + 0 + 0 + 1.875) = 2.187$$

and so on. If this process is continued, we obtain the uncrossed values shown in Figure 14.12 after five iterations. After 10 iterations (not shown in Figure 14.12), we obtain

$$V_1 = 10.04, \quad V_2 = 4.956, \quad V_3 = 15.22, \quad V_4 = 9.786$$

$$V_5 = 21.05, \quad V_6 = 18.97, \quad V_7 = 15.06, \quad V_8 = 11.26$$

Method 2 (Band Matrix Method): This method reveals the sparse structure of the problem. We apply eq. (14.16) to each free node and keep the known terms (prescribed

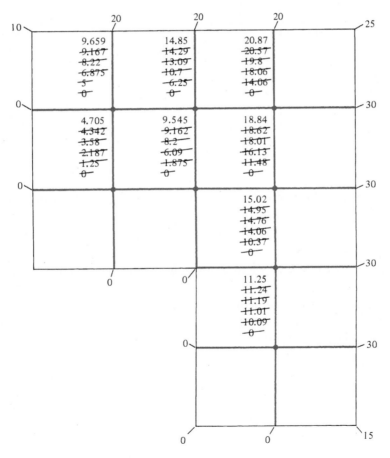

FIGURE 14.12 For Example 14.3; the values not crossed out are the solutions after five iterations.

potentials at the fixed nodes) on the right side; the unknown terms (potentials at free nodes) are on the left side of the resulting system of simultaneous equations, which will be expressed in matrix form as $[A][V] = [B]$.

For node 1,

$$-4V_1 + V_2 + V_3 = -20 - 0$$

For node 2,

$$V_1 + 4V_2 + V_4 = -0 - 0$$

For node 3,

$$V_1 - 4V_3 + V_4 + V_5 = -20$$

For node 4,

$$V_2 + V_3 - 4V_4 + V_6 = -0$$

For node 5,

$$V_3 - 4V_5 + V_6 = -20 - 30$$

For node 6,

$$V_4 + V_5 - 4V_6 + V_7 = -30$$

For node 7,

$$V_6 - 4V_7 + V_8 = -30 - 0$$

For node 8,

$$V_7 - 4V_8 = -0 - 0 - 30$$

Note that since we are using a five-node molecule, we have five terms at each node. The eight equations obtained are put in matrix form as

$$
\begin{bmatrix}
-4 & 1 & 1 & 0 & 0 & 0 & 0 & 0 \\
1 & -4 & 0 & 1 & 0 & 0 & 0 & 0 \\
1 & 0 & -4 & 1 & 1 & 0 & 0 & 0 \\
0 & 1 & 1 & -4 & 0 & 1 & 0 & 0 \\
0 & 0 & 1 & 0 & -4 & 1 & 0 & 0 \\
0 & 0 & 0 & 1 & 1 & -4 & 1 & 0 \\
0 & 0 & 0 & 0 & 0 & 1 & -4 & 1 \\
0 & 0 & 0 & 0 & 0 & 0 & 1 & -4
\end{bmatrix}
\begin{bmatrix}
V_1 \\ V_2 \\ V_3 \\ V_4 \\ V_5 \\ V_6 \\ V_7 \\ V_8
\end{bmatrix}
=
\begin{bmatrix}
-20 \\ 0 \\ -20 \\ 0 \\ -50 \\ -30 \\ -30 \\ -30
\end{bmatrix}
$$

or

$$[A][V] = [B]$$

where $[A]$ is the band, sparse matrix, $[V]$ is the column matrix consisting of the unknown potentials at the free nodes, and $[B]$ is the column matrix formed by the potential at the fixed nodes. The "band" nature of $[A]$ is shown by the dotted loop.

Notice that matrix $[A]$ could have been obtained directly from Figure 14.11 without writing down eq. (14.16) at each free node. To do this, we simply set the diagonal (or self) terms $A_{ii} = -4$ and set $A_{ij} = 1$ if i and j nodes are connected or $A_{ij} = 0$ if i and j nodes are not directly connected. For example, $A_{23} = A_{32} = 0$ because nodes 2 and 3 are not connected, whereas $A_{46} = A_{64} = 1$ because nodes 4 and 6 are connected. Similarly, matrix $[B]$ is obtained directly from Figure 14.11 by setting B_i equal to minus the sum of the potentials at fixed nodes connected to node i. For example, $B_5 = -(20 + 30)$ because node 5 is connected to two fixed nodes with potentials 20 V and 30 V. If node i is not connected to any fixed node, $B_i = 0$.

By using MATLAB to invert matrix $[A]$, we obtain

$$[V] = [A]^{-1}[B]$$

or

$$V_1 = 10.04, \quad V_2 = 4.958, \quad V_3 = 15.22, \quad V_4 = 9.788$$
$$V_5 = 21.05, \quad V_6 = 18.97, \quad V_7 = 15.06, \quad V_8 = 11.26$$

which compares well with the result obtained by means of the iteration method.

PRACTICE EXERCISE 14.3

Use the iteration method to find the finite difference approximation to the potentials at points a and b of the system in Figure 14.13.

Answer: $V_a = 10.135$ V, $V_b = 28.378$ V.

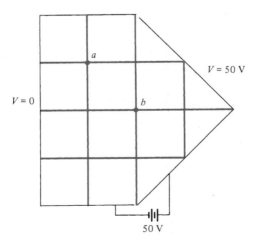

FIGURE 14.13 For Practice Exercise 14.3.

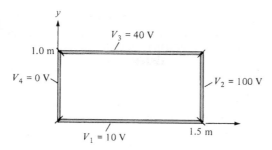

FIGURE 14.14 For Example 14.4.

Obtain the solution of Laplace's equation for an infinitely long trough whose rectangular cross section is shown in Figure 14.14. Let $V_1 = 10$ V, $V_2 = 100$ V, $V_3 = 40$ V, and $V_4 = 0$ V.

Solution:

We shall solve this problem by using the iteration method. In this case, the solution region has a regular boundary. We can easily write a program to determine the potentials at the grid points within the trough. We divide the region into square meshes. If we decide to use a 15×10 grid, the number of grid points along x is $15 + 1 = 16$ and the number of grid points along y is $10 + 1 = 11$. The mesh size $h = 1.5/15 = 0.1$ m. The 15×10 grid is illustrated in Figure 14.15. The grid points are numbered (i, j) starting from the lower left-hand corner of the trough. The computer program in Figure 14.16, for determining the potential at the free nodes, was developed by applying eq. (14.15) and using the iteration method. At points $(x, y) = (0.5, 0.5)$, $(0.8, 0.8)$, $(1.0, 0.5)$, and $(0.8, 0.2)$ corresponding to $(i, j) = (5, 5)$, $(8, 8)$, $(10, 5)$, and $(8, 2)$, respectively, the potentials after 50, 100, and 200 iterations are shown in Table 14.1. The exact values obtained by using the method of separation of variables and a program similar to that of Figure 6.11 are also shown. It should be noted that the degree of accuracy depends on the mesh size h. It is always desirable to make h as small as possible. Also note that the potentials at the fixed nodes are held constant throughout the calculations.

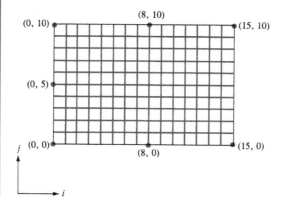

FIGURE 14.15 For Example 14.4; a 15×10 grid.

```
%     USING FINITE DIFFERENCE (ITERATION) METHOD
%     THIS PROGRAM SOLVES THE TWO-DIMENSIONAL BOUNDARY-VALUE
%     PROBLEM (LAPLACE'S EQUATION) SHOWN IN FIG. 14.14.
%     ni = NO. OF ITERATIONS
%     nx = NO. OF X GRID POINTS
%     ny = NO. OF Y GRID POINTS
%     v(i,j) = POTENTIAL AT GRID POINT (i,j) OR (x,y) WITH
%     NODE NUMBERING STARTING FROM THE LOWER LEFT-HAND
%     CORNER OF THE TROUGH

v1 = 10.0;
v2 = 100.0;
v3 = 40.0;
v4 = 0.0;
ni = 200;
nx = 16;
ny = 11;
% SET INITIAL VALUES EQUAL TO ZEROES
v = zeros(nx,ny);
% FIX POTENTIALS ARE FIXED NODES
for i=2:nx-1
    v(i,1) = v1;
    v(i,ny) = v3;
end
for j=2:ny-1
    v(1,j) = v4;
    v(nx,j) = v2;
end
v(1,1) = 0.5*(v1 + v4);
v(nx,1) = 0.5*(v1 + v2);
v(1,ny) = 0.5*(v3 + v4);
v(nx,ny) = 0.5*(v2 + v3);
% NOW FIND v(i,j) USING EQ. (14.15) AFTER ni ITERATIONS
for k=1:ni
    for i=2:nx-1
        for j=2:ny-1
            v(i,j) = 0.25*( v(i+1,j) + v(i-1,j) + v(i,j+1) + v(i,j-1) );
        end
    end
end
diary a:test1.out
[v(6,6), v(9,9), v(11,6), v(9,3)]
[ [1:nx, 1:ny] v(i,j) ]
diary off
```

FIGURE 14.16 Computer program for Example 14.4.

TABLE 14.1 Solution of Example 14.4 (Iteration Method) at Selected Points

Coordinates (x, y)	Number of Iterations			
	50	100	200	Exact Value
(0.5, 0.5)	20.91	22.44	22.49	22.44
(0.8, 0.8)	37.7	38.56	38.59	38.55
(1.0, 0.5)	41.83	43.18	43.2	43.22
(0.8, 0.2)	19.87	20.94	20.97	20.89

PRACTICE EXERCISE 14.4

Consider the trough of Figure 14.17. Use a five-node finite difference scheme to find the potential at the center of the trough using (a) a 4 × 8 grid, and (b) a 12 × 24 grid.

Answer: (a) 31.08 V, (b) 42.86 V.

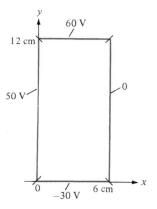

FIGURE 14.17 For Practice Exercise 14.4.

14.4 THE MOMENT METHOD

Like the finite difference method, the moment method,[2] or the method of moments (MOM), has the advantage of being conceptually simple. While the finite difference method is used in solving differential equations, the moment method is commonly used in solving integral equations.

---

[2] The term "moment method" was first used in the Western literature by Harrington. For further exposition on the method, see R. F. Harrington, *Field Computation by Moment Methods*. New York: IEEE Press, 1993.

For example, suppose we want to apply the moment method to solve Poisson's equation in eq. (14.9a). It can be shown that an integral solution to Poisson's equation is

$$V = \int_v \frac{\rho_v\, dv}{4\pi\varepsilon r} \tag{14.26}$$

We recall from Chapter 4 that eq. (14.26) can be derived from Coulomb's law. We also recall that given the charge distribution $\rho_v(x, y, z)$, we can always find the potential $V(x, y, z)$, the electric field $\mathbf{E}(x, y, z)$, and the total charge Q. If, on the other hand, the potential V is known and the charge distribution is unknown, how do we determine ρ_v from eq. (14.26)? In that situation, eq. (14.26) becomes what is called an *integral equation*.

An **integral equation** is one involving the unknown function under the integral sign.

It has the general form of

$$V(x) = \int_a^b K(x, t)\, \rho(t)\, dt \tag{14.27}$$

where the functions $K(x, t)$ and $V(t)$ and the limits a and b are known. The unknown function $\rho(t)$ is to be determined; the function $K(x, t)$ is called the *kernel* of the equation. The moment method is a common numerical technique used in solving integral equations such as in eq. (14.27). The method is probably best explained with an example.

Consider a thin conducting wire of radius a, length $L(L \gg a)$ located in free space as shown in Figure 14.18. Let the wire be maintained at a potential of V_o. Our goal is to determine the charge density ρ_L along the wire by using the moment method. Once we have determined ρ_L, related field quantities can be found. At any point on the wire, eq. (14.26) reduces to an integral equation of the form

$$V_o = \int_0^L \frac{\rho_L\, dl}{4\pi\varepsilon_o r} \tag{14.28}$$

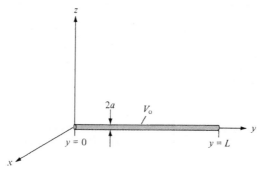

FIGURE 14.18 Thin conducting wire held at a constant potential.

Since eq. (14.28) applies for observation points everywhere on the wire, at a fixed point y_k known as the *match point*.

$$V_o = \frac{1}{4\pi\varepsilon_o} \int_0^L \frac{\rho_L(y)\,dy}{|y_k - y|} \qquad (14.29)$$

We recall from calculus that integration is essentially finding the area under a curve. If Δy is small, the integration of $f(y)$ over $0 < y < L$ is given by

$$\int_0^L f(y)\,dy \simeq f(y_1)\,\Delta y + f(y_2)\,\Delta y + \cdots + f(y_N)\Delta y$$

$$= \sum_{k=1}^N f(y_k)\Delta y \qquad (14.30)$$

where the interval L has been divided into N units, each having length Δy. With the wire divided into N segments of equal length Δ as shown in Figure 14.19, eq. (14.29) becomes

$$4\pi\varepsilon_o V_o \simeq \frac{\rho_1\,\Delta}{|y_k - y_1|} + \frac{\rho_2\,\Delta}{|y_k - y_2|} + \cdots + \frac{\rho_N\,\Delta}{|y_k - y_N|} \qquad (14.31)$$

where $\Delta = L/N = \Delta y$. The assumption in eq. (14.31) is that the unknown charge density ρ_k on the kth segment is constant on that segment. The kth term in eq. (14.31) has $|y_k - y_k|$ in the denominator and causes numerical problems. We shall soon circumvent this problem by modeling the line segment by means of a cylindrical surface charge. Thus in eq. (14.31), we have unknown constants $\rho_1, \rho_2, \ldots, \rho_N$. Since eq. (14.31) must hold at all points on the wire, we obtain N similar equations by choosing N match points at $y_1, y_2, \ldots, y_k, \ldots, y_N$ on the wire. Thus we obtain

$$4\pi\varepsilon_o V_o = \frac{\rho_1\,\Delta}{|y_1 - y_1|} + \frac{\rho_2\,\Delta}{|y_1 - y_2|} + \cdots + \frac{\rho_N\Delta}{|y_1 - y_N|} \qquad (14.32a)$$

$$4\pi\varepsilon_o V_o = \frac{\rho_1\,\Delta}{|y_2 - y_1|} + \frac{\rho_2\,\Delta}{|y_2 - y_2|} + \cdots + \frac{\rho_N\Delta}{|y_2 - y_N|} \qquad (14.32b)$$

$$\vdots$$

$$4\pi\varepsilon_o V_o = \frac{\rho_1\,\Delta}{|y_N - y_1|} + \frac{\rho_2\,\Delta}{|y_N - y_2|} + \cdots + \frac{\rho_N\Delta}{|y_N - y_N|} \qquad (14.32c)$$

The idea of matching the left-hand side of eq. (14.29) with the right-hand side of the equation at the match points is similar to the concept of taking moments in mechanics. Here lies the reason this technique is called the moment method. Notice from Figure 14.19 that the match points $y_1, y_2, \ldots, y_N$ are placed at the center of each segment. Equation (14.32) can be put in matrix form as

$$[B] = [A][\rho] \qquad (14.33)$$

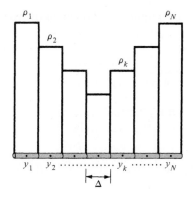

FIGURE 14.19 Division of the wire into N segments.

where

$$[B] = 4\pi\varepsilon_o V_o \begin{bmatrix} 1 \\ 1 \\ \cdot \\ \cdot \\ \cdot \\ 1 \end{bmatrix} \tag{14.34}$$

$$[A] = \begin{bmatrix} A_{11} & A_{12} & \ldots & A_{1N} \\ A_{21} & A_{22} & \ldots & A_{2N} \\ \cdot & & & \cdot \\ \cdot & & & \cdot \\ \cdot & & & \cdot \\ A_{N1} & A_{N2} & \ldots & A_{NN} \end{bmatrix} \tag{14.35a}$$

$$A_{mn} = \frac{\Delta}{|y_m - y_n|}, \quad m \neq n \tag{14.35b}$$

$$[\rho] = \begin{bmatrix} \rho_1 \\ \rho_2 \\ \cdot \\ \cdot \\ \cdot \\ \rho_N \end{bmatrix} \tag{14.36}$$

In eq. (14.33), $[\rho]$ is the matrix whose elements are unknown. We can determine $[\rho]$ from eq. (14.33) by using Cramer's rule, matrix inversion, or the Gaussian elimination technique. With matrix inversion,

$$\boxed{[\rho] = [A]^{-1}[B]} \tag{14.37}$$

where $[A]^{-1}$ is the inverse of matrix $[A]$. In evaluating the diagonal elements (or self terms) of matrix $[A]$ in eq. (14.35), caution must be exercised. Since the wire is conducting, a surface charge density ρ_S is expected over the wire surface. Hence at the center of each segment,

$$V(\text{center}) = \frac{1}{4\pi\varepsilon_o} \int_0^{2\pi} \int_{-\Delta/2}^{\Delta/2} \frac{\rho_S a \, d\phi \, dy}{[a^2 + y^2]^{1/2}}$$

$$= \frac{2\pi a \rho_S}{4\pi\varepsilon_o} \ln \left\{ \frac{\Delta/2 + [(\Delta/2)^2 + a^2]^{1/2}}{-\Delta/2 + [(\Delta/2)^2 + a^2]^{1/2}} \right\}$$

Assuming $\Delta \gg a$,

$$V(\text{center}) = \frac{2\pi a \rho_S}{4\pi\varepsilon_o} 2 \ln \left(\frac{\Delta}{a} \right)$$

$$= \frac{2\rho_L}{4\pi\varepsilon_o} \ln \left(\frac{\Delta}{a} \right).$$

(14.38)

where $\rho_L = 2\pi a \rho_S$. Thus, the self terms $(m = n)$ are

$$A_{nn} = 2 \ln \left(\frac{\Delta}{a} \right)$$

(14.39)

Equation (14.33) now becomes

$$\begin{bmatrix} 2\ln\left(\dfrac{\Delta}{a}\right) & \dfrac{\Delta}{|y_1 - y_2|} & \cdots & \dfrac{\Delta}{|y_1 - y_N|} \\[2ex] \dfrac{\Delta}{|y_2 - y_1|} & 2\ln\left(\dfrac{\Delta}{a}\right) & \cdots & \dfrac{\Delta}{|y_2 - y_N|} \\ \vdots & & & \vdots \\ \dfrac{\Delta}{|y_N - y_1|} & \dfrac{\Delta}{|y_N - y_2|} & \cdots & 2\ln\left(\dfrac{\Delta}{a}\right) \end{bmatrix} \begin{bmatrix} \rho_1 \\ \rho_2 \\ \vdots \\ \vdots \\ \rho_N \end{bmatrix} = 4\pi\varepsilon_o V_o \begin{bmatrix} 1 \\ 1 \\ \cdot \\ \cdot \\ 1 \end{bmatrix}$$

(14.40)

Using eq. (14.37) with eq. (14.40) and letting $V_o = 1\,\text{V}$, $L = 1\,\text{m}$, $a = 1\,\text{mm}$, and $N = 20$ ($\Delta = L/N$), a MATLAB code such as in Figure 14.20 can be developed. The program in Figure 14.20 is self-explanatory. It inverts matrix $[A]$ and plots ρ_L against y. The plot is shown in Figure 14.21. The program also determines the total charge on the wire using

$$Q = \int \rho_L \, dl$$

(14.41)

```
%   THIS PROGRAM DETERMINES THE CHARGE DISTRIBUTION
%   ON A CONDUCTING THIN WIRE, OF RADIUS AA AND
%   LENGTH L, MAINTAINED AT VO VOLT
%   THE WIRE IS LOCATED AT 0 < Y < L
%   ALL DIMENSIONS ARE IN S.I. UNITS


%   MOMENT METHOD IS USED
%   N IS THE NO. OF SEGMENTS INTO WHICH THE WIRE IS DIVIDED
%   RHO IS THE LINE CHARGE DENSITY, RHO = INV(A)*B


%   FIRST, SPECIFY PROBLEM PARAMETERS
ER = 1.0;
EO = 8.8541e-12;
VO = 1.0;
AA = 0.001;
L = 1.0;
N = 20;
DELTA = L/N;
%   SECOND, CALCULATE THE ELEMENTS OF THE COEFFICIENT
%   MATRIX A
I=1:N;
Y=DELTA*(I-0.5);
for i=1:N
    for j=1:N
        if(i ~=j)
            A(i,j)=DELTA/abs(Y(i)-Y(j));
        else
            A(i,j)=2.0*log(DELTA/AA);
        end
    end
end
%   NOW DETERMINE THE MATRIX OF CONSTANT VECTOR B
%   AND FIND Q
B = 4.0*pi*EO*ER*VO*ones(N,1);
C = inv(A);
RHO = C*B;
SUM = 0.0;
for I=1:N
    SUM = SUM + RHO(I);
end
Q=SUM*DELTA;
diary   a:exam145a.out
[EO,Q]
[ [1:N]' Y' RHO ]
diary off
%   FINALLY PLOT RHO AGAINST Y
plot(Y,RHO)
xlabel('y (m)'), ylabel('rho_L (pC/m)')
```

FIGURE 14.20 MATLAB code for calculating the charge distribution on the wire in Figure 14.18.

which can be written in discrete form as

$$Q = \sum_{k=1}^{N} \rho_k \Delta \tag{14.42}$$

With the chosen parameters, the value of the total charge was found to be $Q = 8.5793$ pC. If desired, the electric field at any point can be calculated by using

$$\mathbf{E} = \int \frac{\rho_L \, dl}{4\pi\varepsilon_o R^2} \mathbf{a}_R \tag{14.43}$$

which can be written as

$$\mathbf{E} = \sum_{k=1}^{N} \frac{\rho_k \, \Delta \, \mathbf{R}}{4\pi\varepsilon_o R^3} \tag{14.44}$$

where $R = |\mathbf{R}|$ and

$$\mathbf{R} = \mathbf{r} - \mathbf{r}_k = (x - x_k)\mathbf{a}_x + (y - y_k)\mathbf{a}_y + (z - z_k)\mathbf{a}_z$$

$\mathbf{r} = (x, y, z)$ is the position vector of the observation point, and $\mathbf{r}_k = (x_k, y_k, z_k)$ is that of the source point.

Notice that to obtain the charge distribution in Figure 14.21, we have taken $N = 20$. It should be expected that a smaller value of N would give a less accurate result and a larger value of N would yield a more accurate result. However, if N is too large, we may have the computation problem of inverting the square matrix $[A]$. The capacity of the computing facilities at our disposal can limit the accuracy of the numerical experiment.

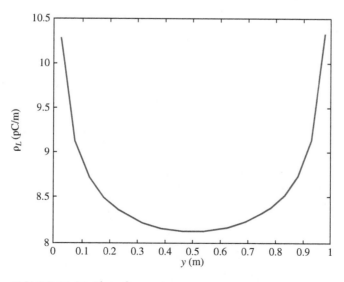

FIGURE 14.21 Plot of ρ_L against y.

EXAMPLE 14.5
Use the moment method to find the capacitance of the parallel-plate capacitor of Figure 14.22. Take $a = 1$ m, $b = 1$ m, $d = 1$ m, and $\varepsilon_r = 1.0$.

Solution:

Let the potential difference between the plates be $V_o = 2$ V so that the top plate P_1 is maintained at $+1$ V while the bottom plate P_2 is at -1 V. We would like to determine the surface charge density ρ_S on the plates so that the total charge on each plate can be found as

$$Q = \int \rho_S \, dS$$

Once Q is known, we can calculate the capacitance as

$$C = \frac{Q}{V_o} = \frac{Q}{2}$$

To determine ρ_S by means of the moment method, we divide P_1 into n subsections: ΔS_1, $\Delta S_2, \ldots, \Delta S_n$ and P_2 into n subsections: $\Delta S_{n+1}, \Delta S_{n+2}, \ldots, \Delta S_{2n}$. The potential V_i at the center of a typical subsection ΔS_i is

$$V_i = \int_S \frac{\rho_S \, dS}{4\pi\varepsilon_o R} \simeq \sum_{j=1}^{2n} \frac{1}{4\pi\varepsilon_o} \int_{\Delta S_i} \frac{\rho_j \, dS}{R_{ij}}$$

$$= \sum_{j=1}^{2n} \rho_j \frac{1}{4\pi\varepsilon_o} \int_{\Delta S_i} \frac{dS}{R_{ij}}$$

It has been assumed that there is uniform charge distribution on each subsection. The last equation can be written as

$$V_i = \sum_{j=1}^{2n} \rho_j A_{ij}$$

where

$$A_{ij} = \frac{1}{4\pi\varepsilon_o} \int_{\Delta S_i} \frac{dS}{R_{ij}}$$

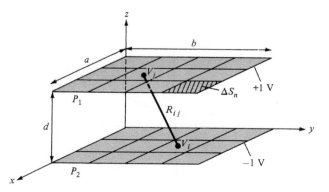

FIGURE 14.22 Parallel-plate capacitor; for Example 14.5.

Thus

$$V_1 = \sum_{j=1}^{2n} \rho_j A_{1j} = 1$$

$$V_2 = \sum_{j=1}^{2n} \rho_j A_{2j} = 1$$

$$\vdots$$

$$V_n = \sum_{j=1}^{2n} \rho_j A_{nj} = 1$$

$$V_{n+1} = \sum_{j=1}^{2n} \rho_j A_{n+1,j} = -1$$

$$\vdots$$

$$V_{2n} = \sum_{j=1}^{2n} \rho_j A_{2n,j} = -1$$

yielding a set of $2n$ simultaneous equations with $2n$ unknown charge densities ρ_j. In matrix form,

$$
\begin{bmatrix}
A_{11} & A_{12} & \cdots & A_{1,2n} \\
A_{21} & A_{22} & \cdots & A_{2,2n} \\
\cdot & & & \cdot \\
\cdot & & & \cdot \\
\cdot & & & \cdot \\
A_{2n,1} & A_{2n,2} & \cdots & A_{2n,2n}
\end{bmatrix}
\begin{bmatrix}
\rho_1 \\
\rho_2 \\
\cdot \\
\cdot \\
\cdot \\
\rho_{2n}
\end{bmatrix}
=
\begin{bmatrix}
1 \\
1 \\
\cdot \\
\cdot \\
-1 \\
-1
\end{bmatrix}
$$

or

$$[A][\rho] = [B]$$

Hence,

$$[\rho] = [A]^{-1}[B]$$

where $[B]$ is the column matrix defining the potentials and $[A]$ is a square matrix containing elements A_{ij}. To determine A_{ij}, consider the two subsections i and j shown in Figure 14.23 where the subsections could be on different plates or on the same plate.

$$A_{ij} = \frac{1}{4\pi\varepsilon_o} \int_{y=y_1}^{y_2} \int_{x=x_1}^{x_2} \frac{dx\,dy}{R_{ij}}$$

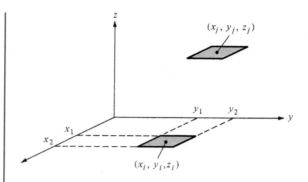

FIGURE 14.23 Subsections i and j; for Example 14.5.

where

$$R_{ij} = [(x_j - x_i)^2 + (y_j - y_i)^2 + (z_j - z_i)^2]^{1/2}$$

For the sake of convenience, if we assume that the subsections are squares,

$$x_2 - x_1 = \Delta\ell = y_2 - y_1$$

it can be shown that

$$A_{ij} = \frac{\Delta S_i}{4\pi\varepsilon_o R_{ij}} = \frac{(\Delta\ell)^2}{4\pi\varepsilon_o R_{ij}} \quad i \neq j$$

and

$$A_{ii} = \frac{\Delta\ell}{\pi\varepsilon_o} \ln(1 + \sqrt{2}) = \frac{\Delta\ell}{\pi\varepsilon_o} (0.8814)$$

With these formulas, the MATLAB code in Figure 14.24 was developed. With $n = 9$, $C = 26.52$ pF, with $n = 16$, $C = 27.27$ pF, and with $n = 25$, $C = 27.74$ pF.

```
%    USING THE METHOD OF MOMENT,
%    THIS PROGRAM DETERMINES THE CAPACITANCE OF A
%    PARALLEL-PLATE CAPACITOR CONSISTING OF TWO CONDUCTING
%    PLATES, EACH OF DIMENSION AA x BB, SEPARATED BY A
%    DISTANCE D, AND MAINTAINED AT 1 VOLT AND -1 VOLT

%    ONE PLATE IS LOCATED ON THE Z=0 PLANE WHILE THE OTHER
%    IS LOCATED ON THE Z=D PLANE

%    ALL DIMENSIONS ARE IN S.I. UNITS
%     N IS THE NUMBER IS SUBSECTIONS INTO WHICH EACH PLATE IS
DIVIDED
```

FIGURE 14.24 MATLAB program for Example 14.5.

```
%    FIRST, SPECIFY THE PARAMETERS

ER = 1.0;
EO = 8.8541e-12;
AA = 1.0;
BB = 1.0;
D = 1.0;
N = 9;
NT = 2*N;
M = sqrt(N);
DX = AA/M;
DY = BB/M;
DL = DX;
%    SECOND, CALCULATE THE ELEMENTS OF THE COEFFICIENT
%    MATRIX A
K = 0;
for K1=1:2
    for K2=1:M
        for K3=1:M
            K = K + 1;
            X(K) = DX*(K2 - 0.5);
            Y(K) = DY*(K3 - 0.5);
        end
    end
end
for K1=1:N
    Z(K1) = 0.0;
    Z(K1+N) = D;
end
for I=1:NT
    for J=1:NT
        if(I==J)
            A(I,J) = DL*0.8814/(pi*EO);
        else
        R = sqrt( (X(I)-X(J))^2 + (Y(I)-Y(J))^2 + (Z(I)-Z(J))^2 );
            A(I,J) = DL^2/(4.*pi*EO*R);
        end
    end
end
%  NOW DETERMINE THE MATRIX OF CONSTANT VECTOR B
for K=1:N
    B(K) = 1.0;
    B(K+N) = -1.0;
end
%   INVERT A AND CALCULATE RHO CONSISTING OF
%   THE UNKNOWN ELEMENTS
%   ALSO CALCULATE THE TOTAL CHARGE Q AND CAPACITANCE C
F = inv(A);
```

FIGURE 14.24 (Continued)

```
RHO = F*B';
SUM = 0.0;
for I=1:N
    SUM = SUM + RHO(I);
end
Q = SUM*(DL^2);
VO = 2.0;
C = abs(Q)/VO;
diary
[C]
[ [1:NT]'   X    Y'   Z'   RHO ]
diary off
```

PRACTICE EXERCISE 14.5

Use the moment method to write a program to determine the capacitance of two identical parallel conducting wires separated at a distance y_o and displaced by x_o as shown in Figure 14.25. If each wire is of length L and radius a, find the capacitance for cases $x_o = 0, 0.2, 0.4, \ldots, 1.0$ m. Take $y_o = 0.5$ m, $L = 1$ m, $a = 1$ mm, $\varepsilon_r = 1$.

Answer: For $N = 10 =$ number of segments per wire, see Table 14.2.

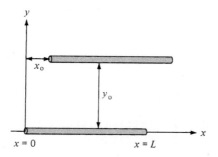

FIGURE 14.25 Parallel conducting wires for Practice Exercise 14.5.

TABLE 14.2 Capacitance for Practice Exercise 14.5

x_o (m)	C (pF)
0.0	4.91
0.2	4.891
0.4	4.853
0.6	4.789
0.8	4.71
1.0	4.643

14.5 THE FINITE ELEMENT METHOD

The finite element method (FEM) has its origin in the field of structural analysis. The method was not applied to EM problems until 1968.[3] Like the finite difference method, the finite element method is useful in solving differential equations. As noticed in Section 14.3, the finite difference method represents the solution region by an array of grid points; its application becomes difficult with problems having irregularly shaped boundaries. Such problems can be handled more easily by using the finite element method.

The finite element analysis of any problem involves basically four steps: (a) discretizing the solution region into a finite number of subregions or *elements,* (b) deriving governing equations for a typical element, (c) assembling all the elements in the solution region, and (d) solving the system of equations obtained.

A. Finite Element Discretization

We divide the solution region into a number of *finite elements* as illustrated in Figure 14.26, where the region is subdivided into four nonoverlapping elements (two triangular and two quadrilateral) and seven nodes. We will assume only triangular elements in this section. We seek an approximation for the potential V_e within an element e and then interrelate the potential distributions in various elements such that the potential is continuous across interelement boundaries. The approximate solution for the whole region is

$$V(x, y) \simeq \sum_{e=1}^{N} V_e(x, y) \tag{14.45}$$

where N is the number of triangular or quadrilateral elements into which the solution region is divided.

The most common form of approximation for V_e within an element is polynomial approximation, namely,

$$V_e(x, y) = a + bx + cy \tag{14.46}$$

for a triangular element and

$$V_e(x, y) = a + bx + cy + dxy \tag{14.47}$$

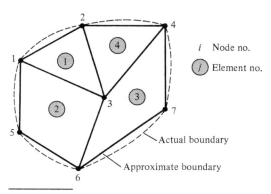

i   Node no.

$\widehat{j}$ Element no.

Actual boundary

Approximate boundary

FIGURE 14.26 A typical finite element subdivision of an irregular domain.

[3] See P. P. Silvester and R. L. Ferrari, *Finite Elements for Electrical Engineers,* 3rd ed. Cambridge, U.K.: Cambridge Univ. Press, 1996.

for a quadrilateral element. The potential V_e in general is nonzero within element e but zero outside e. It is difficult to approximate the boundary of the solution region with quadrilateral elements; such elements are useful for problems whose boundaries are sufficiently regular. In view of this, we prefer to use triangular elements throughout our analysis in this section. Notice that our assumption of linear variation of potential within the triangular element as in eq. (14.46) is the same as assuming that the electric field is uniform within the element; that is,

$$\mathbf{E}_e = -\nabla V_e = -(b\,\mathbf{a}_x + c\,\mathbf{a}_y) \tag{14.48}$$

B. Element-Governing Equations

Consider a typical triangular element, as shown in Figure 14.27. The potential V_{e1}, V_{e2}, and V_{e3} at nodes 1, 2, and 3, respectively, are obtained by using eq. (14.46); that is,

$$\begin{bmatrix} V_{e1} \\ V_{e2} \\ V_{e3} \end{bmatrix} = \begin{bmatrix} 1 & x_1 & y_1 \\ 1 & x_2 & y_2 \\ 1 & x_3 & y_3 \end{bmatrix} \begin{bmatrix} a \\ b \\ c \end{bmatrix} \tag{14.49}$$

The coefficients a, b, and c are determined from eq. (14.49) as

$$\begin{bmatrix} a \\ b \\ c \end{bmatrix} = \begin{bmatrix} 1 & x_1 & y_1 \\ 1 & x_2 & y_2 \\ 1 & x_3 & y_3 \end{bmatrix}^{-1} \begin{bmatrix} V_{e1} \\ V_{e2} \\ V_{e3} \end{bmatrix} \tag{14.50}$$

Substituting this into eq. (14.46) gives

$$V_e = \begin{bmatrix} 1 & x & y \end{bmatrix} \frac{1}{2A} \begin{bmatrix} (x_2 y_3 - x_3 y_2) & (x_3 y_1 - x_1 y_3) & (x_1 y_2 - x_2 y_1) \\ (y_2 - y_3) & (y_3 - y_1) & (y_1 - y_2) \\ (x_3 - x_2) & (x_1 - x_3) & (x_2 - x_1) \end{bmatrix} \begin{bmatrix} V_{e1} \\ V_{e2} \\ V_{e3} \end{bmatrix}$$

or

$$\boxed{V_e = \sum_{i=1}^{3} \alpha_i(x, y)\, V_{ei}} \tag{14.51}$$

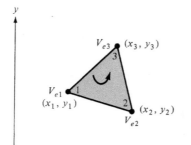

FIGURE 14.27 Typical triangular element; the local node numbering 1-2-3 must be counterclockwise as indicated by the arrow.

where

$$\alpha_1 = \frac{1}{2A}[(x_2y_3 - x_3y_2) + (y_2 - y_3)x + (x_3 - x_2)y] \qquad (14.52a)$$

$$\alpha_2 = \frac{1}{2A}[(x_3y_1 - x_1y_3) + (y_3 - y_1)x + (x_1 - x_3)y] \qquad (14.52b)$$

$$\alpha_3 = \frac{1}{2A}[(x_1y_2 - x_2y_1) + (y_1 - y_2)x + (x_2 - x_1)y] \qquad (14.52c)$$

and A is the area of the element e; that is,

$$2A = \begin{vmatrix} 1 & x_1 & y_1 \\ 1 & x_2 & y_2 \\ 1 & x_3 & y_3 \end{vmatrix}$$

$$= (x_1y_2 - x_2y_1) + (x_3y_1 - x_1y_3) + (x_2y_3 - x_3y_2)$$

or

$$A = 1/2[(x_2 - x_1)(y_3 - y_1) - (x_3 - x_1)(y_2 - y_1)] \qquad (14.53)$$

The value of A is positive if the nodes are numbered counterclockwise (starting from any node) as shown by the arrow in Figure 14.27. Note that eq. (14.51) gives the potential at any point (x, y) within the element, provided the potentials at the vertices are known. This is unlike the situation in finite difference analysis, where the potential is known at the grid points only. Also note that α_i are linear interpolation functions. They are called the *element shape functions,* and they have the following properties:

$$\alpha_i(x_j, y_j) = \begin{cases} 1, & i = j \\ 0, & i \neq j \end{cases} \qquad (14.54a)$$

$$\sum_{i=1}^{3} \alpha_i(x, y) = 1 \qquad (14.54b)$$

The shape functions α_1 and α_2, for example, are illustrated in Figure 14.28.

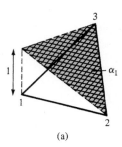
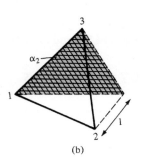

(a)   (b)

FIGURE 14.28 Shape functions α_1 and α_2 for a triangular element.

The energy per unit length associated with the element e is given by eq. (4.96); that is,

$$W_e = \frac{1}{2} \int_S \varepsilon |\mathbf{E}|^2 \, dS = \frac{1}{2} \int_S \varepsilon |\nabla V_e|^2 \, dS \tag{14.55}$$

where a two-dimensional solution region free of charge $(\rho_V = 0)$ is assumed. But from eq. (14.51),

$$\nabla V_e = \sum_{i=1}^{3} V_{ei} \nabla \alpha_i \tag{14.56}$$

Substituting eq. (14.56) into eq. (14.55) gives

$$W_e = \frac{1}{2} \sum_{i=1}^{3} \sum_{j=1}^{3} \varepsilon V_{ei} \left[\int_S \nabla \alpha_i \cdot \nabla \alpha_j \, dS \right] V_{ej} \tag{14.57}$$

If we define the term in brackets as

$$C_{ij}^{(e)} = \int_S \nabla \alpha_i \cdot \nabla \alpha_j \, dS \tag{14.58}$$

we may write eq. (14.57) in matrix form as

$$W_e = \frac{1}{2} \varepsilon [V_e]^T [C^{(e)}] [V_e] \tag{14.59}$$

where the superscript T denotes the transpose of the matrix

$$[V_e] = \begin{bmatrix} V_{e1} \\ V_{e2} \\ V_{e3} \end{bmatrix} \tag{14.60a}$$

and

$$[C^{(e)}] = \begin{bmatrix} C_{11}^{(e)} & C_{12}^{(e)} & C_{13}^{(e)} \\ C_{21}^{(e)} & C_{22}^{(e)} & C_{23}^{(e)} \\ C_{31}^{(e)} & C_{32}^{(e)} & C_{33}^{(e)} \end{bmatrix} \tag{14.60b}$$

The matrix $[C^{(e)}]$ is usually called the *element coefficient matrix*. The matrix element $C_{ij}^{(e)}$ of the coefficient matrix may be regarded as the coupling between nodes i and j; its value is obtained from eqs. (14.52) and (14.58). For example,

$$C_{12}^{(e)} = \int \nabla \alpha_1 \cdot \nabla \alpha_2 \, dS$$

$$= \frac{1}{4A^2} [(y_2 - y_3)(y_3 - y_1) + (x_3 - x_2)(x_1 - x_3)] \int_S dS \tag{14.61a}$$

$$= \frac{1}{4A} [(y_2 - y_3)(y_3 - y_1) + (x_3 - x_2)(x_1 - x_3)]$$

Similarly:

$$C_{11}^{(e)} = \frac{1}{4A} [(y_2 - y_3)^2 + (x_3 - x_2)^2] \tag{14.61b}$$

$$C_{13}^{(e)} = \frac{1}{4A} [(y_2 - y_3)(y_1 - y_2) + (x_3 - x_2)(x_2 - x_1)] \tag{14.61c}$$

$$C_{22}^{(e)} = \frac{1}{4A} [(y_3 - y_1)^2 + (x_1 - x_3)^2] \tag{14.61d}$$

$$C_{23}^{(e)} = \frac{1}{4A} [(y_3 - y_1)(y_1 - y_2) + (x_1 - x_3)(x_2 - x_1)] \tag{14.61e}$$

$$C_{33}^{(e)} = \frac{1}{4A} [(y_1 - y_2)^2 + (x_2 - x_1)^2] \tag{14.61f}$$

Also

$$C_{21}^{(e)} = C_{12}^{(e)}, \quad C_{31}^{(e)} = C_{13}^{(e)}, \quad C_{32}^{(e)} = C_{23}^{(e)} \tag{14.61g}$$

However, our calculations will be easier if we define

$$P_1 = (y_2 - y_3), \quad P_2 = (y_3 - y_1), \quad P_3 = (y_1 - y_2) \tag{14.62a}$$

$$Q_1 = (x_3 - x_2), \quad Q_2 = (x_1 - x_3), \quad Q_3 = (x_2 - x_1)$$

With P_i and Q_i ($i = 1, 2, 3$ are the local node numbers), each term in the element coefficient matrix is found as

$$\boxed{C_{ij}^{(e)} = \frac{1}{4A} [P_i P_j + Q_i Q_j]} \tag{14.62b}$$

where

$$\boxed{A = \frac{1}{2} (P_2 Q_3 - P_3 Q_2)} \tag{14.62c}$$

Note that $P_1 + P_2 + P_3 = 0 = Q_1 + Q_2 + Q_3$ and hence $\sum_{i=1}^{3} C_{ij}^{(e)} = 0 = \sum_{j=1}^{3} C_{ij}^{(e)}$. This may be used in checking our calculations.

C. Assembling All the Elements

Having considered a typical element, the next step is to assemble all such elements in the solution region. The energy associated with the assemblage of all elements in the mesh is

$$W = \sum_{e=1}^{N} W_e = \frac{1}{2} \varepsilon [V]^T [C] [V] \tag{14.63}$$

where

$$[V] = \begin{bmatrix} V_1 \\ V_2 \\ \cdot \\ \cdot \\ \cdot \\ V_n \end{bmatrix}$$

(14.64)

n is the number of nodes, N is the number of elements, and $[C]$ is called the *overall* or *global coefficient matrix,* which is the assemblage of individual element coefficient matrices. The major problem now is obtaining $[C]$ from $[C^{(e)}]$.

The process by which individual element coefficient matrices are assembled to obtain the global coefficient matrix is best illustrated with an example. Consider the finite element mesh consisting of three finite elements as shown in Figure 14.29. Observe the numberings of the nodes. The numbering of nodes as 1, 2, 3, 4, and 5 is called *global* numbering. The numbering *i-j-k* is called *local* numbering, and it corresponds with 1-2-3 of the element in Figure 14.27. For example, for element 3 in Figure 14.29, the global numbering 3-5-4 corresponds to local numbering 1-2-3 of the element in Figure 14.27. Note that the local numbering must be in counterclockwise sequence starting from any node of the element. For element 3, for example, we could choose 4-3-5 or 5-4-3 instead of 3-5-4 to correspond with 1-2-3 of the element in Figure 14.27. Thus the numbering in Figure 14.29 is not unique. However, we obtain the same $[C]$ whichever numbering is used. Assuming the particular numbering in Figure 14.29, the global coefficient matrix is expected to have the form

$$[C] = \begin{bmatrix} C_{11} & C_{12} & C_{13} & C_{14} & C_{15} \\ C_{21} & C_{22} & C_{23} & C_{24} & C_{25} \\ C_{31} & C_{32} & C_{33} & C_{34} & C_{35} \\ C_{41} & C_{42} & C_{43} & C_{44} & C_{45} \\ C_{51} & C_{52} & C_{53} & C_{54} & C_{55} \end{bmatrix}$$

(14.65)

which is a 5×5 matrix, since five nodes $(n = 5)$ are involved. Again, C_{ij} is the coupling between nodes i and j. We obtain C_{ij} by utilizing the fact that the potential distribution

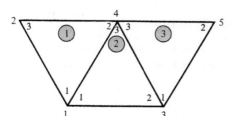

FIGURE 14.29 Assembly of three elements: *i-j-k* corresponds to local numbering 1-2-3 of the element in Figure 14.27.

must be continuous across interelement boundaries. The contribution to the i, j position in $[C]$ comes from all elements containing nodes i and j. To find C_{11}, for example, we observe from Figure 14.29 that global node 1 belongs to elements 1 and 2 and it is local node 1 in both; hence,

$$C_{11} = C_{11}^{(1)} + C_{11}^{(2)} \tag{14.66a}$$

For C_{22}, global node 2 belongs to element 1 only and is the same as local node 3; hence,

$$C_{22} = C_{33}^{(1)} \tag{14.66b}$$

For C_{44}, global node 4 is the same as local nodes 2, 3, and 3 in elements 1, 2, and 3, respectively; hence,

$$C_{44} = C_{22}^{(1)} + C_{33}^{(2)} + C_{33}^{(3)} \tag{14.66c}$$

For C_{14}, global link 14 is the same as the local links 12 and 13 in elements 1 and 2, respectively; hence,

$$C_{14} = C_{12}^{(1)} + C_{13}^{(2)} \tag{14.66d}$$

Since there is no coupling (or direct link) between nodes 2 and 3,

$$C_{23} = C_{32} = 0 \tag{14.66e}$$

Continuing in this manner, we obtain all the terms in the global coefficient matrix by inspection of Figure 14.29 as

$$[C] = \begin{bmatrix} C_{11}^{(1)} + C_{11}^{(2)} & C_{13}^{(1)} & C_{12}^{(2)} & C_{12}^{(1)} + C_{13}^{(2)} & 0 \\ C_{31}^{(1)} & C_{33}^{(1)} & 0 & C_{32}^{(1)} & 0 \\ C_{21}^{(2)} & 0 & C_{22}^{(2)} + C_{11}^{(3)} & C_{23}^{(2)} + C_{13}^{(3)} & C_{12}^{(3)} \\ C_{21}^{(1)} + C_{31}^{(2)} & C_{23}^{(1)} & C_{32}^{(2)} + C_{31}^{(3)} & C_{22}^{(1)} + C_{33}^{(2)} + C_{33}^{(3)} & C_{32}^{(3)} \\ 0 & 0 & C_{21}^{(3)} & C_{23}^{(3)} & C_{22}^{(3)} \end{bmatrix} \tag{14.67}$$

Note that element coefficient matrices overlap at nodes shared by elements and that there are 27 terms (nine for each of the three elements) in the global coefficient matrix $[C]$. Also note the following properties of the matrix $[C]$:

1. It is symmetric $(C_{ij} = C_{ji})$ just like the element coefficient matrix.
2. Since $C_{ij} = 0$ if no coupling exists between nodes i and j, it is evident that for a large number of elements $[C]$ becomes sparse and banded.
3. It is singular. Although this is less obvious, it can be shown by using the element coefficient matrix of eq. (14.60b).

D. Solving the Resulting Equations

From variational calculus, it is known that Laplace's (or Poisson's) equation is satisfied when the total energy in the solution region is minimum. Thus we require that the partial derivatives of W with respect to each nodal value of the potential be zero; that is,

$$\frac{\partial W}{\partial V_1} = \frac{\partial W}{\partial V_2} = \cdots = \frac{\partial W}{\partial V_n} = 0$$

or

$$\frac{\partial W}{\partial V_k} = 0, \quad k = 1, 2, \ldots, n \tag{14.68}$$

For example, to get $\partial W/\partial V_1 = 0$ for the finite element mesh of Figure 14.29, we substitute eq. (14.65) into eq. (14.63) and take the partial derivative of W with respect to V_1. We obtain

$$0 = \frac{\partial W}{\partial V_1} = 2V_1C_{11} + V_2C_{12} + V_3C_{13} + V_4C_{14} + V_5C_{15}$$

$$+ V_2C_{21} + V_3C_{31} + V_4C_{41} + V_5C_{51}$$

or

$$0 = V_1C_{11} + V_2C_{12} + V_3C_{13} + V_4C_{14} + V_5C_{15} \tag{14.69}$$

In general, $\partial W/\partial V_k = 0$ leads to

$$0 = \sum_{i=1}^{n} V_i C_{ik} \tag{14.70}$$

where n is the number of nodes in the mesh. By writing eq. (14.70) for all nodes $k = 1, 2, \ldots, n$, we obtain a set of simultaneous equations from which the solution of $[V]^T = [V_1, V_2, \ldots, V_n]$ can be found. This can be done in two ways similar to those used in solving finite difference equations obtained from Laplace's (or Poisson's) equation.

Iteration Method

The iterative approach is similar to that used in the finite difference method. Let us assume that node 1 in Figure 14.29, for example, is a free node. The potential at node 1 can be obtained from eq. (14.69) as

$$V_1 = -\frac{1}{C_{11}} \sum_{i=2}^{5} V_i C_{1i} \tag{14.71}$$

In general, the potential at a free node k is obtained from eq. (14.70) as

$$V_k = -\frac{1}{C_{kk}} \sum_{i=1, i \neq k}^{n} V_i C_{ik} \tag{14.72}$$

This is applied iteratively to all the free nodes in the mesh with n nodes. Since $C_{ki} = 0$ if node k is not directly connected to node i, only nodes that are directly linked to node k contribute to V_k in eq. (14.72).

Thus if the potentials at nodes connected to node k are known, we can determine V_k by using eq. (14.72). The iteration process begins by setting the potentials at the free nodes equal to zero or to the average potential.

$$V_{ave} = \frac{1}{2}(V_{min} + V_{max}) \tag{14.73}$$

where V_{min} and V_{max} are the minimum and maximum values of the prescribed potentials at the fixed nodes. With those initial values, the potentials at the free nodes are calculated by using eq. (14.72). At the end of the first iteration, when the new values have been calculated for all the free nodes, these values become the old values for the second iteration. The procedure is repeated until the change between subsequent iterations is negligible.

Band Matrix Method

If all free nodes are numbered first and the fixed nodes last, eq. (14.63) can be written such that

$$W = \frac{1}{2}\varepsilon\, [V_f\ V_p] \begin{bmatrix} C_{ff} & C_{fp} \\ C_{pf} & C_{pp} \end{bmatrix} \begin{bmatrix} V_f \\ V_p \end{bmatrix} \tag{14.74}$$

where subscripts f and p, respectively, refer to nodes with free and fixed (or prescribed) potentials. Since V_p is constant (it consists of known, fixed values), we differentiate only with respect to V_f, so that applying eq. (14.68) to eq. (14.74) yields

$$C_{ff}V_f + C_{fp}V_p = 0$$

or

$$[C_{ff}]\,[V_f] = -[C_{fp}]\,[V_p] \tag{14.75}$$

This equation can be written as

$$[A]\,[V] = [B] \tag{14.76a}$$

or

$$\boxed{[V] = [A]^{-1}\,[B]} \tag{14.76b}$$

where $[V] = [V_f]$, $[A] = [C_{ff}]$, and $[B] = -[C_{fp}]\,[V_p]$. Since $[A]$ is, in general, non-singular, the potential at the free nodes can be found by using eq. (14.75). We can solve for $[V]$ in eq. (14.76a) by using the Gaussian elimination technique. We can also use matrix inversion to solve for $[V]$ in eq. (14.76b) if the size of the matrix to be inverted is not large.

Notice that as from eq. (14.55) onward, our solution has been restricted to a two-dimensional problem involving Laplace's equation, $\nabla^2 V = 0$. The basic concepts developed in this section can be extended to finite element analysis of problems involving Poisson's equation ($\nabla^2 V = -\rho_v/\varepsilon$, $\nabla^2 \mathbf{A} = -\mu \mathbf{J}$) or the wave equation ($\nabla^2 \phi - \gamma^2 \phi = 0$). A major problem associated with finite element analysis is the relatively large amount of computer memory required for storing the matrix elements, as well as the associated computational time. However, several algorithms have been developed to alleviate the problem to some degree.

The finite element method has a number of advantages over the finite difference method and the method of moments. First, the FEM can easily handle the complex solution region. Second, the generality of the FEM makes it possible to construct a general-purpose program for solving a wide range of problems. A single program can be used to solve different problems (described by the same partial differential equations) with different solution regions and different boundary conditions; only the input data to the problem need be changed. However, the FEM has its own drawbacks. It is harder to understand and program than the other methods (FDM and MOM). It also requires preparing input data, a process that could be tedious.

EXAMPLE 14.6

Consider the two-element mesh shown in Figure 14.30(a). Using the finite element method, determine the potentials within the mesh.

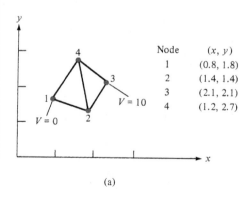

Node	(x, y)
1	(0.8, 1.8)
2	(1.4, 1.4)
3	(2.1, 2.1)
4	(1.2, 2.7)

(a)

FIGURE 14.30 For Example 14.6: (**a**) two-element mesh, (**b**) local and global numbering of the elements.

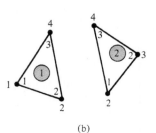

(b)

Solution:

The element coefficient matrices can be calculated by using one of the relations of eq. (14.62). For element 1, consisting of nodes 1-2-4 corresponding to the local numbering 1-2-3 as in Figure 14.30(b),

$$P_1 = -1.3, \quad P_2 = 0.9, \quad P_3 = 0.4$$

$$Q_1 = -0.2, \quad Q_2 = -0.4, \quad Q_3 = 0.6$$

$$A = 1/2\,(0.54 + 0.16) = 0.35$$

Substituting all these into eq. (14.62b) gives

$$[C^{(1)}] = \begin{bmatrix} 1.236 & -0.7786 & -0.4571 \\ -0.7786 & 0.6929 & 0.0857 \\ -0.4571 & 0.0857 & 0.3714 \end{bmatrix} \tag{14.6.1}$$

Similarly, for element 2, consisting of nodes 2-3-4 corresponding to local numbering 1-2-3, as in Figure 14.30(b),

$$P_1 = -0.6, \quad P_2 = 1.3, \quad P_3 = -0.7$$

$$Q_1 = -0.9, \quad Q_2 = 0.2, \quad Q_3 = 0.7$$

$$A = 1/2\,(0.91 + 0.14) = 0.525$$

Hence,

$$[C^{(2)}] = \begin{bmatrix} 0.5571 & -0.4571 & -0.1 \\ -0.4571 & 0.8238 & -0.3667 \\ -0.1 & -0.3667 & 0.4667 \end{bmatrix} \tag{14.6.2}$$

Applying eq. (14.75) gives

$$\begin{bmatrix} C_{22} & C_{24} \\ C_{42} & C_{44} \end{bmatrix} \begin{bmatrix} V_2 \\ V_4 \end{bmatrix} = - \begin{bmatrix} C_{21} & C_{23} \\ C_{41} & C_{43} \end{bmatrix} \begin{bmatrix} V_1 \\ V_3 \end{bmatrix} \tag{14.6.3}$$

This can be written in a more convenient form as

$$\begin{bmatrix} 1 & 0 & 0 & 0 \\ 0 & C_{22} & 0 & C_{24} \\ 0 & 0 & 1 & 0 \\ 0 & C_{42} & 0 & C_{44} \end{bmatrix} \begin{bmatrix} V_1 \\ V_2 \\ V_3 \\ V_4 \end{bmatrix} = \begin{bmatrix} 1 & 0 \\ -C_{21} & -C_{23} \\ 0 & 1 \\ -C_{41} & -C_{43} \end{bmatrix} \begin{bmatrix} V_1 \\ V_3 \end{bmatrix} \tag{14.6.4a}$$

or

$$[C]\,[V] = [B] \tag{14.6.4b}$$

The terms of the global coefficient matrix are obtained as follows:

$$C_{22} = C_{22}^{(1)} + C_{11}^{(2)} = 0.6929 + 0.5571 = 1.25$$

$$C_{42} = C_{24} = C_{23}^{(1)} + C_{13}^{(2)} = 0.0857 - 0.1 = -0.0143$$

$$C_{44} = C_{33}^{(1)} + C_{33}^{(2)} = 0.3714 + 0.4667 = 0.8381$$

$$C_{21} = C_{21}^{(1)} = -0.7786$$

$$C_{23} = C_{12}^{(2)} = -0.4571$$

$$C_{41} = C_{31}^{(1)} = -0.4571$$

$$C_{43} = C_{32}^{(2)} = -0.3667$$

Note that we follow local numbering for the element coefficient matrix and global numbering for the global coefficient matrix. Thus the square matrix $[C]$ is obtained as

$$[C] = \begin{bmatrix} 1 & 0 & 0 & 0 \\ 0 & 1.25 & 0 & -0.0143 \\ 0 & 0 & 1 & 0 \\ 0 & -0.0143 & 0 & 0.8381 \end{bmatrix} \qquad (14.6.5)$$

and the matrix $[B]$ on the right-hand side of eq. (14.6.4a) is obtained as

$$[B] = \begin{bmatrix} 0 \\ 4.571 \\ 10.0 \\ 3.667 \end{bmatrix} \qquad (14.6.6)$$

By inverting matrix $[C]$ in eq. (14.6.5), we obtain

$$[V] = [C]^{-1}[B] = \begin{bmatrix} 0 \\ 3.708 \\ 10.0 \\ 4.438 \end{bmatrix}$$

Thus $V_1 = 0$, $V_2 = 3.708$, $V_3 = 10$, and $V_4 = 4.438$. Once the values of the potentials at the nodes are known, the potential at any point within the mesh can be determined by using eq. (14.51).

PRACTICE EXERCISE 14.6

Calculate the global coefficient matrix for the two-element mesh shown in Figure 14.31 when (a) node 1 is linked with node 3 and the local numbering $(i\text{-}j\text{-}k)$ is as indicated in Figure 14.31(a), (b) node 2 is linked with node 4 with local numbering as in Figure 14.31(b).

Answer: (a)
$$\begin{bmatrix} 0.9964 & 0.05 & -0.2464 & -0.8 \\ 0.05 & 0.7 & -0.75 & 0.0 \\ -0.2464 & -0.75 & 1.5964 & -0.75 \\ -0.8 & 0.0 & -0.75 & 1.4 \end{bmatrix}.$$

(b)
$$\begin{bmatrix} 1.333 & -0.7777 & 0.0 & -1.056 \\ -0.0777 & 0.8192 & -0.98 & 0.2386 \\ 0.0 & -0.98 & 2.04 & -1.06 \\ -1.056 & 0.2386 & -1.06 & 1.877 \end{bmatrix}.$$

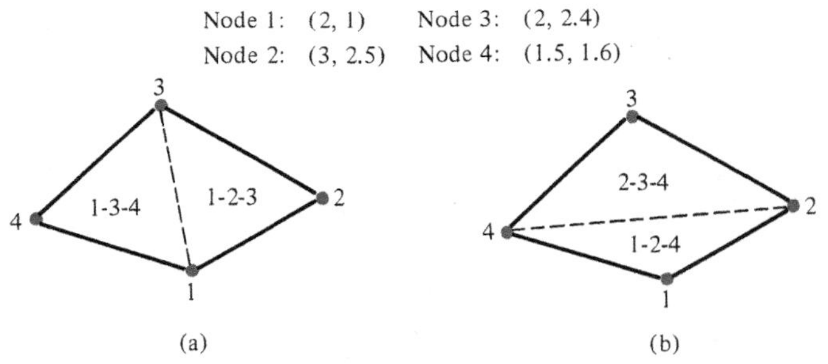

Node 1: (2, 1)   Node 3: (2, 2.4)
Node 2: (3, 2.5)   Node 4: (1.5, 1.6)

FIGURE 14.31 For Practice Exercise 14.6.

EXAMPLE 14.7

Write a program to solve Laplace's equation by means of the finite element method. Apply the program to the two-dimensional problem shown in Figure 14.32(a).

Solution:

The solution region is divided into 25 three-node triangular elements with the total number of nodes being 21, as shown in Figure 14.32(b). This step is necessary to have input data defining the geometry of the problem. Based on our discussions thus far, a general MATLAB program for solving problems involving Laplace's equation by using three-node triangular elements was developed as in Figure 14.33. The development of the program basically involves four steps indicated in the program and explained as follows.

Step 1: This involves inputting the necessary data defining the problem. This is the only step that depends on the geometry of the problem at hand. Through a data file, we input the number of elements, the number of nodes, the number of fixed nodes, the prescribed values of the potentials at the free nodes, the x- and y-coordinates of all nodes, and a list identifying the

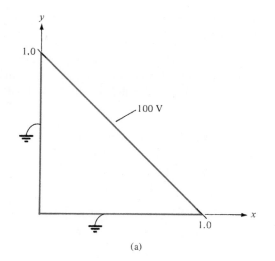

FIGURE 14.32 For Example 14.7: (**a**) two-dimensional electrostatic problem, (**b**) solution region divided into 25 triangular elements.

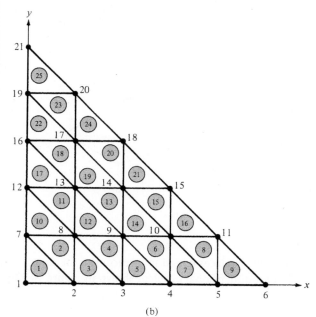

(b)

```
%    FINITE ELEMENT SOLUTION OF LAPLACE'S EQUATION FOR
%    TWO-DIMENSIONAL PROBLEMS
%    TRIANGULAR ELEMENTS ARE USED
%    ND = NO. OF NODES
%    NE = NO. OF ELEMENTS
%    NP = NO. OF FIXED NODES (WHERE POTENTIAL IS PRESCRIBED)
%    NDP(I) = NODE NO. OF PRESCRIBED POTENTIAL, I=1,2,...,NP
%    VAL(I) = VALUE OF PRESCRIBED POTENTIAL AT NODE NDP(I)
%    NL(I,J) = LIST OF NODES FOR EACH ELEMENT I, WHERE
```

FIGURE 14.33 Computer program for Example 14.7.

```
%                  J=1,2,3 REFERS TO THE LOCAL NODE NUMBER
%   CE(I,J) = ELEMENT COEFFICIENT MATRIX
%   C(I,J) = GLOBAL COEFFICIENT MATRIX
%   B(I) = RIGHT-HAND SIDE MATRIX IN THE SYSTEM OF
%   SIMULTANEOUS EQUATIONS; SEE EQ. (14.6.4)
%   X(I), Y(I) = GLOBAL COORDINATES OF NODE I
%   XL(J), YL(J) = LOCAL COORDINATES OF NODE J=1,2,3
%   V(I) = POTENTIAL AT NODE I
%   MATRICES P(I) AND Q (I) ARE DEFINED IN EQ. (14.62a)

%   ******************************************************
%   FIRST STEP - INPUT DATA DEFINING GEOMETRY AND
%                    BOUNDARY CONDITIONS
%   ******************************************************


clear
input('Name of input data file = ')


%   ******************************************************
%   SECOND STEP - EVALUATE COEFFICIENT MATRIX FOR EACH
%                    ELEMENT AND ASSEMBLE GLOBALLY
%   ******************************************************
B = zeros(ND,1);
C = zeros(ND,ND);
for I=1:NE
%   FIND LOCAL COORDINATES XL(J), YL(J) FOR ELEMENT I
    K = NL(I,[1:3]);
    XL = X(K);
    YL = Y(K);
P=zeros(3,1);
Q=zeros(3,1);
    P(1) = YL(2) - YL(3);
    P(2) = YL(3) - YL(1);
    P(3) = YL(1) - YL(2);
    Q(1) = XL(3) - XL(2);
    Q(2) = XL(1) - XL(3);
    Q(3) = XL(2) - XL(1);
    AREA = 0.5*abs( P(2)*Q(3) - Q(2)*P(3) );
%   DETERMINE COEFFICIENT MATRIX FOR ELEMENT I
    CE=(P*P'+Q*Q')/(4.0*AREA);
%   ASSEMBLE GLOBALLY - FIND C(I,J) AND B(I)
    for J=1:3
        IR = NL(I,J);
        IFLAG1=0;
%   CHECK IF ROW CORRESPONDS TO A FIXED NODE
    for K = 1:NP
        if (IR == NDP(K))
```

FIGURE 14.33 (Continued)

```
                    C(IR,IR) = 1.0;
                    B(IR) = VAL(K);
                    IFLAG1=1;
                end
            end % end for K = 1:NP
        if(IFLAG1 == 0)
        for L = 1:3
            IC = NL(I,L);
            IFLAG2=0;
%     CHECK IF COLUMN CORRESPONDS TO A FIXED NODE
            for K=1:NP
                if ( IC == NDP(K) ),
                    B(IR) = B(IR) - CE(J,L)*VAL(K);
                    IFLAG2=1;
                end
            end % end for K=1:NP
        if(IFLAG2 == 0)
            C(IR,IC) = C(IR,IC) + CE(J,L);
            end
        end   % end for L=1:3
    end    %end if(iflag1 == 0)
    end  % end for J=1:3
end % end for I=1:NE
% ****************************************************
%     THIRD STEP - SOLVE THE SYSTEM OF EQUATIONS
% ****************************************************

V = inv(C)*B;
V=V';
% ****************************************************
%     FOURTH STEP - OUTPUT THE RESULTS
% ****************************************************
diary exam147.out
[ND, NE, NP]
[ [1:ND]' X' Y' V']
diary off
```

FIGURE 14.33 (Continued)

nodes belonging to each element in the order of the local numbering 1-2-3. For the problem in Figure 14.32, the three sets of data for coordinates, the element–node relationship, and the prescribed potentials at fixed nodes are shown in Tables 14.3, 14.4, and 14.5, respectively.

Step 2: This step entails finding the element coefficient matrix $[C^{(e)}]$ for each element and the global coefficient matrix $[C]$. The procedure explained in Example 14.6 is applied. Equation (14.6.4) can be written in general form as

$$\begin{bmatrix} 1 & 0 \\ 0 & C_{ff} \end{bmatrix} \begin{bmatrix} V_p \\ V_f \end{bmatrix} = \begin{bmatrix} 1 \\ -C_{fp} \end{bmatrix} [V_p]$$

TABLE 14.3 Nodal Coordinates of the Finite Element Mesh of Figure 14.33

Node	x	y	Node	x	y
1	0.0	0.0	12	0.0	0.4
2	0.2	0.0	13	0.2	0.4
3	0.4	0.0	14	0.4	0.4
4	0.6	0.0	15	0.6	0.4
5	0.8	0.0	16	0.0	0.6
6	1.0	0.0	17	0.2	0.6
7	0.0	0.2	18	0.4	0.6
8	0.2	0.2	19	0.0	0.8
9	0.4	0.2	20	0.2	0.8
10	0.6	0.2	21	0.0	1.0
11	0.8	0.2			

TABLE 14.4 Element–Node Identification

Element No.	Local Node No. 1	2	3	Element No.	Local Node No. 1	2	3
1	1	2	7	14	9	10	14
2	2	8	7	15	10	15	14
3	2	3	8	16	10	11	15
4	3	9	8	17	12	13	16
5	3	4	9	18	13	17	16
6	4	10	9	19	13	14	17
7	4	5	10	20	14	18	17
8	5	11	10	21	14	15	18
9	5	6	11	22	16	17	19
10	7	8	12	23	17	20	19
11	8	13	12	24	17	18	20
12	8	9	13	25	19	20	21
13	9	14	13				

TABLE 14.5 Prescribed Potentials at Fixed Nodes

Node No.	Prescribed Potential	Node No.	Prescribed Potential
1	0.0	18	100.0
2	0.0	20	100.0
3	0.0	21	50.0
4	0.0	19	0.0
5	0.0	16	0.0
6	50.0	12	0.0
11	100.0	7	0.0
15	100.0		

or

$$[C][V] = [B]$$

Both the "global" matrix $[C]$ and matrix $[B]$ are calculated at this stage.

Step 3: The global matrix obtained in step 2 is inverted. The values of the potentials at all nodes are obtained by matrix multiplication as in eq. (14.76b). Instead of inverting the global matrix, it is also possible to solve for the potentials at the nodes by using the Gaussian elimination technique.

Step 4: Finally, the result of the computation is provided. The input and output data are presented in Tables 14.6 and 14.7, respectively.

TABLE 14.6 Input Data for the Finite Element Program in Figure 14.33

```
NE = 25;
ND = 21;
NP = 15;
NL = [  1  2  7
        2  8  7
        2  3  8
           3  9  8
           3  4  9
           4 10  9
           4  5 10
           5 11 10
           5  6 11
           7  8 12
        8 13 12
        8  9 13
        9 14 13
        9 10 14
       10 15 14
       10 11 15
       12 13 16
       13 17 16
       13 14 17
       14 18 17
       14 15 18
       16 17 19
       17 20 19
       17 18 20
       19 20 21];
   X = [0.0   0.2   0.4   0.6   0.8   1.0   0.0 ...
        0.2   0.4   0.6   0.8   0.0   0.2   0.4 ...
        0.6   0.0   0.2   0.4   0.0   0.2   0.0];
   Y = [0.0   0.0   0.0   0.0   0.0   0.0   0.2 ...
        0.2   0.2   0.2   0.2   0.4   0.4   0.4 ...
        0.4   0.6   0.6   0.6   0.8   0.8   1.0];
 NDP = [ 1  2  3  4  5  6 11 15 18 20 21 19 16 12  7];
 VAL = [ 0.0   0.0   0.0   0.0 0.0 ...
             50.0   100.0   100.0   100.0   100.0
             50.0   0.0     0.0     0.0     0.0];
```

TABLE 14.7 Output Data of the Program in Figure 14.33

Node	X	Y	Potential
1	0.00	0.00	0.000
2	0.20	0.00	0.000
3	0.40	0.00	0.000
4	0.60	0.00	0.000
5	0.80	0.00	0.000
6	1.00	0.00	50.000
7	0.00	0.20	0.000
8	0.20	0.20	18.182
9	0.40	0.20	36.364
10	0.60	0.20	59.091
11	0.80	0.20	100.000
12	0.00	0.40	0.000
13	0.20	0.40	36.364
14	0.40	0.40	68.182
15	0.60	0.40	100.000
16	0.00	0.60	0.000
17	0.20	0.60	59.091
18	0.40	0.60	100.000
19	0.00	0.80	0.000
20	0.20	0.80	100.000
21	0.00	1.00	50.000

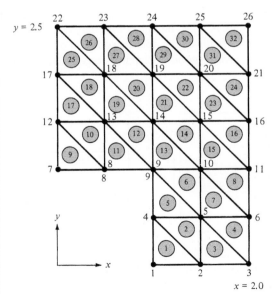

FIGURE 14.34 For Practice Exercise 14.7.

> **PRACTICE EXERCISE 14.7**
>
> Rework Example 14.3 using the finite element method. Divide the solution region into triangular elements as shown in Figure 14.34. Compare the solution with that obtained in Example 14.3 using the finite difference method.
>
> **Answer:** See Example 14.3.

†14.6 APPLICATION NOTE—MICROSTRIP LINES

The numerical methods covered in this chapter have been applied successfully to solve many EM-related problems. Besides the simple examples considered earlier in the chapter, the methods have been applied to diverse problems including transmission line problems, EM penetration and scattering problems, EM pulse (EMP) problems, EM exploration of minerals, and EM energy deposition in human bodies. It is practically impossible to cover all these applications within the limited scope of this text. In this section, we use the finite difference method to consider the relatively easier problem of transmission lines.

The finite difference techniques are suited for computing the characteristic imped-ance, phase velocity, and attenuation of several transmission lines: polygonal lines, shielded strip lines, coupled strip lines, microstrip lines, coaxial lines, and rectangular lines. The knowledge of the basic parameters of these lines is of paramount importance in the design of microwave circuits.

For concreteness, consider the microstrip line shown in Figure 14.35(a). The geometry in Figure 14.35(a) is deliberately selected to illustrate how one uses the finite difference technique to account for discrete inhomogeneities (i.e., homogeneous media separated by interfaces) and lines of symmetry. The techniques presented are equally applicable to other lines. Because the mode is TEM, having components of neither **E** nor **H** fields in the direction of propagation, the fields obey Laplace's equation over the line cross section. The TEM mode assumption provides good approximations if the line dimensions are much

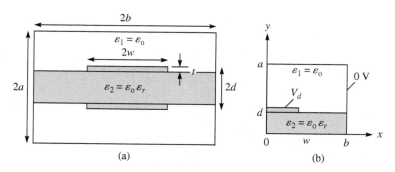

FIGURE 14.35 (**a**) Shielded double strip line with partial dielectric support. (**b**) Problem in (**a**) simplified by taking advantage of symmetry.

smaller than half a wavelength, which means that the operating frequency is far below cut-off frequency for all higher-order modes. Also, owing to biaxial symmetry about the two axes, only one-quarter of the cross section need be considered, as shown in Figure 14.35(b).

The finite difference approximation of Laplace's equation, $\nabla^2 V = 0$, was derived in eq. (14.15), namely

$$V(i,j) = \frac{1}{4}[V(i + 1,j) + V(i - 1,j) + V(i,j + 1) + V(i,j - 1)] \quad (14.77)$$

For the sake of conciseness, let us denote

$$\begin{aligned}
V_o &= V(i, j) \\
V_1 &= V(i, j + 1) \\
V_2 &= V(i - 1, j) \quad\quad (14.78) \\
V_3 &= V(i, j - 1) \\
V_4 &= V(i + 1, j)
\end{aligned}$$

so that eq. (14.77) becomes

$$\boxed{V_o = \frac{1}{4}[V_1 + V_2 + V_3 + V_4]} \quad (14.79)$$

with the computation molecule as shown in Figure 14.36. Equation (14.79) is the general formula to be applied to all free nodes in the free space and dielectric region of Figure 14.35(b). The only limitation on eq. (14.79) is that region is discretely homogeneous.

On the dielectric boundary, the boundary condition

$$D_{1n} = D_{2n} \quad (14.80)$$

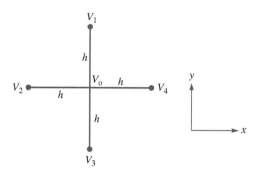

FIGURE 14.36 Computation molecule for Laplace's equation.

must be imposed. We recall that this condition is based on Gauss's law for the electric field; that is,

$$\oint_L \mathbf{D}_n \cdot d\mathbf{l} = \oint_L \mathbf{E}_n \cdot d\mathbf{l} = Q_{enc/m} = 0 \tag{14.81}$$

since no free charge is deliberately placed on the dielectric boundary. Substituting $\mathbf{E} = -\nabla V$ in eq. (14.81) gives

$$0 = \oint_L \varepsilon \nabla V \cdot d\mathbf{l} = \oint_L \varepsilon \frac{\partial V}{\partial n} dl \tag{14.82}$$

where $\partial V/\partial n$ denotes the derivative of V normal to the contour L. Applying eq. (14.82) to the interface in Figure 14.37 yields

$$0 = \varepsilon_1 \frac{(V_1 - V_o)}{h} h + \varepsilon_1 \frac{(V_2 - V_o)h}{h} \frac{h}{2} + \varepsilon_2 \frac{(V_2 - V_o)h}{h} \frac{h}{2}$$

$$+ \varepsilon_2 \frac{(V_3 - V_o)}{h} h + \varepsilon_2 \frac{(V_4 - V_o)h}{h} \frac{h}{2} + \varepsilon_1 \frac{(V_4 - V_o)h}{h} \frac{h}{2}$$

Rearranging the terms, we have

$$2(\varepsilon_1 + \varepsilon_2)V_o = \varepsilon_1 V_1 + \varepsilon_2 V_3 + \frac{(\varepsilon_1 + \varepsilon_2)}{2}(V_2 + V_4)$$

or

$$\boxed{V_o = \frac{\varepsilon_1}{2(\varepsilon_1 + \varepsilon_2)}V_1 + \frac{\varepsilon_2}{2(\varepsilon_1 + \varepsilon_2)}V_3 + \frac{1}{4}V_2 + \frac{1}{4}V_4} \tag{14.83}$$

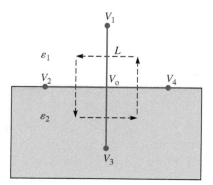

FIGURE 14.37 Interface between media of dielectric permittivities ε_1 and ε_2.

This is the finite difference equivalent of the boundary condition in eq. (14.80). Notice that the discrete inhomogeneity does not affect points 2 and 4 on the boundary but affects points 1 and 3 in proportion to their corresponding permittivities. Also note that when $\varepsilon_2 = \varepsilon_1$, eq. (14.83) reduces to eq. (14.79).

On the line of symmetry, we impose the condition

$$\frac{\partial V}{\partial n} = 0 \tag{14.84}$$

This implies that on the line of symmetry along the y-axis, ($x = 0$ or $i = 0$), $\dfrac{\partial V}{\partial x} = \dfrac{(V_4 - V_2)}{2h} = 0$ or $V_2 = V_4$ so that eq. (14.79) becomes

$$\boxed{V_o = \frac{1}{4}[V_1 + V_3 + 2V_4]} \tag{14.85a}$$

or

$$V(0, j) = \frac{1}{4}[V(0, j+1) + V(0, j-1) + 2V(1, j)] \tag{14.85b}$$

On the line of symmetry along the x-axis ($y = 0$ or $j = 0$), $\dfrac{\partial V}{\partial y} = \dfrac{(V_1 - V_3)}{2h} = 0$, or $V_3 = V_1$, so that

$$\boxed{V_o = \frac{1}{4}[2V_1 + V_2 + V_4]} \tag{14.86a}$$

or

$$V(i, 0) = \frac{1}{4}[2V(i, 1) + V(i-1, 0) + V(i+1, 0)] \tag{14.86b}$$

The computation molecules for eqs. (14.85) and (14.86) are displayed in Figure 14.38.

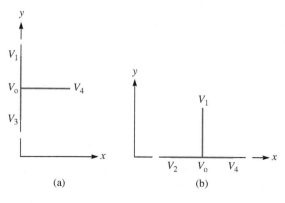

FIGURE 14.38 Computation molecules used for satisfying symmetry conditions: (**a**) $\partial V/dx = 0$ (**b**) $\partial V/dy = 0$

By setting the potential at the fixed nodes equal to their prescribed values and applying eqs. (14.79), (14.83), (14.85), and (14.86) to the free nodes according to the band matrix or iterative methods discussed in Section 14.3, the potential at the free nodes can be determined. Once this is accomplished, the quantities of interest can be calculated.

The characteristic impedance Z_o and phase velocity u of the line are defined as

$$Z_o = \sqrt{\frac{L}{C}} \tag{14.87a}$$

$$u = \frac{1}{\sqrt{LC}} \tag{14.87b}$$

where L and C are the inductance and capacitance per unit length, respectively. If the dielectric medium is nonmagnetic ($\mu = \mu_o$), the characteristic impedance Z_{oo} and phase velocity u_o with the dielectric removed (i.e., the line is air filled) are given by

$$Z_{oo} = \sqrt{\frac{L}{C_o}} \tag{14.88a}$$

$$u_o = \frac{1}{\sqrt{LC_o}} \tag{14.88b}$$

where C_o is the capacitance per unit length without the dielectric. Combining eqs. (14.87) and (14.88) yields

$$Z_o = \frac{1}{u_o\sqrt{CC_o}} = \frac{1}{uC} \tag{14.89a}$$

$$u = u_o\sqrt{\frac{C_o}{C}} = \frac{u_o}{\sqrt{\varepsilon_{eff}}} \tag{14.89b}$$

$$\varepsilon_{eff} = \frac{C}{C_o} \tag{14.89c}$$

where $u_o = c = 3 \times 10^8$ m/s, the speed of light in free space, and ε_{eff} is the effective dielectric constant. Thus to find Z_o and u for an inhomogeneous medium requires calculating the capacitance per unit length of the structure, with and without the dielectric substrate.

If V_d is the potential difference between the inner and the outer conductor,

$$C = \frac{4Q}{V_d} \tag{14.90}$$

so that the problem is reduced to finding the charge per unit length Q. (The factor 4 is needed because we are working on only one-quarter of the cross section.) To find Q, we apply Gauss's law to a closed path L enclosing the inner conductor. We may select L as the rectangular path between two adjacent rectangles as shown in Figure 14.39.

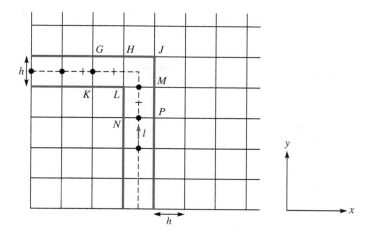

FIGURE 14.39 The rectangular path ℓ used in calculating a charge enclosed.

$$Q = \oint_L \mathbf{D} \cdot d\mathbf{l} = \oint_L \varepsilon \frac{\partial V}{\partial n} dl$$

$$= \varepsilon\left(\frac{V_P - V_N}{\Delta x}\right)\Delta y + \varepsilon\left(\frac{V_M - V_L}{\Delta x}\right)\Delta y + \varepsilon\left(\frac{V_H - V_L}{\Delta y}\right)\Delta x + \varepsilon\left(\frac{V_G - V_K}{\Delta y}\right)\Delta x + \cdots \tag{14.91}$$

Since $\Delta x = \Delta y = h$,

$$Q = (\varepsilon V_P + \varepsilon V_M + \varepsilon V_H + \varepsilon V_G + \cdots) - (\varepsilon V_N + 2\varepsilon V_L + \varepsilon V_K + \cdots) \tag{14.92}$$

or

$$Q = \varepsilon_o[\sum \varepsilon_{ri}V_i \text{ for nodes } i \text{ on external rectangle } GHJMP$$
$$\text{with corners (such as } J\text{) not counted]}$$
$$-\varepsilon_o[\sum \varepsilon_{ri}V_i \text{ for nodes } i \text{ on inner rectangle } KHL$$
$$\text{with corners (such as } L\text{) counted twice].} \tag{14.93}$$

The procedure is outlined as follows:

1. Calculate V (with the dielectric space replaced by free space), using eqs. (14.79), (14.83), (14.85), and (14.86).
2. Determine Q using eq. (14.93).
3. Find $C_o = 4Q_o/V_d$.
4. Repeat steps 1 and 2 (with the dielectric space) and find $C = 4Q/V_d$.
5. Finally, calculate $Z_o = \dfrac{1}{c\sqrt{CC_o}}$, $c = 3 \times 10^8$ m/s.

| EXAMPLE 14.8 | Calculate Z_o for the microstrip transmission line in Figure 14.35 with $a = b = 2.5$ cm, $d = 0.5$ cm, $w = 1$ cm, $t = 0.001$ cm, $\varepsilon_1 = \varepsilon_o$, $\varepsilon_2 = 2.35\varepsilon_o$. |

Solution:

This problem is representative of the various problem types that can be solved by using the concepts developed in this section. A computer program shown in Figure 14.40 was developed based on the five-step procedure just outlined. By specifying the step size h and the number of iterations, the program first sets the potential at all nodes equal to zero. The potential on the outer conductor is also set equal to zero, while that on the inner conductor is set to 100 V so that $V_d = 100$. The program finds C_o when the dielectric slab is removed and C when the slab is in place and finally determines Z_o. For a selected step size h, the number of iterations must be large enough and greater than the number of divisions along the x- or y-direction. Table 14.8 shows some typical results.

```
% Using finite difference,
% this programs finds the characteristic impedance of
% a shielded microstrip line

a=2.5; b=2.5;
d=0.5;
w=1;
h=0.05;
vd=100;
ni=1000;
nx=b/h;
ny=a/h;
nw=w/h;
nd=d/h;
er=2.35;
eo=10^(-9)/(36*pi);
e1=eo;
e2=er*eo;
u =3*10^8;
% Initialization
v=zeros(nx,ny);
for i=1:nw % set the potential on the inner conductor to V_d
     v(i,nd)=vd;
end
% Calculate the potential everywhere
    p1=e1/(2.0*(e1+e2));
   p2=e2/(2.0*(e1+e2));
   for n=1:2
    if n==1
         er=1;
     else
         er=2.35;
     end
```

FIGURE 14.40 MATLAB code for Example 14.8.

```
    for k=1:ni

    for i=2:nx-1
        for j=2:nd-1 % below the interface
            v(i,j)=0.25*( v(i+1,j) + v(i-1,j) + v(i,j+1) +
                v(i,j-1) );
        end
    end
    for i=2:nx-1
        for j=nd+1:ny-1 % above the interface
            v(i,j)=0.25*( v(i+1,j) + v(i-1,j) + v(i,j+1) +
v(i,j-1) ) ;
        end
    end
    j=nd; % on the interface
        for i=nw+1:nx-1
            v(i,j)=0.25*( v(i+1,j) + v(i-1,j)) + p1*v(i,j+1) +
p2*v(i,j-1);
        end
            % on the lines of symmetry
        for i=2:nx-1
            v(i,1)=0.25*( v(i+1,1) + v(i-1,1) + 2*v(i,2) );
        end
        for j=2:nd-1
            v(1,j)=0.25*( 2*v(2,j) + v(1,j+1) + v(1,j-1) );
        end
        for j=nd+1:ny-1
            v(1,j)=0.25*( 2*v(2,j) + v(1,j+1) + v(1,j-1) );
        end
    end
% Now calculate the charge enclosed
% Select two adjacent paths
sum1=0.0;
sum2=0.0;
nm=fix( nd + 0.5*(ny-nd) );
nn=fix( nx/2 );
for i=2:nn
    sum1= sum1 + v(i,nm);
    sum2= sum2 + v(i,nm+1);
end
sum1 = sum1 + 0.5*v(1,nm);
sum2 = sum2 + 0.5*v(1,nm+1);
for j=2:nd-1
    sum1 = sum1 + er*v(nn,j);
    sum2 = sum2 + er*v(nn+1,j);
end
sum1 = sum1 + 0.5*er*v(nn,1);
sum2 = sum2 + 0.5*er*v(nn+1,1);
for j=nd+1:nm
```

FIGURE 14.40 (Continued)

```
      sum1 = sum1 + v(nn,j);
      sum2 = sum2 + v(nn+1,j);
end
      sum1 = sum1 + 0.5*(er+1)*v(nn,nd);
      sum2 = sum2 + 0.5*(er+1)*v(nn+1,nd);
q(n)=eo*abs(sum2 - sum1);
end
% Calculate the characteristic impedance
c1=4*q(1)/vd;
c2=4*q(2)/vd;
zo=1/(u*sqrt(c1*c2))
```

TABLE 14.8 Characteristic Impedance of a Microstrip Line; for Example 14.8

h	Number of Iterations	$Z_o(\Omega)$
0.25	700	72.43
0.1	500	57.56
0.05	500	67.36
0.05	700	66.88
0.05	1000	66.53

MATLAB 14.1

```
% This script allows the user to enter the dimensions and
% dielectric properties of a microstrip line and then use
% the finite-difference algorithm to solve Laplace's
% equation iteratively and obtain the potential as a
% function of space
% The microstrip problem is normally open-aired, meaning
% the upper and side boundaries exist at infinity, but due
% to the limitations of solution space in the numerical
% problem, we have added shielding walls (where E = 0) of
% perfect electric conductor, which should be sufficiently
% far from the microstrip structure a,b >> w,d for accurate
% simulation of free-space and boundaries at infinity.

% The script creates a coarse rectangular grid and then solves

% Prompt user for basic parameters
Vstrip=input('Enter the voltage on the microstrip \n> ');
a=input('Enter the horizontal span of space \n> ');
b=input('Enter the vertical span of space \n> ');
w=input('Enter the microstrip width \n> ');
```

```
d=input('Enter the dielectric thickness \n> ');

% the dielectric boundary
disp('Enter the relative dielectric constant ');
epstop=input(' above the microstrip \n> ');
disp('Enter the relative dielectric constant ');
epsbottom=input(' below the microstrip \n> ');
epsave=(epstop+epsbottom)/2; % the average relative dielectric
                             % along the boundary

% Fill the potential solution space with zeros
P=zeros(b,a);

% set voltage on strip
% the floor rounds odd numbers divided by 2 down to the
% nearest integer
for i=floor(b/2)-(floor(w/2)-1):1:floor(b/2)+floor(w/2),
    P(i,d)=Vstrip;
end

% Begin iterations to solve potential
% --------------------------------
for i=1:600,    % i is the iteration step
    % the larger this number is the more accurate the potential
    % solution
    for j=2:1:a-1,  % sweep each column of y-values
        % if j is not a unit on the microstrip conductor
        if j ~= d
            for i=2:b-1, % sweep each row of x-values
                % this equation solves for the potential by
                % discretizing Laplace's equation on the
                % rectangular grid
                P(i,j)=0.25*(P(i+1,j)+P(i-1,j)+P(i,j+1)+P(i,j-1));
            end
        % else we are at y = d, in the axis of the strip
        else
            for i=2:b-1, % sweep each row of x-values
                if (i < (floor(b/2)-(w/2-1))) | (i >
                (floor(b/2)+w/2))
                    % this equation solves for the potential by
                    % discretizing Laplace's equation on the
                    rectangular
                    % grid
                        P(i,j)=(1/(4*epsave))*(epsave*(P(i+1,j)+...
                        P(i-1,j))+epstop*P(i,j+1)+epsbottom*P(i,j-1));
                end
            end
        % end the if conditional
        end
    end
end
```

```
% ------------------------------------
% Plot potential distribution
% Create the vector of voltage contours
v=[0.005,0.01,0.05,0.1,0.2,0.3,0.4,0.5,0.6,0.7,0.8,0.9,1]*...
  Vstrip;
figure
contour(P',v); % P' is the transposed matrix to meet the
requirements
colorbar   % add the colorbar as a legend for the colors which
                    % define the equipotential lines
xlabel('Horizontal position')
ylabel('Vertical position')
title('Equipotential Curves For Microstrip Line')
```

SUMMARY

1. Electric field lines and equipotential lines due to coplanar point sources can be plotted by using the numerical technique presented in this chapter. The basic concept can be extended to plotting magnetic field lines.

2. An EM problem in the form of a partial differential equation can be solved by using the finite difference method. The finite difference equation that approximates the differential equation is applied at grid points spaced in an ordered manner over the whole solution region. The field quantity at the free points is determined using a suitable method.

3. An EM problem in the form of an integral equation is conveniently solved by using the moment method. The unknown quantity under the integral sign is determined by matching both sides of the integral equation at a finite number of points in the domain of the quantity.

4. While the finite difference method is restricted to problems with regularly shaped solution regions, the finite element method can handle problems with complex geometries. This method involves dividing the solution region into finite elements, deriving equations for a typical element, assembling all elements in the region, and solving the resulting system of equations.

5. The finite difference has been applied to determine the characteristic impedance of a microstrip transmission line.

   Typical examples on how to apply each method to some practical problems have been shown. Computer programs for solving the problems are provided wherever needed.

REVIEW QUESTIONS

14.1 At the point $(1, 2, 0)$ in an electric field due to coplanar point charges, $\mathbf{E} = 0.3\,\mathbf{a}_x - 0.4\,\mathbf{a}_y$ V/m. A differential displacement of 0.05 m on an equipotential line at that point will lead to point

(a) $(1.04, 2.03, 0)$

(b) $(0.96, 1.97, 0)$

(c) $(1.04, 1.97, 0)$

(d) $(0.96, 2.03, 0)$

14.2 Which of the following is *not* a correct finite difference approximation to dV/dx at x_o if $h = \Delta x$?[4]

(a) $\dfrac{V(x_o + h) - V(x_o)}{h}$

(d) $\dfrac{V(x_o + h) - V(x_o - h)}{2h}$

(b) $\dfrac{V(x_o) - V(x_o - h)}{h}$

(e) $\dfrac{V(x_o + h/2) - V(x_o - h/2)}{h}$

(c) $\dfrac{V(x_o + h) - V(x_o - h)}{h}$

14.3 The triangular element of Figure 14.41 is in free space. The approximate value of the potential at the center of the triangle is

(a) 10 V

(c) 5 V

(b) 7.5 V

(d) 0 V

14.4 For finite difference analysis, a rectangular plate measuring 10 by 20 cm is divided into eight subregions by lines 5 cm apart parallel to the edges of the plates. How many free nodes are there if the edges are connected to some source?

(a) 15

(c) 9

(e) 3

(b) 12

(d) 6

14.5 Using the difference equation $V_n = V_{n-1} + V_{n+1}$ with $V_o = V_5 = 1$ and starting with initial values $V_n = 0$ for $1 \le n \le 4$, the value of V_2 after the third iteration is

(a) 1

(c) 9

(e) 25

(b) 3

(d) 15

14.6 The coefficient matrix $[A]$ obtained in the moment method does *not* have one of these properties:

(a) It is dense (i.e., has many nonzero terms).

(b) It is banded.

(c) It is square and symmetric.

(d) It depends on the geometry of the given problem.

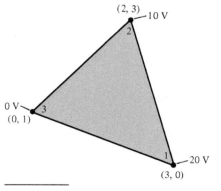

(2, 3) — 10 V

0 V — (0, 1)

— 20 V

(3, 0)

FIGURE 14.41 For Review Questions 14.3 and 14.10.

[4] The formula in (a) is known as a forward-difference formula, that in (b) as a backward-difference formula, and that in (d) or (e) as a central-difference formula.

14.7 A major difference between the finite difference and the finite element methods is that

(a) Using one, a sparse matrix results in the solution.

(b) In one, the solution is known at all points in the domain.

(c) One applies to solving partial differential equation.

(d) One is limited to time-invariant problems.

14.8 If the plate of Review Question 14.4 is to be discretized for finite element analysis such that we have the same number of grid points, how many triangular elements are there?

(a) 32   (c) 12

(b) 16   (d) 9

14.9 Which of these statements is *not* true about shape functions?

(a) They are interpolatory in nature.

(b) They must be continuous across the elements.

(c) Their sum is identically equal to unity at every point within the element.

(d) The shape function associated with a given node vanishes at any other node.

(e) The shape function associated with a node is zero at that node.

14.10 The area of the element in Figure 14.41 is

(a) 14   (c) 7

(b) 8   (d) 4

Answers: 14.1a, 14.2c, 14.3a, 14.4e, 14.5c, 14.6b, 14.7a 14.8b, 14.9e, 14.10d.

PROBLEMS

Section 14.2—Field Plotting

14.1 Use the program developed in Example 14.1 or your own equivalent code to plot the electric field lines and equipotential lines for the following cases:

(a) Three point charges of -1 C, 2 C, and 1 C placed at $(-1, 0)$, $(0, 2)$, and $(1, 0)$, respectively.

(b) Five identical point charges of 1 C located at $(-1, -1)$, $(-1, 1)$, $(1, -1)$, $(1, 1)$, and $(0, 0)$, respectively.

Section 14.3—The Finite Difference Method

14.2 A boundary-value problem is defined by

$$\frac{d^2V}{dx^2} = x + 1, \quad 0 < x < 1$$

Subject to $V(0) = 0$ and $V(1) = 1$. Use the finite difference method to find $V(0.5)$. You may take $\Delta = 0.25$ and perform 5 iterations. Compare your result with the exact solution.

14.3 (a) Obtain $\dfrac{dV}{dx}$ and $\dfrac{d^2V}{dx^2}$ at $x = 0.15$ from the following table.

x	0.1	0.15	0.2	0.25	0.3
V	1.0017	1.5056	2.0134	2.5261	3.0452

  (b) The data in the table are obtained from $V = 10 \sinh x$. Compare your result in part (a) with the exact values.

14.4 Show that the finite difference equation for Laplace's equation in cylindrical coordinates, $V = V(\rho, z)$, is

$$V(\rho_o, z_o) = \frac{1}{4}\left[V(\rho_o, z_o + h) + V(\rho_o, z_o - h) + \left(1 + \frac{h}{2\rho_o}\right) \right.$$
$$\left. V(\rho_o + h, z_o) + \left(1 - \frac{h}{2\rho_o}\right) V(\rho_o - h, z_o) \right]$$

where $h = \Delta z = \Delta\rho$.

14.5 Using the finite difference representation in cylindrical coordinates (ρ, ϕ) at a grid point P shown in Figure 14.42, let $\rho = m\,\Delta\rho$ and $\phi = n\,\Delta\phi$ so that $V(\rho, \phi)|_P = V(m\Delta\rho, n\Delta\phi) = V_m^n$. Show that

$$\nabla^2 V\big|_{m,n} = \frac{1}{\Delta\rho^2}\left[\left(1 - \frac{1}{2m}\right)V_{m-1}^n - 2V_m^n + \left(1 + \frac{1}{2m}\right)V_{m+1}^n \right.$$
$$\left. + \frac{1}{(m\,\Delta\phi)^2}\left(V_m^{n-1} - 2\,V_m^n + V_m^{n+1}\right) \right]$$

14.6 The four sides of a square trough are maintained at potentials 10 V, −40 V, 50 V, and 80 V. Determine the potential at the center of the trough.

14.7 For the potential problem in Figure 14.43, use the finite difference method to determine V_1 to V_4. Five iterations are enough.

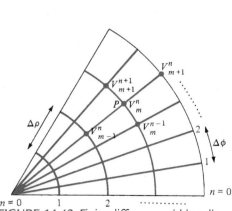

V_{m+1}^n $V_m^n{}_{+1}$ V_{m+1}^{n+1} P V_m^n V_m^{n-1} V_{m-1}^n $\Delta\rho$ $\Delta\phi$ $n = 0$ $m = 0$ 1   2

FIGURE 14.42 Finite difference grid in cylindrical coordinates; for Problem 14.5.

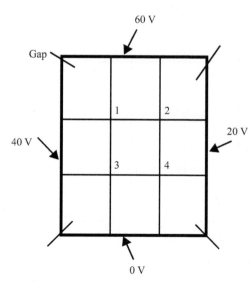

60 V   Gap   40 V   20 V   0 V   1   2   3   4

FIGURE 14.43 For Problem 14.7.

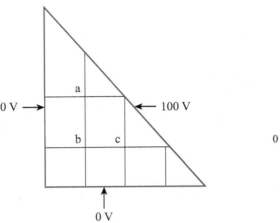

FIGURE 14.44 For Problem 14.8.

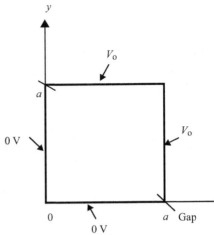

FIGURE 14.45 For Problem 14.9.

14.8 Use the finite difference method to obtain the potential at points a, b, and c in Figure 14.44. Five iterations are enough.

14.9 The cross section of a long conducting pipe is shown in Figure 14.45. For the indicated boundary conditions:
(a) Find the potential distribution $V(x, y)$ by solving Laplace's equation.
(b) Find the potential at the center of the region using the finite difference method. Take $a = 1$ m and $V_0 = 50$ V.

14.10 For the rectangular region shown in Figure 14.46, the electric potential is zero on the boundaries and the charge distribution ρ_v is 50 nC/m³. Although there are six free nodes, there are only four unknown potentials ($V_1 - V_4$) because of symmetry. Solve for the unknown potentials.

14.11 Apply the band matrix technique to set up a system of simultaneous difference equations for each of the problems in Figure 14.47. Obtain matrices [A] and [B].

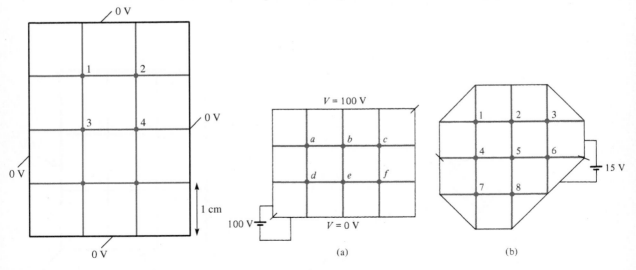

FIGURE 14.46 For Problem 14.10.   FIGURE 14.47 For Problem 14.11.

14.12 (a) How would you modify matrices $[A]$ and $[B]$ of Example 14.3 if the solution region had charge density ρ_v?

   (b) Write a program to solve for the potentials at the grid points shown in Figure 14.48 assuming a charge density $\rho_v = x(y - 1)$ nC/m^3. Use the iterative finite difference method and take $\varepsilon_r = 1.0$.

14.13 Use the finite difference method to find the potentials at nodes 1 to 7 in the grid shown in Figure 14.49.

14.14 The two-dimensional wave equation is given by

$$\frac{1}{c^2} \frac{\partial^2 \Phi}{\partial t^2} = \frac{\partial^2 \Phi}{\partial x^2} + \frac{\partial^2 \Phi}{\partial z^2}$$

By letting $\Phi^j_{m,n}$ denote the finite difference approximation of $\Phi(x_m, z_n, t_j)$, show that the finite difference scheme for the wave equation is

$$\Phi^{j+1}_{\mu,n} = 2\,\Phi^j_{m,n} - \Phi^{j-1}_{\mu,n} + \alpha\,(\Phi^j_{m+1,n} + \Phi^j_{\mu-1,n} - 2\,\Phi^j_{m,n}) +$$
$$\alpha\,(\Phi^j_{m,n+1} + \Phi^j_{\mu,n-1} - 2\,\Phi^j_{m,n})$$

   where $h = \Delta x = \Delta z$ and $\alpha = (c\Delta t/h)^2$.

14.15 Write a program that uses the finite difference scheme to solve the one-dimensional wave equation

$$\frac{\partial^2 V}{\partial x^2} = \frac{\partial^2 V}{\partial t^2}, \quad 0 \le x \le 1, \quad t > 0$$

given boundary conditions $V(0, t) = 0$, $V(1, t) = 0$, $t > 0$ and the initial condition $\partial V/\partial t\,(x, 0) = 0$, $V(x, 0) = \sin \pi x$, $0 < x < 1$. Take $\Delta x = \Delta t = 0.1$. Compare your solution with the exact solution $V(x, t) = \sin \pi x \cos \pi t$ for $0 < t < 4$.

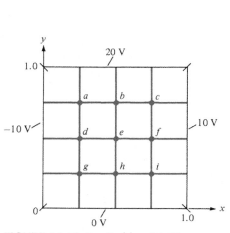

FIGURE 14.48 For Problem 14.12.

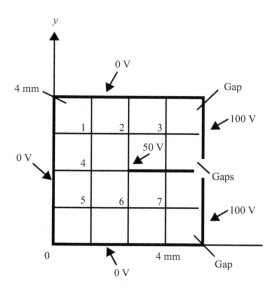

FIGURE 14.49 For Problem 14.13.

Section 14.4—The Moment Method

14.16 The electric field due to a circular loop placed at the origin with charge ρ per unit length at a distance x from origin is given by

$$E_x = \frac{\rho}{2\pi\varepsilon_o a}\int_0^\pi \frac{(u - \cos\theta)d\theta}{(1 + u^2 - 2u\cos\theta)^{3/2}} = \frac{\rho}{2\pi\varepsilon_o a}F(u)$$

where

$$F(u) = \int_0^\pi \frac{(u - \cos\theta)d\theta}{(1 + u^2 - 2u\cos\theta)^{3/2}}$$

and $u = x/a$. Use MATLAB to plot $F(u)$ for $0 < u < 2$.

14.17 A transmission line consists of two identical wires of radius a, separated by distance d as shown in Figure 14.50. Maintain one wire at 1 V and the other at -1 V and use the method of moments to find the capacitance per unit length. Compare your result with exact formula for C in Table 11.1. Take $a = 5$ mm, $d = 5$ cm, $\ell = 5$ m, and $\varepsilon = \varepsilon_o$.

14.18 Two conducting wires of equal length L and radius a are separated by a small gap and inclined at an angle θ as shown in Figure 14.51. Find the capacitance between the wires by using the method of moments for cases $\theta = 10°$, $20°$, . . . , $180°$. Take the gap as 2 mm, $a = 1$ mm, $L = 2$ m, $\varepsilon_r = 1$.

14.19 Given an infinitely long thin strip transmission line shown in Figure 14.52(a), use the moment method to determine the characteristic impedance of the line. We divide each strip into N subareas as in Figure 14.52(b) so that on subarea i,

$$V_i = \sum_{j=1}^{2N} A_{ij}\rho_j$$

where

$$A_{ij} = \begin{cases} \dfrac{-\Delta\ell}{2\pi\varepsilon_o}\ln R_{ij}, & i \neq j \\[2ex] \dfrac{-\Delta\ell}{2\pi\varepsilon_o}[\ln\Delta\ell - 1.5], & i = j \end{cases}$$

R_{ij} is the distance between the ith and jth subareas, and $V_i = 1$ or -1 depending on whether the ith subarea is on strip 1 or 2, respectively. Write a program to find the characteristic impedance of the line using the fact that

$$Z_o = \frac{\sqrt{\mu_o\varepsilon_o}}{C}$$

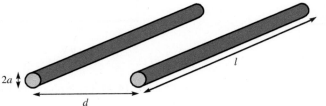

FIGURE 14.50 For Problem 14.17.

$2a$

d

l

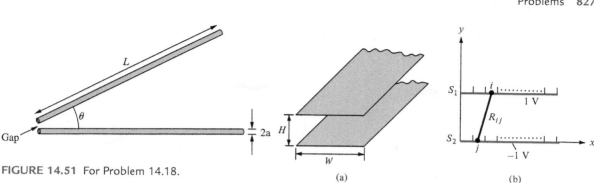

FIGURE 14.51 For Problem 14.18.

FIGURE 14.52 Analysis of strip transmission line using moment method. For Problem 14.19.

where C is the capacitance per unit length and

$$C = \frac{Q}{V_d} = \frac{\sum\limits_{i=1}^{N} \rho_i \, \Delta\ell}{V_d}$$

and $V_d = 2$ V is the potential difference between strips. Take $H = 2$ m, $W = 5$ m, and $N = 20$.

14.20 Consider an L-shaped thin wire of radius 1 mm as shown in Figure 14.53. If the wire is held at a potential $V = 10$ V, use the method of moments to find the charge distribution on the wire. Take $\Delta = 0.1$.

14.21 Consider the coaxial line of the arbitrary cross section shown in Figure 14.54(a). Using the moment method to find the capacitance C per length involves dividing each conductor into N strips so that the potential on the jth strip is given by

$$V_j = \sum_{i=1}^{2N} \rho_i A_{ij}$$

where

$$A_{ij} = \begin{cases} \dfrac{-\Delta\ell}{2\pi\varepsilon} \ln \dfrac{R_{ij}}{r_o}, & i \neq j \\[2ex] \dfrac{-\Delta\ell}{2\pi\varepsilon} \left[\ln \dfrac{\Delta\ell_i}{r_o} - 1.5 \right], & i = j \end{cases}$$

and $V_j = -1$ or 1 depending on whether $\Delta\ell_i$ lies on the inner or outer conductor, respectively. Write a MATLAB program to determine the total charge per length on a coaxial cable of elliptical cylindrical cross section shown in Figure 14.54(b) by using

$$Q = \sum_{i=1}^{N} \rho_i$$

and the capacitance per unit length with $C = Q/2$.

(a) As a way of checking your program, take $A = B = 2$ cm and $a = b = 1$ cm (coaxial line with circular cross section), and compare your result with the exact value of $C = 2\pi\varepsilon/\ln(A/a)$.

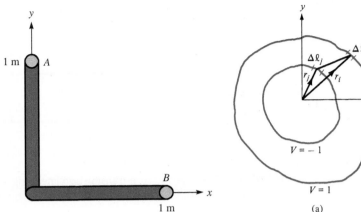

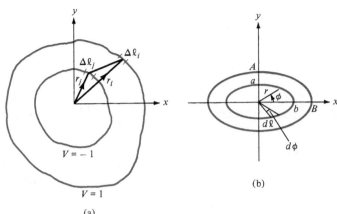

FIGURE 14.53 For Problem 14.20.

FIGURE 14.54 For Problem 14.21; coaxial line of (**a**) arbitrary cross section, (**b**) elliptical cylindrical cross section.

(b) Take $A = 2$ cm, $B = 4$ cm, $a = 1$ cm, and $b = 2$ cm.
Hint: For the inner ellipse of Figure 14.54(b), for example,

$$r = \frac{a}{\sqrt{\sin^2 \phi + v^2\cos^2 \phi}}$$

where $v = a/b$, $d\ell = r\, d\phi$. Take $r_o = 1$ cm.

14.22   A conducting bar of rectangular cross section is shown in Figure 14.55. By dividing the bar into N equal segments, we obtain the potential at the jth segment as

$$V_j = \sum_{i=1}^{N} q_i A_{ij}$$

where

$$A_{ij} = \begin{cases} \dfrac{1}{4\pi\varepsilon_o R_{ij}}, & i \neq j \\[3mm] \dfrac{1}{2\varepsilon_o\sqrt{\pi h\Delta}}, & i = j \end{cases}$$

and Δ is the length of the segment. If we maintain the bar at 10 V, we obtain

$$[A][q] = 10[I]$$

where $[I] = [1\ 1\ 1 \cdots 1]^T$ and $q_i = \rho_v th\Delta$.

(a) Write a program to find the charge distribution ρ_v on the bar and take $\ell = 2$ m, $h = 2$ cm, $t = 1$ cm, and $N = 20$.

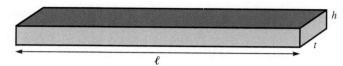

h FIGURE 14.55 For Problem 14.22.

(b) Compute the capacitance of the isolated conductor by using

$$C = \frac{Q}{V} = (q_1 + q_2 + \cdots + q_N)/10$$

Section 14.5—The Finite Element Method

14.23 Another way of defining the shape functions at an arbitrary point (x, y) in a finite element is using the areas A_1, A_2, and A_3 shown in Figure 14.56. Show that

$$\alpha_k = \frac{A_k}{A}, \quad k = 1, 2, 3$$

where $A = A_1 + A_2 + A_3$ is the total area of the triangular element.

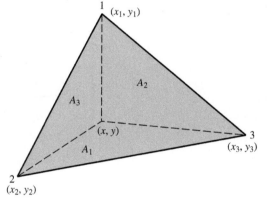

FIGURE 14.56 For Problem 14.23.

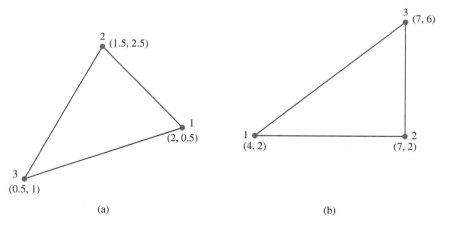

(a) (b)

FIGURE 14.57 For Problem 14.24.

14.24 Determine the element coefficient matrices of the triangular elements in Figure 14.57.

14.25 The nodal potential values for the triangular element of Figure 14.58 are $V_1 = 100$ V, $V_2 = 50$ V, and $V_3 = 30$ V. (a) Determine where the 80 V equipotential line intersects the boundaries of the element. (b) Calculate the potential at $(2, 1)$.

14.26 The triangular element shown in Figure 14.59 is part of a finite element mesh. If $V_1 = 8$ V, $V_2 = 12$ V, and $V_3 = 10$ V, find the potential at (a) (1, 2) and (b) the center of the element.

14.27 Determine the global coefficient matrix for the two-element region shown in Figure 14.60.

14.28 Calculate the global coefficient matrix for the two-element region shown in Figure 14.61.

14.29 Find the global coefficient matrix of the two-element mesh of Figure 14.62.

14.30 For the two-element mesh of Figure 14.62, let $V_1 = 10$ V and $V_3 = 30$ V. Find V_2 and V_4.

14.31 Use the MATLAB code in Figure 14.33 to determine the potentials at node 5 of the system shown in Figure 14.63.

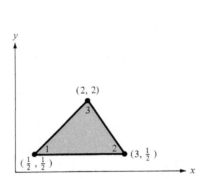

FIGURE 14.58 For Problem 14.25.

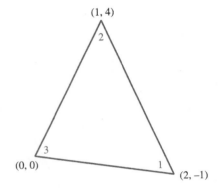

FIGURE 14.59 For Problem 14.26.

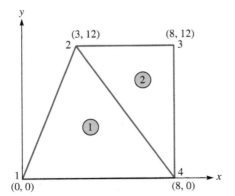

FIGURE 14.60 For Problem 14.27.

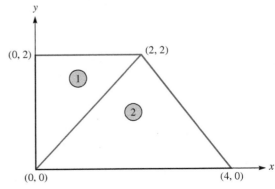

FIGURE 14.61 For Problem 14.28.

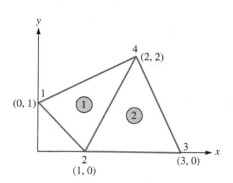

FIGURE 14.62 For Problems 14.29 and 14.30.

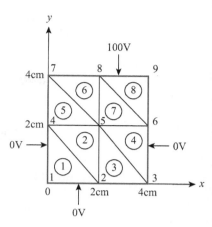

FIGURE 14.63 For Problem 14.31.

14.32 Use the program in Figure 14.33 to solve Laplace's equation in the problem shown in Figure 14.64, where $V_o = 100$ V. Compare the finite element solution to the exact solution in Example 6.5; that is,

$$V(x, y) = \frac{4V_o}{\pi} \sum_{k=0}^{\infty} \frac{\sin n\pi x \sinh n\pi y}{n \sinh n\pi}, \quad n = 2k + 1$$

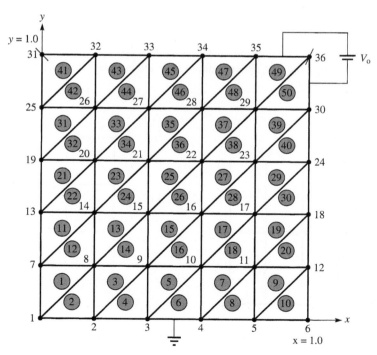

FIGURE 14.64 For Problem 14.32.

14.33 Repeat Problem 14.32 for $V_o = 100 \sin \pi x$. Compare the finite element solution with the theoretical solution [similar to Example 6.6(a)]; that is,

$$V(x, y) = \frac{100 \sin \pi x \sinh \pi y}{\sinh \pi}$$

14.34 Show that when a square mesh is used in the finite difference method, we obtain the same result in the finite element method when the squares are cut into triangles.

Section 14.6—Application Note—Microstrip Lines

14.35 Consider the shielded microstrip problem shown in Figure 14.65. Use the finite difference method to find the potential at points 1 to 6. Five iterations are sufficient.

14.36 Refer to the square mesh in Figure 14.66. By setting the potential values at the free nodes equal to zero, find (by hand calculation) the potentials at nodes 1 to 4 for five or more iterations.

14.37 Determine the characteristic impedance of the microstrip line shown in Figure 14.67. Take $a = 2.02$, $b = 7.0$, $h = 1.0 = w$, $t = 0.01$.

14.38 The cross section of a transmission line is shown in Figure 14.68. Use the finite difference method to compute the characteristic impedance of the line.

14.39 Half a solution region is shown in Figure 14.69 so that the y-axis is a line of symmetry. Use finite difference to find the potential at nodes 1 to 9. Five iterations are sufficient if you use an iterative method.

14.40 The potential system in Figure 14.70 is symmetric about the y-axis. Set the initial values at the free nodes equal to zero and calculate the potentials at nodes 1 to 5 for five iterations.

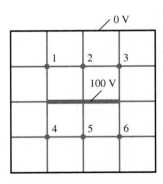

FIGURE 14.65 For Problem 14.35.

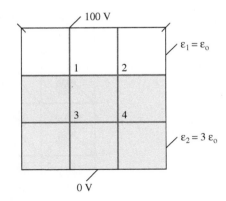

FIGURE 14.66 For Problem 14.36.

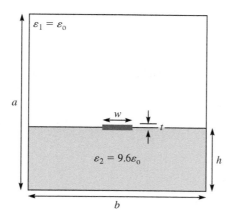

FIGURE 14.67 For Problem 14.37.

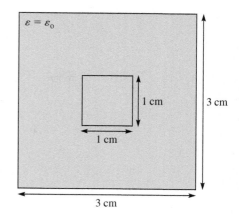

FIGURE 14.68 For Problem 14.38.

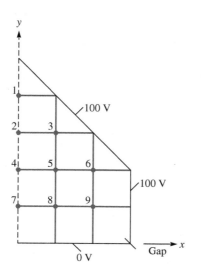

FIGURE 14.69 For Problem 14.39.

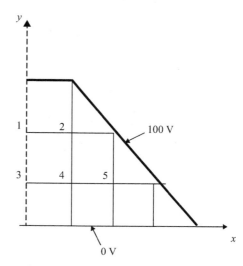

FIGURE 14.70 For Problem 14.40.

MATHEMATICAL FORMULAS

A.1 TRIGONOMETRIC IDENTITIES

$$\tan A = \frac{\sin A}{\cos A}, \quad \cot A = \frac{1}{\tan A}$$

$$\sec A = \frac{1}{\cos A}, \quad \csc A = \frac{1}{\sin A}$$

$$\sin^2 A + \cos^2 A = 1, \quad 1 + \tan^2 A = \sec^2 A$$

$$1 + \cot^2 A = \csc^2 A$$

$$\sin (A \pm B) = \sin A \cos B \pm \cos A \sin B$$

$$\cos (A \pm B) = \cos A \cos B \mp \sin A \sin B$$

$$2 \sin A \sin B = \cos (A - B) - \cos (A + B)$$

$$2 \sin A \cos B = \sin (A + B) + \sin (A - B)$$

$$2 \cos A \cos B = \cos (A + B) + \cos (A - B)$$

$$\sin A + \sin B = 2 \sin \frac{A + B}{2} \cos \frac{A - B}{2}$$

$$\sin A - \sin B = 2 \cos \frac{A + B}{2} \sin \frac{A - B}{2}$$

$$\cos A + \cos B = 2 \cos \frac{A + B}{2} \cos \frac{A - B}{2}$$

$$\cos A - \cos B = -2 \sin \frac{A + B}{2} \sin \frac{A - B}{2}$$

$$\cos (A \pm 90°) = \mp \sin A$$

$$\sin (A \pm 90°) = \pm \cos A$$

$$\tan (A \pm 90°) = -\cot A$$

$$\cos (A \pm 180°) = -\cos A$$

$$\sin (A \pm 180°) = -\sin A$$

$$\tan\left(A \pm 180°\right) = \tan A$$

$$\sin 2A = 2 \sin A \cos A$$

$$\cos 2A = \cos^2 A - \sin^2 A = 2 \cos^2 A - 1 = 1 - 2 \sin^2 A$$

$$\tan\left(A \pm B\right) = \frac{\tan A \pm B}{1 \mp \tan A \tan B}$$

$$\tan 2A = \frac{2 \tan A}{1 - \tan^2 A}$$

$$\sin A = \frac{e^{jA} - e^{-jA}}{2j}, \quad \cos A = \frac{e^{jA} + e^{-jA}}{2}$$

$$e^{jA} = \cos A + j \sin A \quad \text{(Euler's identity)}$$

$$\pi \approx 3.1416$$

$$1 \text{ rad} \approx 57.296°$$

A.2 COMPLEX VARIABLES

A complex number may be represented as

$$z = x + jy = r\underline{/\theta} = re^{j\theta} = r\left(\cos\theta + j\sin\theta\right)$$

$$\text{where } x = \text{Re } z = r\cos\theta, \quad y = \text{Im } z = r\sin\theta$$

$$r = |z| = \sqrt{x^2 + y^2}, \quad \theta = \tan^{-1}\frac{y}{x}$$

$$j = \sqrt{-1}, \quad \frac{1}{j} = -j, \quad j^2 = -1$$

The complex conjugate of $z = z^* = x - jy = r\underline{/-\theta} = re^{-j\theta}$

$$= r\left(\cos\theta - j\sin\theta\right)$$

$$\left(e^{j\theta}\right)^n = e^{jn\theta} = \cos n\theta + j\sin n\theta \quad \text{(de Moivre's theorem)}$$

If $z_1 = x_1 + jy_1$ and $z_2 = x_2 + jy_2$, then $z_1 = z_2$ only if $x_1 = x_2$ and $y_1 = y_2$.

$$z_1 \pm z_2 = \left(x_1 + x_2\right) \pm j\left(y_1 + y_2\right)$$

$$z_1 z_2 = \left(x_1 x_2 - y_1 y_2\right) + j\left(x_1 y_2 + x_2 y_1\right)$$

or

$$z_1 z_2 = r_1 r_2\, e^{j\left(\theta_1 + \theta_2\right)} = r_1 r_2 \underline{/\theta_1 + \theta_2}$$

$$\frac{z_1}{z_2} = \frac{(x_1 + jy_1)}{(x_2 + jy_2)} \cdot \frac{(x_2 - jy_2)}{(x_2 - jy_2)} = \frac{x_1 x_2 + y_1 y_2}{x_2^2 + y_2^2} + j\frac{x_2 y_1 - x_1 y_2}{x_2^2 + y_2^2}$$

or

$$\frac{z_1}{z_2} = \frac{r_1}{r_2} e^{j(\theta_1 - \theta_2)} = \frac{r_1}{r_2} \underline{/\theta_1 - \theta_2}$$

$$\sqrt{z} = \sqrt{x + jy} = \sqrt{r}\, e^{j\theta/2} = \sqrt{r}\, \underline{/\theta/2}$$

$$z^n = (x + jy)^n = r^n e^{jn\theta} = r^n \underline{/n\theta} \qquad (n = \text{integer})$$

$$z^{1/n} = (x + jy)^{1/n} = r^{1/n} e^{j\theta/n} = r^{1/n} \underline{/\theta/n + 2\pi k/n} \quad (k = 0, 1, 2, \ldots, n - 1)$$

$$\ln(re^{j\theta}) = \ln r + \ln e^{j\theta} = \ln r + j\theta + j2k\pi \quad (k = \text{integer})$$

A.3 HYPERBOLIC FUNCTIONS

$$\sinh x = \frac{e^x - e^{-x}}{2}, \quad \cosh x = \frac{e^x + e^{-x}}{2}$$

$$\tanh x = \frac{\sinh x}{\cosh x}, \quad \coth x = \frac{1}{\tanh x}$$

$$\operatorname{csch} x = \frac{1}{\sinh x}, \quad \operatorname{sech} x = \frac{1}{\cosh x}$$

$$\sin jx = j \sinh x, \quad \cos jx = \cosh x$$

$$\sinh jx = j \sin x, \quad \cosh jx = \cos x$$

$$\sinh (x \pm y) = \sinh x \cosh y \pm \cosh x \sinh y$$

$$\cosh (x \pm y) = \cosh x \cosh y \pm \sinh x \sinh y$$

$$\sinh (x \pm jy) = \sinh x \cos y \pm j \cosh x \sin y$$

$$\cosh (x \pm jy) = \cosh x \cos y \pm j \sinh x \sin y$$

$$\tanh (x \pm jy) = \frac{\sinh 2x}{\cosh 2x + \cos 2y} \pm j\frac{\sin 2y}{\cosh 2x + \cos 2y}$$

$$\cosh^2 x - \sinh^2 x = 1$$

$$\operatorname{sech}^2 x + \tanh^2 x = 1$$

$$\sin (x \pm jy) = \sin x \cosh y \pm j \cos x \sinh y$$

$$\cos (x \pm jy) = \cos x \cosh y \mp j \sin x \sinh y$$

A.4 LOGARITHMIC IDENTITIES

$$\log xy = \log x + \log y$$

$$\log \frac{x}{y} = \log x - \log y$$

$$\log x^n = n \log x$$

$$\log_{10} x = \log x \, (\text{common logarithm})$$

$$\log_e x = \ln x \, (\text{natural logarithm})$$

If $|x| \ll 1, \ln(1 + x) \simeq x$

A.5 EXPONENTIAL IDENTITIES

$$e^x = 1 + x + \frac{x^2}{2!} + \frac{x^3}{3!} + \frac{x^4}{4!} + \cdots$$

where $e \approx 2.7183$

$$e^x e^y = e^{x+y}$$

$$[e^x]^n = e^{nx}$$

$$\ln e^x = x$$

A.6 APPROXIMATIONS FOR SMALL QUANTITIES

If $|x| \ll 1$,

$$(1 \pm x)^n \simeq 1 \pm nx$$

$$e^x \simeq 1 + x$$

$$\ln(1 + x) \simeq x$$

$$\sin x \simeq x \text{ or } \lim_{x \to 0} \frac{\sin x}{x} = 1$$

$$\cos x \simeq 1$$

$$\tan x \simeq x$$

If $a \gg b$,

$$(a + b)^n \simeq a^n + na^{n-1}b$$

A.7 DERIVATIVES

If $U = U(x)$, $V = V(x)$, and $a =$ constant,

$$\frac{d}{dx}(aU) = a\frac{dU}{dx}$$

$$\frac{d}{dx}(UV) = U\frac{dV}{dx} + V\frac{dU}{dx}$$

$$\frac{d}{dx}\left[\frac{U}{V}\right] = \frac{V\dfrac{dU}{dx} - U\dfrac{dV}{dx}}{V^2}$$

$$\frac{d}{dx}(aU^n) = naU^{n-1}$$

$$\frac{d}{dx}\log_a U = \frac{\log_a e}{U}\frac{dU}{dx}$$

$$\frac{d}{dx}\ln U = \frac{1}{U}\frac{dU}{dx}$$

$$\frac{d}{dx}a^U = a^U\ln a\frac{dU}{dx}$$

$$\frac{d}{dx}e^U = e^U\frac{dU}{dx}$$

$$\frac{d}{dx}U^V = VU^{v-1}\frac{dU}{dx} + U^V\ln U\frac{dV}{dx}$$

$$\frac{d}{dx}\sin U = \cos U\frac{dU}{dx}$$

$$\frac{d}{dx}\cos U = -\sin U\frac{dU}{dx}$$

$$\frac{d}{dx}\tan U = \sec^2 U\frac{dU}{dx}$$

$$\frac{d}{dx}\sinh U = \cosh U\frac{dU}{dx}$$

$$\frac{d}{dx}\cosh U = \sinh U\frac{dU}{dx}$$

$$\frac{d}{dx}\tanh U = \operatorname{sech}^2 U\frac{dU}{dx}$$

A.8 INDEFINITE INTEGRALS

If $U = U(x)$, $V = V(x)$, a and b are constants, and C is an arbitrary constant,

$$\int a\, dx = ax + C$$

$$\int U\, dV = UV - \int V\, dU \quad \text{(integration by parts)}$$

$$\int U^n\, dU = \frac{U^{n+1}}{n+1} + C, \quad n \neq -1$$

$$\int \frac{dU}{U} = \ln U + C$$

$$\int a^U\, dU = \frac{a^U}{\ln a} + C, \quad a > 0, a \neq 1$$

$$\int e^U\, dU = e^U + C$$

$$\int e^{ax}\, dx = \frac{1}{a} e^{ax} + C$$

$$\int x e^{ax}\, dx = \frac{e^{ax}}{a^2}(ax - 1) + C$$

$$\int x^2 e^{ax}\, dx = \frac{e^{ax}}{a^3}(a^2 x^2 - 2ax + 2) + C$$

$$\int \ln x\, dx = x \ln x - x + C$$

$$\int \sin ax\, dx = -\frac{1}{a} \cos ax + C$$

$$\int \cos ax\, dx = \frac{1}{a} \sin ax + C$$

$$\int \tan ax\, dx = \frac{1}{a} \ln \sec ax + C = -\frac{1}{a} \ln \cos ax + C$$

$$\int \sec ax\, dx = \frac{1}{a} \ln (\sec ax + \tan ax) + C$$

$$\int \sin^2 ax\, dx = \frac{x}{2} - \frac{\sin 2ax}{4a} + C$$

$$\int \cos^2 ax\, dx = \frac{x}{2} + \frac{\sin 2ax}{4a} + C$$

$$\int x \sin ax\, dx = \frac{1}{a^2}(\sin ax - ax \cos ax) + C$$

$$\int x \cos ax \, dx = \frac{1}{a^2} (\cos ax + ax \sin ax) + C$$

$$\int e^{ax} \sin bx \, dx = \frac{e^{ax}}{a^2 + b^2} (a \sin bx - b \cos bx) + C$$

$$\int e^{ax} \cos bx \, dx = \frac{e^{ax}}{a^2 + b^2} (a \cos bx + b \sin bx) + C$$

$$\int \sin ax \sin bx \, dx = \frac{\sin (a - b)x}{2(a - b)} - \frac{\sin (a + b)x}{2(a + b)} + C, \quad a^2 \neq b^2$$

$$\int \sin ax \cos bx \, dx = -\frac{\cos (a - b)x}{2(a - b)} - \frac{\cos (a + b)x}{2(a + b)} + C, \quad a^2 \neq b^2$$

$$\int \cos ax \cos bx \, dx = \frac{\sin (a - b)x}{2(a - b)} + \frac{\sin (a + b)x}{2(a + b)} + C, \quad a^2 \neq b^2$$

$$\int \sinh ax \, dx = \frac{1}{a} \cosh ax + C$$

$$\int \cosh ax \, dx = \frac{1}{a} \sinh ax + C$$

$$\int \tanh ax \, dx = \frac{1}{a} \ln \cosh ax + C$$

$$\int \frac{dx}{ax + b} = \frac{1}{a} \ln |ax + b|$$

$$\int \frac{dx}{x^2 + a^2} = \frac{1}{a} \tan^{-1} \frac{x}{a} + C$$

$$\int \frac{x \, dx}{x^2 + a^2} = \frac{1}{2} \ln (x^2 + a^2) + C$$

$$\int \frac{x^2 \, dx}{x^2 + a^2} = x - a \tan^{-1} \frac{x}{a} + C$$

$$\int \frac{dx}{x^2 - a^2} = \begin{cases} \frac{1}{2a} \ln \dfrac{x - a}{x + a} + C, & x^2 > a^2 \\ \frac{1}{2a} \ln \dfrac{a - x}{a + x} + C, & x^2 < a^2 \end{cases}$$

$$\int \frac{dx}{\sqrt{a^2 - x^2}} = \sin^{-1} \frac{x}{a} + C$$

$$\int \frac{dx}{\sqrt{x^2 \pm a^2}} = \ln (x + \sqrt{x^2 \pm a^2}) + C$$

$$\int \frac{x \, dx}{\sqrt{x^2 + a^2}} = \sqrt{x^2 + a^2} + C$$

$$\int \frac{dx}{(x^2 + a^2)^{3/2}} = \frac{x/a^2}{\sqrt{x^2 + a^2}} + C$$

$$\int \frac{xdx}{(x^2 + a^2)^{3/2}} = -\frac{1}{\sqrt{x^2 + a^2}} + C$$

$$\int \frac{x^2dx}{(x^2 + a^2)^{3/2}} = \ln\left(\frac{\sqrt{x^2 + a^2}}{a} + \frac{x}{a}\right) - \frac{x}{\sqrt{x^2 + a^2}} + C$$

$$\int \frac{dx}{(x^2 + a^2)^2} = \frac{1}{2a^2}\left(\frac{x}{x^2 + a^2} + \frac{1}{a}\tan^{-1}\frac{x}{a}\right) + C$$

A.9 DEFINITE INTEGRALS

If m and n are integers, a, b, and c are constants,

$$\int_0^\pi \sin mx \sin nx \, dx = \int_0^\pi \cos mx \cos nx \, dx = \begin{cases} 0, & m \neq n \\ \pi/2, & m = n \end{cases}$$

$$\int_0^\pi \sin mx \cos nx \, dx = \begin{cases} 0, & m + n = \text{even} \\ \dfrac{2m}{m^2 - n^2}, & m + n = \text{odd} \end{cases}$$

$$\int_0^{2\pi} \sin mx \sin nx \, dx = \int_{-\pi}^\pi \sin mx \sin nx \, dx = \begin{cases} 0, & m \neq n \\ \pi, & m = n \end{cases}$$

$$\int_0^\infty \frac{\sin ax}{x} \, dx = \begin{cases} \pi/2, & a > 0 \\ 0, & a = 0 \\ -\pi/2, & a < 0 \end{cases}$$

$$\int_0^\infty \frac{\sin^2 x}{x^2} \, dx = \frac{\pi}{2}$$

$$\int_0^\infty \frac{\sin^2 ax}{x^2} \, dx = |a|\frac{\pi}{2}, \quad a > 0$$

In the following, $a > 0$ for the integrals to converge.

$$\int_0^\infty x^n e^{-ax} \, dx = \frac{n!}{a^{n+1}}$$

$$\int_0^\infty e^{-ax^2} \, dx = \frac{1}{2}\sqrt{\frac{\pi}{a}}$$

$$\int_{-\infty}^\infty e^{-ax^2} \, dx = \sqrt{\frac{\pi}{a}}$$

$$\int_{-\infty}^\infty e^{-(ax^2 + bx + c)} \, dx = \sqrt{\frac{\pi}{a}} \, e^{(b^2 - 4ac)/4a}$$

$$\int_0^\infty e^{-ax} \cos bx \, dx = \frac{a}{a^2 + b^2}$$

$$\int_0^\infty e^{-ax} \sin bx \, dx = \frac{b}{a^2 + b^2}$$

A.10 VECTOR IDENTITIES

If **A** and **B** are vector fields while U and V are scalar fields, then

$$\nabla (U + V) = \nabla U + \nabla V$$

$$\nabla (UV) = U \nabla V + V \nabla U$$

$$\nabla \left[\frac{U}{V}\right] = \frac{V(\nabla U) - U(\nabla V)}{V^2}$$

$$\nabla V^n = n \, V^{n-1} \, \nabla V \quad (n = \text{integer})$$

$$\nabla (\mathbf{A} \cdot \mathbf{B}) = (\mathbf{A} \cdot \nabla) \mathbf{B} + (\mathbf{B} \cdot \nabla) \mathbf{A} + \mathbf{A} \times (\nabla \times \mathbf{B}) + \mathbf{B} \times (\nabla \times \mathbf{A})$$

$$\nabla \cdot (\mathbf{A} + \mathbf{B}) = \nabla \cdot \mathbf{A} + \nabla \cdot \mathbf{B}$$

$$\nabla \cdot (\mathbf{A} \times \mathbf{B}) = \mathbf{B} \cdot (\nabla \times \mathbf{A}) - \mathbf{A} \cdot (\nabla \times \mathbf{B})$$

$$\nabla \cdot (V\mathbf{A}) = V \nabla \cdot \mathbf{A} + \mathbf{A} \cdot \nabla V$$

$$\nabla \cdot (\nabla V) = \nabla^2 V$$

$$\nabla \cdot (\nabla \times \mathbf{A}) = 0$$

$$\nabla \times (\mathbf{A} + \mathbf{B}) = \nabla \times \mathbf{A} + \nabla \times \mathbf{B}$$

$$\nabla \times (\mathbf{A} \times \mathbf{B}) = \mathbf{A} (\nabla \cdot \mathbf{B}) - \mathbf{B} (\nabla \cdot \mathbf{A}) + (\mathbf{B} \cdot \nabla)\mathbf{A} - (\mathbf{A} \cdot \nabla)\mathbf{B}$$

$$\nabla \times (V\mathbf{A}) = \nabla V \times \mathbf{A} + V(\nabla \times \mathbf{A})$$

$$\nabla \times (\nabla V) = \mathbf{0}$$

$$\nabla \times (\nabla \times \mathbf{A}) = \nabla(\nabla \cdot \mathbf{A}) - \nabla^2 \mathbf{A}$$

$$\oint_L \mathbf{A} \cdot d\mathbf{l} = \int_S \nabla \times \mathbf{A} \cdot d\mathbf{S}$$

$$\oint_L V d\mathbf{l} = -\int_S \nabla V \times d\mathbf{S}$$

$$\oint_S \mathbf{A} \cdot d\mathbf{S} = \int_v \nabla \cdot \mathbf{A} \, dv$$

$$\oint_S V d\mathbf{S} = \int_v \nabla V \, dv$$

$$\oint_S \mathbf{A} \times d\mathbf{S} = -\int_v \nabla \times \mathbf{A} \, dv$$

MATERIAL CONSTANTS

TABLE B.1 Approximate Conductivity* of Some
Common Materials at 20°C

Material	Conductivity (siemens/meter)
Conductors	
Silver	6.1×10^7
Copper (standard annealed)	5.8×10^7
Gold	4.1×10^7
Aluminum	3.5×10^7
Tungsten	1.8×10^7
Zinc	1.7×10^7
Brass	1.1×10^7
Iron (pure)	10^7
Lead	5×10^6
Mercury	10^6
Carbon	3×10^4
Water (sea)	4
Semiconductors	
Germanium (pure)	2.2
Silicon (pure)	4.4×10^{-4}
Insulators	
Water (distilled)	10^{-4}
Earth (dry)	10^{-5}
Bakelite	10^{-10}
Paper	10^{-11}
Glass	10^{-12}
Porcelain	10^{-12}
Mica	10^{-15}
Paraffin	10^{-15}
Rubber (hard)	10^{-15}
Quartz (fused)	10^{-17}
Wax	10^{-17}

* The values vary from one published source to another because there
are many varieties of most materials, and conductivity is sensitive to
temperature, moisture content, impurities, and the like.

TABLE B.2 Approximate Dielectric Constant or Relative Permittivity (ε_r) and Strength of Some Common Materials*

Material	Dielectric Constant ε_r (Dimensionless)	Dielectric Strength E (V/m)
Barium titanate	1200	7.5×10^6
Water (sea)	80	
Water (distilled)	81	
Nylon	8	
Paper	7	12×10^6
Glass	5–10	35×10^6
Mica	6	70×10^6
Porcelain	6	
Bakelite	5	20×10^6
Quartz (fused)	5	30×10^6
Rubber (hard)	3.1	25×10^6
Wood	2.5–8.0	
Polystyrene	2.55	
Polypropylene	2.25	
Paraffin	2.2	30×10^6
Petroleum oil	2.1	12×10^6
Air (1 atm)	1	3×10^6

* The values given are only typical; they vary from one published source to another because of the different varieties of most materials and the dependence of ε_r on temperature, humidity, and the like.

TABLE B.3 Relative Permeability (μ_r) of Some Materials*

Material	μ_r
Diamagnetic	
Bismuth	0.999833
Mercury	0.999968
Silver	0.9999736
Lead	0.9999831
Copper	0.9999906
Water	0.9999912
Hydrogen (STP)	$\simeq 1.0$
Paramagnetic	
Oxygen (STP)	0.999998
Air	1.00000037
Aluminum	1.000021
Tungsten	1.00008
Platinum	1.0003
Manganese	1.001
Ferromagnetic	
Cobalt	250
Nickel	600
Soft iron	5000
Silicon–iron	7000

* The values given are only typical; they vary from one published source to another owing to different varieties of most materials.

MATLAB

MATLAB has become a powerful tool of technical professionals worldwide. The term MATLAB is an abbreviation for **Mat**rix **Lab**oratory, implying that MATLAB is a computational tool that employs matrices and vectors/arrays to carry out numerical analysis, signal processing, and scientific visualization tasks. Because MATLAB uses matrices as its fundamental building blocks, one can write mathematical expressions involving matrices just as easily as one would on paper. MATLAB is available for Macintosh, Unix, and Windows operating systems. A student version of MATLAB is available for PCs. A copy of MATLAB can be obtained from

The Mathworks, Inc.
3 Apple Hill Drive
Natick, MA 01760-2098
Phone: (508) 647-7000
Website: http://www.mathworks.com

The brief introduction on MATLAB presented in this appendix is sufficient for solving problems in this book. Additional information about MATLAB can be found in MATLAB books and from online help. The best way to learn MATLAB is to learn the basics and start working with the program right away.

C.1 MATLAB FUNDAMENTALS

The Command window is the primary area in which the user interacts with MATLAB. A little later, we will learn how to use the text editor to create M-files, which allow executing sequences of commands. For now, we focus on how to work in the Command window. We will first learn how to use MATLAB as a calculator. We do so by using the algebraic operators in Table C.1.

TABLE C.1 Basic Operations

Operation	MATLAB Formula	
Addition	a+b	
Division (right)	a/b	(means $a \div b$)
Division (left)	a\b	(means $b \div a$)
Multiplication	a*b	
Power	a^b	
Subtraction	a—b	

To begin to use MATLAB, we use these operators. Type commands to the MATLAB prompt ">>" in the Command window (correct any mistakes by backspacing) and press the <Enter> key. For example,

```
» a=2; b=4; c=-6;
» dat = b^2 - 4*a*c
dat =
    64
» e=sqrt(dat)/10
e =
   0.8000
```

The first command assigns the values 2, 4, and –6 to the variables a, b, and c, respectively. MATLAB does not respond because this line ends with a semicolon. The second command sets dat to $b^2 - 4ac$, and MATLAB returns the answer as 64. Finally, the third line sets e equal to the square root of dat and divides by 10. MATLAB prints the answer as 0.8. As function $sqrt$ is used here, other mathematical functions listed in Table C.2 can be used. Table C.2 provides just a small sample of MATLAB functions. Others can be obtained from the online help. To get help, type

```
>> help
```

[a long list of topics comes up]

TABLE C.2 Typical Elementary Math Functions

Function	Remark
abs(x)	Absolute value or complex magnitude of x
acos, acosh(x)	Inverse cosine and inverse hyperbolic cosine of x in radians
acot, acoth(x)	Inverse cotangent and inverse hyperbolic cotangent of x in radians
angle(x)	Phase angle (in radians) of a complex number x
asin, asinh(x)	Inverse sine and inverse hyperbolic sine of x in radians
atan, atanh(x)	Inverse tangent and inverse hyperbolic tangent of x in radians
conj(x)	Complex conjugate of x
cos, cosh(x)	Cosine and hyperbolic cosine of x in radians
cot, coth(x)	Cotangent and hyperbolic cotangent of x in radians
exp(x)	Exponential of x
fix	Round toward zero
imag(x)	Imaginary part of a complex number x
log(x)	Natural logarithm of x
log2(x)	Logarithm of x to base 2
log10(x)	Common logarithms (base 10) of x
real(x)	Real part of a complex number x
sin, sinh(x)	Sine and hyperbolic sine of x in radians
sqrt(x)	Square root of x
tan, tanh(x)	Tangent and hyperbolic tangent of x in radians

and for a specific topic, type the command name. For example, to get help on *log to base 2*, type

```
>> help log2
```

[a help message on the log function follows]

Note that MATLAB is case sensitive, so that sin(a) is not the same as sin(A).

Try the following examples:

```
>> 3^(log10(25.6))
>> y=2* sin(pi/3)
>>exp(y+4-1)
```

In addition to operating on mathematical functions, MATLAB easily allows one to work with vectors and matrices. A vector (or array) is a special matrix with one row or one column. For example,

```
>> a = [1 -3 6 10 -8 11 14];
```

is a row vector. Defining a matrix is similar to defining a vector. For example, a 3×3 matrix can be entered as

```
>> A = [1 2 3; 4 5 6; 7 8 9]
```

or as

```
>> A = [1 2 3
         4 5 6
         7 8 9]
```

In addition to the arithmetic operations that can be performed on a matrix, the operations in Table C.3 can be implemented.

We can use the operations in Table C.3 to manipulate matrices as follows.

```
» B = A'
B =
     1    4    7
     2    5    8
     3    6    9
» C = A + B
```

TABLE C.3 Matrix Operations

Operation	Remark
A'	Finds the transpose of matrix A
det(**A**)	Evaluates the determinant of matrix A
inv(**A**)	Calculates the inverse of matrix A
eig(**A**)	Determines the eigenvalues of matrix A
diag(**A**)	Finds the diagonal elements of matrix A
exp(**A**)	Exponential of matrix A

```
C =
    2    6   10
    6   10   14
   10   14   18
» D = A^3 - B*C
D =
      372    432    492
      948   1131   1314
     1524   1830   2136
» e= [1 2; 3 4]
e =
    1    2
    3    4
» f=det(e)
f =
   -2
» g = inv(e)
g =
   -2.0000    1.0000
    1.5000   -0.5000
» H = eig(g)
H =
   -2.6861
    0.1861
```

Note that not all matrices can be inverted. A matrix can be inverted if and only if its determinant is nonzero. Special matrices, variables, and constants are listed in Table C.4. For example, type

```
>> eye(3)
ans=
    1   0   0
    0   1   0
    0   0   1
```

to get a 3 × 3 identity matrix.

TABLE C.4 Special Matrices, Variables, and Constants

Matrix/Variable/Constant	Remark
eye	Identity matrix
ones	An array of ones
zeros	An array of zeros
i or j	Imaginary unit or sqrt(−1)
pi	3.1416
NaN	Not a number
inf	Infinity
eps	A very small number, $2.2e^{-16}$
rand	Random element

C.2 USING MATLAB TO PLOT

To plot using MATLAB is easy. For a two-dimensional plot, use the **plot** command with two arguments as

```
>> plot(xdata,ydata)
```

where *xdata* and *ydata* are vectors of the same length containing the data to be plotted.

For example, suppose we want to plot `y=10*sin(2*pi*x)` from `0` to `5*pi`, we will proceed with the following commands:

```
>> x = 0:pi/100:5*pi;      % x is a vector, 0 <= x <= 5*pi,
                             increments of pi/100
>> y = 10*sin(2*pi*x);     % create a vector y
>> plot(x,y);              % create the plot
```

With this, MATLAB responds with the plot in Figure C.1.

MATLAB will let you graph multiple plots together and distinguish them with different colors. This capability is obtained with the command **plot**(*x*data, *y*data, 'color'), where the color is indicated by using a character string from the options listed in Table C.5.

For example,

```
>> plot(x1, y1, 'r', x2,y2, 'b', x3,y3, '--');
```

will graph data (*x*1, *y*1) in red, data (*x*2, *y*2) in blue, and data (*x*3, *y*3) in dashed lines, all on the same plot.

MATLAB also allows for logarithm scaling. Rather than the **plot** command, we use:

```
loglog           log(y) versus log(x)
semilogx         y versus log(x)
semilogy         log(y) versus x
```

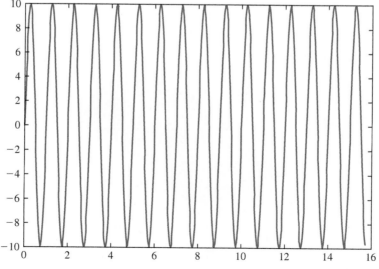

FIGURE C.1 MATLAB plot of `y=10*sin(2*pi*x)`.

TABLE C.5 Various Color and Line Types

y	Yellow	.	Point
m	Magenta	o	Circle
c	Cyan	x	×-mark
r	Red	+	Plus
g	Green	–	Solid
b	Blue	*	Star
w	White	:	Dotted
k	Black	–.	Dashdot
		– –	Dashed

Three-dimensional plots are drawn by using the functions *mesh* and *meshgrid* (mesh domain). For example, to draw the graph of z = x*exp(-x^2-y^2) over the domain -1 <x, y<1, we type the following commands:

```
>> xx = -1:.1:1;
» yy = xx;
» [x,y] = meshgrid(xx,yy);
» z=x.*exp(-x.^2 -y.^2);
» mesh(z);
```

(The dot symbol used in x. and y. allows element-by-element multiplication.) The result is shown in Figure C.2.

Other plotting commands in MATLAB are listed in Table C.6. The **help** command can be used to find out how each of these is used.

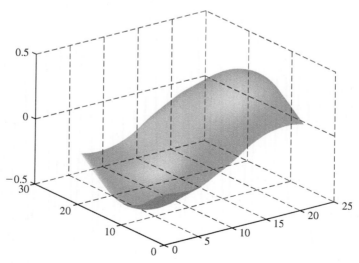

FIGURE C.2 A three-dimensional plot.

TABLE C.6 Other Plotting Commands

Command	Comments
bar(x,y)	A bar graph
contour(z)	A contour plot
errorbar (x,y,l,u)	A plot with error bars
hist(x)	A histogram of the data
plot3(x,y,z)	A three-dimensional version of plot ($\cdots$)
polar(r, angle)	A polar coordinate plot
stairs(x,y)	A stairstep plot
stem(x)	Plots the data sequence as stems
subplot(m,n,p)	Multiple ($m \times n$) plots per window
surf(x,y,x,c)	A plot of three-dimensional colored surface

C.3 PROGRAMMING WITH MATLAB

So far MATLAB has been used as a calculator. You can also use MATLAB to create your own program. The command line editing in MATLAB can be inconvenient if one has several lines to execute. To avoid this problem, one creates a program that is a sequence of statements to be executed. If you are in the Command window, click **File/New/M-files** to open a new file in the MATLAB Editor/Debugger or simple text editor. Type the program and save the program in a file with an extension .m, say filename.m (it is for this reason such files are called M-files). Once a program has been saved as an M-file, exit the Debugger window. You are now back in the Command window. Type the file without the extension .m to get results. For example, the plot that was made earlier in Figure C.1 can be improved by adding a title and labels and typing as an M-file called example1.m as follows:

```
x = 0:pi/100:5*pi;          % x is a vector, 0 <= x <= 5*pi,
                              increments of pi/100
y = 10*sin(2*pi*x);         % create a vector y
plot(x,y);                  % create the plot
xlabel('x (in radians)');   % label the x-axis
ylabel('10*sin(2*pi*x)');   % label the y-axis
title('A sine function');   % title the plot
grid                        % add grid
```

Once the file is saved as example1.m and we exit text editor, we type

```
>> example1
```

in the Command window, and hit <Enter> to obtain the result shown in Figure C.3.

To allow flow control in a program, certain relational and logical operators are necessary. These are shown in Table C.7. Perhaps the most commonly used flow control statements are *for* and *if*. The *for* statement is used to create a loop or a repetitive procedure and has the general form

for x = array
  [commands]
end

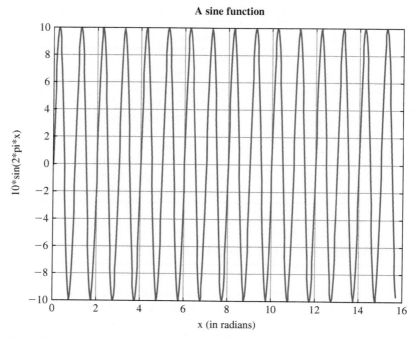

FIGURE C.3 MATLAB plot of `y=10*sin(2*pi*x)` with title and labels.

The *if* statement is used when certain conditions must be met before an expression is executed. It has the general form

if expression
   [commands if expression is True]
else
   [commands if expression is False]
end

TABLE C.7 Relational and Logical Operators

Operator	Remark
<	less than
<=	less than or equal
>	greater than
>=	greater than or equal
==	equal
~=	not equal
&	and
|	or
~	not

For example, suppose we have an array $y(x)$ and we want to determine the minimum value of y and its corresponding index x. This can be done by creating an M-file as follows.

```
% example2.m
% This program finds the minimum y value
% and its corresponding index
x = [1 2 3  4 5 6   7 8 9 10]; %the nth term in y
y = [3 9 15 8 1 0 -2 4 12 5];
min1 = y(1);
for k=1:10
  min2=y(k);
  if(min2 < min1)
   min1 = min2;
   xo = x(k);
  else
   min1 = min1;
  end
end
diary
min1, xo
diary off
```

Note the use of *for* and *if* statements. When this program is saved as `example2.m`, we execute it in the Command window and obtain the minimum value of y as –2 and the corresponding value of x as 7, as expected:

```
» example2
min1 =
  -2
xo =
   7
```

If we are not interested in the corresponding index, we could do the same thing using the command

```
>> min(y)
```

The following tips are helpful in working effectively with MATLAB:

- Comment your M-file by adding lines beginning with a % character.
- To suppress output, end each command with a semicolon (;). You may remove the semicolon when debugging the file.
- Press up and down arrow keys to retrieve previously executed commands.
- If your expression does not fit on one line, use an ellipse (...) at the end of the line and continue on the next line. For example, MATLAB considers
  ```
  y = sin(x + log10(2x + 3)) + cos(x + ...
  log10(2x+3));
  ```
  as one line of expression.
- Keep in mind that variable and function names are case sensitive.

C.4 SOLVING EQUATIONS

Consider the general system of n simultaneous equations as

$$a_{11}x_1 + a_{12}x_2 + \cdots + a_{1n}x_n = b_1$$

$$a_{21}x_1 + a_{22}x_2 + \cdots + a_{2n}x_n = b_2$$

$$\vdots$$

$$a_{n1}x_1 + a_{n2}x_2 + \cdots + a_{nn}x_n = b_n$$

or in matrix form

$$AX = B$$

where

$$A = \begin{bmatrix} a_{11} & a_{12} & \cdots & a_{1n} \\ a_{21} & a_{22} & \cdots & a_{2n} \\ \cdots & \cdots & \cdots & \cdots \\ a_{n1} & a_{n2} & a_{n3} & a_{nn} \end{bmatrix}, \quad X = \begin{bmatrix} x_1 \\ x_2 \\ \cdots \\ x_n \end{bmatrix}, \quad B = \begin{bmatrix} b_1 \\ b_2 \\ \cdots \\ b_n \end{bmatrix}$$

The square matrix A is known as the coefficient matrix, while X and B are vectors. We are seeking X, the solution vector. There are two ways to solve for X in MATLAB. First, we can use the backslash operator (\\) so that

$$X = A \backslash B$$

Second, we can solve for X as

$$X = A^{-1}B$$

which in MATLAB is the same as

$$X = \text{inv}(A)*B$$

We can also solve equations by using the command **solve**. For example, given the quadratic equation $x^2 + 2x - 3 = 0$, we obtain the solution by using the MATLAB command

```
>> [x]=solve('x^2 + 2*x - 3 =0')
x =
[-3]
[ 1]
```

indicating that the solutions are $x = -3$ and $x = 1$. Of course, we can use the command **solve** for a case involving two or more variables. We will see that in the following example.

EXAMPLE C.1

Use MATLAB to solve the following simultaneous equations:

$$25x_1 - 5x_2 - 20x_3 = 50$$
$$-5x_1 + 10x_2 - 4x_3 = 0$$
$$-5x_1 - 4x_2 + 9x_3 = 0$$

Solution:

We can use MATLAB to solve this in two ways:

Method 1:

The given set of simultaneous equations could be written as

$$\begin{bmatrix} 25 & -5 & -20 \\ -5 & 10 & -4 \\ -5 & -4 & 9 \end{bmatrix} \begin{bmatrix} x_1 \\ x_2 \\ x_3 \end{bmatrix} = \begin{bmatrix} 50 \\ 0 \\ 0 \end{bmatrix} \quad \text{or} \quad AX = B$$

We obtain matrix A and vector B and enter them in MATLAB as follows:

```
»  A = [25 -5 -20; -5 10 -4; -5 -4 9]
A =
   25   -5   -20
   -5   10   -4
   -5   -4    9
»  B = [50 0 0]'
B =
   50
    0
    0
»  X = inv(A)*B
X =
   29.6000
   26.0000
   28.0000
»  X=A\B
X =
   29.6000
   26.0000
   28.0000
```

Thus, $x_1 = 29.6$, $x_2 = 26$, and $x_3 = 28$.

Method 2:

Since the equations are not many in this case, we can use the command **solve** to obtain the solution of the simultaneous equations as follows:

```
[x1,x2,x3]=solve('25*x1 - 5*x2 - 20*x3=50', '-5*x1 + 10*x2 - 4*x3
=0', '-5*x1 - 4*x2 + 9*x3=0')
x1 =
```

```
148/5
x2 =
26
x3 =
28
```

The same result obtained by Method 1 appears again.

PRACTICE PROBLEM C.1

Use MATLAB to solve the problem in the following simultaneous equations.

$$3x_1 - x_2 - 2x_3 = 1$$
$$-x_1 + 6x_2 - 3x_3 = 0$$
$$-2x_1 - 3x_2 + 6x_3 = 6$$

Answer: $x_1 = 3 = x_3$, $x_2 = 2$.

C.5 PROGRAMMING HINTS

A good program should be well documented, of reasonable size, and capable of performing computations with reasonable accuracy within a reasonable amount of time. The following hints may make writing and running MATLAB programs easier.

- Use the fewest commands possible and avoid execution of extra commands. This is particularly applicable for loops.
- Use matrix operations directly as much as possible and avoid *for*, *do*, and/or *while* loops if possible.
- Make effective use of functions for executing a series of commands over several times in a program.
- When unsure about a command, take advantage of the help capabilities of the software.
- Start each file with comments to help you remember what it is all about.
- When writing a long program, save frequently. If possible, avoid a long program; break it down into smaller subroutines.

C.6 OTHER USEFUL MATLAB COMMANDS

Table C.8 lists some common useful MATLAB commands that may be used in this book.

TABLE C.8 Other Useful MATLAB Commands

Command	Explanation
diary	Saves screen display output in text format
mean	Mean value of a vector
min(max)	Minimum (maximum) of a vector
grid	Adds a grid mark to the graphic window
poly	Converts a collection of roots into a polynomial
roots	Finds the roots of a polynomial
sort	Sorts the elements of a vector
sound	Plays vector as sound
std	Standard deviation of a data collection
sum	Sum of elements of a vector

THE COMPLETE SMITH CHART

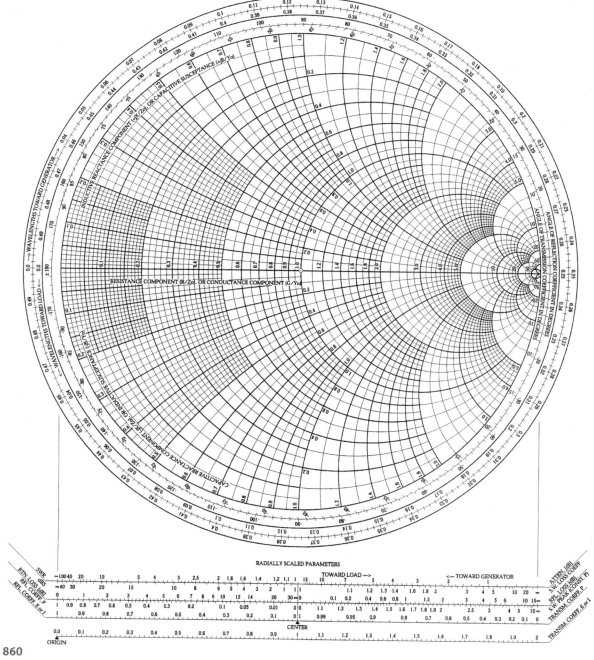

RADIALLY SCALED PARAMETERS

ANSWERS TO ODD-NUMBERED PROBLEMS

CHAPTER 1

1.1 $0.8704\mathbf{a}_x - 0.3482\mathbf{a}_y - 0.3482\mathbf{a}_z$

1.3 $2\mathbf{a}_x + 9\mathbf{a}_y + \mathbf{a}_z$

1.5   (a) $\mathbf{a}_x - \mathbf{a}_y + 8\mathbf{a}_z$
     (b) -32
     (c) $86.69°$

1.7   (a) $T(4,6,-1)$ and $S(10,12,8)$
     (b) $6\mathbf{a}_x + 6\mathbf{a}_y + 9\mathbf{a}_z$
     (c) 12.37

1.9   (a) $2\mathbf{a}_x + 10\mathbf{a}_z$
     (b) -2
     (c) $100.14°$

1.11   (a) 7.3485
     (b) 4
     (c) -4
     (d) 206
     (e) $16\mathbf{a}_x + 12\mathbf{a}_y + 8\mathbf{a}_z$
     (f) $162.3°$
     (g) $86.45°$

1.13 $-13, -21\mathbf{a}_x - 2\mathbf{a}_y + 6\mathbf{a}_z, 120.66°$

1.15   (a) $(\mathbf{B} \cdot \mathbf{A})\mathbf{A} - (\mathbf{A} \cdot \mathbf{A})\mathbf{B}$
     (b) $-A^2(\mathbf{A} \times \mathbf{B})$

1.17   (a) 7.6811,   (b) $-2\mathbf{a}_y - 5\mathbf{a}_z$,   (c) $42.57°$,   (d) 11.023,   (e) 17.31

1.19 8.646

1.21 $-18\mathbf{a}_x - 5\mathbf{a}_y + 4\mathbf{a}_z$

1.23   (a) -2.8577
     (b) $-0.2857\mathbf{a}_x + 0.8571\mathbf{a}_y + 0.4286\mathbf{a}_z$
     (c) $65.91°$

1.25   (a) 10, (b) $15\mathbf{a}_x - 30\mathbf{a}_y - 15\mathbf{a}_z$, (c) $5\mathbf{a}_x + 5\mathbf{a}_z$

1.27   (a) $0.8242\mathbf{a}_x + 0.1374\mathbf{a}_y + 0.5494\mathbf{a}_z$

(b) $100 = 4x^2y^2 + x^2 + 2xz + z^2 + z^4$

1.29   (a) $G = 6.403, H = 7.348$

(b) -18

(c) $\theta_{GH} = 112.5°$

1.31   (a) 6.403

(b) $1.286\mathbf{a}_x - 2.571\mathbf{a}_y + 3.857\mathbf{a}_z$

(c) $\pm\mathbf{a}_x$

CHAPTER 2

2.1   (a) $P(5.3852, 68.2°, 1)$, $P(5.477, 79.48°, 68.2°)$

(b) $Q(5, 306.88°, 0)$, $Q(5.90°, 306.88°)$

(c) $R(6.325, 18.43°, -4)$, $R(7.483, 122.31°, 18.43°)$

2.3   (a) $(6.324, 71.56°, -4)$

(b) $(7.483, 147.69°, 71.56°)$

2.5 $T(x, y, z) = (7.5, 4.33, 5)$

$T(\rho, \phi, z) = (8.66, 30°, 5)$

2.7   (a) $\dfrac{1}{\sqrt{\rho^2 + z^2}}(\rho\mathbf{a}_\rho + 4\mathbf{a}_z)$, $\left(\sin^2\theta + \dfrac{4}{r}\sin\theta\right)\mathbf{a}_r + \sin\theta\left(\cos\theta - \dfrac{4}{r}\right)\mathbf{a}_\theta$

(b) $\dfrac{\rho^2}{\sqrt{\rho^2 + z^2}}(\rho\mathbf{a}_\rho + z\mathbf{a}_z)$, $r^2 \sin^2\theta\,\mathbf{a}_r$

2.9 $-3\mathbf{a}_x + 2\mathbf{a}_y + 4\mathbf{a}_z$

2.11 $\dfrac{4}{(x^2 + y^2 + z^2)}[x\mathbf{a}_x + y\mathbf{a}_y + z\mathbf{a}_z]$

2.13   (a) $\mathbf{B} = \rho\cos\phi\,\mathbf{a}_z$

(b) $\mathbf{B} = 0.5r\sin(2\theta)\cos\phi\,\mathbf{a}_r - r\sin^2\theta\cos\phi\,\mathbf{a}_\theta$

2.15   Proof

2.17   (a) 1, (b) 1, (c) -0.4472

2.19   (a) $8\mathbf{a}_\rho + 2\mathbf{a}_\phi - 7\mathbf{a}_z$

(b) 7

(c) $-16\mathbf{a}_\rho + 29\mathbf{a}_\phi - 10\mathbf{a}_z$

(d) $78.56°$

2.21 $\rho z\sin 2\phi\,\mathbf{a}_\rho + \rho z\cos 2\phi\,\mathbf{a}_\phi + 0.5\rho^2\sin 2\phi\,\mathbf{a}_z$

2.23   (a)

$r\sin\theta[\sin\phi\cos\theta(r\sin\theta + \cos\phi)\mathbf{a}_r + \sin\phi(r\cos^2\theta - \sin\theta\cos\phi)\mathbf{a}_\theta + 3\cos\phi\,\mathbf{a}_\phi]$,

$5\mathbf{a}_\theta - 21.21\mathbf{a}_\phi$

(b) $\sqrt{\rho^2 + z^2}\left(\rho\mathbf{a}_\rho + \dfrac{\rho}{\rho^2 + z^2}\mathbf{a}_\phi + z\mathbf{a}_z\right)$, $4.472\mathbf{a}_\rho + 0.8944\mathbf{a}_\phi + 2.236\mathbf{a}_z$

2.25   64

2.27   (a) 50.07°, (b) $0.9428\mathbf{a}_\rho + 0.2357\mathbf{a}_\phi - 0.2357\mathbf{a}_z$

2.29   9

2.31   (a) An infinite line parallel to the z-axis.
       (b) Point $(2, -1, 10)$.
       (c) A circle of radius $r\sin\theta = 5$, i.e., the intersection of a cone and a sphere.
       (d) An infinite line parallel to the z-axis.
       (e) A semi-infinite line parallel to the $x-y$ plane.
       (f) A semi-circle of radius 5 in the $y-z$ plane.

2.33   2.5

CHAPTER 3

3.1   (a) 2.356
       (b) 0.5236
       (c) 4.189

3.3   (a) 6
       (b) 110
       (c) 4.538

3.5   11.502

3.7   2047.5

3.9   (a) -50
       (b) -39.5

3.11 -1.5

3.13   2

3.15   484.58

3.17   (a) $(6y - 2z)\mathbf{a}_x + 6x\mathbf{a}_y + (1 - 2x)\mathbf{a}_z$
       (b) $(10\cos\phi - z)\mathbf{a}_\rho - 10\sin\phi\mathbf{a}_\phi - \rho\mathbf{a}_z$
       (c) $-\dfrac{2}{r^2}\cos\phi\mathbf{a}_r - \dfrac{2\sin\phi}{r^2\sin\theta}\mathbf{a}_\phi$

3.19   Proof

3.21 $0.4082\mathbf{a}_x - 0.8165\mathbf{a}_y + 0.4082\mathbf{a}_z$

3.23 -40.42

3.25   2.36, $0.5413\mathbf{a}_x + 1.624\mathbf{a}_y + 1.624\mathbf{a}_z$

3.27   (a) $3y - x$
       (b) $2z^2 + \sin 2\phi + 2\rho\sin^2\phi$
       (c) 3

3.29   Proof

3.31   (a) $6YZ\mathbf{a}_x + 3XY^2\mathbf{a}_y + 3X^2YZ\mathbf{a}_z$
         (b) $4YZ\mathbf{a}_x + 3XY^2\mathbf{a}_y + {}_4X^2YZ\mathbf{a}_z$
         (c) $6XYZ + 3XY^3 + 3X^2YZ^2$
         (d) $2(X^2 + Y^2 + Z^2) = 2r^2$

3.33   (a) 209.44
         (b) 209.44

3.35   24

3.37   37.7

3.39   198.97

3.41   (a) $-y^2\mathbf{a}_X + 2z\mathbf{a}_y - x^2\mathbf{a}_Z, \quad 0$
         (b) $(\rho^2 - 3z^2)\mathbf{a}_\phi + 4\rho^2\mathbf{a}_z, \quad 0$
         (c) $-\dfrac{\sin\phi}{r^3\sin\theta}\mathbf{a}_r + \dfrac{\cos\phi}{r^3\sin\theta}\mathbf{a}_\theta + \dfrac{\cos\phi}{r^3}\mathbf{a}_\phi, \quad 0$

3.43   Proof

3.45 -9.4956

3.47   0

3.49   (a) 4π
         (b) 4π
         (c) -4π
         (d) 2.2767
         (e) 7.2552
         (f) 9.532

3.51   (a) $2z + 5x + 8$,   (b) $8\mathbf{a}_x + 2x\mathbf{a}_y + 5y\mathbf{a}_z$

3.53   Proof

3.55   (a) $6(x + y + z)$
         (b) $\left(\dfrac{-3z^2}{\rho} + 2\rho\right)\sin 2\phi$
         (c) $6 + 4\cos\theta\sin\phi - \dfrac{\cos\theta\sin\phi}{\sin^2\theta}$

3.57   (a) $\dfrac{\mathbf{r}}{r^2}$,   (b) $\dfrac{1}{r^2}$

3.59 $2\rho z\cos\phi\mathbf{a}_\rho - \rho z\sin\phi\mathbf{a}_\phi + \rho^2\cos\phi\mathbf{a}_z, \quad 3z\cos\phi$

3.61   Proof

3.63   Proof

3.65   (a) $\mathbf{G}$ is irrotational, (b) 8, (c) -6

3.67   Proof

3.69   Proof

CHAPTER 4

4.1 $-1.8291\mathbf{a}_x - 0.5226\mathbf{a}_y + 1.3065\mathbf{a}_z$ mN

4.3   (a) $\dfrac{Q}{2\pi\varepsilon_0 a^2}\mathbf{a}_x$

         (b) $\dfrac{-Q}{4\sqrt{2}\pi\varepsilon_0 a^2}\mathbf{a}_x$

         (c) $\dfrac{-Q}{10\sqrt{5}\pi\varepsilon_0 a^2}\mathbf{a}_x + \dfrac{Q}{4\pi\varepsilon_0 a^2}\left[1 - \dfrac{1}{5\sqrt{5}}\right]\mathbf{a}_z$

4.5   (a) 0.5 C, (b) 1.206 μC, (c) 1579.1 C

4.7   10.472 mC

4.9   288 mC

4.11   (a) 4 nC,   (b) $3.6 \times 1.0^{-5}\mathbf{a}_z$ V/m

4.13 $\dfrac{\rho_s h}{4\pi\varepsilon_0}\left[\dfrac{1}{\sqrt{a^2 + h^2}} - \dfrac{1}{\sqrt{b^2 + h^2}}\right]\mathbf{a}_z$

4.15 $-151.7\mathbf{a}_x - 303.5\mathbf{a}_y$

4.17   (a) $565.5\mathbf{a}_x - 1131\mathbf{a}_y - 1696.5\mathbf{a}_z$ kV/m
         (b) $-565.5\mathbf{a}_x - 1131\mathbf{a}_y - 1696.5\mathbf{a}_z$ kV/m
         (c) $565.5\mathbf{a}_x + 1131\mathbf{a}_y + 1696.5\mathbf{a}_z$ kV/m

4.19   Proof

4.21   (a) $0.32\mathbf{a}_x$ μC/m^2
         (b) -51.182 μC

4.23 -1.326 nC

4.25   (a) $8y$ C/m^2
         (b) $6\sin\phi + 4z$ C/m^3
         (c) 0

4.27 $-2.7\mathbf{a}_r$ mC/m^2

4.29   (a) $2y$ C/m^3
         (b) 1/3 C
         (c) 1 C

4.31   Proof

4.33   (a) 125.7 mC, 377 mC, (b) $\mathbf{0}, 1.2\mathbf{a}_r$ mC/m^2

4.35   1.325 V

4.37   (a) 135 kV, (b) 135 kV, (c) 135 kV

4.39 $-15.38\mathbf{a}_\rho - 7.688\mathbf{a}_\phi + 30.75\mathbf{a}_z$ V/m

4.41 (a) $\dfrac{8\pi}{15}a^3\rho_o$

(b) $\dfrac{2a^3\rho_o}{15\varepsilon_o r^2}\mathbf{a}_r,\ \dfrac{2a^3\rho_o}{15\varepsilon_o r}$

(c) $\dfrac{\rho_o}{\varepsilon_o}\left(\dfrac{r}{3} - \dfrac{r^3}{5a^2}\right)\mathbf{a}_r,\quad \dfrac{\rho_o}{\varepsilon_o}\left(\dfrac{r^4}{20a^2} - \dfrac{r^2}{6}\right) + \dfrac{a^2\rho_o}{4\varepsilon_o}$

(d) Proof

4.43 $-30\ \text{C/m}^3$

4.45 $\dfrac{2\varepsilon_o E_o}{a},\ \ 0 < \rho < a$

4.47 (a) $0.1592\ \text{nC/m}^2$,   (b) $480\ \mu\text{J}$

4.49 $-8\ \text{V}$

4.51 $45.24\ \text{mJ}$

4.53 (a) $-e^{-z}\sin\phi\,\mathbf{a}_\rho - e^{-z}\cos\phi\,\mathbf{a}_\phi + \rho e^{-z}\sin\phi\,\mathbf{a}_z$

(b) Proof

4.55 $-\dfrac{Q}{4\pi\varepsilon_o}\left(\dfrac{1}{a} - \dfrac{1}{b}\right)$

4.57 Not a genuine EM field;   0.5858 C

4.59 (a) $-1.136\mathbf{a}_y\ \text{kV/m}$

(b) $(\mathbf{a}_x + 0.2\mathbf{a}_y) \times 10^7\ \text{m/s}$

4.61 Proof, $\dfrac{Qd}{4\pi\varepsilon_o r^3}(2\sin\theta\sin\phi\,\mathbf{a}_r - \cos\theta\sin\phi\,\mathbf{a}_\theta - \cos\phi\,\mathbf{a}_\phi)$

4.63 $54.74°,\ 125.26°$

4.65 $1.886\ \text{nJ}$

4.67 $\dfrac{Q^2}{8\pi\varepsilon_o a}$

4.69 (a) $-e^{-z}\sin\phi\,\mathbf{a}_\rho - e^{-z}\cos\phi\,\mathbf{a}_\phi + \rho e^{-z}\sin\phi\,\mathbf{a}_z$

(b) $8.512\ \text{pJ}$

CHAPTER 5

5.1 $-0.1172\ \text{A}$

5.3 $100\ \text{A}$

5.5 $25.13\ \text{A}$

5.7 (a) $33.95\ \text{m}\Omega$

(b) $265.1\ \text{A}$

(c) $2.386\ \text{kW}$

5.9 The silver wire is longer.

5.11   (a) 0.27 mΩ

(b) 50.3 A (copper), 9.7 A   (steel)

(c) 0.322 mΩ

5.13   130.86 A

5.15 $3.6 \times 10^{-8}\,\mathbf{a}_x$ C/m^2,   1.0407

5.17 $2.261\mathbf{a}_\rho$kV/m, $70\mathbf{a}_\rho$nC/m^2

5.19   (a) -86.86×10^{-18} C

(b) 883.5×10^{-18} C

(c) -796.61×10^{-18} C

5.21   2.16 kV,

$2.16\mathbf{a}_x + 0.432\mathbf{a}_y + 2.16\mathbf{a}_z$ kV/m, $133.69\mathbf{a}_x + 26.74\mathbf{a}_y + 133.69\mathbf{a}_z$ nC/m^2

5.23   For $r < a$, $\mathbf{E} = 0 = \mathbf{D} = \mathbf{P}$

For $a < r < b$, $\mathbf{D} = \dfrac{\mathbf{a}_r}{\pi r^2}$, $\mathbf{E} = \dfrac{\mathbf{a}_r}{2\varepsilon_o \pi r^2}$, $\mathbf{P} = \dfrac{\mathbf{a}_r}{2\pi r^2}$

For $b < r < c$, $\mathbf{D} = \dfrac{-\mathbf{a}_r}{2\pi r^2}$, $\mathbf{E} = \dfrac{-\mathbf{a}_r}{10\varepsilon_o \pi r^2}$, $\mathbf{P} = \dfrac{-4\mathbf{a}_r}{10\pi r^2}$

For $r > c$, $\mathbf{D} = \dfrac{2\mathbf{a}_r}{\pi r^2}$, $\mathbf{E} = \dfrac{2\mathbf{a}_r}{\varepsilon_o \pi r^2}$, $\mathbf{P} = 0$

5.25   1.733

5.27

(a) $\dfrac{\rho_o a^2 (2\varepsilon_r + 1)}{6\varepsilon_o \varepsilon_r}$

(b) $\dfrac{\rho_o a^2}{3\varepsilon_o}$

5.29   (a) Possible
(b) Not possible
(c) Not possible
(d) Possible

5.31   (a) $\dfrac{100}{\rho^3}$ C/m^3.s

(b) 314.16 A

5.33   Proof

5.35   (a) 2.741×10^4s,   (b) 5.305×10^4s,   (c) 7.07 μS

5.37   Proof

5.39   (a) $79.6\mathbf{a}_x - 265.3\mathbf{a}_y + 212.2\mathbf{a}_z$ pC/m^2, $53.05\mathbf{a}_x - 88.42\mathbf{a}_y + 70.74\mathbf{a}_z$ pC/m^2,

(b) 3.0593 nJ/m^3,   1.7684 nJ/m^3

5.41   (a) $6.667\mathbf{a}_x - 13.33\mathbf{a}_y + 6.667\mathbf{a}_z$ V/m, $13.3\mathbf{a}_x - 23.3\mathbf{a}_y + 33.3\mathbf{a}_z$ V/m
        (b) $16\mathbf{a}_x - 18\mathbf{a}_y + 36\mathbf{a}_z$ V/m

5.43   (a) $387.8\mathbf{a}_x - 452.4\mathbf{a}_y + 678.6\mathbf{a}_z$ V/m, $12\mathbf{a}_x - 14\mathbf{a}_y + 21\mathbf{a}_z$ nC/m^2
        (b) $4\mathbf{a}_x - 2\mathbf{a}_y + 3\mathbf{a}_z$ nC/m^2, 0
        (c) 12.62 μJ/m^2, 9.839 μJ/m^2

5.45 $30 \sin\theta \mathbf{a}_r + 5 \cos\theta \mathbf{a}_\theta$, $\varepsilon_o(60 \sin\theta \mathbf{a}_r + 10 \cos\theta \mathbf{a}_\theta)$

5.47 0.476 pC/m^2

5.49 $49.11°, 60°, 30°$

CHAPTER 6

6.1   (a) $-270\mathbf{a}_x + 540\mathbf{a}_y - 135\mathbf{a}_z$ V/m
        (b) 14.324 nC/m^3

6.3 27.25 V

6.5 $157.08y^4 - 942.5y^2 + 30.374$ kV

6.7 $\left(0.3142 - \dfrac{66.51}{\rho}\right)\mathbf{a}_\rho$

6.9 127.58 V

6.11 $-3xy$

6.13 436.14 nC/m^2

6.15 36.91 V, $9.102\mathbf{a}_\rho$ kV/m, $161\mathbf{a}_\rho$ nC/m^2, 322 nC/m^2, -107.3 nC/m^2

6.17   (a) Proof
        (b) $-V_o \sin\phi \mathbf{a}_\rho - V_o \cos\phi \mathbf{a}_\phi$

6.19 $V = -\dfrac{100}{r} + 150,$ $E = -\dfrac{100}{r^2}\mathbf{a}_r$ V/m

6.21   (a) $V(\rho = 15\text{ mm}) = 12.4$ V, (b) $u = 8.93 \times 10^6$ m/s

6.23   (a) $\dfrac{4V_o}{\pi} \displaystyle\sum_{n=\text{odd}}^{\infty} \dfrac{\sin(n\pi x/b)\sinh(n\pi(a-y)/b)}{n\sinh(n\pi a/b)}$

        (b) $\dfrac{4V_o}{\pi} \displaystyle\sum_{n=\text{odd}}^{\infty} \dfrac{\sin(n\pi y/a)\sinh(n\pi x/a)}{n\sinh(n\pi b/a)}$

        (c) $\dfrac{4V_o}{\pi} \displaystyle\sum_{n=\text{odd}}^{\infty} \dfrac{\sin(n\pi y/a)\sinh(n\pi(b-x)/a)}{n\sinh(n\pi b/a)}$

6.25 $\dfrac{4V_o}{a} \displaystyle\sum_{n=\text{odd}}^{\infty} \exp(-n\pi x/a)[\sin(n\pi y/a)\mathbf{a}_x - \cos(n\pi y/a)\mathbf{a}_y]$

6.27   Proof

6.29   Proof

6.31   Proof

6.33   Derivations

6.35 $\dfrac{2\pi\sigma V^2}{\ln\dfrac{b}{a}}$

6.37   2.122 pF

6.39   6 pF

6.41   Proof

6.43   1.75

6.45 $1/3\ C_o,\ 1/3\ Q_o,\ 1/3\ E_o,\ 1/3\ W_o$

6.47   (a) 25 pF, (b) 63.66 nC/m^2

6.49 $\dfrac{1}{\dfrac{1}{4\pi\varepsilon_o c}+\dfrac{1}{4\pi\varepsilon_o}\left(\dfrac{1}{a}-\dfrac{1}{b}\right)}$

6.51   0.8665 μF

6.53   Proof

6.55 $\dfrac{2\pi\varepsilon_1\varepsilon_2}{\varepsilon_2\ln(b/a)+\varepsilon_1\ln(c/b)}$

6.57   Proof

6.59   (a) $\dfrac{V_o}{\ln 2}\ln\dfrac{x+d}{d},\ -\dfrac{V_o}{(x+d)\ln 2}\mathbf{a}_x$

      (b) $-\dfrac{\varepsilon_o x V_o}{d(x+d)\ln 2}\,\mathbf{a}_x$

      (c) $0,\ -\dfrac{\varepsilon_o V_o}{2d\ln 2}$

      (d) $\dfrac{\varepsilon_o S}{d\ln 2}$

6.61   Proof

6.63   0.326 nF

6.65   (a) -12.107 pF/m^2, (b) -10 nC

6.67 $-0.1092(\mathbf{a}_x+\mathbf{a}_y+\mathbf{a}_z)$ N

6.69   (a) $-138.2\mathbf{a}_x-184.3\mathbf{a}_y$ V/m, (b) -1.018 nc/m^2

CHAPTER 7

7.1   (a) See text.

       (b) $0.7433\mathbf{a_x} + 0.382\mathbf{a_y} + 0.1404\mathbf{a_z}$ A/m

7.3 $-0.1592\mathbf{a_x}$

7.5 $-0.234\mathbf{a_x} + 0.382\mathbf{a_y} + 0.1404\mathbf{a_z}$ A/m

7.7   (a) $28.471\mathbf{a_y}$ mA/m

       (b) $13(-\mathbf{a_x} + \mathbf{a_y})$ mA/m

       (c) $-5.1\mathbf{a_x} + 1.7\mathbf{a_y}$ mA/m

       (d) $5.1\mathbf{a_x} + 1.7\mathbf{a_y}$ mA/m

7.9   (a) $-0.6792\mathbf{a_z}$ A/m

       (b) $0.1989\mathbf{a_z}$ A/m

       (c) $0.1989(\mathbf{a_x} + \mathbf{a_y})$ A/m

7.11   (a) $1.964\mathbf{a_z}$ A/m

       (b) $1.78\mathbf{a_z}$ A/m

       (c) $-0.1178\mathbf{a_z}$ A/m

       (d) $-0.3457\mathbf{a_x} - 0.3165\mathbf{a_y} + 0.1798\mathbf{a_z}$ A/m

7.13   Proof

7.15 $\dfrac{I\phi_o}{4\pi}\left(\dfrac{1}{\rho_1} - \dfrac{1}{\rho_2}\right) O_z$

7.17   (a) 69.63 A/m, (b) 36.77 A/m

7.19   (a) See text.

$$
\text{(b)} \ \ H_\phi = \begin{cases} 0, & \rho < a \\ \dfrac{I}{2\pi\rho}\left(\dfrac{\rho^2 - a^2}{b^2 - a^2}\right), & a < \rho < b \\ \dfrac{I}{2\pi\rho}, & \rho > b \end{cases}
$$

7.21   94.25 A

$$
\text{7.23} \quad H_\phi = \begin{cases} J_o, & 0 < \rho < a \\ \dfrac{J_o a}{\rho}, & \rho > a \end{cases}
$$

7.25   (a) $\dfrac{2k_o}{a}\mathbf{a_z}, \quad \rho < a$

       (b) $k_o\left(\dfrac{a}{\rho}\right)\mathbf{a_\phi}, \quad \rho > a$

7.27   (a) $3\rho \times 10^3\mathbf{a_z}$ A/m^2

       (b) 50.265 kA

7.29   (a) $80\mathbf{a}_\phi$ nWb/m^2, (b) 1.756 μWb

7.31   (a) $\frac{1}{2}\pi a^2 J_o$,

$$(b)\ H_\phi = \begin{cases} \dfrac{J_o\rho}{4}\left(2 - \dfrac{\rho^2}{a^2}\right), & \rho < a \\ \dfrac{aJ_o}{4\rho}, & \rho > a \end{cases}$$

7.33 $\dfrac{\mu_o I}{4a}\mathbf{a}_z$

7.35   Proof, 1.37×10^{-8} Wb

7.37   2.854 Wb

7.39   12.53 Wb/m^2

7.41   (a) magnetostatic field, (b) magnetostatic field, (c) neither electrostatic nor magnetostatic field.

7.43 $\dfrac{\pi}{\mu_o}\left(\sin \pi x \mathbf{a}_y - 10\cos \pi y \mathbf{a}_z\right)$ A/m, $\dfrac{\pi^2}{\mu_o}(10\sin \pi y \mathbf{a}_x + \cos \pi x \mathbf{a}_z)$ A/m^2

7.45   (a) $\mathbf{B} = (-6xz + 4x^2y + 3xz^2)\mathbf{a}_x + (y + 6yz - 4xy^2)\mathbf{a}_y + (y^2 - z^3 - 2x^2 - z)\mathbf{a}_z$ Wb/m^2
(b) 8 Wb
(c) Proof

7.47   1.011 Wb

7.49   Proof

7.51 $\dfrac{A_o}{r^3}(2\cos \theta \mathbf{a}_r + \sin \theta \mathbf{a}_\theta)$ Wb/m^2

7.53   (a) $-\dfrac{20\rho}{\mu_o}\mathbf{a}_\phi$ μA/m, $\quad -\dfrac{40}{\mu_o}\mathbf{a}_z$ μA/m^2
(b) -400 A

7.55   1A

7.57   Proof

CHAPTER 8

8.1 $-152.8\mathbf{a}_x - 170.4\mathbf{a}_y - 189.6\mathbf{a}_z$ mN

8.3   (a) $-0.05\mathbf{a}_x - 0.25\mathbf{a}_y$ N
(b) $50\mathbf{a}_x + 250\mathbf{a}_y$ V/m

8.5 $(0.2419, 1, 1.923)$, $4.071\mathbf{a}_x - 2.903\mathbf{a}_z$ m/s. The particle gyrates in a circle in the $y = 1$ plane with center at $(0,1,19/12)$.

8.7   26.67 nC

8.9   85.714 Wb/m^2

8.11   100 μN

8.13	(a) $2.197\mathbf{a}_z$, (b) $0.575\mathbf{a}_\rho$
8.15	$15.71\mathbf{a}_z\ \mu\text{N/m}$
8.17	$540(\mathbf{a}_y + \mathbf{a}_z)$ N.m
8.19	627.5 MA
8.21	24 mN.m
8.23	Proof
8.25	$7.5, 1.477\mathbf{a}_z$ mA/m, $0.1477\mathbf{a}_z\ \text{A/m}^2$
8.27	4.54
8.29	(a) Proof, (b) Proof
8.31	$1.6\mathbf{a}_x - 4\mathbf{a}_y + 12\mathbf{a}_z\ \text{mWb/m}^2, 32.34\ \text{J/m}^3$
8.33	$5\mathbf{a}_x - 6\mathbf{a}_z$ A/m
8.35	(a) $4\mathbf{a}_\rho + 15\mathbf{a}_\phi - 8\mathbf{a}_z\ \text{mWb/m}^2$
	(b) $60.68\ \text{J/m}^3, 57.7\ \text{J/m}^3$
8.37	$\mu_o(22\mathbf{a}_\rho + 0.05625\mathbf{a}_\phi)\ \text{Wb/m}^2$
8.39	(a) $-5\mathbf{a}_y$ A/m, $-6.28\mathbf{a}_y\ \mu\ \text{Wb/m}^2$
	(b) $-35\mathbf{a}_y$ A/m, $-110\mathbf{a}_y\ \mu\text{Wb/m}^2$
	(c) $5\mathbf{a}_y$ A/m, $6.283\mathbf{a}_y\ \mu\ \text{Wb/m}^2$
8.41	Proof
8.43	$2\ \mu\text{H}$
8.45	$11.82\ \mu\text{H}$
8.47	Proof
8.49	304.1 pJ
8.51	101.86 J
8.53	190.8 A.t, 19080 A/m
8.55	88.5 mWb/m^2
8.57	7.2643×10^4 A.t/Wb
8.59	(a) 37 mN
	(b) $1.885\ \mu\text{N}$
8.61	Proof

CHAPTER 9

9.1	$0.4738 \sin 377t$ V
9.3	$-12.57 \cos 10^4 t$ A
9.5	$\dfrac{20 \sin 1.6\beta}{\beta}$
9.7	$9.888\ \mu\text{V}$, point A at higher potential
9.9	-1.8 V

9.11   0.97 mV

9.13   6.33 A, counterclockwise

9.15   277.8 A/m^2, 77.78 mA

9.17   (a) 0.444×10^{-3}
        (b) 5.555
        (c) 7.2×10^{-4}

9.19   600 kHz

9.21 $\beta H_o \sin(\omega t - \beta z)\mathbf{a}_y$

9.23   (a) $\nabla \times \mathbf{E}_s = j\omega\mu\mathbf{H}_s, \quad \nabla \times \mathbf{H}_s = (\sigma - j\omega\varepsilon)\mathbf{E}_s$

        (b) $\nabla \cdot \mathbf{D} = \rho_v \rightarrow \dfrac{\partial D_x}{\partial x} + \dfrac{\partial D_y}{\partial y} + \dfrac{\partial D_z}{\partial z} = \rho_v$

        $\nabla \cdot \mathbf{B} = 0 \rightarrow \dfrac{\partial B_x}{\partial x} + \dfrac{\partial B_y}{\partial y} + \dfrac{\partial B_z}{\partial z} = 0$

        $\nabla \times \mathbf{E} = -\dfrac{\partial \mathbf{B}}{\partial t} \rightarrow \dfrac{\partial E_z}{\partial y} - \dfrac{\partial E_y}{\partial z} = -\dfrac{\partial B_x}{\partial t}$

        $\dfrac{\partial E_x}{\partial z} - \dfrac{\partial E_z}{\partial x} = -\dfrac{\partial B_y}{\partial t}$

        $\dfrac{\partial E_y}{\partial x} - \dfrac{\partial E_x}{\partial y} = -\dfrac{\partial B_z}{\partial t}$

        $\nabla \times \mathbf{H} = \mathbf{J} + \dfrac{\partial \mathbf{D}}{\partial t} \rightarrow \dfrac{\partial H_z}{\partial y} - \dfrac{\partial H_y}{\partial z} = J_x + \dfrac{\partial D_x}{\partial t}$

        $\dfrac{\partial H_x}{\partial z} - \dfrac{\partial H_z}{\partial x} = J_y + \dfrac{\partial D_y}{\partial t}$

        $\dfrac{\partial H_y}{\partial x} - \dfrac{\partial H_x}{\partial y} = J_z + \dfrac{\partial D_z}{\partial t}$

9.25   Proof

9.27 $0.3z^2 \cos 10^4 t \text{ mC/m}^3$

9.29 $\dfrac{40\beta}{\omega\varepsilon_o} \sin(\omega t + \beta x)\mathbf{a}_z$ V/m,   0.333 rad/m

9.31 $\dfrac{10\beta}{\omega\mu_o r} \sin\theta \cos(\omega t - \beta r)\mathbf{a}_\phi$ A/m

9.33   Derivation

9.35   (a) Yes, (b) yes, (c) no, (d) no.

9.37 $2.636 \cos(12\pi x) \sin(10^{10}t - 25.66y)\mathbf{a}_x + 3 \sin(12\pi x) \cos(10^{10}t - 25.66y)\mathbf{a}_y$ mA/m, 33/2

9.39 $-\dfrac{12\beta \sin\theta}{\omega\varepsilon_o r} \cos(\omega t - \beta r)\mathbf{a}_\phi, \quad \omega = 2\pi \times 10^8$

9.41   Derivation

9.43   (a) V = constant, $\mathbf{E} = -A_o\omega \cos(\omega t - \beta z)\mathbf{a}_x$
        (b) $\beta = \omega\sqrt{\mu_o\varepsilon_o}$

9.45   (a) $-10e^{j\pi/3}\mathbf{a}_x$

         (b) $4\cos(4y)e^{-j2x}\mathbf{a}_z$

         (c) $5e^{-j\pi/6}\mathbf{a}_x - 8e^{-j\pi/4}\mathbf{a}_y$

9.47   (a) $\dfrac{1}{\rho}e^{-j3z}\mathbf{a}_\phi$

         (b) $\dfrac{3}{\rho\varepsilon_o\omega}\cos(\omega t - 3z)\mathbf{a}_\rho$

         (c) $9 \times 10^8\,\text{rad/s}$

9.49 $3.333, -4.533e^{j3.33x}\,\mathbf{a}_y$

9.51   Proof

CHAPTER 10

10.1   (a) along $\mathbf{a}_x$

         (b) 1 μs, 1.047 m, 1.047×10^6 m/s

         (c) See Figure E.1.

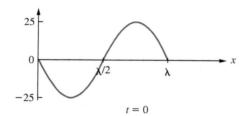

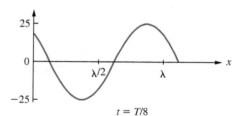

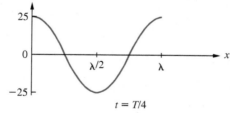

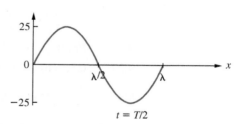

FIGURE E.1 For Problem 10.1(c).

10.3   (a) 10^8 rad/s

         (b) 0.333 rad/m

         (c) 18.85 m

         (d) Along $-\mathbf{a}_y$

         (e) -0.1665 A/m

10.5   Derivations

10.7   (a) $5.41 + j6.129$ /m, (b) 1.025 m, (c) 5.125×10^7 m/s, (d) 101.4

         (e) $-59.16e^{-j41.44°}\,e^{-\gamma}{}_z\,\mathbf{a}_y$ mA/m

10.9   (a) 6.85 m, (b) 0.73×10^{-3} Np

10.11   Proof

10.13   (a) 20 ns, (b) 3 m, (c) $2.257e^{-0.1y}\sin(\pi \times 10^8 t - 2.088y + 2.74°)a_z$ V/m, (d) 2.74°

10.15   (a) 168.8 Np/m
         (b) 210.5 rad/m
         (c) $263.38\angle 38.73°$ Ω,
         (d) 2.985×10^7 m/s

10.17   5.224

10.19   (a) lossless
         (b) 12.83 rad/m, 0.49 m
         (c) 25.66 rad
         (d) 4.62 kΩ

10.21   14.175, 1.0995

10.23   (a) 5.662 V/m
         (b) 5.212 dB
         (c) $36.896\angle 3.365°$ Ω

10.25   (a) No, nonconducting
         (b) No, nonconducting
         (c) Yes, conducting

10.27   (a) 2.287 Ω
         (b) 207.61 Ω
         (c) 12.137 kHz

10.29   1.038 kHz

10.31   2.723 Np/m, 2.723 rad/m, 2.306 m, 2.768×10^7 m/s

10.33 2.94×10^{-6} m

10.35   (a) Linearly polarized along a_z
         (b) 10 MHz
         (c) 205.18
         (d) $-0.456\sin(2\pi \times 10^7 t - 3y)a_x$ A/m

10.37   (a) Circular polarization
         (b) Elliptical polarization

10.39   (a) Elliptically polarized.
         (b) $-159.2\sin(\omega t - \beta z)a_z + 106\cos(\omega t - \beta z)a_y$ mA/m

10.41   (a) Linearly polarized
         (b) Elliptically polarized
         (c) Linearly polarized

10.43   (a) 5.76
         (b) 157.1 Ω
         (c) 1.25×10^8 m/s
         (d) $0.955\cos(10^9 t + 8x)a_y$ A/m
         (e) $-143.25\cos^2(10^9 t + 8x)a_x$ W/m^2

10.45 (a) $\dfrac{1}{12\pi r}\sin\theta e^{-j3r}\mathbf{a}_\phi$ A/m

(b) 7.145 mW

10.47 (a) 2.828×10^8 rad/s, $\dfrac{0.225}{\rho}\sin(\omega t - 2z)\mathbf{a}_\phi$ A/m

(b) $\dfrac{9}{\rho^2}\sin^2(\omega t - 2z)\mathbf{a}_z$ W/m^2

(c) 11.46 W

10.49 12×10^9 rad/s, $-7.539\sin(\omega t - 40x)\mathbf{a}_y - 3.775\sin(\omega t - 40x)\mathbf{a}_z$ kV/m, $94.23\mathbf{a}_x$ kW/m^2

10.51 Proof

10.53 (a) $\dfrac{V_o I_o}{4\pi\rho^2 \ln(b/a)}\mathbf{a}_z$

(b) $\dfrac{1}{2}V_o I_o$

10.55 $-20\sin(\omega t + 10z)\mathbf{a}_x - 10\sin(\omega t + 10z + \pi/6)\mathbf{a}_y$ V/m, $40\sin(\omega t - 10z)\mathbf{a}_x + 20\sin(\omega t - 10z + \pi/6)\mathbf{a}_y$ V/m

10.57 (a) $-1.508\sin(\omega t - 5x)\mathbf{a}_z + 0.503\sin(\omega t + 5x)\mathbf{a}_z$ kV/m

(b) $2.68\mathbf{a}_x$ kW/m^2

(c) 2

10.59 $p_1 = 38.197\cos^2(\omega t - \beta_1 z)\mathbf{a}_x + 13.75\cos^2(\omega t + \beta_1 z)\mathbf{a}_y$ W/m^2, $\beta_1 = 1200$ rad/m, $p_2 = 24.46\cos^2(\omega t - \beta_2 z)\mathbf{a}_x$ W/m^2, $\beta_2 = 300$ rad/m

10.61 89.51%, 10.84%

10.63 (a) 9×10^8 rad/s

(b) 2.094 m

(c) 6.288, $16.71\angle 40.74°$ Ω

(d) $E_r = 9.35\sin(\omega t - 3z + 179.7°)a_x$ V/m
$E_r = 0.857\,e^{43.94z}\sin(9 \times 10^8 t + 51.48z + 38.89°)a_x$ V/m

10.65 $\theta_{t1} = 19.47°$, $\theta_{t2} = 28.13°$

10.67 $0.8146\angle 174.4°$

10.69 Proof

10.71 1.235, 258 MHz, $0.491[(12.6 + j16.8)\mathbf{a}_x + (10.2 + j13.6)\mathbf{a}_y - 8.59\mathbf{a}_z]e^{-j3.4x + 4.2y}$ mA/m

10.73 Proof

10.75 (a) $\theta_i = 36.87°$

(b) $106.1\mathbf{a}_y + 79.58\mathbf{a}_z$ mW/m^2

(c) $E_r = -(1.518\mathbf{a}_y + 2.02\mathbf{a}_z)\sin(\omega t + 4y - 3z)$,
$E_t = (1.879\mathbf{a}_y - 5.968\mathbf{a}_z)\sin(\omega t - 9.539y - 3z)$ V/m

10.77 (a) $5\cos(\omega t - 0.5\pi x - 0.866\pi z)\mathbf{a}_y$, $(4\mathbf{a}_x - 3\mathbf{a}_z)\cos(\omega t - 0.5\pi x - 0.866\pi z)$

(b) 36.87°

10.79 (a) 1.45, (b) 9, (c) 1.643

10.81 (a) $T_{11} = 1/S_{21}$, $T_{12} = -S_{22}/S_{21}$, $T_{21} = S_{11}/S_{21}$, $T_{22} = S_{12} - S_{11}S_{22}/S_{21}$

   (b) $T = \begin{bmatrix} 2.5 & -0.5 \\ 0.5 & 0.3 \end{bmatrix}$

10.83 The microwave wavelengths are of the same magnitude as the circuit components. The physical dimension of the lumped element must be in this range, which is not physically realizable.

CHAPTER 11

11.1 0.0354 Ω/m, 50.26 nH/m, 221 pF/m, 0 S/m

11.3 Proof

11.5 0.2342 pF, $1.4 \times 10^{-2}\ \Omega$

11.7 $\dfrac{\partial V}{\partial z} = -L\dfrac{\partial I}{\partial t}$, $\dfrac{\partial I}{\partial z} = -C\dfrac{\partial V}{\partial t}$

11.9 (a) $8.159 \times 10^{-2} + j2.384 \times 10^3$ /m, (b) $8.159 \times 10^{-2} + j2.384 \times 10^3$

11.11 0.2305 Ω/m, 0.3316 μH/m, 0.1326 nF/m, 92.2 μS/m

11.13 (a) $644.3 - j97\ \Omega$, $(5.415 + j33.96) \times 10^{-3}$ /mi
   (b) 1.85×10^5 mi/s
   (c) 185.02 mi

11.15 0.46 mm

11.17 0.655 μH/m, 59.4 pF/m, 105 Ω

11.19 (a) $0.2898 + j3.278$ /m, 2.3×10^8 m/s, 69Ω
   (b) 5.554 m
   (c) 0.3051m

11.21 $(1 + j10) \times 10^{-3}$ /m, 6.283×10^6 m/s

11.23 $55.12 + j45.85\ \Omega$, 0.1783 μs

11.25 $57.44 + j48.82\ \Omega$

11.27 (a) Proof, (b) $62.6 + j82.93\ \Omega$

11.29 0.0136 Np/m, 38.94 rad/m, $129.1 - j0.045\ \Omega$, 6.452×10^7 m/s

11.31 $0.7071\angle 45°$

11.33 (a) $j29.375\ \Omega$, $5.75\angle 102°$ V
   (b) $j40\ \Omega$, $12.60\angle 0°$ V
   (c) $-j3471.88$, $22.74\angle 0°$ V
   (d) $-j18.2\ \Omega$, $6.607\angle 180°$ V

11.35 (a) 1.6
   (b) $1.2 + j0.8$
   (c) $0.6 - j2.4$
   See Figure E.2 for their locations on the Smith chart.

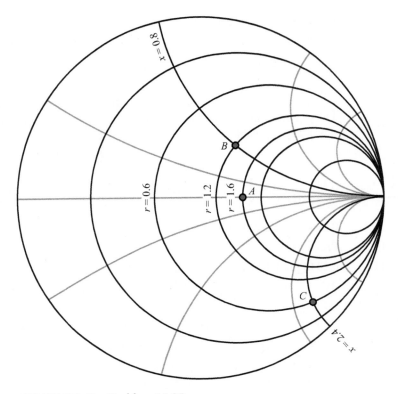

FIGURE E.2 For Problem 11.35.

11.37 $(0.2 - j0.4)Z_o$

11.39 $10.25 + j12.3$, 5.19

11.41 $46 + j26\ \Omega$, $0.4167\angle144°$

11.43   (a) $0.475\angle42°$
        (b) 2.8
        (c) $27.5 + j32.5\Omega$
        (d) At 0.05833λ
        (e) Same as in (d), i.e., 0.05833λ

11.45 $7.4 - j4.4$ mS

11.47 $42\ \Omega$

11.49   (a) $0.5543\angle25°$, $296\ \Omega$, $21.622\ \Omega$
        (b) $184 + j124\ \Omega$, 3.7, $38.4 + j60.4\ \Omega$
        (c) $3Z_{in,\max}$ and $2Z_{in,\min}$

11.51   (a) 0.4714λ
        (b) $s = \infty$, $\Gamma_L = 1\angle106.26°$

11.53   (a) $24.5\ \Omega$
        (b) $55.33\ \Omega$, $67.74\ \Omega$

11.55   10.25 W

11.57 $20 + j15$ mS, $-j10$ mS, $6.408 + j5.189$ mS, $j10$ mS, $2.461 + j5.691$ mS

11.59   (a) $34.2 + j41.4\ \Omega$
       (b) 0.38λ, 0.473λ
       (c) 2.65

11.61 0.0723λ, 0.4093λ

11.63   2.11, 1.764 GHz, $0.357\angle -44.5°$, $70 - j40\ \Omega$

11.65   See Figure E.3.

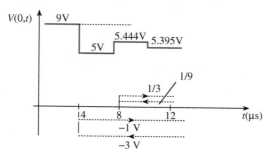

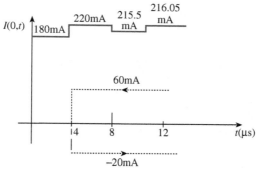

FIGURE E.3 **For Problem 11.65.**

11.67   9.57 V

11.69   17.6 V, 600m

11.71 $\dfrac{V_g}{2}$, where $V_o = \dfrac{Z_o}{Z_o + Z_g} V_g$

11.73   (a) 14.9 mm, (b) 2.52 mm

11.75   127.64 Ω

11.77   13.98 dB

CHAPTER 12

12.1 (a) 2.5 GHz
(b) 7 TE modes and 2 TM modes

12.3 See Table E.1

TABLE E.1 **For Problem 12.3**

Mode	f_c (GHz)
TE_{01}	9.901
TE_{10}	4.386
TE_{11}	10.829
TE_{02}	19.802
TE_{22}	21.66
TM_{11}	10.829
TM_{12}	20.282
TM_{21}	13.228

12.5 A design could be $a = 9$ mm, $b = 3$ mm

12.7 (a) 3 cm, 1.25 cm
(b) 10, 12, 13 GHz
(c) 8.67 GHz

12.9 254.15 rad/m, 3.09×10^8, 388.3 Ω

12.11 2.325 μs

12.13 (a) 5.391 GHz
(b) 2.62×10^8 m/s
(c) 4.368 cm

12.15 375.1 Ω, 1.5 mW

12.17 (a) TE_{23}
(b) $j400.7$ /m
(c) 985.3 Ω

12.19 Proof

12.21 (a) 122.32 rad/m, 3.185×10^8 m/s, 2.826×10^8 m/s, 400.21 Ω,
(b) 189.83 rad/m, 2.052×10^8 m/s, 1.949×10^8 m/s, 257.88 Ω,

12.23 $3u_p$

12.25 4.927 cm

12.27 $\dfrac{\omega\mu\beta a^2}{2\pi^2}H_o^2\sin^2(\pi y/b)\mathbf{a}_z$

12.29 (a) 0.012682 Np/m, 0.0153 Np/m
(b) 0.02344 Np/m, 0.0441 Np/m

12.31 6.5445 m

12.33 (a) 2.165×10^{-2} Np/m
(b) 4.818×10^{-3} Np/m

12.35   0.1339 dB/m

12.37   199.94 rad/m, 1.481×10^{-2} Np/m, 2.183×10^{-2} Np/m, 2.357×10^{8} m/s, 1.689×10^{8} m/s, 5 cm

12.39   Proof

12.41   Proof, $\frac{1}{h^2}(n\pi/b)(\rho\pi/c)E_o \sin(m\pi x/a)\cos(n\pi y/b)\cos(\rho\pi z/c)$ V/m

12.43   (a) 2.606 GHz, (b) 4727.7

12.45   5.701 GHz (TE_{011}), 7.906 GHz (TE_{012}), 10 GHz (TE_{101} and TE_{021}), 10.61 GHz (TE_{013} or TM_{110}), 11.07 GHz (TE_{111} or TM_{111})

12.47   (a) 16.77 GHz, (b) 6589.51

12.49   (a) $a = b = c = 3.788$ cm, (b) $a = b = c = 2.646$ cm

12.51   (a) $a = b = c = 1.77$ cm, (b) 9767.61

12.53   1.4286

12.55   (a) 0.2271, (b) 13.13°, (c) 6 modes

12.57   (a) 29.23°, (b) 63.1%

12.59 $\alpha_{12} = 8686\alpha_{10}$

12.61   3.1698 mW

CHAPTER 13

13.1 $\dfrac{50\eta\beta}{\mu r} \sin(\omega t - \beta r)(\sin\phi\mathbf{a}_\phi + \cos\theta\cos\phi\mathbf{a}_\theta)$ V/m,

$\dfrac{-50}{\mu r}\beta\sin(\omega t - \beta r)(\sin\phi\mathbf{a}_\phi - \cos\theta\cos\phi\mathbf{a}_\theta)$ A/m

13.3 $\mathbf{H}_s = \dfrac{j0.1667}{r}\sin\theta e^{-j\pi r/3}\mathbf{a}_\phi, \quad \mathbf{E}_s = \dfrac{j62.83}{r}\sin\theta e^{-j\pi r/3}\mathbf{a}_\theta$

13.5   (a) Proof
       (b) $\ell = 0.05\lambda$

13.7   1.447 W

13.9   (a) Proof

       (b) For $l = \lambda, f(\theta) = \dfrac{\cos(\pi\cos\theta) + 1}{\sin\theta}$

       See Figure E.4.

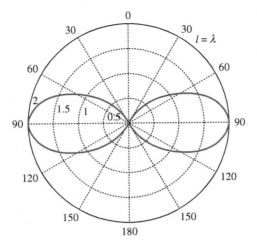

$$\text{For } l = \frac{3\lambda}{2}, f(\theta) = \frac{\cos\left(\dfrac{3\pi}{2}\cos\theta\right)}{\sin\theta}$$

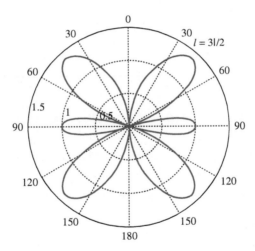

$$\text{For } l = 2\lambda, f(\theta) = \frac{\cos\theta\,\sin(2\pi\,\cos\theta)}{\sin\theta}$$

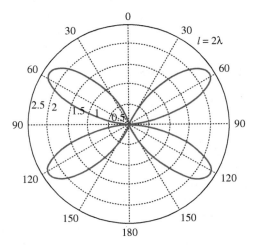

FIGURE E.4 **For Problem 13.9.**

13.11 (a) 65.22 m
  (b) 0.8333 m
  (c) 0.9375 m
  (d) 0.125 m

13.13 (a) 9.071 mA
  (b) 0.25 mW

13.15 (a) 1.26 mΩ
  (b) 1.575 W
  (c) 1.933%

13.17 33.3%

13.19 (a) $\mathbf{H}_s = \dfrac{\cos 2\theta}{120\pi r} e^{-j\beta r} \mathbf{a}_\phi$ A/m, (b) 7.778 mW, (c) 76.78%

13.21 (a) $12.73\mathbf{a}_r \, \mu\text{W/m}_2$, (b) 0.098 V/m

13.23 See Figure E.5.

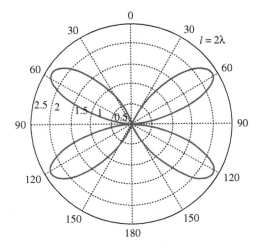

FIGURE E.5 **For Problem 13.23.**

13.25   (a) $1.5 \sin^2\theta$, (b) 1.5, (c) $\dfrac{1.5\lambda^2 \sin^2\theta}{4\pi}$, (d) $3.084\ \Omega$

13.27   (a) 62.5 m, (b) $36.5\ \Omega$, (c) 3.282

13.29   (a) $2.546 \sin\theta \sin^2\phi$, 2.546
   (b) $2.25\pi \sin^2\theta \sin^3\phi$, 7.069
   (c) $0.75(1 + \sin^2\theta \sin^2\phi)$, 1.5

13.31 2.121 A

13.33 $\dfrac{\alpha^2}{30}$

13.35 $\dfrac{j\eta\beta I_o dl}{2\pi} \sin\theta \, \cos\!\left(\dfrac{1}{2}\beta d \cos\theta\right)\mathbf{a}_\theta$

13.37   See Figure E.6.

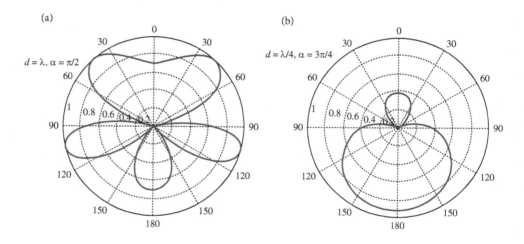

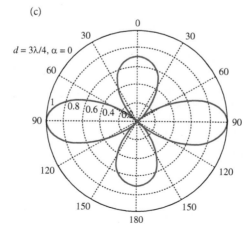

FIGURE E.6 **For Problem 13.37.**

13.39   See Figure E.7.

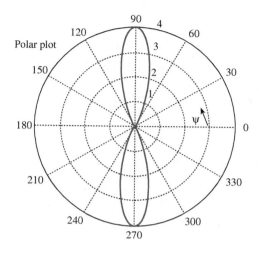

FIGURE E.6 For Problem 13.39.

13.41 $7.162 \times 10^{-3} \, \text{m}^2$

13.43   115.384 dB

13.45   21.29 pW

13.47   19 dB

13.49   272.1 pW

13.51   The transmitted power must be increased 16 times.

13.53   (a) 8.164 dB, (b) 55.05 dB

13.55   (a) 17.1 mΩ/km, (b) 51.93 Ω

CHAPTER 14

14.1   See Figure E.8

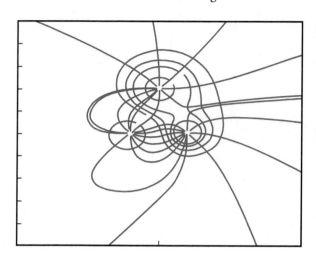 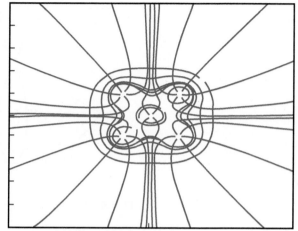

FIGURE E.8 For Problem 14.1.

14.3   (a) 10.117, 1.56
        (b) 10.113, 1.506

14.5   Proof

14.7 $V_1 = 39.93$, $V_2 = 34.97$, $V_3 = 24.96$, $V_4 = 19.98$ V

14.9   (a) $\dfrac{4V_o}{\pi} \displaystyle\sum_{n=\text{odd}}^{\infty} \dfrac{\sin\left(\dfrac{n\pi x}{a}\right)\sinh\left(\dfrac{n\pi y}{a}\right)}{n\sinh(n\pi)} + \dfrac{4V_o}{\pi} \displaystyle\sum_{n=\text{odd}}^{\infty} \dfrac{sin\left(\dfrac{n\pi y}{a}\right)\sinh\left(\dfrac{n\pi x}{a}\right)}{n\sinh(n\pi)}$

        (b) 25 V

14.11   (a)

$$
\begin{bmatrix}
-4 & 1 & 0 & 1 & 0 & 0 \\
1 & -4 & 1 & 0 & 1 & 0 \\
0 & 1 & -4 & 0 & 0 & 1 \\
1 & 0 & 0 & -4 & 1 & 0 \\
0 & 1 & 0 & 1 & -4 & 1 \\
0 & 0 & 1 & 0 & 1 & -4
\end{bmatrix}
\begin{bmatrix}
V_a \\ V_b \\ V_c \\ V_d \\ V_e \\ V_f
\end{bmatrix}
=
\begin{bmatrix}
-200 \\ -100 \\ -100 \\ -100 \\ 0 \\ 0
\end{bmatrix}
$$
$$\qquad\qquad\quad [A] \qquad\qquad\qquad\qquad [B]$$

(b)

$$
\begin{bmatrix}
-4 & 1 & 0 & 1 & 0 & 0 & 0 & 0 \\
1 & -4 & 1 & 0 & 1 & 0 & 0 & 0 \\
0 & 1 & -4 & 0 & 0 & 1 & 0 & 0 \\
1 & 0 & 0 & -4 & 1 & 0 & 1 & 0 \\
0 & 1 & 0 & 1 & -4 & 1 & 0 & 1 \\
0 & 0 & 1 & 0 & 1 & -4 & 0 & 0 \\
0 & 0 & 0 & 1 & 0 & 0 & -4 & 1 \\
0 & 0 & 0 & 0 & 1 & 0 & 1 & -4
\end{bmatrix}
\begin{bmatrix}
V_1 \\ V_2 \\ V_3 \\ V_4 \\ V_5 \\ V_6 \\ V_7 \\ V_8
\end{bmatrix}
=
\begin{bmatrix}
-30 \\ -15 \\ -30 \\ -7.5 \\ 0 \\ -7.5 \\ 0 \\ 0
\end{bmatrix}
$$

$$[A] \qquad\qquad\qquad\qquad [B]$$

14.13 $V_1 = 10.97$, $V_2 = 26.25$, $V_3 = 44.06$, $V_4 = 17.97$, $V_5 = 11.05$, $V_6 = 26.28$, $V_7 = 44.07$ V

14.15   The numerical and analytical solutions are plotted in Figure E.9.

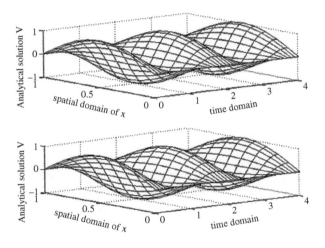

FIGURE E.9 For Problem 14.15.

14.17   12.77 pF/m

14.19   100 Ω

14.21   (a) Exact $C = 80.26$ pF/m, $Z_o = 41.56$ Ω; for numerical solution, see Table E.2.

TABLE E.2 For Problem 14.21(a)

N	C (pF/m)	Z_o (Ω)
10	82.386	40.486
20	80.966	41.197
40	80.438	41.467
100	80.025	41.562

(b) For numerical result, see Table E.3

TABLE E.3   For Problem 14.21(b)

N	C (pF/m)	Z_o (Ω)
10	109.51	30.458
20	108.71	30.681
40	108.27	30.807
100	107.93	30.905

14.23   Proof

14.25   (a) (1.5, 0.5) along 12 and (0.9286, 0.9286) along 13
        (b) 56.67 V

14.27
$$\begin{bmatrix} 0.8802 & -0.2083 & 0 & -0.6719 \\ -0.2083 & 1.5333 & -1.2 & -0.125 \\ 0 & -1.2 & 1.4083 & -0.2083 \\ -0.6719 & -0.125 & -0.2083 & 1.0052 \end{bmatrix}$$

14.29
$$\begin{bmatrix} 0.8333 & -0.667 & 0 & -0.1667 \\ -0.6667 & 1.4583 & -0.375 & -0.4167 \\ 0 & -0.375 & 0.625 & -0.25 \\ -0.1667 & -0.4167 & -0.25 & 0.833 \end{bmatrix}$$

14.31   25 V

14.33   See Table E.4.

TABLE E.4   For Problem 14.33

Node no.	FEM Solution	Exact Solution
8	3.635	3.412
9	5.882	5.521
10	5.882	5.521
11	3.635	3.412
14	8.659	8.217
15	14.01	13.30
16	14.01	13.30
17	8.659	8.217
20	16.99	16.37
21	27.49	26.49
22	27.49	26.49
23	16.99	16.37
26	31.81	31.21
27	51.47	50.5
28	51.47	50.5
29	31.81	31.21

14.35 $V_1 = V_3 = V_4 = V_6 = 35.71$ V, $V_2 = V_5 = 42.85$

14.37   40.587 Ω

14.39   92.01, 74.31, 82.87, 53.72, 61.78, 78.6, 30.194, 36.153, 53.69 V